Advances in
Magnetic Materials

Processing, Properties,
and Performance

Advances in Materials Science and Engineering

Series Editor
Sam Zhang

Advances in Magnetic Materials

Processing, Properties, and Performance

Edited by

Sam Zhang • Dongliang Zhao

CRC Press
Taylor & Francis Group
Boca Raton London New York

CRC Press is an imprint of the
Taylor & Francis Group, an **informa** business

CRC Press
Taylor & Francis Group
6000 Broken Sound Parkway NW, Suite 300
Boca Raton, FL 33487-2742

First issued in paperback 2019

ISBN-13: 978-1-4987-0671-1 (hbk)
ISBN-13: 978-0-367-87182-6 (pbk)

Visit the Taylor & Francis Web site at
http://www.taylorandfrancis.com

and the CRC Press Web site at
http://www.crcpress.com

Contents

Editors

Professor Sam Zhang Shanyong, better known as Sam Zhang, earned his PhD in ceramics in 1991 from the University of Wisconsin-Madison, Wisconsin, USA, and is a tenured full professor (since 2006) at the School of Mechanical and Aerospace Engineering, Nanyang Technological University. Professor Zhang was the founding editor-in-chief for *Nanoscience and Nanotechnology Letters* (2008–2015) and principal editor of the *Journal of Materials Research* (USA) responsible for the thin films and coating field (since 2003). Professor Zhang has been involved in the processing and characterization of nanocomposite thin films and coatings for more than 25 years and has authored/coauthored more than 300 peer-reviewed international journal papers, 12 books, and 23 book chapters, and guest edited 13 journal volumes special issues. From the Scopus webpage as of June 29, 2016, Professor Zhang's total articles in publication are 298; the sum of the times cited, 6586; h-index, 45.

Professor Dongliang Zhao has been the vice president at the Central Iron and Steel Research Institute (CISRI, Beijing, China) since 2012. CISRI plays a leading role in China's R&D in superalloys. Professor Zhao's research interests include computational material sciences, magnetic materials, energy materials, and superalloys. He has been the leading principal investigator of or participated in more than 20 Chinese national research projects. Professor Zhao has published 50 journal papers and was granted 6 patents. In 2003, Professor Zhao received the title of "Beijing Outstanding Young Engineer" by the Beijing City Government. In 2006, he was recognized by the State Department as one of the National Star Researchers and in 2008, he was awarded the title of "National Defense Science and Technology Innovation Leader."

Contributors

Shampa Aich
Department of Metallurgical and
  Materials Engineering
Indian Institute of Technology
  Kharagpur
Kharagpur, India

Animesh Kumar Basak
Adelaide Microscopy
The University of Adelaide
Adelaide, South Australia, Australia

Somnath Bhattacharyya
Indian Institute of Technology Madras
Chennai, India

Prasanta Chowdhury
National Aerospace Laboratories
Council of Scientific and Industrial
  Research
Bengaluru, India

Zhaofu Du
Research Institute of Functional Materials
Central Iron and Steel Research Institute
Beijing, China

Richard YongQing Fu
Faculty of Engineering and
  Environment
University of Northumbria
Newcastle upon Tyne, United Kingdom

M. R. Hossain
Department of Materials and
  Metallurgical Engineering
Bangladesh University of Engineering
  and Technology
Dhaka, Bangladesh

Shaoying Huang
Singapore University of Technology and
  Design
Singapore

Prabhanjan D. Kulkarni
National Aerospace Laboratories
Council of Scientific and Industrial
  Research
Bengaluru, India

Wen Siang Lew
School of Physical and Mathematical
  Sciences
Nanyang Technological University
Singapore

Zhongwu Liu
School of Materials Science and
  Engineering
South China University of
  Technology
Guangzhou, China

Elizaveta Motovilova
Singapore University of Technology and
  Design
Singapore

Chandrasekhar Murapaka
School of Physical and Mathematical
  Sciences
Nanyang Technological University
Singapore

Alokesh Pramanik
Department of Mechanical
  Engineering
Curtin University
Perth, Australia

Indra Purnama
School of Physical and Mathematical
    Sciences
Nanyang Technological University
Singapore

Al Mahmudur Rahman
Institute of Engineering Mechanics
Karlsruhe Institute of Technology
    (KIT)
Karlsruhe, Germany

Rajdeep Singh Rawat
NSSE, National Institute of Education
Nanyang Technological University
Singapore

D. K. Satapathy
Department of Metallurgical and
    Materials Engineering
Indian Institute of Technology
    Kharagpur
Kharagpur, India

Samarpita Senapati
Department of Chemistry
Indian Institute of Technology Kharagpur
Kharagpur, India

J. E. Shield
Department of Mechanical and
    Materials Engineering
College of Engineering
University of Nebraska
Lincoln, Nebraska

Suneel Kumar Srivastava
Department of Chemistry
Indian Institute of Technology Kharagpur
Kharagpur, India

Tat Joo Teo
Mechatronics Group
Singapore Institute of Manufacturing
    Technology
Singapore

S. F. Wang
School of Physical Electronics
University of Electronic Science and
    Technology of China
Chengdu, China
and
Science and Technology on
    Vacuum Technology and Physics
    Laboratory
Lanzhou Institute of Physics
Lanzhou, China

Ying Wang
NSSE, National Institute of Education
Nanyang Technological University
Singapore

Sam Zhang
School of Mechanical and Aerospace
    Engineering
Nanyang Technological University
Singapore

Dongliang Zhao
Research Institute of Functional
    Materials
Central Iron and Steel Research Institute
Beijing, China

Lizhong Zhao
School of Materials Science and
    Engineering
South China University of Technology
Guangzhou, China

Yinghuai Zhu
Institute of Chemical and Engineering
    Sciences
Singapore

X. T. Zu
School of Physical Electronics
University of Electronic Science and
    Technology of China
Chengdu, China

1 Magnetic Behavior of Nanostructured Materials

M. R. Hossain, Animesh Kumar Basak,
Alokesh Pramanik, and Al Mahmudur Rahman

CONTENTS

ABSTRACT

Over the past couple of decades, nanocrystalline materials, commonly noted as nanomaterials, are receiving wide attention among scientists due to unusual material properties at the nanoscale that are not foreseen in the micro-/macroscale. This field of nanomaterials is still very young, requires a long quest, and upholds prospects beyond imagination. With this in view, this chapter compiles the advancements of magnetic behaviors of such nanomaterials. Starting from the origin of magnetism in nanomaterials, we have systematically addressed their functionality, fabrication, and characteristics. This chapter ends with concluding remarks and the future prospects of magnetic nanomaterials.

1.1   INTRODUCTION

The unique properties of materials at the nanometer scale have been the subject of enormous research interest among researchers over the past decades. Incremental shifts in "product-performance" using these materials, for example, as fillers in plastics [1], as coatings on surfaces [2], and as ultraviolet protectants in cosmetics [3], have already been achieved. Nanostructured materials hold even more promise for the future as they could infuse almost any industry including chemistry, medicine/medical, plastics, energy, electronics, aerospace, etc. As a result of recent improvement in technologies to produce, characterize, and manipulate these types of materials, this field has seen a huge surge in endowment as well as research. Originating from the Greek word for dwarf, "nano" signifies one billionth (10^{-9}) of any scale. The word "nano" has gained great importance in science and evolved as a multidisciplinary research topic over the decades. The potential recognition of nanotechnology and research on nanostructured materials resulted from the famous speech of Nobel laureate Richard Feynman in 1959, "There's plenty of room at the bottom..." in which he emphasizes that there were no fundamental physical reasons why materials could not be fabricated by maneuvring individual atoms [4]. The term nanotechnology was first introduced in 1971 by Norio Taniguchi [5] for ultraprecision machining and the main breakthrough occurred with the invention of the scanning tunneling microscope (STM) by Binning and Rohrer in 1981 [6]. The scientific principles and properties exhibited by materials at the nanometer scale are not normally seen at the micron scale. Nonetheless, the well-known quantum effects arise in the case where the size of the system is commensurable with the de Broglie wavelengths of the electrons, phonons, and excitations propagating in them. Such fascinating and useful properties manifested by nanostructured materials can be explored for a variety of structural and nonstructural applications. Atoms and molecules are the essential building blocks of all materials and the way they are "constructed" with these building blocks directly dictate their properties. Accordingly, nanotechnology refers to the maneuvring of individual atoms, molecules, or molecular clusters into structures to create materials and devices with new or vastly different properties. Conventional materials are composed of grains varying in size anywhere from 100's of micrometers (μm) up to millimeters (mm) and in contrast, nanostructured materials, grains/crystals are in the order of 20–30 nm. However, in such cases, a grain of 1 nm size may contain only three to five atoms, depending on the atomic radii and have a well-defined effect of its physical properties. For example, the radius of crack-tip growth and propagation in bulk material is likely to be different from crack proliferation in a nanostructured material where crack and particle size are analogous. Due to these reasons, the fundamental electronic, magnetic, optical, chemical, and biological processes are also different at this level. For example, where proteins are 10–1000 nm in size, and cell walls 1–100 nm thick, their behavior on encountering a nanomaterial may be quite different from that seen in relation to larger-scale materials. Thus, nanocapsules and nanodevices may present new possibilities for drug delivery, gene therapy, and medical diagnostics. Fascinatingly, such nanostructured materials are of the same order of magnitude as the range of the various interactions that define magnetism/magnetic properties in materials and newer magnetic effects can be created. Thus, the objective of this present chapter is to explore magnetic nanostructured materials in terms of their origin, fabrication, characterization, and functionality.

1.2 COMMON TERMINOLOGY IN THE WORLD OF MAGNETISM

1.2.1 Magnetic Moment

A magnet's magnetic moment (μ_m) is a vector that characterizes the magnet's overall magnetic properties. For a bar magnet, the direction of the magnetic moment points from the magnet's south pole to its north pole [7] and the magnitude represents how strong and how far apart these poles are. In SI units, the magnetic moment is $A.m^2$. A magnet produces its own magnetic field and also responds to external magnetic fields as well. The strength of the magnetic field is proportional to the magnitude of its magnetic moment. In addition, when the magnet is put into an external magnetic field, it is subject to a torque tending to orient the magnetic moment parallel to the field [8]. The amount of this torque is proportional to the magnetic moment and the external field as well. According to the positions and orientations of the magnet and source, a magnet may also experience a force driving it in one direction or another. The magnetic field parameters at a given point in space are defined to be the magnetic field strength (H), the magnetic flux density or magnetic induction (B), and the magnetization of the material (M) [9]:

$$B = \mu_0(H + M) \tag{1.1}$$

where $\mu_0 = 4\pi \times 10^{-7}$ H/m is the permeability of free space (H = Henry, Vs/A). We can also define the magnetic field parameters as follows:

$$B = \mu_0 H(1 + \chi) = \mu H \tag{1.2}$$

where $\mu = \mu_0(1 + \chi)$ is the permeability of the material and $\chi = M/H$ is the magnetic susceptibility of the material. And the magnetic moment, μ_m, for a given volume V:

$$M = \mu_m / V \tag{1.3}$$

According to the above equations, the well-known B–H curve is shown in Figure 1.1. Based on these curves, a different class of magnetism can be explained as described hereafter.

If the external field strength increased (in a ferromagnetic solid), then at first magnetization rises slowly and then more rapidly. Finally, it levels off and reaches a constant value, for example, saturation magnetization. When H is reduced to zero, the magnetization retains a positive value, for example, remanent magnetization or remanence. It is this retained magnetization which has been explored in permanent magnets. The remanent magnetization can be removed by reversing the magnetic field strength to a value Hc (coercive field). Without an external magnetic field, orbiting electrons magnetic moments are randomly oriented and thus they mutually cancel one another. As a result, the net magnetization is zero as pointed out earlier. However, when an external field is applied, the individual magnetic vectors tend to turn into the field direction, which is counteracted by thermal agitation alignment. Based on that, different magnetic behavior can be explained as follows:

- Diamagnetism occurs due to the external magnetic field that induces a change in the magnitude of inner atomic currents. Specifically, the external field accelerates or decelerates the orbiting electrons, so that their magnetic moment is in the opposite direction from the external magnetic field.
- Antiferromagnetic materials exhibit a spontaneous alignment of moments below a critical temperature. However, the responsible neighboring atoms in

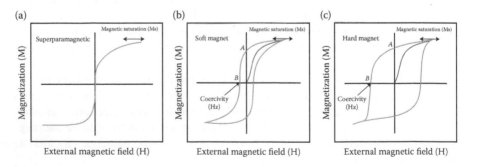

FIGURE 1.1 B–H curve: (a) nonmagnets, (b) soft magnets, and (c) hard magnets. (Adapted from V. V. Mody, A. Singh, B. Wesley, *Eur. J. Nanomed* 5(1), (2013), 11–21.)

antiferromagnetics are aligned in an antiparallel fashion. Antiferromagnetic materials are paramagnetic above Neel temperature (T_N).

- Ferrimagnets are similar to antiferromagnets in that two sublattices exist that couple through a super exchange mechanism to create an antiparallel alignment. However the magnetic moments on the ions of the sublattices are not equal and hence they do not cancel; rather a finite difference remains to leave a net magnetization. This spontaneous magnetization is defeated by the thermal energy above the Curie temperature.

1.2.1.1 Magnetic Moment in Atom

A moving electric charge is the source of a magnetic field and a magnetic moment can be associated with it. At atomic scale, magnetism results from electron motion [9]. There are two contributions to the magnetic moment from electrons:

- The orbital moment m_l, due to the motion of the electron in its orbit in the form of $m_l = -\mu_B l$, where l is the orbital angular momentum and μ_B is the Bohr magneton. The values l = 0, 1, 2, and 3 are attached to electrons in the s, p, d, and f shells, respectively.
- The spin moment $m_s = -2 \mu_B S$, where S is the spin of purely quantum origin. It can explain the spinning motion of the electron about its own axis. The spin is characterized by the value of s, which can only take two possible values, +1/2 and −1/2.

A different set of electrons rotating about the nucleus is associated with each element of the periodic table. In a given electron shell, there are 2l + 1 states available for each spin state, making a total of 2(2l + 1) states (10 for a d shell, 14 for an f shell) as shown in Figure 1.2. The distribution of the electrons over the available orbits for a given shell aims to minimize the energy associated with their mutual electrostatic repulsion. This energy contains the "exchange term," which depends on the spin and is responsible for magnetism in the matter. The electron shells fill up progressively according to the Hund rules. The first of these rules says that the total spin S associated with all the electrons in the same electron shell is maximal by taking consideration of Pauli's exclusion principle. Thus all the electron spins tend to be parallel to one another. The second Hund rule states that the total orbital angular momentum L is maximum, with the restriction that the first rule takes precedence. The motion of the nucleus creates a magnetic field acting on the spin moment. This field is the

I =	-3	-2	-1	0	1	2	3
S = +½	↑	↑	↑	↑	↑	↑	↑
S = -½	↓	↓					

FIGURE 1.2 Itinerant electron system at dysprosium. (Redrawn based on D. Givord, *Nanomaterials and Nanochemistry*, C. Bréchignac, P. Houdy, and M. Lahmani (eds.), Springer-Verlag, Berlin Heidelberg, Germany, 2007, pp. 101–134.)

source of the spin–orbit coupling between the orbital and spin moments. The energy of the spin–orbit coupling is given by

$$E_{spin-orbit} = \lambda LS \tag{1.4}$$

where λ is the spin–orbit coupling constant. The orbital and spin contributions associated with all the electrons in a full (closed) electron shell tend to balance one another so that the resulting magnetic moment is zero.

In solid state, the mixing of orbitals that result from covalence or the formation of energy bands often leads to the disappearance of the atomic magnetic moment. There are only two series of elements in the periodic table that retain magnetism in solid state due to the internal structure of an unfilled electron shell that is responsible for preserving magnetic effects in the solid state:

1. The first series (commonly known as rare earth elements) goes from cerium to lutetium, and corresponds to the progressive filling of the 4f shell. The electrons, localized in their atomic orbits, are essentially subject to the same exchange interactions as in an isolated atom and the magnetic moment as defined by the Hund rules.
2. The second series contains iron, cobalt, and nickel, and corresponds to the filling up of the 3d shell. In insulating systems such as oxides, the 3d electrons are localized and the magnetic moment is still defined by the Hund rules. In metals, the 3d electrons are said to be itinerant, forming an energy band with width of the order of 5 eV. The magnetic moment is no longer produced strictly on the atomic scale.

Stoner showed that an alternative approach can be used to describe most of the observed magnetic behavior [9]. The whole set of electrons is considered, distributed over two half-bands, each containing 5N states, where N is the total number of atoms (Figure 1.2), and characterized by the value of the electron spins +1/2 ($\uparrow$) and –1/2 ($\downarrow$). In the absence of exchange interactions, the minimum energy corresponds to the equal filling of the two half-bands and the system is nonmagnetic. Under the effect of exchange interactions, one half-band will be favored because of the tendency for the electron spins to line up. The state of the system is defined by competition between these two terms. In Cr, Mn, Fe, Co, and Ni, the minimum energy configuration is magnetic.

Despite the itinerant nature of the 3d electrons, they remain essentially localized at each atomic site and at the end of the day many properties of 3d metals can be described by treating the magnetic moments as atomic. However, the distribution of moments considered at two different times will not be strictly the same, since the electrons can hop from one atom to another. The nonintegral values of the magnetic moments per atom (2.2 μ_B for Fe, 1.74 μ_B for Co, 0.60 μ_B for Ni) basically reflect the fact that they correspond to an average value taken over all the atoms making up the given system [9]. A schematic of the orbital and spin magnetic moment is shown in Figure 1.3.

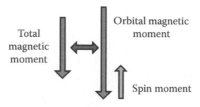

FIGURE 1.3 Schematic of orbital and spin magnetic moment. (Redrawn based on from High-accuracy magnetic property measurement method by separating spin and orbital magnetic moments. 04 Mar, 2013 BL08W (High Energy Inelastic Scattering) Japan Synchrotron Radiation Research Institute (JASRI), University of Hyogo.)

1.2.1.2 Magnetic Moments in Clusters

The first measurements of the magnetic moments of free clusters of 3d ferromagnetic metals, such as iron and cobalt, were made on clusters consisting of 100–500 atoms (diameter 1–2 nm) [7,8]. These clusters were obtained by laser ablation and subsequent condensation of vaporized atoms in the inert atmosphere of helium. The magnetic moments of the clusters were calculated from the extent to which they are deflected when passed through a region of magnetic field gradient similar to the Stern-Gerlach experiment [9]. In order to analyze the results, it should be taken into account that the clusters are superparamagnetic. For iron, the results suggest an average atomic moment of 2.2 μ_B, equal to the value in the bulk solid. For cobalt, the experiments lead to a value of 2.08 μ_B compared to 1.72 μ_B in the bulk. The theoretical band structure of a cluster comprising 15 iron atoms, separated into the successive contributions of the various atomic layers or shells has been calculated by Pastor et al. [12] as shown in Figure 1.4. Qualitatively, the energy distribution of the states in each half-band has two lobes, as in the bulk.

The lower energy lobe is associated with delocalized bonding states, while the higher energy lobe comes from localized antibonding states. However, the band width is less than in the clusters. Indeed, the band states are formed from linear combinations of atomic states. The more states involved, the broader is the energy distribution of the states. When a magnetic moment is formed, a certain number of electrons are transferred from one half-band to another, thus the energy loss is smaller in the case of a narrow band. The narrowing of the d band in clusters is responsible for the 3d moment to increase. The value of the magnetic moment of atoms in a cluster is also affected further by secondary mechanisms [9]. There is a reduction in interatomic distances which leads to an increase in the itinerancy of the electrons and hence to an increased mixing of orbitals and result in a reduction of the atomic magnetic moment. On the other hand, the reduced symmetry of the environment of the atoms causes a reduction in band width that favors an increased magnetic moment.

Measurements of circular dichroism using x-rays synchrotron facilities make it possible to examine the spin and orbital moments separately. Apart from the enhancement of the spin moment, a similar enhancement of the orbital moment has been detected for small iron clusters [8]. In thin films, Bruno [13] has shown that the enhancement of the orbital moment and the increase in magneto-crystalline

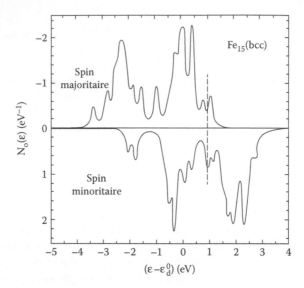

FIGURE 1.4 Theoretical band structure of a cluster comprising 15 iron atoms. (Adapted from G. M. Pastor, J. Dorantes-Davila, K. H. Bennemann, *Phys. Rev. Lett.* 40, 1989, 7642–7451.)

anisotropy which are related to the local reduction of symmetry. However, in clusters, all moments are coupled in parallel by exchange. The value of the orbital moment depends on the orientation of the moments relative to the local easy axis of magnetization. Assuming the clusters to be spherical, the increase in orbital moment should vanish by symmetry. Two phenomena explain the existence of strong orbital moments revealed experimentally: First, there is an anisotropic relaxation which breaks the spherical symmetry of the clusters and second, the cluster atoms are arranged into successive shells. However, the cancelling effects only come fully into effect for systems comprising filled (closed) atomic shells.

1.2.2 MAGNETIC ORDER

The main interactions among electrons in matter are due to electrostatic repulsion. In quantum mechanics the "exchange term," results from the indistinguishability of the electrons. As Heisenberg showed, magnetism then arises naturally as soon as the Pauli exclusion principle is taken into account. This states that two electrons cannot occupy the same quantum state defined by the space and spin variables, and thus requires the wave function for the electron ensemble to be antisymmetric in those variables. This in turn implies that the interaction energy among electrons depends on their spin states. The electrons in different atomic sites affect one another by exchange interactions, a magnetic coupling exists between the atomic moments of different atoms, and this is the source of magnetic order in a material. The exact nature of the coupling between moments depends a great deal on the elements that are present, and also on the crystallographic arrangement of the atoms. When all

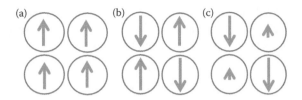

FIGURE 1.5 Schematic of magnetic structures: (a) ferromagnetic, (b) antiferromagnetic and (c) ferrimagnetic structure. (Drawn based on N. Taniguchi, *Precision Engineering*, 16(1), 1994, 5–24; V. V. Mody, A. Singh, B. Wesley, *Eur. J. Nanomed.* 5(1), (2013), 11–21; High-accuracy magnetic property measurement method by separating spin and orbital magnetic moments. 04 Mar, 2013 BL08W (High Energy Inelastic Scattering) Japan Synchrotron Radiation Research Institute (JASRI), University of Hyogo.)

the moments are parallel, the structure is ferromagnetic as shown in Figure 1.5a. In this case, the moment per unit volume of matter is spontaneous magnetization (Ms). In an antiferromagnetic arrangement (Figure 1.5b), the moments are organized into two groups. Within the same group (sublattice), the moments are all parallel to one another. However, the moments of the two sublattices couple in an antiparallel manner and the resulting magnetization is zero. Thus in ferrimagnetic systems, the numbers of atoms or the value of the magnetic moments are not the same in each sublattice. This time, cancellation is not complete and there remains some spontaneous magnetization (Figure 1.5c).

The exchange energy as described earlier can be expressed as

$$E_{exch} = -\frac{1}{2}\sum_{i,j\neq i}J_{ij}S_iS_j \tag{1.5}$$

where J_{ij} is the exchange integral, representing the coupling force between spins, and S_i and S_j are the spins carried by atoms i and j. The exchange interactions are assumed to be very short range. If all the atoms are identical and considering only nearest neighbors, J_{ij} can be replaced by J, and S_i and S_j by S. The exchange energy per atom then becomes

$$e_{exch} = -\frac{1}{2}zJsS^2 \tag{1.6}$$

where z is the number of nearest neighbors of a given atom. When J is positive, that is when the minimal energy configuration corresponds to parallel coupling between all spins, it gives rise to ferromagnetism. More generally, the exchange interactions J_{ij} can vary for one atom I depending on its neighbor's j and from one atom i to another. When there are negative terms J_{ij}, they then favor antiparallel coupling between moments i and j which are required to explain nonferromagnetic structures. Equations 1.5 and 1.6 assumed a temperature of 0 K. At nonzero temperatures, thermal vibrations tend to destroy magnetic order. At a certain temperature (Curie temperature, T_C, in ferromagnetic and Neel temperature, T_N, in ferromagnetic and antiferromagnetic materials) a transition occurs from the ordered magnetic state.

The effects of thermal vibrations can be described using the molecular field model. Let us consider a ferromagnetic material. Since the spin is in fact proportional to the magnetic moment, valid at 0 K, it can be rewritten as

$$e_{exch} = -\frac{1}{2}\mu_{at}H_{m,0} \tag{1.7}$$

where μ_{at} is the atomic magnetic moment, proportional to the spin, and $H_{m,0}$ is a fictitious magnetic field representing the exchange interactions and called the molecular field, first introduced by Weiss:

$$H_{m,0} = nM_{s,0} \tag{1.8}$$

In this expression, n is the molecular field coefficient and $M_{s,0}$ the spontaneous magnetization at 0 K, that is magnetization at absolute saturation. A central assumption in the molecular field model is that, when the temperature increases, the properties of the matter remain strictly homogeneous. The molecular field at temperature T can be expressed as a function of spontaneous magnetization at this temperature:

$$H_m = nM_s \tag{1.9}$$

It can then be shown that the spontaneous magnetization decreases steadily as the temperature increases (Figure 1.6) and disappears at Curie temperature, T_C:

$$T_c = \frac{zJS^2}{3k} = \frac{n\mu_{at}^2}{3k} \tag{1.10}$$

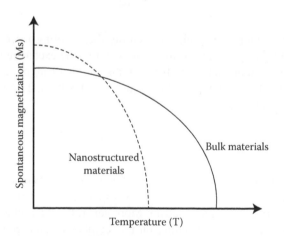

FIGURE 1.6 Schematic of temperature dependence on spontaneous magnetization for different materials. (Drawn based on N. Taniguchi, *Precision Engineering*, 16(1), 1994, 5–24; V. V. Mody, A. Singh, B. Wesley, *Eur. J. Nanomed.* 5(1), (2013), 11–21; High-accuracy magnetic property measurement method by separating spin and orbital magnetic moments. 04 Mar, 2013 BL08W (High Energy Inelastic Scattering) Japan Synchrotron Radiation Research Institute (JASRI), University of Hyogo.)

where k is the Boltzmann constant. The value of the Curie temperature provides a measure of the strength of the exchange interactions. As originally explained by D. Givord [9], the molecular field model can be utilized to describe the behavior of antiferromagnetic and ferrimagnetic materials. To go beyond this model, the disorder of the moments is considered to be produced at finite temperatures by collective excitations known as spin waves. As far as magnetic moments are concerned, these are perfectly analogous to the collective vibrations of atoms known as phonons. For the same entropy as in the molecular field model, the loss of exchange energy is lower. When the wavelength λ of a spin wave increases, the angle between consecutive magnetic moments within the wave reduced. In a ferromagnetic material, the energy E_{SW} of the spin waves can be written as

$$E_{SW} = D_q^2 \tag{1.11}$$

where D is the rigidity constant of the spin waves and $q = 2\pi/\lambda$ is the wave vector.

1.2.2.1 Magnetic Moment in Rare Earth Elements

In rare earth elements, localized 4f electrons are located in the inner shell that is well protected from the effects of the environment by the presence of the outer 5s, 5d, and 6s shells. Interactions between 4f electrons located on different atoms are therefore negligible. In fact, an indirect form of exchange coupling does exist in these systems, involving electrons in the outer 5d and 5s shells as intermediaries. This coupling, known as RKKY (Ruderman-Kittel-Kasuya-Yoshida) coupling, is relatively weak and only extends to 1–2 nm. In contrast, in other exchange coupling mechanisms, the range of the interactions never exceeds one or two times the interatomic distance. The magnetic order temperatures are always lower than room temperature. In 3d transition metal systems, the exchange coupling is related to the itinerant nature of the electrons. When they hop from atom 1 to atom 2, the electrons conserve their spin. They then interact with electrons present on atom 2, their spin tending to lie parallel with these, according to the first Hund rule. The resulting ferromagnetic coupling force is explained by the intra-atomic nature of the exchange terms. In iron, cobalt, and nickel, the Curie temperature is much higher than room temperature, reaching 1380 K in cobalt.

1.2.2.2 Magnetic Moment in Transition Metal Compounds

In insulating transition metal compounds, the exchange mechanism involves a mixture of the 3d and p wave functions of the anions, such as oxygen. The hybridization of the wave functions depends significantly on crystallographic orientation. Hence, the mechanism known as superexchange gives rise to the coupling of different signs and the magnetic arrangements are very often antiferromagnetic (MnO, CoO, NiO) or ferrimagnetic (γ–Fe_2O_3, Fe_3O_4). In some cases, several coupling mechanisms compete with each other and nonaligned magnetic arrangements of moments can exist.

1.2.2.3 Magnetic Moment in Ferromagnetic Materials

The ferromagnetic nature of the bulk solid is conserved in clusters of Fe, Co, or Ni. This result can be related to the fact that, in the 3d shells of metals, the interaction mechanism between magnetic moments is insensitive to minor changes in the atomic environment. In nanoparticles of transition metal oxides, measurements reveal a reduction in the average magnetization. This happens for magnetite nanoparticles (γ–Fe_2O_3), a collinear ferrimagnetic material in the bulk. For an insulating system, such a reduction in magnetization cannot be attributed to a lower magnetic moment, since the latter does not depend on the size of the system. In fact, Mössbauer spectroscopy on the Fe57 nucleus reveals noncollinear arrangements of the moments [12]. Such arrangements are characteristic of atoms located in a low-symmetry environment and subject to magnetic interactions of various signs and supported by numerical simulation. Noncollinear arrangements of the same origin also occur in nanoparticles of systems that are antiferromagnetic in the bulk state, such as NiO [13] and carry a small magnetic moment.

1.2.3 MAGNETIC ANISOTROPY

If the magnetic susceptibility of a material depends on the direction in which it is measured, then the situation is known as magnetic anisotropy [13]. When magnetic anisotropy exists, the total magnetization of a ferromagnet M will prefer to lie along a special direction widely known as the "easy axis." The energy associated with this alignment is "anisotropy energy":

$$E_a = K\sin^2\theta \qquad\qquad (1.12)$$

where θ is the angle between M and "easy axis," and K is the anisotropy constant. Most materials contain some type of anisotropy affecting magnetization behavior. The most common forms are: (i) crystal anisotropy, (ii) shape anisotropy, (iii) stress anisotropy, (iv) externally induced anisotropy, and (v) exchange anisotropy. Among these, crystalline and shape anisotropy play the most important role on the magnetic behavior of nanostructured materials.

1.2.3.1 Crystalline Anisotropy

In crystalline anisotropy, the ease of obtaining saturation magnetization is different for different crystallographic directions. For example, in a single crystal of iron Ms can most easily be obtained in the [100] direction compared to the [110] direction and are most difficult to attain for [111] directions as shown in Figure 1.7. The [100] direction is the easy direction (easy axis). Both iron and nickel are cubic and have three different axes, whereas cobalt is hexagonal with a single easy axis perpendicular to the hexagonal symmetry (Figure 1.7). The physical origin of crystalline anisotropy is the coupling of the electron spins which carry the magnetic moment to the electronic orbit and couple it to the lattice. A polycrystalline sample with no preferred grain orientation has no net crystal anisotropy due to averaging over all orientations.

The magnetic anisotropy can be one or two orders of magnitude stronger in materials composed of 4f elements than in those made from 3d elements as it is much less sensitive to the background crystalline field. In addition, the spin–orbit coupling force, approximately proportional to the square of the atomic number, is much

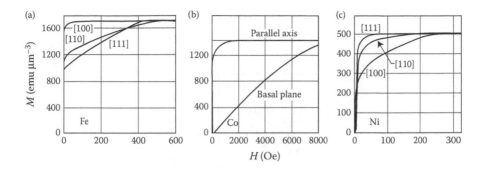

FIGURE 1.7 Magnetization curves for single crystals of (a) iron, (b) cobalt and (c) nickel along different directions. (Adapted from K. J. Klabunde, *Nanoscale Materials in Chemistry*, Wiley-Interscience, 2001.)

stronger in rare earth elements than in 3d transition metals. Thus both spin–orbit coupling and the crystalline field influence the overall performance. In rare earth materials, the spin–orbit coupling dominates the crystalline field, at maximum value of the orbital moment, aligned with the spin moment. Obviously a strong orbital moment indicates a highly asymmetric electron orbit. The energy of an orbit placed in a given environment depends more on its orientation as it becomes more asymmetric. This is basically why rare earth compounds and alloys often exhibit strong crystalline anisotropy. A schematic representation of the hard and easy axis with the corresponding crystallographic structure is shown in Figure 1.8.

For materials made from transition elements, the crystalline field must be considered before the spin–orbit coupling. Electron orbits adopt shapes that follow from their surrounding symmetry and the associated orbital moment is often small, for example, zero for cubic symmetry. The spin–orbit coupling tends to induce an orbital moment in the direction of the spin moment, but the associated distortion of the electron orbit is not favorable for the crystalline field. The orbital moment induced by this mechanism is thus small, for example, a few percent of the spin moment in the metals Fe, Co, and Ni, and the magnetic anisotropy, which is roughly proportional to it, is likewise small.

1.2.3.2 Shape Anisotropy

It is easier to induce a magnetization along a long direction of a nonspherical piece of material than along a short direction as the demagnetizing field is less in the long direction, because of further apart induced poles. Thus, a smaller applied field will negate the internal demagnetizing field. For a pro-late spheroid with major axis c greater than the other two (equal) axes of length a, the shape anisotropy constant is

$$K_s = \frac{1}{2}(N_a - N_c)M^2 \tag{1.13}$$

where N_a and N_c are demagnetization factors. For spheres, $N_a = N_c$ as a = c. It can be shown that $2N_a + N_c = 4\pi$; then the limit c ≫ a, that is, a long rod, $K_s = 2\pi M^2$. Thus a long rod of iron with Ms = 1714 A m^{-1} would have a shape anisotropy constant of

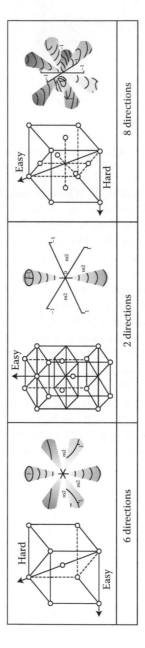

FIGURE 1.8 Schematic representation of hard and easy axis with corresponding crystallographic structure. (Adapted from B. D. Cullity, *Introduction to Magnetic Materials*, Addison-Wesley Publishing Company, 1972, pp. 214–250.)

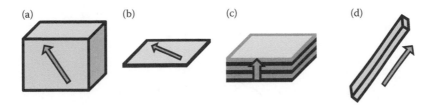

FIGURE 1.9 Schematic of different forms of shape anisotropy. (Redrawn based on K. M. Krishnan et al., *J. Mater. Sci.* 41, 2006, 793–815.)

$K_s = 1.85 \times 10^7$ erg cm^{-3}. This is significantly greater than the crystal anisotropy so we see that shape anisotropy can be very important for nonspherical particles. A schematic of the different forms of shape anisotropy is shown in Figure 1.9.

1.2.3.3 Other Anisotropy

Stress anisotropy results from internal or external stresses that occur due to rapid cooling, application of external pressure, etc. Anisotropy may also be induced by annealing in a magnetic field, plastic deformation, or by ion beam irradiation. Exchange anisotropy occurs when a ferromagnet is in close proximity to an antiferromagnet or ferromagnet. Magnetic coupling at the interface of the two materials can create a preferential direction in the ferromagnetic phase, which takes the form of a unidirectional anisotropy. This type of anisotropy is most commonly observed when an antiferromagnetic or ferromagnetic oxide forms around a ferromagnetic core.

The value of the magnetic anisotropy for nanoparticles can be deduced from the so-called blocking temperature. In cobalt nanoparticles, the anisotropy obtained is equal to 3×10^7 J/m^3, which are almost two orders of magnitude greater than in the bulk [15]. In spherical symmetry, even if the magnetic anisotropy of a surface atom is high, the total anisotropy should disappear due to symmetry. However, the strong anisotropy in nanoparticles must be associated with anisotropic relaxation and the fact that the outer atomic shell of the nanoparticles is not filled.

1.2.4 Coercivity

Coercivity is the magnetic field strength necessary to demagnetize a ferromagnetic material that is magnetized to saturation. It is measured in ampere per meter (A/m), or traditionally in oersted (1 Oe = 79.578 A/m). Let us imagine that a new field is applied to a sample in the saturated magnetization state, but in the opposite direction to the initially applied external field.

The magnetization should grow in the opposite direction in order to minimize the Zeeman energy as shown in Figure 1.10 [9] where the Y axis represents energy and X axis represents θ, the angle between the applied field and the Z axis.

However, the initial nucleation of a new domain in the saturated state requires the system to overcome an energy barrier and a moment taking part in this process to go through a situation in which it lies perpendicular to the easy axis of magnetization. As the situation is unfavorable, the anisotropy energy is maximum. Thus coercivity

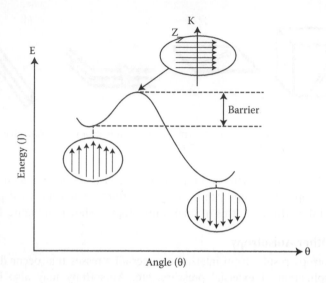

FIGURE 1.10 Relation of coercivity with Zeeman energy. (Redrawn based on E. Kneller, R. Hawig, *IEEE Trans. Mag.* 27, 1991, 3588–3592.)

of a material is its capacity to resist the effect of an applied field. When the applied field is exactly antiparallel to the initial magnetization direction z, the coercive field H_c is equal to the anisotropy field:

$$H_c = \frac{2K}{M_s} = H_A \qquad (1.14)$$

As soon as nucleation occurs, magnetic moments reverse in n as known as coherent rotation according to Stoner and Wohlfarth [7]. In general coercive field is a function of the angle θ between the applied field and the z axis (with the convention that $\theta = 0$ corresponds to a field antiparallel with the initial direction of the moments):

$$H_c(\theta) = H_{SW}(\theta) = H_A/(\sin^{2/3}\theta + \cos^{2/3}\theta)^{3/2} \qquad (1.15)$$

As evident from Equations 1.14 and 1.15, magnetic anisotropy underlies the coercivity. However, in reality the coercive field is much weaker than that of the anisotropy field due to the presence of structural defects. As anisotropy is reduced at defect sites, nucleation is facilitated and accompanied by the formation of a wall that separates the nucleus from the rest of the material with opposite magnetization. Complete reversal of the magnetization takes place by the propagation of the wall through the material. In this case, the angular variation of the coercive field is very different from what would correspond to coherent rotation and is expressed as

$$H_c(\theta) \approx \frac{H_c(0)}{\cos\theta} \qquad (1.16)$$

where $H_c(0)$ is the coercive field at $\theta = 0$. Thus the projection of the applied field in the initial magnetization direction plays an active role in the reversal process. As the coercive field is much weaker than the anisotropy field, the reversible rotation of the moments toward the applied field can be neglected. Some nanostructured alloys are heterogeneous, that it, they contain both magnetically soft and hard particles [18].

Due to the exchange coupling between soft and hard particles, the reversal of magnetization in the soft particles requires a magnetization configuration characterized by progressive rotation, similar to what happens in a domain wall. However, this can only occur over the size of these particles (10–20 nm), well below the equilibrium thickness of a domain wall. As already explained, the energy of a confined wall is greater than the energy of a wall with equilibrium thickness. Due to this reason the soft phase therefore resists the formation of a confined wall and hence becomes coercive. The experimental value of the coercive field is in qualitative agreement with the theoretical value [17,19].

1.3   CONCEPT OF NANOSTRUCTURED MAGNETIC MATERIALS

On October 18, 2011, the European Commission adopted the following definition of a nanomaterial [20]: "A natural, incidental or manufactured material containing particles, in an unbound state or as an aggregate or as an agglomerate and where, for 50% or more of the particles in the number size distribution, one or more external dimensions is in the size range 1–100 nm. In specific cases and where warranted by concerns for the environment, health, safety or competitiveness the number size distribution threshold of 50% may be replaced by a threshold between 1 and 50%." However, this differs from the definition adopted by the International Organization for Standardization (ISO): "Material with any external dimension in the nanoscale or having internal structure in the nanoscale." Nanoscale is, in turn, defined as size range from approximately 1–100 nm [21].

Nanostructured materials are not just another step toward miniaturization. This is mostly due to the fact that, the nanometer scale sits somewhere midway between the atomic scale and microscale, where quantum phenomena are prevalent compared to bulk materials. A bulk material should have constant physical properties in contrast to the nanometer scale. Size-dependent properties, such as quantum confinement in semiconductor particles, surface plasmon resonance in some metal particles, and superparamagnetism in magnetic materials are observed. In this level, some material properties are governed by the laws of atomic physics and do not behave like bulk materials. Research based on nanostructured materials takes a materials science-based approach to nanotechnology through a "top-down" or "bottom-up" approach to fabricate them. Nanostructured materials can be metals, ceramics, polymeric materials, or composite materials.

Nanostructured magnetic materials are a particular class of nanostructured materials which can be manipulated by the application of magnetic field. Such particles commonly consist of magnetic elements in it such as iron, nickel, cobalt, and their chemical compounds. The magnetic nanoparticles have been the focus of recent research activity as they possess attractive properties, which could be of potential use in the field of catalysis including nanomaterial-based catalysts [22], medicine

[23], magnetic resonance imaging [24], magnetic particle imaging [25], data storage [26], environmental remediation [27], nanofluids [28], optical filters [29], defect sensors [30], and cation sensors [31] to mention a few. One of the most important magnetic properties at nanometer scale is superparamagnetism which can lead to particles with much higher magnetic susceptibilities than in traditional paramagnets.

1.3.1 BRIEF HISTORY OF NANOSTRUCTURED MAGNETIC MATERIALS

Nanostructured materials have been used by humans for hundreds of years without knowing of their existence. For example, the beautiful ruby red color of glass, the decorative glaze (luster) found on some medieval glass and pottery contain metallic spherical nanoparticles dispersed in a complex way and exhibit characteristic optical properties. The techniques used to produce these materials were considered trade secrets at the time, and are not fully understood even now. The development of nanostructured materials and nanotechnology as a whole has been pushed forward by the development of sophisticated transmission electron microscopy (TEM) and scanning tunneling microscopy (STM). Such techniques enable researchers to carry out the characterization of such materials. Hollow carbon spheres, commonly known as buckyballs or fullerenes (C_{60}), were discovered in the mid-1980s [32]. The C_{60} (60 carbon atoms chemically bonded together in a ball-shaped molecule) buckyballs inspired research that led to the fabrication of carbon nanofibers, with diameters under 100 nm. Scientists at IBM had managed to position individual xenon atoms on a nickel surface to spell out the company logo, using scanning tunneling microscopy probes, as a demonstration of the extraordinary new technology [33]. In 1991, S. Iijima of NEC in Japan reported the first observation of carbon nanotubes, which are now produced by a number of companies in commercial quantities [34].

1.3.2 COMMON NANOSTRUCTURED MAGNETIC MATERIALS

Magnetic nanostructured materials are extremely small particles of materials that respond to magnetic fields and are classified into three broad categories:

1. Systems with nanometric dimensions such as nanoparticles, nanorods, nanotubes, etc.
2. Systems with macroscopic dimensions, but made up of crystallites with nanometric dimensions such as nanostructured/nanopatterned materials, core/shell nanoparticles, nanorods, nanotubes, etc.
3. A combination of both categories above.

As stated earlier, nanoparticles in general exist on scales larger than the atom but still smaller than traditional bulk materials. Thus, a combination of both classical and quantum mechanics must be used to analyze this transitional region of scale and difference from bulk materials by:

- Intrinsic properties of isolated clusters
- Coercively and remnant magnetization dependence on structure

- Specific properties resulting from coupling between constitutive nanocrystallites

Three different kinds of magnetic nanoparticles, namely, (i) oxides and carbides, (ii) metallic, and (iii) core-shell are mostly available, as described next.

1.3.2.1 Oxides and Carbides Nanoparticles

The most common magnetic nanoparticles are ferrites (Fe_2O_3), including both binary and complex oxides of iron. Among several crystalline modifications of ferrites, there are two magnetic phases: rhombohedral α–Fe_2O_3 (hematite) and cubic γ–Fe_2O_3 (magnetite) that are more prominent. In α–Fe_2O_3 all Fe^{3+} ions have an octahedral coordination, whereas in γ–Fe_2O_3 the structure is cation-deficient AB_2O_4-type spinel, where metal atoms A and B occur in tetrahedral and octahedral environments, respectively. Fe_2O_3 is a reddish brown, inorganic compound which is paramagnetic in nature and also one of the three main oxides of iron, while the other two are FeO and Fe_3O_4. The cubic spinel Fe_3O_4 (magnetite) is ferrimagnetic at temperatures below 858 K and the average size of 3.5 nm have been prepared by thermal decomposition of $Fe_2(C_2O_4)_3 \cdot 5H_2O$ at T > 400°C [35]. The controlled reduction of ultradispersed α –Fe_2O_3 in a hydrogen stream at 723 K (15 min) is a more reliable method of synthesis of Fe_3O_4 nanoparticles with particle size in the range of 13 nm and were prepared in this way [36]. FeO (wustite) is antiferromagnetic (T_C = 185 K) in the bulk state and mainly prepared by comilling of Fe and Fe_2O_3 powders in a definite ratio which gave nanoparticles (5–10 nm) consisting of FeO and Fe [37]. The Fe_3O_4, which also occurs naturally as mineral magnetite, is also superparamagnetic in nature. Among the known hydroxides, orthorhombic α–FeO(OH) (goethite) is antiferromagnetic in the bulk state and has T_C = 393 K [38]. β–FeO(OH) (akagenite) is paramagnetic at 300 K [39], γ–FeO(OH) (lipidocrokite) is paramagnetic at 300 K, and δ–FeO(OH) (ferroxyhite) is ferromagnetic [40]. Iron carbide (Fe_xC_y) is often present in Fe-containing nanoparticles. It has been shown by Mössbauer spectroscopy [41] that thermal decomposition of $Fe(CO)_5$ (at 353 K) on a carbon support forms $Fe_{78}C_{22}$ nanoparticles with average size of 3.9 nm. Ferro-fluids or so-called magnetic liquids are suspensions of colloid magnetic particles stabilized by surfactants in liquid media. The magnetic phase in ferrofluids can be represented by magnetite [42], ferrites [43], and Fe_xC_y particles resulting from the thermal decomposition of $Fe(CO)_5$ [44]. Decalin or silicone oil is the usual liquid phase. The dimensions of magnetic particles are in the range of 5–10 nm. The commercial magnetic liquids most often contain magnetite [45,46]. Magnetic liquids with a Curie point below the boiling point has been reported [43,47,48]. Apart from the "classical" ferroliquids, ferrofluid emulsions in which oil drops containing a magnetic phase are dispersed in water by means of surfactants are also reported in the literature [49]. A TEM micrograph of such materials is shown in Figure 1.11 [50]. The preparation of lyotropic ferronematics stable for several months with a high content (up to 1 vol.%) of the magnetic component, γ–Fe_2O_3 nanoparticles with average size of 6 nm was reported [50]. Once the ferrite particles become smaller than 128 nm [51] they become superparamagnetic which prevents self-agglomeration as they exhibit their magnetic behavior only when an external magnetic field is applied. In the absence of an external magnetic field, the

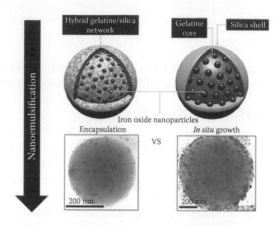

FIGURE 1.11   TEM image of nanoparticles: nanoparticles in vial (top–right corner insert) and STEM image of nanoparticles (bottom-left corner). (Adapted from J. Allouche et al., *Nanomaterials* 4(3), 2014, 612–627.)

remanence falls back to zero. Just like nonmagnetic oxide nanoparticles, the surface of ferrite nanoparticles is often modified by surfactants, silicones, or phosphoric acid derivatives to increase their stability in solution [52].

Cubic cobalt oxide (CoO) is antiferromagnetic with $T_N = 291$ K that has played an important role in the discovery of "exchange shift" of the hysteresis curve [53,54]. This was first found for samples with oxidized Co nanoparticles [55]. The dependency of Neel temperature (T_N) on the particle size was obtained during a study of CoO nanoparticles dispersed in LiF matrix [56]. As the particle size decreased from 3 to 2 nm, T_N decreased from 170 to 55 K, which can be explained satisfactorily in terms of the mean-field theory [57]. Apparently, the presence of an oxide layer on cobalt was identified. The Co_3O_4 nanoparticles (cubic spinel) with sizes of 15–19 nm dispersed in an amorphous silicon matrix exhibited ferromagnetic properties at temperatures below 33 K (for bulk samples, $T_N = 30$ K) [58]. A method for controlled synthesis of Co_3O_4 cubic nanocrystallites (10–100 nm) has been developed [59].

The amount of research devoted to Ni nanoparticles is small, unlike Co and Fe nanoparticles. In addition to the conventional metal evaporation methods [60] thermal decomposition of organonickel compounds like $Ni(CO)_4$, $Ni(C_5H_5)_2$ [61], and $Ni(COD)_2$ (COD is cyclo-octadiene) [62], and the reduction of $NiBr_2$ in the presence of PPh_3 (triphenylphosphine) are some common ways to get Ni nanoparticles [63]. In the latter case, Ni nanoparticles with average 3 nm thin oxide film are formed.

The advantage of oxide nanoparticles is that they can be easily functionalized through the attachment of organic or biological molecules to the particle surface which can increase the selectivity and bonding strength of the particle and target.

1.3.2.2   Metallic Nanoparticles

Metallic nanoparticles are pyrophoric and reactive to oxidizing agents to various degrees, which makes their handling difficult as unwanted side reactions occur. The preparation of nanoparticles consisting of pure iron is a complicated task, because they always contain oxides, carbides, and other impurities. The α–Fe nanoparticles with a body-centered cubic (BCC) lattice could attain an average size of about 10 nm. In the phase diagram of bulk Fe, γ–Fe (FCC) exists (at the ambient pressure) in the temperature range of 1183–1663 K, that is, above the Curie point (1096 K). In some special alloys, this phase, which exhibits antiferromagnetic properties ($T_N = 40$–67 K), was observed at room temperature [64–66]. Therefore, most of the metallic nanoparticles are actually metallic alloys. Some of the most common metallic alloyed nanoparticles are as follows:

1. *Fe–Co alloys:* The saturation magnetization of Fe–Co alloys reaches a maximum at Co content of 35%. In addition, other magnetic characteristics of these metals also increase when they are alloyed. Therefore, Fe_xCo_y nanoparticles attract considerable attention. A TEM micrograph of Co nanoparticles with a graphene layer is shown in Figure 1.12.

   Thus Fe, Co, and Fe–Co (20 at.%, 40 at.%, 60 at.%, 80 at.%) nanoparticles (40–51 nm) [67] with a structure similar to the corresponding bulk phases could be prepared in a stream of hydrogen plasma. The Fe–Co particles could reach a maximum saturation magnetization (61 cm^3 g/L) at 40 at.% of Co and a maximum coercive force (860 Œ) is attained at 80 at.% of Co.

2. *Fe–Ni alloys:* The bulk samples of the iron–nickel alloys are either nonmagnetic or soft ferromagnets. For example, permalloys contain more than 30% of Ni and various doping additives. When the content of nickel is about 30%, its magnetic properties approach the properties of Invar (36% of Ni, 64% of Fe, about 0.05% of C). The Fe–Ni nanoparticles have a much lower saturation magnetization than the corresponding bulk samples over the whole concentration range [68] whereas an alloy containing 37% of Ni has a low T_C and FCC structure. The alloy consists of super paramagnetic nanoparticles (12–80 nm) over a broad temperature range [69]. Theoretical prediction provides a complex magnetic structure for these Fe–Ni particles (clusters) [70].

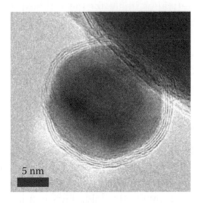

FIGURE 1.12 Cobalt nanoparticle with graphene layers. (Adapted from R. N. Grass et al., *Angew. Chem. Int. Ed.* 46(26), 2007, 4909–4912.)

3. *Fe–Pt alloys:* Fe–Pt nanoparticles have received much attention in recent years due to the substantial increase in data storage density [71]. The Fe–Pt nanoparticles (6 nm) with a narrow size distribution could be prepared by joint thermolysis of $Fe(CO)_5$ and Pt in the presence of oleic acid and oleylamine and Hexade-cane-1,2-diol as the reducing aid for Pt^{2+}. Further heating resulted in the formation of a protective film, from the products of thermal decomposition, which does not significantly change the particle size. These particles can be arranged to form regular films called "colloid crystals" [72]. Nonetheless, unusual magnetic characteristics was observed for the FePt/Fe_3Pt nanocomposite, after the reaction of FePt nanoparticles with Fe_3O_4 followed by heating of the samples at 650°C in an Ar^+ and 5% H_2 stream [73].

1.3.2.3 Core-Shell Nanoparticles

The metallic core of magnetic nanoparticles may be passivated by gentle oxidation, surfactants, polymers, and precious metals [51]. In oxygen-rich environment, Co nanoparticles form an antiferromagnetic CoO layer on the surface of the Co nanoparticle. Recently, research has explored the synthesis and exchange bias effect in these Co–CoO core-shell nanoparticles with a gold outer shell [74]. In general, nanoparticles with a magnetic core consisting either of elementary iron or cobalt with a nonreactive shell made of graphene offers enhanced magnetic properties compared to ferrite or elemental nanoparticles in terms of:

- Higher magnetization
- Higher stability in acidic and basic solution as well as organic solvents
- Chemistry on the graphene surface via methods already known for carbon nanotubes [75–77]

In addition, core-shell nanoparticles may also form external functional groups. The unique magnetic properties are usually inherent in the particles with a core size of 2–30 nm. For most magnetic nanoparticles, this value coincides in the order of magnitude with the theoretical estimate for the smallest dimensions of a magnetic domain. The soft magnetic core provides a high saturation magnetization and the relative hard magnetic shell ensures a high coercive force. Figure 1.13 show a typical example of core-shell magnetic nanoparticles, where the soft magnetic core provides a high saturation magnetization and the relative hard magnetic shell ensures a high coercive force.

1.3.3 Different Types of Magnetism

A magnet is a material that produces a magnetic field, which is invisible but responsible for the most notable property of a magnet. The term magnet is typically reserved for objects that produce their own persistent magnetic field even in the absence of an applied magnetic field. This is a kind of force that pulls on other ferromagnetic materials, such as iron, and attracts or repels other magnets. Only certain classes of materials can produce a magnetic field in response to an

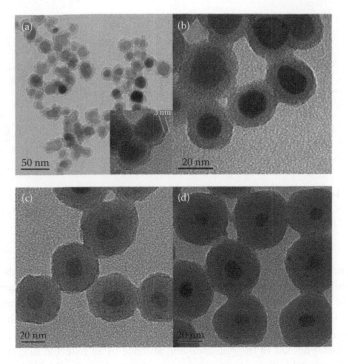

FIGURE 1.13 TEM image of core–shell nanoparticles with a TEOS content of (a) 0.2 mL, (b) 0.4 mL, (c) 0.8 mL, (d) 1.6 mL. (Adapted from J. Zou, Y.-G. Peng, Y.-Y. Tanga, *Rsc. Adv.* 4, 2014, 9693–9700.)

applied magnetic field—a phenomenon known as magnetism. There are several types of magnetism, and the overall magnetic behavior of a material can vary widely, depending on the structure of the material, particularly on its electron configuration, as discussed next.

1.3.3.1 Ferromagnetism

Ferromagnetic and ferrimagnetic materials are the most common as they are attracted to other magnets so strongly that it can be felt physically [78]. It has unpaired electrons in its atomic structure. In addition to the electrons' intrinsic magnetic moment's tendency to be parallel to an applied field, there is also a tendency for these magnetic moments to orient parallel to each other to maintain a lowered-energy state. Thus, even in the absence of an applied field, the magnetic moments of the electrons in the material spontaneously line up parallel to one another. Ferrimagnetic materials, that include ferrites and the oldest magnetic materials magnetite and lodestone, are similar to but weaker than ferromagnetic materials. The difference between ferro- and ferrimagnetic materials is related to their microscopic structure. Every ferromagnetic substance has its own individual temperature, Curie temperature, or Curie point (T_C), above which it loses its ferromagnetic properties. This is because the thermal tendency to disorder overwhelms the energy-lowering due to ferromagnetic order.

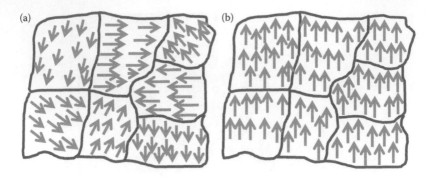

FIGURE 1.14 Schematic representation of ferromagnetism: (a) in presence of magnetic field and (b) in absence of magnetic field. (Drawn based on S. V. Vonsovskii, R. Hardin, *Magnetism*, John Wiley & Sons, 1974, Vol. 1.)

Ferromagnetism only occurs in a few substances: iron, nickel, cobalt, their alloys, and some alloys of rare earth metals. Iron comes with two forms of magnetism: hard and soft. In hard iron, the domains will not shift back to their starting points when the field is taken away and is used in permanent magnets. In soft iron, the domains return to being randomly aligned when the field is removed. A schematic representation of ferromagnetism in the presence and absence of an external magnetic field is shown in Figure 1.14. To make a permanent magnet, a piece of hard iron is placed in a magnetic field. The domains align with the field, and retain a good deal of that alignment when the field is removed, resulting in a magnet. An electromagnet, in contrast, uses soft iron. This allows the field to be turned on and off. One method is to coil a wire around a nail (made of iron or steel) and connect the two ends of the wire to a battery. When the domains in the nail align with the field produced by the current, the magnetic field is magnified by a large factor, typically by 100–1000s.

1.3.3.2 Paramagnetism

There are unpaired electrons in paramagnetic material, that is, atomic or molecular orbital with exactly one electron in them [78]. While paired electrons are required according to Pauli's exclusion principle to have their intrinsic magnetic moments pointing in opposite directions, causing their magnetic fields to cancel out, an unpaired electron is free to align its magnetic moment in any direction. When an external magnetic field is applied, these magnetic moments will tend to align themselves in the same direction as the applied field, thus reinforcing it. Paramagnetic materials, such as platinum, aluminum, and oxygen, are weakly attracted to either pole of a magnet. This attraction is hundreds of thousands of times weaker than that of ferromagnetic materials, so it can only be detected by using sensitive instruments or using extremely strong magnets. Magnetic ferrofluids, although they are made of tiny ferromagnetic particles suspended in liquid, are sometimes considered paramagnetic since they cannot be magnetized. A schematic representation of paramagnetism in the presence and absence of an external magnetic field is shown in Figure 1.15.

(a)   (b)

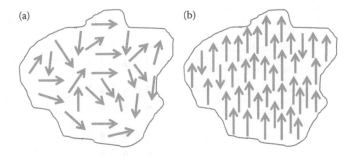

FIGURE 1.15 Schematic representation of paramagnetism: (a) in presence of magnetic field and (b) in absence of magnetic field. (Drawn based on S. V. Vonsovskii, R. Hardin, *Magnetism*, John Wiley & Sons, 1974, Vol. 1.)

1.3.3.3 Diamagnetism

Diamagnetic means repelled by both poles. Diamagnetism appears in all materials and is the tendency of a material to oppose an applied magnetic field and therefore, to be repelled by a magnetic field [80]. Despite its universal occurrence, diamagnetic behavior is observed only in a purely diamagnetic material. In a diamagnetic material, there are no unpaired electrons, so the intrinsic electron magnetic moments cannot produce any bulk effect. In these cases, the magnetization arises from the electrons' orbital motions. Compared to paramagnetic and ferromagnetic materials, diamagnetic materials, such as carbon, copper, water, and plastic, are even more weakly repelled by a magnet. The permeability of diamagnetic materials is less than the permeability of a vacuum. All substances not possessing one of the other types of magnetism are diamagnetic. Although force on a diamagnetic object from an ordinary magnet is far too weak to be felt, using extremely strong superconducting magnets, diamagnetic objects such as pieces of lead and even mice can be levitated, so they float in midair. Superconductors repel magnetic fields from their interior and are strongly diamagnetic. A schematic representation of diamagnetism in the presence and absence of an external magnetic field is shown in Figure 1.16.

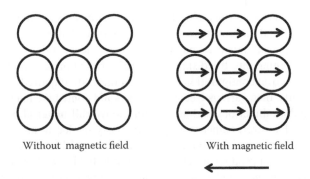

Without magnetic field   With magnetic field

FIGURE 1.16 Schematic representation of diamagnetism. (Drawn based on J. W. McClure, *Phys. Rev.* 104, 1956, 666–669.)

Type	Spin alignment	Spin in simplified plot	Examples
Ferromagnetic	All spins align parallel to one another: spontaneous magnetization: $M = a + b$	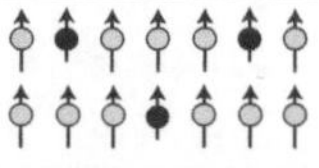	Fe, Co, Ni, Gd, Dy, SmCo$_5$, Sm$_2$Co$_{17}$, Nd$_2$Fe$_{14}$B
Ferrimagnetic	Most spins parallel to one another, some spins antiparallel: spontaneous magnetization: $M = a - b > 0$	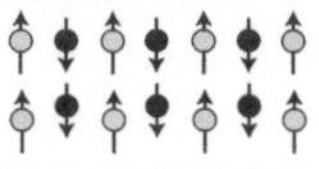	Magnetite (Fe$_3$O$_4$), yttrium iron garnet (YIG), GdCo$_5$
Antiferromagnetic	Periodic parallel-antiparallel spin distribution: $M = a - b = 0$	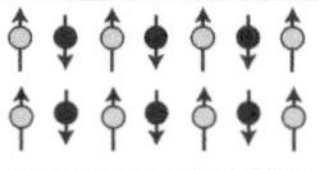	Chromium, FeMn, NiO
Paramagnetic	Spins tend to align parallel to an external magnetic field: $M = 0 @ H = 0, M > 0 @ H > 0$	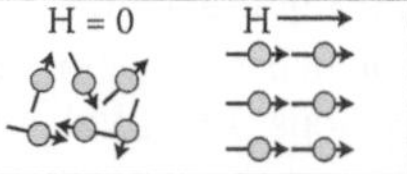 $H = 0$ $H \longrightarrow$	Oxygen, sodium, aluminum, calcium, uranium
Diamagnetic	Spins tend to align antiparallel to an external magnetic field: $M = 0 @ H = 0, M < 0 @ H > 0$	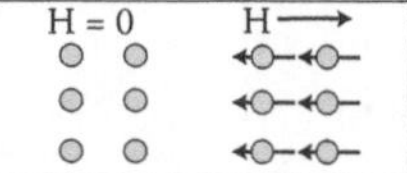 $H = 0$ $H \longrightarrow$	Superconductors, nitrogen, copper, silver, gold, water, organic compounds

FIGURE 1.17 Classification of different types of magnetism. (Adapted from A. G. Kolhatkar et al., *Int. J. Mol. Sci.* 14(8), 2013, 15977–16009.)

The general classification of different types of magnetism with corresponding spin plot is summarized in Figure 1.17.

1.4 ORIGIN OF MAGNETISM

Basically, magnetism in materials arises from two sources:

1. Electric current.
2. Nuclear magnetic moments of atomic nuclei. These kinds of magnetism are typically thousands of times less significant than the electrons' magnetic moments, so they are negligible in the context of the magnetization of materials.

The abundances of electrons in a material are arranged such that their magnetic moments (both orbital and intrinsic) cancel out. This is due to electrons combining into pairs with opposite intrinsic magnetic moments according to Pauli's exclusion principle or combining into filled subshells with zero net orbital motion. Moreover, even when the electron configuration is such that there are unpaired electrons and/or nonfilled subshells, then various electrons in the solid contribute magnetic moments in different random directions and as a result the overall material does not exhibit any magnetism. However, sometimes spontaneously or due to an applied external magnetic field, each of the electron magnetic moments will be lined up, on average. Then the material can produce a net total magnetic field, which can potentially be quite strong. The magnetic behavior of a material depends on its structure, particularly its electron configuration, for the reasons mentioned above, and also on the temperature. At high temperatures, random thermal motion makes it more difficult for the electrons to maintain alignment. When a material is placed in a magnetic field, the electrons circling the nucleus will experience a Coulomb attraction to the nucleus and Lorentz force from the magnetic field. Depending on which direction the electron is orbiting, this force may increase the centrifugal force on the electrons, pulling them in toward the nucleus, or it may decrease the force, pulling them away from the nucleus. This effect systematically increases the orbital magnetic moments that were aligned opposite the field, and decreases the ones aligned parallel to the field. This results in a small bulk magnetic moment, with an opposite direction to the applied field. Note that all materials undergo this orbital response. However, in paramagnetic and ferromagnetic substances, the diamagnetic effect is overwhelmed by much stronger effects caused by the unpaired electrons. There are various other types of magnetism, such as spin glass, superparamagnetism, superdiamagnetism, and metamagnetism. Three main phenomena [82] that characterize magnetism in matter:

- The formation of the magnetic moment on the atomic scale
- The occurrence of magnetic order, resulting from the strong interactions existing between atomic moments
- The alignment of moments along some favored crystallographic axis, leading to the phenomenon of magnetic anisotropy

Iron, cobalt, and nickel belong to the first series of magnetic elements, the 3d series, whilst the second series comprises the 14 rare earth elements. In the following sections, the properties of materials composed of elements belonging to one or the other of these two series have been described. Behavior specific to each category of material is related to different characteristic properties of the 3d and 4f electronic shells [82]. Besides these fundamental explanations regarding the origin of magnetism in matter, the quantum-mechanical theory on magnetism has received much attention as described hereafter.

1.4.1 QUANTUM-MECHANICAL ORIGIN OF MAGNETISM

In principle, all kinds of magnetism originate from specific quantum-mechanical phenomena. A successful model was developed in 1927, by Walter Heitler and Fritz London, who quantum-mechanically derived how hydrogen molecules are formed from hydrogen atoms, that is, from the atomic hydrogen orbitals u_A and u_B centered at the nuclei A and B [83]. According to their theory, the resulting orbital is

$$\psi(r_1, r_2) = \frac{1}{\sqrt{2}}(u_A(r_1)u_B(r_2) + u_B(r_1)u_A(r_2)) \qquad (1.17)$$

Here the last product means that a first electron r_1 is in an atomic hydrogen-orbital centered at the second nucleus, whereas the second electron runs around the first nucleus. This "exchange" phenomenon is an expression for the quantum-mechanical property that particles with identical properties cannot be distinguished. This holds not only for the formation of chemical bonds but also for magnetism. The term exchange interaction arises for the origin of magnetism and is stronger roughly by factors of 100 and even by 1000, than the energies arising from the electrodynamic dipole–dipole interaction.

As for spin function $\chi(s_1, s_2)$ which is responsible for the magnetism, according to Pauli's principle, a symmetric orbital (i.e., with the + sign) must be multiplied with an antisymmetric spin function (i.e., with a − sign), and vice versa. Thus,

$$\chi(s_1, s_2) = \frac{1}{\sqrt{2}} (\alpha(s_1)\beta(s_2) + \beta(s_1)\alpha(s_2)) \tag{1.18}$$

that is, not only u_A and u_B must be substituted by α and β, respectively (the first entity means "spin up," the second one "spin down"), but also the + sign by the − sign, and finally r_1 by the discrete values $s_i = \pm 1/2$:

$$\alpha\left(+\frac{1}{2}\right) = \beta\left(-\frac{1}{2}\right) = 1 \tag{1.19}$$

and

$$\alpha\left(-\frac{1}{2}\right) = \beta\left(+\frac{1}{2}\right) = 0 \tag{1.20}$$

The "singlet state," that is, the − sign, means the spins are antiparallel, that is, for the solid we have antiferromagnetism, and for two-atomic molecules one has diamagnetism. The tendency to form a (homopolar) chemical bond results through the Pauli principle automatically in an antisymmetric spin state (i.e., with the − sign). In contrast, the Coulomb repulsion of the electrons, that is, the tendency that they try to avoid each other by this repulsion, would lead to an antisymmetric orbital function (i.e., with the − sign) of these two particles, and complementary to asymmetric spin function (i.e., with the + sign). Thus, now the spins would be parallel (ferromagnetism in a solid, paramagnetism in two-atomic gases).

The last-mentioned tendency dominates in the metals (iron, cobalt, and nickel) and in some rare earth elements, which are ferromagnetic. Most of the other metals, where the first-mentioned tendency dominates, are nonmagnetic (e.g., sodium, aluminum, and magnesium) or antiferromagnetic (e.g., manganese). Diatomic gases are also almost exclusively diamagnetic, and not paramagnetic. However, the oxygen molecule, because of the involvement of π-orbitals, is an exception important for the life sciences.

The Heitler–London considerations can be generalized to the Heisenberg model of magnetism (Heisenberg 1928). The explanation of the phenomena is thus essentially based on all subtleties of quantum mechanics, whereas the electrodynamics covers mainly the phenomenology. In summary, electron spins are in one of two states, up or down. This is another way of stating that the magnetic quantum number can be +1/2 or −1/2. Electrons are arranged in shells and orbitals in an atom. If they fill the orbitals so that there are more spins pointing up than down (or vice versa), each atom will act like a tiny magnet. In nonmagnetic materials such as aluminum, neighboring atoms do not align themselves with each other or with an external magnetic field. In ferromagnetic materials, the spins of neighboring atoms do align (through a quantum effect known as exchange coupling), resulting in small (a tenth of a millimeter, or less) neighborhoods called domains where all the spins are aligned. When a piece of demagnetized iron (or other ferromagnetic material) is exposed to an external magnetic field, two things happen: First, the direction of magnetization (the way the spins point) of each domain will tend to shift toward the direction of the field. Second, domains which are aligned with the field will expand to take over regions occupied by domains aligned opposite to the field. This is what is meant by magnetizing a piece of iron [84,85].

Magnetic effects are sensitive to temperature. It is much easier to keep permanent magnets magnetized at low temperatures, because at higher temperatures the atoms tend to move around much more, throwing the spins out of alignment. Above a critical temperature known as the Curie temperature (T_C), ferromagnets lose their ferromagnetic properties. Behavior of magnetic nanoparticles in a magnetic field can depend on the size and composition of the particles, the strength of the field, and any obstacles that may be in the way.

1.5   MAGNETISM IN NANOSTRUCTURED MATERIALS

Magnetism in nanostructured materials originates from the magnetic moments in clusters. Since in such materials large numbers of atoms of the clusters are on the surface, one expects the clusters of these elements to exhibit larger magnetic moments. Being small particles, it is possible to measure the magnetic moment directly by measuring the interaction with a magnetic field in terms of their deflection from the original trajectories [86]. First measurements were reported on iron and cobalt clusters of 1–2 nm in size range. For iron, the atomic moment was 2.2 μ_B equal to that of the bulk solid. For Co, the atomic moment was 2.08 μ_B whereas for bulk it was 1.72 μ_B. The change in properties due to nanostructuring is not always in favor. For example, ferroelectric materials smaller than 10 nm can switch their magnetization direction using room temperature thermal energy, thus making them useless for memory storage. Suspensions of nanoparticles are possible because the interaction of the particle surface with the solvent is strong enough to overcome differences in density, which may result in a material either sinking or floating in a liquid. Nanoparticles often have unexpected visual properties as they are small enough to confine their electrons and produce quantum effects. For example gold nanoparticles appear deep red to black in solution. The enhanced interest of the researchers in nanostructured materials are due to the quantum size effects. These arise in the case where the size of the system is comparable to the de Broglie wavelengths of the electrons, phonons, or excitons propagating in them. The change in the physical and chemical properties of small particles depends on their size, conditions (coordination number, symmetry of the local environment, etc.) differing from those of the bulk atoms. From the energy stand point, a decrease in the particle size results in an increase in the fraction of the surface energy in its chemical potential.

Special attention has been paid to the investigation of magnetic properties in which the difference between bulk and nanostructured material is pronounced. In particular, it was revealed that magnetization (per atom) and the magnetic anisotropy of nanoparticles can be much higher than those of the bulk with differences in the Curie (T_C) or Neel (T_N) temperatures. In addition, magnetic nanomaterials were found to possess a number of unusual properties: giant magnetoresistance, abnormally high magnetocaloric effect, and so on. The magnetic properties of nanoparticles are determined by many factors including chemical composition, type, and degree of defectiveness of the crystal lattice, particle size and shape, morphology (for structurally inhomogeneous particles), interaction of the particles with the surrounding matrix and neighboring particles.

In 1930, on the basis of energy considerations, Frenkel and Dorfman showed that particles of a sufficiently small size should be single domain. In the mid-twentieth century, the theory of single-domain particles started to develop actively [87–92] and the related phenomena were investigated experimentally [93–104]. These studies identified a substantial increase in the coercive force of a ferromagnet changing the structure from multidomain to single domain, which is important for the creation of permanent magnets. The results of calculations of the characteristic particle size (for different magnetic materials) where the particle becomes single domain are presented in Table 1.1. The critical diameters corresponding to particle transition from multidomain to the single-domain state were calculated for spherical particles with an axial magnetic anisotropy.

It is important to note that, a single-domain particle is not necessarily "small" [105]. It is considered that significant changes in the main physical characteristics of a bulk material appear when the dimensions of its particles decrease to an extent where the ratio of the number of surface atoms N_S to the total number N of atoms in the particle approaches 0.5. Assuming that in a surface layer of thickness Δr (defectiveness parameter), the number of exchange bonds is twice as low as that in the particle

TABLE 1.1

Critical Diameter (for Room Temperature) of a Single-Domain Spherical Particle with the Axial Magnetic Anisotropy

	d_{crit} (nm)		
Material	From Reference (109)	From Reference (110)	From Reference (111)
Co	70	70	68
Ni	–	55	32
Fe	30	14	12
$BaFe_{12}O_{19}$	–	–	580
Fe_3O_4	–	128	–
α-Fe_2O_3	–	166	–
Nd-Fe-B	200	–	214
$SmCo_5$	1500	55	1528

bulk and that the Curie temperature is directly proportional to the bulk density of the exchange bonds. Nikolaev and Shipilin [102] have analyzed the dependence of T_C on the magnetite particle size. The defectiveness parameter Δr was found to depend on the particle radius r. In particular, for magnetite $\Delta r \to 0$ for $r \to 2.5$ nm (the radius for the single-domain state of magnetite is ~70 nm, see Table 1.1). As the particle radius decreases, the parameter Δr substantially increases and for $r = 2.5$ nm, it amounts to 0.5 nm. Thus, the smaller the magnetic particle size, the greater the effective depth to which the violation of the regular structure extends. TEM micrographs of Fe_5C_2 magnetic nanoparticles are shown in Figure 1.18 [106].

One more remarkable property of the nanoparticles, which allowed their experimental discovery in the mid-twentieth century, is their superparamagnetism. The higher the magnetic moment of the particle, the lower the magnetic field (H_S) required for observing the magnetization saturation. In a rough approximation, the H_S can be estimated from Equation 1.21:

$$\mu_{ef} \cdot H_s \approx k_B \cdot T \tag{1.21}$$

where μ_{ef} is the effective magnetic moment of the particle. For paramagnetic $Gd(SO_4)_3 \cdot H_2O$, the effective magnetic moment of Gd^{3+} ion is 7 μ_B. Thus, the H_S value for this paramagnet at room temperature would be $H_s \approx 300\,k_B/7 \approx 10^6$ Œ H_S. For a particle with an effective magnetic moment of $10^4 \mu_B$, the saturation field would decrease to 10^3 Œ. The phenomenon of saturation of the magnetization curve in low fields of ~1 kŒ has been called "superparamagnetism," while a material exhibiting such properties is called a "superparamagnetic." The model of an ideal superparamagnetic was mainly worked out by the early 1960s [107], but now it seems to get further attention [104,108]. The simplest variant of this model considers a system of

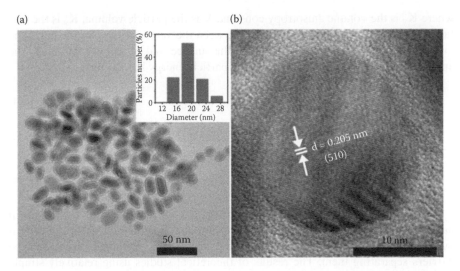

FIGURE 1.18 TEM micrographs of Fe_5C_2 magnetic nanoparticles. (a) Overall particle shape and size distribution (inset) and (b) hi-resolution TEM image showing atomic fringes. (Adapted from G. Huang et al., *Nanoscale* 6, 2014, 726–730.)

N noninteracting identical particles with the magnetic moment μ_{ef}. Since the magnetic moment of the particle is assumed to be large, its interaction with the magnetic field H is calculated without taking account of the quantum effects. In the case of isotropic particles, the equilibrium magnetization, <M >, described by the Langevin equation:

$$< M >= N\mu_{ef}\left[coth\left(\frac{\mu_{ef}H}{k_BT} \right) - k_B \frac{T}{\mu_{ef}} H \right] \tag{1.22}$$

Equation 1.21 has been derived with the assumption that single particles are magnetically isotropic, that is, all directions for their magnetic moments are energetically equivalent, but this condition is hardly ever fulfilled. If the particles are magnetically anisotropic, the calculation of the equilibrium magnetization becomes more complicated. According to the nature of factors giving rise to the nonequivalence of the directions of magnetic moments, one can distinguish the magnetically crystalline anisotropy: shape anisotropy, anisotropy associated with the internal stress and external impact, exchange anisotropy and so on [109].

For nanoparticles, the surface magnetic anisotropy plays a special role. Unlike other kinds of magnetic anisotropy, the surface anisotropy is proportional to the surface area of the particle S rather than to its volume V. The surface anisotropy appears due to the violation of the symmetry of the local environment and the change in the crystal field, the equation for uniaxial magnetic anisotropy in terms of its energy is as follows:

$$E(\theta) = (K_V V + K_S S)\sin^2\theta \tag{1.23}$$

where K_V is the volume anisotropy constant, V is the particle volume, K_S is the surface anisotropy constant, θ is the angle between the vector of the particle magnetic moment (m) and anisotropy axis. When the surface makes no contribution to the anisotropy, the angular dependence of the particle energy has the form

$$E(\theta) = K_V V \sin^2\theta \tag{1.24}$$

In the presence of an external magnetic field (H), at an angle θ to the anisotropy axis, the particle energy is

$$E(\theta) = K_V V \sin^2\theta - M_S VH\cos(\theta - \Psi) \tag{1.25}$$

The difficulty of the theoretical investigation of the magnetic hysteresis in nanoparticles lies on nonlinear, nonequilibrium, and nonlocal phenomenon caused by the existence of energy minima resulting from the magnetic anisotropy and the barriers separating them. The results of theoretical studies using relatively simple models seldom provide a plausible description for real magnetic nanomaterials as they do not take into account their microstructure, in particular, the effect of boundaries and defects on the local magnetization [110].

An important role of the microstructure in the formation of magnetic characteristics is indicated by the studies of nanocomposite materials, for example, the Nd–Fe–B/α–Fe system [111] or Cu–Ni–Fe system which represent a magnetically soft material with dispersed nanosized grains [112]. In these materials, the magnetically hard phase ensures a high coercive force, while the magnetically soft phase provides a high saturation magnetization. In addition, the substantial exchange interaction between the grains ensures a relative residual magnetization higher than 0.7. Therefore, materials of this type are called exchange-coupled magnets. In recent years attempts have been undertaken to study the effect of the internal structure (microstructure) of a nanoparticle on the magnetic characteristics of real nanomaterials. The success was the use of numerical calculations within the framework of the micromagnetism theory [113–116]. Even when a nanoparticle has a defect-free crystal structure, the different local environments of atoms at the particle boundary and inside the particle result in a nonuniform magnetization in the particle and distortion of the perfect collinear magnetic structure [117,118]. The calculations show that at the final temperature, magnetization decreases along the direction from the particle center toward the boundary [119], and the magnetic moment of each particular surface atom can be greater than that of the bulk atoms [120]. The decrease in the magnetization on the particle surface compared to that in the bulk is due to a lower energy of the surface spin wave excitations [108]. In other words, there is more pronounced action of the thermal fluctuations on the surface. The increase in the magnetic moment of surface atoms can be attributed to the decrease in the coordination number and as a consequence, to narrowing the corresponding energy band and an increase in the density of states. Apparently, this also accounts for the rare cases of magnetic order appearing in metal nanoparticles whose bulk analogs are nonmagnetic [121–123].

The qualitative isothermal dependence of the coercive force H_C on the characteristic size of magnetic particles is shown in Figure 1.19. The increase in H_C with decreasing particle size follows from the Stoner-Wohlfarth theory according to which the spins of atoms forming a nanoparticle rotate coherently. It is known from

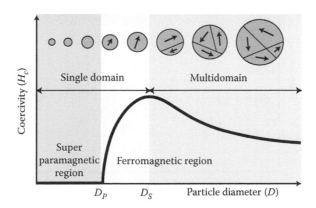

FIGURE 1.19 Qualitative dependences of the coercive force H_C on the particle diameter. (Adapted from J. S. Lee et al., *Sci. Rep.* 5, 2015, Article number: 12135.)

experiments that the coercive force in real magnetic materials (including nanomaterials) is much lower than the limiting values predicted by the theory even at very low temperatures. One reason is that under the action of an external magnetic field, the spins of the atoms forming the nanoparticle can rotate not only coherently but also in a more complex manner to form spin modes: swirls, fans, etc. [124]. The appearance of noncoherent spin modes is facilitated if the nanoparticles form agglomerates. The coherent rotation can, apparently, take place only in absolutely defect-free uniform particles with a zero surface anisotropy. In a multidomain particle, this rotation can be additionally associated with the displacement of domain boundaries. As the particle size decreases, the number of domains decreases, and the role of interdomain boundaries in magnetization reversal becomes less distinct. Therefore, up to the critical particle size (d_{crt}, Figure 1.19), the coercive force increases with a decrease in d. However, further decrease in the particle size and transition to single-domain particles entails an increase in the role of thermal fluctuations [108,114,125]. For a system with identical magnetic nanoparticles, the equilibrium magnetization upon the change in the magnetic field with relaxation time τ is described by Equation 1.15 and schematically shown in Figure 1.20 [124].

$$M(t) = M_0 \cdot e^{-t/\tau} \tag{1.26}$$

It is quite interesting to see the effect of interparticle interactions on the blocking temperature [126–128]. Two models were proposed: one predicting an increase in T_b [127] and a decrease in T_b [128] following an enhancement of the interparticle interactions. The interactions change the height of the energy barrier separating two states of a particle with the opposite directions of the magnetic moment. If

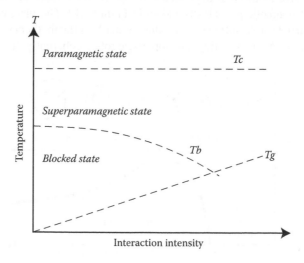

FIGURE 1.20 Scheme of the possible transitions in a system of magnetic nanoparticles arranged randomly in space taking into account interparticle interactions. (Redrawn based on S. P. Gubin et al., *Russian Chemical Reviews* 74(6), 2005, 489–520.)

the barrier grows, T_b increases and vice versa. The influence of the change in the distances between the maghemite nanoparticles with a diameter of 6–7 nm on the blocking temperature has been studied experimentally [129]. The distances between the particles were changed by compacting the sample. The maximum increase in the sample density was 55%; simultaneously, T_b increased from 50 to 80 K. Assuming that for noncharged nanoparticles, the predominant magnetic dipole–dipole interactions are inversely proportional to the cubed distance between the particles, one can expect a linear dependence of T_b on the sample density [129].

In the presence of interparticle interactions, the qualitative picture of the behavior of a system of magnetic nanoparticles following a decrease in temperature may become more complicated than the mere transition to the blocked state [8] shown in Figure 1.20. If the particles are arranged irregularly in space, the interparticle interactions should transfer the system into the spin glass type state at some temperature T_g [8]. These temperatures, either T_g or the average blocking temperature T_b would be higher for the given type of particles and depends on the particle size and on the average distance between them. Often it is assumed that the interparticle interactions can be neglected if the particle concentration in the matrix is low and, hence, the average distance between them is rather high (1.5 times higher than the average particle diameter) [126]. In 2008 [130], an interesting approach to taking account of the effect of interparticle interactions on the magnetization curves of nanoparticles was proposed, on the basis of the monotonic decrease in μ_{ef} with a decrease in temperature [86]. It was assumed that the magnetic dipole–dipole interactions, which predominate in the systems under consideration, act as a random factor and prevent magnetization (ordering) of the system. It is worth noting that, although the proposed approach is strictly applicable only to equilibrium systems, it describes satisfactorily the magnetization curves at any temperatures, including those below the blocking temperature T_b. The applicability of this model for $T < T_b$ is apparently due to the predominance of interparticle interactions over one-particle effects for all the samples studied.

The magnetic moment of transition metals decreases in general with an increase in the number of atoms (size) in the clusters as shown in Figure 1.21. However, the moment of different transition metal clusters are found to depend differently on the number of atoms per cluster (N).

The enclosed data points show the sizes of the respective clusters at which they attain bulk moments. Among the ferromagnetic 3d transition metals, nickel clusters attain the bulk moment at $N = 150$, whereas cobalt and iron clusters reach at the corresponding bulk limit at $N = 450$ and $N = 550$, respectively (Figure 1.21). In icosahedral configuration these numbers correspond to 3, 4, and 5 shells, respectively. The results show that in the lower size limit the clusters have high spin majority configuration and the behavior is more like an atom, but with the increase of size an overlap of the 3d-band with the Fermi level occurs, which reduces the moment toward that of the bulk with slow oscillations. In analogy with the layer-by-layer magnetic moments, the clusters are assumed to be composed of spherical layers of atoms forming shells and the value of moment of an atom in a particular shell is considered to be independent of the size of the cluster. The magnetic moment values of the different shells are optimized to fit with the total moment per atom

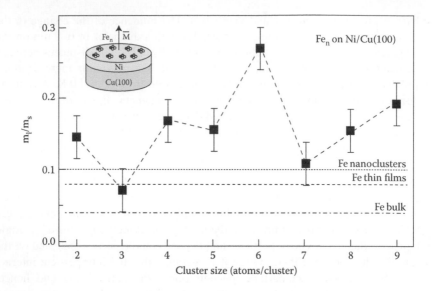

FIGURE 1.21 Spin and orbital magnetic moments of deposited small iron clusters studied by x-ray magnetic circular dichroism spectroscopy. (Adapted from J. T. Lau et al., *New J. Phys.* 4, 2002, 98.)

maintaining the overall trend of decreased size. It is generally observed that the moment per atom for the surface atoms in the first layer are similar to that of an atom in deep layers that corresponds to the bulk. The intermediate layers show an overall decreasing trend. For both nickel and cobalt, there are oscillations in the values at least up to the fifth shell. For nickel, the second layer is found to have a negative coupling with the surface atoms and the third/fourth layers in iron are magnetically dead. The model provides close agreement with the experimental results for cobalt and nickel. But in a more realistic model both the geometric shell closing as well as the electronic structure should be taken into account. However this model provides a good first order picture to understand the magnetic behavior of 3d transition metal clusters. Though the 4d transition metals are nonmagnetic, the small clusters of rhodium are found to be magnetic following the calculations. Within the size range N = 10–20, the moment varies between 0.8 and 0.1 μ_B per atom but the clusters become nonmagnetic within N = 100. Due to the reduced size, the individual moments on the atoms in clusters are aligned. First-principles calculations show that the ground state of Rh6 is nonferromagnetic, while Rh9, Rh13, and Rh19 clusters have nonzero magnetic moments. Rh43 is found to be nonmagnetic as the bulk. Accordingly, ruthenium, palladium, chromium, vanadium, and aluminum clusters are found to be nonmagnetic.

Since the ferromagnetic state requires the moments to remain mutually aligned even at relatively high temperatures, it is important to study the thermal behavior of the magnetic clusters. Temperature-dependent studies show that up to 300 K, the moment remains constant for nickel clusters and then it decreases at higher temperatures closely resembling the bulk behavior. This indicates that the interactions affecting the mutual alignment of moments with temperature are in the same order of magnitude as in the bulk. Cobalt clusters also follow the same trend but the moment increases slightly within 300–500 K, which might be due to a structural phase transition corresponding to the HCP–FCC phase transition in bulk at 670 K, where the moment increases by 1.5%. Clusters of different size ranges of iron behave differently. Fe50–60 shows a gradual reduction of moments from 3 μ_B at 120 K to 1.53 μ_B at 800 K. Up to 400 K, Fe120–140 remains with a constant moment of 3 μ_B and then decreases to finally level off at 700 K to 0.7 μ_B. The size range N = 82–92 behaves like Fe120–140 below 400 K and like Fe50–60 above 400 K. The moment for the higher size ranges decreases steadily with the increase of temperature but Fe250–290 levels off at 700 K whereas Fe500–600 levels off between 500 and 600 K to 0.4 μ_B. Comparing with the thermal behavior of bulk iron, the level-off temperature can be termed as the critical temperature (T_C) for clusters of a particular size and it decreases with the increase of size of the clusters. But, since the bulk Curie temperature is 1043 K, one realizes clearly that this trend must reverse at higher sizes to reach the bulk. It is also interesting to note that for Fe120–140 the moment decreases more rapidly within 600–700 K. The specific heat measurement for both Fe120–140 and Fe250–290 also show a peak at 650 K. This might be because iron clusters undergo a structural phase transition in this temperature range similar to the bulk iron which undergoes BCC–FCC–BCC phase transition, but only beyond the Curie temperature [131,132].

1.5.1 MAGNETIZATION PROCESS

On the basis of the ideas in the aforementioned discussion, the moments in a ferromagnetic material should all turn out to be parallel in order to minimize the exchange energy and lined up along the easy axis of magnetization in order to minimize the anisotropy energy. In reality, materials are divided into so-called magnetic domains as shown in Figure 1.22 [9]. Within magnetic domains, the moments adopt the expected configurations; however, from one domain to another, the direction of

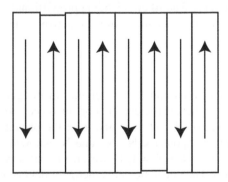

FIGURE 1.22 Magnetic domains and walls. (Redrawn based on D. Givord, *Nanomaterials and Nanochemistry*, C. Bréchignac, P. Houdy, and M. Lahmani (eds.), Springer-Verlag, Berlin Heidelberg, Germany, 2007, pp. 101–134.)

the magnetic moments actually alternates. This division into domains allows the system to minimize energy, namely, the demagnetizing field energy which results from the action of the magnetized matter on itself. When the magnetic domains are formed, magnetic domain walls also come into being among them.

The formation of a wall has a certain cost in anisotropy and exchange energy. The magnetization process describes the action of a magnetic field on magnetized matter. In magnetic soft materials, the magnetization follows the field easily, whereas in hard magnetic materials, it tends to resist the effects of the field. In materials used for magnetic recording, the saturated magnetization state constitutes one bit of information which can be reversed from state 1 to state 0 or the opposite. The energy of the demagnetizing field is given by Equation 1.27:

$$E_D = -\frac{1}{2}\mu_0 \int_V M \cdot H_D dV \tag{1.27}$$

where M is the local magnetization, H_D is the demagnetizing field at the relevant point, and V is the volume of magnetic matter. It can be shown that E_D can also be written in the form

$$E_D = \frac{1}{2}\mu_0 \int_{R^3} (H_D)^2 d^3 R \tag{1.28}$$

where the symbol $\int_{R^3}$ indicates that the integral is taken over the whole of space. Equation 1.28 shows that E_D is always positive. The configuration of moments that minimizes E_D is such that $H_D = 0$. The energy of the demagnetizing field, although less than the exchange energy is not involved in establishing magnetic order. However, the associated interactions are long range and impose the division of the matter into domains with dimensions of micrometric order. On the scale of a material element big enough to contain several domains, the magnetization is practically zero and the demagnetizing field likewise, being proportional to it. The energy density of the demagnetizing field in a uniformly magnetized object can be expressed as

$$E_D = \frac{1}{2}\mu_0 N M^2 V \tag{1.29}$$

where M is the magnetization and V the volume. In the general case, it is intricate matter to calculate the coefficient N. For an object with ellipsoidal shape, one may introduce a tensor **N**. Let x, y, and z be the principal axes of the ellipsoid. Then **N** has three components, N_x, N_y, and N_z, known as the demagnetizing factors, which satisfy $N_x + N_y + N_z = 1$. However, in consideration of a sphere, we have $N_x = N_y = N_z = 1/3$. The magnetic demagnetizing field is uniform and given by

$$H_D = -NM \tag{1.30}$$

In an object with arbitrary shape, it is often justifiable to treat the demagnetizing field as homogeneous to the first order of approximation.

1.5.1.1 Domain Walls

The region of transition between two neighboring domains is called a domain wall, or Bloch wall. In this region, the moments gradually rotate round from an initial position in which they are parallel to the direction of the moments in one domain, to a final position in which they are parallel to the direction of the magnetization in the adjacent domain. To a first approximation, the domain walls form planes running right across the sample. At this condition the energy of the demagnetizing field resulting from the formation of these walls is actually very small. However, the moments within the walls are neither strictly parallel to one another, nor aligned with any easy axis of magnetization. The energy lost in wall formation, called the wall energy, is thus determined purely by the competition between the anisotropy energy and exchange energy. For a uniaxial system with anisotropy constant K, the wall energy expressed per unit area of the wall is equal to $\gamma = 4\sqrt{AK}$, and the thickness of the wall is given by $\delta = \pi\sqrt{A/K}$ where A is the exchange constant and representing exchange interactions and proportional to the Curie temperature: $A \propto Nk\xi^2 T_C$, where N is the number of atoms per unit volume and ξ is the distance between nearest-neighbor atoms. In iron, $A \approx 18 \times 10^{-11}$ J/m and $K \approx 50$ kJ/m^3, whence $\gamma \approx 3 \times 10^{-3}$ J/m^2 and $\delta \approx 100$ nm. The Curie temperatures of all magnetic materials used in applications are higher than room temperature, generally lying in the range 500–1000 K. The value of A is therefore always of the order of 10^{-11} J/m. In contrast, K may vary over several orders of magnitude, so that $K \approx 10^3$ J/m^3 in ultrasoft magnetic materials, whereas $K \approx 10^7$ J/m^3 in hard magnetic materials. The wall energy thus has values between one-tenth and 10 times the value for iron, with wall thicknesses between 300 nm in soft materials and 5 nm in hard materials. Note that the energy of a wall increases rapidly when its thickness cannot take the equilibrium value δ. The most important case is that of a confined wall. The increase in energy of the wall is due to the loss of exchange energy resulting from the fact that the magnetic moments must rotate rapidly from one atomic plane to the next in order to complete the 180° rotation over the available distance. When a magnetic field H_{app} is applied to a ferromagnetic material, the Zeeman energy of coupling with the field is given per unit volume by (Figure 1.23):

$$E_Z = -\mu_0 M H_{app} \tag{1.31}$$

According to Equation 1.31 magnetization tends to occur along the field direction, starting from the initial situation resulting from the division into domains and characterized by $M = 0$. The magnetization increases by displacement of the walls in such a way that domains with magnetization along the field grow larger to the detriment of the others. In a homogeneous material, this could be expressed mathematically as

$$E_T = (1/2)NM^2 - \mu_0 M H_{app} \tag{1.32}$$

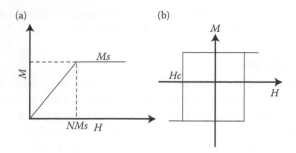

FIGURE 1.23 Magnetization process: (a) Variation of the magnetization due to wall displacement and (b) magnetic field. (Redrawn based on D. Givord, *Nanomaterials and Nanochemistry*, C. Bréchignac, P. Houdy, and M. Lahmani (eds.), Springer-Verlag, Berlin Heidelberg, Germany, 2007, pp. 101–134.)

The wall energy does not depend on the position of the wall and this is why it does not enter in Equation 1.33. Minimizing, one deduces that the magnetization has the linear dependence:

$$M = (1/N)H \tag{1.33}$$

The slope of the functional dependence here, given by $1/N$, is known as the demagnetizing gradient. It depends only on the shape of the sample and is independent of temperature. Wall displacement proceeds until a state is reached in which all moments are aligned with the field. This happens for $H_{app} = NM_s$. In stronger fields, the magnetization remains constant and is said to be saturated as shown schematically in Figure 1.23.

The way materials divide into domains is determined by competition between the wall energy and the energy of the demagnetizing field. The walls are quasi-2D objects and the wall energy is therefore a surface term. When the particle volume is reduced, this term becomes greater than the energy of the demagnetizing field. For very small dimensions, the most stable magnetization state is one in which there is just one domain. To find the corresponding critical volume, Kittel [127] considered a spherical particle of radius R, characterized by a uniaxial anisotropy constant K. Its energy E_1 in the single-domain state contains only one term, due to the demagnetizing field:

$$E_1 = (1/6)\mu_0 M^2 (4/3)\pi R^3 \tag{1.34}$$

When this same particle is divided into two equal domains, Kittel assumes that the energy of the demagnetizing field is divided by two. The corresponding energy E_2 is

$$E_2 = 4\sqrt{AK\pi R^2} + (1/12)\mu_0 M^2 (4/3)\pi R^3 \tag{1.35}$$

where the term in R^2 represents the wall energy. The critical radius R_c is obtained for $E_1 = E_2$ as

$$R_c = 36\sqrt{AK/\mu_0 M_s^2} \tag{1.36}$$

In metallic iron, the critical radius R_c obtained in this way is of the order of 7 nm. Below this size, the nanoparticles are spontaneously magnetized. Such particles are typical objects used in magnetic recording.

Let us consider a magnetic particle with uniaxial anisotropy and radius well below the critical value R_c. In zero field, the two orientations of the magnetization along the easy axis, denoted by $\uparrow$ and $\downarrow$, have equal energy. An energy barrier of height $E = KV$ separates these two states. At low temperatures, the magnetization of each particle remains blocked in one of the two directions, imposed either by chance as the particle cooled from higher temperatures, or by application of some magnetic field stronger than the coercive field. As the temperature increases, thermal activation must be taken into account. The characteristic time required to overcome a barrier of height E is given by the Arrhenius law:

$$\tau = \tau_0 e^{E/kT} \tag{1.37}$$

where τ_0 is the intrinsic magnetization reversal time, of the order of 10^{-9} s. On macroscopic scales, $E/kT \gg 1$ and τ is extremely long. Thermal activation will have no effect on the magnetization state of the particle. But on very small scales, the energy barrier E, with height proportional to the particle volume, is much lower. Above the temperature T_B (blocking temperature), the particle magnetization fluctuates between its two possible orientations. When a physical measurement is made at a temperature above T_B, the particle behaves as though it is no longer ferromagnetic, although in reality the atomic magnetic moments are still rigidly coupled together. This is the phenomenon known as superparamagnetism. There is a characteristic acquisition time t_c associated with each measurement technique. For example, $t_c \approx 1$ s for magnetization measurements, and $t_c \approx 10 - 8$ s for Mössbauer spectroscopy. For an intermediate value of τ, the particle is superparamagnetic as far as magnetization measurements are concerned, but blocked for Mössbauer spectroscopy. More precisely, the value obtained for T_B depends on the experimental technique used according to

$$E = KV \approx kT_B \ln (t_c/\tau_0) \tag{1.38}$$

The superparamagnetism phenomenon characterizes objects of very small dimensions. As far as magnetization measurements are concerned, the blocking temperature of a 3 nm cobalt nanoparticle is of the order of 30 K, but it approaches room temperature for a diameter of 6 nm. For nanoparticles of known volume the magnetic anisotropy can be deduced from the value of the blocking temperature using Equation 1.38.

The remanent magnetization of a particle is zero above the blocking temperature and cannot be used as a magnetic recording medium. The superparamagnetism phenomenon specifies a physical lower limit on the size of particles that can be used for magnetic recording. The corresponding ultimate limit on the recording density is of the order of 50 Gbits/cm^2, close to the recording densities achieved today. A significant increase in the recording density achievable with a magnetic medium will probably involve some way of overcoming or circumventing the superparamagnetic limit.

1.5.1.2 Coherent Rotation

Let us consider a spherical particle of radius R with saturated magnetization in the direction ↑, subject to a magnetic field applied in the direction ↓. Then on the macroscopic scale, magnetization reversal involves the nucleation of a wall on a defect, followed by propagation of this wall. After nucleation, the wall surface increases until it reaches the maximal value πR^2. An energy barrier thus comes into play in the propagation mechanism:

$$E_p = \gamma \pi R^2 \tag{1.39}$$

By a similar argument to the one used in the previous section, this barrier can be compared with the energy barrier $E_{SW} = KV$ characterizing the process of coherent rotation. For $R < 3\sqrt{/4K}$, $E_p > E_{SW}$ and nucleation at defects becomes inoperative due to the propagation barrier. Magnetization reversal must then occur via coherent rotation. Although the coherent rotation theory of magnetization reversal dates from 1948, it could not be confirmed experimentally until it became possible to measure the individual magnetic properties of very small objects, in which the nucleation–propagation scenario is impossible. Magnetization reversal has been studied in 4 nm cobalt nanoparticles using a high-sensitivity magnetometry technique (Figure 1.24) based on micro-SQUIDs [130]. The nature of the reversal process was deduced by

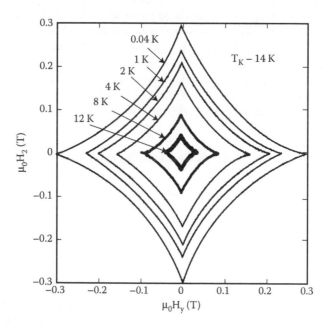

FIGURE 1.24 Angular dependence of the coercive field for a single Co nanoparticle of diameter 3 nm, measured at different temperatures. In this polar diagram, the amplitude of the coercive field at angle θ is represented by the length of the vector making angle. (Adapted from W. Wernsdorfer, *Adv. Chem. Phys.* 118, 2001, 99–103.)

measuring the angular dependence of the coercive field, which corresponds exactly to the prediction of the Stoner–Wohlfarth theory.

1.5.2 ROLE OF PARTICLE SIZE ON MAGNETISM

Among different magnetic materials that have found broad practical applications, ferromagnets deserve the most consideration [77]. An important characteristic of a ferromagnet is the coercive force (H_c), that is, the magnetic field strength H corresponding to the point with B = 0 on the symmetric hysteresis loop B(H) where B is the magnetic field induction in a ferromagnetic material with zero demagnetizing factors. One more characteristic, apart from H_c, is the intrinsic coercive force (H_{ci}), defined as the magnetic field strength at the point M = 0 on the symmetric M(H) hysteresis loop, where M is the magnetization of a ferromagnetic material with zero demagnetizing factor. The coercive force and the intrinsic coercive force normally do not differ much in magnitude, but these are different physical quantities. When designing new magnetic materials, it is often a goal to attain the highest H_c. Most modern magnetic materials have H_c in the range of 2–3 Œ [77]. In terms of the coercive force, ferromagnets are subdivided into soft magnetic ($H_C < 12.6$ Œ) and hard magnetic ($H_C > 126$ Œ) ones. The magnets with intermediate coercive force values are referred to as semihard. Table 1.2 presents data on the dependence of the magnetic properties of ferromagnets on the dimensions of the constituent particles. Apart from the dimensions, the magnetic properties depend on a number of external

---

TABLE 1.2
Change of Magnetic Properties of a Ferromagnet with a Decrease in Its Dimensions from Macroscopic to Atomic

Macroscopic (bulk)	<1 μm	Spontaneous magnetization below T_C. The appearance of a nonzero magnetic moment is suppressed by the formation of domain structure.
Microscopic	50–1000 nm	Magnetic characteristics strongly depend on the sample pre-history, preparation, and processing method.
Single-domain magnetic particles	1–30 nm	The presence of a blocking temperature, $T_b < T_C$, below which the magnetic moment of the particle retains orientation in space, while the particle ensemble demonstrates a magnetic hysteresis. At a temperature higher than T_b, the particle transfers into the superparamagnetic state. In $T_b < T < T_C$ region, the particle has a spontaneous magnetization and a nonzero total magnetic moment, which easily changes the orientation in the external field.
Single atom (ion)	~0.2 nm	Usual paramagnetic properties.

Source: D. Givord, *Nanomaterials and Nanochemistry*, C. Bréchignac, P. Houdy, and M. Lahmani (eds.), Springer-Verlag, Berlin Heidelberg, Germany, 2007, pp. 101–134; High-accuracy magnetic property measurement method by separating spin and orbital magnetic moments. 04 Mar, 2013 BL08W (High Energy Inelastic Scattering) Japan Synchrotron Radiation Research Institute (JASRI), University of Hyogo; S. Mørup, *Europhys. Lett.* 28, 1994, 671–677.

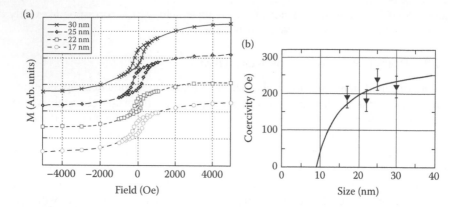

FIGURE 1.25 (a) Magnetization curves measured at 300 K for various sample sizes. (b) The coercivity plotted as a function of particle size. The solid line is the fitting according to the single-domain Stoner and Wohlfarth model. (Adapted from J. Carvell et al., *J. Appl. Phys.* 107, 2010, 103913.)

conditions: temperature, pressure, structure of the materials (crystalline/amorphous), local crystal environment (for a single atom) or the substrate (for a film) [131,132].

For iron alloys the coercivity has been shown to depend on the size of the structure. Reduction in particle size is accompanied by an increase in overall coercivity. Reduction in size below 1 um can produce coercivities, $\mu oHc = 0.5T$. However an extremely fast drop in coercivity occurs at nanometric dimensions (10 nm). Since coercivity depends on anisotropy, then the result may be considered as a signature of a significant drop in effective anisotropy. Indeed the coercivity drops as a function of D6 (Figure 1.25).

1.6 FABRICATION OF MAGNETIC NANOSTRUCTURED MATERIALS

1.6.1 DIFFERENT FABRICATION PROCESSES

Although some nanostructured materials require rather exotic approaches to synthesis, processing of many polymer matrix nanocomposites is quite straight forward. One such approach is the processing of clay/polymer nanocomposites where pretreated clay such as montmorillonite is mixed with polymer melts. There are also a number of other ways to process such materials, including reactive processes involving *in situ* polymerization. The low volume fraction of reinforcement particles allows the use of well-established and well-understood processing methods, such as extrusion and injection molding. The synthetic method should exhibit control of size so that the required property can be attained. Often the methods are divided into two main types: (1) bottom-up, and (2) top-down approach.

1.6.1.1 Bottom-Up Approach

Bottom-up methods involve the assembly of atoms or molecules into nanostructured arrays. In these methods the raw material sources can be in the form of gases, liquids, or solids. The latter require some sort of disassembly prior to their incorporation into a nanostructure. Bottom-up methods generally fall into two categories: (1) chaotic and (2) controlled.

1. *Chaotic processes*: Chaotic processes involve elevating the constituent atoms or molecules to a chaotic state and then suddenly changing the conditions so as to make that state unstable. Through the clever manipulation of any number of parameters, products form largely as a result of the insuring kinetics. The collapse from the chaotic state can be difficult or impossible to control and so ensemble statistics often govern the resulting size distribution and average size. Examples of chaotic processes are: laser ablation, exploding wire, arc, flame pyrolysis, combustion, and precipitation synthesis techniques [131].

2. *Controlled processes*: Controlled processes involve the controlled delivery of the constituent atoms or molecules to the site(s) of nanoparticle formation such that the nanoparticle can grow to a prescribed size in a controlled manner. Generally, the states of the constituent atoms or molecules are never far from that needed for nanoparticle formation. Accordingly, nanoparticle formation is controlled through the control of the state of the reactants. Examples of controlled processes are self-limiting growth solution, self-limiting chemical vapor precipitation, shaped pulse femtosecond laser techniques, and molecular beam epitaxy [1].

1.6.1.2   Top-Down Approach

The knowledge of processes for bottom-up assembly of structures remains in its infancy in comparison to traditional manufacturing techniques. As a result, the most mature products of nanotechnology rely heavily on top-down processes to define structures. The traditional example of a top-down technique for fabrication is lithography which is used to scale a macroscopic plan to the nanoscale. The most commonly used methods for the fabrication of magnetic nanostructured materials are as follows:

1. *Coprecipitation*: Converging advances in the understanding of the molecular biology of various diseases recommended the need of homogeneous and targeted imaging probes along with a narrow size distribution in between 10 and 250 nm in diameter. Developing magnetic nanoparticles in this diameter range is a complex process and various chemical routes for their synthesis have been proposed including microemulsions, sol-gel syntheses, sono-chemical reactions, hydrothermal reactions, hydrolysis and thermolysis of precursors, flow injection syntheses, and electrospray syntheses [132–140]. However, the most common method for the production of magnetite nanoparticles is the chemical coprecipitation of various iron salts [141–144]. The main advantage of the coprecipitation is that a large amount of nanoparticles can be synthesized though control on size distribution is limited. This is mainly due to kinetic factors that control the growth of the crystal. Coprecipitation is the most simplistic and

convenient way to synthesize iron oxides ($Fe_3O_4/\gamma-Fe_2O_3$) from aqueous Fe^{2+}/Fe^{3+} containing salt solutions by the addition of a base under inert atmosphere at room/elevated temperature. The size, shape, and composition of the formed magnetic nanoparticles depends very much on the type of salts used (e.g., chlorides, sulfates, nitrates), the Fe^{2+}/Fe^{3+} ratio, the reaction temperature, the pH value, ionic strength of the media [145] and the mixing rate with the base solution used to provoke the precipitation [146]. The coprecipitation approach has been used extensively to produce ferrite nanoparticles of controlled sizes and magnetic properties [51,147–149]. A variety of experimental arrangements have been reported to facilitate continuous and large-scale co-precipitation of magnetic particles by rapid mixing [150,151]. Recently, the growth rate of the magnetic nanoparticles was measured in real time during the precipitation of magnetite nanoparticles by an integrated AC magnetic susceptometer within the mixing zone of the reactants [152]. This method may be the most promising one because of its simplicity and productivity [155]. It is widely used for biomedical applications because of the ease of implementation and need for less hazardous materials and procedure. The reaction principle is simply [153]

$$Fe^{2+} + 2Fe^{3+} + 8\ OH^- \Leftrightarrow Fe(OH)_2 + Fe(OH)_3 \rightarrow Fe_3O_4 + 2H_2O \qquad (1.40)$$

At first, ferrous hydroxide suspensions are partially oxidized with different oxidizing agents [154]. For example, spherical magnetite particles of narrow size distribution with mean diameters between 30 and 100 nm can be obtained from Fe(II) salt, a base, and a mid-oxidant (nitrate ions). The other method consists in aging stoichiometric mixtures of ferrous and ferric hydroxides in aqueous media, yielding spherical magnetite particles homogeneous in size [155].

2. *Thermal decomposition*: Magnetic nanocrystals with smaller size can essentially be synthesized through the thermal decomposition of organometallic compounds in high-boiling organic solvents containing stabilizing surfactants [151]. For example, the decomposition of iron precursors in the presence of hot organic surfactants has yielded markedly improved samples with good size control, narrow size distribution, and good crystallinity of individual and dispersible magnetic iron oxide nanoparticles [156].

3. *Microemulsion*: Using the microemulsion technique, metallic cobalt, cobalt/platinum alloys, and gold-coated cobalt/platinum nanoparticles have been synthesized in reverse micelles of cetyltrimethlyammonium bromide, using 1-butanol as the cosurfactant and octane as the oil phase [151,156–158].

4. *Flame spray synthesis*: Using flame spray pyrolysis and varying the reaction conditions, oxides, metal, or carbon-coated nanoparticles are produced at a rate of >30 g/h [67] as shown schematically in Figure 1.26.

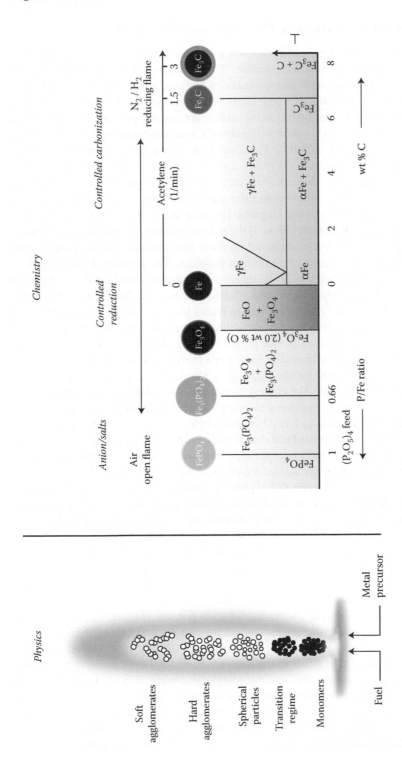

FIGURE 1.26 Various flame spray conditions and their impact on the resulting nanoparticles. (Redrawn based on R. N. Grass et al., *Angew. Chem. Int. Ed.* 46(26), 2007, 4909–4912.)

1.6.2 STABILITY OF MAGNETIC NANOSTRUCTURED MATERIALS

Nanoparticles of some metals are known to be pyrophoric, that is, they spontaneously ignite in air at room temperature. Thus encapsulation of nanoparticles and the preparation of magnetic nanoparticles in a protective coating is a widely used protection and stabilization method. Carbon is often used as the protective coating. The carbon layers formed on the metal surface are usually graphite-like and, hence, conductive. In those cases, where an electrically insulating coating is required, boron nitride layers are used [159,160]. Encapsulation of magnetic nanoparticles makes them stable against oxidation, corrosion, and spontaneous aggregation, which allows them to retain the single-domain structure. The magnetic particles coated by a protective shell can find application in information recording media, for example, as magnetic toners in xerography, magnetic ink, contrasting agents for magnetic resonance images, ferrofluids, and so on. If the nanosized magnetic particles are retained after compaction, the materials based on them can serve as excellent initial components for the preparation of permanent magnets. The coating of metal particles by carbon was first observed in the research of heterogeneous catalysis almost 50 years ago. Subsequently, this process (deleterious to oil refining and other industries) was comprehensively studied and, in recent years, it has been used deliberately to stabilize nanoparticles. The first structurally characterized carbon-encapsulated nanoparticles were obtained as side products in the electric arc synthesis of fullerenes. Subsequently, special studies have been carried out to identify the possibilities of using this method for the targeted synthesis of encapsulated nanoparticles, especially magnetic ones. The coating of magnetic nanoparticles by a thin layer of a nonmagnetic metal is considered to be a promising method for their stabilization, for example, the synthesis of Fe_3O_4 nanoparticles (5 nm) coated by metallic gold [161,162]. Much attention has been devoted in recent years to methods for the formation of thin polymeric coatings (especially those based on biocompatible and readily biodegradable polymers) on the surface of magnetic nanoparticles [163].

Self-assembled monolayers (SAM) on the nanoparticle surfaces are represented by monomolecular layers of amphiphilic molecules, which protect the particles from aggregation and simultaneously stabilize their suspensions (solutions) in certain solvents. A typical example is the self-assembly of a monolayer of amphiphilic molecules of fatty acids on the Fe_3O_4 nanoparticle surface [164]. Freshly prepared nanoparticles (obtained by the standard procedure by treatment of a mixture of Fe^{2+} and Fe^{3+} chlorides with aqueous NH_3) were washed, separated into fractions by centrifuging, and treated with an excess of lauric or decanoic acid, whose molecules were adsorbed onto the surface of each particle (Figure 1.27). The protective role of amphiphilic molecules was manifested most clearly in the synthesis of $CoPt_3$ nanoparticles of a strictly specified size (1.5–7.2 nm) [165]. The success of the synthesis is related first of all to the use of a new stabilizing agent, 1-adamantanecarboxylic acid. The effect of cationic (cetyltrimethylammonium bromide, CTAB) and anionic (sodium didecylbenzenesulfonate, DBS) surfactants on the stabilization of γ–Fe_2O_3 nanoparticles (4–5 nm) has been studied [166]. A nanoparticle having a large number of defects and dangling bonds on the surface is considered to interact with the surfactant rather strongly. This interaction has a pronounced influence

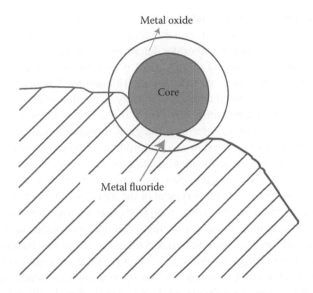

FIGURE 1.27 Model of the structure of a metallic nanoparticle stabilized by PTFE. (Redrawn based on S. P. Gubin et al., *Russian Chemical Reviews* 74(6), 2005, 489–520.)

on the electronic structure of the particle surface [167]. Self-assembled monolayers are needed to create aqueous dispersions of magnetic nanoparticles [168]. Diverse magnetic colloid isotropic (magnetic emulsions and vesicles), anisotropic (steric and electrostatic ferrosmectics), and lyotropic ferronematics have been obtained on the basis of nanoparticles coated by a surfactant monolayer [169].

1.7 CHARACTERIZATION OF MAGNETIC NANOSTRUCTURED MATERIALS

The first observations and size measurements of nanoparticles were made during the first decade of the twentieth century. Zsigmondy made detailed studies of gold sols and other nanomaterials with sizes down to 10 nm and beyond as reported in his book in 1914 [172]. He used an ultramicroscope that employs a dark field method for seeing particles with sizes much less than light wavelength. There are traditional techniques developed during the twentieth century in interface and colloid science for characterizing nanomaterials. These methods include several different techniques for characterizing particle size distribution. This characterization is imperative because many materials that are expected to be nanosized are actually aggregated in solutions. Some of the methods are based on light scattering. Others apply ultrasound, such as ultrasound attenuation spectroscopy for testing concentrated nanodispersions and microemulsions [171]. There are also other groups of traditional techniques for characterizing surface charge or zeta potential of nanoparticles in solutions. This information is required for proper system stabilization, preventing its aggregation or flocculation. These methods include microelectrophoresis, electrophoretic light scattering, and electroacoustics. The last one, for instance, the colloid vibration current method, is suitable for characterizing concentrated systems.

However, there is no unique method for determination of the nanoparticle composition and dimensions. As a rule of thumb, TEM is used for size, shape, and structure; XRD for phase as well as size; particle size distribution by dynamic laser scattering analyzer (DLS) [172]; chemical composition analysis by atomic absorption spectroscopy; magnetite content determination by TGA, ICP-AES; and DSC to analyze structure of the copolymer block to determine the glass transition temperature (T_g). X-ray diffraction analysis of nanomaterials seldom produces diffraction patterns with a set of narrow reflections adequate for identification of the composition of the particles they contain. Some x-ray diffraction patterns exhibit only two or three broad-ended peaks of the whole set of reflections typical of the given phase. First of all, this is the case for freshly prepared samples containing nanoparticles with dimensions of several nanometers. To obtain more reliable information on the composition of these samples, they are "hardened" on heating, which makes the x-ray diffraction pattern more informative. In the case of larger particles, it is often possible not only to determine the phase composition but also to estimate the size of coherent x-ray scattering areas, corresponding to the average crystalline size. This is usually done by the Scherer formula:

$$d = \frac{0.9\lambda(54.7)}{b\cos\theta} \qquad (1.41)$$

where λ is the x-ray wavelength, b is the width at half-height of the reflection after correction for the instrumental broadening, and 2θ is the diffraction angle. In some cases, it is possible to identify fine structural effects, for example, phase transitions in Co metal particles. In most cases, the synthesis of cobalt nanoparticles at moderate temperatures gives the high-temperature FCC β-Co phase whose heating and cooling (down to 28 K) do not result in phase transformations [173]. It was found that long-term (for more than a week) maintenance of samples containing β-Co nanoparticles in air did not cause any significant changes in their x-ray diffraction patterns. It was also shown that even in thin Co films alternating with thicker Cu layers, the FCC structure of the β-phase is retained. The modern possibilities of the use of synchrotron radiation for determination of the structures of magnetic materials have been discussed in the literature [155]. The nanoparticle dimensions are determined most often using TEM, which directly shows the presence of nanoparticles in the material under examination and their arrangement relative to one another (Figures 1.28 and 1.29). The phase composition of nanoparticles can be derived from electron diffraction patterns recorded for the same sample during the investigation. Note that in some cases, TEM investigations of dynamic processes are also possible. For example, the development of dislocations and declinations in the Mössbauer spectroscopy provides data on the phase composition of nanoparticles, especially, magnetic phases. The method is widely used to determine the structure of Fe-containing nanoparticles [174–176]. For example, Mössbauer spectroscopy has been used to establish the composition of magnetic particles formed upon thermal decomposition of $Fe(CO)_5$ in a polyethylene matrix [177]. It was shown that the particles consist of α–Fe, γ–Fe_2O_3,

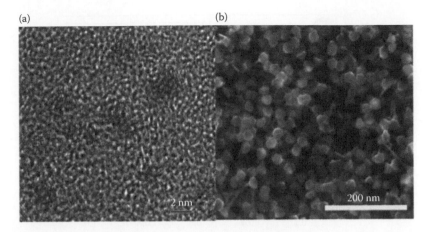

FIGURE 1.28 Micrograph of a sample containing Co nanoparticles: (a) TEM image and (b) SEM image. (Adapted from Q. Liu et al., *Rsc Adv.* 5, 2015, 4861–4871.)

and iron carbides in ratios depending on the nanoparticle concentration in the matrix. Magnetic density of ferromagnet is measured by SJZP-1 Gauss/Tesla Meters and the magnetic property using a vibrating sample magnetometer [178].

Researchers use polarized neutrons or those with a magnetic spin in one direction, to determine the magnetic qualities of materials. Polarized neutron reflectometry is being used to study the magnetism at the interfaces between different materials in epitaxial heterostructures based on lanthanum manganese oxide ($LaMnO_3$) as shown in Figure 1.30.

The first experimental data on the properties of magnetic nanoparticles were obtained in experiments with cluster beams [179–181]. Despite the difficulty of interpretation, these experiments provided the unique possibility of determining the

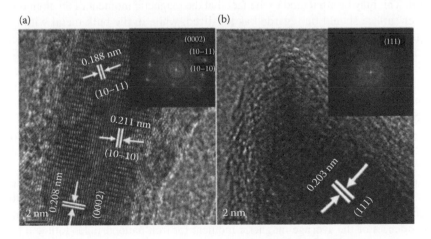

FIGURE 1.29 Typical HRTEM micrograph of a cobalt nanoparticle showing atomic fringes at (a) (0002) and (b) (111) crystallographic orientation. (Adapted from Q. Liu et al., *Rsc Adv.* 5, 2015, 4861–4871.)

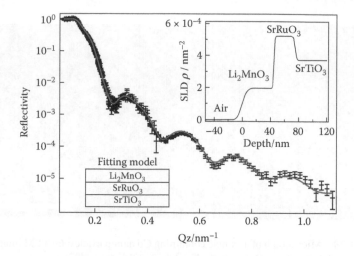

FIGURE 1.30 Neutron reflectometry results for an epitaxial Li_2MnO_3(001) film (inset show the fitting model and the SLD profile calculated using the fitting parameters). (Adapted from S. Taminato et al., *Chem. Commun.* 51, 2015, 1673–1676.)

dependence of magnetic parameters on the number of atoms in the nanoparticle. Notably, the specific magnetic moment of a nanoparticle increases with a decrease in particle size. This trend is most pronounced for Ni, which may be due to higher density of the valence electrons [180]. The Co nanoparticles with a number of atoms n = 56–215 are superparamagnetic at 97 K having average atomic magnetic moment of 2.24 μ_B which is higher than the values for macroscopic samples of cobalt [179–181]. Thus, the experimental data imply that the effective magnetic moment of atoms in 3d-metal nanoparticles can be greater than its magnetic moment in bulk metal. This may be attributed to the fact that the magnetic moment of an atom on the cluster surface should be regarded as localized, while in the bulk metal with band magnetism as delocalized. It is noteworthy that for the smallest Ni nanoparticles, the magnetic moment barely changes with temperature over the whole temperature range studied, due to the large number of surface atoms [180]. For larger clusters, a nonzero contribution of surface atoms into the magnetic moment is retained even at temperatures above 631 K (the Curie temperature of the bulk phase). For Ni550–600 nanoparticles, the magnetic moment at 631 K accounts to 25% of the low-temperature value 0.6 mB. Thus, it shows that magnetic order is retained in nanoparticles at higher temperatures than in the bulk samples [180,181], which is manifested as an increase in the Curie temperature with respect to bulk phases. For cobalt (T_C = 1400 K), the atomic magnetic moment in nanoparticles with the numbers of atoms n = 50–600 changes only slightly with an increase in the temperature to about 1000 K and always remains greater than the bulk value. The temperature dependences of the average magnetic moment for iron nanoparticles follow a more complex pattern. A possible reason is the structural transitions, which complicate the pattern of magnetic behavior. No transitions to the conventional paramagnetic state were detected up to the highest temperatures (about 1000 K).

1.8 APPLICATION OF MAGNETIC NANOSTRUCTURED MATERIALS

A wide variety of applications have been envisaged for magnetic nanoparticles as summarized hereafter:

1. *Medical diagnostics and treatments*: Magnetic nanoparticles are used in an experimental cancer treatment called magnetic hyperthermia [182] in which the fact that nanoparticles heat when they are placed in an alternative magnetic field is used. Another potential treatment of cancer includes attaching magnetic nanoparticles to free-floating cancer cells, allowing them to be captured and carried out of the body. The treatment has been tested in the laboratory on mice and will be looked at in survival studies [183,184]. Moreover, magnetic nanoparticles coated with antibodies targeting cancer cells or proteins are also under investigation. The magnetic nanoparticles can be recovered and the attached cancer-associated molecules can be assayed to test for their existence. Magnetic nanoparticles can be conjugated with carbohydrates and used for the detection of bacteria. Iron oxide particles have been used for the detection of Gram-negative bacteria such as *Escherichia coli* and Gram-positive bacteria such as *Streptococcus suis* [185–190]. In a recent article from Harvard Medical School [190]: "Researchers from Harvard medical school and Massachusetts general hospital have developed a magnetic nanoparticle-based MRI technique for predicting whether-and when-subjects with a genetic predisposition for diabetes will develop the disease. While done initially in mice, preliminary data show that the platform can be used in people as well, so far to distinguish patients that do or do not have pancreas inflammation." "This research is about predicting Type-1 diabetes, and using that predictive power to figure out what is different between those who get it and those who don't get it," according to Diane Mathis, professor of immunohematology in the department of microbiology and immunobiology [*Nature Immunology* Feb. 26, 2012]. This shows that the progression of the disease, at least in this animal model, is determined very early in life and that diabetes does not require an additional trigger such as a secondary infection or environmental …' [187].

2. *Magnetic immunoassay*: Magnetic immunoassay (MIA) [188] is a novel type of diagnostic method utilizing magnetic beads as labels instead of conventional enzymes, radioisotopes, or fluorescent moieties. This assay involves specific binding of an antibody to its antigen, where a magnetic label is conjugated to one element of the pair. The presence of magnetic beads is then detected by a magnetic reader (magnetometer) which measures the magnetic field change induced by the beads. The signal measured by the magnetometer is proportional to the analyte (virus, toxin, bacteria, cardiac marker, etc.) quantity in the initial sample. One issue with the use of magnetic nanoparticles in medical treatment is the risk of allergic reactions. While these materials are not known to be dangerous inside the human body, they might be able to trigger an immune response in the concentrations needed

for medical therapy. This could make the patient more ill with a cascading series of inflammatory reactions as the body attempts to fight the invader. Clinical trials can determine the level of risk and may uncover various approaches to mitigating the risk of such reactions.

3. *Biomedical imaging*: There are many applications for iron-oxide based nanoparticles in concert with magnetic resonance imaging [189]. Magnetic CoPt nanoparticles are being used as an MRI contrast agent for transplanted neural stem cell detection [190]. Magnetic nanoparticles have a number of applications outside of biological systems where their high surface area and ability to disperse in solution can provide unique benefits.

4. *Genetic engineering*: Magnetic nanoparticles can be used for a variety of genetics applications. One application is the isolation of mRNA. This can be done quickly—usually within 15 minutes. In this particular application, the magnetic bead is attached to a poly T tail. When mixed with mRNA, the poly A tail of the mRNA will attach to the bead's poly T tail and the isolation takes place simply by placing a magnet on the side of the tube and pouring out the liquid. Magnetic beads have also been used in plasmid assembly. Rapid genetic circuit construction has been achieved by the sequential addition of genes onto a growing genetic chain, using magnetic nanobeads as an anchor (Figure 1.31). This method has been proven much faster than previous methods, taking less than an hour to create functional multigene constructs *in vitro* [191].

5. *Waste water treatment*: Due to easy separation by applying a magnetic field and very large surface to volume ratio, magnetic nanoparticles have a good potential for the treatment of contaminated water [193]. In this method, attachment of an EDTA-like chelating agent to carbon-coated metal nanomagnets results in a magnetic reagent for the rapid removal of heavy metals

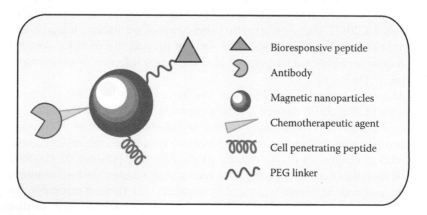

FIGURE 1.31 Schematic diagram representing the functionalization of magnetic nanoparticles with bio-responsive peptide, PEG linker, chemotherapeutic agent, antibody, and cellpenetrating peptide. (Adapted from V. V. Mody et al., *J. Pharm. Bioallied Sci.* 2(4), 2010, 282–289.)

from solutions or contaminated water by three orders of magnitude to concentrations as low as micrograms per liter. Such particles have been shown to serve as recyclable supports in heterogeneous catalysis allowing easy separation of the catalyst post reaction with an induced magnetic field. A similar process has been shown to be effective for the removal of contaminants from water using functionalized magnetic nanoparticles.

6. *Chemistry*: Magnetic nanoparticles are being used or have the potential to be used as a catalyst or catalyst supports [193]. In chemistry, a catalyst support is the material, usually a solid with a high surface area, to which a catalyst is affixed. The reactivity of heterogeneous catalysts occurs at the surface atoms. Consequently, great effort is made to maximize the surface area of a catalyst by distributing it over the support. The support may be inert or participate in the catalytic reactions. Typical supports include various kinds of carbon, alumina, and silica.

7. *Information storage*: Research is taking place toward the use of magnetic nanoparticles for magnetic recording media. The most promising candidate for high-density storage is the face-centered tetragonal FePt alloy with grain sizes as small as 3 nm. If it is possible to modify the magnetic nanoparticles at this small scale, the information density that can be achieved with this media could easily surpass 1 terabyte per square inch [194]. They can also be used in information storage, with a potential for very high capacity storage through the use of magnetic nanoparticles. The development of very small, dense storage media is critical for a number of industries as the demands for storage space increase while consumers expect to see decreases in size for devices like external hard drives and thumb drives. For materials used in magnetic recording, the coercivity must be great enough to ensure the stability of the stored data, but it must not be so strong that the magnetization cannot be reversed, as required during the writing process, when a field of moderate strength is applied. Most of the materials used here are cobalt alloys which exhibit coercivity in the 3d series [195].

In the case of data recording on a disc, one bit requires about 109 atoms, whereas with the use of nanomaterials (particles 10 nm in diameter), not more than 105 atoms are required. Thus, transition to magnetic nanomaterials increases the information recording density by a factor of 10^3–10^4. In order to achieve independent recording of data on each bit, materials used for magnetic recording are made from decoupled nanoparticles. It is not known how to control the orientations of the particles, and the direction of the moments fluctuates from one nanoparticle to another. To reduce read noise, one bit of information comprises some 1000 particles, and the average magnetization direction is then much better defined than that of the individual particles. At today's densities of 20 Gbits/cm^2, the bit size is around 0.2×0.1 μm^2 and the volume of the particles is around 100 nm^3. The very small size of the particles raises the question of stability for the recorded data. For a characteristic fluctuation time of 100 years and assuming that $\tau_0 \approx 10^{-9}$ s, the barrier height must be of the order of 40 kT and it follows that K = 10^6 J/m^3, a very high level of anisotropy.

1.9 CONCLUDING REMARKS

The field of nanoscience has blossomed over the last two decades and the need for nanotechnology to explore beyond cell walls has become more important. Nanoparticles have successfully come to aid various disease states, but advances in biomedical imaging depend largely on the shape, size, and selectivity of the nanoparticle to the target. Moreover, the type of the particle synthesized also governs the imaging modality to be used and thus the cost of diagnosis. Even though current investigations have demonstrated that multivalent composite materials can provide significant advantages, the ambiguity in developing them for a particular target with high specificity is still challenging. Fortunately, the field of nanotechnology continues to grow in interest with major discoveries as well as new scientific challenges. Nevertheless, the future studies should also aim to address safety and biocompatibility of these nanoparticles, in particular long-term toxicities. Additional clinical studies on humans and on animal models should be performed to substantiate their use especially in biomedical imaging using MRI, computed tomography, ultrasound, positron emission tomography, surface-enhanced Raman scattering (SERS), and optical imaging.

REFERENCES

1. S. H. Ng, Nanostructured materials for electrodes in lithium-ion batteries, PhD thesis, University of Wollongong, 2007.
2. A. M. Thayer, Nanomaterials. *Chem. Eng. News*, 13(84), 2006, 47–52, ISSN: 0009–2357.
3. A. M. Thayer, Nanomaterials. *Chem. Eng. News*, 35(81), 2003, 15–22, ISSN: 0009-2347.
4. J. Gribbin, M. Gribbinn, *Richard Feynman: A Life in Science*, Dutton Adult, 1997, ISBN: 0-525-94124-4.
5. N. Taniguchi, The state of the art of nanotechnology for processing of ultraprecision and ultrafine products. *Precis. Eng.*, 16(1), 1994, 5–24.
6. G. Binning, H. Rohrer, Scanning tunneling microscopy—from birth to adolescence. *Rev. Mod. Phys.*, 59, 1987, 615–623.
7. E. Stoner, E. Wohlfarth, A mechanism of magnetic hysteresis in heterogeneous alloys. *Philos. Trans. R. Soc. London*, A240, 1948, 599–610.
8. R. Coehoorn, D. B. de Mooij, D. de Waard, Meltspun permanent magnet materials containing Fe_3B as the main phase. *J. Magn. Magn. Mat.* 80, 1989, 101–105.
9. D. Givord, Magnetism in nanomaterials, in *Nanomaterials and Nanochemistry*, C. Bréchignac, P. Houdy, and M. Lahmani (eds.), Springer-Verlag, Berlin Heidelberg, Germany, 2007, pp. 101–134, ISBN: 978-3-540-72992-1 (print)/978-3-540-72993-8 (online).
10. V. V. Mody, A. Singh, B. Wesley, Basics of magnetic nanoparticles for their application in the field of magnetic fluid hyperthermia. *Eur. J. Nanomed.* 5(1), 2013, 11–21.
11. High-accuracy magnetic property measurement method by separating spin and orbital magnetic moments. 04 Mar, 2013 BL08W (High Energy Inelastic Scattering) Japan Synchrotron Radiation Research Institute (JASRI), University of Hyogo.
12. G. M. Pastor, J. Dorantes-Davila, K. H. Bennemann, Size and structural dependence of the magnetic properties of small 3D-transition-metal clusters. *Phys. Rev. Lett.* 40, 1989, 7642–7451.
13. P. Bruno, Physical origins and theoretical models of magnetic anisotropy, in *Magnetismus von Festkörpern und grenzflächen*, P. H. Dederichs, P. Grünberg, W. Zinn (eds.), 24 IFF-Ferienkurs, Forschungszentrum Jülich 1993, 24–28.

14. K. J. Klabunde, *Nanoscale Materials in Chemistry*, Wiley-Interscience, New York, 2001.
15. B. D. Cullity, *Introduction to Magnetic Materials*, John Wiley & Sons, New Jersey, 1972, pp. 214–250.
16. K. M. Krishnan, A. B. Pakhomov, Y. Bao, P. Blomqvist, Y. Chun, M. Gonzales, K. Griffin, X. Ji, B. K. Roberts, Nanomagnetism and spin electronics: Materials, microstructure and novel properties. *J. Mater. Sci.* 41, 2006, 793–815.
17. E. Kneller, R. Hawig, The exchange-spring magnet: A new material principle for permanent magnets. *IEEE Trans. Mag.* 27, 1991, 3588–2592.
18. B. J. Hickey, M. A. Howson, D. Greig, N. Wiser, Enhanced magnetic anisotropy energy density for superparamagnetic particles of cobalt. *Phys. Rev.* 53(1), 1996, 32–33.
19. R. Coehoorn, D. B. de Mooij, J. P. W. B. Duchateau, K. H. J. Buschow: Novel permanent magnetic materials made by rapid quenching. *J. Phys. (Paris), Colloq.* 49, 1988, C8–669.
20. F. F. Lange, M. Metcalf, Processing-related fracture origins: II, agglomerate motion and cracklike internal surfaces caused by differential sintering. *J. Am. Ceram. Soc.* 66(6), 1983, 398–401.
21. Nanotechnologies vocabulary—Part 1: Core terms. Geneva 2011 ISO/TS 800004-1.
22. T. Fukui, K. Murata, S. Ohara, H. Abe, M. Naito, K. Kiyoshi, Morphology control of Ni–YSZ cermet anode for lower temperature operation of SOFCs. *J. Power Sources* 125(1), 2004, 17–21.
23. P. Barbaro, F. Liguori (eds.), *Heterogenized Homogeneous Catalysts for Fine Chemicals Production: Materials and Processes*. Springer, Dordrecht, 2010, ISBN: 978-90-481-3695-7.
24. S. Zalesskiy, V. Ananikov, $Pd_2(dba)_3$ as a precursor of soluble metal complexes and nanoparticles: Determination of palladium active species for catalysis and synthesis. *Organometallics*, 31(6), 2012, 2302–2309.
25. A. Roucoux, J. Schulz, H. Patin, Reduced transition metal colloids: A novel family of reusable catalysts? *Chem. Rev.* 102(10), 2002, 3757–3778.
26. L. M. Buil, A. M. Esteruelas, S. Niembro, M. Oliván, L. Orzechowski, C. Pelayo, A. Vallribera, Dehalogenation and hydrogenation of aromatic compounds catalyzed by nanoparticles generated from rhodium bis(imino)pyridine complexes. *Organometallics* 29(19), 2010, 4375–4383.
27. W. Yu, H. Liu, M. Liu, Z. Liu, Selective hydrogenation of citronellal to citronellol over polymer-stabilized noble metal colloids. *React. Funct. Polym.* 44(1), 2000, 21–29.
28. W. Yu, M. Liu, H. Liu, X. An, Z. Liu, X. Ma, Immobilization of polymer-stabilized metal colloids by a modified coordination capture: Preparation of supported metal colloids with singular catalytic properties. *J. Mol. Catal. A Chem.* 142(2), 1999, 201–211.
29. W. Yu, M. Liu, H. Liu, X. Ma, Z. Liu, Preparation, characterization, and catalytic properties of polymer-stabilized ruthenium colloids. *J. Colloid Interface Sci.* 208(2), 1998, 439–444.
30. M. Tamura, H. Fujihara, Chiral bisphosphine BINAP-stabilized gold and palladium nanoparticles with small size and their palladium nanoparticle-catalyzed asymmetric reaction. *J. Am. Chem. Soc.* 125(51), 2003, 15742–15743.
31. P. W. N. M. van Leeuwen, J. C. Chadwick, *Homogeneous Catalysts: Activity, Stability, Deactivation*, Wiley-VCH, Weinheim, Germany, ISBN: 978-3-527-32329-6.
32. B. C. Yadav, Nanotechnology–An introduction and its application. *Lucknow J. Sci.*, 4(2), 2007, 45–53.
33. Sarrajusneha, Nano-materials and Nano-technology. *Adv. Electronic Electric Eng.* 3(5), 2013, 575–578.

34. M. F. Thorpe, D. Tománek, R. J. Enbody (eds.), *Science and Application of Nanotubes*, Springer US, New York, ISBN: 978-0-306-46372-3 (Print) 978-0-306-47098-1 (Online).

35. F. K. Yu, I. P. Suzdalev, Size effects in small particles in Fe_3O_4. *Zh. Eksp. Teor. Fiz.* 67, 1974, 736–741.

36. R. N. Panda, N. S. Gajbhiye, G. Balaji, Magnetic properties of interacting single domain Fe_3O_4 particles. *J. Alloys Compd.* 326(1), 2001, 50–53.

37. J. Ding, W. F. Miao, E. Pirault, R. Street, P. G. McCormick, Mechanical alloying of iron-hematite powders. *J. Alloys Compd.* 267, 1998, 199–205.

38. C. J. W. Koch, M. B. Madsenm, S. Mùrup, Decoupling of magnetically interacting crystallites of goethite. *Hyperfine Interact.* 28, 1986, 549–452.

39. S. Mùrup, T. M. Meaz, C. B. Koch, H. C. B. Hansen, Nanocrystallinity induced by heating. *Z. Phys. D* 40, 1997, 167–172.

40. T. Meaz, C. B. Koch, S. Mùrup, An investigation of trivalent substituted Mtype hexagonal ferrite using x-ray and Mössbauer, in *Proceedings of the Conference ICAME-95*, Bologna, 50, 1996, pp. 525–530.

41. J. van Wonterghem, S. Mùrup, Preparation of ultrafine amorphous iron-carbon alloy particles on a carbon support. *J. Phys. Chem.* 92, 1988, 1013–1017.

42. V. I. Nikolaev, A. M. Shipilin, E. N. Shkolnikov, I. N. Zaharova, Elastic properties of magnetic fluids. *J. Appl. Phys.* 86, 1999, 576–581.

43. T. Upadhyay, R. V. Upadhyay, R. V. Mehta, V. K. Aswal, P. S. Goyal, Characterization of a temperature-sensitive magnetic fluid. *Phys. Rev. B* 55, 1997, 5585–5591.

44. J. van Wonterghem, S. Mùrup, S. W. Charles, S. Wells, Formation of a metallic glass by thermal decomposition of $Fe(CO)_5$. *J. Colloid Interface Sci.* 121, 1998, 558–562.

45. R. Rosensweig, Magnetic fluids. *Ann. Rev. Fluid Mech.* 19, 1987, 437–442.

46. K. Raj, R. F. Moskowitz, A review of damping applications of ferrofluids. *IEEE Trans. Magn.* 16, 1980, 358–363.

47. H. Matsuki, K. Murakami, Performance of an automatic cooling device using a temperature-sensitive magnetic fluid. *J. Magn. Magn. Mater.* 65, 1987, 363–367.

48. K. Nakatsuka, Y. Hama, J. Takahashi, Heat transfer in temperature-sensitive magnetic fluids. *J. Magn. Magn. Mater.* 85, 1990, 207–212.

49. J. Liu, E. M. Lawrence, A. Wu, M. L. Ivey, G. A. Flores, K. Javier, J. Bibette, Field-induced structures in ferrofluid emulsions. *J. Richard Phys. Rev. Lett.* 74, 1995, 2828–2832.

50. J. Allouche, C. Chanéac, R. Brayner, M. Boissière, T. Coradin, Design of magnetic gelatine/silica nanocomposites by nanoemulsification: Encapsulation versus in situ growth of iron oxide colloids. *Nanomaterials* 4(3), 2014, 612–627.

51. A.-H. Lu, E. L. Salabas, F. Schüth, Magnetic nanoparticles: Synthesis, protection, functionalization, and application. *Angew. Chem. Int. Ed.* 46(8), 2007, 1222–1244.

52. D. K. Kim, M. Mikhaylova, Protective coating of superparamagnetic iron oxide nanoparticles. *Chem. Mater.* 15(8), 2003, 1617–1627.

53. M. Kiwi, Exchange bias theory. *J. Magn. Mater.* 234, 2001, 584–591.

54. A. E. Berkowitz, K. Takano, Exchange anisotropy—A review. *J. Magn. Mater.* 200, 1999, 552–557.

55. W. H. Meiklejohn, C. P. Bean, New magnetic anisotropy. *Phys. Rev. B* 102, 1956, 1413–1417.

56. S. Sako, K. Ohshima, M. Sakai, S. Bandow, Magnetic property of CoO ultrafine particle. *Surf. Rev. Lett.* 3, 1996, 109–115.

57. L. D. Bianco, A. Hernando, M. Multigner, C. Prados, J. C. Sanchez-Lopez, A. Fernandez, C. F. Conde, A. Conde, Evidence of spin disorder at the surface–core interface of oxygen passivated Fe nanoparticles. *J. Appl. Phys.* 84, 1998, 2189–2195.

58. M. Sato, S. Kohiki, Y. Hayakawa, Y. Sonda, T. Babasaki, H. Deguchi, M. Mitome, Dilution effect on magnetic properties of Co_3O_4 nanocrystals. *J. Appl. Phys.* 88, 2000, 2771–2775.

59. J. Feng, H. C. Zeng, Size-controlled growth of Co_3O_4 nanocubes. *Chem. Mater.* 15, 2003, 2829–2832.

60. S. Sun, H. Zeng, Size-controlled synthesis of magnetite nanoparticles. *J. Am. Chem. Soc.* 124, 2002, 82048210.

61. S. R. Hoon, M. Kilner, G. J. Russel, B. K. Tanner, Preparation and properties of nickel ferrofluids. *J. Magn. Mater.* 39, 1983, 107–112.

62. D. deCaro, J. S. Bradley, Surface spectroscopic study of carbon monoxide adsorption on nanoscale nickel colloids prepared from a zerovalent organometallic precursor. *Langmuir* 13, 1997, 3067–3072.

63. K.-L. Tsai, J.-L. Dye, Synthesis, properties, and characterization of nanometer-size metal particles by homogeneous reduction with alkalides and electrides in aprotic solvents. *Chem. Mater.* 5, 1993, 540–543.

64. U. Gonser, C. J. Meechan, A. H. Muir, H. Wiedersich, Determination of Néel temperatures in fcc iron. *J. Appl. Phys.* 34, 1963, 2373–2375.

65. S. C. Abrahams, L. Guttman, J. S. Kasper, Neutron diffraction determination of antiferromagnetism in face-centered cubic (γ) iron. *Phys. Rev.* 127, 1962, 2052–2057.

66. U. Gonser, H. G. Wagner, Some recent developments in the applications of Mössbauer spectroscopy to physical metallurgy. *Hyperfine Interact.* 24–26, 1985, 769–772.

67. R. N. Grass, N. Robert, E. K. Athanassiou, W. J. Stark, Covalently functionalized cobalt nanoparticles as a platform for magnetic separations in organic synthesis. *Angew. Chem. Int. Ed.* 46(26), 2007, 4909–4912.

68. X. G. Li, A. Chiba, S. Takahashi, Preparation and magnetic properties of ultrafine particles of Fe-Ni alloys. *J. Magn. Mater.* 170, 1997, 339–342.

69. A. M. Afanas'ev, I. P. Suzdalev, M. Y. Gen, V. I. Gol'danskii, V. P. Korneev, E. A. Manykin, Magnetic structure of small weakly nonspherical ferromagnetic particles. *Zh. Eksp. Teor. Fiz.* 58, 1970, 115–221.

70. B. K. Rao, S. R. de Debiaggi, P. Jena, Structure and magnetic properties of Fe-Ni clusters. *Phys. Rev. B* 64, 2001, 024418.

71. S. Sun, C. B. Murray, D. Weller, L. Folks, A. Moser, Monodisperse FePt nanoparticles and ferromagnetic FePt nanocrystal superlattices. *Science* 287, 2000, 1989–1992.

72. E. Shevchenko, D. Talapin, A. Kornowski, F. Wiekhorst, J. Kltzler, M. Haase, A. Rogach, H. Weller, Gold (core)–iron oxide (hollow shell) nanoparticles. *Adv. Mater.* 14, 2002, 287–293.

73. H. Zeng, J. Li, J. P. Liu, Z. L. Wang, S. Sun, Exchange-coupled nanocomposite magnets by nanoparticle self-assembly. *Nature* 420, 2002, 395–399.

74. S. H. Johnson, C. L. Johnson, S. J. May, S. Hirsch, M. W. Cole, J. E. Spanier, Co@CoO@Au core-multi-shell nanocrystals. *J. Mater. Chem.* 20(3), 2010, 439–442.

75. R. N. Grass, N. Rober, W. J. Stark, Co nanoparticles by reducing flame pyrolysis. *J. Mater. Chem.* 16, 2006, 1825–1832.

76. S. Chikazumi, *Physics of Ferromagnetism*, Oxford University Press, UK, 2009, 118–135.

77. R. Jackson, John Tyndall and the early history of diamagnetism. *Ann. Sci.* 72(4), 2015, 435–489.

78. J. Zou, Y.-G. Peng, Y.-Y. Tanga, A facile bi-phase synthesis of Fe_3O_4@SiO_2 core–shell nanoparticles with tunable film thicknesses. *Rsc. Adv.* 4, 2014, 9693–9700.

79. S. V. Vonsovskii, R. Hardin, *Magnetism*, John Wiley & Sons, New York, 1974, Vol. 1.

80. J. W. McClure, Diamagnetism of graphite. *Phys. Rev.* 104, 1956, 666–669.

81. A. G. Kolhatkar, A. C. Jamison, D. Litvinov, R. C. Willson, T. R. Lee, Tuning the magnetic properties of nanoparticles. *Int. J. Mol. Sci.* 14(8), 2013, 15977–16009.

82. G. I. Likhtenshtein, J. Yamauchi, S. Nakatsuji, A. I. Smirnov, R. Tamura, Fundamentals of magnetism, in *Nitroxides: Applications in Chemistry, Biomedicine, and Materials Science*, G. I. Likhtenshtein, J. Yamauchi, S. Nakatsuji, A. I. Smirnov, R. Tamura (eds.), Wiley-VCH Verlag GmbH & Co. KGaA, Weinheim, Germany, 2008.

83. D. J. Griffiths, *Introduction to Electrodynamics,* 3rd ed., Prentice Hall, New Jersey, 1999, pp. 255–258.

84. C. Kittel, Theory of the structure of ferromagnetic domains in films and small particles. *Phys. Rev.* 70, 1946, 965–969.

85. W. Brown Jr., Rigorous approach to the theory of ferromagnetic microstructure. *Appl. Phys.* 29, 1958, 470.

86. W. Brown Jr., Relaxational behavior of fine magnetic particles. *J. Appl. Phys.* 30, 1959, 130–135.

87. W. Brown Jr., Thermal fluctuations of a single-domain particle. *J. Appl. Phys.* 34, 1963, 1319–1322.

88. W. Brown Jr., Thermal fluctuations of a single-domain particle. *Phys. Rev. B* 130, 1963, 1677–1681.

89. V. Gottschalk, The coercive force of magnetite powders. *Physics* 6, 1935, 127–132.

90. W. Elmor, Ferromagnetic colloid for studying magnetic structures. *Phys. Rev.* 54, 1938, 309–312.

91. W. Elmor, The magnetization of ferromagnetic colloids. *Phys. Rev.* 54, 1938, 1092–1101.

92. F. Bitter, A. Kaufmann, C. Starr, S. Pan, Magnetic studies of solid solutions II. The properties of quenched copper-iron alloys. *Phys. Rev.* 60, 1941, 134–139.

93. A. Mayer, E. Vogt, Magnetische messungen an eisenamalgam zur frage: Ferromagnetismus und korngröße. *Z. Naturforsch. A* 7, 1952, 334–339.

94. W. Heukolom, J. Broeder, L. Van Reijen, Hard magnetic materials. *J. Chim. Phys.* 51, 1954, 51–55.

95. C. Bean, Hysteresis loops of mixtures of ferromagnetic micropowders. *J. Appl. Phys.* 26, 1955, 1381–1385.

96. C. Bean, I. Jacobs, Magnetic granulometry and super-paramagnetism. *J. Appl. Phys.* 27, 1955, 1448–1452.

97. E. Kneller, F. Luborsky, Particle size dependence of coercivity and remanence of single-domain particles. *J. Appl. Phys.* 34, 1963, 656–659.

98. J. Dormann, D. Fiorani, E. Tronc, Magnetic relaxation in fine-particle systems. *Adv. Chem. Phys.* 98, 1997, 283–285.

99. D. Leslie-Pelecky, R. Rieke, Chemical synthesis of nanostructured cobalt at elevated temperature. *Chem. Mater.* 8, 1996, 1770–1776.

100. S. Skomski, Nanomagnetics. *J. Phys. Condens. Matter* 15, 2003, R841–845.

101. B. Sadeh, M. Doi, T. Shimizu, M. Matsui, Dependence of the Curie temperature on the diameter of Fe_3O_4 ultra fine particles. *J. Magn. Soc. Jpn.* 24, 2000, 511–517.

102. V. Nikolaev, A. Shipilin, On the thermal expansion of nanoparticles. *Fiz. Tv. Tela* 45, 2003, 1029–1033.

103. I. Jacobs, C. Bean, Fine particles, thin films and exchange anisotropy, in *Magnetism*, G.T. Rado, H. Suhl (eds.), Academic Press, New York, 3, 1963, pp. 271–276.

104. X. Batlle, A. Labarta, Finite-size effects in fine particles: Magnetic and transport properties. *J. Phys. D* 35, 2002, R15.

105. S. Chikazumi, *Physics of Ferromagnetism, Magnetic Characteristics and Engineering Applications*, Shokabo Publ. Co., Tokyo, Japan, 1984.

106. G. Huang, J. Hu, H. Zhang, Z. Zhou, X. Chi, J. Gao, Highly magnetic iron carbide nanoparticles as effective T2 contrast agents. *Nanoscale* 6, 2014, 726–730.

107. G. Hadjipanayis, Nanophase hard magnets. *J. Magn. Magn. Mater.* 200, 1999, 373–377.

108. J. Fidler, T. Schrefl, Modelling of exchange-spring permanent magnets. *J. Magn. Mater.* 177, 1998, 970–975.

109. R. Fischer, H. Kronmuller, The role of grain boundaries in nanoscaled high-performance permanent magnets. *J. Magn. Mater.* 184, 1998, 166–171.
110. R. Kodama, A. Berkowitz, E. McNiff, S. Foner, Surface spin disorder in $NiFe_2O_4$ nanoparticles. *Phys. Rev. Lett.* 77, 1996, 394–397.
111. R. Kodama, A. Berkowitz, E. McNiff, S. Foner, Surface spin disorder in ferrite nanoparticles. *J. Appl. Phys.* 81, 1997, 5552–5557.
112. O. Iglesias, A. Labarta, Finite-size and surface effects in maghemite nanoparticles: Monte Carlo simulations. *Phys. Rev. B* 63, 2001, 184416.
113. H. Kachkachi, M. Noguès, E. Tronc, D. Garanin, Finite-size versus surface effects in nanoparticles. *J. Magn. Magn. Mater.* 221, 2000, 158–162.
114. P. Hendriksen, S. Linderoth, P.-A. Lindgård, Finite-size modifications of the magnetic properties of clusters. *Phys. Rev. B* 48, 1993, 7259–7261.
115. F. Liu, M. Press, S. Khanna, P. Jena, Magnetism and local order: Ab initio tight-binding theory. *Phys. Rev. B* 39, 1989, 6914–6919.
116. S.-K. Ma, J. Lue, Spin-glass states exhibited by silver nano-particles prepared by sol-gel method. *Solid State Commun.* 97, 1996, 979–983.
117. T. Taniyama, E. Ohta, T. Sato, Ferromagnetism of Pd fine particles. *Physica B* 237, 1997, 286–291.
118. T. Nakano, Y. Ikemoto, Y. Nozue, Ferromagnetic properties of rubidium clusters in zeolite LTA. *J. Magn. Mater.* 226, 2001, 238–342.
119. B. Reddy, S. Khanna, B. Dunlap, Giant magnetic moments in 4d clusters. *Phys. Rev. Lett.* 70, 1993, 3323–3329.
120. K. O'Grady, R. Chantrell, Remanence curves of fine particle systems I: Experimental studies, in *Magnetic Properties of Fine Particles*, J. L. Dormann, D. Fiorani (eds.), Elsevier, Amsterdam, 1992, pp. 93–101.
121. J. Dai, J.-Q. Wang, C. Sangregorio, J. Fang, E. Carpenter, J. J. Tang, Magnetic coupling induced increase in the blocking temperature of γ-Fe_2O_3 nanoparticles. *J. Appl. Phys.* 87, 2000, 7397–7401.
122. J. S. Lee, J. M. Cha, H. Y. Yoon, J.-K. Lee, Y. K. Kim, Magnetic multi-granule nano-clusters: A model system that exhibits universal size effect of magnetic coercivity. *Sci. Rep.* 5, 2015, Article number: 12135.
123. Q. Liu, X. Cao, T. Wang, C. Wang, Q. Zhang, L. Ma, Synthesis of shape-controllable cobalt nanoparticles and their shape-dependent performance in glycerol hydrogenolysis. *Rsc. Adv.* 5, 2015, 4861–4871.
124. S. P. Gubin, Y. A. Koksharov, G. B. Khomutov, G. Y. Yurkov, Magnetic nanoparticles: Preparation, structure and properties, *Russ. Chem. Rev.* 74(6), 2005, 489–520.
125. S. Mørup, Superparamagnetism and spin glass ordering in magnetic nanocomposites, *Europhys. Lett.* 28, 1994, 671–677.
126. P. Allia, M. Coisson, P. Tiberto, F. Vinai, M. Knobel, M. Novak, W. Nunes, Granular Cu-Co alloys as interacting superparamagnets. *Phys. Rev. V* 64, 2001, 144420.
127. C. Kittel, Physical theory of ferromagnetic domains. *Rev. Mod. Phys.* 21, 1949, 541–547.
128. T. K. Indira, P. K. Lakshmi, Magnetic nanoparticles—A review. *Int. J. Pharm. Sci. Nanotechnol.* 3, 2010, 1035–1042.
129. D. Lee, J. S. Hilton, S. Liu, Y. Zhang, G. C. Hadjipanayis, C. H. Chen, Hot-pressed and hot-deformed nanocomposite $(Nd,Pr,Dy)_2Fe_{14}B/\alpha$-Fe-based magnets. *IEEE Trans. Mag.* 39, 2003, 2947–2952.
130. W. Wernsdorfer, E. Bonet Orozco, K. Hasselbach, A. Benoit, B. Barbara, N. Demoncy, A. Loiseau, D. Boivin, H. Pascard, D. Mailly, Experimental evidence of the Néel-Brown model of magnetization reversal. *Phys. Rev. Lett.* 78, 1997, 1791–1797.
131. J. T. Lau, A. Föhlisch, M. Martins, R. Nietubyc, M. Reif, W. Wurth, Spin and orbital magnetic moments of deposited small iron clusters studied by x-ray magnetic circular dichroism spectroscopy. *New J. Phys.* 4, 2002, 98.

132. A. Elaissari, J. Chatterjee, M. Hamoudeh, H. Fessi, Advances in the preparation and biomedical applications of magnetic colloids, in *Structure and Functional Properties of Colloidal Systems*, CRC Press, New York, 2010, pp. 315–337.

133. W. Wernsdorfer, Classical and quantum magnetization reversal studied in nanometer-sized particles and clusters. *Adv. Chem. Phys.* 118, 2001, 99–103.

134. J. Carvell, E. Ayieta, A. Gavrin, R. Cheng, V. R. Shah, P. Sokol, Magnetic properties of iron nanoparticle. *J. Appl. Phys.* 107, 2010, 103913.

135. A. B. Chin, Yaacob II, Synthesis and characterization of magnetic iron oxide nanoparticles via w/o microemulsion and Massart's procedure. *J. Mater. Process Technol.* 191, 2007, 235–239.

136. C. Albornoz, S. E. Jacobo, Preparation of a biocompatible magnetic film from an aqueous ferrofluid. *J. Magn. Mater.* 305, 2006, 12–19.

137. E. H. Kim, H. S. Lee, B. K. Kwak, B. K. Kim, Synthesis of ferrofluid with magnetic nanoparticles by sonochemical method for MRI contrast agent. *J. Magn. Mater.* 289, 2005, 328–335.

138. J. Wan, X. Chen, Z. Wang, X. Yang, Y. Qian, A soft-template-assisted hydrothermal approach to single-crystal Fe_3O_4 nanorods. *J. Cryst. Growth.* 276, 2005, 571–577.

139. M. Kimata, D. Nakagawa, M. Hasegawa, Preparation of monodisperse magnetic particles byhydrolisis of iron alkoxide. *Powder Technol.* 132, 2003, 112–117.

140. G. S. Alvarez, M. Muhammed, A. A. Zagorodni, Novel flow injection synthesis of iron oxide nanoparticles with narrow size distribution. *Chem. Eng. Sci.* 61, 2006, 4625–4629.

141. S. Basak, D. R. Chen, P. Biswas, Electrospray of ionic precursor solutions to synthesize iron oxide nanoparticles: Modified scaling law. *Chem. Eng. Sci.* 62, 2007, 1263–1267.

142. I. Martinez-Mera, M. E. Espinosa, R. Perez-Hernandez, Synthesis of magnetite (Fe_3O_4) nanoparticles without surfactants at room temperature. *J. Mater. Lett.* 61, 2007, 4447–1152.

143. S. A. Morrison, C. L. Cahill, E. E. Carpenter, S. Calvin, V. G. Harris, Atomic engineering of mixed ferrite and core-shell nanoparticles. *J. Nanosci. Nanotechnol.* 5, 2005, 1323–1344.

144. S. Sun S, H. Zeng, D. B. Robinson, S. Raoux, P. M. Rice, S. X. Wang, G. Li, Monodisperse MFe_2O_4 (M = Fe, Co, Mn) nanoparticles. *J. Am. Chem. Soc.* 126, 2004, 273–279.

145. J. Qiu, R. Yang, M. Li, N. Jiang, Preparation and characterization of porous ultrafine Fe_2O_3 particles. *Mater. Res. Bull.* 40, 2005, 1968–1972.

146. S. J. Lee, J. R. Jeong, S. C. Shin, J. C. Kim, J. D. Kim, Synthesis and characterization of superparamagnetic maghemite nanoparticles prepared by coprecipitation technique. *J. Magn. Magn. Mater.* 282, 2004, 147–152.

147. M. Fang, V. Ström, R. T. Olsson, L. Belova, K. V. Rao, Rapid mixing: A route to synthesize magnetite nanoparticles with high moment. *Appl. Phys. Lett.* 99, 2011, 222501.

148. G. Gnanaprakash, S. Ayyappan, T. Jayakumar, J. Philip, B. Raj, A simple method to produce magnetic nanoparticles with enhanced alpha to gamma-Fe_2O_3 phase transition temperature. *Nanotechnology* 17, 2006, 5851–5857.

149. G. Gnanaprakash, J. Philip, T. Jayakumar, B. Raj, Effect of digestion time and alkali addition rate on physical properties of magnetite nanoparticles. *J. Phys. Chem. B* 111, 2007, 7978–7986.

150. S. Ayyappan, J. Philip, B. Raj, Effect of digestion time on size and magnetic properties of spinel $CoFe_2O_4$ nanoparticles. *J. Phys. Chem.* C113, 2009, 590–596.

151. S. Ayyappan, S. Mahadevan, P. Chandramohan, M. P. Srinivasan, J. Philip, B. Raj, Influence of Co2+ ion concentration on the size, magnetic properties, and purity of $CoFe_2O_4$ spinel ferrite nanoparticles. *J. Phys. Chem. C* 114, 2010, 6334–6341.

152. S. F. Chin, K. S. Iyer, C. L. Raston, M. Saunders, Size selective synthesis of superparamagnetic nanoparticles in thin fluids under continuous flow conditions. *Adv. Funct. Mater.* 18, 2008, 922–927.

153. N. Smith, C. L. Raston, M. Saunders, R. Woodward, Synthesis of magnetic nanoparticles using spinning disc processing, *NSTI-Nanotech.* 1, 2006, 343–346.

154. V. Ström, R. T. Olsson, K. V. Rao, Real-time monitoring of the evolution of magnetism during precipitation of superparamagnetic nanoparticles for bioscience applications. *J. Mater. Chem.* 20, 2010, 4168–4175.

155. Y. Zhao, Z. Qiu, J. Huamg, Preparation and analysis of Fe_3O_4 magnetic nanoparticles used as targeted drug carriers. *Chin. J. Chem. Eng.* 16(3), 2008, 451–455.

156. T. Sugimoto, Formation of uniform spherical magnetite particles by crystallization from ferrous hydroxide gels. *J. Colloid Interface Sci.* 74, 1980, 227–231.

157. R. Massart, J. Roger, V. Cabuil, New trends in chemistry of magnetic colloids: Polar and nonpolar magnetic fluids, emulsions, capsules and vesicles, *Bra. J. Phy.* 25(2), 1995, 135–141.

158. M. Seenuvasan, N. Balaji, M. A. Kumar, Review on enzyme loaded magnetic nanoparticles. *Asian J. Pharm. Tech.* 3(4), 2013, 200–208.

159. S. S. Rana, J. Philip, B. Raj, Micelle based synthesis of cobalt ferrite nanoparticles and its characterization using Fourier transform infrared transmission spectrometry and thermogravimetry. *Mater. Chem. Phys.* 124, 2010, 264–269.

160. S. P. Gubin, Y. I. Spichkin, G. Y. Yurkov, A. M. Tish, Nanomaterial for high-density magnetic data storage. *Russ. J. Inorg. Chem.* 47(1), 2002, 32–47.

161. E. K. Athanassiou, K. Evagelos, R. N. Grass, W. J. Stark, Chemical aerosol engineering as a novel tool for material science: From oxides to salt and metal nanoparticles. *Aerosol. Sci. Tech.* 44(2), 2010, 161–172.

162. H. Kitahara, T. Oku, T. Hirano, K. Suganuma, Synthesis and characterization of cobalt nanoparticles encapsulated in boron nitride nanocages. *Diamond Relat. Mater.* 10, 2001, 1210.

163. M. E. McHenry, S. Subramoney, Synthesis, structure, and properties of carbon encapsulated metal nanoparticles, in *Fullerenes: Chemistry, Physics and Technology*, K. M. Kadish, R. S. Ruoff (eds.), Wiley-Interscience, New York, 2000, pp. 839–845.

164. T. Kinoshita, S. Seino, K. Okitsu, T. Nakayama, T. Nakagawa, T. A. Yamamoto, Magnetic evaluation of nanostructure of gold–iron composite particles synthesized by a reverse micelle method. *J. Alloys Compd.* 359, 2003, 46–52.

165. K. Landfester, L. P. Ramirez, Encapsulated magnetite particles for biomedical application. *J. Phys. Condens. Matter* 15, 2003, 1345–1351.

166. L. Fu, V. P. Dravid, D. L. Johnson, Self-assembled (SA) bilayer molecular coating on magnetic nanoparticles. *J. Appl. Surf. Sci.* 181, 2001, 173–177.

167. E. V. Shevchenko, D. V. Talapin, A. L. Rogach, A. Kornowski, M. Haase, H. Weller, Colloidal synthesis and self-assembly of CoPt(3) nanocrystals. *J. Am. Chem. Soc.* 124, 2002, 11480.

168. L. Guo, Z. Wu, T. Liu, S. Yang, The effect of surface modification on the microstructure and properties of γ-Fe_2O_3 nanoparticles. *Physica E (Amsterdam)* 8, 2000, 199–205.

169. T. Liu, L. Guo, Y. Tao, T. D. Hu, Y. N. Xie, J. Zhang, Bondlength alternation of nanoparticles Fe_2O_3 coated with organic surfactants probed by EXAFS. *Nanostruct. Mater.* 11, 1999, 1329.

170. J.-C. Bacri, R. Perzynski, D. Salin, V. Cabuil, R. Massart, Ionic ferrofluids: A crossing of chemistry and physics. *J. Magn. Magn. Mater.* 85, 1990, 27–35.

171. V. Berejnov, Y. Raikher, V. Cabuil, J.-C. Bacri, R. Perzynski, Synthesis of stable lyotropic ferronematics with high magnetic content. *J. Colloid Interface Sci.* 199, 1998, 215–219.

172. R. Zsigmondy, *Colloids and the Ultramicroscope,* John Wiley & Sons, New York, 1914.
173. A. S. Dukhin, P. J. Goetz, *Ultrasound for Characterizing Colloids,* Elsevier, 2002.
174. P. Vavassori, E. Angeli, D. Bisero, F. Spizzo, F. Ronconi, Role of particle size distribution on the temperature dependence of coercive field in sputtered Co/Cu granular films. *Appl. Phys. Lett.* 79, 2001, 2225–2229.
175. J. B. Kortright, D. D. Awschalom, J. Stlhr, S. D. Bader, Y. U. Idzerda, S. S. P. Parkin, I. K. Schuller, H.-C. Siegmann, Research frontiers in magnetic materials at soft X-ray synchrotron radiation facilities. *J. Magn. Magn.* 7, 1999, 207–212.
176. S. Muörup, Mössbauer effect studies of microcrystalline materials, in *Mössbauer Spectroscopy Applied to Inorganic Chemistry,* G. J. Long (ed.), Plenum, New York, 1987, Vol. 2, pp. 89–115.
177. S. Muörup, J. A. Dumestic, H. Topsuöe, Magnetic microcrystals, in *Applications of Mössbauer Spectroscopy,* R. L. Cohen (ed.), Academic Press, New York, 1980, pp. 1285–1291.
178. J. L. Dormann, D. Fiorani, Effect of the applied field on the relaxation time of the magnetic moment of uniaxial small particles with easy axis in random position. Influence on Mössbauer spectra, *Hyperfine Interact.* 70(1), 1992, 1109–1112.
179. S. Taminato, M. Hirayama, K. Suzuki, N. L. Yamada, M. Yonemura, J. Y. Son, R. Kanno, Highly reversible capacity at the surface of a lithium-rich manganese oxide: A model study using an epitaxial film system. *Chem. Commun.* 51, 2015, 1673–1676.
180. P. Hua, S. Zhanga, H. Wangb, D. Pana, J. Tiana, Z. Tanga, A. A. Volinsky, Heat treatment effects on Fe_3O_4 nanoparticles structure and magnetic properties prepared by carbothermal reduction. *J. Alloys Compounds* 509, 2011, 2316–2319.
181. I. M. L. Billas, A. Chãtelain, W. A. de Heer, Magnetism of Fe, Co and Ni clusters in molecular beams. *J. Magn. Magn. Mater.* 168, 1997, 64–75.
182. I. M. L. Billas, A. Chãtelain, W. A. de Heer, Magnetism in transition-metal clusters from the atom to the bulk. *Surf. Rev. Lett.* 3, 1996, 429–435.
183. I. Rabias, Rapid magnetic heating treatment by highly charged maghemite nanoparticles on Wistar rats exocranial glioma tumors at microliter volume. *Biomicrofluidics* 4, 2010, 024111.
184. K. E. Scarberry, E. B. Dickerson, J. F. McDonald, Z. J. Zhang, Magnetic nanoparticle-peptide conjugates for in vitro and in vivo targeting and extraction of cancer cells. *J. Am. Chem. Soc.* 130(31), 2008, 10258–10262.
185. Using magnetic nanoparticles to combat cancer, *Newswise,* 2008.
186. N. Perara, A. Kouki, J. Finne, R. J. Pieters, Detection of pathogenic *Streptococcus suis* bacteria using magnetic glycoparticles. *Organic Biomol. Chem.* 8(10), 2010, 2425–2429.
187. Highlights in chemical biology, *Rsc.org* 12, 2010, 2041–5842.
188. S. E. Turvey, E. Swart, M. C. Denis, U. Mahmood, C. Benoist, R. Weissleder, D. Mathis, Noninvasive imaging of pancreatic inflammation and its reversal in type 1 diabetes. *J. Clin. Invest.* 115(9), 2005, 2454–2461.
189. L. Lenglet, P. Nikitin, C. Péquignot, Magnetic immunoassays: A new paradigm in POCT, *IVD Technology* 2008, 43–49.
190. M. Colombo, Biological applications of magnetic nanoparticles. *Chem. Soc. Rev.* 41(11), 2012, 4306–4334.
191. M. Xiaoting, H. C. Seton, L. T. Lu, I. A. Prior, T. K. T. Nguyen, B. Song, Magnetic CoPt nanoparticles as MRI contrast agent for transplanted neural stem cells detection. *Nanoscale* 3(3), 2011, 977–984.
192. V. V. Mody, R. Siwale, A. Singh, H. R. Mody, Introduction to metallic nanoparticles, *J. Pharm. Bioallied Sci.* 2(4), 2010, 282–289.

193. F. M. Koehler, M. Fabian, M. Rossier, M. Waelle, E. K. Athanassiou, L. K. Limbach, R. N. Grass, D. Günther, W. J. Stark, Magnetic EDTA: Coupling heavy metal chelators to metal nanomagnets for rapid removal of cadmium, lead and copper from contaminated water. *Chem. Commun.* 32(32), 2009, 4862–4864.

194. A. Schätz, O. Reiser, W. J. Stark, Nanoparticles as semi-heterogeneous catalyst supports. *Chem. Eur. J.* 16(30), 2010, 8950–8967.

195. A. F. Natalie, S. Shouheng, Magnetic nanoparticle for information storage, in *Applications Inorganic Nanoparticles: Synthesis, Applications, and Perspectives*, C. Altavilla, E. Ciliberto (eds.), 2010, pp. 33–68, ISBN 9781439817612, CRC Press, New York.

2 Elemental Ferromagnetic Nanomaterials

Their Preparation, Properties, and Applications

Suneel Kumar Srivastava and Samarpita Senapati

CONTENTS

ABSTRACT

In recent years, research and development focusing on the synthesis of magnetic nanoparticles have shown wide possibilities for investigation in harnessing their unique properties for numerous potential applications in multifaceted fields. This chapter describes the preparation of zero-, one-, two-, and three-dimensional Fe, Co, Ni, the corresponding core–shell materials. Subsequently, their chemical, electromagnetic, and magnetic properties followed by their magnetic/electrical, catalytic, biomedical, and environmental applications are also described in detail.

2.1   MAGNETIC NANOMATERIALS

In recent years, nanomaterials have been receiving a considerable amount of research interest due to their unique properties that are significantly different from materials at a larger scale. This could be attributed to the presence of a large fraction of atoms on the surface and hence a high surface energy, the presence of reduced imperfections, and the quantum confinement effect. These materials find applications in the field of electrochemical, drug delivery, gas sensing, electronics, and optoelectronics, etc. [1].

Depending on the dimensionality, these materials could be classified as zero, one, two, and three dimensional (3D). When all the dimensions are in nanorange, the material is termed as zero dimensional, for example, nanoparticles, quantum dots, etc. The symmetric shapes, easy chemistry of synthesis and stabilization, and improved optical, magnetic, and catalytic properties make them very promising functional materials. A material is termed as one-dimensional (1D) when two of its dimensions are in the nanorange and the third dimension extends to the micro or millimeter range (aspect ratio: >5). These materials act as functional units as well as interconnect in thermoelectric, electronic, and opto-electronic devices [2]. A material is termed as two dimensional (2D) when one of the dimensions is in the nano range. Some of the examples under this category include nanoplatelets, nanosheets, nanodisks, thin films, etc. If the pore diameters are in the range of 2–30 nm in a nanomaterial, it is referred to as 3D, for example, hierarchical architecture, nano-flower, alumina, silica, zeolites, hybrid materials, etc.

Interest in magnetic nanomaterials has increased enormously over the past two decades due to the possibility of its unique interaction with magnetic fields and field gradient [3]. This enables the controlling of the location and motion of these nano-materials in the presence of an applied (external) magnetic field [3]. The magnetization (per atom) and the magnetic anisotropy of particles in magnetic nanomaterials are found to be greater than in the corresponding bulk specimens, while differences in the Curie (T_C) or Neel (T_N) temperatures between nanoparticles and those of the microscopic phases reach hundreds of degrees [4]. In addition, fundamental research elucidating the structure, physical, and magnetic properties, and the toxicity of nanoparticles, also involved use of magnetic nanoparticles in industrial and biomedical applications.

These magnetic nanomaterials find a wide range of applications in the generation and distribution of electricity, for the storage of data in audiotapes, videotapes, and computer disks, biotechnology/biomedicine, magnetic resonance contrast media and therapeutic agents in cancer treatment, catalysis, magnetic resonance imaging, and environmental remediation [4–15]. All these features clearly demonstrate that magnetic nanomaterials could be one of the key components in the future nanotechnology revolution.

2.2 CLASSIFICATION

Depending on the response of a material to an externally applied magnetic field, the following five basic types of magnetic materials are reported and are summarized in Table 2.1.

2.2.1 PARAMAGNETIC MATERIALS

These materials possess magnetic dipoles due to the presence of unpaired electrons in atoms, ions, or molecules and are attracted by magnetic fields. However, they lose their magnetism in the absence of a magnetic field.

2.2.2 DIAMAGNETIC MATERIALS

The atomic current loops created by the orbital motion of electrons oppose the externally applied magnetic field. The materials exhibiting this type of weak repulsion to a magnetic field are called diamagnetic. This type of magnetism is observed in materials consisting of filled electronic subshells, where the magnetic moments are paired and overall cancel one another.

TABLE 2.1

Classification of Magnetic Materials Based on the Response to Magnetic Field

Sample Type	Magnetic Behavior	Example
Paramagnetic materials	↑↗↙↖	O_2, NO, Na, VO_2, and Ti_2O_3 [16]
Diamagnetic materials		Quartz, NaCl, ZrO_2, and SiO_2 [16]
Antiferromagnetic materials	↑↓↑↓↑↓	FeS [16]
Ferrimagnetic materials	↑↓↑ ↓↑↓↑	Fe_3O_4 and Fe_3S_4 [16]
Ferromagnetic materials	↑↑↑↑↑↑	Fe, Co, and Ni

2.2.3 ANTIFERROMAGNETIC MATERIALS

Antiferromagntic materials are similar to ferromagnetic materials, but the exchange interaction between neighboring leads to the antiparallel alignment of the atomic magnetic moments. Therefore, the magnetic field cancels out to give a net zero moment. These materials also exhibit paramagnetic behavior above a temperature known as the Neel temperature. Chromium is the only element which exhibits antiferromagnetic properties at room temperature.

2.2.4 FERRIMAGNETIC MATERIALS

Ferrimagnetisms are only observed in compounds having complex crystal structures rather than in the pure element. When magnetic moments are aligned in parallel and antiparallel directions in a magnetic material, the unequal numbers result in a net moment.

2.2.5 FERROMAGNETIC MATERIALS

A ferromagnetic material exhibits permanent magnetism even after the removal of the magnetic field due to aligned atomic magnetic moments of equal magnitude. They find more practical applications due to their high magnetic moment [17]. The example of room-temperature elemental ferromagnetic material includes Fe, Co, and Ni. At room temperature, Fe crystalizes in hexagonal (hcp) and cubic (bcc) forms, whereas Co and Ni exist more commonly in hexagonal (hcp) and cubic (fcc) forms. According to Deraz and Fouda, the fcc structure of Co is thermodynamically preferred above 450°C, while the formation of hcp phase is favored at lower temperatures. For small particles, however, the fcc structure appears to be preferred even below room temperature [18]. More recently, the synthesis of a new metastable phase of cobalt (ε phase) was also reported [12].

Elemental ferromagnetic nanomaterials have been extensively studied in recent years due to their morphology-dependent magnetic/electronic properties and potential applications in the field of catalysis, magnetic separation, sensors, high-density magnetic recording, environmental remediation, and in the internal electrodes of multilayer ceramic capacitors, etc. [6,19–22]. These ferromagnetic materials undergo aerial oxidation due to their high sensitivity toward oxygen. As a result, the loss of magnetization as well as reactivity restrict their potential applications in many fields [23]. Additionally, they also tend to agglomerate in order to reduce the energy associated with the high surface area-to-volume ratio [24]. This causes the instability of the materials over longer periods of time and limits their applications. Interestingly, the formation of core (magnetic)–shell (inorganic or organic) materials by suitable chemical or physical means could overcome many of these limitations. In addition, these materials also exhibit some novel properties compared to the individual components. It may be noted that the inorganic shell may be fabricated by nonsilica material (Ag, Au, TiO_2, Fe_3O_4, etc.) or silica [24–27]. These magnetic core–inorganic shell nanomaterials show enhaced optical property and biocompatibility [24,25]. On the other hand magnetic core–organic shell nanomaterials find applications in the field of EM shielding [24,28–31], magnetic separation of cells, biochemical compounds, and controlled drug release within the body [24].

2.3 SYNTHETIC PROCEDURES OF ELEMENTAL FERROMAGNETIC NANOMATERIALS

The properties of elemental magnetic nanoparticles (Fe, Co, and Ni) can be tuned for multifaceted applications depending on their size/morphology [10,21]. In view of this, these materials have been synthesized by various physical and chemical methods.

2.3.1 ZERO DIMENSIONAL

2.3.1.1 Iron Nanoparticles

Fe nanoparticles have been invariably synthesized by chemical reduction [32–49] and thermal decomposition of the iron precursors [50–52]. Klacanova et al. [32] prepared iron nanoparticles by the reduction of central Fe(II) ion in the coordination compounds with amino acid ligands. Here the anion of the amino acid also acted as the reducing agent under this mild reduction condition at a temperature below 52°C and the optimum pH between 9.5 and 9.7. Fe nanoparticles can be produced at room temperature by borohydride mediated reduction of Fe^{3+} solution [33,34], $FeCl_2 \cdot 4H_2O$ [35], $FeSO_4 \cdot 7H_2O$, or $FeCl_3$ [36] and the naturally occurring Fe-containing bioprecursors hemoglobin and myoglobin [37]. Borohydride can also be used for reducing $FeSO_4$ in a reverse micelle medium using cetyltrimethylammonium bromide as surfactant and 1-butanol as cosurfactant [38]. Accordingly, 2.4 mL 0.8 M $FeSO_4$ (aq.) and 2.4 mL $NaBH_4$ 1 M were mixed together under magnetic stirring and inert atmosphere for 1 h. The nanoparticles of Fe can be produced by reduction of Fe^{2+} in a microemulsion medium using KBH_4 [39,40] and $NaBH_4$ [41]. Green tea extract [42–44], aqueous sorghum bran extract [45], and the aqueous extracts of plants (e.g., mango leaves, clove buds, rose leaves, neem leaves, green tea, black tea, coffee seeds, carom seeds, the joy perfume tree, champa leaves, and curry leaves)

[46] have also been harnessed in the reduction of iron salt. Other methods included the reduction of bis(ditrimethylsilyl) amido iron complex with amine–borane complexes [47] and hydrogen in presence of stabilizers [48,49]. In addition, decomposition of $Fe(CO)_5$ in presence of amine surfactant and alkane (dodecane, tetradecane) solvent at 170–220°C and 1–12 bar of CO pressure produces Fe nanoparticles [50]. Depending on the reaction conditions, the particle size varied between 3 and 12 nm. However, nearly monodisperse particles with an average diameter of 7.4 nm were obtained at 200°C with 1 bar decomposition pressure.

The preparation of Fe nanoparticles from the condensation of gases vaporized from $Fe(CO)_5$ precursors is reported [51,52]. Figure 2.1 shows a transmission electron microscopy (TEM) micrograph of iron nanoparticles (insertion shows high resolution transmission electron microscopy (HRTEM) image of oxide–metal phase interface) produced by condensing vapors generated from decomposition of $Fe(CO)_5$ [52]. It is noted that the particles consist of dark cores and light shells; the particle shape is nearly spherical. The core is metallic and the shell (3–4 nm) is composed of metal oxides. The electrochemical routes [53,54] and laser-induction complex heating method [55] are a few other additional modes of preparation for Fe nanoparticles.

2.3.1.2 Cobalt Nanoparticles

Cobalt nanoparticles have been prepared under ambient conditions by the $NaBH_4$ mediated reduction of cobalt sulfate heptahydrate [56] and $Co(NO_3)_2 \cdot 6H_2O$ [57] in the presence of triblock copolymer surfactant/organic solvents and functionalized SBA-15 mesoporous silica, respectively. Dendrimers have also been used as templates/stabilizers for the synthesis of metal nanoparticles. Therefore, Kavas et al. [58] synthesized Co nanoparticles by the addition of $NaBH_4$ solution to Co^{2+} solution in the presence of poly-amidoamine dendrimers. Liang and Zhao [59] synthesized Co nanoparticles by aqueous phase reduction of $CoCl_2 \cdot 6H_2O$ with $NaBH_4$ at ambient temperature. Alternatively, hydrazine hydrate reduced aqueous $Co(NO_3)_2 \cdot 6H_2O$

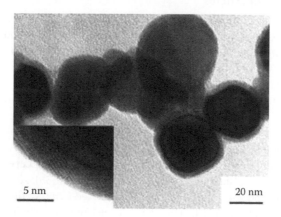

5 nm   20 nm

FIGURE 2.1   TEM micrographs of iron nanoparticles. Insertion shows HRTEM image of oxide–metal phase interface. (Reprinted from *Mater. Lett.*, 56, C. J. Choi, O. Tolochko, and B. K. Kim, Preparation of iron nanoparticles by chemical vapor condensation, 289–294, Copyright 2002, with permission from Elsevier.)

under hydrothermal conditions at 350°C/5 h in the presence of capping ligands followed by the annealing of the product at 150°C which formed cobalt nanoparticles [60]. The formation of Co nanoparticles is also reported using the reducing properties of polyol (polyol process) [61]. In this method, $AgNO_3$ or K_2PtCl_4 was added as the nucleating agent to obtain heterogeneous nucleation to the earlier prepared solution of cobalt acetate tetrahydrate and sodium hydroxide dissolved in 1,2-propanediol. Finally, the overall solution was heated to ~190°C under stirring for 3 h. Chandra and Kumar [62] synthesized Co nanoparticles by reducing aqueous $Co(NO_3)_2 \cdot 6H_2O$ by adding hot ethanolic solution of benzildiethylenetriamine followed by stirring and refluxing for 3–4 h at moderate temperature for overnight.

Kamal et al. [63] synthesized monodispersed cobalt nanoparticles by a modified polyol method using cobalt hydrazine complex. They noted that this method required only 30 min to complete the reaction compared to 3 h for the conventional method [62]. Yang et al. [64] prepared hydrophilic hollow nanospheres and hemispheres of cobalt by the ethylene glycol mediated reduction of cobalt acetate in the presence of PVP and Pd seeds. The phototrophic eukaryotes/biocompatible agents [65] and femtosecond laser [66] have been employed in liquid medium to prepare Co nanoparticles. In addition, the sonochemical [67,68], microemulsion [69,70], and pyrolytic decomposition of $Co(CO)_5$ [71], $Co_2(CO)_8$ [72], $Co(\eta^3-C_8H_{13})(\eta^4-C_8H_{12})$ [73], and $CoC_2O_4 \cdot 2H_2O$ [74,75], [Bis(2-hydroxyacetophenato) cobalt(II)] [76] have also been reported in the literature. Abu-Much and Gedanken [67] explored the feasibility of the sonochemical fabrication of cobalt (and nickel) particles, variation of their size, and formation of colloidal solutions with different solvents.

2.3.1.3 Nickel Nanoparticles

Several methods have been reported in the literature for the preparation of Ni nanoparticles. Hydrazine hydrate mediated chemical reduction of Ni^{2+} in the absence of any protective agent produced Ni nanoparticles [77–81]. It has also been synthesized in the presence of different protecting agents, such as polyethyleneimine [82], starch [83], hydroxyethyl carboxymethyl cellulose [84,85], poly(N-vinyl-2-pyrrolidone) [86–88], and organic modifiers with different functional groups, such as citric acid, Tween 40, CTAB, and D-sorbitol [89]. Sodium borohydride can be used in the reduction of aqueous $NiCl_2 \cdot 6H_2O$ [90–94] and $Ni(NO_3)_2 \cdot 6H_2O$ [95]. In addition, borohydride mediated reduction of $Ni(NO_3)_2 \cdot 6H_2O$ through foam based protocol [96] and $Ni(acac)_2$ in the presence of hexadecylamine (HAD) stabilizer are also reported [97]. The reduction of $Ni(acac)_2$ has been carried out in the presence of HAD and trioctylphosphineoxide (TOPO) to control the Ni particle size/dispersion and prevent their surface oxidation [98]. Hou et al. [98] developed a facile reduction approach, where, $Ni(acac)_2$ (dissolved in o-dichlorobenzene at 100°C) was quickly injected during vigorously stirring into the mixture of dichlorobenzene, TOPO, HDA, and sodium borohydride at 120–160°C. When this mixture was heated to 180°C for 30 min under Ar, monodisperse nickel nanoparticles are formed. The synthetic procedures involving reduction of $Ni(CH_3COO)_2 \cdot 6H_2O$ and $Ni(acac)_2$ with a mixture of oleic acid, n-trioctylamine, and n-trioctylphosphine at high temperature are also described by Winnischofer et al. [99]. The green chemical reduction route has been reported using d-(+) glucose as the stabilizer as well as the reducing agent [100]. Veena Gopalan et al. [101] prepared self-protected Ni nanoparticles through the inter matrix preparation of Ni nanoparticles by cation exchange reduction in two types of resins. Ni nanoparticles have also been produced by hydrolysis of Mg_2Ni in water [102], wet chemical reduction of $NiCl_2$ by $Na/NH_3(liq)$ [103], and hydrogenation [104,105]/thermal decomposition [106–114] of nickel precursors, reductive annealing of NiO films [115,116], and the modified polyol method [117,118].

Single phase, highly crystalline Ni nanoparticles have been synthesized in supercritical water and H_2 through superrapid heating of $Ni(HCOO)_2 \cdot 2H_2O$ solution [119]. In another report, Ni nanoparticles have been produced under hydrothermal conditions from the aqueous solution of nickel formate in supercritical water at 673 K and 30 MPa for 3–30 min. Here hydrogen is produced by the thermal decomposition of formate anion and forms a homogeneous mixture with supercritical water and easily reduces Ni^{2+} to Ni [120]. Monodisperse nickel particles with narrow particle size distributions have successfully been synthesized via a simple solvothermal method. In this procedure, ultrasonicated $Ni(CH_3COO)_2.4H_2O$ (dissolved in ethanol) mixture was transferred in an autoclave and maintained at 85°C for 3 h followed by air cooling to room temperature. Subsequently, hydrazine hydrate (reducing agent) and NaOH was added to it and again placed inside an autoclave at 100°C for 3 h [121]. It was established that the particle size of Ni increased with increasing reaction time in absence of any surfactants. In addition, reverse micelle [23,122], reverse microemulsion [123], nonaqueous sol–gel method [124], pulsed laser [125], electrochemical deposition [126–129], microwave irradiation techniques [130–132], hydrogen plasma method [133], and plasma-induced cathodic discharge electrolysis [128] are other useful techniques for the synthesis of Ni nanoparticles.

Recently, Senapati, Srivastava, and Singh [11] reported the synthesis of Ni nanoparticles through hydrazine hydrate mediated reduction of nickel chloride hexahydrate under solvothermal conditions in the absence of any external $^-$OH medium at 140°C. Figure 2.2 shows TEM and higher magnification TEM images of nickel nanoparticles prepared under hydrothermal conditions at 140°C for 12 h [inset shows the corresponding selected area electron diffraction (SAED) pattern]. The formation of nickel nanoparticles with a spherical morphology is clearly visible from here. The high-magnification TEM image showed the lattice separation of 0.205 nm corresponding to the (111) plane of fcc nickel. The SAED pattern (in the inset) confirmed the polycrystalline nature of nickel, due to the presence of concentric rings corresponding to the (111), (200), (220), and (311) fcc planes. In this procedure, the alkaline medium is generated by the reversible dissociation of water-soluble hydrazine hydrate [134]. Subsequently, ammonia and nitrogen are formed through the disproportionate reaction of hydrazine hydrate [135] and finally Ni^{2+} is reduced to Ni under this condition according to the following reactions [11]:

$$3N_2H_4 \rightarrow N_2 + 4NH_3 \qquad\qquad E^0 = 1.05\,V$$
$$Ni^{2+} + 4OH^- + N_2H_4 \rightarrow Ni^0 + 4H_2O + N_2 \qquad E^0 = 0.49\,V$$

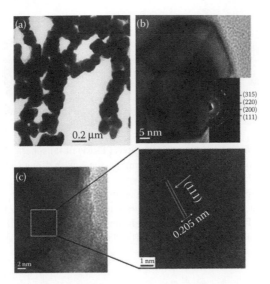

FIGURE 2.2 TEM image of (a and b) nickel nanoparticles prepared under hydrothermal conditions at 140°C for 12 h and (c) higher magnification TEM image of this nickel nanoparticle. (Inset shows the corresponding SAED pattern.) (S. Senapati, S. K. Srivastava, and S. B. Singh, Synthesis, characterization and photocatalytic activity of magnetically separable hexagonal Ni/ZnO nanostructure, *Nanoscale* 4, 2012, 6604–6612. Reproduced by permission of The Royal Society of Chemistry.)

2.3.2 ONE DIMENSIONAL

2.3.2.1 Iron Nanorods

Single-crystalline Fe nanorods are grown directly from submicron-sized Fe grains on Si substrates at ~650°C [136]. Fe nanorods have been fabricated by hydrogen reduction of α and β-FeOOH [137–140]. Fe nanorods are also prepared by polyethylene glycol (PEG) mediated chemical reduction of $FeSO_4$ followed by annealing at 400°C for 10 h [141]. Hydrazine hydrate reduction of $FeCl_3 \cdot 6H_2O$ [142] and $NH_4Fe(SO_4) \cdot 12H_2O$ [143] in the presence of CTAB under hydrothermal conditions produced rod-like Fe nanostructures. Chemical reduction of $FeCl_3 \cdot 6H_2O$ in water-in-oil microemulsions with KBH_4 [144] and $[Fe(NO_3)_3 \cdot xH_2O]$ under microwave irradiation with PEG [145] also facilitated the formation of Fe nanorods. Park et al. [146] prepared it by heating monodispersed spherical Fe nanoparticles with a mixture of trioctylphosphine oxide, $Fe(CO)_5$, and trioctylphosphine at 320°C followed by aging for 30 min and recrystallization. The TEM image in Figure 2.3 shows spherical iron nanoparticles with diameters of 2 nm, rod-shaped iron nanoparticles with dimensions of 2 nm × 11 nm (inset: high-resolution electron micrograph of a single nanorod), and the electron microdiffraction pattern of these nanorods exhibiting a body center cubic (bcc) structure of α-Fe. Fe nanorods have also been fabricated by using anodic alumina membrane [147]. Recently, Senapati et al. [14] prepared Fe nanorods by reducing $FeCl_3$ by $NaBH_4$ in a glycerol–water mixture (1:4) solvent under an ice bath condition and subsequently annealed it at 500°C for 4 h under H_2 atmosphere. It is observed that the ratio of

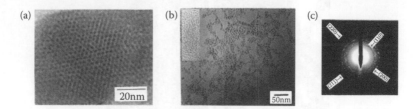

FIGURE 2.3   Transmission electron micrographs (TEM) of (a) spherical iron nanoparticles with diameters of 2 nm, (b) rod-shaped iron nanoparticles with dimensions of 2 nm × 11 nm (inset shows high-resolution electron micrograph of a single nanorod), and (c) the electron microdiffraction pattern of these nanorods. The images were obtained with a JEOL JEM-2000EX II instrument. (Reprinted with permission from S.-J. Park, S. Kim, S. Lee, Z. G. Khim, K. Char, and T. Hyeon, Synthesis and magnetic studies of uniform iron nanorods and nanospheres, *J. Am. Chem. Soc.* 122, 8581–8582. Copyright 2000, American Chemical Society.)

glycerol to water in the reaction medium affect the morphology of the final product. The morphology of the product annealed at 500°C has also been carried out by TEM and HRTEM as displayed in Figure 2.4. The formation of nanorods like the structure of iron (length ~100 nm; diameter ~15 nm) and concentric rings in the SAED pattern confirming the polycrystalline nature of iron are clearly evident from TEM and SEAD, respectively. The corresponding HRTEM image indicates the lattice planes in

FIGURE 2.4   (a) TEM image, (b) SAED pattern, and (c) HRTEM image of iron nanorods. (Reprinted from *Environ. Res.*, 135, S. Senapati et al., SERS active Ag encapsulated Fe@SiO$_2$ nanorods in electromagnetic wave absorption and crystal violet detection, 95–104, Copyright 2014, with permission from Elsevier.)

parallel without any noticeable defects. In addition, the interplanar distance between the lattice planes is also in good agreement with the d-value of (110) planes of bcc Fe.

2.3.2.2 Iron Nanotubes

Fe nanotubes have successfully been electrodeposited from H_3BO_3 and $FeSO_4$ electrolyte in the pores of anodic alumina membranes [148,149] and polycarbonate [150,151].

2.3.2.3 Iron Nanowires

The synthesis of iron nanowires has been carried out by borohydrate reduction of $FeCl_3 \cdot 6H_2O$ in an aqueous phase in a high magnetic field [152–154]. He et al. [155] used the hydrogen reduction of $Fe(OH)_3/NaCl$ at 400°C to produce Fe nanowires. The formation of Fe nanowires is also reported by chemical vapor condensation [156–158] and thermal decomposition [159] of iron pentacarbonyl. The nanoporous polycarbonate membranes [150,151,160] or anodic alumina membranes [147,161–172] are used as templates for the electrochemical deposition of Fe nanowires in their pores. Figures 2.5 and 2.6 show transmission electron micrographs of as-grown iron nanowires prepared by different methods [163,171].

2.3.2.4 Cobalt Nanorods

Co nanorods can be prepared under hydrothermal conditions by the reduction of $CoCl_2 \cdot 6H_2O$ using hydrazine hydrate in the presence of dimethylglyoxime as a complexing agent (120°C/12 h) [173] and $NaH_2PO_2 \cdot H_2O$ in the absence of any complexing agent (140°C/72 h) [174]. Hydrazine hydrate mediated reduction of $CoCl_2 \cdot 6H_2O$ has been successfully carried out in a microemulsion medium (90°C/15 h) [175]. Alternatively, the decomposition of organometallic precursors $Co[N(SiMe_3)_2]_2$ [176], $[Co(\eta^3\text{-}C_8H_{13})(\eta^4\text{-}C_8H_{12})]$ [177], and $Co_2(CO)_8$ [178] also produced Co nanorods in the presence of HAD/oleylamine/H_2, HAD/lauric acid/H_2, and o-dichlorobenzene oleic acid/Ar, respectively. Soumare et al. [179] reduced cobalt laurate with 1,2-butanediol in the presence of a nucleating agent ($RuCl_2$). In another pulsed current method, the reduction of $CoSO_4$ solution is carried out in the presence of sodium sulfate and

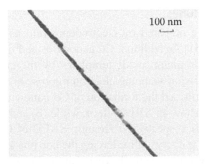

FIGURE 2.5 Morphology of a Fe nanowire. (Reprinted from *J. Magn. Magn. Mater.*, 222, S. Yang et al., Preparation and magnetic property of Fe nanowire array, 97–100, Copyright 2000, with permission from Elsevier.)

FIGURE 2.6 Transmission electron micrographs of as-grown iron nws liberated from the template (a) and after their storage in the air for 5 months (b). (Reprinted from *Mater. Chem. Phys.,* 109, A. Jagminas et al., Template synthesis, characterization and transformations of iron nanowires while aging, 82–86, Copyright 2008, with permission from Elsevier.)

polyvinyl pyrrolidone (PVP) as the ionic strength adjuster and structure director, respectively [180].

2.3.2.5 Cobalt Nanotubes

Cobalt nanotubes can be prepared by electrodeposition using polycarbonate track etched templates [150,151]. In addition, Co nanowires and nanotubes have also been deposited in anodized aluminum oxide templates by the electrodeposition method [181–187]. The field emission scanning electron microscopy (FESEM) and HRTEM images in Figure 2.7 confirmed the formation of Co nanowires and nanotubes [185]. The insets of (a) and (c) show an XRD pattern, while (b) and (d) show a SAED of Co nanowires and nanotubes, respectively. Though, SAED of Co NWs/NTs confirmed that most of the NWs are polycrystalline having the hcp phase, some small grains with the fcc structure were also observed. However, the fcc phase is found to be a minor phase in the 50 nm thick Co NWs. On the contrary, the SAED pattern of Co NTs shows diffuse rings depicting that the Co NTs are not entirely amorphous. Atomic layer deposition is another useful method for the synthesis of cobalt nanotubes [188,189].

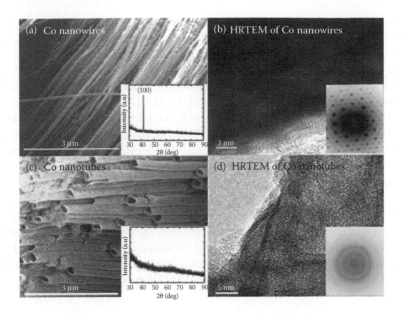

FIGURE 2.7 (a and c) FESEM images of Co nanowires and nanotubes. (b and d) HRTEM of Co nanowires and nanotubes. The insets of (a) and (c) show XRD pattern, while (b) and (d) show SAED of Co nanowires and nanotubes, respectively. (Reprinted with permission from N. Ahmad et al., Temperature dependent magnetic properties of Co nanowires and nanotubes prepared by electrodeposition method, *J. Appl. Phys.* 109, 07A331/1–07A331/3. Copyright 2011, American Institute of Physics.)

2.3.2.6 Cobalt Nanowires

Several groups prepared Co nanowires by magnetic field induced chemical reduction of cobalt salt in the presence of $NaBH_4$ [190] and $N_2H_4 \cdot H_2O$ [191–194]. Athanassiou et al. [195] prepared them by keeping magnets and steel cubes within the hot exhaust stream containing cobalt nanoparticles prepared earlier. The reduction of cobalt carboxylate (solvent: 1,2 butanediol) [196] and cobalt acetate (propanediol: solvent and stearic acid: surfactant) [197] in the presence of a heterogeneous nucleating agent ($RuCl_2$) has also been reported. In addition, decomposition of cobalt precursors ($Co(\eta^3$-$C_8H_{13})(\eta^4$-$C_8H_{12})$, cobalt stearate) produced cobalt nanowire [198]. The electrochemical deposition using different templates, that is, anodic aluminium oxide [163,164,183,185,199–209] and ion track polycarbonate [148,149,210–212], is other commonly used preparative method for Co nanowire. Template-free one-step electrosynthesis of cobalt nanowires has also been reported from aqueous [$Co(NH_3)_6$] Cl_3 solution [213].

2.3.2.7 Nickel Nanorods

According to Guo et al. [214] nanorod-like nickel is formed by microwave irradiated liquid-phase reduction of $Ni(CH_3COO)_2 \cdot 4H_2O$ at 85°C under hydrothermal conditions in the presence of hydrazine hydrate and polyvinyl alcohol. Hydrazine hydrate mediated reduction of $NiCl_2 \cdot 6H_2O$ (ethanolic solution) in the presence of trimethylamine acting as the morphology directing agent under solvothermal conditions at 120°C for 8–12 h also leads to the production of Ni nanorods [215]. Ghosh et al. [216] prepared nickel nanorods by solvothermal decomposition of nickel acetate in n-octylamine medium at 250°C for 3 h. Hydrazine reduction of nickel chloride in microemulsion medium consisting of water/butanol/potassium oleate/kerosene at ~90°C also produced Ni nanorods [217]. Ghosal et al. [218] applied green synthetic route by reducing $NiCl_2 \cdot 6H_2O$ with natural polyol castor oil. In addition, plasma-enhanced atomic layer deposition [219], electrodeposition in the nanopores of polycarbonate membrane [220–222], and anodic aluminium oxide membrane [187,223–228] remian other interesting modes of fabrication.

A novel method has been developed for the growth of nickel nanorod arrays employing electrochemical deposition on anodized aluminium oxide templates [221]. Figure 2.8 shows FESEM top-view images of the nickel nanorods before and after the AAO template is resolved. It is clearly evident from the image that the AAO template wraps the nickel nanorods tightly. Generally, the nanorods are well ordered and parallel to each other, being the same size of the AAO nanopores. The XRD pattern (inset) revealed a face-centered cubic structure (a = 0.3523 nm) of the nickel nanorods.

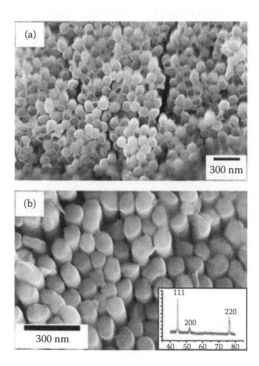

FIGURE 2.8 FE-SEM top-view images of the nickel nanorods (a) before and (b) after the AAO template is resolved. The inset XRD spectrum shows that the nickel nanorods prefer a fcc structure. (Reprinted from *Mater. Lett.*, 62, Z. F. Zhou et al., Growth of the nickel nanorod arrays fabricated using electrochemical deposition on anodized Al templates, 3419–3421, Copyright 2008, with permission from Elsevier.)

2.3.2.8 Nickel Nanotubes

Similar to Ni nanorods, the electrodeposition techniques have also been adopted to prepare Ni nanotubes using anodized alumina membrane [182,184,186,229,230] and polycarbonate membrane [221] as the templating agents. In another report, atomic layer deposition of Ni nanotubes is reported inside the pores of anodic alumina membranes using nickelocene and H_2O/O_3 as the precursor [231].

2.3.2.9 Nickel Nanowires

Ni nanowires can be prepared by hydrazine hydrate mediated reduction of $NiCl_2$ $6H_2O$ in the presence of TX-100 [232]/L-methionine [233] surfactants. In another experiment, Ni nanowire has been synthesized by refluxing aqueous $NiCl_2 \cdot 6H_2O$ at 190°C in the presence of NaOH, ethylene glycol, ethylenediamine (reducing agent), and silicione nanowire (template) [234]. Wang et al. [235] reported the formation of self-assembled Ni nanowires by hydrazine hydrate mediated reduction of $NiSO_4 \cdot 6H_2O$ in ethylene glycol. A microemulsion template-assisted hydrazine hydrate mediated wet chemical reduction of $NiCl_2 \cdot 6H_2O$ at 70°C has been used for the fabrication of prickly nanowires [13]. Figure 2.9 shows TEM images of Ni nanowires consisting of a network structure of interconnecting nickel nanowires of diameters 200–500 nm. Further, it is clearly evident that the surfaces of the nanowires are prickly in nature consisting of nanocones with their diameters lying in the range of 80–120 nm and growing radially from a rod-like axis. The presence of concentric rings in the SAED pattern of nickel confirmed the formation of the fcc structure of Ni. The HRTEM image shows the equispaced lattice fringes of spacing 0.203 nm corresponding to the (111) plane of the fcc nickel.

External magnetic field assisted preparation of nickel nanowire has also been carried out by the hydrazine hydrate mediated reduction of $NiCl_2 \cdot 6H_2O$ in presence of ethylene glycol [236] and ethanol [237,238]. In addition, Ni nanowires could be prepared by hydrazine hydrate reduction of $NiCl_2 \cdot 6H_2O$ in the absence of a magnetic field [239,240]. The electrochemical deposititon on different templates, for example, amine functionalized SBA-15 [241], halloysite [129], anodic aluminium oxide [164,165,206,242–250], and polymer [251] are other efficient strategies for the fabrication of nickel nanowires.

2.3.3 Two Dimensional

Thin plate-like shaped elemental ferromagnetic nanomaterials have drawn a lot of attention, specially due to their high-frequency EM properties [252].

2.3.3.1 Iron

Guan et al. [253] synthesized amorphous iron nanoplatelets by reducing concentrated $FeSO_4$ solution with $NaBH_4$ under an applied magnetic field at room temperature and nitrogen atmosphere. The preparation of Fe nanoflakes is also reported by a hydrazine hydrate mediated solvothermal reduction of $FeCl_3 \cdot 6H_2O$ in the presence of sodium dodecyl benzene sulfonate (SDBS) at 120°C [142], pyrolysis of $Fe(CO)_5$ under Ar [254], and ball milling of micron scale powders [255], and carbonyl iron powder [256].

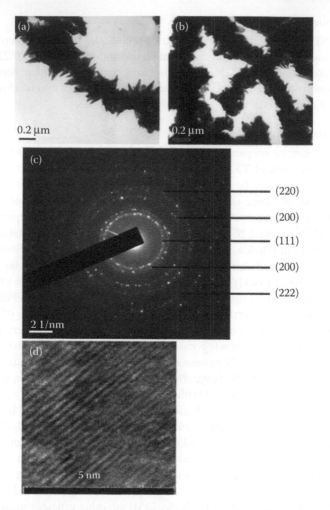

FIGURE 2.9 TEM images of (a) the prickly Ni nanowires, (b) network-like structure of Ni nanowires, (c) SAED pattern of a prickly nanowire, and (d) high-resolution image of a prickly structure. (S. Senapati et al., Magnetic Ni/Ag core-shell nanostructure from prickly Ni nanowire precursor and its catalytic and antibacterial activity, *J. Mater. Chem.* 22, 2012, 6899–6906. Reproduced by permission of The Royal Society of Chemistry.)

2.3.3.2   Cobalt

Co nanoflakes have been synthesized by surfactant assisted hydrazine hydrate mediated reduction of $CoCl_2$ at 180°C under hydrothermal conditions [257]. The reduction of $Co(\eta^3-C_8H_{13})(\eta^4-C_8H_{12})$ in the presence of rhodamine B and hexadecyl amine at 150°C under H_2 provides an alternative method [258]. Co nanodiscs are formed by the decomposition of $Co_2(CO)_8$ in the presence of surfactant [73,259]. Huang et al. [260] carried out transformation of Co nanodisks to Co caterpillars. Figure 2.10 depicts TEM images of Co nanodisks prepared in this method taking

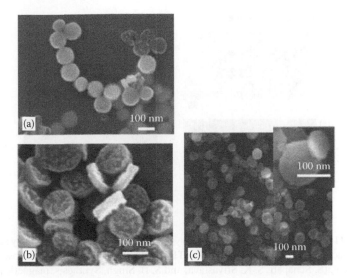

FIGURE 2.10 TEM images of cobalt nanodisks (a) seedless, (b) with seeds added, (c) with capping agent PVP. (Reprinted from *J. Magn. Magn. Mater.*, 304, F. T. Huang, R. S. Liu, and S. F. Hu, Transformation of Co nanodisks to Co caterpillars, e19–e21. Copyright 2006, with permission from Elsevier.)

into consideration seedless, seeds added, and with PVP as a capping agent. It is inevitabe that cobalt nanodisks synthesized without the addition of seeds are smaller than those synthesized in the presence of seeds. Interestingly, thick cobalt disks were composed of many thinner disks when synthesized in the presence of PVP due to mutual magnetic attraction of thin Co disks.

2.3.4 Three Dimensional

3D nanomaterials derived from the self-assembly of low dimensional inorganic building blocks yield excellent structural units. They are important in terms of stability, spatial uniformity, and physical properties depending on their size, spacing, and high-order structure [21,261,262].

2.3.4.1 Cobalt Nanoflowers

Liu et al. [263] fabricated hierarchical cobalt nanoflowers by a simple solvothermal reduction of $Co(OAc)_2 \cdot 4H_2O$ in polyol using Ru as the heterogeneous nucleation agent and HAD as the structure-directing agent. The solid cobalt alkoxide produced in the initial stage mediates the growth rate of Co nanoflowers, which follow a hierarchical growth mode. The core initially formed is transformed to a wheat fringe, which further grows to a nanoflower. The size of these nanoflowers and the petals could be simply adjusted by varying the concentration of HAD. Hydrothermal reduction of $CoCl_2 \cdot 6H_2O$ by $NaH_2PO_2 \cdot H_2O$ [262] and $N_2H_4 \cdot H_2O$ [264,265] in the presence of dodecyl benzenesulfonate [262], polyvinylpyrrolidone [264], and dimethylglyoxime [265] produced Co nanoflowers. The formation of Co nanoflowers

FIGURE 2.11 SEM images of cobalt nanocrystal prepared at 90°C catalyzed by silver using 5 mL of reducing agent in the presence of (a) TEA, (b) DEA, (c) EG, and (d) in absence of any capping agent. (S. Senapati, S. K. Srivastava, and S. B. Singh, Synthesis, magnetic properties and catalytic activity of hierarchical cobalt microflowers, *J. Nanosci. Nanotech.* 12, 2012, 3048–3058. Reprinted with permission from American Scientific Publishers.)

has also been reported under refluxing conditions, when $N_2H_4 \cdot H_2O$ mediated reduction of $CoCl_2 \cdot 6H_2O$ takes place in the presence of polyvinylpyrrolidone [266] and 2-hydroxy-4-(1-methylheptyl) benzophenone oxime [267].

Recently, the seed-mediated synthesis method has been used to overcome the problem of uncontrolled nucleation leading to polydispersity. We have prepared Co microflowers by the hydrazine hydrate mediated reduction of cobalt acetate tetrahydrate in the presence of TEA, DEA, and EG acting as capping agents and an Ag seed catalyst at 90°C within 10 min [12]. Interestingly, no Co was formed in the absence of the Ag catalyst keeping other experimental conditions unaltered. The corresponding scanning electron microscopy (SEM) images of the product are displayed in Figure 2.11. These images show the formation of porous microspherical flowery architectures of diameter 1–3 μm depending on the capping agents used. It is inferred from these images that these are composed of densely packed uniform nanoflakes with thickness of nearly 5–15 nm. Such nanoflakes like petals consisting of wavy rough edges are actually interlinked with each other leading to the formation of porous flower-like architectures.

2.3.4.2 Ni Nanoflowers

Ni nanoflowers can be prepared by template-free hydrazine hydrate mediated reduction of $NiCl_2 \cdot 6H_2O$ at 60°C in ethylene glycol [21] and ethanol [268] medium. The procedures involving hydrazine reduction of $NiCl_2 \cdot 6H_2O$ in the absence [261] and presence [269] of a nucleating agent are also carried out hydrothermally at 120°C and 160°C, respectively. Recently, Senapati et al. [10] described the silver-catalyzed growth of the nickel nanoflower by hydrazine hydrate mediated reduction of nickel(II) chloride hexahydrate employing a simple heating method in the presence and absence of any capping agents. Most importantly, the reaction is completed within 10 min at 60°C in absence of any external hydroxide force. Based on these findings, the following mechanism for the formation of Ni has been proposed:

$$N_2H_4.H_2O \Leftrightarrow N_2H_5^+ + OH^-$$

$$Ag^+ + H_2N-NH_2 \rightarrow H_2N-NH_2-Ag^+$$

$$H_2N-NH_2-Ag^+ + OH^- \rightarrow Ag + \dot{N}_2H_3 + H_2O$$

$$\dot{N}_2H_3 \xrightarrow{\text{dimerization}} H_2NHN-NHNH_2 \xrightarrow{\text{Fast}} N_2 + NH_3$$

$$2Ni^{2+} + N_2H_4 + 4OH^- \rightarrow 2Ni + N_2 + 4H_2O$$

SEM images of the product obtained by adding hydrazine hydrate to aqueous $NiCl_2$ and TEA solution at room temperature and thereafter heating this mixture at 60°C for 5, 7, 8, and 10 min are displayed in Figure 2.12. According to this, the

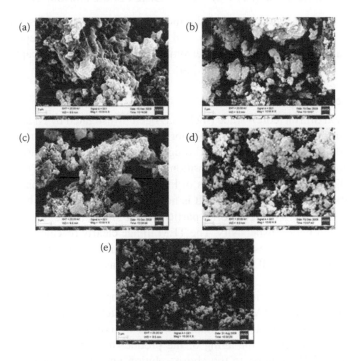

FIGURE 2.12 SEM images of the product obtained by adding hydrazine hydrate to the aqueous $NiCl_2$ and TEA solution at (a) room temperature, and thereafter heating this reaction at 60°C for (b) 5 min, (c) 7 min, (d) 8 min, and (e) 10 min. (Reprinted with permission from S. Senapati et al., Capping agent assisted and Ag-catalyzed growth of Ni nanoflowers, *Cryst. Growth Des.* 10, 4068–4075. Copyright 2010, American Chemical Society.)

product exhibits irregular morphology after 5 min of the reaction. After 7 and 8 min, some spiky surfaces are formed due to the formation of more nickel particles obtained by the silver-catalyzed reduction of nickel hydrazine complex. After the completion of the reaction (10 min), flower-like particles are formed.

Ni et al. [270] employed a complex mixture of $Ni(N_2H_4)_3^{2+}$ and nickel dimethyl-glyoximate for the synthesis of flowery shaped nickel at 110°C under hydrothermal conditions. Ethylenediamine tetraacetic acid sodium as a complexant has also been utilized in fabricating flower-like Ni crystals through a facile chemical reduction route [271]. Jia et al. [272] followed the nonaqueous sol–gel synthesis route for the reduction of nickel acetylacetonate with benzyl alcohol in the presence of varying magnetic fields at 200°C. Xiong et al. [273] developed a template and surfactant-free strategy to prepare porous hierarchical Ni nanostructures by directly calcining the nickel-based flower-like precursor in Ar. The precursor is preformed by refluxing the solution of nickel nitrate and the co-precipitators of hexamethylenetetramine and oxalic acid at 100°C for 6 h.

2.4 SYNTHESIS OF CORE (Fe, Co, OR Ni)–SHELL NANOMATERIALS

2.4.1 MAGNETIC CORE (Fe, Co, OR Ni)–SILICA SHELL NANOMATERIALS

The Stöber process remains one of the most widely accepted routes for the preparation of silica-encapsulated Fe, Co, and Ni nanoparticles [24]. Reports are also available on the preparation of these ferromagnetic nanomaterials encapsulated with silica using Na_2SiO_3 and SiO_2 powder [24].

2.4.1.1 Fe@SiO$_2$

The formation of iron (core)–silica (shell) has been investigated by Ni and coworkers [142,274]. In one such approach, they fabricated bcc phase iron nanocrystals with granular, rod-like, and flaky shapes through a simple surfactant-controlled chemical reduction route followed by deposition of thin silica coating through a Stöber process [142]. SEM and TEM images in Figure 2.13 show the morphology of iron particles after surface silica coating. It is noted that compared with that of the pure Fe counterparts, the silica-coated Fe particles well maintain the original morphology but showed better dispersivity. The TEM image clearly revealed the core–shell structure of these particles with different shapes. This is also evident from these images that the outer shell and inner dark core were ascribed to silica and metallic iron, respectively.

Leng et al. [275] prepared iron nanoparticles capped with PVP molecules for silica coating by adding PVP dispersed in ethanol to tetraethoxysilane (TEOS) and ammonia solution. In another method, TEOS is introduced to the ethanolic solution of $FeCl_3$ followed by the aging of the product for 0.5 h while stirring [276]. On injecting KBH_4 solution to the above solution and stirring vigorously, Fe^{3+} is reduced to Fe particles. This is subsequently coated by silica due to its condensation through the aerosol process. Alternatively, the mixture of $Fe(NO_3)_2$ and TEOS in alcoholic solution has also been used for the preparation of silica-coated Fe [277]. Fernandez-Pacheco et al. [278] prepared Fe nanorods coated with SiO_2 using a

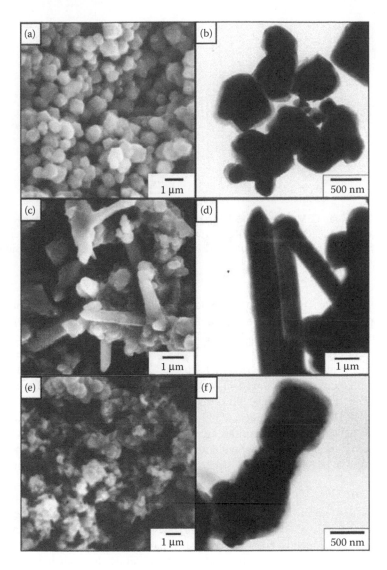

FIGURE 2.13 SEM and TEM images of the three silica-coated iron samples: (a and b) sample d; (c and d) sample e; (e and f) sample f. (Reprinted from *Mater. Chem. Phys.*, 120, X. Ni et al., Silica-coated iron nanoparticles: Shape-controlled synthesis, magnetism and microwave absorption properties, 206–212, Copyright 2010, with permission from Elsevier.)

modified arc-discharge method. Recently, Senapati et al. [14] fabricated SiO_2-coated Fe nanorods employing the modified Stöber method.

2.4.1.2 Co@SiO$_2$

Park et al. [279] coated silica on cobalt nanoparticles by hydrogen reduction of early prepared $CoO@SiO_2$ core–shell precursor without using any etching step. Low-temperature calcination of SiO_2-coated Co_3O_4 nanoparticles also produced $Co@SiO_2$

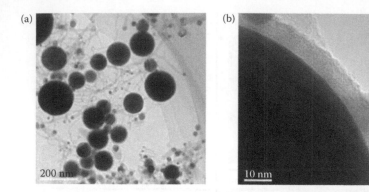

FIGURE 2.14 (a) TEM and (b) HRTEM images of the Co/SiO2 nanocapsules. (Reprinted from *Acta Mater.*, 55, X. F. Zhang et al., Synthesis, growth mechanism and magnetic properties of SiO2-coated Co nanocapsules, 3727–3733, Copyright 2007, with permission from Elsevier.)

core–shell nanostructures [280]. In another method, an ethanolic solution containing 3-aminopropyl-trimethoxysilane and tetraethyl orthosilicate (molar ratio 1:9) and Co colloid produced silica-coated cobalt [281]. Arc-discharge [282] and electrochemical methods [283] have also been employed in the preparation of Co@SiO$_2$ nanostructures. Zhang et al. [282] synthesized SiO$_2$-coated Co nanocapsules by a modified arc-discharge method, using Co and SiO$_2$ micropowders as raw materials. The prepared Co nanoparticles are found to be uniformly encapsulated with amorphous SiO$_2$ shells. TEM and HRTEM images of the Co/SiO$_2$ nanocapsules are displayed in Figure 2.14. It is noted that Co/SiO$_2$ nanocapsules are spherical in shape with diameters from 80 to 100 nm and the SiO$_2$ shell is about 8 nm thick. The HRTEM image indicated that the nanocapsule has a "core–shell" structure with a clearly distinguishable interface.

2.4.1.3 Ni@SiO$_2$

Park et al. [284] applied the microemulsion method consisiting of the dispersion of nickel particles in cyclohexane and tetramethyl ortho silicate and octadecyltrime-thoxysilane. Subsequently, Ni@SiO$_2$ core–shell particles were precipitated by adding methanol at room temperature. In another method, silica shell has been fabricated via decomposition of TEOS on Ni nanoparticle cores prepared by evaporation of pure metal in an electric arc [285].

2.4.2 Magnetic Core (Fe, Co, and Ni)–Inorganic (Nonsilica) Shell

2.4.2.1 Fe@Au, Fe@Ag, and Ag-Coated Fe@SiO$_2$

Xu et al. [286] reported the formation of T4 virus supported Fe@Au nanostructures by NaBH$_4$ mediated chemical reduction of aqueous AuCl$_3$ in the presence of the T4 virus supported Fe nanoparticles at room temperature. They also provided a speculated general growing mechanism for preparation of gold-coated iron ternary core–shell nanostructure. NaBH$_4$ mediated reduction of HAuCl$_4$ in reverse micelle has also been used to deposit Au on Fe [287–291]. Dong and coworkers [292] fabricated Fe/Au core–shell nanostructure by adding $HAuCl_4$ to a mixture of ascorbic acid and $Fe(NO_3)_3$ in a nitrogen atmosphere and maintaining the pH at 4. Laser beam irradiation ($\lambda = 532$ nm) of a mixture of Fe and Au nanoparticles also produced Fe@Au nanostructures [293]. In another study, Carroll et al. [294] investigated a one-step method for preparing core–shell nanoparticles of Fe and Ag by aqueous reduction using sodium borohydride and sodium citrate. These findings show that the formation of different core–shell structures are determined by the various introduction times of $AgNO_3$ to the reaction vessel after the $NaBH_4$ addition. Scheme 2.1 depicts the proposed reaction scheme for the reduction of sodium citrate, $FeCl_2 \cdot 7H_2O$, and $NaBH_4$, and the addition of $AgNO_3$ at different times after the addition of $NaBH_4$ in creating various core–shell structures.

Wang et al. [295] synthesized core shell Fe/Ag nanoparticles by aqueous phase reduction of $AgNO_3$ with sodium borohydride. The synthesis procedure first included the reduction of iron nanoparticles followed by the addition of silver nitrate. Ag-coated Fe nanostructures have been prepared by the reduction of $AgNO_3$ in organic media [296], vitamin C [297], and ultrahigh vacuum evaporation deposition [298]. Senapati et al. [14] used Tollen's reagent to deposit silver on the surface of Fe@ SiO_2 prepared by the Stöber method.

Transmetalation [299] and laser ablation [300] methods have been utilized for the preparation of Au-coated Co nanostructures. Bao et al. [301] prepared Co/Au

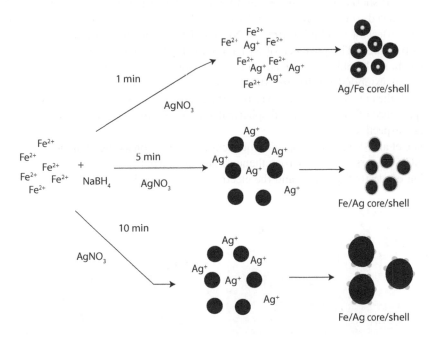

SCHEME 2.1 Proposed reaction scheme for the reduction of sodium citrate, $FeCl_2 \cdot 7H_2O$ and $NaBH_4$; addition of $AgNO_3$ at various times after the addition of $NaBH_4$ produced various core–shell morphologies. (Reprinted with permission from K. J. Carroll et al., One-pot aqueous synthesis of Fe and Ag core/shell nanoparticles, *Chem. Mater.* 22, 6291–6296. Copyright 2010 American Chemical Society.)

core–shell nanoparticles by reduction of organo-Au compound on Co seeds in a non-polar solvent. The fabrication of Au-coated Co nanostructures has also been successfully carried out by galvanic replacement reaction of magnetic cobalt nanoparticles in presence of $HAuCl_4$ [302,303].

The preparation of Ag-coated cobalt has been carried out by many researchers. Garcia-Torres et al. [69] reported the $N_2H_4 \cdot H_2O$ mediated reduction of $AgClO_4$ in the microemulsion medium. Co@Ag nanostructures can also be fabricated by electrodeposition from an electrolyte containing aqueous solution of $AgNO_3$, $CoCl_2$, and NaCl [304–307]. Yamauchi et al. [308] prepared them by chemical leaching of Al atoms from various ternary Al–Ag–Co alloys. In addition ZnO-coated Co nanoparticles have been reported by a chemical precipitation method at a lower temperature [309].

2.4.2.2 Ni@Au, Ni@Ag, Ni@TiO₂, and Ni@ZnO

Ni/Au core–shell nanostructures can be obtained by a transmetalation reaction involving $HAuCl_4$/Ni in the aqueous phase [232,240]. Core–shell Ni–Au nanoparticles were chemically synthesized through a redox-transmetalation method in a reverse microemulsion [310]. The core–shell structure could be clearly established by the transmission electron micrograph as shown in Figure 2.15. The cores of the nanoparticles are found to be in the range from 5 to 10 nm in diameter, the thickness of the shells are 5 to 10 nm, and the diameter of the integrated nanoparticles are estimated to range from 15 to 30 nm. Sarkar et al. [232] also thoroughly studied the effect of incubation time of Ni and $HAuCl_4$ on the morphology of Ni/Au nanostructures. The presence of Ni nanowires together with a prickly surface and nanospherical gold was noticed after incubation for 5 min. When the incubation time was extended for 10 min, the growth of gold clusters in colonial fashion all through the nanowires surface was observed. The incubation for 4 and 12 h showed the presence of densely packed spherical superstructures and dendritic nanostructures, respectively.

She et al. [311] used an injection-quenching process consisting of quickly injecting $HAuCl_4$ precursor in dichloromethane to Ni nanoparticles in oleylamine. A combination of electrodeposition and the electroplating method has also been reported to

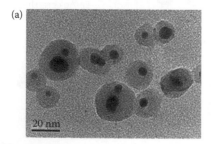
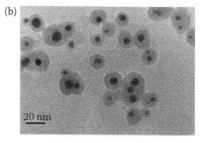

(a)   (b)

20 nm   20 nm

FIGURE 2.15 TEM micrograph of core–shell Ni–Au nanoparticles (a) 1 day and (b) 1 week after synthesis. (Reprinted with permission from D. Chen et al., Properties of core-shell Ni-Au nanoparticles synthesized through a redox-transmetalation method in reverse microemulsion, *Chem. Mater.* 19, 3399–3405. Copyright 2007, American Chemical Society.)

prepare Ni/Au core–shell nanowires [312]. Kim et al. [313] synthesized Ni/Au nano-composite by x-ray exposure to a mixed electroless solution of Ni and Au.

Ni/Ag nanoparticles can be successfully prepared by $N_2H_4 \cdot H_2O$ mediated reduction in the presence and absence of a protective agent [79,82,314]. A few reports are also available by redox transmetallation on the formation of Ni/Ag nanostructures in aqueous [25,90,96] or nonaqueous phase [315]. A nonaqueous reduction route involving the mixing of $AgNO_3$ (in octadecane) to Ni nanoparticles at 160°C has been employed [316]. Senapati et al. [13] reported the preparation of Ni/Ag nano-structure through a redox-transmetalation method using freshly prepared Tollens' reagent to the dispersion of earlier prepared Ni nanowire. Low magnification FESEM images show the formation of core–shell Ni/Ag nanostructures. However, high-magnification FESEM and TEM images show the deposition of spherical silver particles on the prickly surface of nickel nanowires. The formaton of TiO_2-coated Ni nanoparticles [26], ZnO-coated Ni nanoparticles [317], and nanowires [239] has also been reported. In another method, the synthesis of hexagonal zinc oxide-coated nickel (Ni/ZnO) nanostructures has also been reported [11]. The procedure involved reduction of nickel chloride hexahydrate using hydrazine hydrate through the solvo-thermal process at 140°C followed by the surface modification of the product by the reflux method at 110°C for 1 h. The TEM image of Ni/ZnO hybrid in Figure 2.16 clearly shows that a flake-like ZnO is coated around the spherical nanoparticles of

FIGURE 2.16 TEM images of the Ni/ZnO nanostructure prepared using (a) 0.063 M [Zn^{2+}] and (b and c) 0.25 M [Zn^{2+}] at 110°C; (d) high-resolution TEM image of Ni/ZnO nanostructure prepared using 0.25 M [Zn^{2+}]. Inset shows the SAED pattern of Ni/ZnO prepared using 0.25 M [Zn^{2+}]. (S. Senapati, S. K. Srivastava, and S. B. Singh, Synthesis, characterization and photocatalytic activity of magnetically separable hexagonal Ni/ZnO nanostructure, *Nanoscale* 4, 2012, 6604–6612. Reproduced by permission of The Royal Society of Chemistry.)

nickel at the lower concentration of (0.063 M) Zn^{2+}. However, nickel nanoparticles at a relatively higher concentration of zinc salt solution (0.25 M [Zn^{2+}]) are coated with "nut like" stepped hexagonal nanodisk-shaped zinc oxide. The lattice fringes exhibit parallel alignment at a distance of 0.193 nm corresponding to the (102) plane of ZnO.

2.4.3   MAGNETIC CORE (Fe, Co, AND Ni)–ORGANIC SHELL

A very limited amount of work has been reported so far on the formation of magnetic core–organic shell nanostructures. Burke and coworkers [318] reported the preparation of polyisobutylene, polyethylene, or polystyrene-coated iron nanoparticles by the thermal decomposition of $Fe(CO)_5$ in presence of ammonia and the respective polymers. Molday and Molday [319] prepared ferromagnetic dextran-coated iron particles by reacting the mixture of ferrous chloride/ferric chloride and dextran polymers under alkaline conditions. According to Feng et al. [320] polypyrrole (PPy)-coated Co nanostructures can be fabricated through chemical oxidation with ferric chloride as an oxidant in the absence and presence of dopants (Na dodecylbenzene sulfonate, Na paratoluene sulfonate, and Na benzene sulfonate). Xu et al. [28] prepared Ni/PPy core–shell composites by *in situ* chemical oxidative polymerization of pyrrole monomer in the presence of Ni powder. Senapati [15] reported room-temperature synthesis of magnetic Ni/PPy nanostructures via an *in situ* oxidative polymerization of pyrrole (Py) monomer in the presence of $FeCl_3$ oxidant in an aqueous suspension of Ni nanoflowers.

2.5   PROPERTIES OF Fe, Co, AND Ni NANOMATERIALS

Nanomaterials of Fe, Co, and Ni exhibit extreme chemical reactivity, and unique electrical and magnetic properties. All these properties have been investigated by many workers and are reviewed next.

2.5.1   CHEMICAL PROPERTIES

These elemental ferromagnetic nanomaterials are highly sensitive to air and undergo aerial oxidation which reduces their magnetic moment and limit their applications in different fields. Therefore, liquid or solid dispersants are often used to slow down the diffusion of oxygen on their surface [6]. Alternative procedures can also be used to overcome this problem without compromising their magnetic properties. Preoxidation of the metal surface is one such strategy to prevent further oxidation, for example, Ni@NiO [321] and $Fe@Fe_2O_3$ [6]. The surface oxidation may be overcome by alloying the nanomaterials with suitable agents [12] or by the formation of core–shell nanostructures [13,24]. These techniques not only prevent the aerial oxidation but also extend their applications in a wider range.

2.5.2   ELECTROMAGNETIC PROPERTIES

Materials possessing both magnetic as well as electrically conducting properties are termed as electromagnetic (EM) materials [322]. They find application in EM shielding, molecular electronics, nonlinear optics, microwave absorption, and catalysis [28]. These materials are characterized by EM parameters, such as relative complex permittivity (ϵ) and relative complex permeability (μ). The real parts of relative complex permittivity (ϵ') and permeability (μ') symbolize the storage ability of EM energy and the imaginary parts (ϵ'' and μ'') represent the loss ability [323]. It is well known that the permittivity and permeability in a material mainly originate from different types of polarizations and magnetic properties depending on the crystal structure, size, and special geometrical morphology [29,323–325]. The anisotropic magnetic particles may exhibit higher resonance frequency above Snoek's limit in the GHz frequency range due to their low eddy current loss coming from the particle shape effect [154]. Interestingly, Fe, Co, and Ni exhibit excellent EM properties because of the large saturation magnetization and higher Snoek's limit [154,326].

2.5.3 Magnetic Properties

A ferromagnetic material becomes single domain when the size is reduced below a critical value. The single domain sizes of Fe, Co, and Ni spherical particles correspond to 14, 70, and 55 nm, respectively [18]. If the size of a single-domain particle further decreases below a critical diameter, it exhibits superparamagnetism [16]. Figure 2.17 depicts the magnetic behavior of ferromagnetic and superparamagnetic nanoparticles under an external magnetic field. It shows that the magnetic moment of ferromagnetic as well as single-domain superparamagnetic nanoparticles is aligned under the applied magnetic field. However, the ferromagnetic nanoparticles retain the net magnetization in absence of the external field. In contrary, no net magnetization is observed in the superparamagnetic nanoparticles due to rapid reversal of the magnetic moment [16].

The susceptibility (κ) of a material is the measure of effectivenesss of an applied field in inducing magnetic dipole moment. It can be expressed as: $\kappa = M/H$; where, M is the magnetization of the material and H is the applied magnetic field [18].

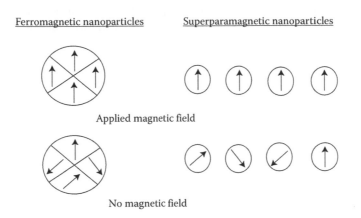

FIGURE 2.17 Magnetization behavior of ferromagnetic and superparamagnetic nanoparticles under an external magnetic field.

The magnetization is a measure of magnetic moment per unit volume of material and it is called a specific magnetization when expressed as magnetic moment per unit mass. In ordered materials, the susceptibility is found to be dependent on temperature and the applied field. When H increases in ferromagnetic materials, M increases initially, and finally attains a saturation limit called saturation magnetization (M_s) at large values of H. However, the magnetization does not follow the initial curve with decreasing H. Such irreversibility in the M–H curve is referred to as hysteresis, where a nonzero remnant magnetization (M_r) exists even in the absence of the applied field (H = 0) when the applied field is reversed. When the field is applied in the opposite direction (a negative field), M becomes zero at a field magnitude referred as coercivity (H_c). These characteristics determine the practical applications of magnetic material such as hard magnets ($H_c \geq 100$ Oe) and soft magnets ($H_c \leq 100$ Oe) [18]. Further increase in the negative field leads to the saturation of magnetization (M_s) in the negative direction. An example of such hysteresis in a magnetic material illustrating all these features is displayed in Figure 2.18.

Temperature has a pronounced effect on the magnetic properties of a material. It induces a random rotation of the magnetic moment as well as realigns the spins in a magnetic material [6]. The net magnetic moment becomes zero at high enough temperatures in the absence of an EM field [7]. In addition, the magnetization of a ferromagnetic material is extremely sensitive to the morphology, crystallinity, and magnetization direction [327]. Room-temperature magnetic properties (M_s, M_r, and H_c) of Fe, Co, and Ni with diverse morphology have also been measured. The saturation magnetization of iron nanoparticles are lower than the corresponding bulk counterpart [142,328–330] due to high surface/volume ratio. The dependence of magnetic properties on the particle size (10–100 nm) of Fe shows that the magnetization decreases with decreasing particle size [51]. This could be attributed to the increasing volume of the amorphous layer of surface Fe_3O_4 [51]. Fan et al. [139] reported that the smaller size, bigger shape anisotropy, and stronger surface pinning effect of iron nanorods account for the higher H_c values compared to the bulk material. Ni et al. [142] observed larger H_c in rod-like Fe compared to spherical and flaky morphology, though the flaky Fe showed the higher M_s. The measurement of

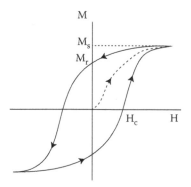

FIGURE 2.18   Typical ferromagnetic hysteresis curve.

H_c values along the easy axis in Fe nanowires exhibits higher value compared to that measured along the perpendicular direction [163].

Co nanoparticles exhibit lower M_s due to the presence of an oxidized surface layer compared to its bulk counterpart [12,331,332]. Abu-Much and Gedanken [67] reported that Co nanoparticles, when prepared with and without application of magnetic field, show different M_s values due to spin disorder and surface oxidation. Senapati et al. [12] prepared Co microflowers, which exhibit lower M_s value compared to the bulk Co mainly due to the presence of trace amounts of antiferromagnetic substance on the surface as detected by EDX and surface spin disorder. In addition, the reduction in the total magnetic moment and the increased magnetic interaction between the nanoflake-like petals of the microflowers cannot be ruled out. Li et al. [264] also noticed that H_c is enhanced in hierarchical Co nanostructures compared to the bulk cobalt [333]/spherical Co nanoparticles [334]. Such higher H_c values might be attributed to the increased surface/volume ratio and enhanced shape anisotropy. Co nanorods [116,174,177] and nanoflowers [12,265,267] also exhibit higher H_c values due to shape anisotropy. In addition, the higher H_c values are also observed, when a magnetic field is applied parallel to the easy axis of 1D nanomaterial [185,200,335]. Baber et al. [336] noted that the coercivity of Co nanowires decreases with increasing wire diameter (>30 nm). The magnetic properties of self-assembled Co nanodisk arrays [337] and cobalt flowerlike architectures [262] have also been studied.

The presence of spin disorder on the surface and surface oxidation significantly reduce the M_s value of Ni nanoparticles compared with bulk material [214]. Eluri and Paul [80] observed a decrease in M_s (and M_r) values of Ni nanoparticles with reduction in the particle size. Wang et al. [112] also noted the decrease in M_s values with reduction in the particle size of Ni nanoparticles. This is attributed to high surface/volume ratio and the corresponding surface effects. Senapati et al. [10] prepared Ni nanoflowers in the presence/absence of different capping agents, which also exhibited lower value of M_s compared to the bulk Ni. This reduction may be due to the increase in surface area and the magnetic interaction between the petals in the nickel nanoflowers. As a consequence, the total magnetic moment at a given field as well as M_s also reduced in comparison to the bulk counterpart. Ni nanostructures, similar to Fe and Co, also exhibit enhanced H_c value compared to the bulk counterpart [214,215,217,219,270]. The coercivity of 1D Ni is found to be larger when the applied field is parallel than in the perpendicular direction [206,222,229,243,246]. The coercivity and remnant magnetization variation of Ni nanotubes [230,231] and nanorods [222] have also been measured as a function of the diameter in the applied field along the axis direction. It is noted that coercivity and remnant magnetization increase with decreasing diameter. However, Ni nanoflowers exhibit lower H_c in comparison to the nanorods, nanotubes, and nanowires. In all probability, radially oriented petals of the Ni nanoflower, unlike in 1D nanomaterial, could not align themselves simultaneously along the direction of the external magnetic field [270]. Tables 2.2 through 2.4 record the room temperature M_s, M_r, and H_c data of different morphology of Fe, Co, and Ni.

Wu et al. [277] noted how saturation magnetization increased with increasing calcination temperature in SiO_2-coated Fe nanoparticles, whereas they coercively

TABLE 2.2
Room Temperature Magnetic Properties of Iron of Different Morphology

Type of Sample	M_s(emu g^{-1})	M_r(emu g^{-1})	H_c(Oe)	Reference
Iron bulk	225		0.9	[51,142]
Nanoparticles	200	–	~1100	[51]
	67.5	–	193	[142]
	–	–	198	[329]
Nanorods	120	32	538	[139]
	142	14	189	
	150 ‖, 115 ⊥	50	900	[140]
	–	–	183	[143]
	70.6	–	224	[142]
	77.7	0.63	346	[14]
Nanowires	145	36	206	[155]
	–	–	1832 ⊥, 118 ‖	[163]
	157.93	–	9.74	[328]
Nanodisks	78.1	–	176	[142]
	193	–	23	[330]
Fe@SiO$_2$	71.2	0.61	344	[14]
Ag@@Fe@SiO$_2$	46.1	0.60	275	[4]

TABLE 2.3
Room Temperature Magnetic Property of Cobalt of Different Morphology

Type of Sample	M_s(emu g^{-1})	M_r(emu g^{-1})	H_c(Oe)	Reference
Bulk Co	168	–	10	[116]
Nanoparticles	163	–	100	[64]
	–	–	–	[67]
Nanorods	150	–	340	[116]
	–	–	695	[174]
	113	–	4500	[179]
Nanotubes	–	–	691,175	[181]
	–	–	233 ‖, 43 ⊥	[185]
Nanowires	–	–	930	[196]
	–	–	30	[199]
	–	–	1848 ⊥, 507‖	[200]
	133.1	–	166.8	[332]
	–	–	1549 ⊥, 711‖	[335]
	–	–	430 ‖, 470 ⊥	[336]
Nanodisks	48.5	–	100	[258]
	–	–	900	[337]
Nanoflowers	95.5	1.1	261	[12]
	108.7	1.4	258	
	131	1.3	250	
	129	1.5	454.3	
	90	5.7	207	[265]
	128	–	308	[265]
	90	33	360	[267]
	97	28.3	407	[332]
	–	–	197	[262]

TABLE 2.4

Room Temperature Magnetic Property of Nickel of Different Morphology

Type of Sample	M_s(emu g^{-1})	M_r(emu g^{-1})	H_c(Oe)	Reference
Bulk Ni	57	–	0.7	[21]
	34.9	6.1	175	[11]
	19.6	5.4	73	[80]
	17.2	4.0	81	[82]
	45	–	89 (15 nm); 148 (58 nm)	[84]
	43	–	–	[85]
	37.17	4.5	54	[90]
	–	0.458, 0.693	148, 289	[92]
	42	–	130	[111]
	27.9	–	27.3	[112]
	23	10.1	445	[117]
	50	–	300	[120]
	54.2	–	19.4	[272]
Nanorods	44.2	19.4	240	[214]
	–	–	332	[217]
	3	–		[218]
	–	–	692 ∥, 359 ⊥	[222]
	54, 53	–	197, 204	[223]
	37.5	–	68.5	[338]
Nanotubes	–	–	610	[229]
	–	–	700–100	[230]
	–	–	190 ∥, 90 ⊥	[231]
Nanowires	25.63	1.89	472.56	[13]
	–	–	180	[203]
	39.6	6.1	245	[232]
	49.5	–	300	[242]
	–	–	830 ∥, 165 ⊥	[243]
	–	–	300, 250	[246]
	–	–	315	[235,247]
	54.8		186.2	[272]
Nanoflowers	52.3	–	238	[21]
	47	–	160	[269]
	30.8	9.9	173.2	[270]
	43.2	–	185.45	[272]
	40.2, 51.9	16.5, 13.2	192.9, 142.4	[273]
	29.8	1.143	259.44	[10]
	46.5	20.10	404	
	48.55	1.512	322	
	56.45	12.86	275	
Ni/Ag core–shell	20.50	0.60	391.88	[13]
Ni/ZnO (0.063 M [Zn^{2+}])	16.5	1.25	187	
Ni/ZnO (0.125 M [Zn^{2+}])	15.44	4.04	528	
Ni/ZnO (0.195 M [Zn^{2+}])	14.42	1.14	532	
Ni/ZnO (0.25 M [Zn^{2+}])	9.6	1.6	529	

decreased with increasing calcination temperature. Senapati et al. [14] also investigated the variation of magnetization as a function of the applied magnetic field of Fe, Fe@SiO$_2$, and Ag-encapsulated Fe@SiO$_2$ nanostructures at room temperature; the corresponding M$_s$, M$_r$, and H$_c$ data are also presented in Table 2.2. According to this, the M$_s$ values of the samples follow the order: Fe nanorods (77.7 emu g^{-1}) > Fe@SiO$_2$ (71.2 emu g^{-1}) > Ag-encapsulated Fe@SiO$_2$ (46.1 emu g^{-1}). The observed decrease in the M$_s$ value of Fe@SiO$_2$ nanostructures could be ascribed to the coating of the amorphous SiO$_2$ on the surface of Fe nanorods. A further decrease in M$_s$ value of Ag-coated Fe@SiO$_2$ could be attributed to the slight increase in size and mass due to the silver coating.

Senapati et al. [13] investigated the room-temperature magnetic properties of Ni and Ni/Ag nanostructures that have also been studied and the variation of the magnetization versus the applied magnetic field. The appearance of hysteresis loop in both the cases confirmed the ferromagnetic behavior of the bare Ni wires and its transmetalated product. The coercivity, saturation magnetization, and remanent magnetization values of Ni nanowire and its transmetalated product (Ni/Ag) are recorded in Table 2.4. This clearly demonstrates that the magnetization is retained even in the core–shell Ni/Ag, though there exists a slight loss in its saturation magnetization compared to bare nickel. This in all probability is due to the increase in the size and mass of silver-coated nickel. Optical and magnetic analyses of Ni@Ag nanoparticles exhibited the characteristic absorption band of Ag nanoparticles at 430 nm and the nearly superparamagnetic property of Ni nanoparticles [82].

Rivase et al. [338] investigated temperature dependence of magnetization in cobalt particles coated with silver and observed a strong dependency of magnetic properties on temperature. Fe@Au magnetic core–shell nanoparticles are superparamagnetic at room temperature (300 K) with a blocking temperature, T$_b$, of ≈170 K [293].

The Co$_{core}$Pt$_{shell}$ supermagnetic nanoalloy showed block temperature of 15 K and coercive value of 330 Oe (5 K) [364]. When annealed at 700°C, it exhibited ferromagnetic behavior of core–shell particles (H$_c$ = 5300 Oe). Paulus et al. [365] reported that magnetic anisotropy of Co@Au was highly reduced and corresponded closer to that of the bulk. Xu and Wang [366] observed that Co@Au core–shell nanoparticles retain the properties of core cobalt as evident from low and room-temperature hysteresis. The saturation magnetization, remanent magnetization, and coercive value were found to be 17.2 emu g^{-1}, 4.0 emu g^{-1}, and 81 Oe respectively. M$_s$ and M$_r$ values decrease with coating of Ag shells suggesting that interface of Ni@Ag plays a quenching role on the magnetic moment. Temperature dependent magnetic property measurements of Co/Au core–shell nanoparticles confirmed that the particles are superparamagnetic with a blocking temperature, T$_b$ ~ 55 K, consistent with a magnetic diameter ~6 nm [301]. The ultraviolet–visible absorption spectra of these nanoparticles showed a red shift (relative to pure gold nanoparticles) in agreement with an Au–shell morphology. The giant magnetoresistance values in Co/Ag magnetic nanostructures were found to be as high as 0.5% at room temperature [304]. Wang et al. [367] recently reported theoretical and experimental studies that explain the observed strong enhancement of the magnetooptical (MO) Faraday rotation in all-metal core–shell Co/Ag nanoparticles attributed to localized surface plasmon resonance (LSPR). They established direct correlation between the strong LSPR induced EM fields and the enhanced MO activity of these nanoparticles.

Based on room-temperature magnetization plots for nickel nanoparticles and Ni/ZnO hybrid nanostructures, the corresponding data are presented in Table 2.4 [11]. Though ferromagnetic behavior is retained in both, M_s of the nickel is significantly higher compared to Ni/ZnO hybrid nanostructures and gradually decreases with increasing $[Zn^{2+}]$ used in preparation of Ni/ZnO hybrids. Moreover, the H_c of Ni/ZnO nanostructures ($[Zn^{2+}] > 0.063$ M) increased approximately by two and half folds compared to pure Ni. This could be attributed either due to the "surface anisotropy effect" or "exchange bias effect" at the interface between the ferromagnetic Ni and nonmagnetic ZnO. Bala et al. [309] reported that the Co/ZnO with a core–shell nanocomposite exhibited both magnetic and optical bifunctionalities simultaneously.

2.6 APPLICATIONS OF Fe, Co, AND Ni NANOMATERIALS

2.6.1 MAGNETIC AND ELECTRICAL APPLICATIONS

The applications of Fe, Co, and Ni as magnetic material are influenced by their magnetic properties [1]. The material with high values of H_c, M_s, and M_r finds application as permanent magnets, for example, motors, loudspeakers, refrigerators, generators, and high-density recording media [177]. In addition, Fe-coated magnetic nanocomposites [368] and FePt nanostructures [369,370] have also been used as permanent magnets in most of these applications. However, the lower H_c, M_s, and M_r values are desirable in soft magnetic material for their various applications, for example, electronic circuit, transformer cores [18,19], etc.

The area of magnetic recording media constitutes another important application of ferromagnetic nanomaterials. For example, iron nanoparticles have been used in many advanced magnetic tapes, such as those used in computer backup and camcorders of very high capacity [6]. Recently, 1D ferromagnetic Fe nanorods [137] and Ni nanowire arrays [371], and CoPt and FePt [369,370] nanoparticles with high aspect ratio have found application in high-density magnetic recording media.

2.6.2 CATALYTIC APPLICATIONS

The superior catalytic activity of ferromagnetic materials compared to bulk counterparts could be attributed to their extremely high surface area per unit volume [6]. Additional advantages in these materials lie with their magnetic field induced separation and recovery of the catalysts. All these features make the ferromagnetic materials better catalysts in terms of their recyclability and cost.

According to the available literature, Fe nanoparticles act as a catalyst in the electrophilic amination of functionalized organo copper and organo zinc reagents [372], asymmetric hydrogenation of ketones [373], and nitrite reduction [374]. Fe and Co nanoparticles find catalytic application in Fischer–Tropsch synthesis [45,375]. Ni and Co nanoparticles and Co microflowers have also been used in the hydrogenation of benzene [376], p-nitrophenol [12,377–379], and CO_2 [380,381], respectively. The catalytic decomposition of hydrocarbon in presence of ferromagnetic nanomaterials produced disordered carbon nanostructures [382] and multiwalled carbon nanotubes (MWCNT) [382–384].

Acid-functionalized iron nanoparticles are found to be good catalysts in the ring opening of epoxidized methyl oleate [385]. Hierarchical Fe-anchored Zeolite Socony Mobil (ZSM) acts as an efficient catalyst in selective oxidation of benzene [386] and benzylation [387]. Zeolite framework stabilized Ni nanoparticles showed high catalytic activity for hydrogen generation [388]. The Fischer–Tropsch reaction has also been catalyzed by ZSM-5 supported Co [389], CNT supported Co [390,391], and Fe/Co [392]. Nickel nanoparticles embedded in the pores of a mesoporous metal-organic framework also act as an efficient catalyst for the hydrogenolysis of nitrobenzene [393].

The shell invariably acts as a catalyst, whereas the core enables its easy separation for recycling. These features help in finding out the economical use of precious noble metals as active catalysts on the surface of ferromagnetic matels [6]. For example, Au-coated Ni [232], Ag-coated Ni [13], and Ni@SiO$_2$ [394] have successfully been used in catalytic hydrogenation of p-nitrophenol. It is also reported that Ni/Ag is a better catalyst compared to bare Ni due to the ligand effect. Hydrogenation of benzene has also been investigated in the presence of Ni@Pd [395]. In addition, nano Co@SiO$_2$ and Au/Ni-fiber act as good catalysts in the phenoxycarbonylation of iodobenzene [279] and low-temperature gas-phase oxidation of alcohol, respectively [396]. Reports are also available on the catalytic activity of nanoshell carbon-supported Co in the aerobic oxidation of alcohols [397] and hydrogen generation [398].

2.6.3 BIOMEDICAL APPLICATIONS

Magnetic nanoparticles offer some attractive possibilities in medicine. Though cells in living organisms are in the range of 10 μm in diameter, the cell parts are much smaller (submicron size domain). Therefore, nanoparticles can place themselves at dimensions that are smaller than those of a cell (10–100 μm), or comparable to the size of a virus (20–450 nm), a protein (5–50 nm), or a gene (2 nm wide and 10–100 nm length), and they can become close to the biological entity of interest. In addition, magnetic nanoparticles can be manipulated by an external magnetic field gradient. These properties open up many applications involving the transport and immobilization of magnetic nanoparticles, or of magnetically tagged biological entities. In this way, they can be made to deliver a package, such as an anticancer drug, to a targeted region of the body, such as a tumor. Magnetic nanomaterials having higher magnetization values are beneficial for various biomedical applications [6]. However, the toxicity of nanostructured materials remains an important consideration for most of these applications [399]. In this regard, Fe nanowires find several biomedical applications due to their biocompatibility and low cytotoxicity [399]. Ni [400] and FePt [401] nanoparticles find applications in cancer diagnosis and chemotherapy.

Functionalized Ni nanoparticles could also enhance the permeability of cell membrane facilitating the cellular uptake of outer target molecules into cancer cells [402]. Fe-based nanoparticles are also useful in drug delivery systems supplying essential nutrients to the human body [160,403]. The high magnetization, antihemolysis property, and cell compatibility of TiO_2 [404] and Au-coated Fe nanoparticles [405] make them very attractive materials in biomedical fields. Carboxylated SiO_2-coated α-Fe nanoparticles are a versatile material used in biomedical applications due to their enhanced corrosion resistivity, excellent aqueous dispersibility, and low cytotoxicity [406].

Carbon-encapsulated Fe nanoparticles [407], Fe filled MWCNT [408], Au-coated Fe nanoclusters [409], and Au-coated Ni nanowires [410] have been considered as potential candidates for applications in drug delivery. Au-coated Co [411], CoPt [412], C-encapsulated Fe [408] and Fe/MWCNT [408], and functionalized Fe nanoparticles [27,413] are widely applied as MRI contrast agents. Reports are also available on using Ni/NiO nanoparticles stabilized with the amphipathic peptide H_2N–Cys–Leu–Pro–Phe–Phe–Asp–NH_2 in biomedical applications [414]. Co–Au [302] Ni–Au nanowire/Ni–Au nanorods [415] core–shell nanomaterials are found to be very efficient for gene delivery.

2.6.4 ENVIRONMENTAL APPLICATIONS

Water is continuously polluted by a wide range of biological and chemical contaminants [9]. Biological contaminants can be remediated by various methods if detected before consumption by humans. In this regard, ferromagnetic nanomaterials are found to be very effective due to their enhanced catalytic activity, reusability, and cost effectiveness [26]. In order to remove the biological pollutant, Fe nanoparticles (cornlike) have been employed as the antibacterial agent against four bacteria that cause digestive problems [416]. Senapati et al. [13] studied the excellent antibacterial activity of Ni/Ag nanowire against *E. coli* and Ni/Ag is found to be superior compared to commercially available antibacterial beads. Most importantly, they can be easily magnetically separated below the WHO specified toxicity limits of Ni and Ag.

Chemical contaminants are divided into organic and inorganic contaminants. Water is also polluted by organic contaminants, such as nitrophenols, dyes, pesticides, industrial solvents, fuels, etc. Their toxicity, stability to natural decomposition, and persistence in the environment is the cause of much concern to societies and regulation authorities around the world. Therefore, the removal of organic pollutants from water is very important in environmental protection. Among the processes developed for the destruction of organic contaminants, biodegradation has received the greatest attention. However, many organic chemicals, especially those that are toxic or refractory, are not amenable to microbial degradation. Therefore, in recent years considerable attention has been focused on advanced oxidation processes, which are considered as the alternative green techniques to replace existing conventional methods for the degradation of dyes and pesticides. In this regard, maximum interest has been focused on semiconductor-based photocatalysis [11]. However, the difficulty in removing suspended nanosized photocatalysts from the water medium is a serious drawback. On the other hand, the recyclability of the catalyst is also important for its multiple usage and cost effectiveness. Commonly used separation steps, such as filtration or centrifugation are tedious. Fortunately, the problem has been overcome by selecting magnetically separable photocatalysts consisting of ferromagnetic nanoparticles and a semiconductor oxide, either in the core–shell structure or in the form of a composite. Therefore, the removal of aromatic compounds, methylene green, arsenazo (III) dye, congo red, and gaseous pollutants has been investigated using hierarchical Co [59], Ni nanoparticles functionalized graphene sheets, carbon–cobalt nanocomposites, nickel nanoparticles loaded on activated carbon, and nickel and Ni impregnated activated carbon fibers as adsorbents [59,417–421], respectively. Alternatively, the photodegradation of organic pollutants in the presence of stepped hexagon-shaped Ni/ZnO [11], Fe nanoparticles [44], Fe/TiO$_2$ [422] Si-doped iron–iron oxide [423], Pd-doped Co nanofibers [339], and TiO$_2$-coated Ni nanoparticles [26,340] have also been successfully used to purify contaminated water. It has been noted that Ni/ZnO nanostructures exhibit better photocatalytic activity compared to pure ZnO. This may be due to interfacial charge transfer between the metal semiconductor interface which prevents electron–hole recombination and improves photocatalytic activity. Room-temperature degradation of nitrophenols has also been reported in the presence of Co [12], pure Ni [13,373–375], Ni/Ag [13], and Ni/Au [232] nanostructures acting as the catalysts. Interestingly, inorganic pollutants, for example, heavy metals, metalloids, and nonmetal pollutants can also be removed by these ferromagnetic nanomaterials and their core–shell nanostructures [33,276,341–363,424,425] as referred to in Table 2.5.

Theoretical calculations show that a lightening-rod effect in the EM mechanism can enhance the surface-enhanced Raman spectroscopy (SERS) activity of Ni nanowires [426]. Yao et al. [427] reported surface Raman spectra of pyridine adsorbed at their surface and studied the SERS mechanism by probing the SERS effect fusing Ni nano-wire arrays with different aspect ratios. Saur et al. [428] observed plasmon-enhanced absorption at around 400 nm in hexagonally arranged monodisperse Ni nanowires. They were also observed in typically SERS inactive metals like nickel, a strong, but locally strongly inhomogeneous SERS signal during *in situ* Raman microspectroscopy. Pradeep et al. [239] investigated SERS activity of Ni nanowires using crystal violet (CV) as an analyte and collected the Raman spectra of adsorbed CV molecules at different concentrations. Their study showed that Ni nanowires give significant signal intensity up to a concentration of 10^{-6} M.

Silver is regarded as an ideal shell material, which could provide easy control of the deposition process, controllable thickness, and excellent oxidation resistance [13]. In addition, the choice of silver as a shell could also lead to the SERS phenomena [41] resulting in its excellent sensitivity and selectivity, when the analyte is close to the rough surface of the silver (<20 nm) [429,430]. However, the nanosized silver is easily agglomerated limiting its application in SERS. It is reported that the formation of SiO$_2$ shell on Fe could prevent the agglomeration of Ag [431]. Senpati et al. [14] reported that Ag-encapsulated Fe@SiO$_2$ nanostructures also exhibited the SERS phenomena. The presence of a broad peak (200–400 cm^{-1}) corresponds to the central C$^+$-phenyl bond vibrations [432]. In addition, the peaks in the range of 1370–1390 and 1510–1628 cm^{-1} in the spectra correspond to the vibrations of N-phenyl stretching and ring C–C stretching vibration, respectively [433]. It is also noted that the overall intensity of the peak decreases with the concentration of the CV. These observations clearly demonstrate that the CV can be well detected with the Ag-encapsulated Fe@SiO$_2$ SERS substrate, when the concentration is as low as

TABLE 2.5
Metal Ions Removed by Fe Nanomaterial and Core (Fe or Ni)–Shell Nanostructure as Adsorbents

Metal Ions Removed	Magnetic Nanomaterial Used as Adsorbent	Reference
Cr(VI)	Organo-montmorillonite supported iron	[33]
	Silica-coated Fe	[276]
	Fe	[339–344]
	Graphene-coated Fe	[345]
	Bentonite-supported nanoscale zero-valent iron	[346]
As(III)	Fe	[347,348]
As(V)	Fe	[348]
Nitrate	Fe	[349,350]
Pb(II)	Fe	[351,352]
	Amine functionalized Fe nanoparticles	[353]
Cd(II)	Fe	[351]
Pb(II),Cd(II), and Cr(VI)	Fe	[354]
Pb(II), Zn(II), Cd(II), Ni(II), Cu(II), and Ag(I)	Fe	[355]
Cr(VI), Cu(II), Cd(II), and Pb(II)	Chitosan beads-supported Fe	[356]
As(III) and As(V)	Iron-modified activated carbon	[357]
	Iron-coated pottery granules	[358]
Cu(II)	Fe@C	[359]
Cr(VI), Ni(II), Cd(II), Pb(II), Co(II), and Mn(II)	Carbon-coated Fe nanoparticles	[360]
Cr(VI), Pb(II), Cd(II), and Hg(II)	Fe@Fe$_3$O$_4$ matrix in graphene	[361]
Cu(II), As(III), Cr(VI), Pb(II), and Hg(II)	Iron nanoparticle-immobilized hybrid electrospun polymer nanofibrous mats	[362]
Pb(II), Cu(II), and Cd(II)	Honeycomb-like Ni@C composite	[363]

1×10^{-10} M. Ray et al. [434] discussed the possible mechanism and operating principle for the targeted separation, label free SERS imaging, and photothermal destruction of multidrug resistance bacteria from a food sample.

The rapid growth of wireless communication, information technology, high-frequency circuit devices (in the GHz range), and radar stealth systems also contribute toward environmental pollution. This is due to the emission of EM irradiations causing the malfunctioning of electronic devices as well as an adverse effect on biological systems [139,435]. Therefore, the soft metallic magnets remain the most preferred choice for EM wave absorption at high frequency over the gigahertz range due to high saturation magnetization and Snoek's limit [436]. As a consequence, nanoparticles of Fe [14,139,254,437–440], Co [323], Ni [441], Fe@C [442–446], Fe@ZnO [447,448], Ni–P-coated Fe [449], Fe/SiO$_2$ [14,142,274], Ag-coated Fe/SiO$_2$ [14], Fe/SnO$_2$ [450], C-coated Co [451], polymer-protected Co [31], C-coated Ni [446,452], CuO/Cu$_2$O-coated Ni [453], Ni/polypyrrole [28], Ni/graphene [29,30], Fe/Fe$_2$O$_3$, and Ni/NiO [326] have been investigated as EM wave absorbing materials.

The impedance mismatch between the free space and shielding material accounts for reflection loss (SE$_R$). Its magnitude under plane wave (far field conditions) can be expressed as [454,455]:

$$SE_R(dB) = -10\log_{10}\left(\frac{\sigma_T}{16\omega\varepsilon_1\varepsilon_0\mu_r}\right)$$

where σ_T, ω, ε_1, ε_0, and μ_r refer to total conductivity, angular frequency, relative permittivity, and relative permeability referred to as free space, respectively. Table 2.6 records the reflection loss data of Fe, Co, and Ni nanostructures. The EM property of the urchin-like Ni chain is found to be relatively better compared to Ni with spherical morphology [325]. The absorbing mechanism can be explained in terms of geometrical effect quantitatively as well as by high initial permeability, point discharge

TABLE 2.6
EM Data Reported on Fe, Co, and Ni Nanostructures

Sample Type	Frequency (GHz)	Reflection Loss (dB)	Reference
Iron			
Nanoparticles	10	−8	[142]
Nanorods	9	−9.5	[142]
Nanorods	10	−13.1	[14]
Nanoflakes	13	−15.5	[142]
Cobalt			
Spherical	5.5	−8	[323]
Flower with dendritic petal	9	−13.5	[323]
Flower with sharp petal	5.5	−12.5	[323]
Nickel			
Sting like	9	−15	[324]
Chain like	7.5	−27	[324]
Spherical	7.5	−12	[324]
Smooth chain like	5	−5	[325]
Ring like	10.5	−5	[325]
Urchin like chain	9.5	−25	[325]
SiO$_2$-Coated Fe			
Nanoparticles	11	−8.5	[142]
Nanorods	10	−15	[142]
Nanorods	10	−15.6	[14]
Nanoflakes	11	−16.5	[142]

effect, and multiple absorption qualitatively. Zhao et al. [324] established that the EM absorption ability of sting-like nickel arises from the point discharge effect, while the absorption property of chain-like nickel is ascribed to the geometrical effect. The hierarchical Co exhibits better EM properties compared to the spherical Co [323]. Wang et al. [323] observed that flowerlike structures (flowers with dendritic petals and flowers with sharp petals) of cobalt exhibit enhanced EM absorbing ability compared to cobalt spheres. The measurements on silica-coated iron nanoparticles, nanorods, and nanoflakes indicated the larger reflection loss in a wider frequency range [142]. Senapati et al. [14] recently studied the EM property of Fe nanorods and SiO_2-coated Fe nanorods. These results show the improved EM property of SiO_2-coated Fe nanorods compared to pure Fe of identical thickness. This improvement may be due to the increase in surface anisotropy by the reduction of the magnetic coupling effect through dispersivity of the magnetic particles into the shell.

2.7 SUMMARY AND CONCLUSION

This chapter is focused on the synthesis of elemental ferromagnetic nanomaterials of variable morphologies due to their morphology dependent properties. These are found to be highly susceptible to aerial oxidation. Therefore, their suitable surface protection stretegies have also been discussed which consists of either the formation of core–shell nanomaterials or some other preoxidation procedure with a slight loss of magnetization. The magnetic as well as the EM properties of these materials in different morphology has also been explained in detail. The possible applications of these nanomaterials either in pure form or its surface modified forms are also discussed.

REFERENCES

1. J. Wang and X. C. Zeng, Core–shell magnetic nanoclusters, in *Nanoscale Magnetic Materials and Applications*, J. P. Liu, E. Fullerton, O. Gutfleisch, and D. J. Sellmayer (eds.), Springer Publications, 2009, pp. 35–62.
2. Y. Xia, P. Yag, Y. Sun, Y. Wu, B. Mayers, B. Gates, Y. Yin, F. Kim, and H. Yan, One-dimensional nanostructures: Synthesis characterization and application, *Adv. Mater.* 15, 2003, 353–389.
3. J. S. Beveridge, J. R. Stephens, and M. E. Williams, The use of magnetic nanoparticles in analytical chemistry, *Annu. Rev. Anal. Chem.* 4, 2011, 251–273.
4. S. P. Gubin, Y. A. Koksharov, G. B. Khomutov, and G. Y. Yurkov, Magnetic nanoparticles: Preparation, structure and properties, *Russ. Chem. Rev.* 74, 2005, 489–520.
5. A.-H. Lu, E. L. Salabas, and F. Schüth, Magnetic nanoparticles: Synthesis, protection, functionalization, and application, *Angew. Chem. Int. Ed.* 46, 2007, 1222–1244.
6. D. L. Huber, Synthesis, properties, and applications of iron nanoparticles, *Small* 1, 2005, 482–501.
7. M. Colombo, S. Carregal-Romero, M. F. Casula, L. Gutiérrez, M. P. Morales, I. B. Böhm, J. T. Heverhagen, D. Prosperi, and W. J. Parak, Biological applications of magnetic nanoparticles, *Chem. Soc. Rev.* 41, 2012, 4306–4334.
8. X. W. Teng and H. Yang, Synthesis of face-centered tetragonal FePt nanoparticles and granular films from Pt@Fe$_2$O$_3$ core-shell nanoparticles, *J. Am. Chem. Soc.* 125, 2003, 14559–14563.

9. C. S. S. R. Kumar, *Nanomaterials for the Life Sciences Vol. 4: Magnetic Nanomaterials*, Wiley-VCH Verlag GmbH & Co. KGaA, Weinheim, Germany, 2009.

10. S. Senapati, S. K. Srivastava, S. B. Singh, and K. Biswas, Capping agent assisted and Ag-catalyzed growth of Ni nanoflowers, *Cryst. Growth Des.* 10, 2010, 4068–4075.

11. S. Senapati, S. K. Srivastava, and S. B. Singh, Synthesis, characterization and photocatalytic activity of magnetically separable hexagonal Ni/ZnO nanostructure, *Nanoscale* 4, 2012, 6604–6612.

12. S. Senapati, S. K. Srivastava, and S. B. Singh, Synthesis, magnetic properties and catalytic activity of hierarchical cobalt microflowers, *J. Nanosci. Nanotech.* 12, 2012, 3048–3058.

13. S. Senapati, S. K. Srivastava, S. B. Singh, and H. N. Mishra, Magnetic Ni/Ag core-shell nanostructure from prickly Ni nanowire precursor and its catalytic and antibacterial activity, *J. Mater. Chem.* 22, 2012, 6899–6906.

14. S. Senapati, S. K. Srivastava, S. B. Singh, and A. R. Kulkarni, SERS active Ag encapsulated Fe@SiO$_2$ nanorods in electromagnetic wave absorption and crystal violet detection, *Environ. Res.* 135, 2014, 95–104.

15. S. Senapati, Development and application of ferromagnetic nanomaterials in environmental remediation, PhD thesis, IIT Kharagpur, 2014.

16. A. Akbarzadeh, M. Samiei, and S. Davaran, Magnetic nanoparticles: Preparation, physical properties, and applications in biomedicine, *Nanoscale Res. Lett.* 7, 2012, 144.

17. K. J. Klabunde, *Nanoscale Materials in Chemistry*, John Wiley & Sons, USA, 2001.

18. N. M. Deraz and M. M. G. Fouda, Synthesis, structural, morphological properties of cobalt-aluminum nano-composite, *Int. J. Electrochem. Sci.* 8, 2013, 2756–2767.

19. Y.-J. Zhang, Q. Yao, Y. Zhang, T.-Y. Cui, D. Li, W. Liu, W. Lawrence, and Z.-D. Zhang, Solvothermal synthesis of magnetic chains self-sssembled by flowerlike cobalt submicrospheres, *Cryst. Growth Des.* 8, 2008, 3206–3212.

20. A. S. Lanje, S. J. Sharma, and R. B. Pode, Magnetic and electrical properties of nickel nanoparticles prepared by hydrazine reduction method. *Arch. Phys. Res.* 1, 2010, 49–56.

21. P. Li, N. Wang, and R. Wang, Flower-like nickel nanocrystals: Facile synthesis, shape evolution and their magnetic properties, *Eur. J. Inorg. Chem.* 2010, 2261–2265.

22. L. Bai, F. Yuan, and Q. Tang, Synthesis of nickel nanoparticles with uniform size via a modified hydrazine reduction route, *Mater. Lett.* 62, 2008, 2267–2270.

23. D. Chen, S. Liu, J. Li, N. Zhao, C. Shi, X. Du, and J. Sheng, Nanometre Ni and core/shell Ni/Au nanoparticles with controllable dimensions synthesized in reverse microemulsion, *J. Alloy Compd.* 475, 2009, 494–500.

24. R. Ghosh Chaudhuri and S. Paria, Core/shell nanoparticles: Classes, properties, characterization, and applications, *Chem. Rev.* 112, 2012, 2373–2433.

25. X. Ni, J. Zhang, Y. Zhang, and H. Zheng, Citrate-assisted synthesis of prickly nickel microwires and their surface modification with silver, *J. Colloid Interf. Sci.* 307, 2007, 554–558.

26. H. Pang, Y. Li, L. Guan, Q. Lu, and F. Gao, TiO$_2$/Ni nanocomposites: Biocompatible and recyclable magnetic photocatalysts, *Catal. Commun.* 12, 2011, 611–615.

27. S. Cheong, P. Ferguson, I. F. Hermans, G. N. L. Jameson, S. Prabakar, D. A. J. Herman, and R. D. Tilley, Synthesis and stability of highly crystalline and stable iron/iron oxide core/shell nanoparticles for biomedical applications, *Chem. Plus. Chem.* 77, 2012, 135–140.

28. P. Xu, X. Han, C. Wang, D. Zhou, Z. Lu, A. Wen, X. Wang, and B. Zhang, Synthesis of electromagnetic functionalized nickel/polypyrrole core/shell composites, *J. Phys. Chem. B* 112, 2008, 10443–10448.

29. T. Chen, F. Deng, J. Zhu, C. Chen, G. Sun, S. Ma, and X. Yang, Hexagonal and cubic Ni nanocrystals grown on graphene: Phase-controlled synthesis, characterization and their enhanced microwave absorption properties, *J. Mater. Chem.* 22, 2012, 15190–15197.

30. J. Fang, W. Zha, M. Kang, S. Lu, L. Cui, and S. Li, Microwave absorption response of nickel/graphene nanocomposites prepared by electrodeposition, *J. Mater. Sci.* 48, 2013, 8060–8067.
31. Y. Kato, S. Sugimoto, K. Shinohara, N. Tezuka, T. Kagotani, and K. Inomata, Magnetic properties and microwave absorption properties of polymer-protected cobalt nanoparticles, *Mater. Trans.* 43, 2002, 406–409.
32. K. Klacanova, P. Fodran, P. Simon, P. Rapta, R. Boca, V. Jorik, M. Miglierini, E. Kolek, and L. Caplovic, Formation of Fe(0)-nanoparticles via reduction of Fe(II) compounds by amino acids and their subsequent oxidation to iron oxides, *J. Chem.* 2013, 961629, 10.
33. K.-C. Huang and S. H. Ehrman, Synthesis of iron nanoparticles via chemical reduction with palladium ion seeds, *Langmuir* 23, 2007, 1419–1426.
34. T. I. Yang, R. N. C. Brown, L. C. Kempel, and P. Kofinas, Controlled synthesis of core-shell iron-silica nanoparticles and their magneto-dielectric properties in polymer composites. *Nanotechnology* 22, 2011, 105601/1–105601/8.
35. P. Wu, S. Li, L. Ju, N. Zhu, J. Wu, P. Li, and Z. Dang, Mechanism of the reduction of hexavalent chromium by organo-montmorillonite supported iron nanoparticles, *J. Hazard. Mater.* 219–220, 2012, 283–288.
36. N. Goldstein and L. F. Greenlee, Influence of synthesis parameters on iron nanoparticle size and zeta potential, *J. Nanopart. Res.* 14, 2012, 760/1–760/15.
37. A. S. Sayyad, K. Balakrishnan, L. Ci, A. T. Kabbani, R. Vajtai, and P. M. Ajayan, Synthesis of iron nanoparticles from hemoglobin and myoglobin, *Nanotechnology* 23, 2012, 055602/1–055602/5.
38. C. Leostean, O. Pana, R. Turcu, M. L. Soran, S. Macavei, O. Chauvet, and C. Payen, Comparative study of core-shell iron/iron oxide gold covered magnetic nanoparticles obtained in different conditions, *J. Nanopart. Res.* 13, 2011, 6181–6192.
39. M. Shekarriz, S. Taghipoor, F. Haji-Aliakbari, M. Soleymani-Jamarani, R. Kaveh-Ahangar, and M. Eslamian, Optimal synthesis and nitrate and mercury removal ability of microemulsion-made iron nanoparticles, *Int. J. Nanopart.* 3, 2010, 123–137.
40. X.-F. Cheng, B.-S. Wu, Y. Yang, and Y.-W. Li, Synthesis of iron nanoparticles in water-in-oil microemulsions for liquid-phase Fischer–Tropsch synthesis in polyethylene glycol, *Catal. Commun.* 12, 2011, 431–435.
41. L. Guo, Q. Huang, X. Li, and S. Yang, Iron nanoparticles: Synthesis and applications in surface enhanced Raman scattering and electrocatalysis, *Phys. Chem. Chem. Phys.* 3, 2001, 1661–1665.
42. V. Smuleac, R. Varma, S. Sikdar, and D. Bhattacharyya, Green synthesis of Fe and Fe/Pd bimetallic nanoparticles in membranes for reductive degradation of chlorinated organics, *J. Membrane Sci.* 379, 2011, 131–137.
43. Y. Kuang, Q. Wang, Z. Chen, M. Megharaj, and R. Naidu, Heterogeneous Fenton-like oxidation of monochlorobenzene using green synthesis of iron nanoparticles, *J. Colloid Interf. Sci.* 410, 2013, 67–73.
44. T. Shahwan, S. Abu Sirriah, M. Nairat, E. Boyaci, A. E. Eroglu, T. B. Scott, and K. R. Hallam, Green synthesis of iron nanoparticles and their application as a Fenton-like catalyst for the degradation of aqueous cationic and anionic dyes, *Chem. Eng. J.* 172, 2011, 258–266.
45. E. C. Njagi, H. Huang, L. Stafford, H. Genuino, H. M. Galindo, J. B. Collins, G. E. Hoag, and S. L. Suib, Biosynthesis of iron and silver nanoparticles at room temperature using aqueous sorghum bran extracts, *Langmuir* 27, 2011, 264–271.
46. M. Pattanayak and P. L. Nayak, Ecofriendly green synthesis of iron nano particles from various plants and spices extract, *Int.J. Plant, Animal Envn. Sci.* 3, 2013, 68–78.
47. F. Pelletier, D. Ciuculescu, J.-G. Mattei, P. Lecante, M.-J. Casanove, N. Yaacoub, J.-M. Greneche, C. Schmitz-Antoniak, and C. Amiens, On the use of amine-borane complexes to synthesize iron nanoparticles, *Chemistry* 19, 2013, 6021–6026.

48. O. Margeat, F. Dumestre, C. Amiens, B. Chaudret, P. Lecante, and M. Respaud, Synthesis of iron nanoparticles: Size effects, shape control and organisation, *Prog. Solid State Ch.* 33, 2006, 71–79.

49. L.-M. Lacroix, S. Lachaize, A. Falqui, M. Respaud, and B. Chaudret, Iron nanoparticle growth in organic superstructures, *J. Am. Chem. Soc.* 131, 2009, 549–557.

50. M. Huuppola, Z. Zhu, L.-S. Johansson, K. Kontturi, K. Laasonen, and C. Johans, Anomalous dependence of particle size on supersaturation in the preparation of iron nanoparticles from iron pentacarbonyl, *J. Colloid Interf. Sci.* 386, 2012, 28–33.

51. D. W. Lee, T. S. Jang, D. Kim, O. V. Tolochko, and B. K. Kim, Nanocrystalline iron particles synthesized without chilling by chemical vapor condensation, *Glass Phys. Chem.* 31, 2005, 545–548.

52. C. J. Choi, O. Tolochko, and B. K. Kim, Preparation of iron nanoparticles by chemical vapor condensation, *Mater. Lett.* 56, 2002, 289–294.

53. Y.-L. Zhu, Y. Katayama, and T. Miura, Electrochemical preparation of nickel and iron nanoparticles in a hydrophobic ionic liquid, *Electrochem. Solid St. Lett.* 14, 2011, D110–D115.

54. R. Ray, S. Das, M. Patra, and M. Thakur, Iron nanoparticles from an electrochemical route, *Nanosci. Methods* 1, 2012, 1–8.

55. K. Huang, C. Tan, M. Hu, and C. Xie, Synthesis of iron nanoparticles by a new laser-induction complex heating method and a study of its thermal phenomena, *J. Comput. Theoretical Nanosci.* 9, 2012, 1417–1421.

56. E. Escalera, M. A. Ballem, J. M. Cordoba, M.-L. Antti, and M. Oden, Synthesis of homogeneously dispersed cobalt nanoparticles in the pores of functionalized SBA-15 silica, *Powder Technol.* 221, 2012, 359–364.

57. Y. Chen, K. Y. Liew, and J. Li, Size controlled synthesis of Co nanoparticles by combination of organic solvent and surfactant, *Appl. Surf. Sci.* 255, 2009, 4039–4044.

58. H. Kavas, Z. Durmus, E. Tanriverdi, M. Senel, H. Sozeri, and A. Baykal, Fabrication and characterization of dendrimer-encapsulated monometallic Co nanoaprticles, *J. Alloy. Compd.* 509, 2011, 5341–5348.

59. X. Liang and L. Zhao, Room-temperature synthesis of air-stable cobalt nanoparticles and their highly efficient adsorption ability for Congo red, *RSC Adv.* 2, 2012, 5485–5487.

60. S. Pauline and A. Persis Amaliya, Size and shape control evaluation of cobalt (Co) and cobalt ferrite ($CoFe_2O_4$) magnetic nanoparticles, *Arch. Phys. Res.* 3, 2012, 78–83.

61. F. Fellah, F. Schoenstein, A. Dakhlaoui-Omrani, S. M. Cherif, G. Dirras, and N. Jouini, Nanostructured cobalt powders synthesized by polyol process and consolidated by spark plasma sintering: Microstructure and mechanical properties, *Mater. Charact.* 69, 2012, 1–8.

62. S. Chandra and A. Kumar, Modulation of synthetic parameters of cobalt nanoparticles: TEM, EDS, spectral and thermal studies, *Spectrochim. Acta. A* 98, 2012, 23–26.

63. S. S. K. Kamal, P. K. Sahoo, M. Premkumar, N. V. Rama Rao, T. J. Kumar, B. Sreedhar, A. K. Singh, S. Ram, and K. Chandra Sekhar, Synthesis of cobalt nanoparticles by a modified polyol process using cobalt hydrazine complex, *J. Alloy. Compd.* 474, 2009, 214–218.

64. H. Yang, C. Shen, N. Song, Y. Wang, T. Yang, H. Gao, and Z. Cheng, Facile synthesis of hollow nano-spheres and hemispheres of cobalt by polyol reduction, *Nanotechnology* 21, 2010, 375602, 5.

65. K. B. Narayanan and N. Sakthivel, Green synthesis of biogenic metal nanoparticles by terrestrial and aquatic phototrophic and heterotrophic eukaryotes and biocompatible agents, *Adv. Coll. Interf. Sci.* 169, 2011, 59–79.

66. P. Boyer and M. Meunier, Modeling solvent influence on growth mechanism of nanoparticles (Au, Co) synthesized by surfactant free laser processes, *J. Phys. Chem. C* 116, 2012, 8014–8019.

67. R. Abu-Much and A. Gedanken, Sonochemical synthesis under a magnetic field: Fabrication of nickel and cobalt particles and variation of their physical properties, *Chemistry*, 2008, 14, 10115–10122.
68. W. J. Erasmus and S. E. Van, Some insights in the sonochemical preparation of cobalt nano-particles, *Ultrason. Sonochem.* 14, 2007, 732–738.
69. J. Garcia-Torres, E. Vallés, and E. Gómez, Synthesis and characterization of Co@Ag core–shell nanoparticles, *J. Nanopart. Res.* 12, 2010, 2189–2199.
70. J. Ahmed, S. Sharma, K. V. Ramanujachary, S. E. Lofland, and A. K. Ganguli, Microemulsion-mediated synthesis of cobalt (pure fcc and hexagonal phases) and cobalt-nickel alloy nanoparticles, *J. Colloid Interf. Sci.* 336, 2009, 814–819.
71. E. Redel, J. Kraemer, R. Thomann, and C. Janiak, Synthesis of Co, Rh and Ir nanoparticles from metal carbonyls in ionic liquids and their use as biphasic liquid-liquid hydrogenation nanocatalysts for cyclohexene, *J. Organometallic Chem.* 694, 2009, 1069–1075.
72. H. Taeghwan, Chemical synthesis of magnetic nanoparticles, *Chem. Commun.* 8, 2003, 927–934.
73. D. Wostek-Wojciechowska, J. K. Jeszka, C. Amiens, B. Chaudret, and P. Lecante, The solid-state synthesis of metal nanoparticles from organometallic precursors, *J. Colloid Interf. Sci.* 287, 2005, 107–113.
74. X. Fu, Preparation of metallic cobalt nanoparticles by the thermal decomposition of $CoC_2O_4 \cdot 2H_2O$ precursor in the argon gas, *Appl. Mech. Mater.* 127, 2012, 85–88.
75. J. Ahmed, T. Ahmad, K. V. Ramanujachary, S. E. Lofland, and A. K. Ganguli, Development of a microemulsion-based process for synthesis of cobalt (Co) and cobalt oxide (Co_3O_4), *J. Colloid Interf. Sci.* 321, 2008, 434–441.
76. M. Salavati-Niasari and F. Davar, Synthesis of cobalt and cobalt oxide nanoparticles and their magnetic properties, *Int. J. Nanosci.* 8, 2009, 273–276.
77. Z. G. Wu, M. Munoz, and O. Montero, The synthesis of nickel nanoparticles by hydrazine reduction, *Adv. Powder Technol.* 21, 2010, 165–168.
78. Z. Libor and Q. Zhang, The synthesis of nickel nanoparticles with controlled morphology and SiO_2/Ni core-shell structures, *Mater. Chem. Phys.* 114, 2009, 902–907.
79. D.-H. Chen and S.-R. Wang, Protective agent-free synthesis of Ni-Ag core-shell nanoparticles, *Mater. Chem. Phys.* 100, 2006, 468–471.
80. R. Eluri and B. Paul, Synthesis of nickel nanoparticles by hydrazine reduction: Mechanistic study and continuous flow synthesis, *J. Nanopart. Res.* 14, 2012, 800/1–800/14.
81. M. Edrissi, H. A. Hosseinabadi, and S. Z. Hajibagher, Synthesis of coral-like and spherical nickel nanoparticles using Taguchi L8 experimental design, *Powder Metall.* 54, 2011, 572–576.
82. C.-C. Lee and D.-H. Chen, Large-scale synthesis of Ni-Ag core-shell nanoparticles with magnetic, optical and anti-oxidation properties, *Nanotechnology* 17, 2006, 3094–3099.
83. N. V. Suramwar, S. R. Thakare, and N. T. Khaty, One-pot green synthesis of Ni nanoparticles and study of its catalytic activity in the hydrothermal reduction of *p*-nitrophenol synthesis and reactivity in inorganic, metal-organic, and nano-metal, *Chemistry* 43, 2013, 57–62.
84. H. Wang, X. Kou, L. Zhang, and J. Li, Size-controlled synthesis, microstructure and magnetic properties of Ni nanoparticles, *Mater. Res. Bull.* 43, 2008, 3529–3536.
85. H. Wang, X. Kou, J. Zhang, and J. Li, Large scale synthesis and characterization of Ni nanoparticles by solution reduction method, *Bull. Mater. Sci.* 31, 2008, 97–100.
86. H.-J. Song, X.-H. Jia, X.-F. Yang, H. Tang, Y. Li, and Y.-T. Su, Controllable synthesis of monodisperse polyhedral nickel nanocrystals, *Cryst. Eng. Comm* 14, 2012, 405–410.
87. M. Singh, M. Kumar, F. Stepanek, P. Ulbrich, P. Svoboda, E. Santava, and M. L. Singla, Liquid-phase synthesis of nickel nanoparticles stabilized by PVP and study of their structural and magnetic properties, *Adv. Mater. Lett.* 2, 2011, 409–414.

88. X. Liu, M. Guo, M. Zhang, X. Wang, X. Guo, and K. Chou, Effects of PVP on the preparation and growth mechanism of monodispersed Ni nanoparticles, *Rare Metals* 27, 2008, 642–647.

89. A. Wang, H. Yin, H. Lu, J. Xue, M. Ren, and T. Jiang, Effect of organic modifiers on the structure of nickel nanoaprticles and catalytic activity in the hydrogenation of *p*-nitrophenol to *p*-aminophenol, *Langmuir* 25, 2009, 12736–12741.

90. A. Roy, V. Srinivas, S. Ram, and T. V. C. Rao, The effect of silver coating on magnetic properties of oxygen-stabilized tetragonal Ni nanoparticles prepared by chemical reduction, *J. Phys.: Cond. Matt.* 19, 2007, 346220/1–346220/16.

91. A. Roy, V. Srinivas, S. Ram, J. A. De Toro, and J. P. Goff, A comprehensive structural and magnetic study of Ni nanoparticles prepared by the borohydride reduction of $NiCl_2$ solution of different concentrations, *J. Appl. Phys.* 100, 2006, 094307/1–094307/8.

92. A. Roy, V. Srinivas, S. Ram, J. A. De Toro, and U. Mizutani, Structure and magnetic properties of oxygen-stabilized tetragonal Ni nanoparticles prepared by borohydride reduction method, *Phys. Rev. B* 71, 2005, 184443/1–184443/10.

93. J.-M. Yan, X.-B. Zhang, S. Han, H. Shioyama, and Q. Xu, Synthesis of longtime water/air-stable Ni nanoparticles and their high catalytic activity for hydrolysis of ammonia-borane for hydrogen generation, *Inorg. Chem.* 48, 2009, 7389–7393.

94. P. K. Khanna, P. V. More, J. P. Jawalkar, and B. G. Bharate, Effect of reducing agent on the synthesis of nickel nanoparticles, *Mater. Lett.* 63, 2009, 1384–1386.

95. T. Bala, B. Joshi, N. Iyer, M. Sastry, and B. L. V. Prasad, Assembly of phase transferred nickel nanoparticles at air-water interface using Langmuir-Blodgett technique, *J. Nanosci. Nanotechnol.* 2006, 6, 3736–3745.

96. T. Bala, S. D. Bhame, P. A. Joy, B. L. V. Prasad, and M. Sastry, A facile liquid foam based synthesis of nickel nanoparticles and their subsequent conversion to $Ni_{core}Ag_{shell}$ particles: Structural characterization and investigation of magnetic properties, *J. Mater. Chem.* 14, 2004, 2941–2945.

97. Y. Hou and S. Gao, Monodisperse nickel nanoparticles prepared from a monosurfactant system and their magnetic properties, *J. Mater. Chem.* 13, 2003, 1510–1512.

98. Y. Hou, H. Kondoh, T. Ohta, and S. Gao, Size-controlled synthesis of nickel nanoparticles, *Appl. Surf. Sci.* 241, 2005, 218–222.

99. H. Winnischofer, T. C. R. Rocha, W. C. Nunes, L. M. Socolovsky, M. Knobel, and D. Zanchet, Chemical synthesis and structural characterization of highly disordered Ni colloidal nanoparticles, *ACS Nano* 2, 2008, 1313–1319.

100. M. Vaseem, N. Tripathy, G. Khang, and Y.-B. Hahn, Green chemistry of glucose-capped ferromagnetic hcp-nickel nanoparticles and their reduced toxicity, *RSC Adv.* 3, 2013, 9698–9704.

101. E. Veena Gopalan, K. A. Malini, G. Santhoshkumar, T. N. Narayanan, P. A. Joy, I. A. Al-Omari, D. Sakthi Kumar, Y. Yoshida, and M. R. Anantharaman, Template-assisted synthesis and characterization of passivated nickel nanoaprticles, *Nanoscale Res. Lett.* 5, 2010, 889–897.

102. H. Wang and D. O. Northwood, Synthesis of Ni nanoparticles by hydrolysis of Mg_2Ni, *J. Mater. Sci.* 43, 2008, 1050–1056.

103. Z. Han and H. Zhu, A novel method for the preparation of nano-sized nickle powders, *Adv. Mater. Res.* 512–515, 2012, 1849–1853.

104. P. Migowski, S. R. Teixeira, G. Machado, M. C. M. Alves, J. Geshev, and J. Dupont, Structural and magnetic characterization of Ni nanoparticles synthesized in ionic liquids, *J. Electron Spectrosc.* 156–158, 2007, 195–199.

105. M. A. Dominguez-Crespo, E. Ramirez-Meneses, A. M. Torres-Huerta, V. Garibay-Febles, and K. Philippot, Kinetics of hydrogen evolution reaction on stabilized Ni, Pt and Ni-Pt nanoparticles obtained by an organometallic approach, *Int. J. Hydrogen Energ.* 37, 2012, 4798–4811.

106. M. H. G. Prechtl, P. S. Campbell, J. D. Scholten, G. B. Fraser, G. Machado, C. C. Santini, J. Dupont, and Y. Chauvin, Imidazolium ionic liquids as promoters and stabilizing agents for the preparation of metal(0) nanoparticles by reduction and decomposition of organometallic complexes, *Nanoscale* 2, 2010, 2601–2606.

107. S. Carenco, C. Boissiere, L. Nicole, C. Sanchez, P. Le Floch, and N. Mezailles, Controlled design of size-tunable monodisperse nickel nanoparticles, *Chem. Mater.* 22, 2010, 1340–1349.

108. S. Carenco, S. Labouille, S. Bouchonnet, C. Boissiere, X.-F. Le Goff, C. Sanchez, and N. Mezailles, Revisiting the molecular roots of a ubiquitously successful synthesis: Nickel (0) nanoparticles by reduction of [Ni(acetylacetonate)$_2$], *Chem. Eur. J.* 18, 2012, 14165–14173.

109. K. P. Donegan, J. F. Godsell, D. J. Otway, M. A. Morris, S. Roy, and J. D. Holmes, Size-tuneable synthesis of nickel nanoparticles, *J. Nanoparti. Res.* 14, 2012, 670/1–670/10.

110. D. E. Zhang, X. K. Dou, H. J. Mao, X. S. Ma, S. H. Cai, X. Q. Liu, and Z. W. Tong, Controllable synthesis of size-tunable h-nickel nanoparticles, *Cryst. Eng. Comm.* 15, 2013, 6923–6927.

111. X. Luo, Y. Chen, G.-H. Yue, D.-L. Peng, and X. Luo, Preparation of hexagonal close-packed nickel nanoparticles via a thermal decomposition approach using nickel acetate tetrahydrate as a precursor, *J. Alloy. Compd.* 476, 2009, 864–868.

112. H. Wang, X. Jiao, and D. Chen, Monodispersed nickel nanoparticles with tunable phase and size: Synthesis, characterization, and magnetic properties, *J. Phys. Chem. C* 112, 2008, 18793–18797.

113. F. Davar, Z. Fereshteh, and M. Salavati-Niasari, Nanoparticles Ni and NiO: Synthesis, characterization and magnetic properties, *J. Alloy. Compd.* 476, 2009, 797–801.

114. Y. He, X. Li, and M. T. Swihart, Laser-driven aerosol synthesis of nickel nanoparticles, *Chem. Mater.* 17, 2005, 1017–1026.

115. T. Peng, X. Xiao, W. Wu, L. Fan, X. Zhou, F. Ren, and C. Jiang, Size control and magnetic properties of single layer monodisperse Ni nanoparticles prepared by magnetron sputtering, *J. Mater. Sci.* 47, 2012, 508–513.

116. K. Sue, A. Suzuki, M. Suzuki, K. Arai, T. Ohashi, K. Matsui, Y. Hakuta, H. Hayashi, and T. Hiaki, Synthesis of Ni nanoparticles by reduction of NiO prepared with a flow-through supercritical water method, *Chem. Lett.* 35, 2006, 960–961.

117. A. D. Omrani, M. A. Bousnina, L. S. Smiri, M. Taibi, P. Leone, F. Schoenstein, and N. Jouini, Elaboration of nickel nanoparticles by modified polyol process and their spark plasma sintering, characterization and magnetic properties of the nanoparticles and the dense nano-structured material, *Mater. Chem. Phys.* 123, 2010, 821–828.

118. T. G. Altincekic, I. Boz, A. C. Basaran, B. Aktas, and S. Kazan, Syhthesis and characterization of ferromagnetic nickel nanoparticles, *J. Supercond. Novel Magn.* 25, 2012, 2771–2775.

119. K. Sue, A. Suzuki, Y. Hakuta, H. Hayashi, K. Arai, Y. Takebayashi, S. Yoda, and T. Furuya, Hydrothermal-reduction synthesis of Ni nanoparticles by superrapid heating using a micromixer, *Chem. Lett.* 38, 2009, 1018–1019.

120. K. Sue, A. Suzuki, M. Suzuki, K. Arai, Y. Hakuta, H. Hayashi, and T. Hiaki, One-pot synthesis of nickel particles in supercritical water, *Ind. Eng. Chem. Res.* 45, 2006, 623–626.

121. X.-J. Shen, J.-P. Yang, Y. Liu, Y.-S. Luo, and S.-Y. Fu, Facile surfactant-free synthesis of monodisperse Ni particles via a simple solvothermal method and their superior catalytic effect on thermal decomposition of ammonium perchlorate, *New J. Chem.* 35, 2011, 1403–1409.

122. P. Calandra, Synthesis of Ni nanoparticles by reduction of NiCl$_2$ ionic clusters in the confined space of AOT reversed micelles, *Mater. Lett.* 63, 2009, 2416–2418.

123. Y.-I. Lee, J.-W. Joung, and K.-J. Lee, Manufacturing nickel nanoparticles by reducing nickel-hydrazine complex using sodium borohydride in microemulsion. *U.S. Pat. Appl. Publ.* US 20070237669 A1 20071011, 2007.

124. C. T. G. Petit, M. S. A. Alsulaiman, R. Lan, and S. Tao, Direct synthesis of Ni nanoparticles by a non-aqueous sol-gel process, *Nanosci. Nanotechnol. Lett.* 4, 2012, 136–141.

125. B. Jaleh, M. J. Torkamany, R. Golbedaghi, M. Noroozi, S. Habibi, F. Samavat, V. J. Hamedan, and L. Albeheshti, Preparation of nickel nanoparticles via laser ablation in liquid and simultaneously spectroscopy, *Adv. Mater. Res.* 403–408, 2012, 4440–4444.

126. Z. Wu, J. Chen, Q. Di, and M. Zhang, Size-controlled synthesis of a supported Ni nanoparticle catalyst for selective hydrogenation of p-nitrophenol to p-aminophenol, *Catal. Commun.* 18, 2012, 55–59.

127. Z. Cheng, J. Xu, H. Zhong, and J. Song, A modified electroless route to monodisperse and uniform nickel nanoparticles, *Mater. Chem. Phys.* 131, 2011, 4–7.

128. M. Tokushige, T. Nishikiori, and Y. Ito, Synthesis of Ni nanoparticles by plasma-induced cathodic discharge electrolysis, *J. Appl. Electrochem.* 39, 2009, 1665–1670.

129. Y. Fu and L. Zhang, Simultaneous deposition of Ni nanoparticles and wires on a tubular halloysite template: A novel metallized ceramic microstructure, *J. Solid St. Chem.* 178, 2005, 3595–3600.

130. K. Yamada, S. Inoue, H. Nomoto, T. Yamauchi, Y. Wada, and Y. Tsukahara, Process for production of nickel nanoparticles, *PCT Int. Appl.* WO 2011115213 A1 20110922, 2011.

131. T. Yamauchi, Y. Tsukahara, T. Sakamoto, T. Kono, M. Yasuda, A. Baba, and Y. Wada, Microwave-assisted synthesis of monodisperse nickel nanoparticles using a complex of nickel formate with long-chain amine ligands, *B. Chem. Soc. Jpn.* 82, 2009, 1044–1051.

132. D. Li and S. Komarneni, Microwave-assisted polyol process for synthesis of Ni nanoparticles, *J. Am. Ceram. Soc.* 89, 2006, 1510–1517.

133. H. Duan, X. Lin, G. Liu, L. Xu, F. Li, Synthesis of Ni nanoparticles and their catalytic effect on the decomposition of ammonium perchlorate, *J. Mater. Process. Technol.* 208, 2008, 494–498.

134. P. L. Freund and M. Spiro, Colloidal catalysis: The effect of sol size and concentration, *J. Phys. Chem.* 89, 1985, 1074–1077.

135. Q. Liao, R. Tannenbaum, and Z. L. Wang, Synthesis of FeNi$_3$ alloyed nanoparticles by hydrothermal reduction, *J. Phys. Chem. B* 110, 2006, 14262–14265.

136. C. Pan, Z. Zhang, X. Su, Y. Zhao, and J. Liu, Characterization of Fe nanorods grown directly from submicron-sized iron grains by thermal evaporation, *Phys. Rev. B* 70, 2004, 233404-1–233404-4.

137. K.-S. Lin, Z.-P. Wang, S. Chowdhury, and A. K. Adhikari, Preparation and characterization of aligned iron nanorod using aqueous chemical method, *Thin Solid Films* 517, 2009, 5192–5196.

138. M. Chen, B. Tang, and D. E. Nikles, Preparation of iron nanoparticles by reduction of acicular β-FeOOH particles, *IEEE Trans. Mag.* 34, 1998, 1141–1143.

139. X. Fan, J. Guan, W. Wang, and G. Tong, Morphology evolution, magnetic and microwave absorption properties of nano/submicrometre iron particles obtained at different reduced temperatures, *J. Phys. D: Appl. Phys.* 42, 2009, 075006 (7 pp).

140. L. Vayssieres, L. Rabenberg, and A. Manthiram, Aqueous chemical route to ferromagnetic 3-D arrays of iron nanorods, *Nano Lett.* 2, 2002, 1393–1395.

141. T. Ling, H. Yu, X. Liu, Z. Shen, and J. Zhu, Five-fold twinned nanorods of FCC Fe: Synthesis and characterization, *Cryst. Growth Des.* 8, 2008, 4340–4342.

142. X. Ni, Z. Zheng, X. Xiao, L. Huang, and L. He, Silica-coated iron nanoparticles: Shape-controlled synthesis, magnetism and microwave absorption properties, *Mater. Chem. Phys.* 120, 2010, 206–212.

143. D. Zhang, X. Ni, and H. Zheng, Surfactant-controlled synthesis of Fe nanorods in solution, *J. Colloid Interf. Sci.* 292, 2005, 410–412.

144. X. Cheng, B. Wu, Y. Yang, and Y. Li, Synthesis of iron nanoparticles in water-in-oil microemulsions for liquid-phase Fischer–Tropsch synthesis in polyethylene glycol, *Catal. Commun.* 12, 2011, 431–435.

145. M. N. Nadagouda, and R. S. Varma, Microwave-assisted shape-controlled bulk synthesis of Ag and Fe nanorods in poly(ethylene glycol) solutions, *Cryst. Growth Des.* 8, 2008, 291–295.

146. S.-J. Park, S. Kim, S. Lee, Z. G. Khim, K. Char, and T. Hyeon, Synthesis and magnetic studies of uniform iron nanorods and nanospheres, *J. Am. Chem. Soc.* 122, 2000, 8581–8582.

147. J.-H. Lim, W.-S. Chae, H.-O. Lee, L. Malkinski, S.-G. Min, J. B. Wiley, J.-H. Jun, S.-H. Lee, and J.-S. Jung, Fabrication and magnetic properties of Fe nanostructures in anodic alumina membrane, *J. Appl. Phys.* 107, 2010, 09A334/1–09A334/3.

148. X. Xu, J. Huang, M. Shao, and P. Wang, Synthetic control of large-area, ordered Fe nanotubes and their nanotube-core/alumina-sheath nanocables, *Mater. Chem. Phys.* 135, 2012, 6–9.

149. X. Cao and Y. Liang, Controlled fabrication of branched Fe nanotubes, *Mater. Lett.* 63, 2009, 2215–2217.

150. J. Verbeeck, O. I. Lebedev, G. Van Tendeloo, L. Cagnon, C. Bougerol, and G. Tourillon, Fe and Co nanowires and nanotubes synthesized by template electrodeposition, *J. Electrochem. Soc.* 150, 2003, E468–E471.

151. G. Tourillon, L. Pontonnier, J. P. Levy, and V. Langlais, Electrochemically synthesized Co and Fe nanowires and nanotubes, *Electrochem. Solid St.* 3, 2000, 20–23.

152. W.-S. Lin, Z.-J. Jian, H.-M. Lin, L.-C. Lai, W.-A. Chiou, Y.-K. Hwu, S.-H. Wu, W.-C. Chen, and Y. D. Yao, Synthesis and characterization of iron nanowires, *J. Chin. Chem. Soc.* 60, 2013, 85–91.

153. W. F. Liang, R. B. Yang, W. S. Lin, Z. J. Jian, C. Y. Tsay, S. H. Wu, H. M. Lin, S. T. Choi, and C. K. Lin, Electromagnetic characteristics of surface modified iron nanowires at x-band frequencies, *J. Appl. Phys.* 111, 2012, 07B545/1–07B545/3.

154. R.-B. Yang, W.-F. Liang, W.-S. Lin, H.-M. Lin, C.-Y. Tsay, and C.-K. Lin, Microwave absorbing properties of iron nanowire at x-band frequencies, *J. Appl. Phys.* 109, 2011, 07B527/1–07B527/3.

155. C. N. He, F. Tian, and S. J. Liu, A simple method for fabricating single-crystalline Fe nanowires, *Mater. Lett.* 63, 2009, 1252–1254.

156. J.-K. Ha, H.-J. Ahn, K.-W. Kim, T.-H. Nam, and K.-K. Cho, Effect of precursor supply on structural and morphological characteristics of Fe nanomaterials synthesized via chemical vapor condensation method, *J. Nanosci. Nanotechnol.* 12, 2012, 531–538.

157. J.-K. Ha, K.-K. Cho, K.-W. Kim, T.-H. Nam, H.-J. Ahn, and G.-B. Cho, Consideration of Fe nanoparticles and nanowires synthesized by chemical vapor condensation process, *Mater. Sci. Forum* 534–536, 2007, 29–32.

158. J. Liu, M. Itoh, M. Terada, T. Horikawa, and K. Machida, Enhanced electromagnetic wave absorption properties of Fe nanowires in gigaherz range, *Appl. Phys. Lett.* 91, 2007, 093101/1–093101/3.

159. G. H. Lee, S. H. Huh, J. W. Jeong, S. H. Kim, B. J. Choi, B. Kim, and J. Park, Processing of ferromagnetic iron nanowire arrays, *Scripta Mater.* 49, 2003, 1151–1155.

160. S. Song, G. Bohuslav, A. Capitano, J. Du, K. Taniguchi, Z. Cai, and L. Sun, Experimental characterization of electrochemical synthesized Fe nanowires for biomedical applications, *J. Appl. Phys.* 111, 2012, 056103/1–056103/3.

161. J. B. Wang, X. Z. Zhou, Q. F. Liu, D. S. Xue, F. S. Li, B. Li, H. P. Kunkel, and G. Williams, Magnetic texture in iron nanowire arrays, *Nanotechnology* 15, 2004, 485–489.

162. J. Stankiewicz, F. Luis, A. Camon, M. Kroell, J. Bartolome, and W. Blau, Magnetization switching of Fe nanowires at very low temperatures, *J. Magn. Magn. Mater.* 272–276, 2004, 1637–1639.

163. S. Yang, H. Zhu, D. Yu, Z. Jin, S. Tang, and Y. Du, Preparation and magnetic property of Fe nanowire array, *J. Magn. Magn. Mater.* 222, 2000, 97–100.

164. S. Thongmee, Y. W. Ma, J. Ding, J. B. Yi, and G. Sharma, Synthesis and characterization of ferromagnetic nanowires using AAO templates, *Surf. Rev. Lett.* 15, 2008, 91–96.

165. M. Kroll, W. J. Blau, D. Grandjean, R. E. Benfield, F. Luis, P. M. Paulus, and L. J. de Jongh, Magnetic properties of ferromagnetic nanowires embedded in nanoporous alumina membranes, *J. Magn. Magn. Mater.* 249, 2002, 241–245.

166. M. A. Zeeshan, S. Pane, S. K. Youn, E. Pellicer, S. Schuerle, J. Sort, S. Fusco, A. M. Lindo, H. G. Park, and B. J. Nelson, Iron nanowires: Graphite coating of iron nanowires for nanorobotic applications: Synthesis, characterization and magnetic wireless manipulation, *Adv. Funct. Mater.* 23, 2013, 823–831.

167. T. Ghaffary, M. Ebrahimzadeh, M. M. Gharahbeigi, and L. Shahmandi, Fabrication of iron nanowire arrays using nanoporous anodic alumina template, *Asian J. Chem.* 24, 2012, 3237–3239.

168. J. Azevedo, C. T. Sousa, A. Mendes, and J. P. Araujo, Influence of the rest pulse duration in pulsed electrodeposition of Fe nanowires, *J. Nanosci. Nanotechnol.* 12, 2012, 9112–9117.

169. X. F. Qin, C. H. Deng, C. Y. Liu, X. J. Meng, J. Q. Zhang, F. Wang, and X. H. Xu, Magnetization reversal of high aspect ratio iron nanowires grown by electrodeposition, *IEEE Trans. Magn.* 48, 2012, 3136–3139.

170. X. Wang, C. Li, G. Chen, C. Peng, L. He, and L. Yang, Synthesis and characterization of Fe nanowire arrays by AC electrodeposition in PAMs, *Surf. Rev. Lett.* 17, 2010, 419–423.

171. A. Jagminas, K. Mazeika, J. Reklaitis, M. Kurtinaitiene, and D. Baltrunas, Template synthesis, characterization and transformations of iron nanowires while aging, *Mater. Chem. Phys.* 109, 2008, 82–86.

172. B.-Y. Yoo, S. C. Hernandez, B. Koo, Y. Rheem, and N. V. Myung, Electrochemically fabricated zero-valent iron, iron-nickel, and iron-palladium nanowires for environmental remediation applications, *Water Sci. Technol.* 55, 2007, 149–156.

173. M. Alagiri and C. Muthamizhchelvan, Solvothermal preparation of cobalt nanorods, *J. Mater. Sci.* 24, 2013, 1112–1115.

174. C. Jiang, L. Wang, and K. Kuwabara, Selective-precursor reducing route to cobalt nanocrystals and ferromagnetic property, *J. Solid St. Chem.* 180, 2007, 3146–3151.

175. W. Liu, W. Zhong, X. Wu, N. Tang, and Y. Du, Hydrothermal microemulsion synthesis of cobalt nanorods and self-assembly into square-shaped nanostructures, *J. Cryst. Growth* 284, 2005, 446–452.

176. F. Wetz, K. Soulantica, M. Respaud, A. Falqui, and B. Chaudret, Synthesis and magnetic properties of Co nanorod superlattices, *Mater. Sci. Eng. C* 27, 2007, 1162–1166.

177. F. Dumestre, B. Chaudret, C. Amiens, M.-C. Fromen, M.-J. Casanove, P. Renaud, and P. Zurcher, Shape control of thermodynamically stable cobalt nanorods through organometallic chemistry, *Angew. Chem. Int. Ed.* 41, 2002, 4286–4289.

178. W.-W. Ma, Y. Yang, C.-T. Chong, A. Eggeman, S. N. Piramanayagam, T.-J. Zhou, T. Song, and J.-P. Wang, Synthesis and magnetic behavior of self-assembled Co nanorods and nanoballs, *J. Appl. Phys.* 95, 2004, 6801–6803.

179. Y. Soumare, C. Garcia, T. Maurer, G. Chaboussant, F. Ott, F. Fievet, J.-Y. Piquemal, and G. Viau, Kinetically controlled synthesis of hexagonally close-packed cobalt nanorods with high magnetic coercivity, *Adv. Func. Mater.* 19, 2009, 1971–1977.

180. H. Karami and E. Mohammadzadeh, Synthesis of cobalt nanorods by the pulsed current electrochemical method, *Int. J. Electrochem. Sci.* 5, 2010, 1032–1045.

181. H. Zhang, X. Zhang, T. Wu, Z. Zhang, J. Zheng, and H. Sun, Template-based synthesis and discontinuous hysteresis loops of cobalt nanotube arrays, *J. Mater. Sci.* 48, 2013, 7392–7398.

182. M. P. Proenca, C. T. Sousa, J. Ventura, J. P. Araujo, J. Escrig, and M. Vazquez, Crossover between magnetic reversal modes in ordered Ni and Co nanotube arrays, *SPIN* 2, 2012, 1250014/1–1250014/7.

183. M. P. Proenca, C. T. Sousa, J. Escrig, J. Ventura, M. Vazquez, and J. P. Araujo, Magnetic interactions and reversal mechanisms in Co nanowire and nanotube arrays, *J. Appl. Phys.* 113, 2013, 093907/1–093907/7.

184. H. Zhang, X. Zhang, J. Zhang, Z. Li, and H. Sun, Template-based electrodeposition growth mechanism of metal nanotubes, *J. Electrochem. Soc.* 160, 2013, D41–D45.

185. N. Ahmad, J.-Y. Chen, J. Iqbal, W.-X. Wang, W.-P. Zhou, and X.-F. Han, Temperature dependent magnetic properties of Co nanowires and nanotubes prepared by electrodeposition method, *J. Appl. Phys.* 109, 2011, 07A331/1–07A331/3.

186. W. Wang, N. Li, X. Li, W. Geng, and S. Qiu, Synthesis of metallic nanotube arrays in porous anodic aluminum oxide template through electroless deposition, *Mater. Res. Bull.* 41, 2006, 1417–1423.

187. K. Kant, D. Losic, and R. Sabzi, Emamali template synthesis of nickel, cobalt, and nickel hexacyanoferrate nanodot, nanorod, and nanotube arrays, *Int. J. Nanosci.* 10, 2011, 1–6.

188. R. Zierold and K. Nielsch, Tailor-made, magnetic nanotubes by template-directed atomic layer deposition, *ECS Trans.* 41, 2011, 111–121.

189. M. Daub, M. Knez, U. Goesele, and K. Nielsch, Ferromagnetic nanotubes by atomic layer deposition in anodic alumina membranes, *J. Appl. Phys.* 101, 2007, 09J111/1–09J111/3.

190. A. K. Srivastava, S. Madhavi, and R. V. Ramanujan, Directed magnetic field induced assembly of high magnetic moment cobalt nanowires, *Appl. Phys. A: Mater.* 98, 2010, 821–830.

191. M. D. L. Balela, S. Yagi, and E.-I. Matsubara, Fabrication of cobalt nanowires by electroless deposition under external magnetic field, *J. Electrochem. Soc.* 158, 2011, D210–D216.

192. L. Zhang, T. Lan, J. Wang, L. Wei, Z. Yang, and Y. Zhang, Template-free synthesis of one-dimensional cobalt nanostructures by hydrazine reduction route, *Nanoscale Res. Lett.* 6, 2011, 1–5.

193. M. Li, K. Xie, Y. Wu, Q. Yang, and L. Liao, Synthesis of cobalt nanowires by template-free method, *Mater. Lett.* 111, 2013, 185–187.

194. M. D. L. Balela, S. Yagi, and E. Matsubara, Electroless deposition of cobalt nanowires in an aqueous solution under external magnetic field, *Electrochem. Solid St.* 14, 2011, D68–D71.

195. E. K. Athanassiou, P. Grossmann, R. N. Grass, and W. J. Stark, Template free, large scale synthesis of cobalt nanowires using magnetic fields for alignment, *Nanotechnology* 18, 2007, 165606/1–165606/7.

196. G. Viau, C. Garcia, T. Maurer, G. Chaboussant, F. Ott, Y. Soumare, and J.-Y. Piquemal, Highly crystalline cobalt nanowires with high coercivity prepared by soft chemistry, *Phys. Status Solidi A* 206, 2009, 663–666.

197. Q. Liu, X. Guo, J. Chen, J. Li, W. Song, and W. Shen, Cobalt nanowires prepared by heterogeneous nucleation in propanediol and their catalytic properties, *Nanotechnology* 19, 2008, 365608/1–365608/9.

198. D. Ciuculescu, F. Dumestre, M. Comesana-Hermo, B. Chaudret, M. Spasova, M. Farle, and C. Amiens, Single-crystalline Co nanowires: Synthesis, thermal stability, and carbon coating, *Chem. Mater.* 21, 2009, 3987–3995.

199. D. Saini, R. P. Chauhan, and S. Kumar, Effects of annealing on structural and magnetic properties of template synthesized cobalt nanowires useful as data storage and nano devices, *J. Mater. Sci.: Mater. Electron.* 25, 2014, 124–127.

200. S. G. Yang, H. Zhu, G. Ni, D. L. Yu, S. L. Tang, and Y. W. Du, A study of cobalt nanowire arrays, *J. Phys. D: Appl. Phys.* 33, 2000, 2388–2390.

201. T. Ohgai, L. Gravier, X. Hoffer, M. Lindeberg, K. Hjort, R. Spohr, and J.-Ph. Ansermet, Template synthesis and magnetoresistance property of Ni and Co single nanowires electrodeposited into nanopores with a wide range of aspect ratios, *J. Phys. D: Appl. Phys.* 36, 2003, 3109–3114.

202. C. Li, C. Ni, W. Zhou, X. Duan, and X. Jin, Phase stability of Co nanowires prepared by electrodeposition via AAO templates, *Mater. Lett.* 106, 2013, 90–93.

203. M. Najafi, S. Soltanian, H. Danyali, R. Hallaj, A. Salimi, S. M. Elahi, and P. Servati, Preparation of cobalt nanowires in porous aluminum oxide: Study of the effect of barrier layer, *J. Mater. Res.* 27, 2012, 2382–2390.

204. P. Wang, L. Gao, Z. Qiu, X. Song, L. Wang, S. Yang, and R.-I. Murakami, A multistep ac electrodeposition method to prepare Co nanowires with high coercivity, *J. Appl. Phys.* 104, 2008, 064304/1–064304/5.

205. P. Yang, M. An, C. Su, and F. Wang, Fabrication of cobalt nanowires from mixture of 1-ethyl-3-methylimidazolium chloride ionic liquid and ethylene glycol using porous anodic alumina template, *Electrochimi. Acta* 54, 2008, 763–767.

206. T. N. Narayanan, M. M. Shaijumon, L. Ci, P. M. Ajayan, and M. R. Anantharaman, On the growth mechanism of nickel and cobalt nanowires and comparison of their magnetic properties, *Nano Res.* 1, 2008, 465–473.

207. X. H. Huang, L. Li, X. Luo, X. G. Zhu, and G. H. Li, Orientation-controlled synthesis and ferromagnetism of single crystalline Co nanowire arrays, *J. Phys. Chem. C* 112, 2008, 1468–1472.

208. J. Xu and Y. Xu, Structural and magnetic properties of Co and Co71Ni29 nanowire arrays prepared by template electrodeposition, *J. Mater. Sci.* 43, 2008, 4163–4166.

209. N. B. Chaure, P. Stamenov, F. M. F. Rhen, and J. M. D. Coey, Oriented cobalt nanowires prepared by electrodeposition in a porous membrane, *J. Magn. Magn. Mater.* 290–291, 2005, 1210–1213.

210. J. Rivas, A. Kazadi Mukenga Bantu, G. Zaragoza, M. C. Blanco, and M. A. Lopez-Quintela, Preparation and magnetic behavior of arrays of electrodeposited Co nanowires, *J. Magn. Magn. Mater.* 249, 2002, 220–227.

211. A. Kazadi Mukenga Bantu, J. Rivas, G. Zaragoza, M. A. Lopez-Quintela, and M. C. Blanco, Influence of the synthesis parameters on the crystallization and magnetic properties of cobalt nanowires, *J. Non-Cryst. Solids* 287, 2001, 5–9.

212. C. Schoenenberger, B. M. I. van der Zande, L. G. J. Fokkink, M. Henny, C. Schmid, M. Krueger, A. Bachtold, R. Huber, H. Birk, and U. Staufer, Template synthesis of nanowires in porous polycarbonate membranes: Electrochemistry and morphology, *J. Phys. Chem. B* 101, 1997, 5497–5505.

213. K. Hoshino and Y. Hitsuoka, One-step template-free electrosynthesis of cobalt nanowires from aqueous [Co(NH$_3$)$_6$]Cl$_3$ solution, *Electrochem. Commun.* 7, 2005, 821–828.

214. Y. Guo, G. Wang, Y. Wang, Y. Huang, and F. Wang, Large-scale and shape-controlled synthesis and characterization of nanorod-like nickel powders under microwave radiation, *Mater. Res. Bull.* 47, 2012, 6–11.

215. M. Alagiri, C. Muthamizhchelvan, S. Ponnusamy, Solvothermal synthesis of nickel nanorods and its magnetic, structural and surface morphological behavior, *Mater. Lett.* 65, 2011, 1565–1568.

216. S. Ghosh, M. Ghosh, and C. N. R. Rao, Nanocrystals, nanorods and other nanostructures of nickel, ruthenium, rhodium and iridium prepared by a simple solvothermal procedure, *J. Cluster Sci.* 18, 2007, 97–111.

217. X.-M. Ni, X.-B. Su, Z.-P. Yang, and H.-G. Zheng, The preparation of nickel nanorods in water-in-oil microemulsion, *J. Cryst. Growth* 252, 2003, 612–617.

218. A. Ghosal, J. Shah, R. K. Kotnala, and S. Ahmad, Facile green synthesis of nickel nanostructures using natural polyol and morphology dependent dye adsorption properties, *J. Mater. Chem. A* 1, 2013, 12868–12878.

219. H.-B.-R. Lee, G. H. Gu, J. Y. Son, C. G. Park, and H. Kim, Spontaneous formation of vertical magnetic-metal-nanorod arrays during plasma-enhanced atomic layer deposition, *Small* 4, 2008, 2247–2254.

220. S. W. Joo and A. N. Banerjee, Field emission characterization of vertically oriented uniformly grown nickel nanorod arrays on metal-coated silicon substrate, *J. Appl. Phys.* 107, 2010, 114317/1–114317/9.

221. Z. F. Zhou, Y. C. Zhou, Y. Pan, and X. G. Wang, Growth of the nickel nanorod arrays fabricated using electrochemical deposition on anodized Al templates, *Mater. Lett.* 62, 2008, 3419–3421.

222. E. Feizi, K. Scott, M. Baxendale, C. Pal, A. K. Ray, W. Wang, Y. Pang, and S. N. B. Hodgson, Synthesis and characterisation of nickel nanorods for cold cathode fluorescent lamps, *Mater. Chem. Phys.* 135, 2012, 832–836.

223. H. Zheng, J. Zhong, Z. Gu, and W. Wang, Preparation and magnetic properties of Ni nanorod arrays, *J. Magn. Magn. Mater.* 320, 2008, 565–570.

224. X. Jin, Y. Hu, Y. Wang, R. Shen, Y. Ye, L. Wu, and S. Wang, Template-based synthesis of Ni nanorods on silicon substrate, *Appl. Surf. Sci.* 258, 2012, 2977–2981.

225. P. Bender, A. Guenther, A. Tschoepe, and R. Birringer, Synthesis and characterization of uniaxial ferrogels with Ni nanorods as magnetic phase, *J. Magn. Magn. Mater.* 323, 2011, 2055–2063.

226. R. Kaur, N. K. Verma, and S. K. Chakarvarti, Morphological, structural and optical characterization of nickel nanostructures fabricated through electrochemical template synthesis, *J. Mater. Sci.* 42, 2007, 8083–8087.

227. S. Xue, C. Cao, and H. Zhu, Electrochemically and template-synthesized nickel nanorod arrays and nanotubes, *J. Mater. Sci.* 41, 2006, 5598–5601.

228. S. H. Xue and Z. D. Wang, Metal nanorod arrays and their magnetic properties, *Mater. Sci. Eng. B-Solid* 135, 2006, 74–77.

229. X. W. Wang, Z. H. Yuan, S. Q. Sun, Y. Q. Duan, and L. J. Bie, Electrochemically synthesis and magnetic properties of Ni nanotube arrays with small diameter, *Mater. Chem. Phys.* 112, 2008, 329–332.

230. W. Lee, R. Scholz, K. Nielsch, and U. Goesele, A template-based electrochemical method for the synthesis of multisegmented metallic nanotubes, *Angew. Chem. Int. Ed.* 44, 2005, 6050–6054.

231. J. Bao, C. Tie, Z. Xu, Q. Zhou, D. Shen, and Q. Ma, Template synthesis of an array of nickel nanotubules and its magnetic behavior, *Adv. Mater.* 13, 2001, 1631–1633.

232. S. Sarkar, A. K. Sinha, M. Pradhan, M. Basu, Y. Negishi, and T. R. Pal, Transmetalation of prickly nickel nanowires for morphology controlled hierarchical synthesis of nickel/gold nanostructures for enhanced catalytic activity and SERS responsive functional material, *J. Phys. Chem. C* 115, 2011, 1659–1673.

233. N. H. S. Kalwar, S. T. H. Sherazi, M. I. Abro, Z. A. Tagar, S. S. Hassan, Y. Junejo, and M. I. Khattak, Synthesis of L-methionine stabilized nickel nanowires and their application for catalytic oxidative transfer hydrogenation of isopropanol, *Appl. Catal. A: Gen.* 400, 2011, 215–220.

234. M. Shao, G. Qian, H. Ban, M. Li, H. Hu, L. Lu, and P. Zhang, Synthesis and magnetic property of quasi one-dimensional Ni nanostructures via Si nanowire template, *Scripta Mater.* 55, 2006, 851–854.

235. D.-P. Wang, D.-B. Sun, H.-Y. Yu, Z.-G. Qiu, and H.-M. Meng, Preparation of one-dimensional nickel nanowires by self-assembly process, *Mater. Chem. Phys.* 113, 2009, 227–232.

236. Q. Ding, H. Liu, L. Yang, and J. Liu, Speedy and surfactant-free *in situ* synthesis of nickel/Ag nanocomposites for reproducible SERS substrates, *J. Mater. Chem.* 22, 2012, 19932–19939.

237. J. Wang, L. Y. Zhang, P. Liu, T. M. Lan, J. Zhang, L. M. Wei, Y. F. Zhang, and C. H. Jiang, Preparation and growth mechanism of nickel nanowires under magnetic field assistance, *Nano-Micro Lett.* 2, 2010, 134–138.

238. L. Y. Zhang, J. Wang, L. M. Wei, P. Liu, H. Wei, and Y. F. Zhang, Synthesis of Ni nanowires via a hydrazine reduction route in aqueous ethanol solutions assisted by external magnetic fields, *Nano-Micro. Lett.* 1, 2009, 49–52.

239. K. R. Krishnadas, P. R. Sajanlal, and T. Pradeep, Pristine and hybrid nickel nanowires: Template-, magnetic field-, and surfactant-free wet chemical synthesis and Raman studies, *J. Phys. Chem. C* 115, 2011, 4483–4490.

240. P. R. Sajanlal and T. Pradeep, Functional hybrid nickel nanostructures as recyclable SERS substrates: Detection of explosives and biowarfare agents, *Nanoscale* 4, 2012, 3427–3437.

241. H. Li, H. Lin, S. Xie, W. Dai, M. Qiao, Y. Lu, and H. Li, Ordered mesoporous Ni nanowires with enhanced hydrogenation activity prepared by electroless plating on functionalized SBA-15, *Chem. Mater.* 20, 2008, 3936–3943.

242. P. C. Pinheiro, C. T. Sousa, J. P. Araujo, A. J. Guiomar, and T. Trindade, Functionalization of nickel nanowires with a fluorophore aiming at new probes for multimodal bioanalysis, *J. Colloid Interf. Sci.* 410, 2013, 21–26.

243. N. Winkler, J. Leuthold, Y. Lei, and G. Wilde, Large-scale highly ordered arrays of freestanding magnetic nanowires, *J. Mater. Chem.* 22, 2012, 16627–16632.

244. W. J. Zheng, G. T. Fei, B. Wang, and D. Z. Li, Preparation of free-standing bamboo-like Ni nanowire arrays, *Chem. Lett.* 38, 2009, 394–395.

245. A. Guenther, S. Monz, A. Tschoepe, R. Birringer, and A. Michels, Angular dependence of coercivity and remanence of Ni nanowire arrays and its relevance to magnetoviscosity, *J. Magn. Magn. Mater.* 320, 2008, 1340–1344.

246. J.-H. Jeong, S.-H. Kim, J. H. Min, Y. K. Kim, and S.-S. Kim, High-frequency noise absorbing properties of nickel nanowire arrays prepared by DC electrodeposition, *Phys. Status Solidi. A* 204, 2007, 4025–4028.

247. J. Joo, S. J. Lee, D. H. Park, Y. S. Kim, Y. Lee, C. J. Lee, and S.-R. Lee, Field emission characteristics of electrochemically synthesized nickel nanowires with oxygen plasma post-treatment, *Nanotechnology* 17, 2006, 3506–3511.

248. K. Jain, S. K. N. Rashmi, and S. T. Lakshmikumar, Preparation of Ni nanowires in porous alumina templates using unipolar pulse electrodeposition technique, *J. Surf. Sci. Technol.* 21, 2005, 139–147.

249. S. Kato, H. Kitazawa, and G. Kido, Magnetic properties of Ni nanowires in porous alumina arrays, *J. Magn. Magn. Mater.* 272–276, 2004, 1666–1667.

250. M. T. Wu, I. C. Leu, J. H. Yen, and M. H. Hon, Preparation of Ni nanodot and nanowire arrays using porous alumina on silicon as a template without a conductive interlayer, *Electrochem. Solid St. Lett.* 7, 2004, C61–C63.

251. E. A. A. Mel, E. Gautron, B. Angleraud, A. Granier, W. Xu, C. H. Choi, K. J. Briston, B. J. Inkson, and P. Y. Tessier, Fabrication of a nickel nanowire mesh electrode suspended on polymer substrate, *Nanotechnology* 23, 2012, 275603/1–275603/7.

252. S.-S. Kim, S.-T. Kim, Y.-C. Yoon, and K.-S. Lee, Magnetic, dielectric, and microwave absorbing properties of iron particles dispersed in rubber matrix in gigahertz frequencies, *J. Appl. Phys.* 97, 2005, http://dx.doi.org/10.1063/1.1852371.

253. J. Guan, G. Yan, W. Wanga, and J. Liuc, External field-assisted solution synthesis and selectively catalytic properties of amorphous iron nanoplatelets, *J. Mater. Chem.* 22, 2012, 3909–3915.

254. Q. Liu, Z. Zi, D. Wu, Y. Sun, and J. Dai, Controllable synthesis and morphology-dependent microwave absorption properties of iron nanocrystals, *J. Mater. Sci.* 47, 2012, 1033–1037.

255. R. M. Walser and W. Kang, Fabrication and properties of microforged ferromagnetic nanoflakes, *IEEE Trans. Magn.* 34, 1998, 1144–1146.

256. L. Yan, J. Wang, X. Han, Y. Ren, Q. Liu, and F. Li, Microwave absorption of Fe nanoflakes after coating with SiO_2 nanoshell, *Nanotechnology* 21, 2010, 095708/1–095708/5.

257. F. Ma, Y. Qin, and Y.-Z. Li, Enhanced microwave performance of cobalt nanoflakes with strong shape anisotropy, *Appl. Phys. Lett.* 96, 2010, 202507/1–202507/3.

258. M. Comesana-Hermo, D. Ciuculescu, Z.-A. Li, S. Stienen, M. Spasova, M. Farle, and C. Amiens, Stable single domain Co nanodisks: Synthesis, structure and magnetism, *J. Mater. Chem.* 22, 2012, 8043–8047.

259. V. F. Puntes, D. Zanchet, C. K. Erdonmez, and A. P. Alivisatos, Synthesis of hcp-Co nanodisks, *J. Am. Chem. Soc.* 124, 2002, 12874–12880.

260. F. T. Huang, R. S. Liu, and S. F. Hu, Transformation of Co nanodisks to Co caterpillars, *J. Magn. Magn. Mater.* 304, 2006, e19–e21.

261. X. Ni, H. Zheng, Q. Yang, K. Tang, and G. Liao, Ammonia-assisted fabrication of flowery nanostructures of metallic nickel assembled from hexagonal platelets, *Eur. J. Inorg. Chem.*, 2009, 677–682.

262. Y. Zhang, Y. Zhang, Z. Wang, D. Li, T. Cui, W. Liu, and Z. Zhang, Controlled synthesis of cobalt flowerlike architectures by a facile hydrothermal route, *Eur. J. Inorg. Chem.* 2008, 2733–2738.

263. Q. Liu, X. Guo, Y. Li, and W. Shen, Hierarchical growth of Co nanoflowers composed of nanorods in polyol, *J. Phys. Chem. C* 113, 2009, 3436–3441.

264. S.-K. Li, F.-Z. Huang, X. Guo, X.-R. Yu, C. Lv, Y.-H. Shen, and A.-J. Xie, Morphology-controlled synthesis of hierarchical ball-flower metallic Co superstructures and their thermal catalytic property, *Mater. Res. Bull.* 47, 2012, 3499–3507.

265. X. Ni, D. Li, Y. Zhang, and H. Zheng, Complexant-assisted fabrication of flowery assembly of hexagonal close-packed cobalt nanoplatelets, *Chem. Lett.* 36, 2007, 908–909.

266. H. Helia, I. Eskandari, N. Sattarahmady, and A. A. Moosavi-Movahedi, Cobalt nanoflowers: Synthesis, characterization and derivatization to cobalt hexacyanoferrate—Electrocatalytic oxidation and determination of sulfite and nitrite, *Electrochim. Acta.* 77, 2012, 294–301.

267. Y. Zhu, Q. Yang, H. Zheng, W. Yu, and Y. Qian, Flower-like cobalt nanocrystals by a complex precursor reaction route, *Mater. Chem. Phys.* 91, 2005, 293–297.

268. A. Mathew, N. Munichandraiah, and G. M. Rao, Synthesis and magnetic studies of flower-like nickel nanocones, *Mater. Sci. Eng. B* 158, 2009, 7–12.

269. G. Zhang, X. Zhao, and L. Zhao, Preparation of single-crystalline nickel nanoflowers and their potential application in sewage treatment, *Mater. Lett.* 66, 2012, 267–269.

270. X. Ni, Q. Zhao, H. Zheng, B. Li, J. Song, D. Zhang, and X. Zhang, A novel chemical reduction route towards the synthesis of crystalline nickel nanoflowers from a mixed source, *Eur. J. Inorg. Chem.* 2005, 4788–4793.

271. H. Li, J. Liao, Z. Jin, X. Zhang, X. Lu, J. Liang, Y. Feng, and S. Yu, Flowery Ni microcrystals consisting of star-shaped nanorods: Facile synthesis, formation mechanism and magnetic properties, *Aus. J. Chem.* 64, 2011, 1494–1500.

272. F. Jia, L. Zhang, X. Shang, and Y. Yang, Non-aqueous sol–gel approach towards the controllable synthesis of nickel nanospheres, nanowires, and nanoflowers, *Adv. Mater.* 20, 2008, 1050–1054.

273. J. Xiong, H. Shen, J. Mao, X. Qin, P. Xiao, X. Wang, Q. Wu, and Z. Hu, Porous hierarchical nickel nanostructures and their application as a magnetically separable catalyst, *J. Mater. Chem.*, 22, 2012, 11927–11932.

274. X. Ni, Z. Zheng, X. Hub, and X. Xiao, Silica-coated iron nanocubes: Preparation, characterization and applicationin microwave absorption, *J. Coll. Interf. Sci.* 341, 2010, 18–22.

275. Y. Leng, K. Sato, J.-G. Li, T. Ishigaki, M. Iijima, H. Kamiya, and T. Yoshida, Iron nanoparticles dispersible in both ethanol and water for direct silica coating, *Powder Technol.* 196, 2009, 80–84.

276. Y. Li, Z. Jin, and T. Li, A novel and simple method to synthesize SiO_2-coated Fe nanocomposites with enhanced Cr (VI) removal under various experimental conditions, *Desalination* 288, 2012, 118–125.

277. M. Wu, Y. D. Zhang, S. Hui, T. D. Xiao, S. Ge, W. A. Hines, J. I. Budnick, and M. J. Yacaman, Magnetic properties of SiO_2-coated Fe nanoparticles, *J. Appl. Phys.* 92, 2002, 6809–6812.

278. R. Fernandez-Pacheco, M. Arruebo, C. Marquina, R. Ibarra, J. Arbiol, and J. Santamaria, Highly magnetic silica-coated iron nanoparticles prepared by the arc-discharge method, *Nanotechnology* 17, 2006, 1188–1192.

279. J. C. Park, H. J. Lee, H. S. Jung, M. Kim, H. J. Kim, K. H. Park, and H. Song, Gram-scale synthesis of magnetically separable and recyclable $Co@SiO_2$ yolk-shell nanocatalysts for phenoxycarbonylation reactions, *Chem. Cat. Chem.* 3, 2011, 755–760.

280. M. Wu, Y. D. Zhang, S. Hui, T. D. Xiao, S. Ge, W. A. Hines, and J. I. Budnick, Temperature dependence of magnetic properties of SiO_2-coated Co nanoparticles, *J. Magn. Magn. Mater.* 268, 2004, 20–23.

281. M. Wu, Y. D. Zhang, S. Hui, T. D. Xiao, S. Ge, W. A. Hines, and J. I. Budnick, Structure and magnetic properties of SiO_2-coated Co nanoparticles, *J. Appl. Phys.* 92, 2002, 491–495.

282. X. F. Zhang, X. L. Dong, H. Huang, B. Lv, X. G. Zhu, J. P. Lei, S. Ma, W. Liu, and Z. D. Zhang, Synthesis, growth mechanism and magnetic properties of SiO_2-coated Co nanocapsules, *Acta Mater.* 55, 2007, 3727–3733.

283. U.-H. Lee, J. B. Park, S. K. Kim, and Y.-U. Kwon, Templated synthesis of nano-structured cobalt thin film for potential terabit magnetic recording, *NANO* 1, 2006, 41–45.

284. J. C. Park, H. J. Lee, J. Y. Kim, K. H. Park, and H. Song, Catalytic hydrogen transfer of ketones over $Ni@SiO_2$ yolk-shell nanocatalysts with tiny metal cores, *J. Phys. Chem. C* 114, 2010, 6381–6388.

285. K. L. Klug, V. P. Dravid, and D. L. Johnson, Silica-encapsulated magnetic nanoparticles formed by a combined arc evaporation/chemical vapor deposition technique, *J. Mater. Res.* 18, 2003, 988–993.

286. Z. Xu, H. Sun, F. Gao, L. Hou, and N. Li, Synthesis and magnetic property of T4 virus-supported gold-coated iron ternary nanocomposite, *J. Nanopart. Res.* 14, 2012, 1267/1–1267/12.

287. M. R. Islam, L. G. Bach, T. T. Nga, and K. T. Lim, Covalent ligation of gold coated iron nanoparticles to the multi-walled carbon nanotubes employing click chemistry, *J. Alloy. Compd.* 561, 2013, 201–205.

288. L. G. Bach, M. Rafiqul Islam, J. H. Kim, H. G. Kim, and K. T. Lim, Synthesis and characterization of poly(2-hydroxyethyl methacrylate)-functionalized Fe-Au/core-shell nanoparticles, *J. Appl. Polym. Sci.* 124, 2012, 4755–4764.

289. Y. Zhang, Z. Wang, and W. Jiang, A sensitive fluorimetric biosensor for detection of DNA hybridization based on Fe/Au core/shell nanoparticles, *Analyst* 136, 2011, 702–707.

290. S. Kayal and R. V. Ramanujan, Anti-cancer drug loaded iron-gold core-shell nanoparticles (Fe@Au) for magnetic drug targeting, *J. Nanosci. Nanotechnol.* 10, 2010, 5527–5539.

291. S.-J. Cho, S. M. Kauzlarich, J. Olamit, K. Liu, F. Grandjean, L. Rebbouh, and G. J. Long, Characterization and magnetic properties of core/shell structured Fe/Au nanoparticles, *J. Appl. Phys.* 95, 2004, 6804–6806.

292. J. Dong, T. Liu, X. Meng, J. Zhu, K. Shang, S. Ai, and S. Cui, Amperometric biosensor based on immobilization of acetylcholinesterase via specific binding on biocompatible boronic acid-functionalized Fe@Au magnetic nanoparticles, *J. Solid St. Electrochem.* 16, 2012, 3783–3790.

293. J. Zhang, M. Post, T. Veres, Z. J. Jakubek, J. Guan, D. Wang, F. Normandin, Y. Deslandes, and B. Simard, Laser-assisted synthesis of superparamagnetic Fe@Au core-shell nanoparticles, *J. Phys. Chem. B* 110, 2006, 7122–7128.

294. K. J. Carroll, D. M. Hudgins, S. Spurgeon, K. M. Kemner, B. Mishra, M. I. Boyanov, L. W. Brown, M. L. Taheri, and E. E. Carpenter, One-pot aqueous synthesis of Fe and Ag core/shell nanoparticles, *Chem. Mater.* 22, 2010, 6291–6296.

295. L. Wang, K. Yang, C. Clavero, A. J. Nelson, K. J. Carroll, E. E. Carpenter, and R. A. Lukaszew, Localized surface plasmon resonance enhanced magneto-optical activity in core-shell Fe-Ag nanoparticles, *J. Appl. Phys.* 107, 2010, 09B303/1–09B303/3.

296. L. Lu, W. Zhang, D. Wang, X. Xu, J. Miao, and Y. Jiang, Fe@Ag core–shell nanoparticles with both sensitive plasmonic properties and tunable magnetism, *Mater. Lett.* 64, 2010, 1732–1734.

297. M. N. Nadagouda and R. S. Varma, A greener synthesis of core (Fe, Cu)-shell (Au, Pt, Pd, and Ag) nanocrystals using aqueous vitamin C, *Cryst. Growth Des.* 7, 2007, 2582–2587.

298. Y. Gu, F. Zeng, F. Lv, Y. Gu, P. Yang, and F. Pan, Epitaxial growth of Fe/Ag single crystal superlattices and their magnetic properties, *Int. J. Miner., Metal. Mater.* 16, 2009, 71–76.

299. W. Lee, M. G. Kim, J. Choi, J.-I. Park, S. J. Ko, S. J. Oh, and J. Cheon, Redox-transmetalation process as a generalized synthetic strategy for core-shell magnetic nanoparticles, *J. Am. Chem. Soc.* 127, 2005, 16090–16097.

300. P. Boyer, D. Menard, and M. Meunier, Nanoclustered Co-Au particles fabricated by femtosecond laser fragmentation in liquids, *J. Phys. Chem. C* 114, 2010, 13497–13500.

301. Y. Bao, H. Calderon, and K. M. Krishnan, Synthesis and characterization of magnetic-optical Co-Au core-shell nanoparticles, *J. Phys. Chem. C* 111, 2007, 1941–1944.

302. Y. Lu, Y. Zhao, L. Yu, L. Dong, C. Shi, M.-J. Hu, Y.-J. Xu, L.-P. Wen, and S.-H. Yu, Hydrophilic Co@Au yolk/shell nanospheres: Synthesis, assembly, and application to gene delivery, *Adv. Mater.* 22, 2010, 1407–1411.

303. S. Mandal and K. M. Krishnan, CocoreAushell nanoparticles: Evolution of magnetic properties in the displacement reaction, *J. Mater. Chem.* 17, 2007, 372–376.

304. J. Garcia-Torres, E. Gomez, and E. Valles, Measurement of the giant magnetoresistance effect in cobalt-silver magnetic nanostructures: Nanowires, *J. Phys. Chem. C* 116, 2012, 12250–12257.

305. J. Garcia-Torres, E. Valles, and E. Gomez, Giant magnetoresistance in electrodeposited Co-Ag granular films, *Mater. Lett.* 65, 2011, 1865–1867.

306. J. Garcia-Torres, E. Valles, and E. Gomez, Relevant GMR in as-deposited Co-Ag electrodeposits: Chronoamperometric preparation, *J. Phys. Chem. C* 114, 2010, 12346–12354.

307. S. Kenane, J. Voiron, N. Benbrahim, E. Chainet, and F. Robaut, Magnetic properties and giant magnetoresistance in electrodeposited Co-Ag granular films, *J. Magn. Magn. Mater.* 297, 2006, 99–106.

308. I. Yamauchi, H. Kawamura, K. Nakano, and T. Tanaka, Formation of fine skeletal Co-Ag by chemical leaching of Al-Co-Ag ternary alloys, *J. Alloy. Compd.* 387, 2005, 187–192.

309. H. Bala, W. Fu, Y. Yu, H. Yang, and Y. Zhang, Preparation, optical properties, magnetic properties and thermal stability of core–shell structure cobalt/zinc oxide nanocomposites, *Appl. Surf. Sci.* 255, 2009, 4050–4055.

310. D. Chen, J. Li, C. Shi, X. Du, N. Zhao, J. Sheng, and S. Liu, Properties of core-shell Ni-Au nanoparticles synthesized through a redox-transmetalation method in reverse microemulsion, *Chem. Mater.* 19, 2007, 3399–3405.

311. H. She, Y. Chen, X. Chen, K. Zhang, Z. Wang, and D.-L. Peng, Structure, optical and magnetic properties of Ni@Au and Au@Ni nanoparticles synthesized via non-aqueous approaches, *J. Mater. Chem.* 22, 2012, 2757–2765.

312. I. T. Jeon, M. K. Cho, J. W. Cho, B. H. An, J. H. Wu, R. Kringel, D. S. Choi, and Y. K. Kim, Ni-Au core-shell nanowires: Synthesis, microstructures, biofunctionalization, and the toxicological effects on pancreatic cancer cells, *J. Mater. Chem.* 21, 2011, 12089–12095.

313. C.-C. Kim, C. Wang, Y.-C. Yang, Y. K. Hwu, S.-K. Seol, Y.-B. Kwon et al. X-ray synthesis of nickel-gold composite nanoparticles, *Mater. Chem. Phys.* 100, 2006, 292–295.

314. J. J. Jing, J. M. Xie, H. R. Qin, W. H. Li, and M. M. Zhang, Preparation and characterization of nickel(Ni)-silver(Ag) core-shell nanoparticles for conductive pastes, *Adv. Mater. Res.* 531, 2012, 211–214.

315. T. Bala, A. Swami, B. L. V. Prasad, and M. Sastry, Phase transfer of oleic acid capped Ni(core)Ag(shell) nanoparticles assisted by the flexibility of oleic acid on the surface of silver, *J. Colloid Interf. Sci.* 283, 2005, 422–431.

316. X. Hou, X. Zhang, S. Chen, H. Kang, and W. Tan, Facile synthesis of Ni/Au, Ni/Ag hybrid magnetic nanoparticles: New active substrates for surface enhanced Raman scattering, *Colloid Surf. A* 403, 2012, 148–154.

317. J. Xu, H. Yang, W. Fu, W. Fan, Q. Zhu, M. Li, and G. Zou, Synthesis and characterization of nickel coated by zinc oxide: Bifunctional magnetic-optical nanocomposites, *J. Alloys Compd.* 458, 2008, 119–122.

318. N. A. D. Burke, H. D. H. Stöver, and F. P. Dawson, Magnetic nanocomposites: Preparation and characterization of polymer-coated iron nanoparticles, *Chem. Mater.* 14, 2002, 752–4761.

319. R. S. Molday and L. L. Molday, Separation of cells labeled with immunospecific iron dextran microspheres using high gradient magnetic chromatography, *FEBS Lett.* 170, 1984, 232–238.

320. W. Feng, H. Li, X. Cheng, T.-C. Jao, F.-B. Weng, A. Su, and Y.-C. Chiang, A comparative study of pyrolyzed and doped cobalt-polypyrrole eletrocatalysts for oxygen reduction reaction, *Appl. Surf. Sci.* 258, 2012, 4048–4053.

321. T. Homma, N. N. Khoi, W. W. Smeltzer, and J. D. Embury, The influence of surface preparation on the structures of nickel oxide formed on the (100) face of nickel, *Oxid. Met.* 3, 1971, 463–473.

322. P. Xu, X. Han, C. Wang, H. Zhao, J. Wang, X. Wang, and B. Zhang, Synthesis of electromagnetic functionalized barium ferrite nanoparticles embedded in polypyrrole, *J. Phys. Chem. B* 112, 2008, 2775–2781.

323. C. Wang, X. Han, X. Zhang, S. Hu, T. Zhang, J. Wang, Y. Du, X. Wang, and P. Xu, Controlled synthesis and morphology-dependent electromagnetic properties of hierarchical cobalt assemblies, *J. Phys. Chem. C* 114, 2010, 14826–14830.

324. H. Zhao, X. Han, L. Zhang, G. Wang, C. Wang, X. Li, and P. Xu, Controlled synthesis and morphology-dependent electromagnetic properties of nickel nanostructures by γ-ray irradiation technique, *Radiat. Phys. Chem.* 80, 2011, 390–393.

325. C. Wang, X. Han, P. Xu, J. Wang, Y. Du, X. Wang, W. Qin, and T. Zhang, Controlled synthesis of hierarchical nickel and morphology-dependent electromagnetic properties, *J. Phys. Chem. C* 114, 2010, 3196–3203.

326. B. Lua, X. L. Donga, H. Huanga, X. F. Zhanga, X. G. Zhua, J. P. Leia, and J. P. Sun, Microwave absorption properties of the core/shell-type iron and nickel nanoparticles, *J. Magn. Magn. Mater.* 320, 2008, 1106–1111.

327. A. Y. Ku, S. T. Taylor, and S. M. Loureiro, Mesoporous silica composites containing multiple regions with distinct pore size and complex pore organization, *J. Am. Chem. Soc.* 127, 2005, 6934–6935.

328. W.-S. Lin, H.-M. Lin, H.-H. Chen, Y.-K. Hwu, and Y.-J. Chiou, Shape effects of iron nanowires on hyperthermia treatment, *J. Nanomater.* 2013, 237439, 6, http://dx.doi.org/10.1155/2013/237439.

329. N. Xiaomin, S. Xiaobo, Z. Huagui, Z. Dongen, Y. Dandan, and Z. Qingbiao, Studies on the one-step preparation of iron nanoparticles in solution, *J. Cryst. Growth* 275, 2005, 548–553.

330. L. Yan, J. Wang, X. Han, Y. Ren, Q. Liu, and F. Li, Enhanced microwave absorption of Fe nanoflakes after coating with SiO_2 nanoshell, *Nanotechnology* 21, 2010, 095708, 5.

331. H. Li and S. Liao, Synthesis of flower-like Co microcrystals composed of Co nanoplates in water/ethanol mixed solvent, *J. Phys. D: Appl. Phys.* 41, 2008, 065004, 7 pp.

332. B.-Q. Xie, Y. Qian, S. Zhang, S. Fu, and W. Yu, A hydrothermal reduction route to single-crystalline hexagonal cobalt nanowires, *Eur. J. Inorg. Chem.* 12, 2006, 2454–2459.

333. H. Q. Cao, Z. Xu, H. Sang, D. Sheng, and C. Y. Tie, Template synthesis and magnetic behavior of an array of cobalt nanowires encapsulated in polyaniline nanotubules, *Adv. Mater.* 13, 2001, 121–123.

334. Y. C. Zhu, H. G. Zheng, Q. Yang, A. L. Pan, Z. P. Yang, and Y. T. Qian, Growth of dendritic cobalt nanocrystals at room temperature, *J. Cryst. Growth* 260, 2004, 427–434.

335. H. R. Khan and K. Petrikowski, Synthesis and properties of the arrays of magnetic nanowires of Co and CoFe, *Mater. Sci. Eng. C* C19, 2002, 345–348.

336. S. Baber, M. Zhou, Q. L. Lin, M. Naalla, Q. X. Jia, Y. Lu, and H. M. Luo, Nanoconfined surfactant template electrodeposition to porous hierarchical nanowires and nanotubes, *Nanotechnology* 21, 2010, 165603, 9.

337. Y. Gao, Y. Bao, A. B. Pakhomov, D. Shindo, and K. M. Krishnan, Spiral spin order of self-assembled Co nanodisk arrays, *Phys. Rev. Lett.* 96, 2006, 137205.

338. J. Rivas, R. D. Sánchez, A. Fondado, C. Izco, A. J. García-Bastida, J. García-Otero, J. Mira, D. Baldomir, A. Gonzáles, I. Lado, M. A. López- Quintela, and S. B. Oseroff, Structural and magnetic characterization of Co partical coated with Ag, *J. Appl. Phys.* 76, 1994, 6564–6566.

339. N. A. M. Barakat, Catalytic and photocatalytic hydrolysis of ammonia borane complex using Pd-doped Co nanofibers, *Appl. Catal. A: Gen.* 451, 2013, 21–27.

340. M. Yoshinaga, K. Yamamoto, N. Sato, K. Aoki, T. Morikawa, and A. Muramatsu, Remarkably enhanced photocatalytic activity by nickel nanoparticle deposition on sulfur-doped titanium dioxide thin film, *Appl. Catal. B: Environ.* 87, 2009, 239–244.

341. K. P. Singh, A. K. Singh, S. Gupta, and S. Sinha, Optimization of Cr(VI) reduction by zero-valent bimetallic nanoparticles using the response surface modeling approach, *Desalination* 270, 2011, 275–284.

342. K.-S. Lin, K. Dehvari, Y.-J. Liu, H. Kuo, and P.-J. Hsu, Synthesis and characterization of porous zero-valent iron nanoparticles for remediation of chromium-contaminated wastewater, *J. Nanosci. Nanotechnol.* 13, 2013, 2675–2681.

343. T. Y. Liu, L. Zhao, and Z. L. Wang, Removal of hexavalent chromium from wastewater by Fe0-nanoparticles-chitosan composite beads: Characterization, kinetics and thermodynamics, *Water Sci. Technol.* 66, 2012, 1044–1051.

344. T. Fan, Z. Shen, M. Jin, J. Zhu, N. Wang, and Z. Zhang, The removal of chromium(VI) by synthesized CMC stabilized zero-valent iron materials, *Adv. Mater. Res.* 534, 2012, 188–191.

345. R. Singh, V. Misra, and R. P. Singh, Removal of hexavalent chromium from contaminated ground water using zero-valent iron nanoparticles, *Environ. Monit. Assess.* 184, 2012, 3643–3651.

346. M. Vemula, V. B. R. Ambavaram, G. R. Kalluru, and M. Gajulapalli, A simple method for the determination of efficiency of stabilized Fe0 nanoparticles for detoxification of chromium(VI) in water, *J. Chem. Pharm. Res.* 4, 2012, 1539–1545.

347. H. Jabeen, V. Chandra, S. Jung, J. W. Lee, K. S. Kim, and S. B. Kim, Enhanced Cr(VI) removal using iron nanoparticle decorated graphene, *Nanoscale* 3, 2011, 3583–3585.

348. L. Shi, X. Zhang, and Z. Chen, Removal of chromium (VI) from wastewater using bentonite-supported nanoscale zero-valent iron, *Water Res.* 45, 2011, 886–892.

349. S. R. Kanel, B. Manning, L. Charlet, and H. Choi, Removal of arsenic(III) from ground water by nanoscale zero-valent iron, *Environ. Sci. Technol.* 39, 2005, 1291–1298.

350. C. Su and R. W. Puls, Arsenate and arsenite removal by zerovalent iron: Effects of phosphate, silicate, carbonate, borate, sulfate, chromate, molybdate, and nitrate, relative to chloride, *Environ. Sci. Technol.* 35, 2001, 4562–4568.

351. T. Li, S. Li, S. Wang, Y. An, and Z. Jin, Preparation of nanoiron by water-in-oil (W/O) microemulsion for reduction of nitrate in groundwater, *J. Water Res. Protection*, 1, 2009, 1–57.

352. M. Z. Kassaee, E. Motamedi, A. Mikhak, and R. Rahnemaie, Nitrate removal from water using iron nanoparticles produced by arc discharge vs. reduction, *Chem. Engn. J.* 166, 2011, 490–495.

353. A. Alqudami, N. A. Alhemiary, and S. Munassar, Removal of Pb(II) and Cd(II) ions from water by Fe and Ag nanoparticles prepared using electro-exploding wire technique, *Environ. Sci. Pollut. Res. Int.* 19, 2012, 2832–2841.

354. Y. Zhang, Y. Su, X. Zhou, C. Dai, and A. A. Keller, A new insight on the core-shell structure of zerovalent iron nanoparticles and its application for Pb(II) sequestration, *J. Hazard. Mater.* 263, 2013, 685–693.

355. Q. Liu, Y. Bei, and F. Zhou, Removal of lead(II) from aqueous solution with amino-functionalized nanoscale zero-valent iron, *Cent. Eur. J. Chem.* 7, 2009, 79–82.

356. N. S. Bharadwaj, J. Sudarshan, K. Prasanna, and N. Singh, Mitigation of heavy metals from different industrial effluents using different nano-particles, *Asian J. Chem.* 25, 2013, S287–S289.

357. X.-Q. Li and W.-X. Zhang, Sequestration of metal cations with zerovalent iron nanoparticles—A study with high resolution X-ray photoelectron spectroscopy (HR-XPS), *J. Phys. Chem. C* 111, 2007, 6939–6946.

358. T. Liu, X. Yang, Z.-L. Wang, and X. Yan, Enhanced chitosan beads-supported Fe(0)-nanoparticles for removal of heavy metals from electroplating wastewater in permeable reactive barriers, *Water Res.* 47, 2013, 6691–6700.

359. W. Chen, R. Parette, J. Zou, F. S. Cannon, and A. Dempsey, Brian arsenic removal by iron-modified activated carbon, *Water Res.* 41, 2007, 1851–1858.

360. L. Donga, P. V. Zininb, J. P. Cowenb, and L. C. Ming, Iron coated pottery granules for arsenic removal from drinking water, *J. Hazard. Mater.* 168, 2009, 626–632.

361. H. Wang, N. Yan, Y. Li, X. Zhou, J. Chen, B. Yu, M. Gong, and Q. Chen, Fe nanoparticle-functionalized multi-walled carbon nanotubes: One-pot synthesis and their applications in magnetic removal of heavy metal ions, *J. Mater. Chem.* 22, 2012, 9230–9236.

362. J. Pang, A. Deng, L. Mao, X. Peng, and J. Zhu, Adsorption of heavy metal ions by carbon-coated iron nanoparticles, *Carbon* 56, 2013, 392–395.

363. P. Bhunia, G. Kim, C. Baik, and H. Lee, A strategically designed porous iron-iron oxide matrix on graphene for heavy metal adsorption, *Chem. Commun.* 48, 2012, 9888–9890.

364. J.-I. Park, M. G. Kim, Y.-W. Jun, J. S. Lee, W.-R. Lee, and J. Cheon, Characterization of superparamagnetic core-shell nanoparticles and monitoring their anisotropic phase transition to ferromagnetic solid solution nanoalloys, *J. Am. Chem. Soc.* 126, 2004, 9072–9078.

365. P. M. Paulus, H. Bönnamann, A. M. van der Kraan, F. Luis, J. Sinzig, and L. J. de Jongh, Magnetic properties of nanosized transition metal colloids: The influence of noble metal coating, *Euro. Phys. J*, D9, 1999, 501–504.

366. Y.-H. Xu and J.-P. Wang, Magnetic properties of heterostructured Co–Au nanoparticles direct-snthesized from gas phase, *IEEE Trans. Magn.* 43, 2007, 3109–3111.

367. L. Wang, C. Clavero, Z. Huba, K. J. Carroll, E. E. Carpenter, D. F. Gu et al. Plasmonics and enhanced magneto-optics in core-shell Co-Ag nanoparticles, *Nano. Lett.* 11, 2011, 1237–1240.

368. M. Marinescu and J. F. Liu, Fe-nanoparticle coated anisotropic magnet powders for composite permanent magnets with enhanced properties, *J. Appl. Phys.* 103, 2008, 07E120–07E123.

369. S. Sun, Recent advances in chemical synthesis, self-assembly, and applications of FePt nanoparticles, *Adv. Mater.* 18, 2006, 393–403.

370. M. Chen, J. Kim, J. P. Liu, H. Fan, and S. Sun, Synthesis of FePt nanocubes and their oriented self-assembly, *J. Am. Chem. Soc.* 128, 2006, 7132–7133.

371. A. S. Samardak, E. V. Sukovatitsina, A. V. Ognev, L. A. Chebotkevich, R. Mahmoodi, S. M. Peighambari, M. G. Hosseini, and F. Nasirpouri, High-density nickel nanowire arrays for data storage applications, *J. Phys: Confer. Ser.* 345, 2012, 012011, 5.

372. S. Bhadra, S. Ahammed, and B. C. Ranu, Iron nanoparticles-catalyzed electrophilic amination of functionalized organocopper and organozinc reagents, *Curr. Org. Chem.* 16, 2012, 1453–1460.

373. J. F. Sonnenberg, N. Coombs, P. A. Dube, and R. H. Morris, Iron nanoparticles catalyzing the asymmetric transfer hydrogenation of ketones, *J. Am. Chem. Soc.* 134, 2012, 5893–5899.

374. Y.-X. Chen, S.-P. Chen, Q.-S. Chen, Z.-Y. Zhou, and S.-G. Sun, Electrochemical preparation of iron cuboid nanoparticles and their catalytic properties for nitrite reduction, *Electrochimi. Acta.* 53, 2008, 6938–6943.

375. Z. Wang, S. Skiles, F. Yang, Z. Yan, and D. W. Goodman, Particle size effects in Fischer–Tropsch synthesis by cobalt, *Catal. Today* 181, 2012, 75–81.

376. M. Chettibi, A.-G. Boudjahem, and M. Bettahar, Synthesis of Ni/SiO$_2$ nanoparticles for catalytic benzene hydrogenation, *Transit. Metal Chem.* 36, 2011, 163–169.

377. A. Wang, H. Yin, M. Ren, H. Lu, J. Xue, and T. Jiang, Preparation of nickel nanoparticles with different sizes and structures and catalytic activity in the hydrogenation of *p*-nitrophenol, *New J. Chem.* 34, 2010, 708–713.

378. Z. Ma, R. Wu, Q. Han, R. Chen, and Z. Gu, Preparation of well-dispersed and antioxidized Ni nanoparticles using polyamioloamine dendrimers as templates and their catalytic activity in the hydrogenation of *p*-nitrophenol to *p*-aminophenol, *Korean J. Chem. Eng.* 28, 2011, 717–722.

379. Z. Zhu, X. Guo, S. Wu, R. Zhang, J. Wang, and L. Li, Preparation of nickel nanoparticles in spherical polyelectrolyte brush nanoreactor and their catalytic activity, *Ind. Eng. Chem. Res.* 50, 2011, 13848–13853.

380. S. He, C. Li, H. Chen, D. Su, B. Zhang, X. Cao, B. Wang, M. Wei, D. G. Evans, and X. Duan, A surface defect-promoted Ni nanocatalyst with simultaneously enhanced activity and stability, *Chem. Mater.* 25, 2013, 1040–1046.

381. V. Iablokov, S. K. Beaumont, S. Alayoglu, V. V. Pushkarev, C. Specht, J. Gao, A. P. Alivisatos, N. Kruse, and G. A. Somorjai, Size-controlled model Co nanoparticle catalysts for CO$_2$ hydrogenation: Synthesis, characterization, and catalytic reactions, *Nano Lett.* 12, 2012, 3091–3096.

382. U. Narkiewicz, M. Podsiadly, R. Jedrzejewski, and I. Pelech, Catalytic decomposition of hydrocarbons on cobalt, nickel and iron catalysts to obtain carbon nanomaterials, *Appl. Catal. A: Gen.* 384, 2010, 27–35.

383. T. Somanathan and A. Pandurangan, Catalytic activity of Fe, Co and Fe-Co-MCM-41 for the growth of carbon nanotubes by chemical vapour deposition method, *Appl. Surf. Sci.* 254, 2008, 5643–5647.

384. J. C. De Jesus, I. Gonzalez, M. Garcia, and C. Urbina, Preparation of nickel nanoparticles and their catalytic activity in the cracking of methane, *J. Vac. Sci. Technol. A* 26, 2008, 913–918.

385. B. Kollbe Ahn, H. Wang, S. Robinson, T. B. Shrestha, D. L. Troyer, S. H. Bossmann, and X. S. Sun, Ring opening of epoxidized methyl oleate using a novel acid-functionalized iron nanoparticle catalyst, *Green Chem.* 14, 2012, 136–142.

386. A. J. J. Koekkoek, W. Kim, V. Degirmenci, H. Xin, R. Ryoo, and E. J. M. Hensen, Catalytic performance of sheet-like Fe/ZSM-5 zeolites for the selective oxidation of benzene with nitrous oxide, *J. Catal.* 299, 2013, 81–89.

387. K. Bachari, R. M. Guerroudj, and M. Lamouchi, Catalytic performance of iron-mesoporous nanomaterials synthesized by a microwave-hydrothermal process, *React. Kinet. Mech. Cataly.* 100, 2010, 205–215.

388. M. Zahmakiran, T. Ayvali, S. Akbayrak, S. Caliskan, D. Celik, and S. Ozkar, Zeolite framework stabilized nickel(0) nanoparticles: Active and long-lived catalyst for hydrogen generation from the hydrolysis of ammonia-borane and sodium borohydride, *Catal. Today* 170, 2011, 76–84.

389. Y. Zhu, Y. Ye, S. Zhang, M. E. Leong, and F. F. Tao, Synthesis and catalysis of location-specific cobalt nanoparticles supported by multiwall carbon nanotubes for Fischer-Tropsch synthesis, *Langmuir* 28, 2012, 8275–8280.

390. A. Jung, C. Kern, and A. Jess, Carbon nanomaterials as supports for Fischer-Tropsch catalysts, *Chem. Ind.* 128, 2010, 17–29.

391. M. Dalil, M. Sohrabi, and S. J. Royaee, Application of nano-sized cobalt on ZSM-5 zeolite as an active catalyst in Fischer-Tropsch synthesis, *J. Ind. Eng. Chem.* 18, 2012, 690–696.

392. Y. M. Mikhailov, A. V. Aleshin, L. V. Zhemchugova, M. V. Kulikova, A. A. Panin, S. A. Sagitov, V. I. Kurkin, A. Yu. Krylova, and S. N. Khadzhiev, Fischer-Tropsch synthesis in presence of composite materials containing iron and cobalt nanoparticles, *Chem. Technol. Fuels Oils* 48, 2012, 253–261.

393. Y. K. Park, S. B. Choi, H. J. Nam, D.-Y. Jung, H. C. Ahn, K. Choi, H. Furukawa, and J. Kim, Catalytic nickel nanoparticles embedded in a mesoporous metal-organic framework, *Chem. Commun.* 46, 2010, 3086–3088.

394. Z. Jiang, J. Xie, D. Jiang, J. Jing, and H. Qin, Facile route fabrication of nano-Ni core mesoporous-silica shell particles with high catalytic activity towards 4-nitrophenol reduction, *Cryst. Eng. Comm.* 14, 2012, 4601–4611.

395. Y. Li, L. Zhu, K. Yan, J. Zheng, B. H. Chen, and W. Wang, A novel modification method for nickel foam support and synthesis of a metal-supported hierarchical monolithic Ni@Pd catalyst for benzene hydrogenation, *Chem. Eng. J.* 226, 2013, 166–170.

396. G. Zhao, J. Huang, Z. Jiang, S. Zhang, L. Chen, and Y. Lu, Microstructured Au/Ni-fiber catalyst for low-temperature gas-phase alcohol oxidation: Evidence of Ni_2O_3-Au$^+$ hybrid active sites, *Appl. Catal. B: Environ.* 140–141, 2013, 249–257.

397. Y. Kuang, Y. Nabae, T. Hayakawa, and M. Kakimoto, Nanoshell carbon-supported cobalt catalyst for the aerobic oxidation of alcohols in the presence of benzaldehyde: An efficient, solvent free protocol, *Appl. Catal. A: Gen.* 423–424, 2012, 52–58.

398. D. Xu, P. Lu, P. Dai, H. Wang, and S. Ji, In situ synthesis of multiwalled carbon nanotubes over $LaNiO_3$ as support of cobalt nanoclusters catalyst for catalytic applications, *J. Phys. Chem. C* 116, 2012, 3405–3413.

399. M.-M. Song, W.-J. Song, H. Bi, J. Wang, W.-L. Wu, J. Sun, and M. Yu, Cytotoxicity and cellular uptake of iron nanowires, *Biomater.* 31, 2010, 1509–1517.

400. D. Guo, C. Wu, H. Hu, X. Wang, X. Li, and B. Chen, Study on the enhanced cellular uptake effect of daunorubicin on leukemia cells mediated via functionalized nickel nanoparticles, *Biomed. Mater.* 4, 2009, 025013.

401. Y. Kitamoto and J.-S. He, Chemical synthesis of FePt nanoparticles with high alternate current magnetic susceptibility for biomedical applications, *Electrochimi. Acta* 54, 2009, 5969–5972.

402. D. Guo, C. Wu, X. Li, H. Jiang, X. Wang, and B. Chen, In vitro cellular uptake and cytotoxic effect of functionalized nickel nanoparticles on leukemia cancer cells, *J. Nanosci. Nanotechnol.* 8, 2008, 2301–2307.

403. A. Amirnasr, G. Emtiazi, S. Abasi, and M. Yaaghoobi, Adsorption of hemoglobin, fatty acid and glucose to iron nanoparticles as a mean for drug delivery, *J. Biochem. Technol.* 3, 2011, 280–283.

404. H. Tokoro, Y. Kaneko, M. Adachi, T. Nakabayashi, and F. Shigeo, Fe fine particles encapsulated by titanium oxides with high magnetization for biomedical application, *J. Magn. Magn. Mater.* 311, 2007, 101–105.

405. M. Chen, S. Yamamuro, D. Farrell, and S. Majetich, Gold-coated iron nanoparticles for biomedical applications, *J. Appl. Phys.* 93, 2003, 7551–7553.

406. K. Kohara, S. Yamamoto, L. Seinberg, T. Murakami, M. Tsujimoto, T. Ogawa, H. Kurata, H. Kageyama, and M. Takano, Carboxylated SiO_2-coated α-Fe nanoparticles: Towards a versatile platform for biomedical applications, *Chem. Commun.* 49, 2013, 2563–2565.

407. I. P. Grudzinski, M. Bystrzejewski, M. A. Cywinska, A. Kosmider, M. Poplawska, A. Cieszanowski, Z. Fijalek, A. Ostrowska, and A. Parzonko, Assessing carbon-encapsulated iron nanoparticles cytotoxicity in Lewis lung carcinoma cells, *J. Appl. Toxicol.* 34, 2013, 380–394.

408. I. Moench, A. Leonhardt, A. Meye, S. Hampel, R. Kozhuharova-Koseva, D. Elefant, M. P. Wirth, and B. Buechner, Synthesis and characteristics of Fe-filled multi-walled carbon nanotubes for biomedical application, *J. Phys.: Conf. Ser.* 61, 2007, 820–824.

409. Q. Sun, A. K. Kandalam, Q. Wang, P. Jena, Y. Kawazoe, and M. Marquez, Effect of Au coating on the magnetic and structural properties of Fe nanoclusters for use in biomedical applications: A density-functional theory study, *Phys. Rev. B: Condens. Matter Mater. Phys.* 73, 2006, 134409/1–134409/6.

410. K. M. Pondman, A. W. Maijenburg, F. B. Celikkol, A. A. Pathan, U. Kishore, B. Haken, and J. E. Elshof, Au coated Ni nanowires with tuneable dimensions for biomedical applications, *J. Mater. Chem. B* 1, 2013, 6129–6136.

411. L.-S. Boucharda, M. S. Anwarb, G. L. Liuc, B. Hannc, Z. H. Xied, J. W. Grayc, X. Wange, A. Pinesf, and F. F. Chen, Picomolar sensitivity MRI and photoacoustic imaging of cobalt nanoparticles, *P. Natl. Acad. Sci. USA* 106, 2009, 4085–4089.

412. X. M. H. C. Seton, L. T. L. Prior, A. Ian, N. T. K. Thanh, and B. Song, Magnetic CoPt nanoparticles as MRI contrast agent for transplanted neural stem cells detection, *Nanoscale* 3, 2011, 977–984.

413. S. Cheong, P. Ferguson, K. W. Feindel, I. F. Hermans, P. T. Callaghan, C. Meyer et al. Simple synthesis and functionalization of iron nanoparticles for magnetic resonance imaging, *Angew. Chem. Int. Ed.* 50, 2011, 4206–4209.

414. S. Rodriguez-Llamazares, J. Merchan, I. Olmedo, H. P. Marambio, J. P. Munoz, P. Jara et al. Ni/Ni oxides nanoparticles with potential biomedical applications obtained by displacement of a nickel-organometallic complex, *J. Nanosci. Nanotechnol.* 8, 2008, 3820–3827.

415. L. A. Bauer, N. S. Birenbaum, and G. J. Meyer, Biological applications of high aspect ratio nanoparticles, *J. Mater. Chem.* 14, 2004, 517–526.

416. O. Mahapatra, S. Ramaswamy, S. V. K. Nune, T. Yadavalli, and C. Gopalakrishnan, Corn flake-like morphology of iron nanoparticles and its antibacterial property, *J. Gen. Appl. Microbiol.* 57, 2011, 59–62.

417. S. Li, Z. Niu, X. Zhong, H. Yang, Y. Lei, F. Zhang, W. Hu, Z. Dong, J. Jin, and J. Ma, Fabrication of magnetic Ni nanoparticles functionalized water-soluble graphene sheets nanocomposites as sorbent for aromatic compounds removal, *J. Hazard. Mater.* 229–230, 2012, 42–47.

418. M. Dai and B. D. Vogt, High capacity magnetic mesoporous carbon-cobalt composite adsorbents for removal of methylene green from aqueous solutions, *J. Colloid. Interf. Sci.* 387, 2012, 127–134.

419. F. Taghizadeh, M. Ghaedi, K. Kamali, E. Sharifpour, R. Sahraie, and M. K. Purkait, Comparison of nickel and/or zinc selenide nanoparticle loaded on activated carbon as efficient adsorbents for kinetic and equilibrium study of removal of Arsenazo (III) dye, *Powder Technol.* 245, 2013, 217–226.

420. G. Zhang and L. Zhao, Synthesis of nickel hierarchical structures and evaluation on their magnetic properties and Congo red removal ability, *Dalton Trans.* 42, 2013, 3660–3666.

421. M. Bikshapathi, G. N. Mathur, A. Sharma, and N. Verma, Surfactant-enhanced multiscale carbon webs including nanofibers and Ni-nanoparticles for the removal of gaseous persistent organic pollutants, *Ind. Eng. Chem. Res.* 51, 2012, 2104–2112.

422. W.-C. Hung, Y.-C. Chen, H. Chu, and T.-K. Tseng, Synthesis and characterization of TiO_2 and Fe/TiO_2 nanoparticles and their performance for photocatalytic degradation of 1,2-dichloroethane, *Appl. Surf. Sci.* 255, 2008, 2205–2213.

423. Y. Jing, S.-H. He, and J.-P. Wang, Magnetic nanoparticles of core-shell structure for recoverable photocatalysts, *Appl. Phys. Lett.* 102, 2013, 253102/1–253102/4.

424. S.-L. Xiao, H. Ma, M.-W. Shen, S.-Y. Wang, Q.-G. Huang, and X.-Y. Shi, Excellent copper(II) removal using zero-valent iron nanoparticle-immobilized hybrid electrospun polymer nanofibrous mats, *Colloid. Surf. A* 381, 2011, 48–54.

425. Y. Ni, L. Jin, L. Zhanga, and J. Hong, Honeycomb-like Ni@C composite nanostructures: Synthesis, properties and applications in the detection of glucose and the removal of heavy-metal ions, *J. Mater. Chem.* 20, 2010, 6430–6436.

426. Y. Zhilin, W. Deyin, Y. Jianlin, H. Jianqiang, R. Bin, Z. Haiguang, and T. Zhongqun, SERS mechanism of nickel electrode, *Ch. Sci. Bull.* 47, 2002, 1983–1986.

427. J. L. Yao, J. Tang, D. Y. Wu, D. M. Sun, K. H. Xue, B. Ren, B. W. Mao, and Z. Q. Tian, Surface enhanced Raman scattering from transition metal nano-wire array and the theoretical consideration, *Surf. Surface* 514, 2002, 108–116.

428. G. Saur, G. Brehm, S. Schneider, H. Graener, G. Seifert, K. Nielsch, J. C. P. Goring, U. Gosele, P. Miclea, and R. B. Wehrspohn, Surface-enhanced Raman spectroscopy employing monodisperse nickel nanowire arrays, *Appl. Phys. Lett.* 88, 2006, 023106/1–023106/3.

429. J. Yin, T. Wu, J. Song, Q. Zhang, S. Liu, R. Xu, and H. Duan, SERS-active nanoparticles for sensitive and selective detection of cadmium ion (Cd^{2+}), *Chem. Mater.* 2011, 23, 4756–4764.

430. M. J. Banholzer, J. E. Millstone, L. D. Qin, and C. A. Mirkin, Rationally designed nanostructures for surface-enhanced Raman spectroscopy, *Chem. Soc. Rev.* 2008, 37, 885–897.

431. X. Wang, Y. Dai, J. Zou, L. Meng, S. Ishikawa, S. Li, M. Abuobeidah, and H. Fu, Characteristics and antibacterial activity of Ag embedded $Fe_3O_4@SiO_2$ magnetic composite as a reusable water disinfectant, *RSC Adv.* 3, 2013, 11751–11758.

432. R. M. Liu, Y. P. Kang, X. F. Zi, M. J. Feng, M. Cheng, and M. Z. Si, The ultratrace detection of crystal violet using surface enhanced Raman scattering on colloidal Ag nanoparticles prepared by electrolysis, *Chin. Chem. Lett.* 20, 2009, 711–715.

433. I. Persaud and W. E. L. Grossman, Surface-enhanced Raman scattering of triphenyl-methane dyes on colloidal silver, *J. Raman Spectrosc.* 24, 1993, 107–112.
434. P. C. Ray, From Abstracts of Papers, 245th ACS National Meeting & Exposition, New Orleans, LA, April 7–11, 2013, AGFD-40.
435. M. Yu, X. Li, R. Gong, Y. He, H. He, and P. Lu, Magnetic properties of carbonyl iron fibers and their microwave absorbing characterization as the filer in polymer foams, *J. Alloy. Compd.* 456, 2008, 452–455.
436. S. Yoshida, M. Sato, E. Sugawara, and Y. Shimadab, Permeability and electromagnetic-interference characteristics of Fe–Si–Al alloy flakes–polymer composite, *J. Appl. Phys.* 85, 1999, 4636–4638.
437. G.-X. Tong, W.-H. Wu, Q. Hu, J.-H. Yuan, R. Qiao, and H.-S. Qian, Enhanced electro-magnetic characteristics of porous iron particles made by a facile corrosion technique, *Mater. Chem. Phys.* 132, 2012, 563–569.
438. G. Tong, J. Ma, W. Wu, Q. Hua, R. Qiao, and H. Qian, Grinding speed dependence of microstructure, conductivity, and microwave electromagnetic and absorbing character-istics of the flaked Fe particles, *J. Mater. Res.* 26, 2011, 682–688.
439. G. Sun, B. Dong, M. Cao, B. Wei, and C. Hu, Hierarchical dendrite-like magnetic mate-rials of Fe_3O_4, γ-Fe_2O_3, and Fe with high performance of microwave absorption, *Chem. Mater.* 23, 2011, 1587–1593.
440. Z. Wang, Y. Zuo, Y. Yao, L. Xi, J. Du, J. Wang, and D. Xue, Microwave absorption properties of amorphous iron nanostructures fabricated by a high-yield method, *J. Phys. D: Appl. Phys.* 46, 2013, 135002 (8 pp).
441. Z. An, S. Pan, and J. Zhang, Facile preparation and electromagnetic properties of core-shell composite spheres composed of aloe-like nickel flowers assembled on hollow glass spheres, *J. Phys. Chem. C*113, 2009, 2715–2721.
442. X. Qi, W. Zhong, C. Deng, C. Au, and Y. Du, Large-scale synthesis, electromagnetic and enhanced microwave absorption properties of low helicity carbon nanotubes/Fe nanoparticles hybrid, *Mater. Lett.* 107, 2013, 374–377.
443. Q. Liu, B. Cao, C. Feng, W. Zhang, S. Zhu, and D. Zhang, High permittivity and micro-wave absorption of porous graphitic carbons encapsulating Fe nanoparticles, *Compos. Sci. Technol.* 72, 2012, 1632–1636.
444. V. Prasad, Low temperature charge transport and microwave absorption of car-bon coated iron nanoparticles-polymer composite films, *Mater. Res. Bull.* 47, 2012, 1529–1532.
445. V. Gupta, M. K. Patra, A. Shukla, L. Saini, S. Songara, R. Jani, S. R. Vadera, and N. Kumar, Synthesis of core-shell iron nanoparticles from decomposition of Fe-Sn nanocomposite and studies on their microwave absorption properties, *J. Nanopart. Res.* 14, 2012, 1271/1–1271/10.
446. H. Huang, X. F. Zhang, B. Lv, J. P. Lei, J. P. Sun, X. L. Dong, and C. J. Choi, Characterization and microwave absorption of "core/shell"-type nanoparticles, *Mater. Sci. Forum* 561–565, 2007, 1097–1100.
447. G.-M. Shi, W.-M. Sun, D.-W. Lu, and S. Lian, Broadband micro-wave absorption prop-erties of Ni-P coated Fe nanoparticles by electroless plating process, *Adv. Mater. Res.* 549, 2012, 665–669.
448. X. G. Liu, D. Y. Geng, H. Meng, P. J. Shang, and Z. D. Zhang, Microwave-absorption properties of ZnO-coated iron nanocapsules, *Appl. Phys. Lett.* 92, 2008, 173117/1–173117/3.
449. X. Liu, D. Geng, P. Shang, H. Meng, F. Yang, B. Li, D. Kang, and Z. Zhang, Fluorescence and microwave-absorption properties of multi-functional ZnO-coated α-Fe solid-solution nanocapsules, *J. Phys. D: Appl. Phys.* 41, 2008, 175006/1–175006/7.

450. X. Qi, Y. Deng, W. Zhong, Y. Yang, C. Qin, C. Au, and Y. Du, Controllable and large-scale synthesis of carbon nanofibers, bamboo-like nanotubes, and chains of nano-spheres over Fe/SnO$_2$ and their microwave-absorption properties, *J. Phys. Chem. C* 114, 2010, 808–814.

451. J. Sui, C. Zhang, J. Li, Z. Yu, and W. Cai, Microwave absorption and catalytic activity of carbon nanotubes decorated with cobalt nanoparticles, *Mater. Lett.* 75, 2012, 158–160.

452. X. F. Zhang, X. L. Dong, H. Huang, Y. Y. Liu, W. N. Wang, X. G. Zhu, B. Lv, J. P. Lei, and C. G. Lee, Microwave absorption properties of the carbon-coated nickel nanocapsules, *Appl. Phys. Lett.* 89, 2006, 053115/1–053115/3.

453. X. Liu, C. Feng, S. W. Or, Y. Sun, C. Jin, W. Li, and Y. Lv, Investigation on microwave absorption properties of CuO/Cu$_2$O-coated Ni nanocapsules as wide-band microwave absorbers, *RSC Adv.* 3, 2013, 14590–14594.

454. N. F. Colaneri and L. W. Shacklette, EMI shielding measurements of conductive poly-mer blends, *IEEE Trans. Instrum. Meas.,*41, 1992, 291–297.

455. P. Saini, Electrical properties and electromagnetic interference shielding response of electrically conducting thermosetting nanocomposites, in *Thermoset Nanocomposites*, V. Mittal (ed.), Wiley-VCH, Weinheim, Germany, 2013, pp. 211–237.

3 Magnetic Materials for Nuclear Magnetic Resonance and Magnetic Resonance Imaging

Elizaveta Motovilova and Shaoying Huang

CONTENTS

ABSTRACT

This chapter presents the applications of magnetic materials for nuclear magnetic resonance (NMR) and magnetic resonance imaging (MRI), including the history and latest developments. Magnetic materials are applied for MRI/NMR mainly in two ways. One is the use of permanent magnets for generating the main magnetic fields B_0 and the other is magnetic nanoparticles (MNPs) injected in the object under scan for contrast enhancement. The physics and techniques of these two main applications will be detailed. Discussions and future perspectives will be provided at the end of the chapter.

In this chapter, the physics of magnetic materials, especially those related to MRI/NMR, are introduced in detail in Section 3.1. It is followed by Section 3.2 which details the working principles of MRI and NMR. In Section 3.3, magnetic materials used to provide the main magnetic field for imaging are presented. Examples of MRI systems using permanent magnets for imaging are demonstrated. In Section 3.4, the other main application of magnetic materials for MRI, contrast enhancement using MNPs, is presented. The history, physics, design criteria, and classifications of MNPs are introduced in detail. Different types of contrast agents (CAs) are systematically introduced and compared.

3.1   INTRODUCTION OF MAGNETIC MATERIALS

Magnetism originates from magnetic moments m of elementary particles, that is, electrons, protons, and neutrons, and from the way these magnetic moments interact with one another. The classical definition of a *magnetic moment* is given in terms of the current magnitude I and the enclosing area S of a planar loop (Figure 3.1) and expressed by the following equation:

$$m = IS \tag{3.1}$$

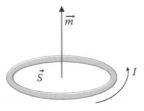

FIGURE 3.1   Magnetic moment of a planar current loop.

Moving charged particles, for example, electrons orbiting around the nucleus in an atom, form electric currents and these currents produce so-called *orbit magnetic moments*.

There is another type of magnetic moment called *spin magnetic moment*, or simply *spin*. It is an intrinsic property of all elementary particles and cannot be understood from a classical point of view. One could imagine an electron as a negatively charged sphere which spins around its own axis and thus produces a magnetic moment. However, this model does not agree with the behaviors of spin obtained from experiments. In 1922, Otto Stern and Walter Gerlach demonstrated the quantum nature of an electron spin, and that experiment gave rise to the further development of the quantum theory which fully explains the nature of spin.

These electric currents produced by moving charged particles and the fundamental magnetic moments all together give rise to a magnetic field. All matter is magnetic to some degree because of the non-cooperative behavior of orbiting electrons in an atom. However, it is the collective interaction of the atomic magnetic moments that gives the variety of different types of magnetic behaviors observed for different materials. In order to classify magnetic materials and understand the nature of magnetism, it is good to start with the interaction of different materials with an external magnetic field.

The relationship between the external magnetic field H and the magnetic induction B inside an object is the following:

$$B = \mu H \tag{3.2}$$

where μ is the *magnetic permeability* of the object. For vacuum, $\mu = \mu_0 = 4\pi \times 10^{-7}$ V·s/(A·m), and it is called the magnetic constant or permeability of free space. However, the situation is different for any other kind of matter. All substances exhibit magnetic properties to some degree and different phenomena can be observed depending on the properties of the substance and the applied magnetic field strength. When a material is subjected to an external field H, it is magnetized. The *magnetization M* is a vector sum of all the magnetic moments m_i within a given volume V

$$M = \frac{\sum m_i}{V} \tag{3.3}$$

The magnetization M and the external magnetic field H are related by the following equation:

$$M = \chi_m H \qquad (3.4)$$

where, χ_m is the *magnetic susceptibility* of the substance. It is dimensionless and shows the degree of magnetization, or in other words, how easily the substance can be magnetized. Magnetic susceptibility is an intrinsic property of a matter. It depends on factors such as the orientation of atoms in a molecule. The sign and value of χ_m determine the magnetic behavior of a matter.

Therefore, for an object placed in an external magnetic field, the actual magnetic induction inside the object is a sum of the external field and magnetization contributions

$$B = \mu_0 H + \mu_0 M = \mu_0 (H + M) \qquad (3.5)$$

Furthermore, by substituting Equations 3.2 and 3.4 into Equation 3.5, the magnetic permeability μ can be expressed in terms of magnetic susceptibility χ_m

$$\mu = \mu_0 (1 + \chi_m) \qquad (3.6)$$

All materials can be classified depending on the sign and the magnitude of the magnetic susceptibility χ_m. This approach is called the *phenomenological classification* of magnetic materials. It has been used for a long time to describe, rather than explain, different types of magnetic behavior naturally observed for pure elements and commonly used compounds. According to this classification there are three types of magnetic behavior: *diamagnetism, paramagnetism*, and *ferromagnetism*.

The most common types of magnetic materials at room temperature are diamagnetic and paramagnetic. Almost all the elements of the periodic table fall into these two categories (Figure 3.2). In our everyday life we usually refer to diamagnetic or paramagnetic materials as nonmagnetic. That is because their response to an external magnetic field is weak, that is, their magnetic susceptibility χ_m has a small magnitude. The difference in diamagnetic and paramagnetic materials is in the direction of the induced magnetization relative to the applied field, that is, positive or negative χ_m.

Beyond the phenomenological classification, there are cases which do not fit in the three aforementioned classes. For this reason, people typically recognize two more forms of magnetism: *antiferromagnetism* and *ferrimagnetism*. In this section, these five types of magnetic behaviors are introduced in detail, especially the aspects related to MRI/NMR.

3.1.1 DIAMAGNETISM

Being placed in an inhomogeneous magnetic field, an object is either pulled into or pushed out of the area of the stronger magnetic field depending on its magnetic susceptibility. If the direction of the induced magnetization and direction of the external

Legend:

- Paramagnetic
- Diamagnetic
- Rare earth elements
- Ferromagnetic
- Antiferromagnetic

1	2											13	14	15	16	17	18
Hydrogen 1 H 1.0079																	Helium 2 He 4.0026
Lithium 3 Li 6.941	Beryllium 4 Be 9.0122											Boron 5 B 10.811	Carbon 6 C 12.011	Nitrogen 7 N 14.007	Oxygen 8 O 15.999	Fluorine 9 F 18.998	Neon 10 Ne 20.180
Sodium 11 Na 22.990	Magnesium 12 Mg 24.305											Aluminium 13 Al 26.982	Silicon 14 Si 28.086	Phosphorus 15 P 30.974	Sulfur 16 S 32.065	Chlorine 17 Cl 35.453	Argon 18 Ar 39.948
Potassium 19 K 39.098	Calcium 20 Ca 40.078	Scandium 21 Sc 44.956	Titanium 22 Ti 47.867	Vanadium 23 V 50.942	Chromium 24 Cr 51.996	Manganese 25 Mn 54.938	Iron 26 Fe 55.845	Cobalt 27 Co 58.933	Nickel 28 Ni 58.693	Copper 29 Cu 63.546	Zinc 30 Zn 65.39	Gallium 31 Ga 69.723	Germanium 32 Ge 72.61	Arsenic 33 As 74.922	Selenium 34 Se 78.96	Bromine 35 Br 79.904	Krypton 36 Kr 83.80
Rubidium 37 Rb 85.468	Strontium 38 Sr 87.62	Yttrium 39 Y 88.906	Zirconium 40 Zr 91.224	Niobium 41 Nb 92.906	Molybdenum 42 Mo 95.94	Technetium 43 Tc [98]	Ruthenium 44 Ru 101.07	Rhodium 45 Rh 102.91	Palladium 46 Pd 106.42	Silver 47 Ag 107.87	Cadmium 48 Cd 112.41	Indium 49 In 114.82	Tin 50 Sn 118.71	Antimony 51 Sb 121.76	Tellurium 52 Te 127.60	Iodine 53 I 126.90	Xenon 54 Xe 131.29
Caesium 55 Cs 132.91	Barium 56 Ba 137.33	57–70 *	Lutetium 71 Lu 174.97	Hafnium 72 Hf 178.49	Tantalum 73 Ta 180.95	Tungsten 74 W 183.84	Rhenium 75 Re 186.21	Osmium 76 Os 190.23	Iridium 77 Ir 192.22	Platinum 78 Pt 195.08	Gold 79 Au 196.97	Mercury 80 Hg 200.59	Thallium 81 Tl 204.38	Lead 82 Pb 207.2	Bismuth 83 Bi 208.98	Polonium 84 Po [209]	Astatine 85 At [210]	Radon 86 Rn [222]
Francium 87 Fr [223]	Radium 88 Ra [226]	89–102 **	Lawrencium 103 Lr [262]	Rutherfordium 104 Rf [261]	Dubnium 105 Db [262]	Seaborgium 106 Sg [266]	Bohrium 107 Bh [264]	Hassium 108 Hs [269]	Meitnerium 109 Mt [268]	Ununnilium 110 Uun [271]	Unununium 111 Uuu [272]	Ununbium 112 Uub [277]		Ununquadium 114 Uuq [289]				

* Lanthanide series

Lanthanum 57 La 138.91	Cerium 58 Ce 140.12	Praseodymium 59 Pr 140.91	Neodymium 60 Nd 144.24	Promethium 61 Pm [145]	Samarium 62 Sm 150.36	Europium 63 Eu 151.96	Gadolinium 64 Gd 157.25	Terbium 65 Tb 158.93	Dysprosium 66 Dy 162.50	Holmium 67 Ho 164.93	Erbium 68 Er 167.26	Thulium 69 Tm 168.93	Ytterbium 70 Yb 173.04

** Actinide series

Actinium 89 Ac [227]	Thorium 90 Th 232.04	Protactinium 91 Pa 231.04	Uranium 92 U 238.03	Neptunium 93 Np [237]	Plutonium 94 Pu [244]	Americium 95 Am [243]	Curium 96 Cm [247]	Berkelium 97 Bk [247]	Californium 98 Cf [251]	Einsteinium 99 Es [252]	Fermium 100 Fm [257]	Mendelevium 101 Md [258]	Nobelium 102 No [259]

FIGURE 3.2 Periodic table of elements.

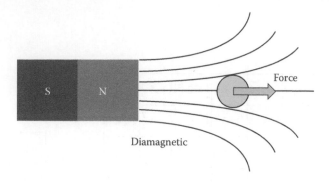

FIGURE 3.3 Schematic illustration of a diamagnetic sample behavior in an external magnetic field.

field are opposite to each other, the effective field will push the object toward the low field region (Figure 3.3), and this behavior is termed *diamagnetism*. Mathematically this tendency of matter to oppose an external magnetic field is expressed by the negative sign of magnetic susceptibility for diamagnetic materials, with the average value of χ_m being of about 10^{-5}.

Diamagnetism is a quantum mechanical effect. Diamagnetism originates from the orbital motion of electrons. All the electrons circulate in orbitals acting like current loops, as shown in Figure 3.1. They are paired in diamagnetic materials and therefore the net magnetic moment is zero. In the presence of an external magnetic field, the applied field aligns the electron paths and meanwhile generates currents in the loops that oppose the change of the field. This results in the repelling phenomenon for diamagnetic behavior. The electrons are rigidly held in orbitals by the charge of the protons and are constrained by the Pauli exclusion principle. Therefore, diamagnetism is generally weak in materials. In short, diamagnetic materials naturally do not have magnetization in the absence of a magnetic field and they are repelled by an externally applied magnetic field.

Generally speaking, all matters possess the diamagnetic property because diamagnetism originates from the orbital motion of electrons. For the same reason, diamagnetism is a property of *every* atom and molecule. However, this effect is so weak that, despite its universal occurrence, diamagnetism is usually masked by other effects, such as paramagnetism or ferromagnetism. It is difficult to observe truly diamagnetic phenomena. Usually substances that mostly display diamagnetic behavior and are generally thought of as nonmagnetic are said to be diamagnetic materials. Practically all organic compounds and the majority of inorganic compounds are examples of diamagnetic materials. The strongest diamagnetic materials are pyrolytic carbon and bismuth. Other notable diamagnetic materials include water, wood, diamond, living tissues (note that the last three examples are carbon-based), and many metals such as copper, gold, and mercury. Magnetic susceptibilities of some diamagnetic materials are shown at Table 3.1. It should be noted here that χ_m is temperature independent.

Probably the most interesting example of diamagnetism application is magnetic levitation. Due to the strong diamagnetism of pyrolytic carbon, it is easy to

TABLE 3.1

Diamagnetic Materials and Their Magnetic Susceptibilities

Material	Water	Copper	Graphite	Lead	Diamond	Silver	Mercury	Bismuth	Pyrolytic Carbon
$\chi_m \times 10^{-5}$	−0.91	−1.0	−1.6	−1.8	−2.1	−2.6	−2.9	−16.6	−40.9

Source: Adapted from C. L. Neave, Magnetic properties of solids. *HyperPhysics.* Retrieved November 9, 2008.

demonstrate the magnetic levitation effect with the help of neodymium (NdFeB) permanent magnets and a thin slice of pyrolytic carbon (Figure 3.4a). The most spectacular part of this experiment is that all the components are at room temperature and no special conditions are required.

As living organisms are diamagnetic, they can also exhibit magnetic levitation. However, because the magnitude of their magnetic susceptibility is much smaller compared to pyrolytic carbon or bismuth, these objects can levitate only in much stronger magnetic fields. In 2010, the Radboud University Nijmegen in the Netherlands demonstrated a live frog levitation (Figure 3.4b) in a 16 Tesla (T) magnetic field inside a bore solenoid (note that so far the maximum strength of a static magnetic field approved by the U.S. Food and Drug Administration (FDA) which can be used in medicine for human beings is 8 T). In 2009, NASA's Jet Propulsion Laboratory in Pasadena, California, demonstrated a live mice levitation. This is a great step forward because mice are biologically closer to human beings than frogs.

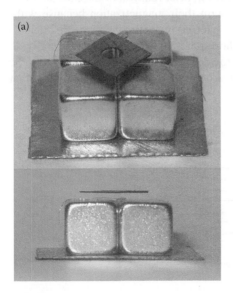
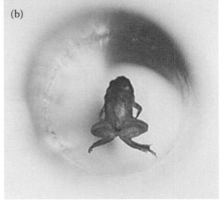

FIGURE 3.4 (a) Pyrolytic carbon levitating above NdFeB magnets and (b) live frog levitation.

However, the experiment required a superconducting magnet that makes the whole experiment more complicated.

3.1.2 PARAMAGNETISM

Looking back at the periodic table in Figure 3.2, many chemical elements at room temperature are paramagnetic. For paramagnetic materials, the directions of the induced magnetization M and the applied magnetic field H are the same. Due to the effective field, the object is pulled toward the area with the higher magnetic field (Figure 3.5). In this case, the susceptibility has a positive sign and the magnitude is of the order of 10^{-3}–10^{-5}, which is comparable to or slightly larger than that of diamagnetic materials.

Multiple theories have been proposed to explain paramagnetism in different types of materials. Some of them explain one specific type of material better, while others are valid for other types. Here we consider the Langevin model of paramagnetism. The origin of paramagnetism comes from the unpaired noninteracting electrons. In fact, in many atoms and in the vast majority of molecules, electrons are combined in pairs with their spins pointing in opposite directions obeying the Pauli exclusion principle, which results in a zero magnetic moment. The only magnetization left is from the orbital motion of such electrons pairs that gives rise to the diamagnetism considered in Section 3.1.1. However, some atoms have unpaired electron spins which results in nonzero permanent magnetic moments. In the absence of an applied magnetic field, these magnetic moments are randomly oriented resulting in a zero net magnetic moment. In the presence of an external magnetic field H, the magnetic moments inside paramagnetic objects align with the field, resulting in an attracting force as shown in Figure 3.5. However, after removal of the external magnetic field, paramagnetic objects do not retain their magnetization because without the alignment by the external force the internal magnetic moments disorient to achieve thermodynamic equilibrium. In short, paramagnetic materials naturally do not have a magnetization and they are attracted by an externally applied magnetic field. Elements from chromium to copper, iron, cobalt, nickel, and rare-earth that

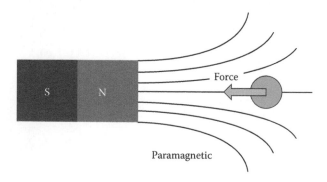

FIGURE 3.5 Schematic illustration of a paramagnetic sample behavior in an external magnetic field.

sequence around gadolinium have this property. Their compounds and alloys are generally paramagnetic or even ferromagnetic.

There is always competition between the diamagnetic contribution from the core electrons and the paramagnetic contribution from the outer shell unpaired electrons, and the resulting magnetic behavior depends on the relative strength of these two. The competition can clearly be seen in s- and p-type metals* where the electrons are delocalized, that is, traveling as an electron gas. This results in weak paramagnetism or even diamagnetism. For example, in the case of gold, the orbital magnetic moments overshadow spin magnetic moments, resulting in the diamagnetic behavior of this metal. In the case of aluminum, the paramagnetic contribution happens to be slightly stronger than the diamagnetic one resulting in the weak attraction of aluminum to an external magnetic field. These small differences are difficult to detect and usually require sensitive analytical devices.

Stronger paramagnetic effects can be observed for d- and f-type elements[†] with strongly localized electrons. The high magnetic moments of lanthanides explains the reason that gadolinium, neodymium, and samarium are typically used for strong magnets.

It should be noted that paramagnetism, unlike diamagnetism, is temperature dependent. The randomizing thermal effect becomes significant at high temperature, making it hard to align the magnetic moments along the external magnetic field. This behavior was experimentally found and named as the Curie law

$$\chi = \frac{C}{T} \tag{3.7}$$

where C is the Curie constant which depends on material and T is the temperature in kelvins.

Figure 3.6 summarizes the behaviors of diamagnetic and paramagnetic materials with the change of an external magnetic field and temperature. As shown in Figure 3.6a and c the relationship between H and M is linear for both of the materials, the difference is only in the sign of the magnetic susceptibility, χ_m. However, the temperature behavior of χ_m for these two types of materials differs significantly, as shown in Figure 3.6b and d. Magnetic susceptibility is a constant for a diamagnetic material, while it obeys the Curie law for a paramagnetic one.

Due to the absence of a strong permanent net magnetic moment, paramagnetic materials are not widely used. However, they have one interesting application: they can be used to achieve extremely low temperatures. The working principle is based on the adiabatic demagnetization effect. When a paramagnet is cooled to liquid-helium temperature (4 K, or –269°C) in the presence of a strong magnetic field, almost all the spins are aligned along the field. If the sample is thermally isolated and the field is gradually decreased, the temperature of the paramagnet

---

* s-elements are those which the outer electronic configuration of ns^1 or ns^2, p-elements are those which the outer electronic configuration of ns^2np^x, where $x = 1...6$.
† The general outer electronic configuration of d-elements is $(n-1)d^xns^y$, where $x = 1...10$, $y = 1,2$; the general outer electronic configuration of f-elements is $(n-2)f^x(n-1)d^xns^2$, where $x = 1...14$, $y = 0,1$.

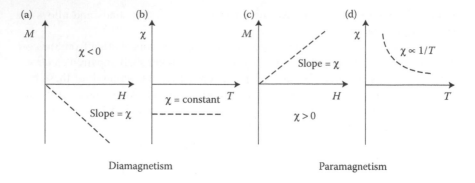

(a) M $\chi < 0$ H Slope = χ

(b) χ χ = constant T

(c) M Slope = χ H $\chi > 0$

(d) χ $\chi \propto 1/T$ T

Diamagnetism   Paramagnetism

FIGURE 3.6   Typical behavior of diamagnetic and paramagnetic materials depending on an external magnetic field and temperature.

will decrease further. The reason is as follows. In the decreasing external field, magnetic moments inside the paramagnetic sample start to reorient back to the random arrangement. This reorientation consumes thermal energy of the system thus, with thermal isolation, the temperature of the system decreases further (<4 K). Praseodymium alloyed with nickel (PrNi5) has an extremely strong adiabatic demagnetization effect and is used to approach a temperature within one thousandth of a degree of absolute zero.

3.1.3   FERROMAGNETISM

The third type of magnetic behavior, according to the phenomenological classification, is called ferromagnetism. In contrast to diamagnetic and paramagnetic materials, ferromagnetic materials have large spontaneous magnetization even in the absence of an external magnetic field. This is the result of the cooperative ordering of spins. It should be noted that ferromagnetism can occur only in materials with paramagnetic properties, because it requires the presence of unpaired electron spins in atoms in order to produce a nonzero net magnetic moment.

The unique feature of ferromagnetic materials is that the relation between M and H is not linear. Moreover, the complicated relation between M and H for ferromagnetic materials is difficult to characterize by simple mathematic functions. Thus, it is usually obtained through a series of experiments by plotting the magnetization M against the strength H of the external magnetic field applied. A magnetic hysteresis loop is shown in Figure 3.7. Ferromagnetic materials can be easily magnetized, and in strong magnetic fields the magnetization reaches a certain limit called *saturation magnetization* M_s, as shown in Figure 3.7. Beyond this limit no further significant increase in magnetization occurs. It should be noted that saturation magnetization is an intrinsic property of a material, which does not depend on the shape or size of the material. Interestingly, with the gradual reduction of the applied field, the magnetization M of a ferromagnetic material does not decrease by its original path, but at a slower rate. When |H| reaches 0 A/m, M is still positive. This positive magnetization is called *remanence* and denoted using M_r. Therefore, unlike dia- or paramagnetic materials, ferromagnetic materials retain a magnetization after the externally

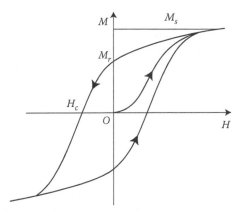

FIGURE 3.7 A magnetic hysteresis loop: magnetization versus magnetic field strength schematic plot of a ferromagnetic material.

applied magnetic field is removed, or in other words, they "remember" their history. This feature of ferromagnetic materials is exploited in magnetic memory devices such as magnetic tape, hard disk drives, and credit cards. When $|H|$ is reduced further (becomes negative and changes its direction), the magnetization becomes zero again at a certain value of magnetic field strength which is called the *point of coercivity* and denoted by H_c. This parameter indicates the strength of the reverse field needed to remove the residual magnetization from the material after the saturation. As the field strength increases further, the magnetization reaches the saturation magnetization in the opposite direction. To complete the entire cycle, $|H|$ along the negative direction can be moved gradually back to 0 A/m and then increased till M reaches the positive saturation magnetization. As is shown in Figure 3.7, by changing the magnetic field strength H and measuring the induced magnetization M for ferromagnetic materials, a loop, called the *magnetic hysteresis* loop is formed, that characterizes the unique behavior of ferromagnetic magnetization.

Different ferromagnetic materials have different shapes and sizes of hysteresis loops. The form of this loop tells the properties of the material. The area inside the hysteresis loop shows the energy losses due to the change of the magnetization. This energy in converted into heat. The parameters M_s, M_r, and H_c indicate important magnetization properties of the material. M_r and H_c indicate how easily the material can be magnetized/demagnetized; M_s shows the amount of magnetization it can store. For example, the coercivity in pure iron is about 0.16 kA/m while in the neodymium magnet (NdFeB) it is more than 800 kA/m. It means that a 5000 times stronger field is required in order to demagnetize NdFeB compared to the amount of field needed to demagnetize iron. This is the reason that the neodymium magnet is the most widely used permanent magnet. Magnetic materials are classified as soft or hard depending on the ease of demagnetization, in other words, the classification is based on the value of their coercivity, or the shape of the hysteresis loop. Hard magnetic materials are those that retain their magnetism and are difficult to demagnetize; they have high H_c and therefore a wide hysteresis loop. One the other hand, soft magnetic materials are those that are easy to magnetize and demagnetize; they have low

H_c and a narrow hysteresis loop. The permanent neodymium magnet belongs to the first category. It has a broad hysteresis loop and a large coercivity. Due to its stability, NdFeB is used as a permanent magnet. Permanent magnets are used to supply the main magnetic field in some MRI systems for imaging. They are introduced in detail in Section 3.3.1. For other applications, sometimes, it is better to reduce energy losses during the hysteresis cycles. One of the examples is the transformer cores in stators and rotors for electrical machinery. In this case soft magnetic materials with a narrow hysteresis loop and small remanence magnetization are more suitable.

Another remarkable feature of ferromagnetic materials is the way magnetic moments are organized inside the material. When cooled below the Curie temperature all ferromagnetic materials consist of microscopic blocks (or the so-called *domains*) of atoms. The magnetic domain is a region within which the magnetization is in a uniform direction, meaning that individual atoms have the same direction of magnetic moment, as shown in Figure 3.8. However, the direction of magnetization of different domains may be different. Therefore, the magnetic field lines from different domains of a ferromagnetic material pass through each other in alternating directions and thus reduce the field outside the material. The domains are separated by thin layers of boundaries (also called walls) where the magnetization vector gradually changes its direction from the direction in one domain to that in the neighboring one, as shown in Figure 3.8d.

The formation of domains, alignment of magnetic moments of atoms in a domain, and formation of flux closure domains where the field lines are allowed to form a closed loop crossing the domains, are the results of energy minimization. The formation of a domain is done in order to reduce magnetostatic energy. Within a domain, when two nearby atoms both have unpaired electrons, it is favorable for the electrons to have their spins aligned because in this way they occupy different orbitals and thus the Coulomb repulsion is smaller and the exchange energy is minimized. The flux closure domains can only be formed when the magnetostatic energy saved is

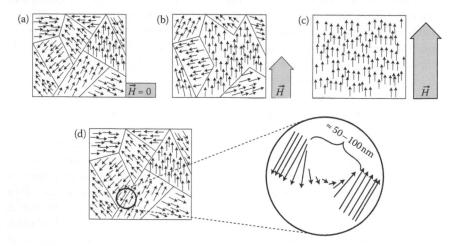

FIGURE 3.8 Domain formation in a ferromagnetic sample (a) without field, (b) with some field applied, (c) with a strong linear field, and (d) a closer look at the domain wall formation.

greater than the energy cost for changing the local net magnetization. It takes two types of energy to form a loop of field lines. One is called *magnetocrystalline anisotropy energy* that is the energy which magnetizes a material in directions other than the favorable "easy axis." The other type is *magnetoelastic anisotropy energy* that is the energy needed due to magnetostriction, the energy for overcoming the mechanical stresses due to the change of the orientation of the molecule in the process of magnetization.

In the presence of an external field H, the favorable direction of magnetization is along H and the magnetization of the material can be increased by the displacement or rotation of the domain walls. The material is so-called magnetized. Typically in the presence of a weak external magnetic field, the increase in the magnetization of the material is due to the displacement of boundaries, as can be seen in Figure 3.8b, c. In the case of a strong magnetic field, the increase is mostly due to the rotation of the domain and aligning along the favorable direction of the external magnetic field (Figure 3.8c). With an external magnetic field, the domain walls are orientated and the domains are aligned, producing a magnetic field. The new orientations of the domain walls and the domains are pinned and not easy to be re-orientated when the external magnetic field is removed. This is the reason that when a piece of ferromagnetic material is magnetized, it becomes a permanent magnet.

Going back to the periodic table in Figure 3.2, only three pure elements Fe, Co, and Ni are examples of ferromagnetic materials at room temperature, and all other ferromagnets are their products, alloys, and combinations. When these materials are heated up, the ordered domain structure is destroyed and they become paramagnetic. The temperature at which such transition occurs is called the Curie temperature, T_c. Different materials have different Curie temperatures (Fe: $T_c = 770°C$, Co: $T_c = 1131°C$, Ni: $T_c = 358°C$). There is a more general law of magnetic susceptibility temperature behavior called the Curie–Weiss law, which is valid for ferromagnetic materials in the paramagnetic state,

$$\chi = \frac{C}{T - T_c} \tag{3.8}$$

This law is only valid for behavior of a ferromagnetic material above Curie temperature when it is paramagnetic and disordered. Below the Curie temperature the ferromagnetic material is ordered into domains. This magnetic ordering temperature is another key feature of ferromagnetic materials.

Moreover, when the size of a ferromagnetic material is very small, for example, a ferromagnetic nanoparticle (NP), ferromagnetism in the material becomes superparamagnetism. In superparamagnetic NPs, magnetization can randomly flip direction under the influence of temperature. Superparamagnetic NPs are one of the important types of NPs applied for CAs for MRI enhancement. More physics on superparamagnetic particles and their application for MRI enhancement are detailed in Section 3.4.6.

The phenomenological approach of classifying materials gives a general idea about different types of magnetic behavior but does not explain the physical mechanisms of the phenomenon. Moreover, there are cases where it is not possible to fit materials to one of the three classes. For this reason, people typically recognize two more forms of magnetism: antiferromagnetism and ferrimagnetism.

3.1.4 ANTIFERROMAGNETISM

Antiferromagnetic materials have properties of both ferro- and paramagnets. Antiferromagnets are similar to ferromagnetic materials in the way magnetic moments are organized: they are also magnetically ordered. However, unlike ferromagnets, in antiferromagnets all magnetic moments are aligned antiparallel to each other, as shown in Figure 3.9, resulting in a zero net magnetic moment like in paramagnets. This complex form of magnetic ordering occurs due to the specific crystal structure. Magnetic oxides are well-known antiferromagnets and they are composed of two interpenetrating and identical magnetic sublattices, typically called sublattice A and sublattice B. The interaction between spins in this system leads to the antiparallel spontaneous magnetization of these two sublattices.

To better understand the origin of this antiparallel alignment of magnetic moments, we consider MnO as an example below. As shown in Figure 3.10, the crystal structure consists of linear chains of Mn^{2+} and O^{2-} ions. Due to the fact that all 3rd orbitals of Mn^{2+} are occupied with spin-up electrons, the only way of covalent bonding with O^{2-} is by donating the spin-down electron of O^{2-} to Mn^{2+}. In this case, a spin-up electron of oxygen is left behind. It can be donated to another Mn^{2+} ion in the chain where this ion must have all spin-down electrons on the 3rd orbital. Such

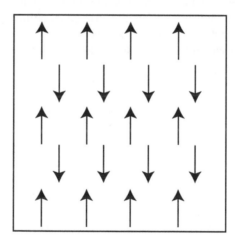

FIGURE 3.9   Spin arrangement in an antiferromagnetic material.

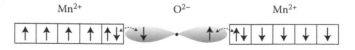

FIGURE 3.10   Super-exchange interaction scheme in MnO.

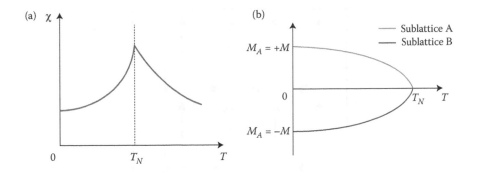

FIGURE 3.11 (a) Susceptibility behavior of an antiferromagnetic material, and (b) spontaneous magnetization of the A and B sublattices in antiferromagnetic materials below the Neel temperature.

type of indirect interaction (mediated by oxygen in this particular case) is called super-exchange interaction.

Like all ferromagnets, antiferromagnets also have the Curie–Weiss dependence but only above a certain critical temperature of magnetic ordering, called the *Néel temperature* (T_N). Figure 3.11a shows the susceptibility of aniferromagnetic materials versus temperature. Above the Néel temperature, an antiferromagnet becomes paramagnetic with randomly oriented magnetic moments and follows the Curie–Weiss law of the following form:

$$\chi = \frac{C}{T + T_N} \tag{3.9}$$

Figure 3.11b shows the spontaneous magnetization of sublattices A and B below T_N. As shown in Figure 3.11b, below the Néel temperature, the sublattices (A and B) have spontaneous magnetizations of the same amount but in opposite directions and thus cancel each other resulting in zero net magnetic moments of the bulk material. The small and positive susceptibility decreases with decreasing temperature. This enables antiferromagnets to respond to an external field in the same manner as paramagnets, and in the meantime, the magnets have a microscopic structure similar to that of ferromagnets.

In their paramagnetic state, antiferromagnets do not have a wide range of applications like ferromagnets. This is because of the absence of spontaneous magnetization. However, they can be a good toy system where theoretical models of more complex ferrimagnets can be tested. The only antiferromagnetic element at room temperature is chromium with the Néel temperature of 37°C.

3.1.5 FERRIMAGNETISM

Ferrimagnets are similar to both ferromagnets and antiferromagnets. They have a spontaneous magnetization below a certain temperature, even in the absence of

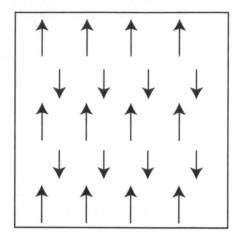

FIGURE 3.12 Spin arrangement in a ferrimagnetic material.

an external magnetic field, like ferromagnets. At the same time, in terms of magnetic ordering, they are related to antiferromagnets because of the super-exchange
mechanism of coupling. This type of coupling exists in both ferrimagnetic and antiferromagnetic materials. Therefore, these two types of magnetic material are both
composed of two sublattices which are antiparallelly aligned. Unlike antiferromagnetic materials, the magnetizations of the sublattices in a ferrimagnetic material are
not identical in magnitude (Figure 3.12, cf. Figure 3.9). Therefore, they do not cancel
each other resulting in the existence of a nonzero spontaneous net magnetization
like that in ferromagnets. Figure 3.12 shows the spin arrangement in ferrimagnetic
materials. In ferrimagnetic materials, the magnetizations of the sublattices are not
identical and they do not necessarily vary monotonically with temperature, making
the net magnetization behavior complicated. Figure 3.13a and b show the magnetization versus temperature curves of $NiO-Cr_2O_3$ and $Li_{0.5}Fe_{1.25}Cr_{1.25}O_4$. In the case
of the $Li_{0.5}Fe_{1.25}Cr_{1.25}O_4$ compound as shown in Figure 3.13b, the net spontaneous
magnetization decreases to zero even before the critical temperature, and changes to

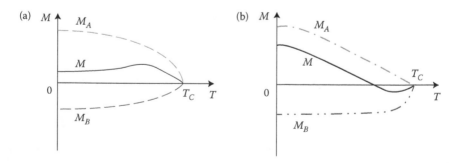

FIGURE 3.13 Magnetization curves of A and B sublattices (dotted lines) and the net magnetization (solid line) in two different ferrimagnets (a) $NiO-Cr_2O_3$ and (b) $Li_{0.5}Fe_{1.25}Cr_{1.25}O_4$.

the opposite direction. The temperature at which the magnetizations are exactly balanced is called the *compensation point*. There are structures which have more than two magnetic sublattices and the magnetization behavior there is even more complex with multiple compensation points.

The applications of ferro- and ferrimagnetic materials have a long history. The term ferromagnetism comes from the most common material which exhibits this property, iron (*ferrum* in Latin). Ferrimagnetism, in turn, originates from the name ferrites, compounds which demonstrate ferrimagnetic behavior. Ferrites are ferromagnetic transition-metal oxides that have been used for centuries. The very first magnetic material which was used for navigation compasses is lodestone and it contains a magnetic mineral, magnetite, which is ferrite. They are now widely used in high-frequency applications due to their high saturation magnetization and low electrical conductivity. Moreover, with the advancement in processing techniques, such as the ceramic processing technique, ferrites can be produced readily with precisely tuned properties for specific applications.

3.2 WORKING PRINCIPLES OF NMR AND MRI

The history of NMR and MRI dates back to the end of the eighteenth century. In 1895, Roentgen discovered the x-ray which enables the visualizing of the interior of the human body without surgical intervention. Today, besides NMR, there are different medical imaging modalities including x-ray radiography, x-ray computerized tomography (CT), ultrasound, and nuclear medicine. Imaging based on NMR is called MRI for short. In chemistry and physics communities, magnetic resonance is usually referred to as NMR while in the imaging community, the world "nuclear" is omitted because of the concerns of public relations.

NMR was discovered in 1946 by Felix Bloch [1] and Edward Purcell [2] independently. The two physicists shared the 1952 Nobel Prize in physics for "their development of new ways and methods for nuclear magnetic precision measurements." In 1973, Paul Lauterbur reported the first MR image [3] using gradient fields in *Nature*. Table 3.2 shows the scientific contributions to MRI before the first image was reported. In the 1970s, most of the work was done in academia. In the 1980s, industry joined in and accelerated the development. As a result, the image quality has been improved dramatically, and MRI and MRI systems have become popular worldwide.

Compared to other modalities, MRI provides advantages of good contrast especially soft tissue contrast, non-invasiveness, no ionizing radiation, and arbitrary scan planes. Figure 3.14 shows MRI images of a human head. Sagittal, coronal, and axial refer to the slice orientation as illustrated. The MRI images in Figure 3.14 are acquired in a 3 T MRI scanner. As shown in the figure, anatomical details can be seen using MRI. More details can be seen with an increase in the main magnetic field of an MRI system, which motivates the research and development of MRI to increase the field to 4 T, 7 T, and so on. Besides anatomical imaging, metabolic information is available with MR, which enables noninvasive *in vivo* physiological studies. Next, the basic physics and imaging methods of MR will be introduced for the purpose of understanding the nature of the MR phenomenon and the imaging modalities.

TABLE 3.2
Scientific Contributions to MRI before the First MRI Image

Year	Contributor	Contribution
1952	Herman Carr (Harvard University)	Produced 1D MRI image
1960	Vladislav Ivanov (Soviet Union)	Filed a document for a MRI device (USSR State Committee for Inventions and Discovery at Leningrad)
1970	Peter Mansfield (University of Nottingham)	Developed a mathematical technique that would allow scans to take seconds rather than hours and produce clearer images than Lauterbur had
1971	Raymond Damadian (State University of New York)	Reported tumors and normal tissue can be distinguished *in vivo* by NMR; this method is not effective and not practical
1972	Raymond Damadian	Created the world's first MRI machine and filed a patent
1973	Paul Lauterbur (State University of New York)	Expended Carr's technique and generated and published the first NMR 2D and 3D images (using gradients)

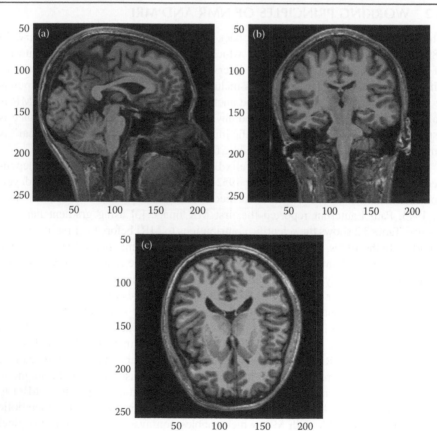

FIGURE 3.14   MRI images of a human head. (a) Sagittal, (b) coronal, and (c) axial refer to the slice orientation as illustrated.

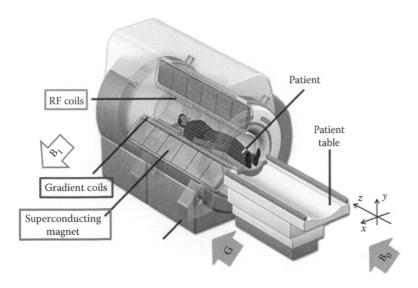

FIGURE 3.15 A cutaway view of a superconducting magnet-based MRI scanner.

3.2.1 SPINS

Atoms with an odd number of protons and/or odd number of neutrons possess nonzero nuclear spin angular momentum. A nonzero spin is associated with a nonzero magnetic moment and exhibits an MR phenomenon. In biological specimens, hydrogen (^{1}H) that has a single proton is the most abundant and the most sensitive. For imaging, polarizations, precession, and relaxations of spins take place, and signals are acquired and processed to construct an image. For the aforementioned processes to take place, three kinds of magnetic fields are needed, namely the main magnetic field ($\boldsymbol{B_0}$), radiofrequency (RF) magnetic field ($\boldsymbol{B_1}$), and linear gradient field ($\boldsymbol{G}$). Figure 3.15 shows a cutaway view of a traditional MRI scanner with a patient lying in the bore. As shown in Figure 3.15, superconducting magnets, RF coils, and gradient coils are the hardware generating $\boldsymbol{B_0}$, $\boldsymbol{B_1}$, and $\boldsymbol{G}$, respectively.

3.2.2 B_0 AND POLARIZATION OF SPINS

The main magnetic field, B_0, is applied for polarization of spins. As shown in Figure 3.16, in the absence of an external magnetic field, the spins are randomly oriented and there is no net magnetic moment macroscopically. When $\boldsymbol{B_0}$ is applied along the z-direction, the spins are aligned and a net magnetic moment is created. This process is called polarization. Moreover, the nuclear spins exhibit resonance at a defined frequency called Larmor frequency, ω. The Larmor frequency is linked to B_0 in the way below,

$$\omega = \gamma B_0 \tag{3.10}$$

No applied field   Applied field

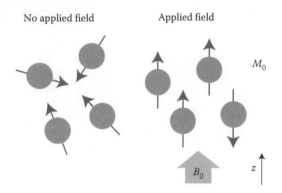

FIGURE 3.16   Spins and the applied field, B_0.

or alternatively

$$f = \frac{\gamma}{2\pi} B_0 \qquad (3.11)$$

where γ is called gyromagnetic ratio which is the ratio of the magnetic dipole moment to the angular momentum of a particle, such as an atom, or a system (SI unit: rad $\cdot$ s^{-1} $\cdot$ T^{-1}). Different atoms have different gyromagnetic ratios. For ^{1}H, $\gamma/2\pi = 42.58$ MHz/T. Therefore, for a system with $B_0 = 1$ T, its Larmor frequency is 42.58 MHz.

The resonant frequency is in the RF frequency range at which RF signals are transmitted and received for imaging. B_0 is required to be homogeneous for imaging using the gradient field that will be introduced shortly. The homogeneity of fields is calculated using Equation 3.12 or Equation 3.13 that follows in parts per million (ppm). The main magnetic field in a MRI scanner in the hospital is generated by superconducting magnets.

$$\text{Homogeneity} = \frac{\sum_i^n \frac{|B_i - B|}{B} 10^6}{n} \qquad (3.12)$$

where B_i is the magnitude of the magnetic field on ith node in the region of interest, B is that of the magnetic field at the center of the region, and n is the total number of nodes in the region.

$$\text{Homogeneity} = \frac{B_{max} - B_{min}}{\bar{B}} \times 10^6 \qquad (3.13)$$

B_{max} and B_{min} are the maximum and minimum field in the region of interest, and $\bar{B}$ is the average field in the region.

3.2.3 B_1 AND PRECESSION

The RF magnetic field, B_1, is the magnetic field of the RF signal that is tuned at Larmor frequency in an MRI system for imaging. Unlike B_0 that is applied in the z-direction, B_1 is applied on the xy-plane to excite the spins out of their equilibrium along the z-direction, as shown in Figure 3.17. In Figure 3.17, the spin is represented using its magnetization vector by a gray arrowed line. B_1-field is generated by RF coils around the bore. It applies a torque which rotates the spins by a prescribed angle dependent on the strength of B_1 and its duration. In a clinical scanner, the strength of B_1 is typically a small fraction of a Gauss and its duration is normally a few milliseconds. There are two ways to describe the behaviors of the spins when they are exposed to both B_0 and B_1, one using the laboratory frame shown in Figure 3.17a and b, the other one using a frame rotating at the Larmor frequency about the z-axis, as shown in Figure 3.17c. As shown in Figure 3.17a, when using a laboratory frame, the magnetization vector is tipped to the xy-plane following a conical spiral trajectory whereas in the rotating frame shown in Figure 3.17c, the spin is tipped with a trajectory on a plane that is perpendicular to the xy-plane. Figure 3.17b shows a top view of the precession in the laboratory frame.

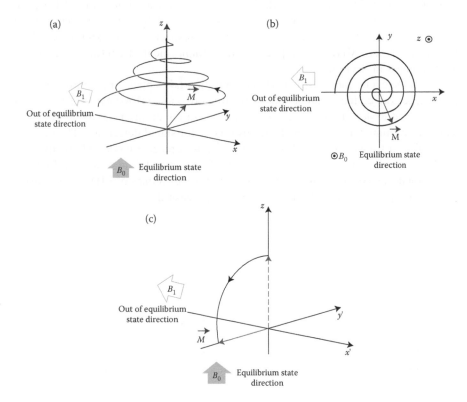

FIGURE 3.17   Effect of B_1 field. (a) 3D view of a spin in a laboratory frame; (b) top view of a spin in a laboratory frame; and (c) 3D view of a spin in a rotating frame.

3.2.4 RELAXATION

For a 90-degree excitation, the spins can be excited and tipped to the xy-plane by B_1 generated by RF coils. When the excitation is turned off, relaxation of the spins back to the equilibrium occurs. During the relaxation, the magnitude of the magnetization of the spins which is the length of the magnetization vector does not remain constant over time. The z-component of the magnetization vector ($|M_z|$) increases while the component of the vector on the xy-plane ($|M_{xy}|$) decreases. Figure 3.18 shows the changes of the different components over time. There are two time constants characterizing the relaxation of the spins, T_1 and T_2. T_1 is called the *longitudinal recovery time constant* and characterizes the recovery of the magnetization vector along the z-axis and T_2 is called the *transverse decay time constant* and characterizes the decay of the vector components on the xy-plane. T_1 is determined by the thermal interactions between the resonating protons and other protons and other magnetic nuclei in the magnetic environment called "lattice." T_2-decay is due to magnetic interactions that occur between spinning protons. T_2-interactions do not involve a transfer of energy but only a change in phase, which leads to a loss of coherence between different spins. In humans, T_1 values of most tissues range from 100 to 1500 ms whereas T_2 values range from 20 to 300 ms. Relaxation time constants have distinct values for different tissues. They are important MR parameters for creating a tissue contrast. Images using T_1- and T_2-relaxation contrasts are called T_1- and T_2-insert spaceweighted images, respectively. T_1- and T_2-weighted images are where MNPs are applied for the enhancement of contrast. The details are presented in Section 3.4.

When the excitation is turned off and relaxation takes place, the rotating magnetization vectors on the xy-plane induce electromotive force (EMF) in an RF receiver coil oriented to detect the change of the magnetization on the xy-plane. Figure 3.19 shows a schematic diagram for RF receiver coils for signal detections. The receiver coil can be a transmission coil that is switched to a receive mode. Alternatively, it can be a separate RF coil only for receiving. The generated time signal is called free induction decay (FID). Figure 3.20 (3rd row) shows examples of FIDs. To construct an MRI image, a set of FIDs are recorded and processed.

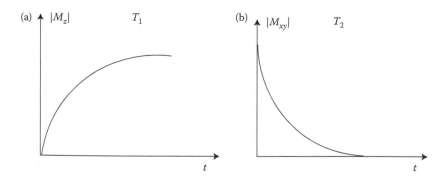

FIGURE 3.18 Relaxation of a spin: (a) longitudinal relaxation characterized by time constant T_1 and (b) transverse relaxation characterized by time constant T_2.

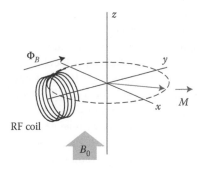

FIGURE 3.19 RF receiver coil and MRI signal detection.

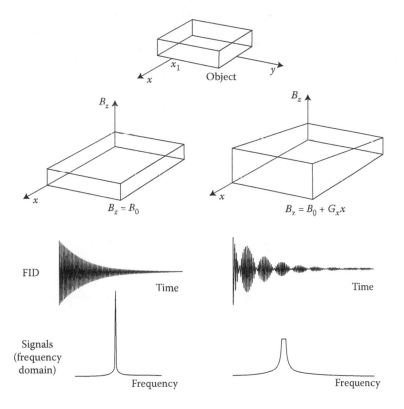

FIGURE 3.20 Comparison of imaging with and without gradient in the x-direction.

3.2.5 LINEAR GRADIENT FIELD, G

For an MRI scan, if the subject under scan is exposed to a homogeneous magnetic field, all spins possess the same resonant frequency. The transmit/receiver RF coil encompasses the whole region of interest. If only B_0 exists, it is impossible to excite a certain portion of the volume or distinguish the signals from different spatial locations. Linear gradient field (G) is added for spatial localization. Taking the gradient in the x-direction (G_x), for example, the gradient field that is applied to the region under imaging is $G_x = G_x x \hat{z}$. This equation means first, the gradient field is in the z-direction and second, its magnitude depends on the locations along the x-direction. Therefore, the total field at a location is $(B_0 + G_x x)\hat{z}$ which is in the z-direction as well. Based on Equation 3.10, the frequency of spins becomes a function of location in the x-direction, $\omega(x) = \gamma(B_0 + G_x x) = \omega_0 + \gamma G_x x$. Therefore, if the dimension of the region under imaging along the x-direction is Δx, the FID contains signals spanning over a frequency range of $\Delta\omega(\Delta x)$. For example, if $G_x = 1$ G/cm, then for an object 5 cm wide in the x-direction, the frequency bandwidth is 21.26 kHz. On a whole body system, the gradient field strength is usually less than 1 Gauss/cm (10 mT/m).

The columns of Figure 3.20 show a comparison of imaging without and with gradient fields along the x-direction, respectively. The FID contains signal of a single frequency for the case without a gradient field, which can be seen clearly in the frequency domain after performing the Fourier transform. On the other hand, with a gradient field, the FID contains signals over a frequency band. After performing the Fourier transform, the signals in the frequency domain show the contributions from each frequency component which can be linearly mapped to a particular location along the x-direction because of the aforementioned way the gradient field is applied along the x-direction.

3.2.6 BLOCH EQUATION

The behavior of the magnetization vector M is described by an empirical equation, the Bloch equation as shown below

$$\frac{dM}{dt} = M \times \gamma B - \frac{M_x x + M_y y}{T_2} - \frac{(M_z - M_0)z}{T_1} \tag{3.14}$$

where M_0 is the equilibrium magnetization arising from the main field, B_0, and B includes B_0, B_1, and G.

If B is static and homogeneous, $B = B_0 \hat{z}$, the solution for Equation 3.14 is as follows:

$$M(t) = \begin{pmatrix} M_x(t) \\ M_y(t) \\ M_z(t) \end{pmatrix} = \begin{pmatrix} e^{-t/T_2} & 0 & 0 \\ 0 & e^{-t/T_2} & 0 \\ 0 & 0 & e^{-t/T_1} \end{pmatrix} \begin{pmatrix} \cos(\omega_0 t) & \sin(\omega_0 t) & 0 \\ -\sin(\omega_0 t) & \cos(\omega_0 t) & 0 \\ 0 & 0 & 1 \end{pmatrix} M^0$$

$$+ \begin{pmatrix} 0 \\ 0 \\ M_0(1 - e^{-t/T_1}) \end{pmatrix} \tag{3.15}$$

The x- and y-components of M contain a factor of e^{-t/T_2} which indicate the decay of the precessing magnetization in the xy-plane with a time constant T_2.

Simultaneously, there is a return to the equilibrium along the z-direction at a time constant of T_1. Equation 3.15 is a mathematical expression for relaxation.

When B changes with time and gradient fields are applied, $B(r,t) = [B_0 + G(t) \cdot r]z$ the solution for the transverse magnetic field ($M = M_x + iM_y$) is

$$M(r,t) = M^0(r)e^{-t/T_2(r)}e^{-i\omega t}\exp\left(-i\gamma\int_0^t G(\tau) \cdot r d\tau\right) \qquad (3.16)$$

3.2.7 SIGNAL AND MR IMAGING

The received time signal $s_r(t)$ is calculated based on the contributions of all processing transverse magnetization in the volume. It is written as follows:

$$s_r(t) = \int_{\text{vol}} M(r,t)dV \qquad (3.17)$$

Based on Equations 3.16 and 3.17, the expression for the signal can be written as follows:

$$s_r(t) = \iiint M^0(x,y,z)e^{-t/T_2(r)}e^{-i\omega t}\exp\left(-i\gamma\int_0^t G(\tau) \cdot r d\tau\right)dxdydz \qquad (3.18)$$

For a 2D image, we are interested in imaging based on the integral over the slide centered at z_0 with a width of Δz. Therefore,

$$m(x,y) \equiv \int_{z_0-\Delta z/2}^{z_0+\Delta z/2} M^0(x,y,z)dz \qquad (3.19)$$

and when the relaxation term e^{-t/T_2} is ignored

$$s_r(t) = \iiint m(x,y)e^{-i\omega t}\exp\left(-i\gamma\int_0^t G(\tau) \cdot r d\tau\right)dxdy \qquad (3.20)$$

Furthermore, to drop the factor $e^{-i\omega_0 t}$, let

$$s(t) = s_r(t)e^{+i\omega t} = \iint m(x,y)\exp\left(-i\gamma\int_0^t G(\tau) \cdot r d\tau\right)dxdy \qquad (3.21)$$

As can be seen in Equation 3.21, $s(t)$ provides information about $m(x, y)$, the transverse magnetization of interest. $m(x, y)$ is a function of the NMR parameters $\rho(x, y)$ (density), $T_1(x, y)$, and $T_2(x, y)$ which is the so-called image in MRI. If the spatial localization is required only in the x- and the y-direction, Equation 3.21 can be further simplified as follows:

$$s(t) = \iint m(x, y) \exp\left[-i\gamma\left(\int_0^t G_x(\tau)d\tau\right)x\right] \exp\left[-i\gamma\left(\int_0^t G_y(\tau)d\tau\right)y\right] dxdy \quad (3.22)$$

or

$$s(t) = \iint m(x, y) e^{-i2\pi[k_x(t)x + k_y(t)y]} dxdy \quad (3.23)$$

where

$$k_x(t) = \frac{\gamma}{2\pi}\int_0^t G_x(\tau)d\tau \quad \text{and} \quad k_y(t) = \frac{\gamma}{2\pi}\int_0^t G_y(\tau)d\tau$$

Equation 3.23 is the signal equation stating the relation between the baseband signal $s(t)$ and the magnetization $m(x, y)$. The signal is a surface integration of the magnetization multiplied by a spatially dependent phase factor. The phase factor is linearly dependent on the spatial position when linear gradient fields are applied. In Equation 3.23, $s(t)$ is the 2D Fourier transform of $m(x, y)$, which can be expressed mathematically as

$$s(t) = M[k_x(t), k_y(t)]$$

With k_x and k_y as variables in the 2D Fourier transform space, the Fourier transform space is called the k-space. Based on Equation 3.23, imaging can be realized by acquiring a set of $\{s(t)\}$ in the k-space and applying the Fourier transform to the data. Figure 3.21 shows raw data in the k-space and the corresponding image in the physical domain before and after the Fourier transform, respectively.

3.3 MAGNETIC MATERIALS USED TO SUPPLY THE MAIN MAGNETIC FIELD IN MRI

As shown in Figure 3.15 earlier, an MRI scan needs a main magnetic field, B_0. For imaging using gradient fields, B_0 is required to be homogeneous. For a human scanner, it is popular to use superconducting magnets for generating homogeneous B_0 over a volume that hosts human parts or a human body. Alternatively, B_0 can be generated using permanent magnets. This is the first application of magnetic materials for MRI. Normally, a scanner using permanent magnets and gradient fields has a small imaging volume, for example, a diameter of a few centimeters. For a human scanner, it requires a large chunk of permanent magnets that can have a weight of a few hundreds of tons. A magnet array can possibly supply B_0 with an increased image volume. However, the homogeneity is difficult to achieve. This problem has

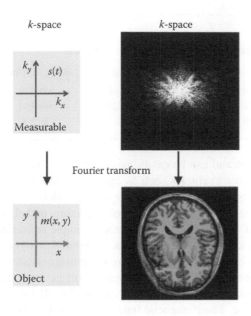

FIGURE 3.21   Raw data and image.

recently been overcome by applying nonlinear imaging methods. In this session, we will review key aspects of permanent magnets that are associated with the application for MR imaging. Moreover, recent work on using permanent magnets to supply B_0 for imaging is presented.

3.3.1   PERMANENT MAGNETS

A magnet is a material or object that produces a magnetic field. A permanent magnet is an object made from a material that is magnetized and creates its own persistent magnetic field even in the absence of an applied magnetic field. It is made of ferromagnetic or ferrimagnetic materials. As we have introduced earlier in Sections 3.1.3 and 3.1.5, these two types of magnetic materials show spontaneous magnetizations, the magnetizations without an external magnetic field below Curie temperature. Examples of such kinds of materials are iron, nickel, cobalt, and alloys of rare-earth metals.

There are different types of permanent magnets. One type is naturally occurring ferromagnets, such as magnetite (or lodestone), nickel, cobalt, and rare-earth metals such as gadolinium and dysprosium (at a very low temperature due to their low Curie temperatures). They are used in the early experiments with magnetism. With the advance of technology, composites based on natural magnetic materials are produced, with improved magnetic field strength and mechanical properties. They are the second type of permanent magnets. The magnet field strength of this type of magnet can reach 1 T. One example is *ceramic magnets* that are made of the sintered composite of powdered iron oxide and barium/strontium carbonate ceramic.

TABLE 3.3
Magnetic Properties of Rare-Earth Magnets

Magnet	B_r (T)	H_c (kA/m)	$(BH)_{max}$ (kJ/m³)	T_c (°C)
$Nd_2Fe_{14}B$ (sintered)	1.0–1.4	750–2000	200–440	310–400
$Nd_2Fe_{14}B$ (bonded)	0.6–0.7	600–1200	60–100	310–400
$SmCo_5$ (sintered)	0.8–1.1	600–2000	120–200	720

They are inexpensive and can be easily mass produced. They are noncorroding but brittle. There are other examples such as *injection-molded magnets* which are the composite of various types of resin and magnetic powders, *alnico magnets* that are made by casting or sintering a combination of aluminum, nickel, and cobalt with iron and a small amount of other elements. Another example is *flexible magnets* that are composed of a high-coercivity ferromagnetic compound mixed with a plastic binder.

The third type is rare-earth magnets. Rare-earth magnets are the strongest type of permanent magnets made from the alloy of rare-earth elements that are the 15 metallic chemical elements with atomic numbers from 57 to 71 (as shown in Figure 3.2) and are ferromagnetic. Their magnetic field can exceed 1 T. The high magnetic field comes from the rare-earth elements (e.g., scandium, $_{21}Sc$, yttrium, $_{39}Y$) that have atoms that retain high magnetic moments in the solid state, which is a consequence of incomplete filling of the *f*-shell allowing up to seven unpaired electrons with aligned spins. The rare-earth elements show low Curie temperature above which the material loses magnetism. However, when they form compounds with transition metals (e.g., iron, nickel, and cobalt), the Curie temperatures of the compounds increase to higher than room temperature. There are mainly two types of rare-earth magnets, neodymium ($Nd_2Fe_{14}B$) and samarium ($SmCo_5$). Table 3.3 shows their magnetic properties. As introduced previously, remanence (B_r) measures the strength of the magnetic field, coercivity (H_c) is the resistance of the material to becoming demagnetized, energy product $(BH)_{max}$ is the density of magnetic energy; and T_c is Curie temperature. To supply the main magnetic field for MRI, high-field strength is preferred because it results in high signal-to-noise ratio (SNR) thus improving image quality. Rare-earth magnets have relatively high magnetic field strength, therefore they are widely used to supply B_0 for MRI.

3.3.2 USING PERMANENT MAGNETS TO GENERATE MAIN MAGNETIC FIELD FOR IMAGING

Multiple MRI systems are built using permanent magnets to provide B_0 field. Field homogeneity, image volume, and weight are crucial parameters for evaluate a magnet/magnet system. There are mainly two categories of a magnet system, those using the magnetic field between two poles and those using a magnet array.

3.3.2.1 Magnet System Using the Magnetic Field between Two Poles

Within this category, the C-shaped permanent magnet (Texas A&M University, 1997) [4–6] is well known. The C-shaped permanent magnet was proposed in Reference 4, detailed in Reference 5, and used in a desktop MRI imaging system in Reference 6. It provides a magnetic field of 0.21 T with homogeneity of 20 ppm within a cylindrical region of 0.5 in. in diameter times 0.75 in. in length.

Figure 3.22a and b shows a photo and the cross-sectional view (with dimensions) of the C-shaped magnet, respectively. As shown in Figure 3.22b, the C-shaped magnet consists of a rectangular C-arm connecting two iron necks, two NdFeB poles, and two round-shaped pole faces with an air gap in between. The size of the air gap is 7 in. × 4 in. Both the poles and the necks are cylindrical, the pole has a diameter of 7 in. and the necks are tapered from a diameter of 7 in. to one of 4 in. The tapered diameter of the neck at the junction to the C-arm is to reduce the iron volume, and,

(a)

(b)

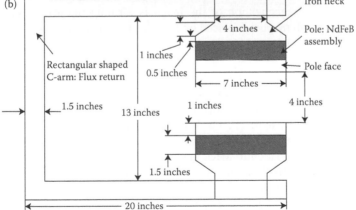

FIGURE 3.22 C-shaped permanent magnet (0.21 T) (a) photograph and (b) drawing with dimensions.

in turn, to reduce the overall weight of the system. Both the dimensions of the neck and the C-arm are optimized so that the iron does not saturate and the reluctance does not increase dramatically. Pole faces are designed to re-focus magnetic field lines toward the C-gap.

The magnet poles were constructed using 330 pieces of NdFeB material in $1 \times 1 \times 0.5$ in. blocks. Each block is numbered, and the energy is measured. The magnets are stacked in groups of three in order to minimize energy variation. This resulted in several groups of magnetization energy, which are placed symmetrically around the pole pieces. The pole pieces are designed using numerical solutions, the 2D Pandira code from Los Alamos National Lab. Shimming is achieved by adjustment of the pole pieces and with four electrical shim coils. 20 ppm in homogeneity at 0.21 T from a cylindrical phantom 0.75 in. long by 0.5 in. in diameter is achieved.

Another example is the magnet built by the Institute of Electrical Engineering of the Chinese Academy of Sciences in Beijing [7]. Figure 3.23a and b shows the photo and the cross-sectional view of the magnet, respectively. The field strength is 0.19 T and the weight is 13 kg. As shown in Figure 3.23b, two iron pole faces, two rare-earth magnets, and an iron case form a magnetic circuit allowing the flow of the magnetic flux. The image volume is between the two pole faces. The distance between the pole faces is 4 cm and the homogeneity is about 50 ppm over 1 cm diameter of spherical volume (DSV). It is used in a tabletop MRI scanner developed in the Martinos Center for Biomedical Imaging, Massachusetts General Hospital [8].

3.3.2.2 Magnet Array

Within the category of magnet array, the Halbach array, especially the Halbach cylinder, is popular in the application for MRI.

3.3.2.2.1 Halbach Cylinder

The Halbach cylinder used widely in MRI is the one that provides a magnetic field pointing in the same direction, as shown in Figure 3.24a. Figure 3.24b shows the directions of the magnetization of the magnets forming the array. As shown in Figure 3.24b, the ith magnet is placed on a circle at an angle $\alpha_i = 2\pi i/n$ and its magnetization is defined by an angle β_i where $\beta_i = 2\alpha_i$, $i = 0, 1, 2, \ldots n - 1$, and n is the number of magnets.

Ideally, the magnetic field inside the cylinder is uniform when the cylinder is infinitely long and the magnetization varies continuously. However, in reality, the length is finite, which introduces nonuniformity at two ends (called end effects). Moreover, continuously varying magnetization is not practical and is implemented by magnet blocks with rotated magnetization. This leads to inhomogeneity of the field inside the bore. For the application to MRI, homogeneity of the magnetic field is required in a desired volume. To eliminate the inhomogeneity, there are different shimming methods proposed in the literature.

In Reference 9, two strategies are applied for magnetic shimming. First, as shown in Figure 3.25a, the magnet bars are split into two parts with a gap in the longitudinal direction (z-direction) in order to increase the homogeneity in that direction. Figure 3.26 shows the simulated field profiles at different gap sizes (*eps*). As shown

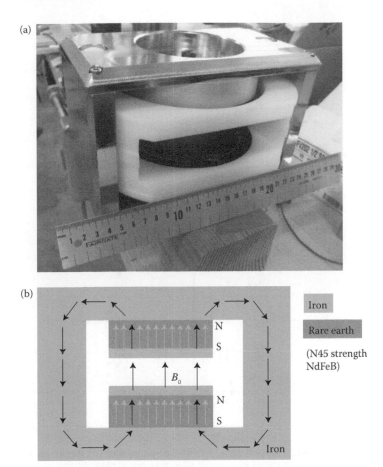

FIGURE 3.23 Magnet built by the Institute of Electrical Engineering of the Chinese Academy of Sciences in Beijing. (a) Photo and (b) side view. (Adapted from Martinos Center for Biomedical Imaging, Massachusetts General Hospital. Available: http://iee.ac.cn/Website/index.php?ChannelID=2195.)

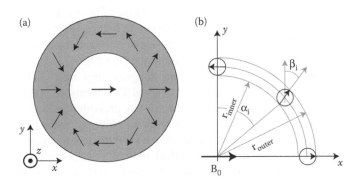

FIGURE 3.24 Halbach cylinder. (a) Side view and (b) labeled side view.

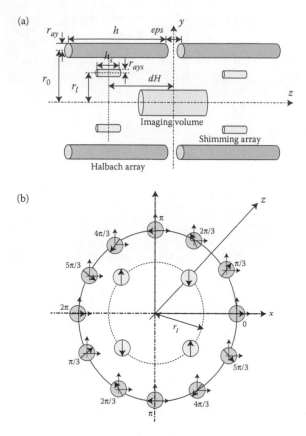

FIGURE 3.25 Proposed Halbach array. (a) The *yz*-plane and (b) the *xy*-plane. (Adapted from H. D. Phuc et al., *Int J Smart Sens Intell Syst*, 7(4), 2014, 1555–1579.)

in Figure 3.26, the homogeneity of field changes considerably with the change of the gap size. Figure 3.27 shows the volume with a field homogeneity of less than 100 ppm, based on Equation 3.13 versus the gap size. The optimal gap size is 0.77 mm where the homogeneous volume is 2640 mm^3. The second strategy is using shimming array inside the Halbach array as shown in Figure 3.25b. The radius of the shimming ring array (r_l) and their location (dH) are optimized for field homogeneity. The homogeneity is improved with the shimming arrays [9].

Shimming the magnetic field using inner magnet rings in a Halbach array is applied to magnet arrays in different portable MRI systems [10,11]. Figure 3.28 shows the 3D and the side view of the Halbach array presented in Reference 11, two cube arrays are inserted inside the Halbach array for eliminating the end effect. The location of the cube arrays is studied in Reference 11. If the location of the cube array is defined by *D* as shown in Figure 3.28b, the bandwidth of the resonance increases as *D* increases, as can be seen from Figure 3.29a, which is the result of an increase in field inhomogeneity. Besides shimming, the cube arrays inside the cylinder increase the average magnetic field strength of the magnet array. Figure 3.29b shows the

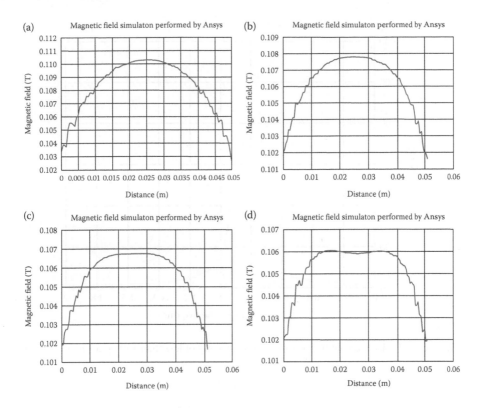

FIGURE 3.26 Field profile along z-direction with different gap sizes (eps). (a) eps = 0 mm, (b) eps = 0.5 mm, (c) eps = 0.9 mm, and (d) eps = 1.3 mm.

simulated magnetic field profiles (Bxy) of the magnet array with and without the shimming cube arrays.

3.3.2.2.2   Other Magnet Arrays

There are other magnet arrays proposed for generating B_0 for MRI/NMR. The array shown in Figure 3.30 is an example. It was invented by G. Aubert in 1991. As shown in Figure 3.30, two magnet arrays with the designed magnetization arranged a distance apart generate homogeneous magnetic fields along the z-direction. The volume where the field is homogeneous is affected by the discretization of the rings [12].

Using permanent magnets to generate B_0 for MR imaging is a low-cost approach compared to using superconducting magnets. However, this approach is limited by low magnetic field strength (thus low SNR), inhomogeneity or small imaging volume if the field is relatively homogeneous, and heavy for those magnet systems used for imaging humans. The advancement of nonlinear signal reconstruction recently relaxed the requirement on field homogeneity and thus, a less bulky permanent magnet array for imaging human or human parts becomes possible [10]. More developments along this direction can be expected in the near future for low cost and portable MRI systems.

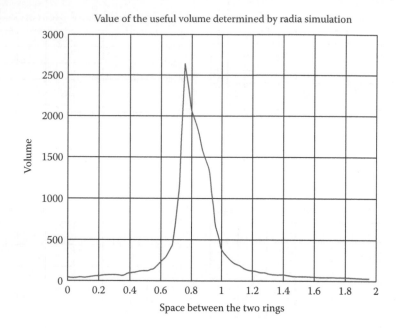

FIGURE 3.27   Image volume with a field homogeneity of less than 100 ppm versus gap size.

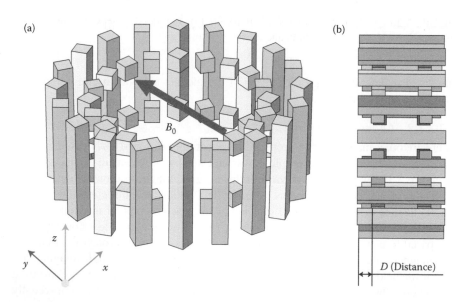

FIGURE 3.28   (a) 3D model and (b) the side view of the Halbach array. (Adapted from Z. H. Ren et al., *RF and Wireless Technologies for Biomedical and Healthcare Applications (IMWS-BIO)*, IEEE MTT-S 2015 International Microwave Workshop Series on, pp. 92–95, Taiwan, 2015. © 2015 IEEE.)

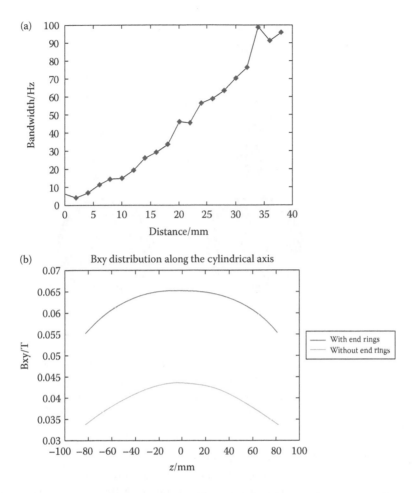

FIGURE 3.29 (a) The Lamor frequency bandwidth in 12 cm × 12 cm area at the center of the bore and (b) the magnetic field profile (Bxy) of the Halbach cylinder in the z-direction. (From Z. H. Ren et al., *RF and Wireless Technologies for Biomedical and Healthcare Applications (IMWS-BIO)*, IEEE MTT-S 2015 International Microwave Workshop Series on, pp. 92–95, Taiwan, 2015. © 2015 IEEE.)

3.4   MRI CONTRAST ENHANCEMENT USING MNPs

Using MNPs for MRI contrast enhancement is another important application of magnetic materials for MRI. Paramagnetic NPs have been used for *in vitro* diagnostics as CAs in MRI for more than four decades [13]. The *in vivo* applications, however, require them to play a more challenging role. They have to be highly specific, efficient, small enough to be transported within the blood stream, and at the same time, they should be able to attach to cells or even enter a cell. In order to fulfill the requirements of contrast enhancement, the materials used must be magnetically active. Paramagnetic materials, like different lanthanide, iron-based

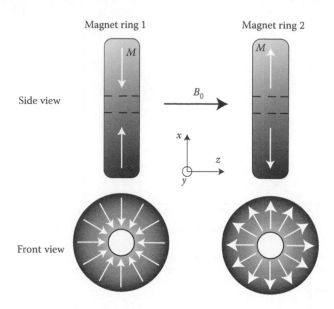

FIGURE 3.30   Aubert's magnet array.

or ion like (gadolinium, manganese, and dysprosium) compounds have unpaired electrons and thus a positive susceptibility [14]. The large magnetic moment of paramagnetic NPs creates large magnetic field heterogeneity and thus shortening the T_1 (MRI positive) or T_2 (MRI negative) relaxation times. Iron oxides are well known and widely used MRI negative CAs, while Gd-based materials are used as MRI positive CAs.

The challenge in their *in vivo* application rises from the following problems: (a) NP agglomeration (NPs tend to agglomerate due to their high surface/volume ratio), (b) short half-life of the particles in blood circulation (if NPs agglomerate or adsorb plasma proteins, they are taken up by the macrophages of the mononuclear phagocyte system and are eliminated from the bloodstream before they can reach the target cells), (c) low efficiency of the intracellular uptake of the NPs, and (d) nonspecific targeting [15,16].

The transport of the CAs is typically done by intravenous administration, which determines the maximum size of the NPs. The reason is that for a successful delivery of NPs, they have to pass through the vascular capillary wall. Gd-based complexes are typically less than 10 nm in diameter which makes it relatively easy to deliver them. Iron oxide-based complexes can be up to 100 nm and their delivery to the targeted tissue is more problematic. Depending on the size, charge, and coating these NPs are metabolized by the reticuloendothelial system (RES), which consist of macrophages and monocytes, and accumulate in the lymph nodes, spleen, and liver. Particles with the size larger than 50 nm are generally taken up by liver cells [17]. If they are not entirely captured by the liver and spleen, they might be used as markers for imaging of inflammatory and degenerative disorders (like plaque or brain ischemia) [18,19]. Smaller particles generally have a longer circulation time and tend to

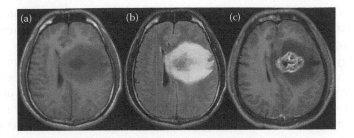

FIGURE 3.31   (a) T_1- and (b) T_2-weighted brain images without contrast and (c) T_1-weighted image after CA administration. (Adapted from X. Z. Peng et al., *Sao Paulo Med J*, 133(5), 2015, 445–449.)

accumulate in the cells of the lymphatic system and bone marrow. The control of the NPs' size, charge, and configuration of the coating gives a possibility to enhance the contrast of the imaging of different parts of the body.

The macrophages in normal tissues and in tumor lesions react differently with the magnetic particles. In healthy tissues, the macrophages uptake the particles and thus darken the image, whereas tumor lesions do not uptake the particles, leaving the lesion tissue bright in the MRI image. This is the reason MNPs are widely used for tumor detections. Figure 3.31 demonstrates the importance of injecting CAs. As shown in Figure 3.31a, from the T_1-weighted image without any contrast, one can notice only a vague area of edema (swelling). In Figure 3.31b in the T_2-weighted image, the edema is much more prominent and shows up bright on the image. However, the best way to see the actual tumor is to look at a T_1-weighted image with a contrast as shown in Figure 3.31 where the cancer tissue is much more obvious.

The following sections will cover the history and physics of MNPs used in clinical practice, their classification and differences, advantages and disadvantages, as well as future perspectives.

3.4.1 HISTORY

MRI CAs have become an indispensable and routine part of modern MRI technology. Currently, about 35% of MRI images are done with the use of CAs, and this percentage is expected to grow further with the development of more effective and specific contrast media. Although MRI provides high-resolution soft tissue contrast by means of noninvasive unenhanced imaging, in many cases CAs offer additional diagnostic information, and improve sensitivity and/or specificity. For example, in many instances of brain metastasis, the properties of the tumorous tissues and the surrounding edema are very similar. The use of CAs helps to differentiate between them, which might be otherwise indistinguishable. Continuous development of MRI hardware and the emergence of nanotechnology in the 1990s have driven the creation of new contrast media designs and led to increased sensitivity and SNR of the MRI images.

From the perspective of the clinical application, MRI CAs can be divided into two groups. Those which shorten the T_1 and T_2 relaxation times are called MRI positive and negative CAs, respectively. In terms of chemical composition and magnetic properties, there are two main classes of contrast media: they are paramagnetic and superparamagnetic agents.

Paramagnetic metals, which include gadolinium, were well known for their relaxation effect in *in vitro* MRI studies for a long time. However, these metals were toxic in their ionic forms which prevented them from being used in humans. In 1981, it was proposed that in order to create a safe agent, the metal ion should be tightly bound by a chelate. The presence of the ligand does not affect the paramagnetic property of gadolinium ion significantly, but the toxicity is limited by achieving rapid and total renal excretion. That is when the forefather of the gadolinium-based contrast agents (GBCAs), Gd^{3+} diethylenetriamine pentaacetate (Gd-DTPA, Magnevist®), was first described. Soon after this, in 1983, the first report of an animal model study using Gd-DTPA as a CA was published. In 1988, Magnevist received U.S. FDA approval for clinical contrast-enhanced imaging of the central nervous system (CNS). Gadolinium chelates are now the major class of CAs used in MRI clinical practice, with the total number of nine FDA-approved GBCAs up to date.

CAs based on Gd-chelates primarily affect T_1 relaxation rates, resulting in positive lesion enhancement. At very high concentrations, they also affect the T_2 relaxation rate, although in most clinical situations, T_2-weighted scans are not appreciable. Administration of the Gd-based contrast agents can significantly improve lesion identification and characterization. Within the CNS, lesion enhancement occurs as a result of the disruption of the blood–brain barrier (BBB). In the case of extra-axial abnormalities and lesions outside the CNS, contrast enhancement is seen in the differences in tissue vascularity.

The research and development of new Gd-based contrast agents have been focused on the improvement of tolerance, the physiochemical properties, and relaxivity. The key safety factors are thermodynamics, solubility, selectivity, and kinetics. The affinity of the chelate for the metal ion must be high, which is associated with the thermodynamic binding constant of the complex (K_{eq}). The CA must be sufficiently soluble in order to prevent potential toxicity of gadolinium ion precipitation. The chelate must have high selectivity for the Gd ion in order to prevent a potential metal exchange with other endogenous ion, such as zinc (Zn) and copper (Cu).

CAs with low osmolality and viscosity are excreted almost completely from the body with a normal renal function, thus allowing a faster administration at higher doses. Other modification steps of Gd-based contrast agents include the development of nonionic (neutral) compounds instead of ionic (charged) ones, and the evolution from linear chelates to macrocyclic chelates.

For a long time CAs that principally affect T_2 relaxation rates have not received much attention. This is explained by the fact that originally T_2-weighted scans required longer imaging time and in most clinical cases, changes on T_2-images had generally little contribution to the diagnosis. However, now with the fast spin echo techniques, T_2-scanning time is no longer the issue. One of the most studied groups of intravenous T_2 CAs is the iron NP group.

Discoveries of new MNPs that shorten T_2 relaxation time push the development of T_2 CAs further. Polymer-coated iron oxide particles have a long history of clinical use. For example, in the 1960s they were used for iron deficiency and anemia treatments [20]. Dextran polysaccharide has been typically used as a polymer coating of the core iron oxide. The reason for this is its well-known antithrombotic, volume expanding properties, and its affinity for iron oxides. Later a new NP termed "dextran magnetite" was developed that exhibited a much stronger magnetism than the paramagnetic particles used for anemia. This unusually strong magnetic behavior was called superparamagnetism. It was soon found that superparamagnetic iron oxides (SPIONs) could shorten the water relaxation time and thus could be used as CAs for MRI. One of the first controlled and reliable demonstrations of the T_2 relaxation time shortening due to the presence of SPIONs particles was performed by Oghushi in 1978 with dextran-magnetite particles [21]. After this discovery and experiment, the field of magnetic particles for T_2-weighted MRI images has continuously been developing, resulting in a range of CAs now being approved and widely used in clinics.

The first generation of MRI negative CAs was polydisperse SPIO (with one, two, or more crystals per NP and a broad size range in solution). After an intravenous injection of SPIO NPs, they can be easily detected by macrophages of the RES of the body and transported to the liver and/or the spleen, because these organs are responsible for blood purification [22–24]. These materials darken normal, but not metastatic liver tissue at T_2-weighted images, and therefore serve as contrast enhancing agents to detect cancerous lesions.

Quite soon it was discovered that the decrease of NP size can extend the circulation time in blood. The reason is that smaller NPs cannot be detected by the RES and in this case macrophages of lymph nodes uptake the NPs. Therefore, smaller MNPs act as a CA for hepatic metastases lymph nodes imaging. The size reduction was achieved by using monodisperse NPs (with only one crystal per NP in solution) [25,26].

The next generations of MNPs for MRI have been modified considerably and achieved significant number of improvements since the first polymer-coated iron oxides. Now they are more sophisticated and can be molecularly targeted to a specific biomolecule via attached antibody, peptide, or polysaccharide; serve as a label for cell tracking; or can be combined with fluorescent components and thus give additional optical information. The aforementioned CAs change the T_1 or T_2 relaxation only, therefore they are also called single mode CAs. However, in modern diagnostics, single mode CAs are not always sufficient and there is a new quickly growing field of dual mode T_1–T_2 CAs [27,28]. MNPs have gone through years of technological improvement, and preclinical and clinical testing. It has given us a diverse range of applications by far. In the future, even more developments can be foreseen in various areas in biology and medicine.

3.4.2   Physics of MRI CAs

MRI is a noninvasive technique used in radiology, which provides information on local biology, anatomy, and physiology with high spatial resolution by detection of the signals coming from proton relaxation in an external magnetic field. With the help of MRI 3D images of different types of tissues can be seen. In principle, it is possible to track the relaxation of different elements, and the reason proton relaxation is typically studied is because of the abundance of protons in the human body and because different tissues have high contrasts in terms of proton density. There are two independent relaxation processes going on during the proton recovery to its original state, namely T_1 and T_2 relaxations. T_1 is the time of magnetic moment $\boldsymbol{m}$ recovery in the direction of the $\boldsymbol{B}_0$ field, and T_2 is the time of the loss of the signal in the transverse plane. These T_1 and T_2 times strongly depend on the tissue type and its physical properties, resulting in high tissue contrast in MRI images. Therefore, MRI scanners are particularly useful at providing highly detailed information about soft tissues. T_1 and T_2 independently provide different information of images and each has its own advantages and disadvantages, and thus are used to better visualize different types of substances. Certain substances have their own magnetic moment and generate local magnetic field ($\boldsymbol{B}_1$) that changes the speed of proton relaxation (T_1 or T_2) significantly and therefore leads to the brightening or darkening of the image. A CA that predominantly affects T_1 relaxation time by reducing it and thus increases signal intensity on a T_1-weighted image is called an *MRI positive CA*. On the other hand, a CA that predominantly affects T_2 relaxation time, reducing it and thus decreasing signal intensity on a T_2-weighted image is called an *MRI negative CA*.

Although a profound review of the relaxivity theory is beyond the scope of this chapter, a basic conceptual understanding will help to appreciate the physics involved in CA enhancement phenomena.

A magnetic moment created by unpaired electrons can interact with surrounding water protons either directly or indirectly via its local magnetic field influence, and thus enhance the T_1 or T_2 relaxation times. Therefore, only magnetic ions with exceptionally slowly relaxing unpaired electrons are effective as MRI CAs, because they give the most profound enhancement. The quantum theory says that the ions with the highest spin quantum number have the most slowly relaxing electrons. Examples of such ions are gadolinium (Gd^{3+}), iron (Fe^{3+}), dysprosium (Dy^{3+}), and manganese (Mn^{3+}), as shown in Figure 3.32. However, this theoretically desirable

	Ion	Configuration		Magnetic moment
		3d	4f	
Transition metal ion	$^{24}Cr^{3+}$	↑ ↑ ↑ — —		3.88
	$^{25}Mn^{2+}$	↑ ↑ ↑ ↑ ↑		5.92
	$^{26}Fe^{3+}$	↑ ↑ ↑ ↑ ↑		5.92
	$^{29}Cu^{2+}$	↑↓ ↑↓ ↑↓ ↑↓ ↑		1.73
Lanthanide metal ion	$^{63}Eu^{3+}$		↑↓ ↑ ↑ ↑ ↑ ↑ ↑	3.4
	$^{64}Gd^{3+}$		↑ ↑ ↑ ↑ ↑ ↑ ↑	7.94
	$^{66}Dy^{3+}$		↑↓ ↑↓ ↑ ↑ ↑ ↑ ↑	10.65

FIGURE 3.32 Configuration and magnetic moment of some of the paramagnetic ions. (Adapted from H. B. Na, I. C. Song, and T. Hyeon, *Adv Mater*, 21, 2009, 2133–2148.)

high spin quantum number is not the only factor determining the efficacy of an MRI CA.

The interaction mechanism between water molecules and a CA is complex and can be divided into two parts, inner-sphere relaxation and outer-sphere relaxation. During the inner-sphere relaxation the formation or dissociation of a coordinate covalent bond between a water molecule and the CA occurs, leading to a chemical exchange and catalyzation of the water protons relaxation. The ability of a CA to bind a large number of water molecules and to perform a rapid exchange is a highly desirable feature because it allows a greater relaxation enhancement. Outer-sphere relaxation, in contrast, does not involve any direct bonding or chemical exchange mechanism, but it is associated with the relative rotational and translational diffusion of water molecules and CAs. The efficiency of the enhancement in this case depends on the mobility of the CA and the ease of approaching interaction with water molecules protons.

All these effects must be taken into consideration while designing CAs, especially those with functional coatings and complexes. For example, if there is a magnetic particle–ligand complex, the system will rotate and translate slower in space, decreasing the number of interactions. Moreover the distance between the magnetic core and the water proton will be increased, reducing the relaxation enhancement effect of the magnetic particle.

There is a mathematical description of the relaxation of water protons in the presence of a magnetic ion, called the Solomon–Bloemberger–Morgan equation:

$$1/t_c = 1/t_r + 1/t_s + 1/t_m \qquad (3.24)$$

where t_c is the total correlation time, t_r is the correlation time of rotation, t_s is the correlation time of electron relaxation, and t_m is the correlation time of chemical exchange. Equation 3.24 describes the probability of contact (correlation time) between the CA and the water molecule proton. It should be noted that the component with the smallest magnitude will affect the total correlation time of interaction the most. A schematic view of the correlation parameters is shown in Figure 3.33. The relaxivities of

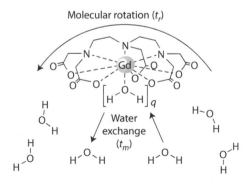

FIGURE 3.33 Schematic of selected key factors that affect proton correlation rate. (Adapted from E. Werner et al., *Angew Chem Int Ed*, 47(45), 2008, 8568–8580.)

current commercial CAs are small compared to what is theoretically possible. The optimization of the correlation times can lead to a dramatic increase in relaxivity, and potentially can result in the values of relaxivities which are about an order of magnitude larger than those of the commercially available ones [30].

3.4.3 Design Criteria

The human body normally contains magnetic substances, for example, degradation products of hemoglobin, or molecular oxygen. However, there should be a significant concentration of such substances in the area of interest in order to get a profound disturbance of the local magnetic field and thus better contrast. That is the reason why the first and foremost criterion of an MRI CA is its ability to influence the parameters responsible for image contrast at low concentration. Second, a contrast media should possess some tissue specificity *in vivo* so that the CA is delivered to an area of a tissue or an organ in a higher concentration than to other locations in the body. Third, the CA must be substantially cleared from the targeted organ or tissue in a reasonable period of time (typically several hours after the imaging) in order to minimize potential toxicity, and eventually excreted from the body via renal or hepatobiliary routes. Fourth, MRI CAs must meet toxicity and tolerance criteria, and pass many other tests for chemical stability *in vivo*, potential mutagenicity, teratogenicity, and carcinogenicity. Finally, a commercial need for *in vitro* stability of the CA has to be satisfied. It must have a shelf life of at least several years.

All the criteria mentioned above stimulate the development of different types of MNPs used as CAs.

3.4.4 Classification of CAs

Although GBCAs are the most common class of MRI CAs to date, many other types of agents are appearing on the market. MRI CAs can be classified by their

- Magnetic properties
- Effect on the image intensity
- Chemical composition
- Administration route
- Applications

In terms of their magnetic properties, CAs are typically divided into two groups: paramagnetic and superparamagnetic. Most of the CAs now in clinical use are based on paramagnetic metal ions, such as gadolinium (Gd^{3+}). Materials used in superparamagnetic CAs are iron-based, such as magnetite (Fe_3O_4) or maghemite (γ-Fe_2O_3). The most common way of describing existing CAs is to classify them into two categories, namely MRI positive CAs and MRI negative CAs. As mentioned previously, the difference between them is that positive CAs reduce T_1 relaxation times (increasing signal intensity and thus appearing bright on the T_1-weighted images), while negative CAs predominantly affect T_2 relaxation times (decreasing T_2 time and appearing dark on MRI images). GBCAs are the most studied and common examples of MRI positive CAs. They are now routinely used in clinics. It should be noted as well that Gd is paramagnetic. SPIO NPs are, on the other hand, an example of MRI negative CAs.

The choice of type of CA, positive or negative, depends on the specific organ or disease suspected, as well as the pulse sequence used. Both types of CAs have certain advantages and disadvantages. For example, for gastrointestinal (GI) MRI positive CAs, ghosting artifacts due to respiratory or peristaltic motion is a problem. One of the solutions is to use breath holding pulse sequences and first order flow compensations. Another solution is to use a pharmaceutical which reduces bowel motion. On the other hand, GI MRI negative CAs do not have this ghosting problem due to the lack of signal in the bowel. However, metallic artifacts are seen when gradient echo sequences are used. Moreover their cost is generally higher and there are limited evaluations of safety on a large number of patients.

Properties, functions, and principles of work of these two types of CAs used now in clinical practice will be covered in the following sections. Some examples of novel CAs will be discussed as well.

3.4.5   MRI POSITIVE CAs

Paramagnetic materials are used as positive CAs due to their ability to develop a magnetic moment in the presence of a magnetic field (inside the bore of the MRI scanner). This large induced magnetic moment enhances the relaxation of the water molecule protons in the vicinity of the agent and creates bright contrast on the T_1-weighted images. For a paramagnetic material to be an effective MRI positive CA, the electron spin-relaxation time must match the Larmor frequency of the protons, and this condition is better met for Gd^{3+}, Mn^{2+}, and Dy. The main problem with paramagnetic metal ions is their toxicity in their native form. Chelating ligands, such as diethylenetriamine pentaacetic acid, DTPA, are bound to the paramagnetic ion in order to prevent the lanthanide from binding to chelates in the body.

3.4.5.1   Gadolinium-Based CAs

The use of GBCAs to enhance the T_1-weighted images has been part of standard clinical practice for over two decades. There are now nine FDA-approved GBCAs, which can be classified into two groups on the basis of their chemical structure: linear and macrocyclic (Figure 3.34 and Table 3.4). In Table 3.4, several evolutional directions in the design of the CAs can be seen as well, from linear to macrocyclic, and from ionic to nonionic.

The most important consideration about GBCAs is their stability. Free Gd^{3+} is toxic and thus the ability of the ligand to tightly bind to the Gd ion is an important safety consideration, especially after 2006 when an association between the development of nephrogenic systemic fibrosis (NSF) and the administration of Gd was noticed.

3.4.5.1.1   Stability of the GBCAs

The thermodynamic stability constant (K_{therm}) is a measure of stability: the higher the K_{therm} constant the more stable the Gd complex. However, the thermodynamic

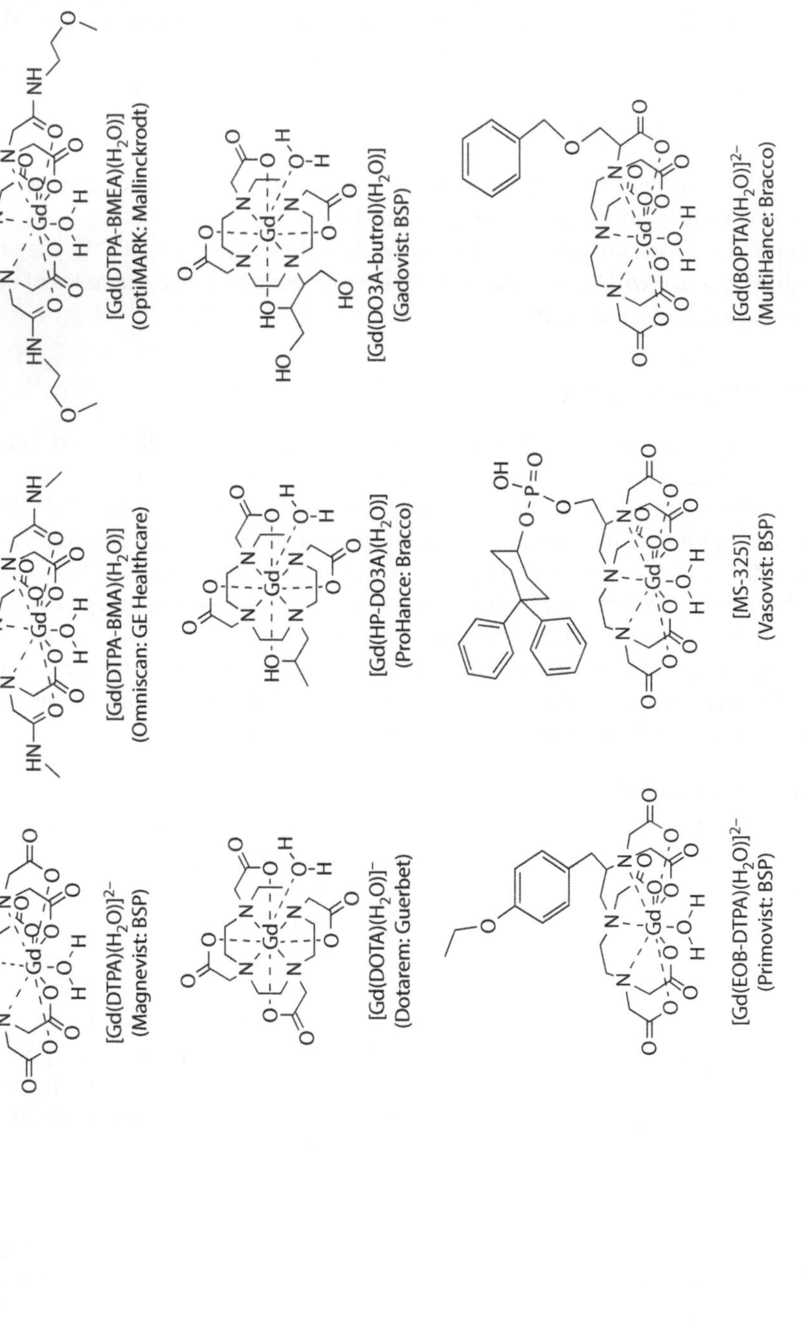

FIGURE 3.34 Commercial Gd-based MRI CAs. (Adapted from E. Werner et al., *Angew Chem Int Ed*, 47(45), 2008, 8568–8580.)

TABLE 3.4
Currently Available GBCAs

Trade Name	Chemical Name	FDA Approval	Structure	Ionicity	Clinical Use	Intra-Venous Injection (mg/mL)
Dotarem	Gadoterate	2013	Macrocyclic	Ionic	Multi-purpose	376.9
Gadavist	Gadobutrol	2011	Macrocyclic	Nonionic	CNS, breast	604.72
Ablavar	Gadofosveset	2008	Linear	Ionic	Blood pool	244
Eovist	Gadoxetate	2008	Linear	Ionic	Liver	181.43
MultiHance	Gadobenate	2004	Linear	Ionic	Multi-purpose	529
OptiMARK	Gadoversetamide	1999	Linear	Nonionic	CNS, liver	330.9
Omniscan	Gadodiamide	1993	Linear	Nonionic	Multi-purpose	287
ProHance	Gadoteridol	1992	Macrocyclic	Nonionic	Multi-purpose	279.3
Magnevist	Gadopentetate	1988	Linear	Ionic	Multi-purpose	469.01

Source: Adapted from M. F. Tweedle, E. Kanal, and R. Muller, *Suppl Appl Radiol*, 43(5), 2014, 1–11.

stability constant does not take the pH of the environment into account. Therefore, the conditional stability of a complex constant (K_{cond}), which is a measure of the stability of a complex at a physiologic pH, is considered a more relevant stability parameter [31]. Table 3.5 presents thermodynamic constants of the FDA-approved GBCAs.

There are a number of factors that affect the stability of the Gd complex *in vivo*. A number of endogenous metals, such as zinc, copper, calcium, and iron normally presented *in vivo* environment can act as destabilizers of the GBCA complex, leading to its dissociation into Gd ion and a ligand. This displacement of the Gd ion from its ligand by other metals via competitive ionic binding is called transmetallation. An *in vitro* comparison analysis of linear and macrocyclic GBCAs demonstrates that in the presence of competitor metals (Zn and Cu) macrocyclic agents (Dotarem and ProHance) remain essentially intact (<1% reaction), while the linear Gd-based complexes are more reactive [32]. A similar analysis *in vivo* also demonstrates a better performance of the macrocyclic GBCAs: up to 14 days after injection of GBCAs,

TABLE 3.5
Thermodynamic Constants of GBCAs

Trade Name	Dotarem	Gada- vist	Ablavar	Eovist	Multi- Hance	Opti- MARK	Omni- scan	Pro- Hance	Magne- vist
Structure	Macro- cyclic	Macro- cyclic	Linear	Linear	Linear	Linear	Linear	Macro- cyclic	Linear
$Log(K_{therm})$	25.6	21.8	22.1	23.5	22.6	16.6	16.9	23.8	22.1
$Log(K_{cond})$	19.3	14.7	18.9	18.7	18.4	15	14.9	17.1	17.7

Source: Adapted from M. F. Tweedle, E. Kanal, and R. Muller, *Suppl Appl Radiol*, 43(5), 2014, 1–11.

macrocyclic agents have the lowest level of residual Gd [33]. *Ex vivo* human data show that dissociation of the nonionic linear GBCAs is much more pronounced. Up to 30% of the nonionic linear agents was released in human plasma 2 weeks after administration, while for the ionic linear agents this number was reduced to about 2%, and for the macrocyclic GBCAs the percent of dissociation processes is 0% (Figure 3.35, adapted from Reference 31). The most credible and important data of *in vivo* human studies support the high stability of macrocyclic GBCAs. The

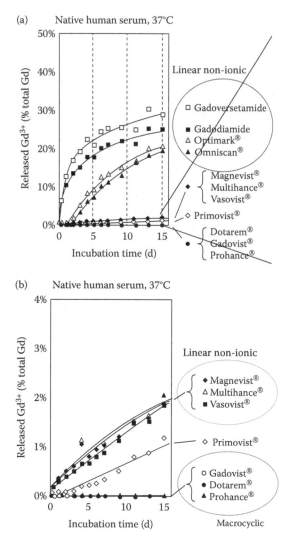

FIGURE 3.35 Amounts of Gadolinium ions released from 1 mM solutions of (a) all commercial GBCAs at 37°C in native human serum from healthy volunteers; (b) an enlarged section of the graph visualizing the data for the ionic linear and macrocyclic GBCAs. (Adapted from M. F. Tweedle, E. Kanal, and R. Muller, *Suppl Appl Radiol*, 43(5), 2014, 1–11.)

TABLE 3.6

In Vivo **Test of Elimination Time-Course of Gadolinium in Skin Tissue of Rats**

	Concentration of Gadolinium (nmolGd/g) in Skin	
Trade Name of GBCA	Day 35 Post-Injection	Day 364 Post-Injection
Omniscan	132 ± 23	72 ± 12
OptiMARK	47 ± 5	18 ± 5
Magnevist	36 ± 6	9 ± 2
MultiHance	7 ± 1	1.4 ± 0.4
Dotarem	2 ± 1	0.22 ± 0.17
Gadovist	2 ± 1	0.06 ± 0.03
ProHance	2 ± 1	0.08 ± 0.02

Source: Adapted from M. F. Tweedle, E. Kanal, and R. Muller, *Suppl Appl Radiol*, 43(5), 2014, 1–11.

amount of Gd^{3+} deposited in the bone of hip replacement patients after administration of Omniscan is four times more than that after ProHance administration [34]. However, recent studies on a sensitive *in vivo* animal model show that even at day 364 after the administration, a very small amount of Gd was still present in the skin of rats administered macrocyclic GBCAs, indicating that there is still a "low-risk" rather than "no-risk" of using GBCAs, especially for patients with a high risk of NSF (Table 3.6, [31]).

3.4.5.1.2   Relaxivity of the GBCAs

Relaxivity rates, another important property of the CAs, can be modified depending on the contrast medium ability to interact with proteins. It was found that interaction with proteins, most notably human serum albumin (HSA), noticeably increases the effective size of the Gd-ligand complex, and meanwhile reduces the molecular motion, thus shortening the T_1 relaxation time and greatly increasing signal intensity enhancement [35]. Figure 3.36 shows the relaxivity profiles of solutions containing 1 mM of either Gadavist or MultiHance in the absence and presence of 4% HAS [36]. For Gadavist, as well as for other nonprotein binding GBCAs, the presence of HSA has no effect on the relaxivity profile, whereas for MultiHance the presence of HSA results in a notable increase in relaxivity.

3.4.5.1.3   GBCAs Safety and NSF

Until 2006 GBCAs were considered as one of the safest CAs used in humans [37]. Over 200 million patients have been exposed to gadolinium since the late 1980s. Worldwide post-marketing surveillance studies have all demonstrated that nearly all drug reactions can be characterized as very mild (<2.5%) or moderately severe (<0.02%) [38]. However in 2006 Grobner et al. reported about the possible association between GBCAs and a new and rare disease, NSF [39]. NSF was first described in the medical literature in 2000. NSF causes fibrosis of the skin, connective tissues

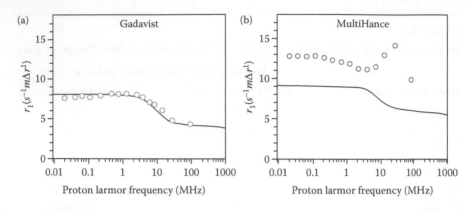

FIGURE 3.36 Relaxivity profiles of solutions containing 1 mM of (a) Gadavist (gadobutol) and (b) MultiHance (gadobenate dimeglumine) in the absence (lines) and presence (dots) of 4% HAS. (Adapted from S. Laurent, L. Elst, and R. Muller, *Contrast Media Mol Imaging*, 1(3), 2006, 128–137.)

like muscles, tendons, ligaments, and blood vessels throughout the body, leading to a thickening of the skin and severe decreasing of joints mobility. However, NSF has been reported only in patients with pre-existing chronic kidney disease and end-stage kidney disease. Due to the weak kidney function, the human body cannot clear itself of GBCAs and the extended presence of gadolinium may lead to irreversible health problems, possible confinement to a wheelchair, and even death. Table 3.7 shows an overview of worldwide unconfounded NSF cases for various GBCAs. A retrospective study with Omnisan in about 370 patients with severe renal insufficiency estimated the risk of NSF to be 4% [40]. It is still unknown what causes NSF, there is no cure for it so far, and skin biopsy is the only true means of diagnosis.

TABLE 3.7
NSF Case and Relative Frequency

Trade Name of GBCA	NSF Cases Global	Contrast Media Examinations Global (in Millions)	NSF Relative Frequency (Cases/1 Million Applications)
Omniscan	438	>47	9.3
Magnevist	135	>115	1.2
OptiMARK	7	>9	0.8
Gadavist	1	>6	0.7
ProHance	1	>14	<0.1
Dotarem	1	>21	<0.1
MultiHance	0	>11	<0.1

Source: Adapted from J. T. Heverhagen, G. A. Krombach, and E. Gizewski, *Rofo*, 186(7), 2014, 661–669.

3.4.5.2 Mn-Based CAs

Although the most of the research involving MRI positive CAs has been carried out by Gd^{3+}-based CAs, Mn^{2+} agents have recently received considerable attention. Manganese, being one of the first reported examples of paramagnetic CAs used in cardiac and hepatic MRI, is especially useful for the detection of anatomical structures and for the mapping of functional brain regions. However, there are several drawbacks which prevent it from being widely used and developed as a contrast media, for example, the toxicity of the Mn ion and the difficulty to find ligands capable of binding to the Mn ion [42].

3.4.5.3 Dy-Based CAs

According to the relaxivity theory of Dy^{3+}, Dy^{3+}-based complexes are the most efficient agents to be encapsulated, due to their high magnetic moment. Incorporation of the amphiphilic Dy^{3+} complexes in the liposome bilayer yields a marked sensitivity enhancement, allowing for the MRI visualization of cellular epitopes, such as membrane transporters present at very low concentrations [43].

3.4.6 MRI NEGATIVE CAs

In terms of MRI negative CAs, the most studied and commonly used magnetic materials are magnetite (Fe_3O_4) and its oxidized form maghemite (γ-Fe_2O_3).

Magnetite is a black ferrimagnetic mineral which has both Fe^{2+} and Fe^{3+} and a saturation magnetization of 4.76×10^5 A/m. Maghemite is a red brown ferrimagnetic mineral isostructural with magnetite, with all or most of the iron being trivalent. It is formed when the magnetite is oxidized and has the saturation magnetization of 4.26×10^5 A/m [44]. Ferrites have an inverse spinel structure, where oxygen atoms form face centered cubic lattices and iron ions occupy tetrahedral (T_d) and octahedral (O_h) interstitial sites, as shown in Figure 3.37. Iron oxides have been intensively

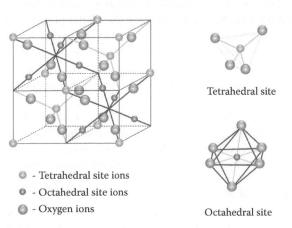

     Tetrahedral site

- Tetrahedral site ions
- Octahedral site ions
- Oxygen ions

     Octahedral site

FIGURE 3.37 Ferrite's crystal structure.

investigated by both chemists and materials scientists for 5 decades, resulting in a number of different synthesis methods [45].

Interestingly, magnetite is found in many bio-systems, from bacteria to human bodies. For example, in 1962 Lowenstam found magnetite in the radula teeth of chitons (marine molluscs) and proved its biological origin [46]. Before his discovery the formation of magnetite was thought to occur only under high temperature and pressure in volcanic or metamorphic rocks. For chitons the magnetite serves to harden the tooth caps, enabling the chitons to extract and eat endolithic algae from within the outer few millimeters of rock substrates. In 1975 Blackmore discovered a bacterium which is sensitive to the Earth's magnetic field and called it magnetotactic bacterium [47]. Its organelles called magnetosomes contain magnetite crystalline particles of size 50–100 nm. These magnetosomes act just like a compass needle and force the bacteria to migrate along oxygen gradients in aquatic environments, under the influence of the Earth's magnetic field. So it means that the bacteria are passively torqued even if they are dead. In both of the above examples the magnetite particles are formed by the organisms by a process called biomineralization. A good review of the formation of magnetite by living organisms is done by Arakaki et al. [48]. Remarkably, the structures formed by the biomineralization process often exhibit excellent physical and/or chemical properties which outperform artificial material, and moreover the conditions required for this are incredibly mild in comparison with common synthetic methods. That is why, in order to understand the key biological and chemical principles of biomineralization, molecular studies including genome sequence, mutagenesis, gene expression, and proteome analysis have been recently performed giving us a simple way to prepare functional protein–magnetic particle complexes and clues for the development of advanced nanomaterials. According to Thomas-Keprta et al. [49] there are six criteria that are unique for biologically produced magnetic crystals, such as a definite size range and width/length ratio, chemical purity, crystallographic perfection, arrangement of crystals in linear chains, unusual crystal morphology, and elongation of crystals in the |111| crystallographic direction. The simultaneous presence of all of them should constitute the evidence of the biological origin of the material. Surprisingly, recent observation of Martian meteorite ALH84001 by NASA researchers shows the presence of magnetite NPs which fulfill all six criteria. These findings lead to the hypothesis that they could be in fact microfossils of former Martian magnetotactic bacteria and that life on Earth could have been brought by meteorites from Mars where conditions could have been more favorable for the creation of life from non-living ingredients in the early history of the solar system [50].

Therefore, the biocompatibility and availability of different synthesis methods of iron oxide nanoparticles make them popular in biomedical studies and they are now considered the gold standard for MRI contrast imaging. They are commercially available and approved by the U.S. FDA for clinical applications.

3.4.6.1   The Effect of Size

It is necessary to keep in mind that some of the magnetic responses are structure-sensitive and some are relatively structure-insensitive. Susceptibility and coercivity are examples of the first category, while saturation magnetization is an example of

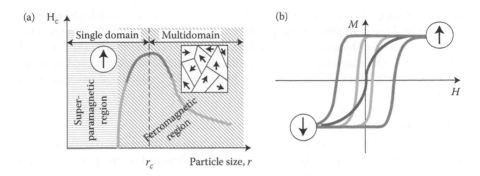

FIGURE 3.38 (a) Variation of coercivity (H_c) of MNPs with size, and (b) theoretical magnetization versus magnetic field curve for SPM and ferro- or ferrimagnetic NPs. (Adapted from A. Figuerola et al., *Pharmacol Res*, 62(2), 2010, 126–143.)

the second one [51]. The reduction of size in magnetic materials leads to profound changes in their intrinsic properties. Nano-sized magnetic materials are governed by laws which are different from those for the material in its bulk form, resulting in phenomena which were never detected before for their macro-size counterparts. For example, in large ferromagnetic particles (with size more than 1 µm) there are still many magnetic domains and the coercivity is relatively small forming a narrow hysteresis loop. However, for smaller particles (with size <1 µm) it becomes more energetically efficient to have a single domain and coercivity of ferromagnetic particles experience a dramatic increase with the reduction of the particle size, leading to the hysteresis loop broadening. If the particle size is reduced further to a certain critical value (to about 20 nm) or the so-called superparamagnetic radius, the particle becomes *superparamagnetic* (SPM) with a zero remanence, forming the *M–H* curve without hysteresis, as shown in Figure 3.38. This superparamagnetic feature of nanoparticles is highly advantageous for biomedical applications because it means that there is no remanent magnetic moment after the removal of the external magnetic field. In other words there will be no aggregation of the magnetic particles in the blood vessels after a diagnostic measurement or a therapy [52].

3.4.6.2 Surface Effects

There are certain disadvantages associated with NP size reduction. With the decrease of the particle size, the surface to volume ratio increases introducing noticeable surface effects such as spin canting, spin glass behavior, and non-collinear spins. These effects result in crystal structure disorder, thus changing the magnetic properties of the particles. Subsequently, this results in an unfavorable saturation magnetization reduction [53]. The association between magnetization and size is quantified by the *magnetic anisotropy constant (Ku)*. It measures the energy to be overcome in order to preserve the direction of the magnetic dipoles of the material. This constant is determined by crystal lattice symmetry, the surface coordination with the core of the NP, and the shape of it. Thus, it is different for different materials [52,54]. The lower the constant, the more size dependent the magnetization is, and therefore, the faster the decrease of magnetization with size will be. Moreover, materials with high magnetic anisotropies have a significant magnetization even at very small particle sizes. This can be used in many biomedical applications [52]. It has been observed [55] that metal alloys have higher magnetization values than their oxide counterparts and thus can act as potentially more efficient CAs with higher image contrast and lower doses of MNPs.

3.4.6.3 Pharmacokinetics

Iron oxide-based NPs are highly biocompatible, they are metabolized in the hepato-renal system, added into body's iron reserves, and finally become part of the red blood cells as hemoglobin [56].

Different surface coatings are used for iron oxide NPs in order to provide chemical stability during and after synthesis of the MNPs, prevent their agglomeration inside the body, and change NP recognition by human immune cells [57]. Typical surface coatings are dextran and its derivatives, albumin, silicon, PEG (poly ethylene glycol), PEI (polyethyleneimine), chitosan, co-polymer, liposomes, and starch. The resulting organic/inorganic complexes have a core–shell nanoarchitecture, which can be functionalized by adding various compounds. The chemistry of the magnetic core and surface of a core–shell SPIO NP dictates its function and working environment [58]. The design of the chemical nature and crystal lattice of the magnetic core enables the control of the magnetic response of the NPs by modifying the core size and chemical composition. On the other hand, the functionalization of the surface allows us to modulate the behavior of particles in solution. With the help of surface coatings homogeneous suspensions can be formed which remain stable in blood and aqueous environments [59]. Moreover, a study [60] shows that polymer (cyclodextran and F127) coatings can reduce NP size and attenuate their cluster behavior consistently from more than 300 nm down to 90 nm. Studies [59,61] show that polymer (PEG) coating of NPs can reduce the recognition and uptake of the NPs by the liver and spleen, and thus increase their circulation times.

3.4.6.4 Superparamagnetic Iron Oxide CAs

It is generally considered that when iron oxide NPs being in an aqueous environment have an overall hydrodynamic size larger than 40 nm they are called SPIO. If their hydrodynamic size is smaller than 40 nm they are called ultra-small superparamagnetic iron oxide (USPIO). Most of the magnetic NPs available on sale are SPIO NPs with the size ranging from 60 nm up to several micrometers.

The first clinical use of SPIO NPs as a contrast media was done for imaging liver tumors. After an intravenous injection of SPIO NPs they can be easily detected by the macrophages of the RES of the body and are therefore accumulated in the liver and in the spleen because these organs are responsible for blood purification. Healthy liver cells can uptake the particles, whereas diseased cells cannot, as schematically shown in Figure 3.39. In the presence of NPs the relaxation time T_2 is reduced, thus on T_2-weighted images only the change in the brightness of the normal cells will be seen. The use of SPIO as contrast media increases the characterization accuracy of lesions in hepatic cellular carcinoma (HCC) and focal nodular hyperplasia (FNH) patients [62].

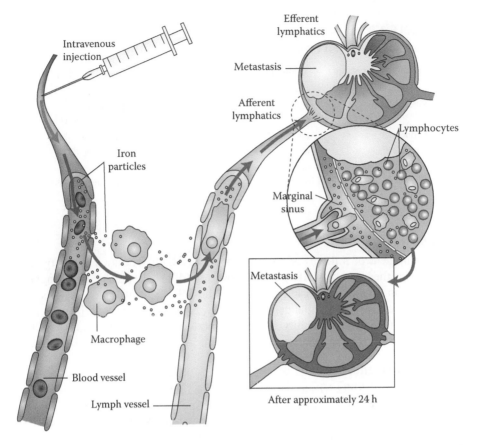

FIGURE 3.39 Schematic illustration of SPIO nanoparticles injection, transport, and agglomeration in body. (Adapted from S. K. Mouli et al., *Nat Rev Urol*, 7.2, 84–93, 2010.)

On the other hand, USPIO NPs do not accumulate in the RES because they are too small for the RES to detect and therefore have a longer circulation time in the blood. They are able to pass across capillary walls, drain via the lymphatic circulation, and localize in lymph nodes independent of size or morphological features of the node. This tendency of USPIO NPs to passively target tumors, due to the enhanced permeation retention (EPR) effect of solid tumors, allows us to identify nodal metastases of less than 2 mm in diameter, which is under the threshold of detection of other imaging modalities [63,64].

For other kinds of tumors functionalized NPs with tumor specific antibodies are utilized. These functionalized complexes are actively targeted to the tumor and darken the tumor cells on T_2-weighted images [65].

3.4.6.5 Clinical Application of Iron Oxides

Although a number of SPIONs and USPIOs have been approved for clinical use in the past, as shown in Table 3.8, only one oral iron oxide CA Lumirem/Gastromark is available to date.

TABLE 3.8
Iron Oxide-Based CAs

Product Trade Name	Clinical Use	Approval	Manufacturing
Lumirem/Gastromark	Gastrointestinal	FDA (1996)	For sale
Sinerem/Combidex	Liver, Lymph nodes	NA	Application withdrawn (2007)
Feridex	Liver	FDA (1996)	Discontinued (2008)
Resovist/Cliavist	Liver	European Market (2001)	Discontinued (2009)
Clariscan	Liver	NA	Discontinued (2009)

3.4.6.6 Toxicity

The requirement of millions of particles with very high saturation magnetization values limits the usage of MNPs as CAs. However, SPIO have much higher magnetic moments and thus require a lower dose for the MRI administration of CAs. For this reason, it reduces potential cellular toxicity. No serious adverse effect has been observed to date. The possibility of mild-to-moderate adverse effects is 3%–28%, with back pain being the most common one [59].

3.4.6.7 Theragnostics

The capability of SPIO NPs to incorporate a broad range of diagnostic and chemo-therapeutic agents allows one to combine diagnostic and therapeutic approaches for complex simultaneous cancer detection and therapy [45]. A study [65] demonstrates multifunctional polyspartic acid nanoparticles (MPAN), containing the SPIO core and chemotherapeutic drug Doxorubicin. They act both as a T_2 MRI CA and an anti-tumor drug delivery system. This combination treatment has a great potential to advance and personalize medicine.

3.4.7 Multimodal ($T_1 + T_2$) Imaging

Today single-mode CAs are not always sufficient, and dual-mode CAs have recently been receiving a great amount of attention. These new contrast media combine the advantages of positive and negative MRI CAs, thus sharpening anatomical details and allowing improved accuracy for diagnosis.

Although the design and preparation of such complexes is a highly challenging task, there have been several techniques reported. For example, one study [55] shows a core–shell-type agent with FeCo core and single graphite shell. It has a relaxation rate r_2 six times higher than that of the commercially available CA Ferridex. In the study [66] silica layer of different thickness was used to separate T_1 ($Gd_2O(CO_3)_2$) and T_2 ($MnFe_2)_4$) contrast modes. A recent report [67] demonstrates three different approaches for dual-mode CAs' preparation based on SPIO NPs, the polymer coating of which is combined with Gd ions. In general, these complex dual-mode contrast media exhibit outstanding relaxivity performance.

REFERENCES

1. F. Bloch, Nuclear induction, *Phys Rev*, 70(7–8), 1946, 460–473.
2. E. M. Purcell, H. C. Torrey, and R. V. Pound, Resonance absorption by nuclear magnetic moments in a solid, *Phys Rev*, 69(1–2), 1946, 37–38.
3. P. C. Lauterbur, Image formation by induced local interactions: Examples employing nuclear magnetic resonance, *Nature*, 242, 1973, 190–191.
4. D. M. Cole, E. Esparza, F. R. Huson, S. M. Wright, and A. Elekes, Design improvements for permanent magnet-based MRI, *Proc Ann Mtg Soc Magn Reson*, Suppl S3, 1994, 1079–1128.
5. E. Esparza-Coss and D. Cole, A low cost MR/permanent magnet prototype, *Second Mexican Symp on Med Physics. American Institute of Physics Conf. Proc.*, Coyoacan, Mexico, February 26–28, 1998.
6. S. M. Wright, D. G. Brown, J. R. Porter, D. C. Spence, E. Esparza, D. C. Cole and F. R. Huson, A desktop magnetic resonance imaging system, *Magn Reson Mater Phys Biol Med*, 13(3), 2002, 177–185.
7. Martinos Center for Biomedical Imaging, Massachusetts General Hospital. Available: http://iee.ac.cn/Website/index.php?ChannelID=2195.
8. Tabletop MRI, [Online]. Available: https://gate.nmr.mgh.harvard.edu/wiki/Tabletop_MRI/index.php/Main_Page.
9. H. D. Phuc, P. Poulichet, T. T. Cong, A. Fakri, C. Delabie, and L. Fakri-Bouchet, Design and construction of light weight portable NMR halbach magnet, *Int J Smart Sens Intell Syst*, 7(4), 2014, 1555–1579.
10. C. Z. Cooley, J. P. Stockmann, B. D. Armstrong, M. Sarracanic, M. H. Lev, M. S. Rosen, and L. L. Wald, Two-dimensional imaging in a lightweight portable MRI scanner without gradient coils, *Mag Resonan Med*, 73(2), 2015, 872–883.
11. Z. H. Ren, W. Luo, J. Su, and S. Y. Huang, Magnet array for a portable magnetic resonance imaging system, in *RF and Wireless Technologies for Biomedical and Healthcare Applications (IMWS-BIO)*, 2015 IEEE MTT-S 2015 International Microwave Workshop Series on, pp. 92–95, Taiwan, 2015.
12. C. Hugon, P. M. Aguiar, G. Aubert, and D. Sakellariou, Design, fabrication and evaluation of a low-cost homogeneous portable permanent magnet for NMR and MRI, *Comptes Rendus Chimie*, 13(4), 2010, 388–393.
13. M. Di Marco et al., Physicochemical characterization of ultrasmall superparamagnetic iron oxide particles (USPIO) for biomedical application as MRI contrast agents, *Int J Nanomedicine*, 2(4), 2007, 609–622.
14. Y. Okuhata, Delivery of diagnostic agents for magnetic resonance imaging, *Adv Drug Deliv Rev*, 37(1–3), 1999, 121–137.
15. C. Alexiou et al., Targeting cancer cells: Magnetic nanoparticles as drug carriers, *Eur Biophys J*, 35(5), 2006, 446–50.
16. B. Stella et al., Design of folic acid-conjugated nanoparticles for drug targeting, *J Pharm Sci*, 89(11), 2000, 1452–1464.
17. K. Saebo, Degradation, metabolism and relaxation properties of iron oxide particles for magnetic resonance imaging, *Disseration at the University of Uppsala*, 2004.
18. C. von zur Muhlen, D. von Elverfeldt, N. Bassler, I. Neudorfer, B. Steitz, A. Petri-Fink, H. Hofmann, C. Bode, and K. Peter, Superparamagnetic iron oxide binding and uptake as imaged by magnetic resonance is mediated by the integrin receptor Mac-1 (CD11b/CD18): Implications on imaging of atherosclerotic plaques, *Atherosclerosis*, 193(1), 2007, 102–111.
19. C. Corot, P. Robert, J. Idee, and M. Port, Recent advances in iron oxide nanocrystal technology for medical imaging, *Adv Drug Deliv Rev*, 58(14), 2006, 1471–1504.
20. S. Callender, Treatment of iron deficiency, *Clin Haematol*, 11(2), 1982, 327–338.

21. M. Ohgushi, K. Nagayama, and A. Wada, Dextran-magnetite: A new relaxation reagent and its application to T2 measurements in gel systems, *J Mag Res*, 29(3), 1978, 599–601.
22. S. Saini et al., Ferrite particles: A superparamagnetic MR contrast agent for the reticuloendothelial system, *Radiology*, 162(1), 1987, 211–216.
23. A. Hemmingsson et al., Relaxation enhancement of the dog liver and spleen by biodegradable superparamagnetic particles in proton magnetic resonance imaging, *Acta Radiol*, 28(6), 1987, 703–705.
24. J. T. Ferrucci and D. D. Stark, Iron oxide-enhanced MR imaging of the liver and spleen: Review of the first 5, *AJR Am J Roentgenol*, 155(5), 1990, 943–950.
25. M. G. Harisinghani et al., Noninvasive detection of clinically occult lymph-node metastases in prostate cancer, *N Engl J Med*, 348(25), 2003, 2491–2499.
26. R. Sigal et al., Lymph node metastases from head and neck squamous cell carcinoma: MR imaging with ultrasmall superparamagnetic iron oxide particles (Sinerem MR)— results of a phase-III multicenter clinical, *Eur Radiol*, 12(5), 2002, 1104–1113.
27. M. De, S. Chou, H. Joshi, and V. Dravid, Hybrid magnetic nanostructures (MNS) for magnetic resonance imaging applications, *Adv Drug Deliv Rev*, 63(14–15), 2011, 1282–1299.
28. Z. Zhou, D. Huang, J. Bao, Q. Chen, G. Liu, Z. Chen, X. Chen, and J. Gao, A synergistically enhanced T1–T2 dual-modal contrast agent, *Ad Mater*, 24(46), 2012, 6223–6228.
29. H. B. Na, I. C. Song, and T. Hyeon, Inorganic nanoparticles for MRI contrast agents, *Adv Mater*, 21, 2009, 2133–2148.
30. E. Werner, A. Datta, C. Jocher, and K. Raymond, High-relaxivity MRI contrast agents: Where coordination chemistry meets medical imaging, *Angew Chem Int Ed*, 47(45), 2008, 8568–8580.
31. M. F. Tweedle, E. Kanal, and R. Muller, Consideration in the selection of a new gadolinium-based contrast agents, *Suppl Appl Radiol*, 43(5), 2014, 1–11.
32. M. Tweedle, J. Hagan, K. Kumar, S. Mantha, and C. Chang, Reaction of gadolinium chelates with endogenously available ions, *Magn Resonan Imag*, 9(3), 1991, 409–415.
33. M. Tweedle, P. Wedeking, and K. Kumar, Biodistribution of radiolabeled, formulated gadopentetate, gadoteridol, gadoterate, and gadodiamide in mice and rats, *Invest Radiol*, 30(6), 1995, 372–380.
34. G. White, W. Gibby, and M. Tweedle, Comparison of Gd(DTPA-BMA) (Omniscan) Versus Gd(HP-DO3A) (ProHance) relative to gadolinium retention in human bone tissue by inductively coupled plasma mass spectroscopy, *Invest Radiol*, 41(3), 2006, 272–278.
35. F. Giesel, H. von Tengg-Kobligk, I. Wilkinson, P. Siegler, C. von der Lieth, M. Frank, K. Lodemann, and M. Essig, Influence of human serum albumin on longitudinal and transverse relaxation rates ($R1$ and $R2$) of magnetic resonance contrast agents, *Invest Radiol*, 41(3), 2006, 222–228.
36. S. Laurent, L. Elst, and R. Muller, Comparative study of the physicochemical properties of six clinical low molecular weight gadolinium contrast agents, *Contrast Media Mol Imaging*, 1(3), 2006, 128–137.
37. P. Carvan et al., Gadolinium (III) chelates as MRI contrast agents: Structure, dynamics, and applications, *Chem Rev*, 99(9), 1999, 2293–2352.
38. A. Li et al., Acute adverse reactions to magnetic resonance contrast media-gadolinium chelates, *Brit J Radiol*, 79(941), 2006, 368–371.
39. Grobner et al., Gadolinium—a specific trigger for the development of nephrogenic fibrosing dermopathy and nephrogenic systemic fibrosis? *Nephrology, Dialysis and Transplantation*, 21(4), 2006, 1104–1108.
40. P. Marckmann, Nephrogenic systemic fibrosis: Suspected causative role of gadodiamide used for contrast-enhanced magnetic resonance imaging, *J Am Soc Nephrol*, 17(9), 2006, 2359–2362.

41. J. T. Heverhagen, G. A. Krombach, and E. Gizewski, Application of extracellular gadolinium-based MRI contrast agents and the risk of nephrogenic systemic fibrosis, *Rofo*, 186(7), 2014, 661–669.
42. E. Terreno, D. Castelli, A. Viale, and S. Aime, Challenges for molecular magnetic resonance imaging, *Chem Rev*, 110(5), 2010, 3019–3042.
43. P. V. Baptista, P. Quaresma, and R. Franco, Nanoparticles in molecular diagnostics, *Prog Mol Biol Transl Sci*, 104, 2011, 427–488.
44. U. Schwertmann and R. M. Cornell, *Iron Oxides in the Laboratory: Preparation and Characterization*, Weinheim, Germany: Wiley-VCH Verlag GmbH, 2000.
45. N. T. Thanh, Chapter 2. Synthesis and characterisation of iron oxide ferrite nanoparticles and ferrite-based aqueous fluids, in *Magnetic Nanoparticles: From Fabrication to Clinical Applications*, Nguyen T.K. Thanh (ed.), CRC Press, USA, 2012.
46. H. A. Lowenstam, Magnetite in denticle capping in recent chitons (Polyplacophora), *Geol. Soc. Am. Bull.*, 73(4), 1962, 435–438.
47. R. Blakemore, Magnetotactic bacteria, *Science*, 190(4212), 1975, 377–379.
48. Atsushi Arakaki et al., Formation of magnetite by bacteria and its application, *J R Soc Interface*, 5(26), 2008, 977–999.
49. Thomas-Keprta et al., Elongated prismatic magnetite crystals in ALH84001 carbonate globules: Potential Martian magnetofossils, *Geochem Et Cosmochem Acta*, 64(23), 2000, 4049–4081.
50. Weiss et al., A low temperature transfer of ALH84001 from Mars to Earth, *Science*, 290(5492), 2000, 791–795.
51. Raju V. Ramanujan, Magnetic particles for biomedical applications, in *Biomedical Materials*, R. Narayan (ed.), Springer, New York, NY, 2009.
52. A. Figuerola, R. Di Corato, L. Manna, and T. Pellegrino, From iron oxide nanoparticles towards advanced iron-based inorganic materials designed for biomedical applications, *Pharmacol Res*, 62(2), 2010, 126–143.
53. Y. Koo, G. Reddy, M. Bhojani, R. Schneider, M. Philbert, A. Rehemtulla, B. Ross, and R. Kopelman, Brain cancer diagnosis and therapy with nanoplatforms, *Adv Drug Deliv Rev*, 58(14), 2006, 1556–1577.
54. E. Pauwels, K. Kairemo, P. Erba, and K. Bergstrom, Nanoparticles in cancer, *Curr Radiopharm*, 1(1), 2008, 30–36.
55. W. S. Seo et al., FeCo/graphitic-shell nanocrystals as advanced magnetic-resonance-imaging and near-infrared agents, *Nat Mater*, 5(12), 2006, 971–976.
56. A. Gupta and M. Gupta, Synthesis and surface engineering of iron oxide nanoparticles for biomedical applications, *Biomaterials*, 26(18), 2005, 3995–4021.
57. S. Benderbous, C. Corot, P. Jacobs, and B. Bonnemain, Superparamagnetic agents: Physicochemical characteristics and preclinical imaging evaluation, *Acad Radiol*, 3(suppl. 2), 1996, s292–s294.
58. O. Veiseh, J. Gunn, and M. Zhang, Design and fabrication of magnetic nanoparticles for targeted drug delivery and imaging, *Adv Drug Deliv Rev*, 62(3), 2010, 284–304.
59. M. Russell and Y. Anzai, Ultrasmall superparamagnetic iron oxide enhanced MR imaging for lymph node metastases, *Radiography*, 13(suppl. 1), 2007, e73–e84.
60. M. Yallapu, S. Othman, E. Curtis, B. Gupta, M. Jaggi, and S. Chauhan, Multifunctional magnetic nanoparticles for magnetic resonance imaging and cancer therapy, *Biomaterials*, 32(7), 2011, 1890–1905.
61. Y. Zhang, N. Kohler, and M. Zhang, Surface modification of superparamagnetic magnetite nanoparticles and their intracellular uptake, *Biomaterials*, 23(7), 2002, 1553–1561.
62. W. Zheng, K. Zhou, Z. Chen, J. Shen, C. Chen, and S. Zhang, Characterization of focal hepatic lesions with SPIO enhanced MRI, *World J Gastroenterol*, 8(1), 2002, 82–86.

63. M. G. Harisinghani et al., Noninvasive detection of clinically occult lymph-node, *N Eng J Med,* 348(25), 2003, 2491–2499.
64. M. Elsabahy and K. Wooley, Design of polymeric nanoparticles for biomedical delivery applications, *Chem Soc Rev,* 41(7), 2012, 2545.
65. W. Cheng, Y. Ping, Y. Zhang, K. Chuang, and Y. Liu, Magnetic resonance imaging (MRI) contrast agents for tumor diagnosis, *J Healthc Eng,* 4(1), 2013, 23–46.
66. J. S. Choi et al., Self-confirming "AND" logic nanoparticles for fault-free MRI, *J Am Chem Soc,* 132(32), 2010, 11015–11017.
67. A. Szpak et al., T1–T2 dual-modal MRI contrast agents based on superparamagnetic iron oxide nanoparticles with surface attached gadolinium complexes, *J Nanopart Res,* 16(11), 2014, 2678–2711.
68. X. Z. Peng, L. H. Hua, S. Z. Qiang, W. Qiang, A case of tumor-like inflammatory demyelinating disease with progressive brain and spinal cord involvement, *Sao Paulo Med J,* 133(5), 2015, 445–449.

4 Plasma Nanotechnology for Nanophase Magnetic Material Synthesis

Rajdeep Singh Rawat and Ying Wang

CONTENTS

ABSTRACT

The plasma, fourth state of matter, is a complex, reactive and non-equilibrium environment that provides excellent avenues for nanoscale magnetic material synthesis and processing at lower reactor zone temperature in relatively shorter time duration with better control on size distribution and material properties. The plasmas used for nanomaterial synthesis exist over enormous range of densities and temperature leading to the possibility of exotic range and combinations of novel nanomaterial synthesis with desired properties. In this chapter, we present the nanophase magnetic material synthesis using three different types of plasmas from (i) pulsed laser ablation deposition (PLAD) facility with high plasma density ($\sim 10^{15-16}$ cm^{-3}) but relatively low plasma temperature ($\sim$few electron volts) at substrate surface, (ii) high energy density dense plasma focus (DPF) device with high plasma density ($\sim 10^{16-18}$ cm^{-3}) and high plasma temperature ($\sim$several tens to hundreds of electron volts), and (iii) atmospheric microplasma (AMP) device with relatively cold plasma with plasma density either lower or almost similar to that of PLAD plasmas. In Section 4.1, basic concepts of plasma nanotechnology and plasma classifications are generally introduced. Then the basic knowledge, challenges and objects for nanophase magnetic material in data storage and biomedical applications are briefly given in Section 4.2. With the supporting knowledge in the above two sections, synthesis mechanisms and magnetic properties of various nanophase magnetic materials by PLAD family, DPF device, and AMP device are discussed in Sections 4.3, 4.4, and 4.5, respectively. The chapter highlights the novelty of plasma as tool for nanophase magnetic material synthesis.

4.1   GENERAL INTRODUCTION

Plasma, widely considered as the fourth state of matter due to its unique properties, is referred to as a complex mixture of electrons, ions (positive as well as negative), neutrals in a ground as well as excited state, and photons. Being composed of charged particles, plasmas are affected by electric and magnetic fields, which allows them to be used as controllable reactive gas. There are very wide varieties of plasmas that exist with densities and temperatures varying over several orders of magnitudes and plasmas are traditionally divided into two general types: (i) industrial plasmas and (ii) fusion plasmas.

Industrial plasmas are typically low-temperature plasmas with ions of masses generally above hydrogen and are usually of two types: (i) nonequilibrium plasmas with electron temperatures much higher than ion temperatures and (ii) equilibrium (thermal) plasmas with electron and ion temperatures approximately equal. Fusion plasmas are plasmas with temperatures much higher than industrial plasmas and are composed of heavier isotopes of hydrogen, that is, deuterium and/or tritium. While hot fusion plasmas are actively pursued with the aim of triggering and controlling thermonuclear fusion reactions to the ignition point to achieve self-sustained fusion reactors in the search for clean and long-term (almost inexhaustible) energy resource, with various worldwide large-scale programs or experiments in place such as the International Thermonuclear Experimental Reactor (ITER) at Cadarache in France [1] and the National Ignition Facility (NIF) at Lawrence Livermore National Laboratory in the United States [2]. Other examples of hot plasmas include relativistic plasmas of very high electron temperatures and quantum plasmas of very high electron densities, typically found in space plasmas. All other plasmas are classified as low-temperature or "cold" plasmas. The low-temperature plasmas have long been studied in laboratories (since the 1920s) and in space (since the 1960s) and have also been successfully harnessed by industry. The wide range of plasma parameters (i.e., plasma density and plasma temperature), even for the cold plasmas, has contributed to the long and expanding list of industrial plasma applications as a result of both scientific and economic drivers.

Industrial plasma applications can be broadly classified into two major areas. First is plasma processing, which typically involves the use of ions and reactive species in plasmas to modify the chemical and physical properties of a material surface. Plasma processing includes plasma activation, plasma etching, ion implantation, or surface modification through plasma functionalization, cleaning, and hardening. The second major area is plasma synthesis, which refers to the use of plasmas to drive or assist chemical reactions either to synthesize complex multicomponent compounds, alloys, or polymers starting from simpler starting precursors or to synthesize simpler material systems through the inverse processes of plasma decomposition of complex multicomponent materials. Plasma processing and synthesis of materials impact several large-scale manufacturing industries in the world, and foremost among them is the electronics industry. The several trillion dollar electronics industry uses plasma-based synthesis techniques for manufacturing very large-scale integrated (VLSI) microelectronic circuits (or chips) [3] and plasma processing for adhesion promotion of encapsulants, adhesives, and sealants by promoting their surface wetting properties. Plasma processing of materials is also a critical technology in the aerospace, automotive, steel, biomedical, and toxic waste management industries [3]. The use of plasma processing at atmospheric conditioning is fast becoming an intense area of research and application in plasma medicine.

4.1.1 Plasma Nanotechnology

Nanotechnology refers to the manipulation of nanostructured materials (with at least one dimension being less than 100 nm) through certain chemical and/or physical processes, to create materials, devices, and systems with fundamentally new properties and functions because of their nanoscale dimensions. The functionalization of nanoscale materials and their integration into nanodevices is leading to a revolution in wide-ranging industries with numerous nanotechnology-enabled products fast becoming part of our daily life. The consumer world is exploding with "nanotechnology-enhanced" or "nanotechnology-enabled" products. Currently, there are numerous products on the market that are the result of nanotechnology [4,5] such as (i) long-lasting tennis balls, stronger tennis and badminton rackets and racket strings, golf balls that fly straighter, harder bowling balls, stain-resistant and water-repelling cloths, shoe inserts that keep athletes cool in summer and warm in winter, non-smelling socks that use the antibacterial properties of silver nanoparticles, and so on for the sporting enthusiast; (ii) nanoceramic coatings on photo-quality picture paper to deliver sharp and high-quality printing on inkjet printers; (iii) nanolithography-enabled products in the electronics industry that have led to the development and manufacture of faster and more powerful computers and smart phones, semiconductor memory-based flash drives and magnetic memory-based hard disk drives, digital cameras and displays, liquid-crystal display (LCDs), light emitting diodes (LEDs), moving pictures experts group (MPEG) audio layer III players (MP3s), electronic displays, thin-film batteries, and flexible electronics to name a few; (iv) biomedical applications of nanotechnology that are developing at a fast pace with numerous applications in the field of medicine such as bandages embedded with silver nanoparticles being used in wound healing, drug delivery via a patch, nanostructured thin-film-coated implants into the human body with better physical properties and bioadaptability, highly sensitive respiration monitors utilizing nanomaterials, use of man-made skin—a nanofabricated network, skin graft applications, diabetic insulin biocapsules, pharmaceuticals utilizing "bucky ball" technology to selectively deliver drugs, cancer therapies using targeted radioactive biocapsules, and so on; and (v) a number of other applications such as catalysts, antifungal and antialgae paints, personal health testing kits (lab-on-a-chip), sensors for security systems, water purification using nanoporous membranes, and so on.

The future impact of high-performance nanotechnology-enabled products is going to be even greater as the size of the global nanotechnology market is expected to exceed $3 trillion by 2020 from about $254 billion in 2009 [5]. Several countries have realized the importance of research and development in nanotechnology for future economic gain and have established national nanotechnology initiatives (with 60 countries starting their nanotechnology initiatives between 2000 and 2012) with an average annual R&D budget ranging from a few tens of millions of dollars to a few billion dollars [6]. This has led to enormous scientific and industrial efforts in evolving and developing various aspects of nanotechnology for discovering new nanoscale materials, new characterization tools, and new novel phenomena and processes leading to new, novel, and efficient products. The conceptualization, development, and optimization of various possible routes and processes for the synthesis of nanoscale materials are some of the most active areas of research in nanotechnology. There are numerous routes for nanoscale material synthesis, which include milling and vapor condensation routes, chemical routes and nature's biological routes. Plasma-based and plasma-assisted processes have become the most versatile tools of nanoscale fabrication as plasmas provide a complex, reactive, and far-from-equilibrium chemical factory. Plasma-based methods have demonstrated (i) higher-throughput, shorter-nanostructure growth time and thus reduced the cost, especially for atmospheric pressure plasma methods; (ii) greater versatility with the capability of synthesizing nanostructured material through bottom-up and top-down production approaches; (iii) better control of size distribution and some properties, particularly for low-pressure plasma methods; (iv) nonthermal synthesis at room temperature; (v) broader range of nanomaterial synthesis, which otherwise were not synthesized using typical physical, chemical, and biological methods; and (vi) higher quality of nanomaterial synthesis, for example, less structural defects of nanotubes, fewer layers of graphene, and so on. This has led to the development of a new field that is popularly referred to as "plasma nanoscience" and/or "plasma nanotechnology" [7–9].

The plasmas used in plasma nanotechnology can be broadly classified into (i) low-temperature or cold plasmas, which have been routinely and conventionally used by industry and the materials research community, and (ii) high-temperature or hot high-energy-density plasmas from fusion, pinch, and intense discharge sources with densities and temperatures many orders of magnitude higher than that of cold plasmas, which were initially not used in materials science research but have recently been gaining interest and momentum largely due to the need for appropriate first wall materials for fusion reactors, which will be subjected to enormous radiation, and particle and heat flux.

Plasmas are typically generated in a gaseous medium, either in vacuum chamber or at atmospheric conditions, by applying a DC, AC, radiofrequency (RF), or microwave-frequency field. Normally, gases are electrical insulators, but there are always a few charge carriers (electrons and ions) present due the impinging UV or high-energy cosmic radiation. The electrons, being lighter, can be accelerated efficiently by the applied electric field, which then collide with neutral particles, producing an avalanche breakdown, thus making the plasma. The electric field needed for breakdown can be made with a potential setup between a pair of electrodes, with an "electrodeless" RF induction coil, with microwave energy coupling to a gaseous medium using waveguide, with shockwaves, with lasers, or with charged or neutral particle beams. The following sections provide the key features of low- and high-temperature plasmas used in plasma nanotechnology.

4.1.2 Low-Temperature Cold Plasmas for Plasma Nanotechnology

Low-temperature plasmas are mainly produced by low-current AC or DC electric gas discharge or by gas discharges initiated by RF or microwave electromagnetic fields or by concentrating intense laser beam on the solid or gaseous targets, and are characterized by low electron kinetic temperatures ranging from fractions to a few tens of electron volts with low degree of gas ionization resulting in a large portion of the gas in neutral state. The low-temperature plasmas used in plasma nanotechnology, or in material synthesis and processing in general, are grouped in the hexagon box in Figure 4.1. Low-temperature plasmas can be further classified into nonequilibrium and equilibrium (thermal) plasmas. Nonequilibrium plasmas are formed (i) when the low operating pressures with lower electrons/ions/neutrals densities results in insufficient electron-ion, electron-neutral, and ion-neutral collisions thus disallowing the thermal equilibrium to be achieved, or (ii) when the high operating pressure plasmas allows higher collision rates among different plasma species but the plasmas are short lived (e.g., pulsed discharge) interrupting the equilibration process

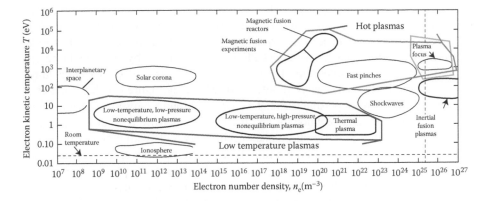

FIGURE 4.1 Plasma parameters of various types of plasmas. The low-temperature plasmas routinely used in plasma nanotechnology are inside the hexagon box while the high-energy-density DPF source is shown in the pentagon and enneagon boxes, which belong to the category of hot plasmas.

as the plasma life-time is less than the equilibration time. They are described by the relation $T_e \gg T_i = T_g$, where T_e, T_i, and T_g are temperatures of electrons, ions, and background gas (neutrals), respectively, and represent their mean kinetic energies. The equilibrium (thermal) plasmas are normally high-pressure gas discharges where electrons, ions, and neutrals have frequent collision between them resulting in equilibration of temperatures among the charged and neutral species, that is, $T_e \sim T_i \sim T_g$. The low-temperature, equilibrium- and nonequilibrium-type, cold plasmas have been used extensively in the synthesis and processing of nanostructured materials due to their interesting combination of electrical, thermal, and chemical properties making them an indispensable and versatile tool for plasma nanotechnology [8–10].

4.1.3 HIGH-TEMPERATURE PLASMAS FOR PLASMA NANOTECHNOLOGY

The plasmas used in thermonuclear fusion research using magnetic, inertial, or hybrid confinement schemes use hot plasmas that have temperatures from a few to several tens of kiloelectron volts (refer to the enneagon box in Figure 4.1). Fusion research schemes often use auxiliary heating mechanisms to achieve extreme plasma temperatures, making matter to be in an almost fully ionized state. In this group of hot fusion plasmas, one can find the dense plasma focus (DPF) device, which was originally envisioned as an alternative magnetic fusion device on almost the highest side of plasma density. The DPF device is one of the plasma devices extensively investigated and presented in this chapter for the processing and synthesis of nanostructured magnetic materials.

High-energy-density plasmas, by definition, refer to plasmas that are heated and compressed to extreme energy densities, exceeding 10^{11} J/m^3 (the energy density of a hydrogen molecule) [11]. The magnitude of plasma and other physical parameters associated with high-energy-density physics is enormous: shockwaves at hundreds of km/s (approaching a million kilometers per hour), temperatures of millions of degrees, and pressures that exceed 100 million atmospheres. Plasmas with energy densities in the range of $1-10 \times 10^{10}$ J/m^3 are also now classified as high-energy-density plasmas. The DPF device is referred to as high-energy-density plasma facility as the energy density of pinch plasma in DPF devices, estimated by dividing the energy stored in the DPF capacitor bank by the volume of the final pinch plasma column, is reported to be in the range of $1.2-9.5 \times 10^{10}$ J/m^3 [12]. Figure 4.1 shows the clear differences, in terms of plasma parameters, between routinely used low-temperature plasmas in plasma nanotechnology and high-energy-density DPF device plasmas, which are emerging as novel plasma nanotechnology tools with both plasma densities and temperatures to be about a minimum of two to three orders of magnitude higher than that of low-temperature plasmas. It is not only the plasma temperature and the plasma density that are different in DPF devices compared to low-temperatures plasma devices but there are also several other features that uniquely belong to DPF devices, which will be discussed later.

4.2 NANOPHASE MAGNETIC MATERIALS FOR DATA STORAGE AND BIOMEDICAL APPLICATIONS

Nanostructured magnetic materials, in the narrow sense, are ferromagnetic materials of nano size. Iron (Fe), nickel (Ni), and cobalt (Co) are three novel magnetic elements in the periodic table. In addition to these elements and their compounds and alloys (such as iron oxides, FePt, CoPt, etc.) many of the rare-earth elements are also magnetic in nature. Nanostructured magnetic materials are of great interest in many applications, including high-density data storage [13,14], biomedicine [15,16], energy storage [17], catalysis [18,19], biosensors [16,20,21], and so on due to their typical magnetic as well as electrochemical properties.

The magnetic hysteresis curve, shown in Figure 4.2, is used for the assessment of ferromagnetic material, and the three important quantities, remanent magnetization M_r, saturation magnetization M_s, and coercivity H_c, are the basic analyzed parameters for a ferromagnetic material. Ferromagnetic material is classified as soft ferromagnetic and hard ferromagnetic depending on the value of coercivity H_c exhibited in hysteresis measurements. The material with a coercivity less than 100 Oe is called

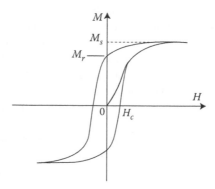

FIGURE 4.2 Typical hysteresis loop of ferromagnetic material.

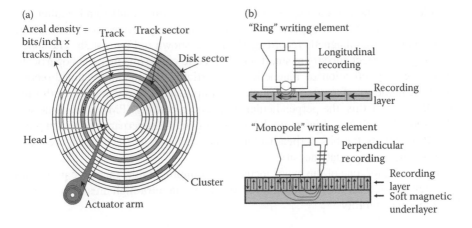

FIGURE 4.3 (a) Typical disk structure. (b) Longitudinal magnetic recording (LMR) and perpendicular magnetic recording (PMR).

soft ferromagnetic, while that with a coercivity larger than 100 Oe is called hard ferromagnetic [22].

4.2.1 APPLICATION IN DATA STORAGE AND RELATED ISSUES

One of the major applications of nanostructured magnetic material is in the field of magnetic data storage, primarily as a magnetic recording medium on hard disk drives (HDD), which is shown in Figure 4.3a. An HDD consists of a recording medium and a read/write head. The recording medium consists of a substrate coated with a ferromagnetic material that can be permanently magnetized and can be later on read repeatedly, and if needed can be rewritten again. The data in HDD are recorded by magnetizing nanosized areas of the ferromagnetic materials of the magnetic recording medium to either one or other direction, also referred to as positive or negative magnetic direction, which represents either a 0 or 1 digital bit. A track is a circular path on the surface of the recording medium on which data bits are recorded. The bits are written closely to form a sector, the minimal physical division of data on an HDD. Each sector stores a fixed amount of user-accessible data, traditionally 512 bytes for HDDs and 2048 bytes for CD-ROMs and DVD-ROMs. Newer HDDs use 4096-byte sectors, which are known as the advanced format (AF). A circular track on the disk is composed of these sectors, and numerous concentric tracks cover the whole surface of the disk. The product of bits per inch along the track and tracks per inch radially on the disk is named areal density in bits per inch square (refer Figure 4.3a), which is one of the most important parameters for researchers to improve for achieving ultra-high data storage in industry [23].

4.2.1.1 Thin-Film Magnetic Recording Media

Magnetic storage technology began with the longitudinal recording format where the magnetization of the recorded bit lies in the plane of the disk as shown in Figure 4.3b. As the continuous improvement of the head design allowed ultra-high areal densities of magnetic recording with magnetic bits getting smaller and smaller, the superparamagnetic effect appears when the areal density arrives at 10–100 Gbits/in² for longitudinal magnetic recording (LMR) as shown in Figure 4.3b. The superparamagnetic effect appears in small ferromagnetic or ferrimagnetic nanoparticles, where the magnetization of magnetic nanoparticles can randomly flip direction under the influence of temperature, making it impossible to store magnetic information. Because of this, the perpendicular magnetic recording (PMR) technique was later studied to achieve a higher areal density without the superparamagnetism problem. PMR aligns the bits perpendicular to the media plane (refer to Figure 4.3b), enabling higher areal data density.

In this chapter, one of the major focuses is on improving the magnetic properties of materials used for the recording layer to achieve an ultra-high areal density with good thermal stability.

4.2.1.2   Progress in High Areal Density Recording and Issue of Thermal Stability

Figure 4.4 shows the areal density versus time since the original IBM RAMAC brought disk storage to computing. The compound growth rate (CGR) kept increasing from 25% in 1980 to 60% in the 1990s, and then increased further to 100% at

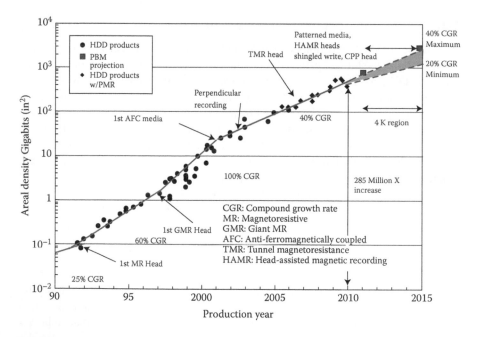

FIGURE 4.4 Trends of historical areal density for HDD products. (Reprinted from T. Coughlin, *Invest in New Technologies or Divest in Market Share*, Coughlin Associates. Copyright 2010, with permission from Coughlin Associates; and E. Grochowski, J. Robert, and E. Fontana, Future technology challenges for NAND flash and HDD products, in: *Flash Memory Summit, Conference ConCepts, Inc.*, Santa Clara, CA. Copyright 2012, with permission from Conference ConCepts. Inc.)

the end of the twentieth century. However, as the areal density increased toward 100 Gbit/in^2 with LMR in 2002, thermal stability limitation and superparamagnetic effects appeared and the incredible growth rate in the areal density of magnetic data storage was no longer possible. The PMR technique was introduced to further increase the areal storage density toward 1 Tbit/in^2 with a growth rate of about 40% per year, but the growing areal density is still not high enough for the requirement of the rapidly improving information industries [26,27].

To increase areal density, the bit size should be decreased. This however brings in the issue of thermal stability of the magnetization state, that is, the magnetic data bit. The thermal stability limit is also known as superparamagnetic limit, which occurs when the magnetic energy stored in the grains becomes comparable to their thermal energy, making them impossible to store magnetic states reliably and stably. The relaxation of the magnetization orientation of each data storage bit is determined by the relaxation time $\tau = \tau_0 e^{K_u V / 2 k_B T}$ of the magnetization state in one of the orientations, where K_u is the storage media's anisotropy constant, V is the volume of the storage bit, k_B is the Boltzmann's constant, and T is temperature. The term $K_u V$ measures the energy barrier between two stable orientations of magnetization states, which is equivalent to the stored magnetic energy. As the size of storage bit decreases, the magnetic energy $K_u V$ will decrease and it can become comparable to the thermal energy kT, which will make the magnetization state fluctuate randomly from one orientation to another (magnetization reversal) due to the short relaxation time. As a result, at this T, the overall magnetic moment of the storage bit is randomized to zero and it becomes impossible to store the magnetic state, and it is said to be superparamagnetic (thermal stability limit) and become the fundamental density limit for magnetic data storage in both longitudinal and perpendicular recording.

The requirement for a magnetic material to maintain thermal stability for about 10 years is given as

$$\frac{K_u V}{k_B T} \geq 40$$

Therefore, magnetic material with higher anisotropy constant (K_u) is required so that smaller bit sizes can be achieved for ultra-high areal density data storage. Table 4.1 lists the properties of certain candidate materials for ultra-high areal density data storage. Using simple grain counting statistics for noise control, the areal density (AD) is scaled according to the relation AD $\propto 1/D_p^2$, where D_p is the minimum stable grain size. As seen in Table 4.1, bimetallic FePt, CoPt, and FePd show a much larger anisotropy constant K_u and smaller minimum grain sizes D_p at which the superparamagnetic behavior is exhibited than recently used magnetic recording media, that is, CoPtCr. Thus, FePd, FePt, and CoPt with atomic percentage of 50:50 are believed to be the most promising candidates for realizing ultra-high-density magnetic data storage with areal densities of Tbits/in^2.

4.2.1.3 Writability Issues of High-K_u Materials

It was estimated that if all microstructural and grain isolation problems could be solved for around 2.8-nm-sized grain of high-K_u FePt to achieve Tbits/in^2 areal

TABLE 4.1

Intrinsic Magnetic Properties of a Number of Potential Alternative Media Alloys [23,27–31]

	Materials	K_u (10^7 erg/cm³)	M_s (emu/cm³)	H_k (kOe)	T_c (K)	D_p (nm)
Today's media	CoPtCr	0.2	300	14	–	10.4
$L1_0$	FePd	1.8	1100	33	760	5.0
$L1_0$	FePt	7	1140	120	750	2.8–3.3
$L1_0$	CoPt	4.9	800	123	840	3.6
Rare earth	$Fe_{14}Nd_2B$	4.6	1270	73	585	3.7
Amorphous	$SmCo_5$	11–20	910	240–400	1000	2.2–2.7

Note: Where K_u is magnitude of the magnetic anisotropy energy density, $K = (1/2)M_sH_k$; M_s is the saturation magnetization of the material; H_k is the anisotropy field, $H_k = 2K/M_s$; T_c is the Curie temperature; and D_p is the minimum stable grain size, $D_P = (60k_BT/K)^{1/3}$.

densities, then write fields of these materials would be somewhere in the region of 50–100 kOe. Thus, the values of the write field required for some of the high-K_u materials listed in Table 4.1 are out of range of the fields that can be reached by recent writing heads.

A contour plot of K versus M_s^2 is shown in Figure 4.5 for a variety of hard magnetic materials with high K_u [31]. The solid stability boundary line [32] represents the optimal media design point for any given combination of K_u and M_s for various hard magnetic materials to achieve storage stability for an areal density of 40 Gbits/in². The low-K_u material below the solid line generates a stable media but large grain size and is therefore not optimal from the noise perspective and large areal density.

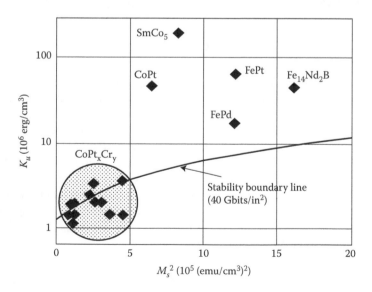

FIGURE 4.5 Stability contour plot of K_u versus M_s^2 for materials listed in Table 4.1.

The high-K_u materials above the optimal design point solid line form smaller grains and quieter media, but they may not be easily writable. Most of the materials in Table 4.1 are over the boundary line, as seen in Figure 4.5, making it difficult to write on them. Thus, to make the media writable, that is, reducing the write field, one would need to operate at higher M_s. The higher M_s, however, will also increase the demagnetization field for the written bits. The increase in the write field of the write heads or temporary reductions of the coercivity during writing, for example, via temperature or other means, might be tried to shift the boundary line upward to move some of the materials into the writable region for improved writability [28].

4.2.1.4   Other Issues for High-K_u Materials

It may be noted from Table 4.1 that the candidate materials for high areal density data storage, such as FePt/FePd/CoPt, need to be in chemically ordered face-centered-tetragonal, *fct*, phase ($L1_0$ phase) to exhibit high magnetic anisotropy constant K_u values, that is, hard magnetic properties. As-grown bimetallic FePt/FePd/CoPt materials normally exist in chemically disordered face-centered-cubic, *fcc*, phase (A1 phase) with low anisotropy constant K_u and hence show a small coercivity value exhibiting magnetically soft behavior. Both these structures (*fcc* and *fct*) will be discussed next, and we will take FePt as an example.

Figure 4.6 illustrates schematically the chemically ordered $L1_0$ and $L1_2$ structures, as well as the disordered A1 structure of FePt. The as-deposited FePt thin films are either amorphous in nature or have the *fcc* (A1) phase with chemically disordered structure with iron and platinum atoms occupying the lattice sites randomly as shown in Figure 4.6c and thus they exhibit soft magnetic behavior as the magnetic anisotropy constant is low for them. The transformation of as-grown chemically disordered (A1) low-K_u soft magnetic phase material to chemically ordered ($L1_0$) high-K_u hard magnetic phase material requires high-temperature thermal annealing, which is discussed later.

If the as-deposited material is with equiatomic compositions, then the crystal structure will transit from the disordered A1 structure to the ordered $L1_0$ structure (shown in Figure 4.6a) after thermal treatment. The cubic symmetry is broken due to the stacking of alternate planes of Fe and Pt atoms along the [001] direction. This periodic structure of layers of two (or more) elements is normally called

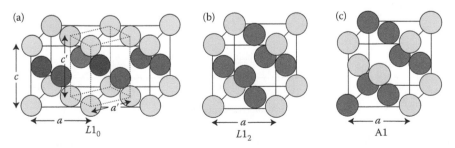

FIGURE 4.6 Schematic illustration of (a) the $L1_0$ structure, (b) the $L1_2$ structure, and (c) the chemically disordered A1 structure. Light spheres represent Fe (Pt) and dark spheres represent Pt (Fe).

a superlattice. However, sometimes, deposited material does not have equiatomic compositions; then the thermal treatment leads to $L1_2$ ordered structure, shown in Figure 4.6b. The $L1_2$ ordered structure is a cubic phase that can form a stoichiometry around 1:3. In Fe_3Pt (or $FePt_3$), the Pt (Fe) atoms occupy the cube corners and the Fe (Pt) atoms occupy the face-centered positions.

The ordered structures in Figure 4.6 represent the case of a perfect long range order (LRO). However, the chemical order may not be perfect and to characterize the degree of LRO, the LRO parameter, S, is induced and defined as

$$S = \gamma_{Pt} + \gamma_{Fe} - 1$$

where γ_{Pt} and γ_{Fe} are the fractions of Pt and Fe sites occupied by a correct atom, Pt or Fe, respectively. When the order is perfect, the order parameter S reaches unity, while for a completely random atom arrangement, S is equal to zero.

The FePt phase diagram, in Figure 4.7, shows the range of annealing temperature needed for fcc to fct phase transition changes with the composition of Fe and Pt in the alloy [33]. At high temperatures, an fcc solid solution of FePt is observed. At

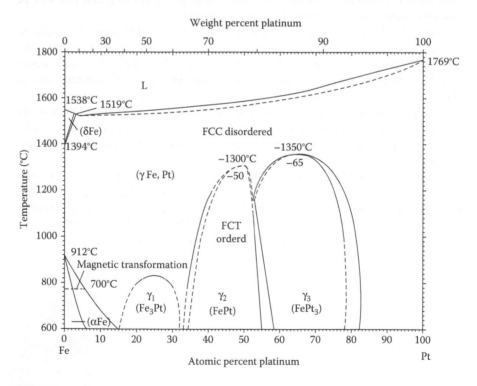

FIGURE 4.7 Equilibrium phase diagram of the FePt system. Shown schematically are crystallographic structures of the phases in the corresponding phase equilibrium regions. (Reprinted from *Acta Mater.*, 46, S. H. Whang, Q. Feng, and Y. Q. Gao, Ordering, deformation and microstructure in L10 type FePt, 6485–6495, Copyright 1998, with permission from Elsevier.)

temperatures below 1300°C, alloys close to the equiatomic composition exhibit transformation from the disordered *fcc* A1 phase to the ordered *fct* $L1_0$ phase. The existence region of the $L1_0$ structure extends from approximately 35 to 55 atomic percent of Pt. In $Fe_{100-x}Pt_x$ alloys deviating from the $35 \leq x \leq 55$ stoichiometry, formation of the stable cubic superstructures $FePt_3$ ($L1_2$) and Fe_3Pt ($L1_2$) is expected at temperatures below 850°C and 1350°C, respectively. An eutectic reaction A1 $\rightarrow L1_0 + L1_2$ is observed to occur at approximately 1160°C at the composition of the A1 phase of about 55 at% Pt.

4.2.1.4.1   Need to Lower the Phase Transition (Ordering) Temperature

As mentioned, the thermal annealing of as-grown material is required to achieve the phase transformation from chemically disordered soft magnetic *fcc* phase to chemically ordered hard magnetic *fct* phase. The heating/annealing temperature, at which the phase transformation happens, is called ordering temperature or phase transition temperature. Take FePt for example. Such ordering temperatures are typically larger than 600°C. Since aluminum is often used as hard disk substrates, such high-temperature treatment may result in substrate damage. More importantly, the high-temperature annealing will lead to grain growth and grain coarsening, which is not favorable for high areal density magnetic data recording. Therefore, techniques that enable ordering at significantly reduced temperatures are critically and urgently needed. Many different schemes such as third elements addition, adding new underlayers or multilayers, and ion irradiation have been tried and tested to achieve low-temperature phase transition to highly ordered $L1_0$ FePt, which will be discussed later in Section 4.3.3.

4.2.1.4.2   Small and Uniformly Isolated Grains of Magnetic Materials

Another important requirement for ultra-high areal density data storage is the formation of small and well-isolated grains of high K_u magnetic material on the substrate surface. One way to achieve this is to use the synthesis techniques that allow the formation of small and isolated magnetic nanoparticles along with the lower-temperature phase transition of $L1_0$ phase; this is one of the focus issues reported in this chapter and will be discussed in Section 4.3.3. Another methodology normally adopted is the use of nonmagnetic matrix materials, such as SiO_2 and Al_2O_3, to isolate FePt nanoparticles to form the so-called nanocomposite structures to reduce grain growth and agglomeration during high-temperature annealing; this is also one of the focuses of this chapter and will be discussed in Section 4.3.2. Piramanayagam [14] reported that several methods such as doping FePt with Ag, Cu, or C, or laminating FePt with Ag, Cu, or C, or depositing underlayers and overlayers have been attempted by several researchers to achieve small and isolated grains. Ko et al. reported that the $L1_0$ ordering and grain growth kinetics of FePt (001) films could be finely controlled by C doping [35].

4.2.2   MAGNETIC NANOPARTICLES FOR BIOMEDICAL APPLICATIONS

Magnetic nanoparticles are finding increasing applications in the field of biomedical sciences. It may be noted that magnetic nanoparticle sizes (from a few to tens of nanometers) are smaller than or comparable to many biological entities such as a cell (10–100 μm), virus (20–450 nm), protein (5–50 nm), and gene (2 nm wide and 10–100 nm long) [36]. This makes them "get close" to a biological entity, rendering them suitable for labeling or tagging, or separation of biological entities. The magnetic nanoparticles can be coated with biocompatible molecules such as dextran [37], polyvinyl alcohol (PVA) [38], and phospholipids [39], so that they can interact with or bind to a biological entity, making entity labeling or tagging possible. Then, a fluid-based magnetic separation device can help to separate out the tagged entities [36].

Another potential application of magnetic nanoparticles is in drug delivery [40,41] as their movement in the human body can be manipulated/controlled by an external magnetic field gradient due to the intrinsic penetrability of magnetic fields into human tissues [36]. If the magnetic nanoparticles are coated with a porous biocompatible polymer, they can be used as drug carriers. The magnetic nanoparticles loaded with drug/carrier complexes are injected via the circulatory system and then can be targeted or localized to a specific site, say a tumor site, via an external magnetic field gradient. The main advantages of using magnetic carriers are (i) the reduced amount of systemic distribution of the cytotoxic drug, which in turn reduces chances of attack/effect on normal healthy cells, and (ii) the reduced amount of drug dosage so it can be effectively targeted to the required site. The targeted delivery of magnetic nanoparticles is also very useful for application in hyperthermia with its main advantage being heating only the intended target cancer tissue. The procedure involves injecting and localizing magnetic nanoparticles to the target tissue and applying an AC magnetic field of sufficient strength and frequency ($H \cdot f$ of a value less than 4.85×10^8 $Am^{-1}s^{-1}$) [42] to cause the magnetic nanoparticles to oscillate resonantly to heat the surrounding. If the temperature of a cancer cell can be maintained above the therapeutic threshold of 42°C for 30 min or more, the cancer cell is killed. Iron oxides magnetite (Fe_3O_4) and maghemite (γ-Fe_2O_3) are the most studied magnetic material for hyperthermia. Other candidate materials can be either ferromagnetic/ferrimagnetic particles or superparamagnetic particles.

Magnetic nanoparticles can also be used in magnetic resonance imaging (MRI) [43]. This relies on the counterbalance between the small magnetic moment on a proton, and the large number of protons present in tissue, leading to a measurable effect in the presence of large magnetic fields. The contrast agents used in MRI can be paramagnetic gadolinium ion complexes or superparamagnetic nanoparticle-based complexes. Iron oxide nanoparticles are the most commonly used superparamagnetic contrast agents.

For the different applications mentioned above, the requirement for the properties of magnetic nanoparticles varies. To control their properties such as particle size, shape, and magnetic properties well, well-controlled synthesis techniques are required. The most common synthesis technique, for iron oxide and PdPt nanoparticles used in biomedical applications, is thermal decomposition. The thermal decomposition method requires a high thermal temperature or long reaction duration. Synthesized PdPt bimetallic nanoparticles of 30 nm size take a reaction duration of around 15 h and a maintained reaction temperature at 80–90°C [44]. Iron oxide nanoparticles of 300 nm size are reported to be synthesized at 210°C for 10 h [40]. The long reaction time and high reaction temperature make it difficult for thermal decomposition to synthesize smaller nanoparticles (around 10 nm) without adding surfactant, which may induce unexpected impurities. A newly developed technique, the atmospheric microplasma-induced chemical method, is found to highly reduce the reaction duration and reaction temperature compared to thermal decomposition and will be presented in this chapter. Nanoparticles with well-controlled size and shape are synthesized using atmospheric microplasmas [45]. In this method, the atmospheric microplasma discharge time of about 10 min at ambient temperature was found to be enough to synthesize bimetallic or iron oxide nanoparticles.

4.3   NANOPHASE MAGNETIC MATERIAL SYNTHESIS BY PULSED LASER ABLATION DEPOSITION

Pulsed laser ablation deposition (PLAD) is one of the forms of physical vapor deposition processes, performed in a vacuum chamber [46]. It is perhaps one of the simplest growth techniques to deposit high-quality smoother and nanostructured thin films of a variety of materials ranging from superconductors to metals to semiconductors to dielectrics and many more under suitable tailored optimized operating conditions. In PLAD, a pulsed laser is focused on a target of the material to be deposited. For sufficiently high laser energy density with laser energy flux above the ablation threshold of the material being irradiated, each laser pulse vaporizes or ablates a small amount of the material. The ablated material is ejected from the laser-irradiated target surface in a highly forward-directed plume. The ablation plume provides the material flux for film and/or nanostructured material growth. Several features that make PLAD particularly attractive include stoichiometric transfer of material from the target to surface, generation of energetic species, hyperthermal reaction between the ablated cations and the background gas in the ablated plasma plume, and capability of operating over a large background pressure ranging from ultrahigh vacuum (UHV) to about several torr [46,47]. Various aspects related to PLAD are detailed in the following sections.

4.3.1   PULSED LASER ABLATION DEPOSITION: EXPERIMENTAL SETUP AND DEPOSITION PROCESS

4.3.1.1   PLAD Experimental Setup

The typical setup of the PLAD system, available in our Laser Technology Lab of the National Institute of Education in Singapore, is shown in Figure 4.8. The PLAD system includes a main deposition chamber where the target (material to be ablated) and substrate (platform for deposition of ablated material) assembly is placed, a vacuum system to set up the background gas pressure in the deposition chamber, and a gas system to set up the ambient medium and pulsed laser system. Two magnetic motors, on diametrically opposite sides, are used to rotate the target and the substrate holder, which helps in irradiating more area on the target surface for more uniform ablation and also to deposit uniform thin films on bigger substrate surface area.

There are two PLAD systems available in our lab. The first is a simple PLAD system, which allows only one target to be used at a time. In this system, both the

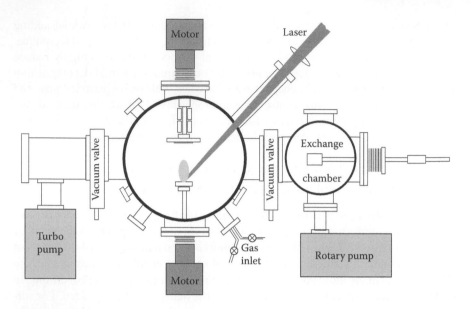

FIGURE 4.8 Schematic of typical PLAD system.

target and the substrate can be rotated. However if the substrate needs to be heated, then it uses a static (nonrotating) substrate heater with a maximum substrate temperature of 500°C. The second PLAD system is an advanced system with a multiple target mounting facility along with the rotatable substrate heater assembly. The heating range of the substrate heater in the second system is from room temperature to 850°C.

The two pulsed Nd:YAG lasers used for these PLAD systems are Continuum Surelite II-10 and Lotis Tii LS-2137U. The active material in Nd:YAG consists of neodymium atoms that are accommodated in a transparent host crystal called yttrium aluminium garnet. The output energy, for the two laser systems mentioned above, range from 100 to 850 mJ at 1064 nm and harmonic options at 532, 355, and 266 nm. For our deposition experiments, either the 1064 or 532 nm Nd:YAG laser with a pulse width of about 10 ns operating at 10 Hz was used. The chambers used in both the PLAD systems have a number of standard ports required for pumping, pressure and temperature gauges, gas inlets, and so on. A dedicated laser light input port sealed with a quartz window is used to ensure maximum transmission of the laser beam. As a routine, the chamber is pumped down to a base vacuum pressure of ~10^{-5}–10^{-6} mbar using a turbo molecular pump, which is helpful to minimize impurities due to residual gases. The laser beam is focused on the target surface using a converging lens of appropriate focal length. The target is continuously rotated and the laser beam is scanned over the target during the ablation process to ensure uniform ablation.

4.3.1.2 Deposition Process in PLAD System

The entire deposition process in any typical PLAD system, in a simplistic way, can be divided into three stages [47,48]: (i) plume generation, (ii) plume propagation, and (iii) ablated species deposition, as shown in Figure 4.9.

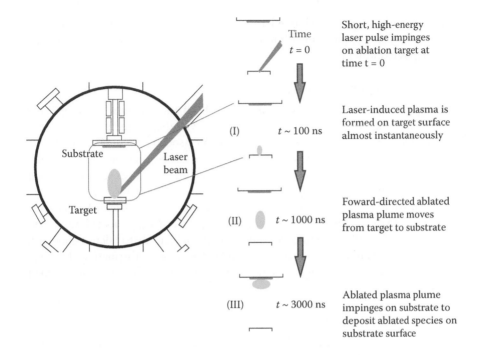

FIGURE 4.9 Typical schematic of the PLAD process.

4.3.1.2.1 First Stage: Plume Generation

A pulsed laser with high energy, with typical energy flux greater than the ablation threshold, is focused on the target of the material, which is to be deposited in thin-film form on the substrate. The laser radiation is absorbed by the surface of the target where the photon energy causes surface heating. The transient temperature of the surface can easily exceed 1500°C. The amount of transient temperature rise on the laser-irradiated substrate surface depends on the optical penetration depth of the material, the thermal conductivity of the target, and the rate at which the photons are supplied [48].

As the pulsed laser continues to irradiate on the target, the molten surface layer is evaporated in the form of a plasma plume. Because the laser energy is supplied to a small volume (10^{-13} m³) in a very short time (typically from femto- to nanosecond range), all the components of the target material are effectively liberated simultaneously, which ensures the same stoichiometry between the evaporated material and the target material [48]. For more details about this process, see the paper by Willmott and Huber [49].

A parameter named laser energy density is normally used to represent the energy received by per unit area of the target surface. The impinging laser energy density above a certain threshold value is required for the ablation of the target material. The ablation threshold for metallic systems is within 1–8 J/cm². High laser fluences on metallic targets lead to considerable droplet contamination and low deposition rates [47]. Furthermore, high laser fluences also result in high energetic plasma species ablated from the target, which bombards the substrate surface and may cause damage to the surface by sputtering off atoms from the surface or cause defect formation in the deposited film [46].

4.3.1.2.2   Second Stage: Plume Propagation

The target surface is almost instantaneously ablated and the laser beam, if it is ns duration as in our case, then continues to interact with the ablated plasma to further break up the plasma plume constituents via photo-dissociation and photo-ionization, and both lead to the expansion of the plasma plume. The expanding plasma plumes move into a vacuum or into an ambient gas, depending on the experimental conditions. The ablated plume is highly cone shaped, and is characterized by a forward peaked distribution with a $\cos^n\theta$ dependence, where $9 < n < 12$ and θ is measured with respect to the target normal. In general, n depends strongly on the laser fluence, and is lower when the plasma propagates into an ambient that is dense enough for multiple collisions to occur, which will broaden the angular distribution. The angular distribution is also governed by parameters such as the target–substrate geometry [48,49].

4.3.1.2.3   Third Stage: Ablated Species Deposition

The moving plasma plume impinges onto the substrate surface and the material begins to grow on the substrate place downstream. At the beginning of the deposition, when the deposited thickness is less than a molecular layer, an initial nucleation process happens first. Here, two atoms mostly form a stable nucleus. If we assume that E_{FS} is the binding energy between the film and the substrate, and E_{FF} is the binding energy between atoms or molecules in the film [48], then

- If $E_{FS} < E_{FF}$, called the weak binding case, most of the nucleation takes place on substrate defects.
- If $E_{FS} > E_{FF}$, called the strong binding case, the nucleation takes place on substrate lattice sites.

In this nucleation process, the nucleation sites are formed as islands, with typical sizes of about 100 Å on the substrate surface. There will be several nucleation sites and several islands are formed over different regions of the substrate surface. When the number of islands reaches a saturation density, the subsequent deposition will lead to the growth of the nuclei until they coalesce. After the initial nucleation process, an island growth mechanism or a layer-by-layer mechanism happens for continuation film growth.

- If $E_{FS} < E_{FF}$, islands of deposited film grow out from these nuclei with different orientations from island to island, which is called an island growth mechanism.
- If $E_{FS} > E_{FF}$, the film grows by taking on the lattice structure and orientation over the entire layer, which is called a layer-by-layer mechanism.

When the film growth is by the strong bonding case, the role of the substrate is more important, and the lattice mismatch (between the film and the substrate), chemical bonding type, and crystal symmetry type must be taken into consideration.

4.3.2 FePt NANOPARTICLES AND NANOCOMPOSITES USING PLAD

The section will provide the details of the FePt nanoparticles and nanocomposites thin films synthesized using the PLAD system and the various physical properties of these thin-film samples. The parameters of the laser used were as follows for most of these experiments (unless stated otherwise): $\lambda = 532$ nm, repetition rate of 10 Hz, laser pulse width of about 10 ns, laser pulse energy of about 75 mJ, and laser focal spot size of about 100 μm. Hence, the laser energy density at the target surface was estimated to be about 955 J/cm^2. High-purity equiatomic FePt target with Fe:Pt ratio of 1:1 was used in these experiments. Different deposition conditions, such as ambient gas pressure, number of laser shots, and *target–substrate geometry*, were explored to study their effects on the formation and magnetic, morphological, and structural properties of FePt nanoparticles.

Two different target–substrate geometries were explored: (i) conventional PLAD geometry where substrates were placed on rotatable substrate holder placed opposite the target holder at a certain target–substrate distance and (ii) a special target–substrate geometry, which was coined as backward plume deposition (BPD) [50,51], with the substrate to be coated placed on the target itself rather than on the substrate holder. Both these geometries are shown in Figure 4.10. As shown in Figure 4.10b, in BPD, the laser beam irradiates the target around the silicon substrate(s) mounted on the target itself along a circular strip on the rotating target, which also helps to obtain homogeneous material ablation and uniform deposition across the entire substrate.

4.3.2.1 FePt Nanoparticles by Conventional PLAD and BPD Target Substrate Geometry

In this section, the use of conventional PLAD and BPD target–substrate geometry-based synthesis of FePt nanoparticles in the form of nanoparticle agglomerates, nanoparticle clusters, or floccule-like nanoparticle networks by varying the ambient gas pressure in the PLAD chamber is discussed.

4.3.2.1.1 Comparison of Surface Morphology

The surface morphologies of FePt samples deposited by conventional PLAD and BPD methods at different argon ambient gas pressures using 6000 laser shots are shown in Figure 4.11. Let us first look at the samples synthesized using conventional PLAD, which are shown on the left in Figure 4.11. For conventional PLAD synthesis, smooth thin films are formed at very low argon ambient gas pressure of ≤1 Pa, which transforms into densely packed nanoparticle agglomerates with increase in ambient pressure to 50 Pa. The population density of nanoparticle agglomerates decreases with further increase in ambient gas pressure to 1 kPa and they are transformed into

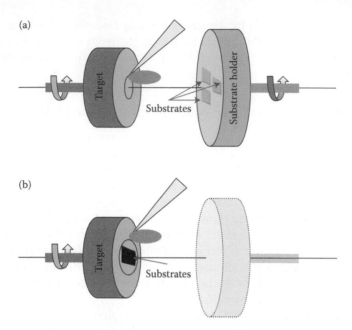

FIGURE 4.10 Target–substrate geometries: (a) conventional geometry, substrates mounted on substrate holder, and (b) backward plume deposition geometry, substrates mounted on target itself.

floccule-like nanoparticle networks. FePt nanoparticles, which exist in the form of nanoparticle agglomerates or floccule-like nanoparticle networks, are found to have uniform particle size and size distribution.

The change in the morphological features with varying ambient gas pressure for conventional PLAD geometry can be explained on the basis of surface mobility of impinging plasma plume species at the substrate surface, which is governed by the kinetic energy these impinging species have when they arrive at the substrate surface. The higher the kinetic energy, the greater will be the surface mobility. The kinetic energy of the ablated material species upon their arrival at the substrate surface is controlled by the collision rate and collision duration of these species as they traverse from the target to the substrate. Thus, they can be tailored through deposition conditions, such as incident laser beam energy and energy coupling of laser with ablated material, ambient gas pressure, and target-to-substrate distance. Smooth and uniform thin films are expected to form if the species impinge on the substrate surface with sufficient surface mobility [52], which is expected in deposition conducted in vacuum or low ambient gas pressure due to much fewer collisions with the ambient gas. As ambient gas pressure increases, the collision rate of the ablated plasma species with ambient gas particles increases, reducing their kinetic energy. Therefore, FePt nanoparticle agglomerates and nanoparticle networks, instead of smooth thin films, are formed at relatively high ambient gas pressure as impinging ablated species on the substrate surface do not have enough surface mobility to form smooth film due to significant reduction in their kinetic energy at higher ambient

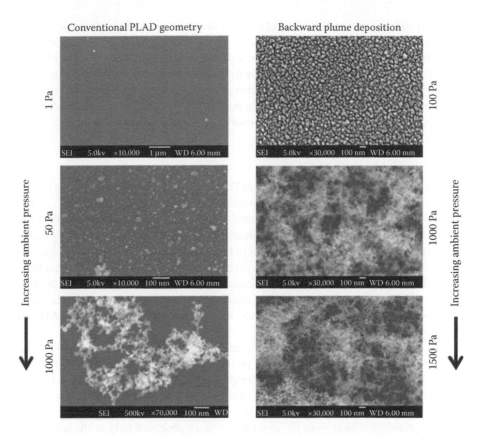

FIGURE 4.11 Comparison of surface morphology of FePt samples deposited at different argon ambient pressures using conventional PLAD and backward plume deposition (BPD) geometries.

gas pressures. Readers are referred to a paper by Happy et al. [53], who conducted a detailed study on the effects of deposition conditions, including deposition shots, ambient gas pressure, and target–substrate distance, on the nanoparticle formation and particle size.

The scanning electron microscope (SEM) images of samples deposited by BPD target–substrate geometry at different argon ambient gas pressures using 6000 shots are shown on the right in Figure 4.11. The basic idea behind the BPD method was that the ablated species can move backward and deposit on the target surface due to their lateral expansion and backscattering by ambient gas. The higher the ambient gas pressure, the greater will be the lateral expansion and backscattering of ablated species but it will also result in greater energy loss due to higher collisions. At ambient gas pressure of 100 Pa, high-density FePt nanoparticle agglomerates are seen to form on the sample surface. FePt nanoparticle agglomerates synthesized by BPD are found to have higher uniformity and density than those deposited by conventional PLAD, which is obviously due to the closer proximity of the substrate surface to the ablated plasma plume. Another notable feature of BPD-based synthesis (refer to Figure 4.4 in Reference 50) is significantly fewer laser droplets compared to that of conventional PLAD synthesis. The laser droplets (large undesirable particulates from the target surface under the thermal stress generated by the focused laser beam) mostly move in the forward direction due to their momentum. The increase in ambient gas pressure will reduce the kinetic energy of the backscattered species, resulting in lesser diffusion due to reduced surface mobility. Thus, they would form relatively more open floccule-like nanoparticle networks as seen for BPD synthesis at an ambient gas pressure of 1000 and 1500 Pa.

4.3.2.1.2 Comparison of Crystalline Structure

The samples deposited with conventional PLAD and BPD geometry were annealed for 2 h at different annealing temperatures of 400°C, 500°C, and 600°C. X-ray diffraction (XRD) patterns of as-deposited sample and sample annealed at 600°C synthesized by (i) conventional PLAD methodology at 50 Pa and (ii) by BPD methodology at 100 Pa 6000 shots are shown in Figure 4.12a and b, respectively. As seen in Figure 4.12, as-deposited samples prepared by both conventional PLAD and BPD methods exhibit a weak *fcc* phase indicated by the weak and broad (111) peak at around 41°. The mean crystallite sizes, estimated using the Scherrer formula, were found to be about 5 and 4 nm, respectively, for as-deposited conventional PLAD and BPD samples.

The samples were then annealed at increasing temperatures of 400°C, 500°C, and 600°C. Please note that the diffraction patterns of samples annealed at 400°C and 500°C are not shown in Figure 4.12. Samples annealed at 500°C, for both target–substrate geometries, exhibited splitting of the fundamental (111) peak indicating the partial phase transition of FePt to high-K_u *fct* phase. However, much better indications of phase transition from *fcc* to *fct* phase were observed only at the annealing temperature of 600°C for both samples, which is indicated by the appearance of the (001) peak at around 24° as shown in Figure 4.11. According to XRD Database@ Socabin (03-065-9121 and 03-065-9122), the intensity ratios $I_{(001)}/I_{(111)}$ and $I_{(110)}/I_{(111)}$ for the *fcc* phase FePt are very low (about 0.001) and they increase significantly to 0.33 and 0.27, respectively, for *fct* phase FePt. Therefore, the presence of the (001) peak at about 24° with intensity ratio $I_{(001)}/I_{(111)}$ of about 0.31 can be used to determine the phase transition of FePt. Compared with 0.18 of those deposited by conventional PLAD and annealed in the same conditions, samples deposited by BPD have a lower

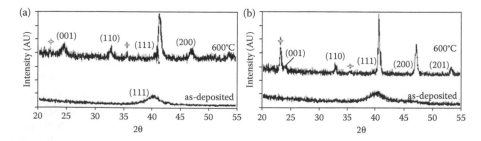

FIGURE 4.12 XRD patterns of as-deposited sample and sample annealed at 600°C for different target–substrate geometries using (a) conventional PLAD and (b) BPD system.

peak intensity ratio $I_{(001)}/I_{(111)}$ of 0.11, after annealing at 600°C, indicating poorer phase transition in the BPD-grown sample.

It may be noted that a few peaks in both XRD patterns taken on samples annealed at 600°C do not match with any of the FePt peaks in the database. These peaks are marked with star signs in Figure 4.12. The XRD pattern of the BPD sample annealed at 600°C shows a very strong unknown peak just before the (001) diffraction peak at 24°. These unknown peaks, as discussed later on exclusively in Section 4.3.4, are due to impurity phase formation, which results in the degradation of magnetic properties.

4.3.2.1.3 Comparison of Magnetic Properties

The hysteresis curves, recorded using a vibration sample magnetometer (VSM), of the as-deposited sample and sample annealed at 600°C for 2 h for both target–substrate geometries of conventional PLAD and BPD are shown in Figure 4.13. These are the same samples whose XRD patterns are shown in Figure 4.12. As seen in Figure 4.13a and c, as-deposited samples prepared by both conventional PLAD and BPD methods exhibit weak ferromagnetic behavior with coercivity values of 67 and 53 Oe, respectively. This validates the formation of a magnetically soft *fcc* phase in as-deposited samples of both target–substrate geometries, confirming their XRD results shown earlier in Figure 4.12. The remanence ratio, $S = M_r/M_s$, where M_r and M_s are remanent, and saturation magnetization are estimated to be 0.45 and 0.33 for

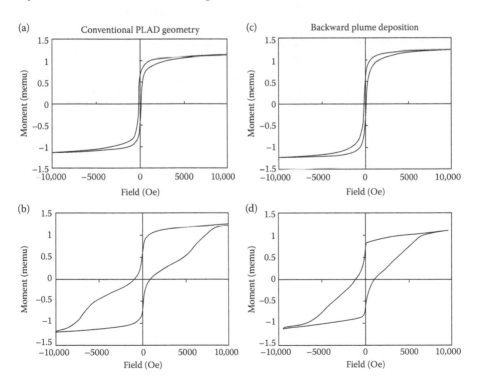

FIGURE 4.13 VSM hysteresis curves of both target–substrate geometries for as-deposited samples (a, c) and samples annealed at 600°C (b, d).

as-deposited conventional PLAD and BPD samples, respectively. It is well known that the remanence ratio is about 0.5 for randomly oriented nanoparticles undergoing coherent magnetic moment rotations without interaction. The remanence ratio decreases when the nanoparticles exhibit magnetostatic interaction and increases when they exhibit exchange coupling interaction. The remanence ratio of both as-deposited samples is less than 0.5, indicating that the major intergranular interaction for these samples is magnetostatic in nature.

It may be noted that the hysteresis loops of samples annealed at 400°C and 500°C are not included in this figure as they are quite similar to the as-deposited samples though with slightly higher coercivities. The variation of coercivities with annealing temperature for both conventional PLAD and BPD samples is shown in Figure 4.14. As seen in Figure 4.14, with the increase in annealing temperature to 400°C and 500°C, the coercivities of samples increase to a few hundred Oe. Only at the annealing temperature of 600°C the coercivity value reaches about a thousand Oe with values being 938 and 1149 Oe for conventional PLAD- and BPD-grown samples, respectively. The increase in coercivity can be attributed to the gradual phase transition from the chemically disordered A1 soft magnetic *fcc* phase to partially chemically ordered $L1_0$ hard magnetic *fct* phase with increasing annealing temperature.

The remanence ratio of the samples annealed at 600°C is found to increase to be greater than 0.5 at values of 0.73 and 0.75, respectively, for conventional PLAD- and BPD-grown samples, indicating the intergranular interaction changing from magnetostatic to exchange coupling. The high remanence ratio after partial phase transition to *fct* phase at 600°C annealing, indicating high exchange coupling effects, can be related to the grain growth and agglomeration of FePt nanoparticles during high-temperature annealing. Weak kinks can also be observed in the hysteresis loop of the sample annealed at 600°C, indicating the coexistence of the magnetically

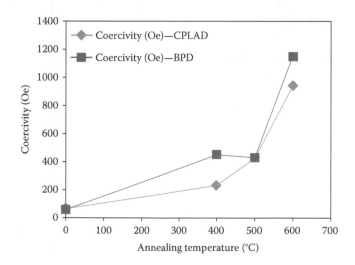

FIGURE 4.14 Variation of coercivities with annealing temperature for FePt nanoparticles samples synthesized with conventional PLAD (CPLAD) and BPD system.

soft *fcc* phase and magnetically hard *fct* phase. The kinks in the hysteresis loops originate from the exchange coupling between the grains of magnetically hard and soft phases. The higher the percentage of the magnetically soft *fcc* phase, the greater will be the exchange coupling interaction between the two phases increasing the magnitude of the kinks.

4.3.2.2 FePt:Al$_2$O$_3$ Nanocomposites: FePt Nanoparticles in Nonmagnetic Al$_2$O$_3$ Matrix

FePt nanoparticles synthesized by conventional PLAD and BPD were in the form of nanoparticle agglomerates and nanoparticle networks with significant exchange coupling effects, which needed to be reduced to improve the magnetic properties of FePt nanoparticles for their potential applications in ultra-high-density magnetic data storage. One of the schemes normally adopted to reduce exchange coupling is to embed magnetic nanoparticles (such as FePt) in nonmagnetic matrix materials (such as Al$_2$O$_3$ and MgO). FePt:Al$_2$O$_3$ nanocomposites were prepared using co-ablation of specially fabricated concentric targets. The nonmagnetic matrix materials are expected to physically isolate FePt nanoparticles, thereby helping in minimizing their annealing effects such as grain growth and agglomeration and in turn reducing the exchange coupling.

In order to prepare FePt:Al$_2$O$_3$, the nanocomposites PLAD facility was used with a specially designed concentric target as shown in Figure 4.15. The special concentric target consisted of a FePt target disk enclosed by a concentric alumina (Al$_2$O$_3$) target disk. The co-ablation of the concentric FePt and alumina target disks used two different laser beams. Two different materials (FePt and alumina) were simultaneously ablated from the two target surfaces by two lasers and were co-deposited onto silicon substrates mounted on a rotating substrate holder (refer to Figure 4.15). The Continuum Nd:YAG laser (532 nm, 10 Hz, 10 ns, 80 mJ, and 1.02×10^3 J/cm^2) was focused on the central 1-inch FePt target disk, and another Lotis Nd:YAG laser (532 nm, 10 Hz, 7 ns, 140 mJ, and 1.47×10^3 J/cm^2) was focused on the concentric Al$_2$O$_3$ target disk. Depositions were done using 36,000 shots in vacuum (with base pressure better than 2×10^{-5} mbar) at room-temperature silicon substrate.

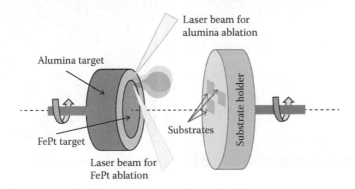

FIGURE 4.15 Concentric alumina and FePt target geometry with their co-ablation by two laser pulses for FePt:Al$_2$O$_3$ nanocomposites synthesis in PLAD facility.

4.3.2.2.1 Morphological and Structural Features of FePt:Al$_2$O$_3$ Nanocomposites

The morphological features of FePt:Al$_2$O$_3$ nanocomposite thin films are shown in Figure 4.16a. A large number of laser droplets can be observed on the surface of the deposited thin films. The point scan energy-dispersive x-ray (EDX) spectrum of big laser droplets, shown in the inset in Figure 4.16a, indicates that most of these laser droplets are that of alumina. The higher laser energy density used on Al$_2$O$_3$ targets coupled with their porous nature might be the reason for greater laser droplets of this material on FePt:Al$_2$O$_3$ nanocomposite thin films. Figure 4.16b shows the typical EDX spectrum of entire FePt:Al$_2$O$_3$ nanocomposite thin films whose SEM image is shown in Figure 4.16a. The composition is determined to be Fe$_{0.48}$Pt$_{0.52}$, which is close to equiatomic percentages of target materials.

Figure 4.17 shows XRD patterns of as-deposited and annealed thin films of FePt:Al$_2$O$_3$ nanocomposite. Thermal annealing of samples was performed in vacuum with samples heated to the desired temperature at 60°C/min and maintained at the desired annealing temperature for 3 h before allowing them to cool down naturally to room temperature. As-deposited FePt:Al$_2$O$_3$ nanocomposite thin films exhibit *fcc* phase FePt with a broad and weak peak of (111) at about 41°. After annealing at 300°C, the intensity of the (111) peak increases slightly, indicating the improved crystallinity. As the annealing temperature increases further to 500°C, the intensities of the fundamental (111) and (002) peaks increase, which indicates the improved crystallinity of the *fcc* phase. The slight splitting of the (111) and (002) peaks indicates the partial phase transition of FePt from low-K_u *fcc* to high-K_u *fct*, since the lattice constants of *fct* phase are different from those of *fcc* phase. No diffraction peak of Al$_2$O$_3$ was observed in any of the XRD patterns, confirming that it is in amorphous form in nanocomposite thin films. As the annealing temperature increases to 600°C, the appearance of a weak diffraction peak at around 24° corresponding to the (001) diffraction plane indicates the partial phase transition of FePt from *fcc* to *fct*. Other diffraction peaks become sharper and narrower, indicating the increase in crystallite size. The mean crystallite size of FePt is estimated to be about 7.9, 14.3,

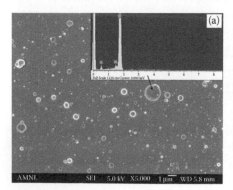
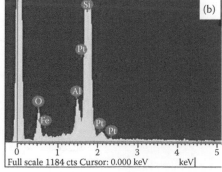

FIGURE 4.16 (a) SEM image of FePt:Al$_2$O$_3$ nanocomposite thin film deposited by conventional PLAD system. Inset in (a) is the point scan EDX spectrum of big laser droplet. (b) EDX spectrum of entire SEM image shown in (a).

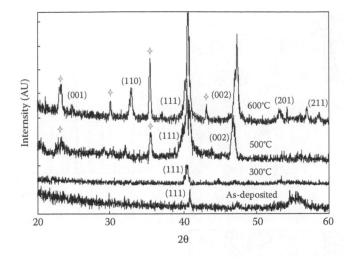

FIGURE 4.17   XRD patterns of as-deposited and annealed FePt:Al$_2$O$_3$ nanocomposite thin films.

and 29.7 nm for as-deposited sample and samples annealed at 500°C and 600°C, respectively, indicating the increase in crystallite size with annealing temperature.

Once again it may be noted that a few diffraction peaks in XRD patterns of FePt:Al$_2$O$_3$ samples annealed at 500°C and 600°C do not match with any of the FcPt peaks in the database. These peaks are marked with star signs in Figure 4.17. These unknown peaks, as discussed later on exclusively in Section 4.3.4, are due to impurity phase formation, which results in the degradation of magnetic properties.

Figure 4.18 shows transition electron microscopy (TEM) images of as-deposited and 500°C-annealed samples of FePt:Al$_2$O$_3$ nanocomposite thin films. The TEM image of as-deposited nanocomposite sample, shown in Figure 4.18a1, exhibits the formation of uniformly sized FePt nanoparticles. The size distribution of as-deposited FePt nanoparticles, estimated from TEM images, is shown in Figure 4.18a2. The average nanoparticle size is estimated to be 7.3 ± 1.1 nm. The corresponding selected area diffraction (SAD) pattern as as-deposited nanocomposite thin film is shown in the inset of Figure 4.18a, indicating that the sample has weak *fcc* phase FePt polycrystallites. As shown in Figure 4.18b1, after annealing at 500°C, FePt nanoparticles grow to slightly bigger size with an average nanoparticle size of about 12.3 ± 2.4 nm with broadened size distribution (refer Figure 4.18b2). It can be noticed that grain growth and agglomeration of FePt nanoparticles is observed after thermal annealing, though reduced, but it cannot be eliminated at high annealing temperature even in the presence of the nonmagnetic Al$_2$O$_3$ matrix. The SAD pattern in the inset of Figure 4.18b1 shows a clearer ring pattern for the sample annealed at 500°C, indicating the improved crystallinity of *fcc* phase FePt nanoparticles.

4.3.2.2.2   Magnetic Properties of FePt:Al$_2$O$_3$ Nanocomposites

Hysteresis loops of as-deposited and annealed samples of FePt:Al$_2$O$_3$ nanocomposite thin films deposited are shown in Figure 4.19. The coercivity and remanence ratio

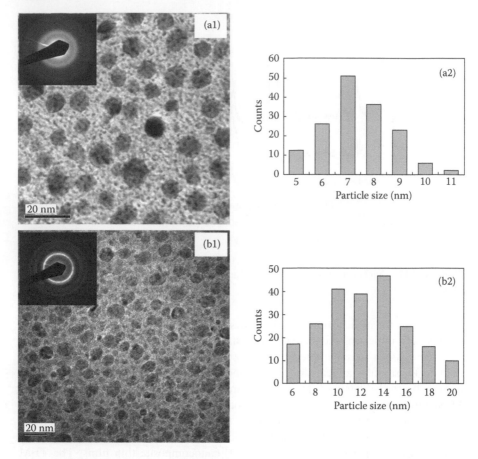

FIGURE 4.18 TEM images of FePt:Al_2O_3 nanocomposite thin films and corresponding particle size distribution of FePt nanoparticles in these nanocomposite thin films (a1, a2) for as-deposited sample and (b1, b2) sample annealed at 500°C. (Reprinted from J. J. Lin et al., FePt:Al_2O_3 nanocomposite thin films synthesized by magnetic trapping assisted pulsed laser deposition with reduced intergranular exchange coupling, *J. Phys. D: Appl. Phys.* 41, 095001, is acknowledged. Copyright 2008, © IOP Publishing. Reproduced with permission. All rights reserved. DOI: http://dx.doi.org/10.1088/0022-3727/41/9/095001.)

of samples were found to increase from low values of 4.19 Oe and 0.12, respectively, for an as-deposited sample to moderate values of 249 Oe and 0.45, respectively, for a sample annealed at 500°C and finally to the highest values of 1207 Oe and 0.72 for a 600°C-annealed sample, respectively. This indicates the change from the soft magnetic phase in as-deposited samples to the hard magnetic phase in annealed samples. Also, the remanence ratio of less than 0.5 in as-deposited and 500°C-annealed samples indicate the presence of magnetostatic interactions among FePt nanoparticles in these nanocomposite samples. The major intergranular interaction changed from magnetostatic to exchange coupling after annealing at 600°C as the remanence

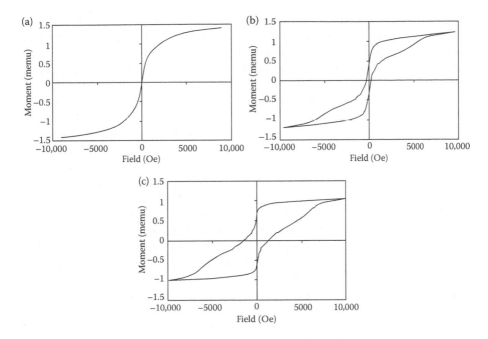

FIGURE 4.19 Hysteresis loops of FePt:Al$_2$O$_3$ nanocomposite thin films deposited by conventional PLAD: (a) as-deposited and annealed at (b) 500°C and (c) 600°C. (Adapted from J. J. Lin, Pulsed laser deposition of nanostructured magnetic materials, PhD thesis, National Institute of Education, Nanyang Technological University, Singapore, 2009, p. 197.)

ratio at 0.72 is more than 0.5. Moreover, the existence of weak kinks in the hysteresis loops of annealed samples indicates the presence of mixed soft and hard magnetic phases.

The synthesis of high anisotropy constant, K_u, FePt nanoparticles, and nanocomposites was motivated with the aims of (i) achieving high coercivities (high magnetic hardness) in FePt nanoparticles for greater stability of stored information and (ii) restricting nanoparticle size to as small as possible for high areal density of magnetic data storage. The reading and writing of magnetic information on storage medium, however, also involves the usage of soft magnetic materials (such as FeCo), which is discussed in the next section.

4.3.3 Schemes for Low-Temperature Phase Transition to Hard Magnetic Phase

As was highlighted earlier, the high-temperature annealing of materials such as FePt and CoPt is required to achieve the phase transformation from the chemically disordered soft magnetic *fcc* phase in as-grown FePt/CoPt to the chemically ordered hard magnetic *fct* phase, for them to be suitable for reliable, thermally stable magnetic data storage. However, high-temperature annealing leads to grain growth and grain coarsening, which is not suitable for high areal density data storage as the particle size is increased, leading to increase in data bit size. Therefore, experimental schemes that enable ordering, phase transition to chemically ordered high-K_u $L1_0$ phase, at significantly reduced temperatures are critically and urgently needed.

Many efforts, like third elements addition [56–62], adding new underlayers or multilayers [63–67], and ion irradiation [68–72], have been devoted to achieve low-temperature phase transition to highly ordered $L1_0$ FePt. Introducing additives, such as Ag [56–58], Cu [56,59], Au [58], Sb [61], Ti [60], and Nb [60], have been tried. Aimuta et al. [56] found that for FePt doped with 3 at% Ag or 7 at% Cu, the ordering temperature is reduced from 500°C to 400°C and they suggested that the movement of Ag to the surface during annealing might have caused the decrease in the ordering temperature. Kang et al. [57,62] also reported that FePt doped with 12 at% Ag decreased the ordering transition temperature by about 100–150°C down from 550°C for the undoped one. According to them, the defects and lattice strain introduced by the Ag and the subsequent segregation of the Ag upon annealing activated the nucleation of the ordered phase. Furthermore, Sb-doped FePt nanoparticles were studied by Yan et al. [61], and the ordering temperature was lowered to 300°C, around 50–100°C more than the lowest reported for Ag- or Au-doped FePt nanoparticles [56,57]. However, Ag- or Au-doped FePt nanoparticles by Wang et al. [58], Cu-doped FePt by Platt et al. [59], and Ti, Nb-doped FePt by Mahalingam et al. [60] were reported without a positive effect on the ordering temperature.

The underlayer/multilayers systems such as FePt/Cu [63], FePt/CrX (X = Ru, Mo, W, Ti) [64], FePt/Ag [65], FePt/Au [66], and [Pt/Fe]$_n$ [67] have also been investigated to reduce the ordering temperature. The decreased ordering temperature for them is attributed to the induced strain resulting from the lattice mismatch between underlayers and FePt thin film. Lai et al. [63] found that FePt with a Cu underlayer obtained an in-plane coercivity as high as 6900 Oe after postannealing at 300°C. Underlayers with CrX (X = Ru, Mo, W, Ti) alloys were studied by Chen et al. [64], and reported that large uniaxial anisotropy, good magnetic squareness, and (001) texture were obtained at an ordering temperature of 250°C and above by using CrRu. The FePt/Ag multilayers reported by Konagai et al. [65] obtained a perpendicular magnetized film around the annealing temperature of 200°C, and a large anisotropy was exhibited at 300°C. Au underlayers were tried by Feng et al. [66], and they found a decrease in the ordering temperature by 150°C, that is at around 450°C, and a considerable increase in H_c. Furthermore, other kinds of multilayers, such as [Pt(1 nm)/Fe(1 nm)]$_3$/Pt(3 nm)/Fe(3 nm) multilayers studied by Ogata et al. [67], were found to reduce the ordering annealing temperature to 500°C with a high value of saturation magnetization around 12 kG.

In addition to the aforementioned methods, we will discuss in detail in this chapter two methods of lowering the phase transition temperature of FePt that were used by our research group. They are (i) magnetic trapping assisted pulsed laser ablation deposition used by Lin et al. [73] discussed in the following section and (ii) high-energy high-flux ion irradiation of PLAD-grown FePt thin films in plasma focus device [68–70], which is discussed later in Section 4.4.3.

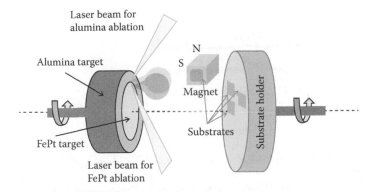

FIGURE 4.20 Experimental setup for magnetic trapping assisted PLAD. Note the magnet placed perpendicular to the target–conventional substrate holder axis.

4.3.3.1 Magnetic Trapping Assisted PLAD for FePt:Al_2O_3 Nanocomposites

The experimental setup for magnetic trapping assisted PLAD, as special target–substrate geometry, to reduce the phase transition temperature of FePt nanoparticles is shown in Figure 4.20. In the target–substrate geometry, silicon substrates were placed on the surface of a strong permanent magnet of magnetic field strength of about 0.4 T at its surface. The magnet, as shown in the schematic of the setup, was placed perpendicular to the target–conventional substrate holder axis. The substrates mounted on the magnet were therefore parallel to the laser-ablated plasma plume for magnetic trapping assisted PLAD.

The setup shown in Figure 4.20 was used to deposit FePt:Al_2O_3 nanocomposite samples using magnetic trapping assisted PLAD, so the rest of the experimental setup, that is, use of a special concentric target, consisted of a FePt target disk enclosed by a concentric alumina (Al_2O_3) target disk, two laser beam-based co-ablations, laser pulse energy densities of two lasers on target surface, and the base pressure of the PLAD chamber (better than 2×10^{-5} mbar) were all similar to the one described in Section 4.3.2.2. The major difference was that the magnet was mounted rigidly so the magnet and hence the substrates were not rotating. Ions and electrons of co-ablated plasma plumes were expected to be trapped and spirally drift along magnetic field lines and move toward the substrate surface for deposition. It was also expected that the laser droplets, which are the large particulates ejected directly from the target by mechanical forces, will not be trapped by magnetic field due to their large momentum and will continue to move forward toward the conventional substrate holder place down the target axis. Magnetic trapping assisted methods were used by Kobayashi et al. [74] to synthesize $SrTiO_3$ films with improved crystallinity as well as Huh et al. [75] to synthesize ferromagnetic nanocluster rods.

4.3.3.2 XRD Results of FePt:Al_2O_3 Nanocomposites Grown by Magnetic Trapping Assisted PLAD

XRD patterns of as-deposited and thermally annealed FePt:Al_2O_3 nanocomposite thin films synthesized by magnetic trapping assisted PLAD are shown in Figure 4.21. As-deposited thin film exhibits the *fcc* phase with a broad, asymmetric, and

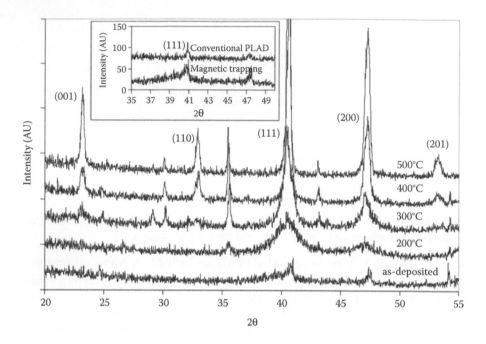

FIGURE 4.21 XRD patterns of FePt: American Institute of Physics. Al_2O_3 nanocomposite thin films deposited by magnetic trapping assisted PLAD and annealed at different temperatures. Inset: (111) peak of as-deposited samples by conventional PLAD and magnetic trapping assisted PLAD. (Reprinted with permission from J. J. Lin et al., Magnetic trapping induced low temperature phase transition from fcc to fct in pulsed laser deposition of FePt:Al_2O_3 nanocomposite thin films, *Appl. Phys. Lett.* 91, 063120. Copyright 2007, American Institute of Physics Publishing LLC.)

weak peak of (111) at about 41°. The asymmetry in the diffraction peak is known to result from a high concentration of crystal defects. The inset in Figure 4.21 shows that the (111) peak of the as-deposited sample grown by magnetic trapping assisted PLAD exhibits significantly higher asymmetry than that of the (111) peak in the as-deposited sample grown by conventional PLAD. Thus, the concentration of crystal defects in the magnetic trapping assisted grown as-deposited sample is much higher. The sample annealed at 200°C exhibits increased crystallinity and decrease in crystal defects concentration (indicated by improved symmetry of the (111) peak), but it still remains in the *fcc* phase. Both as-deposited and 200°C-annealed samples also show weak peaks corresponding to the (200) diffraction plane. The samples annealed at 300°C, 400°C, and 500°C exhibit further consolidation of crystallinity as the (111) and (200) diffraction peak intensities increased and also several new diffraction peaks, such as (001), (110), and (201), of FePt emerged and strengthened with increasing annealing temperature. The partial conversion to chemically ordered *fct* FePt phase is confirmed by the appearance of the (001) peak at around 24° at an annealing temperature of 300°C. Hence, this sample has already been partially converted to the magnetically hard *fct* phase. However, FePt:Al_2O_3 nanocomposite thin films synthesized by conventional PLAD, without magnetic trapping, remain predominantly in the *fcc* phase even after annealing at 500°C. The peaks without labeling are attributed by the nonmagnetic matrix materials as well as Si substrates. As the annealing temperature increases further to 400°C and then to 500°C, the peak around 24° (of *fct* phase FePt) becomes stronger and narrower. It indicates the improved crystallinity in the sample along with greater conversion to chemically ordered $L1_0$ *fct* phase.

The reason for the initiation of phase transition to the $L1_0$ ordered *fct* phase at lower annealing temperature of 300°C can be understood following the explanation given by Wiedwald et al. [76]. According to them [76], one of the ways to lower the phase transition temperature is to reduce the activation energy, E_D, for the diffusion of atom (needed to achieve ordered state) by increasing the number of point defects, such as vacancy and interstitial, in the crystal structures. The phase transition can be related to the characteristic diffusion length λ during postannealing, which is given by $\lambda = \sqrt{Dt_A}$ with $D = D_0 \exp(-E_D/k_B T_A)$, where D is the diffusion coefficient, k_B is the Boltzmann's constant, t_A is the annealing time, and T_A is the annealing temperature. The samples synthesized by magnetic trapping assisted PLAD have higher defect concentration, indicated by large asymmetry in the (111) diffraction peak. The increase in defect concentration for magnetic trapping assisted PLAD synthesis can be related to the increased degree of ionization and activation of neutral ablated species due to the increase in collision with the electrons trapped along the magnetic field lines due to their smaller Larmor radius.

4.3.3.3 TEM Results of FePt:Al₂O₃ Nanocomposites Grown by Magnetic Trapping Assisted PLAD

Figure 4.22 shows the TEM images of as-deposited and thermally annealed FePt:Al_2O_3 nanocomposite thin films synthesized by magnetic trapping assisted PLAD. The as-deposited sample, shown in Figure 4.22a, has well-separated FePt nanoparticles with narrow size distribution with most particles having a size of about 6 nm. The TEM image of sample annealed at 300°C (refer to Figure 4.22b) shows that FePt nanoparticles remain separated but the average particle size increases slightly to about 8 nm and the particle size distribution is broadened. The SAD pattern in the inset in Figure 4.22b shows that FePt nanoparticles have been converted to the *fct* phase after annealing at the temperature of 300°C, which is identified by JEMS@JEOL. The dot-line-like SAD ring pattern in Figure 4.22c for the sample

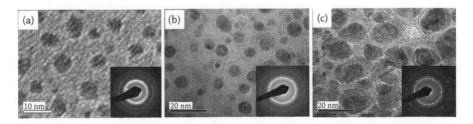

FIGURE 4.22   TEM images of FePt:Al₂O₃ nanocomposite thin films synthesized by magnetic trapping assisted PLAD (a) as-deposited, (b) annealed at 300°C, and (c) annealed at 500°C. Insets are corresponding SAD patterns.

annealed at 500°C indicates improved crystallinity along with an increased average particle size of about 13–15 nm along with some signs of particle agglomeration.

4.3.3.4 Magnetic Properties of FePt:Al$_2$O$_3$ Nanocomposites Grown by Magnetic Trapping Assisted PLAD

The coercivity of FePt:Al$_2$O$_3$ nanocomposite thin films synthesized by conventional PLAD and magnetic trapping assisted PLAD and annealed at different temperatures is shown in Figure 4.23. Typical hysteresis loops for as-deposited and 400°C-annealed samples synthesized by magnetic trapping assisted PLAD are shown in the inset of Figure 4.23. For the samples synthesized by magnetic trapping assisted PLAD, coercivity increases by about 5 times from 130 to 617 Oe after annealing at 300°C, which can be related to the phase transition from low-K_u fcc to high-K_u fct as indicated by XRD and TEM results discussed earlier. As the annealing temperature increases, the coercivity increases due to the improved crystallinity in the fct phase, which enhances the magnetocrystalline anisotropy. The increase in coercivity is, however, slowed down at a higher annealing temperature due to the grain growth and agglomeration, which may change FePt nanoparticles from single domain and single nanocrystallite to multidomains and nanopolycrystallites with random orientation, which will reduce the net magnetocrystalline anisotropy.

Figure 4.24 shows the remanence ratio S and the coercive squareness $S*$ of FePt:Al$_2$O$_3$ nanocomposite thin films synthesized by conventional PLAD and

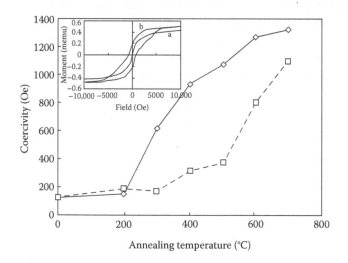

FIGURE 4.23 The coercivity of FePt:Al$_2$O$_3$ nanocomposite synthesized by (□ line) conventional PLAD and (◊ line) magnetic trapping assisted PLAD and annealed at different temperatures. Inset shows the hysteresis loops of sample synthesized by magnetic trapping assisted PLAD: (a) As-deposited and (b) annealed at 400°C. (Reprinted from J. J. Lin et al., FePt: Al$_2$O$_3$ nanocomposite thin films synthesized by magnetic trapping assisted pulsed laser deposition with reduced intergranular exchange coupling, *J. Phys. D: Appl. Phys.* 41, 095001, is acknowledged. Copyright 2008 © IOP Publishing. Reproduced with permission. All rights reserved. DOI: http://dx.doi.org/10.1088/0022-3727/41/9/095001.)

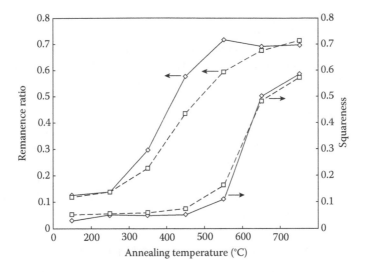

FIGURE 4.24 The remanence ratio S and coercive squareness S* of FePt:Al$_2$O$_3$ nanocomposite synthesized by (□ dash line) conventional PLAD and (◊ solid line) magnetic trapping assisted PLAD and annealed at different temperatures. (Reprinted from J. J. Lin et al., FePt: Al$_2$O$_3$ nanocomposite thin films synthesized by magnetic trapping assisted pulsed laser deposition with reduced intergranular exchange coupling, *J. Phys. D: Appl. Phys.* 41, 095001, is acknowledged. Copyright 2008 © IOP Publishing. Reproduced with permission. All rights reserved. DOI: http://dx.doi.org/10.1088/0022-3727/41/9/095001.)

magnetic trapping assisted PLAD and annealed at different temperatures. While the remanence ratio S (also referred to as remanence squareness) is a measure of the degree to which the shape of hysteresis loop can be approximated by a rectangle, the coercive squareness S* measures how close the tangent to the loop at H_c (coercive field value) is to the vertical. For magnetic trapping assisted PLAD synthesis, the remanence ratio of as-deposited sample and samples annealed at 200°C and 300°C remains smaller than 0.5, indicating that the major intergranular interaction is magnetostatic. The remanence ratio increases and becomes greater than 0.5 at the annealing temperature of 400°C, indicating that the main intergranular interaction has changed from magnetostatic to exchange coupling. It must be pointed out over here that the exchange coupling interaction should be decreased as much as possible to reduce the media noise in order to increase the signal-to-noise ratio, even if an appropriate amount of exchange coupling is indeed required for stabilizing the grain against thermal and demagnetizing effects. Therefore, one of the major advantages of lower phase transition temperature of 300°C in magnetic trapping assisted PLAD is the reduced intergranular exchange coupling in the *fct* phase FePt nanoparticles as the remanence ratio is less than 0.5. The strength of exchange interaction in granular films, the kind of film we have, can be estimated by evaluating coercive squareness, $S^* = 1 - M_r / \alpha H_c$, where α is the hysteresis slope at coercive field value of H_c. It is generally assumed that the small value of coercive squareness indicates weak intergranular exchange interactions. The coercive squareness also has an inverse relationship with the switch field distribution (SFD). The SFD is defined as the reversing field range required to reduce the magnetization of an initially saturated medium from $M_r/2$ to $-M_r/2$. The SFD value is finite for randomly oriented nanoparticles. The magnetostatic interaction can broaden the SFD and hence decrease the coercive squareness, whereas the intergranular exchange coupling can encourage the cooperative switching and hence increase the coercive squareness. The coercive squareness has a significant increase for samples annealed at 600°C and above for both conventional PLAD and magnetic trapping assisted PLAD-based synthesis, indicating that the major intergranular interaction changes from magnetostatic to exchange coupling after annealing at a temperature greater than 500°C, which is almost consistent with the conclusion based on variation in remanence ratio. Hence, it can be concluded that the FePt:Al_2O_3 nanocomposite thin films synthesized by magnetic trapping assisted PLAD can be converted to hard magnetic $L1_0$ fct phase at a lower annealing temperature of 300°C with significantly reduced exchange coupling interaction.

4.3.4 IMPURITY PHASE FORMATION: ISSUES AND SOLUTIONS

Ultra-high magnetic data storage application requires recording media such as FePt nanoparticles to have a particle size as small as possible but continue exhibiting the fct phase and hard magnetic behavior. The degree of phase transformation from soft magnetic phase to hard magnetic phase in FePt thin films [63,70,76,77] has been reported to vary with experimental conditions, including (i) annealing temperature [78], (ii) number and thickness of underlayers [79], (iii) deposition rate [80], and (iv) defects such as antisite defects and doping defects [81]. The impurity phases due to certain defects may affect the magnetic properties of FePt thin films positively or negatively. The relatively small amount of Pt silicide addition is reported to decrease the phase transition temperature for the sample, resulting in an improved phase transformation [82,83]. The formation of Pt silicide on the interface between FePt films and Si substrates is also reported to play an important role in the transformation of FePt films [84]. It is very difficult to quantify the range of impurity phase content to have positive effects on magnetic properties, but it can certainly be stated that if the impurity phase content is too high, then it will definitely have a detrimental effect on the samples' magnetic properties. No clear discussion about impurity phase formation in FePt thin films has been provided in the literature to date, although XRD patterns in some reports [85–87] showed unknown impurity phase formation. In this section, the major focus is on the observation of the impurity phase formation in FePt thin films, their effects on the magnetic properties, and their elimination.

4.3.4.1 Observation of Impurity Phases in PLAD Deposited FePt Thin Films

The FePt thin films are synthesized using PLAD in vacuum with a base pressure better than 5.7×10^{-6} mbar at room temperature. A Lotis-TII (LS-2137U) second-harmonic Nd:YAG laser (532 nm, 10 Hz, 10 ns) is used to ablate the 2.5 cm diameter FePt target (50:50 at.% with 99.99% purity). The formation of impurity phases is observed in ~70 nm FePt films synthesized at room temperature using PLAD with laser energy fluence (LEF) of 182 J/cm² (corresponding to a flash lamp pump energy of 30 J). As shown in the XRD pattern for the annealed FePt sample (Figure 4.25a),

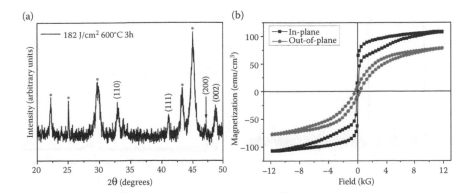

(a)   (b)

FIGURE 4.25   (a) The XRD pattern of PLAD deposited FePt thin films annealed at 600°C for 3 h. The impurity peaks are marked with an asterisk (*). (b) VSM results of the sample; the squares curve for the in-plane hysteresis loop, the circles curve for out-of-plane one. (Reprinted from *Appl. Surf. Sci.*, 288, Y. Wang et al., Elimination of impurity phase formation in FePt magnetic thin films prepared by pulsed laser deposition, 381–391, Copyright 2014, with permission from Elsevier.)

unknown diffraction peaks (labeled as "*") with relatively higher intensities than FePt diffraction peaks (peaks (110), (111), (200) and (002)) are shown in FcPt thin films annealed at 600°C for 3 h in vacuum. Although the diffraction peak of (110) at about 33° and split diffraction peaks of (200) and (002) at about 47° and 49°, respectively, demonstrate the formation of the *fct* phase in the FePt sample, the absence of the typical superlattice peak (001) at about 24° and the relatively smaller intensities of the FePt diffraction peaks indicate the partial phase transformation from *fcc* to *fct* phase in the sample.

The VSM results for the annealed FePt sample (Figure 4.25b) show small values of both out-of-plane coercivity (about 312 G) and in-plane coercivity (about 93 G), revealing the soft magnetic behavior in FePt thin films. This soft magnetic behavior along with the relatively weak FePt diffraction peaks and the absence of the superlattice (001) peak are probably due to the formation of impurity phases in the FePt thin-film sample. Thus, it is necessary to find the reasons for the appearance of impurities, and to eliminate them so that the samples can exhibit hard magnetic properties.

4.3.4.2   Identification of Possible Reasons for Impurity Phase Formation

The possible reasons for impurity phase formation might be (i) impurities from the original target used in PLAD; (ii) contribution of the Si substrate; (iii) incorrect annealing parameters such as annealing temperature, duration, and ambience; and (iv) improper deposition parameters such as deposition environment and LEF. The effects of all the above parameters have been studied, and LEF is found to be the main reason for the impurity phase formation.

The XRD pattern for the FePt target (Figure 4.26a) matches well with the XRD Database@Socabin (PDF #03-065-9121) without exhibiting any impurity phase, indicating zero contribution of the PLAD target to the impurity phase formation.

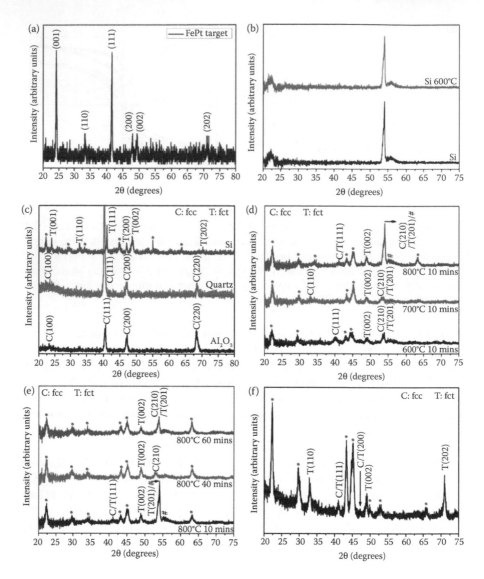

FIGURE 4.26 XRD patterns of (a) the FePt target; (b) bare Si substrate; (c) Si substrate annealed at 600°C for 3 h; (d) FePt thin films annealed at 600°C, 700°C, and 800°C for 10 min; (e) FePt thin films annealed at 800°C for 10, 40, and 60 min in vacuum; (f) FePt thin films annealed at 600°C for 60 min in 95%Ar:5%H$_2$ flowing gas ambience; the impurity peaks are marked with an asterisk (*), and the peaks related to Si substrate are marked with a number sign (#). (Reprinted from *Appl. Surf. Sci.*, 288, Y. Wang et al., Elimination of impurity phase formation in FePt magnetic thin films prepared by pulsed laser deposition, 381–391, Copyright 2014, with permission from Elsevier.)

The XRD patterns for the bare Si substrate and 600°C-annealed Si substrate also do not show any impurity diffraction peaks, revealing that the impurity phases do not originate from the Si substrate signal. However, the XRD patterns for three FePt thin-film samples, which were deposited on different substrate materials (Si, quartz, and alumina) but with the same deposition and postannealing conditions, exhibit obvious difference for impurity diffraction peaks among the three samples. Impurity phase formation is only observed for the sample deposited on Si substrate, demonstrating that the Si substrate must contribute to the impurity phase formation by certain unknown ways during the postannealing process or deposition process. XRD patterns in Figure 4.26d–f show the effect of annealing conditions, including annealing temperature (Figure 4.26d), annealing duration (Figure 4.26e), and annealing ambience (Figure 4.26f) on the impurity formation. Three sets of samples are prepared on Si substrates: (i) three PLAD deposited FePt samples are postannealed at 600°C, 700°C, and 800°C for 10 min in vacuum; (ii) another three PLAD deposited FePt samples are postannealed at the same temperature of 800°C but for different durations of 10, 40, and 60 min in vacuum; (iii) one PLAD deposited FePt sample was postannealed at 600°C for 60 min in 95%Ar:5%H_2 flowing gas ambience instead of in vacuum. No obvious decrease of impurity diffraction peaks has been found with the change of annealing temperature, annealing duration, and annealing ambience, indicating that impurity phases should not be induced during the annealing process.

The effect of deposition parameters on the impurity phase formation is further studied, and the related XRD patterns are shown in Figure 4.27a and b. One sample is prepared in the PLAD chamber, which was flushed with argon (Ar) gas before FePt thin-film deposition. The deposited sample was annealed at 500°C for 60 min in 95%Ar:5%H_2 flowing gas ambience. The XRD pattern of this sample (Figure 4.27a) still shows obvious impurity phase formation with some relatively stronger diffraction peaks than FePt diffraction peaks, indicating that deposition ambience is also not the reason for impurity phase formation.

Another two samples are deposited with different LEFs of 51 and 182 J/cm² (corresponding to flash lamp pump energies of 25 and 30 J, respectively). The thickness of the samples, determined by cross-sectional FESEM imaging (Figure 4.27b), are about 37.2 ± 0.6 and 34.0 ± 1.5 nm, respectively. XRD patterns of the 600°C-annealed samples (Figure 4.27c) show obvious reduction of impurity diffraction peaks with decreasing LEF from 182 to 51 J/cm², demonstrating an important role of LEF in the impurity phase formation. The XRD pattern of the sample synthesized at 51 J/cm² LEF matches well with the XRD Database@Socabin (PDF #03-065-9121). The obvious superlattice (001) peak at about 24° and the splitting of the (200) and (002) diffraction peaks demonstrate the phase transformation from the soft magnetic *fcc* phase to hard magnetic *fct* phase in the sample. The calculated order parameter S of the samples with LEFs of 51 and 182 J/cm² have similar values of 0.95 and 0.94, respectively, which indicates high but not complete *fcc* to *fct* phase transformation in both the samples. However, a large amount of impurity diffraction peaks (labeled as "*") with relatively higher intensity than FePt diffraction peaks and the absence of the superlattice (001) peak are observed for the sample synthesized at higher LEF of 182 J/cm².

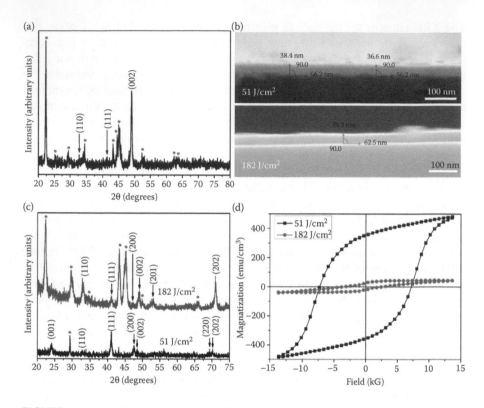

FIGURE 4.27 (a) The XRD pattern of FePt thin films deposited in the argon gas-based vacuum ambience and annealed at 500°C for 60 min. (b) Cross-sectional SEM images, (c) XRD patterns, and (d) in-plane hysteresis loops of FePt thin films deposited with LEFs of 51 and 182 J/cm² and annealed at 500°C for 60 min; the impurity peaks are marked with an asterisk (*). (Reprinted from *Appl. Surf. Sci.*, 288, Y. Wang et al., Elimination of impurity phase formation in FePt magnetic thin films prepared by pulsed laser deposition, 381–391, Copyright 2014, with permission from Elsevier.)

The in-plane hysteresis loops for the two annealed samples deposited with LEF of 51 and 182 J/cm² (Figure 4.27d) exhibit hard magnetic behavior with coercivities of 7242 and 3043 G, respectively, which is coincident with the high phase transformation in both the samples shown in the XRD patterns. Much smaller anisotropy K_u is found for the sample with LEF of 182 J/cm² when compared to that of the 51 J/cm² LEF deposited sample, which might be due to the large amount of impurity phases formed in the sample with LEF of 182 J/cm². The anisotropy K_u is proportional to the closed area of the hysteresis loop.

The definite role played by higher-ablation LEF in the impurity phase formation is evidenced by the XRD and VSM results discussed above. The possible effect of LEF on the formation of the impurity phase can be explained by the snow plow and shockwave model reported by Mahmood et al. [89]. They performed the plasma plume imaging for different incident LEFs and fitting of the experimental data with the model. The fraction of input laser energy deposited to the ablated plasma species is demonstrated to be about 70%, which means that at higher incident LEF, the energy of the ablated plasma species (Fe and Pt in this case) must be higher. Higher ablated plasma species will result in greater kinetic energy of the impinging plasma plume species on the Si substrate. The energetic impinging plasma plume is reported to sputter the substrate surface, resulting in the interaction and re-condensation of substrate materials with impinging plasma species [89]. Hence, the impinging energetic Fe and Pt plasma species formed at higher LEF may ablate the native oxide layer and Si on substrate and form Fe and Pt silicide phases in the thin-film samples.

The effect of the energy of the ablated plasma species on the magnetic and crystalline properties of the PLAD deposited FePt thin films was further investigated by increasing the deposition distance between target and substrates from the initial 2 to 5 cm. The energy and energy flux of the ablated plasma species are predicted to decrease for depositions at longer target-to-substrate distance due to their interaction with ambient gas and expansion of the plume. Three samples are deposited by using LEFs of 51, 136, and 182 J/cm^2 (corresponding to flash lamp pump energies of 25, 28, and 30 J, respectively). The thicknesses of the as-deposited samples are 6.9 ± 1.0, 13.3 ± 0.8, and 13.9 ± 0.9 nm, respectively (Figure 4.28a). The as-deposited samples are postannealed at 600°C for 60 min in 95%Ar:5%H$_2$ flowing gas.

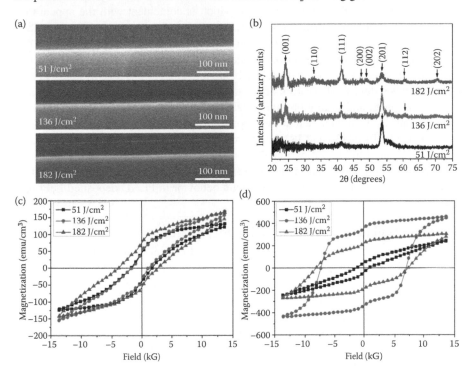

FIGURE 4.28 (a) Cross-sectional SEM images, (b) XRD patterns, (c) in-plane hysteresis loops, and (d) out-of-plane hysteresis loops of the annealed FePt thin films synthesized at 51, 136, and 182 J/cm^2 with target-to-substrate distance of about 5 cm. (Reprinted from *Appl. Surf. Sci.*, 288, Y. Wang et al., Elimination of impurity phase formation in FePt magnetic thin films prepared by pulsed laser deposition, 381–391, Copyright 2014, with permission from Elsevier.)

XRD patterns of all the annealed samples (Figure 4.28b) match well with the XRD Database@Socabin (PDF #03-065-9121). It must be noted that no impurity diffraction peak is observed in XRD patterns for all three samples, especially for the sample deposited at the highest LEF of 182 J/cm^2, the sample that exhibits impurity phase formation when the target-to-substrate distance was 2 cm. This indicates the elimination of impurity phases in the PLAD deposited FePt thin films. The energy of the ablated plasma species at the substrate surface is significantly decreased with increasing target-to-substrate distance from 2 to 5 cm, which results in the elimination of the probability of sputtering the Si substrate and no formation of Pt silicide or other impurity phases. Hence, the energy of the ablated plasmas species at the substrate surface is the most critical factor that controls the impurity phase formation.

VSM results in Figure 4.28c and d show both in-plane and out-of-plane hard magnetic behavior for all three samples. The in-plane coercivities of the annealed samples deposited at the LEFs of 51, 136, and 182 J/cm^2 are about 1478, 1409, and 3313 G, respectively, while their out-of-plane coercivities are about 1077, 7146, and 7712 G, respectively. The higher out-of-plane coercivities than the in-plane ones for the samples deposited at 136 and 182 J/cm^2 LEFs indicates a larger percentage of perpendicular orientation in the samples, which is coincident with the appearance of FePt (001) diffraction peaks in the XRD patterns of the samples (Figure 4.28b). The appearance of FePt (001) diffraction peaks indicates the formation of a (001) crystalline structure that has an out-of-plane-aligned c axis in the samples, which leads to improved out-of-plane magnetic behavior. The out-of-plane hysteresis loops of the samples deposited at LEFs of 136 and 182 J/cm^2 also exhibit higher squareness, remanence, and saturation magnetization when compared to the in-plane hysteresis loops. In the perpendicular direction, the squarenesses (M_r/M_s) of the samples deposited at LEFs of 136 and 182 J/cm^2 are 0.77 and 0.70, respectively. All the XRD and VSM results discussed earlier demonstrate that the elimination of the impurity phase formation results in improved perpendicular orientation crystallization and hard in-plane and out-of-plane magnetic behavior in PLAD deposited FePt thin films.

4.3.4.3 Possible Compositions of Impurity Phases

The formation of Fe and Pt silicide phases in FePt thin-film samples grown on Si substrates can be studied by using the composition depth profile as reported by Sharma et al. [90]. Thus, to find more information about the impurity phases, a comparison of the composition depth profile of the samples with and without impurity formation was carried out by using secondary ion mass spectrometry (SIMS). The depth profiles are shown in Figure 4.29. The composition depth profile in Figure 4.29a belongs to the sample with impurities, whose results are discussed in Figure 4.25. The sputtered depth where Si profile saturates, that is, at the depth of about 70 nm in Figure 4.29a matches well with the real thickness of this sample (about 74.3 ± 1.0 nm) as characterized by cross-sectional SEM. Thus, the interface between FePt thin film and Si substrate is identified at a depth of about 70 nm as marked in Figure 4.29a. Based on the distribution profiles of Fe$^+$, Pt$^-$, and Si$^-$, three different regions (marked as I, II, and III) are identified above the thin-film–substrate interface. In region III, the complex FePt silicide composition is formed immediately above the Si substrate

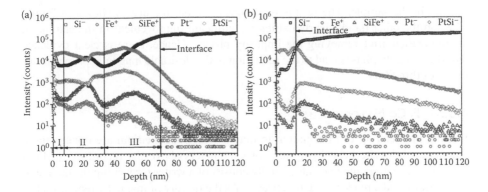

FIGURE 4.29 SIMS depth profiles of the samples (a) with impurity phase formation and (b) without impurity phase formation. (Reprinted from *Appl. Surf. Sci.,* 288, Y. Wang et al., Elimination of impurity phase formation in FePt magnetic thin films prepared by pulsed laser deposition, 381–391, Copyright 2014, with permission from Elsevier.)

as the profiles of Fe^+, Pt^-, $FeSi^+$, and $PtSi^-$ composition plots are similar with peaks in this region. In region II (about 7–33 nm), the profile of Si^- signal clearly indicates the Si diffusion into the FePt thin film. Although region II is significantly rich in Pt, it has similar compositional profiles of Fe^+, Si^-, and $FeSi^+$ with peaks. Region II, therefore, has a stronger Fe silicide phase formation along with FePt and Pt silicide phases. In region I, essentially FePt is found at the top of this sample as a minimum amount of Si is observed. Considering the three regions as a whole, it can be concluded that the FePt phase is formed with low volume fraction and the major part of FePt diffusing with Si leads to the formation of various Pt silicide, Fe silicide, and FePt silicide phases.

Figure 4.29b shows the depth profile of the sample without impurity formation, whose XRD pattern is shown in Figure 4.28b. This sample was grown with the same LEF of 182 J/cm^2 as the sample with impurity formation shown in Figure 4.29a. The Si compositional profile for this sample shows a much sharper and larger drop in Si signal counts when compared to the sample with impurity formation, suggesting significantly lower diffusion of Si into the top thin-film layer. This is further proved by the sharp dip exhibited in compositional profiles of $FeSi^+$ and $PtSi^-$ in the thin-film region. The almost parallel Pt^- and Fe^+ compositional profiles at the top of the sample (0 to ~13 nm) indicate the formation of a FePt thin-film layer on the Si substrate with a thickness of about 13 nm, which agrees with the real thickness of the sample (13.9 ± 0.9 nm) characterized by cross-sectional SEM (Figure 4.28a).

Considering the SIMS results discussed earlier, it can be concluded that impurities in our samples are composed of complex combinations of Pt silicides, Fe silicides, and FePt silicides. Combining the results obtained from SIMS, XRD, and VSM of the two samples with and without impurity formation, it seems that the formation of Pt silicides, Fe silicides, and FePt silicides in FePt thin-film samples resulted in relatively weaker magnetic properties. This is also confirmed by the fact that the lower saturation magnetization for the sample deposited at LEF of 182 J/cm^2 as shown in Figure 4.27d is attributed to the relatively higher amount of impurities formed in the sample. The results of samples without impurities shown in Figure 4.28, on the other hand, clearly exhibit a significant improvement in the saturation magnetization after removing the impurity phases. Hence, there is clear evidence that the impurity phases lead to the reduction in the desirable magnetic properties of the FePt thin films.

4.3.5 FeCo Nanoparticles

Writing and reading of information in magnetic storage medium are done using a magnetic read/write head flying tens of nanometers above the recording medium surface. Current magnetic heads actually consist of a writer for writing the data onto the medium and a reader for data retrieval. As the density of magnetic recording has been increasing rapidly, to maintain the reliability of data recording, the head core material must possess a high resistivity (r) to minimize eddy-current loss. Soft magnetic materials with high r values, such as FeCo, NiFe, NiFeMo, or NiFeP, are great interest for such applications.

The uniformly distributed granular FeCo thin films, containing about 50%–70% iron, have been attractive as one of the best prospective materials among all 3d-transition alloys for a recording head pole tip due to its (i) ultra-high saturation magnetization $\mu_0 M_s \geq 2.4$ T [91] at room temperature and (ii) soft magnetic properties such as low coercivity due to the low magnetocrystalline anisotropy of the body-centered cubic structure [92]. This allows high-speed switching, which is necessary for recording data and is indispensable for meeting the requirements for higher medium coercivity and higher data transfer rate [93].

According to the FeCo phase diagram [94], the melting points of iron and cobalt at the pressure of 1.01×10^5 Pa are 1538°C and 1495°C, respectively. In the FeCo system, at approximately 50 at% Co and below 1000 K, the random body-centered cubic (bcc) $A2$ α phase transforms into the ordered simple cubic $B2$ structure of identical overall composition, where the iron atoms arrange themselves preferentially to the cube corners and the cobalt atoms to the cube centers. This second-order transformation involves a relatively small amount of energy. Ordering into the cubic $B2$ structure is accompanied by an increase in hardness and saturation magnetization compared to the disordered (random) bcc $A2$ alpha phase.

The FeCo thin films have been produced using various techniques mostly based on chemical methods such as sol–gel [95] and chemical reduction and also using physical thin-film deposition techniques such as magnetron sputtering, pulsed laser ablation deposition [53,96], and plasma focus devices [97]. In this section, we will present the synthesis of FeCo nanoparticle using pulsed laser ablation-based methods while the synthesis using a plasma focus device will be discussed later in Section 4.4.4.

4.3.5.1 Diffusion Cloud Chamber-Based PLAD for FeCo Nanoparticles

The PLAD system was modified to incorporate a diffusion cloud chamber with a liquid nitrogen-cooled substrate stage achieving the thermal gradient needed for diffusion. It has been reported [98,99] that PLAD systems with controlled condensation in a diffusion cloud chamber allow better control over morphology and

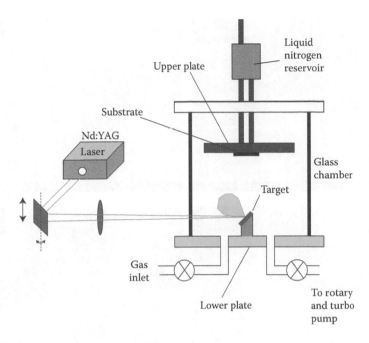

FIGURE 4.30   Schematic of diffusion cloud chamber-based PLAD for FeCo nanoparticle synthesis.

particle size of synthesized material. An equiatomic FeCo disk is used as a target and placed on a tilted target holder at the base of the chamber (refer to Figure 4.30). A liquid nitrogen-filled substrate holder plate, connected to a liquid nitrogen reservoir outside the chamber, was used to keep the substrates at a temperature of about 150 K, while the target disk was placed on the lower plate and remained at room temperature. A glass chamber was used, instead of the normally used metal chamber, whose poor thermal conduction helped to maintain the large temperature gradient between the upper liquid nitrogen-cooled substrate plate and lower plate, generating a steady convection current. The diffusion cloud chamber was evacuated to a base pressure of 10^{-4} Pa. The ablation of target material was accomplished by a Nd:YAG laser (Continuum Surelite; wavelength of 532 nm, 10 ns pulse, and repetition rate of 10 Hz). Deposition was performed at a pressure of 50 kPa using between 10,000 and 50,000 pulses. Deposition was also performed at argon gas pressures ranging from 10 to 90 kPa with the number of laser pulses fixed at 30,000.

4.3.5.2   Morphologies of FeCo Nanoparticles Grown in Diffusion Cloud Chamber-Based PLAD

SEM images (refer to Figure 4.31) show the formation of branched fractal network structures of FeCo nanoparticles synthesized at a fixed ambient argon gas pressure of 50 kPa but using different numbers of laser pulses, namely, 20,000, 30,000, 40,000, and 50,000 pulses. FeCo nanoparticles were found to arrange in

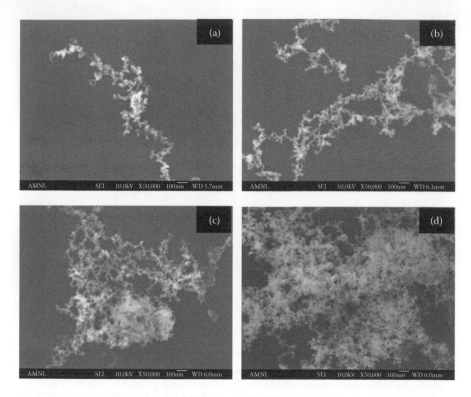

FIGURE 4.31 Morphology of deposited materials at argon ambient gas pressure of 50 kPa. From (a) to (d), the number of laser pulses are 20,000, 30,000, 40,000, and 50,000, respectively. (Reprinted from *Appl. Surf. Sci.*, 254, P. L. Ong et al., Synthesis of FeCo nanoparticles by pulsed laser deposition in a diffusion cloud chamber, 1909–1914, Copyright 2008, with permission from Elsevier.)

interconnected chains due to diffusion-limited aggregation processes. As the number of laser ablation pulses increased, a denser fibrous network structure was seen to form in three dimensions as more material was ablated. Next, the dependence of morphology as a function of ambient argon gas pressure was investigated for a fixed number (30,000) of laser pulses. The combined mix of SEM and TEM images of the FeCo nanoparticle network obtained at 10, 30, 50, 70, and 90 kPa are shown in Figure 4.32. The high-magnification TEM images are included in Figure 4.21 to demonstrate the formation of very uniformly sized nanoparticles aggregates and nanoparticle chains.

It can also be noticed from Figure 4.32e and f that as the ambient gas pressure increases, the morphology changes from loosely attached dense aggregates of nanoparticles (nanonetworks) to chains of nanoparticles. This is due to the increasing number of ambient gas molecules in the diffusion cloud chamber with increasing ambient pressure resulting in increasing collisions between the gas molecules and the nanoparticles that are formed in the gas expansion phase of the ablated plasma plume as they traverse from target to substrate. The increased collisions lower the

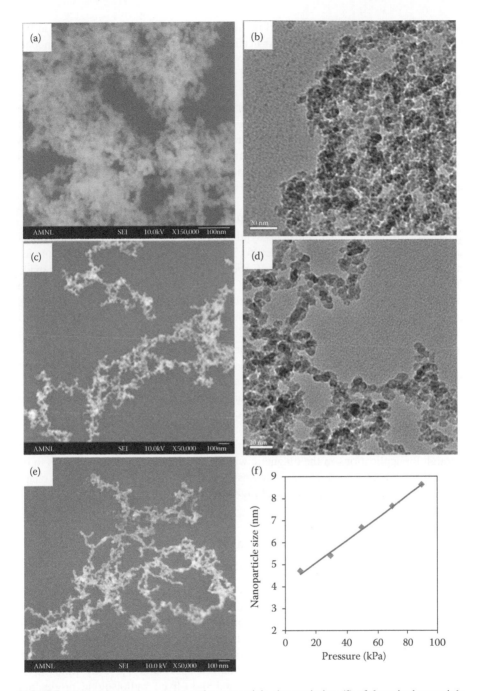

FIGURE 4.32 Morphology (a–e) and nanoparticle size variation (f) of deposited materials with argon ambient gas pressures for fixed 30,000 laser pulses. (a, c, and e) SEM image for 10, 30, and 50 kPa samples; (b, d) TEM image for 30 and 50 kPa samples. Note the scale bar is 20 nm in (b) and (d) and 100 nm in (a), (c), and (e).

kinetic energy as well as the overall number of FeCo nanoparticles arriving at the substrate surface and consequently form chains of nanoparticles rather than nanoparticles aggregates.

The ambient gas pressure also affects the nanoparticle size, as the average size of FeCo nanoparticles was found to increase with increasing ambient gas pressure as shown in Figure 4.32f. This increase in particle size can be related to the plasma expansion volume and the time for which the nanoparticles reside in the nucleation zone. The plasma expansion volume decreases while the residence time increases with increasing ambient gas pressures, which will increase the probability as well as the duration of collision between the ablated atoms and/or ions resulting in higher average size of the nanoparticles, as observed in Figure 4.32f.

4.3.5.3　Structural and Magnetic Properties of FeCo Nanoparticles

XRD patterns of FeCo nanoparticles synthesized at different ambient gas pressures are shown in Figure 7 of Reference 96. The diffraction peaks in all the samples mainly corresponded to cobalt ferrite ($CoFe_2O_4$), iron oxide (Fe_2O_3), and cobalt oxide (Co_3O_4). The peaks corresponding to FeCo were absent and it is probably due to higher residual oxygen in the diffusion cloud chamber as high argon ambient gas pressures were used as well due to exposure to air after synthesis as the nanoparticles were really very small (typically 5–10 nm) and mostly well separated allowing a much larger surface-to-volume ratio for their interaction with residual or atmospheric oxygen leading to conversion to mostly cobalt ferrites. The typical hysteresis loop of FeCo nanoparticles synthesized at the ambient gas pressure of 70 kPa is shown in Figure 4.9 of Reference 96. It can be observed that the as-deposited FeCo nanoparticles exhibited soft ferromagnetic behavior with coercivity in the range of 80–140 Oe.

The soft magnetic behavior (with coercivity values 80–140 Oe) is still not soft enough for applications in the magnetic write head, which desires coercivities to be in single digit number, preferably 1 Oe or smaller [100]. In the later part of this chapter (Section 4.4.4), we report the use of high-energy-density plasma to synthesize low coercivity (less than 10 Oe) FeCo.

4.4　NANOPHASE MAGNETIC MATERIAL SYNTHESIS BY PULSED HIGH-ENERGY-DENSITY PLASMA FOCUS DEVICE

The DPF device was first developed in the mid-1960s and over the last five decades has firmly established itself not only an intense source of fusion relevant neutrons but also as an intrinsic, energetic, and abundant source of relativistic electrons [101–103], soft/hard x-rays [104–108], and fast ions [109–114]. Being a multiple radiations source, DPF has found applications in different fields such as pulsed activation analysis [115,116], production of highly ionized species [117], x-ray lithography [118–124], x-ray radiography [125–133], and short-lived radioisotope production [134–138]. Many of these applications were attempted/developed as DPF was mostly considered and envisioned as a multiple radiation source only. It was only in the late 1980s to early 1990s that another set of applications in the field of material processing and material synthesis were developed using high-energy-density plasma and energetic ion irradiation in the plasma focus device. The first-ever application of the DPF device for the processing of bulk samples AISI 304 stainless steel was reported by Feugeas et al. [139], while Rawat et al. [140] were the first to report the processing of magnetron sputtered-coated lead zirconate titanate thin film. The first-ever deposition of thin film, of carbon, was reported by Kant et al. [141]. After that, more than a hundred research papers have been written on various material processing and synthesis-related work done by various plasma focus groups across the globe, making this the most routinely and actively pursued application of DPF devices. Many of these papers highlight that in both cases, that is, in the case of processing of bulk substrates and thin films and also in the case of direct synthesis of thin films, nanostructured materials are formed.

This section of the chapter will (i) provide the basic details of the DPF device, (ii) highlight several key and distinctive features of the DPF device, which sets it apart from routinely used low-temperature plasma devices for nanostructured material synthesis, and (iii) provide the details of the thematic use of the DPF device for processing and synthesis of magnetic materials to establish DPF as a novel high-energy-density plasma processing and deposition facility, in the fast-emerging field of "plasma nanotechnology," for nanoscale magnetic material fabrications.

4.4.1 DPF AS PULSED HIGH-ENERGY-DENSITY PLASMA SOURCE FOR MATERIAL PROCESSING AND SYNTHESIS: DEVICE DETAILS, PRINCIPLES OF OPERATION, AND KEY CHARACTERISTICS OF DPF DEVICE

4.4.1.1 DPF Device: Basic Details

Figure 4.33 shows the schematic of the DPF device with its main subsystems. The typical DPF device, which has been tailored specifically for use as a processing/deposition facility consists of

1. The DPF vacuum chamber that contains
   a. The coaxial electrode assembly consisting of central electrode (anode) surrounded coaxially by multiple cathode rods—for material processing work, where either bulk or thin film is exposed to energetic ions, a hollow anode is used to minimize anode material ablation to minimize contamination of irradiated material by anode material, whereas for material deposition work, suitable solid material is fitted to anode top as shown in Figure 4.33.
   b. An aperture assembly—normally used only in material processing work.
   c. A shutter—employed during the conditioning shots of DPF.
   d. An axially moveable substrate holder—to adjust the distance of irradiation/deposition.
2. The vacuum pump and gas inlet connections—DPF devices typically working at an operating gas pressure of a few mbar pressure are known to work quite well and reliably even if the system is not very clean or does not have

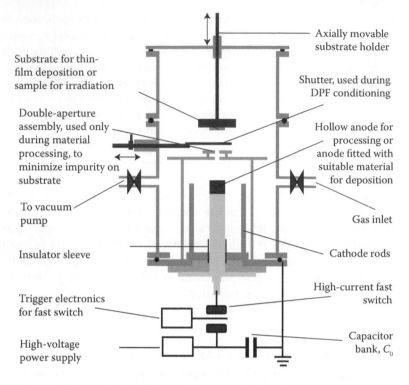

Axially movable
substrate holder

Substrate for thin-
film deposition or
sample for irradiation

Shutter, used during
DPF conditioning

Double-aperture
assembly, used only
during material
processing, to
minimize impurity on
substrate

Hollow anode for
processing or
anode fitted with
suitable material
for deposition

To vacuum
pump

Gas inlet

Insulator sleeve

Cathode rods

High-current fast
switch

Trigger electronics
for fast switch

High-voltage
power supply

Capacitor
bank, C_0

FIGURE 4.33 Schematic of DPF material processing/synthesis facility with its various subsystems.

low-enough base pressure. Hence, normally, a rotary pump is used to set up operating pressures; however, for works related to materials processing and depositions, a turbo pump is used to minimize the impurities due to residual gases.

3. A capacitor bank (which can be a single capacitor such as we have in our UNU-ICTP DPF device [142] or multiple capacitors arranged in multiple modules as we have in our NX2 device that has 48 capacitors arranged in four modules [143]) connected to a high-voltage power supply.

4. Low-inductance high-current fast-discharging switch (or switches) and transmission lines between the capacitor bank and the DPF load.

5. Trigger electronics to activate the high current switches.

6. Various diagnostics as per the requirements of the experiment being conducted.

4.4.1.2 Plasma Dynamics and Key Characteristics of DPF Devices

Excellent reviews about DPF devices have been written by Gribkov et al. [132] and Krishnan [144] and readers are advised to refer them for greater detail, but in the following part of this paragraph, a short description of device operation and plasma dynamics is provided. The electrical energy is first stored in the capacitor bank by charging it to its recommended voltage values (typically between 10 and 30 kV range) using a high-voltage power supply. The electrical energy is then transferred to the DPF chamber across its coaxial electrode assembly by activating low-inductance high-current-carrying fast switches. If the operating gas pressure is adjusted in the required pressure range of a few mbar, then an electric discharge is initiated across the insulator sleeve at the closed end of the electrode assembly. This discharge then evolves into a well-defined sheath of plasma. The amount of time taken to form a well-defined current sheath at the closed end of the electrode assembly depends on the electric parameters of the DPF device; for example, it takes about 100 ns in a low-energy 200 J DPF device, which has a quarter time period of about 500 ns [145], and about 500 ns in a mid-energy 3 kJ UNU-ICTP DPF device, which has a quarter time period of about 3 μs [108]. Driven by $\mathbf{J} \times \mathbf{B}$ force, the well-defined current sheath first moves in inverse pinch manner to form axis-symmetric current sheath at the closed end and then accelerates along the electrode assembly toward its open end, in what is commonly referred to as the axial acceleration phase, to typical maximum current sheath speed of about 4–6 cm/μs. At the open end of the anode, the current sheath starts to roll over the anode and has both axial as well as radially inward motion toward the anode axis. This is commonly referred to as the radial acceleration phase wherein the current sheath collapses on the anode axis, forming the hot and dense pinch plasma column. The radial speed of current sheath in the radial compression phase is about 2–2.5 times the speed in the axial phase, which results in extremely fast-moving shock. The fast-moving shock front ahead of the collapsing current sheath can result in a plasma temperature of about several hundred electron volts, while the reflected shock at the axis, together with the magnetic compression, can finally raise this temperature to around 1–2 keV. Such a hot pinch plasma column radiates efficiently in the soft x-ray region and if the operating gas is deuterium (D), then there is a finite reaction cross section for D–D fusion to take place, resulting in the generation of thermal fusion neutrons. Soon after the formation of pinched plasma column, the cylindrical pinch plasma column breaks up due to $m = 0$ and $m = 1$ instabilities. The $m = 0$ mode instabilities accelerate the ions of the filling gas species to very high energies toward the top of the chamber and accelerate electrons to relativistic energies (100 keV and above) toward the positively charged anode. For deuterium-filled DPF devices, the instability-accelerated fast deuterons also produce fusion neutrons through their interaction with deuterium target atoms/ions through beam–target mechanism. A fast-moving axial ionization wavefront, produced by the ionization caused by energetic ions, later develops into a bubble-like structure [146]. The ionizing front coincides with the beginning of the hard x-ray emission due to the interaction of the energetic electron with the anode tip material and also the neutron pulses for the deuterium-filled DPF device. The pinched plasma column finally breaks up and disintegrates, leading to the decay of hot and dense plasma.

The optimized DPF devices exhibit unique universality as many typical parameters and key characteristics of plasmas, plasma dynamics, and various radiation and energetic charged particles are independent of the device storage energy, discharge current, and electrode dimensions. Typical values of various key parameters of optimized DPF devices are given below:

- Peak current sheath speed in the axial phase is about a few cm/μs (2–10 cm/μs) [147,148].
- Electron and ion temperatures at the end of the axial phase are about 100 and 300 eV, respectively [149].
- Peak current sheath speed in radial phase is about 2–2.5 times that of the peak axial speed [148].
- Electron/ion densities in pinch plasmas are in the range of 5×10^{24}–10^{26} m^{-3} [113,150].
- Electron and ion temperatures of pinch plasmas are in the range of 200 eV–2 keV [146] and 300 eV–1.5 keV [150], respectively.
- Energies of instability-accelerated electrons are in the range of a few tens to few hundreds of keV [101,151].
- Energies of instability-accelerated ions are in the range of tens of keV to a few MeV [109,152]. The ions are mostly forward directed with most of the ions being emitted in a narrow angle of 20° with respect to the anode axis.

The DPF devices are pulsed plasma devices and the durations of pinch plasma, radiation, and energetic particles in this device change significantly from about sub-tens of ns to several tens of ns to about hundred or several hundreds of ns [52,53,56,57] depending on the quarter time period of the discharge, which in turn is governed by the inductance (static system inductance plus plasma inductance) and capacitance of the device. This makes most phenomena of interest in DPF devices highly transient in nature. To summarize, immense compression and heating of pinch plasmas to very high densities and temperatures at the top of the central electrode, very high energy flux of instability-accelerated ions and relativistic electrons, intense energetic radiations such as EUV, x-ray, and neutrons (for deuterium-filled operation), a fast-moving shockwave with hot-dense decaying plasma, combined with their transient nature, offers an altogether different kind of plasma and radiation environment in plasma focus device that is drastically different from low-temperature plasmas conventionally used in plasma nanotechnology.

The following sections under Section 4.4 will deal specifically with the magnetic material synthesis and processing in various plasma focus devices at our Plasma Radiation Source Lab of the National Institute of Education, Nanyang Technological University. In the starting phase of our work on magnetic material, we successfully used a plasma focus device for the synthesis of soft magnetic Fe [153] and FeCo [97,101] thin films and later on concentrated on hard magnetic material such as CoPt and FePt for magnetic data storage and very soft magnetic FeCo for write head application by controlling FeCo nanoparticle size.

The first ever nanostructured Fe magnetic thin films were successfully synthesized using high-energy-density plasmas in 3.3 kJ Mather-type plasma focus by Rawat et al. [153] using an anode fitted with solid Fe top. The deposition was done using different numbers of deposition shots at two different angular positions. The size of the nanophase agglomerate was smaller when either number of deposition shots was low or the angular position of the sample with respect to the anode axis was high. The films deposited at off-center position (18° with respect to the anode axis) were composed of Fe nanoagglomerates with an average size ranging from

60 ± 5 (for 10 shots deposition) to 95 ± 6 nm (for 40 shots deposition); with agglomerates themselves made up of smaller (20–30 nm) sized grains. At the outermost position (18° with respect to anode axis), the average nanoagglomerate size reduced to 44 ± 4 and 79 ± 5 nm for 10 and 40 shots deposition, respectively. The magnetization and coercivity was of intermediate values for Fe thin films deposited at a smaller angular position, which reduced further for samples at the outermost positions. This was one of the first usages of plasma focus devices for magnetic film synthesis and with its success, we concentrated on the synthesis of other magnetic materials such as CoPt and FeCo and also worked on the processing of PLAD-grown FePt thin films. The following sections present details of some of these works.

4.4.2 HARD MAGNETIC NANOPHASE CoPt SYNTHESIS USING DPF DEVICE

Cobalt–platinum (CoPt) alloy has exactly similar solid structures as that of FePt, shown earlier in Figure 4.6, with Fe simply being replaced by Co. So, just like FePt, CoPt also exhibits the chemically disordered face-centered-cubic (*fcc*) structured $A1$ phase and chemically ordered face-centered-tetragonal (*fct*) structured $L1_0$ phase. CoPt with a $L1_0$-ordered structure is also an attractive ultra-high-density magnetic recording media [154] due to its very large magnetic anisotropy constant of about 10^7 erg/cc [31]. In this subsection, the deposition of nanostructured magnetic CoPt thin films at room temperature on Si substrates using an NX2 DPF device operated at sub-kJ storage energy of about 880 are summarized. More details can be found in Reference 155.

4.4.2.1 Nanostructured CoPt Thin Films Using NX2 DPF Facility

A high-performance repetitive DPF, NX2 (Nanyang x-ray source), was used for nanostructured CoPt thin film synthesis with a copper central electrode fitted with high purity (50:50 at%; Kurt J. Lesker, 99.99%) solid CoPt. Hydrogen was used as the filling gas as it was found [101] that the NX2 device produces the highest electron energy flux for hydrogen compared to that of He, Ar, and Ne operations, resulting in the most efficient ablation of the anode target material with hydrogen as the operating gas. Substrates were placed at a distance of 25 cm from the anode top along the anode axis and the NX2 was operated at the lower charging voltage of 8 kV with a storage energy of about 880 J. The effects of hydrogen filling gas pressure (for fixed 25 focus deposition shots) and number of focus deposition shots (for fixed hydrogen ambient pressure of 6 mbar) on the morphological, structural, and magnetic properties as deposited and thermally annealed (at temperatures from 300°C to 700°C for 1 h) CoPt thin films were investigated.

4.4.2.2 Qualitative Understanding of Nanostructured Thin-Film Deposition in DPF Facilities

The nanostructured CoPt thin-film deposition mechanism in the DPF device (which would also be the same for the DPF device-based FeCo thin-film deposition discussed in a later section) can be qualitatively outlined on the basis of our understanding about two key plasma and radiation/charge particles emission characteristics of this device. First, the interaction of instability-accelerated backward-moving relativistic electrons and the hot dense pinch plasma with the anode tip material, which results in the ablation of the anode tip material. Second, the generation and acceleration (generated simultaneously with electrons) of instability-accelerated high-energy ions of filling/operating gas species toward the top of the chamber and the bombardment of the substrate surface on which the ablated anode tip material is to be deposited or is already deposited (for multiple shot deposition experiment). The ablated anode tip material plasma (ablated by energetic electron beam and hot dense pinch plasma) moves toward the substrate and is deposited on to it in the form of a thin film. However, as this ablated anode tip material plasma traverses from the anode top to the substrate placed downstream (either along the anode axis or at certain angle with respect to the anode axis), it may or may not interact with the plasma of the filling gas species depending on whether the filling gas species is reactive or inert. For the deposition of (i) nitrides such as TiN [156,157], WN [158,159], and so on, (ii) carbides such as TiC [160–164], SiC [163,164], and so on, (iii) oxides such as titanium/zinc/zirconium oxides [165–168], and (iv) complex carbo-nitride, oxy-nitride, and other phases [169–173], the DPF device can be operated with nitrogen, methane/acetylene, or oxygen gases. For bimetallic CoPt thin-film synthesis, therefore, an inert gas (such as He, Ne, or Ar) ambience is desired. However, as mentioned in the previous section, we decided to use hydrogen-operated DPF to achieve the maximum ablation efficiency of the anode tip material and also due to the fact that hydrogen is expected to reduce oxide impurities, which might be there due to residual oxygen in the DPF chamber.

It may be noted that the depositions in DPF devices are carried out using multiple deposition shots, which is also the case for CoPt and FeCo synthesis discussed in this chapter. Therefore, the energetic ions and hot dense decaying of the filling gas species from the next DPF shot processes the material deposited in the previous shot deposition in DPF devices. The deposition in the DPF device therefore is essentially an *ion- and plasma-assisted pulsed deposition process*. The energetic ion- and plasma-assisted deposition leads to high packing density in the deposited films and the simultaneous processing of thin films during deposition leads to the direct synthesis of crystalline thin films.

4.4.2.3 Morphological, Structural, and Magnetic Properties of CoPt Thin Films

The CoPt thin films deposited at different hydrogen ambience gas pressures using 25 DPF shots (refer to Figure 4.34) reveal the strong influence of operating pressure on surface morphology. The sample synthesized at 2 mbar (Figure 4.34a) exhibits a top layer having big island-like structures with a sublayer of some agglomerates between these big nanoislands. The rest of the samples showed similar morphological features with surfaces composed of agglomerates of nanoparticles with the average particle agglomerate size decreasing from 35 ± 6.0 nm to 15 ± 3.0 and 10 ± 2.0 nm for increasing operating gas pressure to 4, 6, and 8 mbar, respectively. The cross-sectional SEM images revealed the CoPt film thickness to be about 94.0 ± 4.0, 58.0 ± 3.0, 44.0 ± 2.0, and 22.0 ± 1.0 nm for depositions at 2, 4, 6, and 8 mbar, respectively. In the 2–8 mbar pressure range, focusing (plasma pinching) efficiency was found to decrease with increasing gas pressure. This resulted in a lower amount

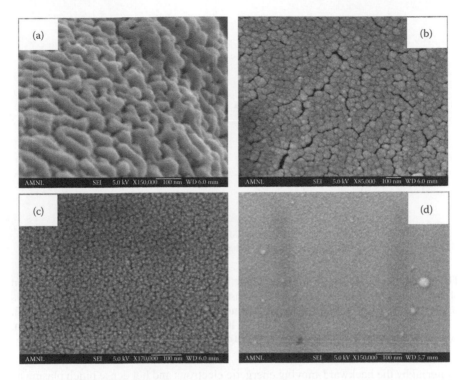

FIGURE 4.34 SEM images of nanostructured CoPt thin films grown using 25 focus shots at hydrogen filling gas pressures of (a–d) 2, 4, 6, and 8 mbar, respectively. (Reprinted from Z. Y. Pan et al., Nanostructured magnetic CoPt thin films synthesis using dense plasma focus device operating at sub-kilojoule range, *J. Phys. D: Appl. Phys.* 42, 175001, is acknowledged. Copyright 2009 © IOP Publishing. Reproduced with permission. All rights reserved. DOI: http://dx.doi.org/10.1088/0022-3727/42/17/175001.)

and energy of ablated anode tip material plasma. Additionally, with increasing ambience pressure the ablated material plasma would not only lose more energy but will also be scattered more during its movement from the anode to the substrate due to the increased collision with background gas. The reduced amount and energy of the ablated material, with increasing gas pressure, is responsible for reduced average particle size and CoPt thin-film thickness.

The surface morphology of CoPt thin-film samples synthesized using different numbers, namely, 25, 50, 100, 150, and 200 DPF deposition shots, but at a fixed operating gas pressure of 6 mbar, was investigated and it was found to change from well-separated small-sized nanoparticles (15 nm @ 25 shots) to nanoparticle agglomerates of increasing size. The thickness of thin-film samples increased with the increased number of deposition shots and is shown in Figure 4.35. The average deposition rate was estimated to be about 1.78 nm/shot for CoPt thin films deposited using an 880 J NX2 DPF device operated at 6 mbar hydrogen. This deposition rate is more than 30 times higher as compared to that of conventional PLAD, which is found to be about 0.50 Å/shot [50]. The comparatively high deposition rate achieved in the DPF deposition facility points to a much stronger ablation of target (anode tip)

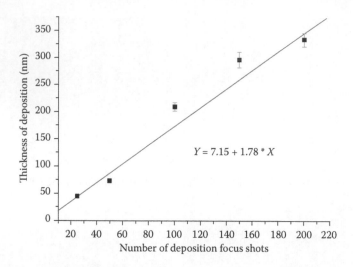

FIGURE 4.35 CoPt film thickness as a function of the number of plasma focus deposition shots. (Reprinted from Z. Y. Pan et al., Nanostructured magnetic CoPt thin films synthesis using dense plasma focus device operating at sub-kilojoule range, *J. Phys. D: Appl. Phys.* 42, 175001, is acknowledged. Copyright 2009 © IOP Publishing. Reproduced with permission. All rights reserved. DOI: http://dx.doi.org/10.1088/0022-3727/42/17/175001.)

material by the backward-moving energetic electrons and hot dense pinch plasma as compared to the ablation achieved by pulsed laser beam in PLAD facilities.

The XRD patterns of as-deposited and thermally annealed CoPt thin films deposited using 25 DPF shots at different hydrogen ambient gas pressures from 2 to 8 mbar are shown in Figure 4.36. All of the as-deposited samples were in the *fcc* phase with a broad and weak peak of (111) plane at about 41°, except for the sample deposited at a low operating pressure of 2 mbar, which exhibited much better crystallinity. One hour annealing at 500°C improved the crystallinity of the samples but they remained in the *fcc* phase. Samples annealed at 650°C exhibited transition to the hard magnetic $L1_0$ *fct* phase; as confirmed by the emergence of the (001) peak about 24°, shifting of (111), and slight splitting of the fundamental (200) peak into the (200) and (002) peaks. As-deposited samples were found to have very small coercivity values, in confirmation with the *fcc* phase observed in the XRD spectrum, while samples annealed at 650°C showed significantly enhanced coercivities values to about 1700, 3014, 618, and 696 Oe for samples deposited at 2, 4, 6, and 8 mbar, respectively. The enhanced coercivity in annealed samples confirmed phase transition from the disordered *fcc* phase to ordered *fct* phase.

There is an additional interesting point to note over here that the sample deposited at the lowest operating pressure of 2 mbar exhibited a very different XRD pattern, which at the time of our first analysis has been identified and labeled as shown in Figure 4.36. However, after our detailed analysis of impurity phase formation in PLAD-grown FePt samples, particularly at high laser energy flux, it can be hypothesized that this might be the case here as well, that is, there might as well be the impurity phase formation in the sample deposited at 2 mbar in the DPF device. As

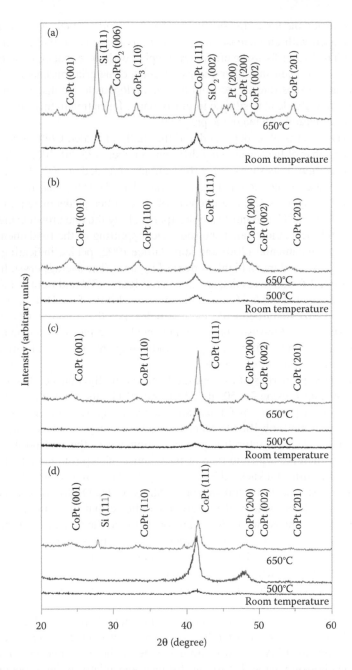

FIGURE 4.36 XRD patterns of CoPt nanostructures synthesized at different gas pressures by 25 plasma focus shots annealed at different temperatures. From (a) to (d): 2, 4, 6, and 8 mbar. (Reprinted from Z. Y. Pan et al., Nanostructured magnetic CoPt thin films synthesis using dense plasma focus device operating at sub-kilojoule range, *J. Phys. D: Appl. Phys.* 42, 175001, is acknowledged. Copyright 2009 © IOP Publishing. Reproduced with permission. All rights reserved. DOI: http://dx.doi.org/10.1088/0022-3727/42/17/175001.)

mentioned earlier, the focusing efficiency of the NX2 DPF device was best at 2 mbar and it decreased with an increase in operating gas pressure, in the 2–8 mbar pressure range. The higher focusing efficiency coupled with lower background operating gas pressure will result in higher energy of ablated material plasma species when they arrive at the substrate surface and hence, analogous to the discussion done for PLAD, this may also lead to silicide impurity phase formation in this sample as well as affect its structural and magnetic properties.

The XRD patterns of as-deposited and thermally annealed CoPt thin films synthesized at 6 mbar of hydrogen using different numbers (25, 50, 100, 150, and 200) of focus deposition shots showed that the samples annealed at $\geq 600°C$ transformed into the fct phase as the fct superlattice peaks of (001), (110), and (201) appeared clearly (refer to Figure 4.37). The intensity of superlattice peaks increased for samples annealed at 700°C, and most samples (particularly the one grown using a larger number of DPF deposition shots) showed a clear splitting of the fundamental (200) peak into the fundamental (200) and superlattice (002) peaks, indicating that the samples had converted into a highly ordered $L1_0$ phase. The average crystallite sizes, estimated from diffraction results, are shown in Figure 4.37f and were found to be in the narrow range of about 6–20 nm, indicating the ability of the DPF device to synthesize narrow-size disturbed CoPt nanoparticles. Another important point to note about XRD patterns of the as-deposited and annealed samples was that none of them exhibited any unknown diffraction peak, indicating the absence of any impurity phase formation in this set of samples.

The magnetic properties of as-deposited and thermally annealed CoPt thin-film samples synthesized using 25, 50, 100, 150, and 200 focus deposition shots at fixed hydrogen ambient pressure of 6 mbar are shown in Figure 4.38. All as-deposited samples showed small sub-100 Oe coercivity values as they were in low-K_u soft magnetic fcc phase. The total saturation magnetic moments of the as-deposited samples was found to increase with the increasing number of deposition shots due to the increase in thin-film thickness. The coercivity of the samples annealed at 500°C was found to increase from 260 Oe for 25 shots to 742 Oe for 200 shots deposition. The intermediate values of coercivities indicate that the samples essentially remained in the fcc phase with slight transformation to the fct phase. At the annealing temperatures of 600°C and 700°C, most of the samples, except for the 25 shots deposition sample, exhibited very hard magnetic properties as coercivities values rose above 4900 Oe, reaching a maximum of 8973 Oe for the 700°C-annealed 200 shots sample. This indicates the achievement of a highly ordered fct structure $L1_0$ phase in these samples.

It is important to note that the amount of magnetic hardness achieved in DPF-grown CoPt samples grown at 6 mbar gas pressure was much better as compared to that achieved for PLAD-grown samples. One of the main reasons for this was the suitable choice of large deposition distance and relatively large operating gas pressure, which reduced the energy of the ablated material species at the substrate surface, resulting in no impurity phase formation. The absence of any impurity phase formation and formation of highly crystalline superlattice peaks of (001), (110), and (201) and well-defined splitting of the fundamental (200) peak into the fundamental (200) and superlattice (002) peaks for the sample annealed at 700°C led to very hard

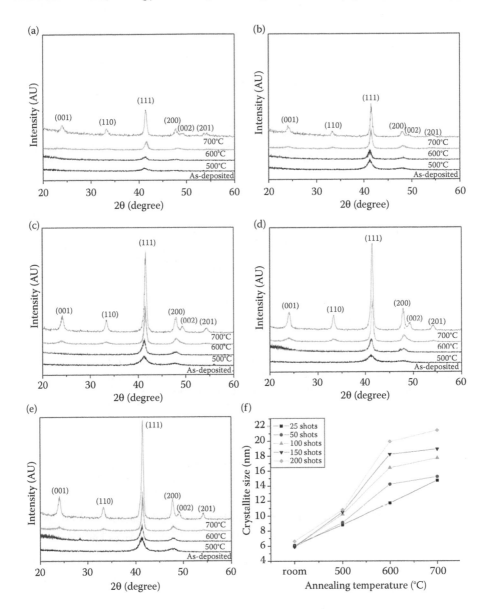

FIGURE 4.37 XRD patterns of CoPt nanostructures synthesized by (a) 25, (b) 50, (c) 100, (d) 150, and (e) 200 plasma focus shots annealed at different temperatures (400°C, 500°C, 600°C, and 700°C), and (f) the average crystallite size variation as a function of the number of plasma focus deposition shots. (Reprinted from Z. Y. Pan et al., Nanostructured magnetic CoPt thin films synthesis using dense plasma focus device operating at sub-kilojoule range, *J. Phys. D: Appl. Phys.* 42, 175001. Copyright 2009 © IOP Publishing. Reproduced with permission. All rights reserved. DOI: http://dx.doi.org/10.1088/0022-3727/42/17/175001.)

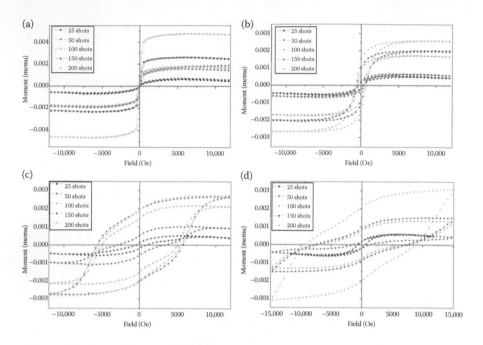

FIGURE 4.38 Hysteresis loops of CoPt nanostructures synthesized by different numbers of plasma focus shots (a) before annealing, and after annealing at (b) 500°C, (c) 600°C, and (d) 700°C. (Reprinted from Z. Y. Pan et al., Nanostructured magnetic CoPt thin films synthesis using dense plasma focus device operating at sub-kilojoule range, *J. Phys. D: Appl. Phys.* 42, 175001, is acknowledged. Copyright 2009 © IOP Publishing. Reproduced with permission. All rights reserved. DOI: http://dx.doi.org/10.1088/0022-3727/42/17/175001.)

magnetic nanostructured CoPt thin films, confirming the DPF facility as a novel high-energy-density pulse plasma device for magnetic data storage media application.

4.4.3 LOWER-PHASE TRANSITION TEMPERATURE IN PLAD-GROWN FePt THIN FILMS USING DPF DEVICE AS ENERGETIC ION IRRADIATION FACILITY

As mentioned in Section 4.3.3, we adopted two different strategies to lower the phase transition temperature in PLAD-grown FePt thin films, one of which was to use magnetic trapping assisted PLAD synthesis of $FePt:Al_2O_3$ nanocomposite thin films, and the second method, which is discussed in this section, was to use the energetic ion exposure of PLAD-grown FePt thin films. It may be recalled here that an enormous amount of research effort has been put into and is still ongoing with regard to devising various schemes to lower the phase transition temperature (the temperature at which the phase changes from the magnetically soft chemically disordered A1 *fcc* phase to magnetically hard chemically ordered $L1_0$ *fct* phase) to control the grain growth and agglomeration to minimize the exchange coupling effects to enhance the signal-to-noise ratio for magnetic data recording. Ion irradiation has been considered as an effective technique to lower the phase transition temperature and simultaneously tune the structural and magnetic properties of films. Devolder et al. [174] reported the mechanism involved in continuous ion beam irradiation. Helium ion irradiation was used by Ravelosona et al. [175,176] to control the degree of chemical ordering in FePt/FePd films. High-energy (350 keV) He+ irradiation was reported by Wiedwald et al. to lower the ordering temperature of FePt film by more than 100°C [76]. It was also reported that the ordered $L1_0$ FePt phase was directly achieved by using continuous 2 MeV He+ irradiation for about 1 h [71]. However, all of these ion irradiation experiments were conducted using continuous ion sources with typically hours of irradiation duration whereas our group reported the lowering of the *fcc* to *fct* phase transition temperature of the PLAD-grown FePt thin films using the transient pulse of energetic ions in the DPF facility, which lasts typically for about 50 to several hundreds of ns.

4.4.3.1 Experiments for Energetic Ion Irradiation of PLAD-Grown FePt Thin Films

The Continuum Surelite Nd:YAG laser (532 nm, 10 Hz, 10 ns and 80 mJ) with energy density of about 1×10^3 J/cm^2 at the rotating FePt (50:50 at%) target disk was used to ablate the target material in the PLAD chamber operated at base pressure better than 3×10^{-5} mbar. The FePt thin films of two different thicknesses of about 67 and 100 nm were grown on room-temperature Si (001) substrates at a target–substrate distance of about 3 cm [68,69]. The PLAD-grown FePt thin films were then irradiated by highly energetic H+ ions in the UNU/ICPT (United Nation University/International Center for Theoretical Physics) DPF device [142] at different distances from the anode top and also using different numbers of irradiation shots [68–70]. It may be noted that for ion irradiation studies, the double-aperture assembly shown in Figure 4.33 was used. DPF-based ion-irradiated FePt thin films were then thermally annealed in a vacuum furnace to various temperatures for 1 h to investigate the effect of ion irradiation on the *fcc* to *fct* phase transition temperature.

A Faraday cup (FC) was used to investigate the ions emitted from the UNU-ICTP plasma focus device. The typical FC signal, along with the voltage probe signal, for the hydrogen-operated UNU-ICTP DPF device is shown in Figure 4.39a, which exhibits the pulsed duration of H+ ions to be in the order of a few hundred ns. To

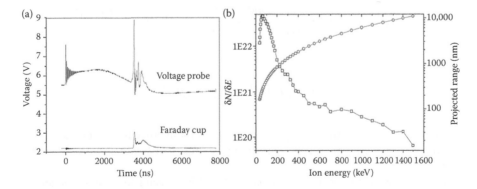

FIGURE 4.39 (a) Typical voltage probe and Faraday cup signal from UNU-ICTP DPF device. (b) Deduced energy spectrum (line) of H+ ions at filling gas pressure of 4 mbar and their corresponding projected range (o line) in FePt.

deduce the energy spectrum, the FC signal is first converted from the voltage signal, captured by digital storage oscilloscope traces, into the current signal by dividing it with corresponding resistance value of 50 Ω across which it is measured. Assuming that charge dQ is associated with the incident ions with energy in the range of E to $E + dE$, which contribute to the FC current signal in the time interval of t to $t + dt$, the number of ions corresponding to dQ is then given as $dN = dQ/eZ(E)$, where e is the electron charge and $Z(E)$ is the effective average charge of ions of energy E. The $Z(E)$ for hydrogen can simply be taken as 1 for hot pinch plasmas. The hydrogen will be fully ionized, whereas for other gases one needs to estimate it from the Corona model.

The interaction of energetic ions with the copper metal surface of the FC results in secondary electron emission. By assuming that the secondary electron emission coefficient of the FC cup metal is g (i.e., g electrons are generated for each of the incident ions and it depends on incident ion energy), the additional charge on FC due to secondary electron emission in time dt for ions in the range of E to $E + dE$ can be given by $dq = gedN = gedQ/eZ(E)$. For simple modeling of the ion emission characteristics, it is assumed that all the ions are emitted at the same time and from the same point above the anode. It is a reasonable assumption as the typical ion flight time of about 1 μs is much larger than the ion production time of about 100 ns and the distance between the FC and ion source point, typically >50 cm, is much longer than the dimension of ion emission source, that is, the pinch column, which is about 1 cm for the UNU-ICTP device. For this assumption, the total charge generated by ions in the energy range E to $E + dE$ in time dt can be written as

$$dQ_T = dQ + kdq = dQ\left(1 + \frac{kg}{Z(E)}\right) = eZ(E)dN\left(1 + \frac{kg}{Z(E)}\right)$$

where k represents the fraction of secondary electrons that are able to escape the FC.

Hence, the collector current can be shown to be

$$I = \frac{dQ_T}{dt} = eZ(E)\left(1 + \frac{kg}{Z(E)}\right)\frac{dN}{dt} = \left(\frac{(2E)^{3/2}}{m^{1/2}l}\right)eZ(E)\left(1 + \frac{kg}{Z(E)}\right)\frac{dN}{dE}$$

$$\frac{dN}{dE} = \left(\frac{m^{1/2}l}{(2E)^{3/2}}\right)\frac{1}{e(kg + Z(E))}I$$

where m is the mass of the hydrogen ion, $Z(E) = 1$ for hydrogen-filled operation, l is the distance of the Faraday cup from the ion source, and $k = 16/450$ (based on FC geometry).

The deduced ion energy of H$^+$ ions produced by plasma focus devices, shown in Figure 4.39b, is found to be in the range from about 35 keV to about 1.5 MeV. Following the method reported by Sanchez et al. [177], the total number of ions passing through a 0.5-mm-diameter pinhole of FC assembly placed at 10 cm above the anode, with energies in the range from 35 keV to 1.5 MeV, was estimated to be about 1.29×10^{11} with a mean energy of about 124 keV per ion [68]. The ion flux at 5 cm is therefore estimated to be 2.63×10^{14} ions/cm^2. Estimation of ion energies and their flux is very useful information that is needed to understand the FePt or any other thin-film processing mechanism in a DPF device.

4.4.3.2   H$^+$ Irradiation of 67 and 100 nm PLAD-Grown FePt Thin Films

The effects of hydrogen ion irradiation on morphological, structural, and magnetic properties of 67 and 100 nm PLAD-grown FePt thin films are presented in this section. The 67-nm-thick FePt films were irradiated at a distance of 5 cm from the anode top using one and two DPF irradiation shots, whereas the 100-nm-thick FePt films were irradiated at a distance of 4 cm from the anode top using different numbers of focus shots and also at different distances of 5, 6, and 7 cm using a single DPF shot.

4.4.3.2.1   Effects of Ion Irradiations on Morphological Properties

The morphology of the 67-nm-thick FePt film samples before and after ion irradiation is shown in Figure 4.40. The as-deposited FePt thin films (refer to Figure 4.40a) were smooth with laser droplets dispersing on the surface. Ion irradiation in the single DPF shot changed the morphology of FePt thin films from smooth uniform film to film with uniform and isolated nanoparticles as seen in Figure 4.40b. The increase in the number of ion irradiation shots to two made isolated nanoparticles to agglomerate as seen in Figure 4.40c. The changes in the surface morphology of 100-nm-thick FePt thin-film samples (SEM images are not shown) after exposure to different numbers of focus shots were similar the one reported for ~67 nm FePt thin-film samples as they all changed from smooth thin films to thin films composed of nanoparticles (for single shot irradiation) and/or nanoparticle agglomerates (for two or more shots irradiation).

It is interesting to note that the SEM results of two independent experiments were very consistent with single shot irradiation of (i) 67-nm-thick FePt samples, resulting in the formation of very small, uniform, and isolated nanoparticles of an average nanoparticle size of about 9.1 ± 2.3 nm and (ii) 100-nm-thick FePt samples, resulting in the formation of very small and uniform nanoparticles of an average nanoparticle size of about 8.0 ± 2.5 nm, along with some bigger-sized (~30–40 nm) nanoparticle agglomerates. The TEM image of a single shot-irradiated 100-nm-thick FePt film sample, irradiated at 4 cm above the anode and later annealed at 400°C, is shown in Figure 4.41. The TEM image in Figure 4.41a depicts the formation of well-separated FePt nanoparticles with an average particle size estimated to be about 11.6 ± 3.4 nm, while the high-resolution image in Figure 4.41b shows the crystalline nature of FePt nanoparticles.

The nanostructurization of single DPF shot-irradiated FePt thin films can be understood on the basis of ion emission characteristics and the corresponding processing of irradiated thin films. As mentioned earlier, the FC signal analysis showed that the mean energy of H$^+$ in the UNU-ICTP DPF device is about 124 keV and using SRIM code [178], it can be estimated that 124 keV H$^+$ ions will have a projected range of 521 nm in FePt and can create about 0.4 vacancies/ion. As the thickness of FePt thin films was only about 67 or 100 nm, most of the H$^+$ ions while losing

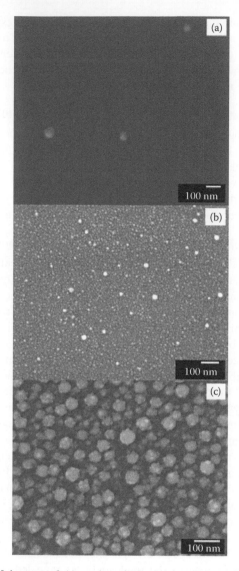

FIGURE 4.40 SEM images of (a) as-deposited sample and samples after (b) one shot and (c) two shots ion irradiation by plasma focus device. (Reprinted from J. J. Lin et al., FePt nanoparticle formation with lower phase transition temperature by single shot plasma focus ion irradiation, *Phys. D: Appl. Phys.* 41, 135213, is acknowledged. Copyright 2008 © IOP Publishing. Reproduced with permission. All rights reserved. DOI: http://dx.doi. org/10.1088/0022-3727/41/13/135213.)

energy continuously as they traverse through thin film mostly deposit the bulk of their energy in the silicon substrate at their stopping depth, heating it to a very high temperature in a very short span of time. The energy deposited in the silicon substrate is also conducted to the FePt thin films and causes the diffusion of metal atoms either through the lattice or along the grain boundaries. The diffusion releases the

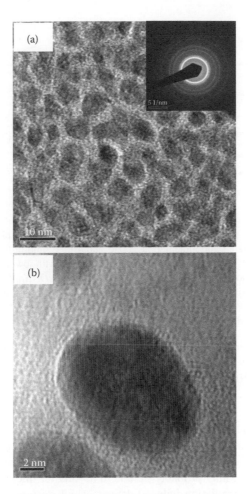

FIGURE 4.41 (a and b) TEM images of FePt nanoparticles induced by single shot ion irradiation and annealed at 400°C. Inset of (a) is the corresponding SAD pattern. (With kind permission from Springer Science+Business Media: *Appl. Phys. A*, Nanostructuring of FePt thin films by plasma focus device: Pulsed ion irradiation dependent phase transition and magnetic properties, 96, 2009, 1027–1033, Z. Y. Pan et al.)

thermal expansion mismatch stresses between the silicon oxide layer and the FePt thin film, leading to the formation of nanoparticles at the surface layer of the FePt thin film in single shot irradiation. The change in morphology with the increase in the number of irradiation shots can be attributed to the increased amount of energy being imparted by energetic ions from the next shot to the nanoparticles created in the first exposure, which causes them to migrate and form bigger agglomerates to reduce their surface energy.

4.4.3.2.2 Effects of Ion Irradiations on Structural Properties

The XRD spectra of as-deposited, single shot-irradiated, and annealed samples of 67- and 100-nm-thick FePt thin-film samples are shown in Figure 4.42a and b,

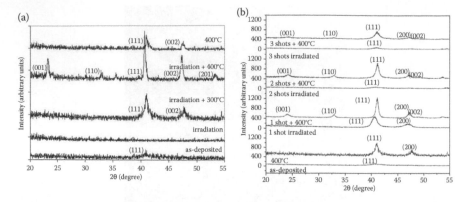

FIGURE 4.42 XRD patterns of (a) 67-nm-thick FePt thin-films samples (as-deposited, one shot irradiated, irradiated and annealed at 300°C and 400°C, and only annealed at 400°C). (Reprinted from J. J. Lin et al., FePt nanoparticle formation with lower phase transition temperature by single shot plasma focus ion irradiation, *Phys. D: Appl. Phys.* 41, 135213, is acknowledged. Copyright 2008 © IOP Publishing. Reproduced with permission. All rights reserved. DOI: http://dx.doi.org/10.1088/0022-3727/41/13/135213) (b) 100-nm-thick FePt thin-film samples (as-deposited and samples irradiated using different numbers of DPF shots at 4 cm from anode top, before and after annealing at 400°C). (With kind permission from Springer Science+Business Media: *Appl. Phys.* A, Nanostructuring of FePt thin films by plasma focus device: Pulsed ion irradiation dependent phase transition and magnetic properties, 96, 2009, 1027–1033, Z. Y. Pan et al.)

respectively. The as-deposited thin films of both 67 and 100 nm thickness exhibited the *fcc* phase with a broad and weak (111) diffraction.

Let us first discuss the XRD results of 67-nm-thick FePt thin films shown in Figure 4.42a. The single shot DPF exposure leads to the amorphization of the irradiated sample, which may probably be due to its smaller thickness combined with either energetic ion irradiation-induced defects, which destabilize the crystal structure, or due to the local melting of the ion-irradiated region and consequent rapid quenching of the liquid phase [179]. The amorphization of the crystalline material such as CdS [180] and Si [181] by energetic ion exposure in the DPF device has also been reported before. The XRD spectrum of the single shot-irradiated sample that was also annealed at 300°C showed the (111) and (002) diffraction peaks of the *fcc* phase. Samples that were single shot irradiated and later annealed at 400°C exhibited the emergence of the superlattice (001) peak at around 24° and the weak splitting of the fundamental (111) and (002) peaks, indicating the phase transition to the $L1_0$ ordered *fct* phase. The nonirradiated sample was found to remain in the *fcc* phase after annealing at 400°C.

Let us now look at the XRD results of 100-nm-thick FePt thin films shown in Figure 4.42b. Interestingly, the single DPF shot exposure this time did not lead to amorphization of the sample. It rather became more crystalline with its diffraction peak intensities becoming almost similar to that of the nonirradiated sample annealed at 400°C, implying therefore that if the thin-film thickness is appropriate, then ion irradiation in plasma focus can provide transient annealing of the sample to a more crystalline phase with the amount of energy almost equal to that offered by conventional thermal annealing for 1 h at 400°C. The two and three shots irradiation, however, reduced the crystallinity of the samples, indicating that the vacancy defects created by ion irradiation dominated over the transient thermal annealing with the increase in irradiation shots. Another noticeable difference in diffraction patterns of 67 and 100 nm FePt thin films is much better crystallinity in 100-nm-thick irradiated and annealed samples with well-defined diffraction peaks of much higher intensities. Moreover, there is a clear shift in the (111) and (200) peak position and the emergence of the (001) superlattice peak at about 24° indicates significant and much better transformation from the *fcc* to *fct* phase in 100-nm-thick samples as compared to that of the 67-nm-thick sample.

The structural properties of 100-nm-thick FePt thin films that were irradiated at distances of 5, 6, and 7 cm and later annealed were also investigated (XRD results are not shown here but the reader may refer to Figure 2 in Reference 70). It was found that the crystallinity of samples after a single DPF shot irradiation at all three distances of exposure increased as compared to that of the as-deposited nonirradiated sample confirming the crucial role played by film thickness on outcome: crystalline or amorphous. Once again, the samples irradiated to a single DPF shot at 5, 6, and 7 cm after annealing showed significant *fcc* to *fct* phase transition. To summarize, the annealing of all ion-irradiated 100-nm-thick FePt thin-film samples, whether irradiated by different numbers of DPF shots at 4 cm or irradiated by a single shot at different distances of 5, 6, and 7 cm, at 400°C lead to the phase transition from the disordered *fcc* structured $A1$ phase to chemically ordered *fct* structured $L1_0$ phase indicated by the appearance of the superlattice tetragonal (001), (110), and (002) peaks. The XRD results were also confirmed by the selected area electron diffraction (SAED) pattern shown in the inset of Figure 4.41a and analyzed by JEMS@ JEOL, which confirmed the FePt nanoparticles to be in the polycrystalline *fct* phase. The low phase transition temperature of ion-irradiated samples can be explained by the impact of energetic ions that promote significant adatom mobility and also decrease the activation energy for atomic ordering by increasing point defects such as vacancies and interstitials.

4.4.3.2.3 Effects of Ion Irradiations on Magnetic Properties

The magnetic properties of 67-nm-thick film samples were investigated through ferromagnetic response of VSM measurements. The in-plane hysteresis loops are not shown here but the results are summarized as follows. The as-deposited 67-nm-thick FePt thin films were soft magnetic in nature with a coercivity of about 75 Oe. The sample irradiated with 1 focus shot, but not annealed, had marginal increase in coercivity to about 96 Oe. The 300°C annealing of the ion-irradiated sample led to the increase in coercivity to about 447 Oe, indicating some transformation to the $L1_0$-ordered *fct* phase. A significant increase in coercivity to about 1563 Oe (with a 16-fold increase) was observed for the sample irradiated by one DPF shot and then annealed at 400°C. The manyfold increase in the coercivity of the last sample indicates the increased conversion of this sample to the *fct* phase. The remanence ratio S $(=M_r/M_s)$ for the single shot-irradiated sample annealed at 400°C was estimated to be about 0.4, which is smaller than 0.5 and hence the main intergranular interaction was the desirable magnetostatic rather than undesirable exchange coupling. It may however be pointed out that none of the 67-nm-thick FePt thin-film samples exhibited very hard magnetic properties, which is the reflection of its crystalline properties as the samples also did not exhibit good crystallinity or stronger evidences of conversion to the desired *fct* phase.

The hysteresis loops of 100-nm-thick FePt thin films (i) irradiated at a distance of 4 cm from the anode top using 1, 2, and 3 focus shots are shown in Figure 4.43a–d and (ii) irradiated at different distances of 5, 6, and 7 cm from the anode top using 1 focus shot are shown in Figure 4.43e and f. The as-deposited 100-nm FePt thin film showed soft ferromagnetic behavior with a coercivity of about 78 Oe, which after annealing at 400°C increased to 177 Oe, as shown in Figure 4.43a. It can be inferred from Figure 4.43b–e that the samples irradiated by different numbers of focus shots and at different distances, but not annealed, exhibited soft magnetic behavior with coercivity values similar to that of the as-deposited sample. The examination of Figure 4.43b and f reveals that all the samples that were irradiated by single DPF shot at all distances of exposures and later annealed at 400°C shows an enormous increase in coercivity to values >5000 Oe, confirming significant degree of transformation of these samples to the magnetically hard ordered $L1_0$ *fct* phase, as observed in the XRD results as well. The increase in the number of irradiation shots to 2 and 3 shots (refer to Figure 4.43c and d), resulted in the sharp decrease of coercivity. This can be explained on the basis of the lower ordering degree of the (001) diffraction peak for these samples. The ordering degree of the (001) diffraction peak was estimated using the texture coefficient of this peak for all the samples. The ordering degree of the (001) diffraction peak was found to be the maximum for the sample that was single focus shot ion-irradiated and later annealed at 400°C, which decreased by 4.4% and 24.7% for two and three shots ion-irradiated and annealed samples, respectively. The (001) plane, according to Lim et al. [182], coincides with the magnetic easy axis of the ordered $L1_0$ *fct* phase. Hence, the decrease in the ordering degree of the (001) peak with an increase in number of shots explains the corresponding decrease in coercivity values.

As a concluding remark to this section, we would like to point out that single shot energetic ion irradiation of PLAD-grown FePt thin films of 100 nm thickness in the DPF device was not only able to convert the samples from smooth thin films to very small, uniform, and isolated nanoparticles but also transformed them into a very hard magnetic $L1_0$ phase with coercivity values in excess of 5000 Oe at a significantly lower phase transition temperature of 400°C. Thus, this beyond doubt established the high-energy-density plasma focus device as a novel material processing tool for plasma nanotechnology.

4.4.4 SOFT MAGNETIC NANOPHASE FeCo SYNTHESIS USING LOW-ENERGY DPF DEVICE

It is known that the large magnetostriction constant of FeCo makes it difficult to achieve soft magnetic properties although this alloy has a large saturation magnetic flux density. However, good soft magnetic properties can still be obtained if (i) the size of the magnetic particles is reduced to less than a characteristic length (the

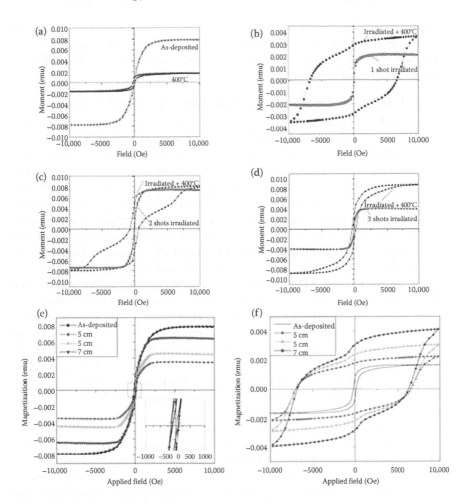

FIGURE 4.43 In-plane hysteresis loops of 100 nm FePt thin-films samples (a) as-deposited and annealed at 400°C (b–d) before and after 400°C annealing the samples irradiated to 1, 2, and 3 focus shots at a distance of 4 cm from anode top sample and finally the samples irradiated at 5, 6, and 7 cm (e) before annealing and (f) after annealing at 400°C. ((a–d): With kind permission from Springer Science+Business Media: *Appl. Phys. A*, Nanostructuring of FePt thin films by plasma focus device: Pulsed ion irradiation dependent phase transition and magnetic properties, 96, 2009, 1027–1033, Z. Y. Pan et al. (e) and (f): Reprinted from *Thin Solid Films*, 517, Z. Y. Pan et al., Lowering of L10 phase transition temperature of FePt thin films by single shot H+ ion exposure using plasma focus device, 2753–2757, Copyright 2009, with permission from Elsevier.)

exchange length), and (ii) the exchange coupling between the neighboring magnetic particles takes place, which leads to a cancellation of magnetic anisotropy and compensation of the demagnetization effect of individual particles, resulting in much better soft magnetic properties. Since our earlier investigations had already established that it is very much possible to make small nanoparticles of the size of about 10–30 nm using a mid-energy kJ-range plasma focus device [155], we hypothesized that a miniature plasma focus with a storage energy of about 120 J will be able to synthesize very small FeCo nanoparticles of about 10 nm size with narrow size distribution with the potential to exhibit very soft magnetic behavior suitable for magnetic recording heads.

4.4.4.1 FMPF-1 200 J Plasma Focus Device for FeCo Thin-Film Synthesis

Previous works of nanostructured thin-film deposition reported in this chapter were done using mid-energy kJ-range DPF devices such as UNU/ICTP [68,69,142] and NX2 [101,155]. The FeCo thin-film deposition reported in this section was however done on the first version of the fast miniature plasma focus device FMPF-1 (200 J @ 12.9 kV, 2.4 μF, 27 nH, T/4 ~400 ns) developed in-house at the Plasma Radiation Source Lab of NIE, Nanyang Technological University, Singapore [183], whose performance has been investigated for a wide range of applications [184,185]. The inherent advantage of a low-energy 200 J miniature plasma focus device compared to that of mid-energy kJ-range DPF devices in material synthesis/deposition applications is that the radiation and energetic particle fluence produced by pinch is less. This minimizes some of the typical problems associated with kJ-range PF-assisted material deposition technique such as (i) formation of droplets due to the splashing of the anode material because of the larger thermal load in the higher-energy plasma focus, (ii) damage of substrate surfaces due to strong shocks produced inside kJ-range plasma focus devices, and (iii) excessive thermal loading on the anode material also causes severe anode damage over the period of use. The low-energy miniature plasma focus devices can also be easily operated at a high repetition rate (up to a few Hz range) since the driving power requirement is low.

The schematic of specially configured coaxial electrode assembly of the FMPF-1 device is shown in Figure 4.44 in which a tapered stainless-steel (SS) anode was fitted with a high-purity (Amstrong Science, 99.99% purity 50:50 at%) solid FeCo target. A custom-designed substrate holder was used to mount silicon substrates, as shown in Figure 4.44, at a fixed axial distance of 10 cm kept at an angle of 70° with respect to the horizontal plane, and the angular position with respect to the anode axis was 8°. The FMPF-1 was operated at 10 kV, and hence with a 2.4 μF capacitor bank, the stored energy was about 120 J. The FeCo thin-film samples were deposited (i) using hydrogen and nitrogen as operating gas on Si (100) substrates to find out which gas results in soft magnetic properties in deposited films and (ii) using three different substrates of Si (100), MgO (100), and amorphous Al_2O_3 with hydrogen as the filling gas to investigate the effects of substrate type on the magnetic properties of FeCo thin films. All depositions were carried out using 200 focus shots at 1 Hz repetition rate.

4.4.4.2 Soft Magnetic Properties of Miniature Plasma
      Focus-Based FeCo Thin Films

In-plane hysteresis loops of FeCo samples deposited on (i) Si using 4 mbar hydrogen and 1 mbar nitrogen (optimized pressures for these two gases) and (ii) Si, MgO, and Al_2O_3 substrates using 4 mbar hydrogen in the FMPF-1 DPF device are shown in Figure 4.45a and b, respectively. The coercivity for the sample synthesized using

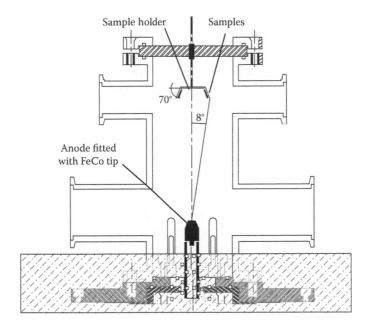

FIGURE 4.44   Schematic of fast miniature plasma focus device (FMPF-1) for FeCo thin-film deposition. (Reprinted from *Phys. Lett. A*, 374, Z. Y. Pan et al., Miniature plasma focus as a novel device for synthesis of soft magnetic FeCo thin films, 1043–1048, Copyright 2010, with permission from Elsevier.)

hydrogen on the Si substrate is only about 6.3 Oe, indicating the formation of very soft magnetic FeCo thin films while the sample deposited using nitrogen-operated FMPF-1 is magnetically hard with a coercivity of about 622.9 Oe (refer to the enlarged view in the inset in Figure 4.45a). According to Zou et al. [187], lower

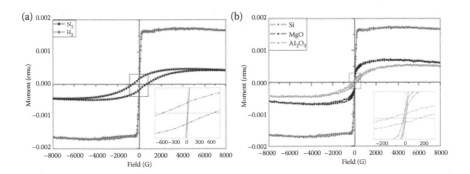

FIGURE 4.45   Hysteresis loops of (a) samples deposited on Si(100) substrate using H2 (-•-) and N2 (-■-) gases (b) samples deposited on Si(100) (-■-), MgO(100) (-•-) and amorphous Al2O3 (-▲-) using H2 as filling gas. (Reprinted from *Phys. Lett. A,* 374, Z. Y. Pan et al., Miniature plasma focus as a novel device for synthesis of soft magnetic FeCo thin films, 1043–1048, Copyright 2010, with permission from Elsevier.)

stress in the film can help achieve lower coercivity and according to them the stress induced in the film is strongly dependent on the filling gas species and pressures as these factors affect the impingement of Fe and Co atoms onto a substrate and hence alter the intrinsic stress. In DPF devices, the deposition process as mentioned is assisted by ion irradiation and hot dense decaying plasma exposure and hence the energetic hydrogen/nitrogen ions may be implanted in the FeCo thin films, resulting in tensile stress in deposited thin films. The tensile stress for the sample deposited in nitrogen is expected be larger than that of the sample deposited in hydrogen since the N^+ ion is bigger and heavier than the H^+ ion. Hence, lower stress in hydrogen ion-assisted deposition might lead to lower coercivity in the corresponding FeCo thin films.

The comparison of in-plane hysteresis loops of samples deposited on Si, MgO, and amorphous Al_2O_3 (shown in Figure 4.45b) shows that (i) the coercivity of the sample deposited on MgO was still very small at about 6.4 Oe but its saturation magnetization reduced significantly as compared to the sample deposited on the Si substrate and (ii) the sample deposited on the amorphous Al_2O_3 exhibited relatively hard magnetic behavior with a coercivity of about 160.6 Oe and its saturation magnetization was lowest among all three samples. The change in the coercivity with substrate type may be attributed to the changes in the grain size of FeCo thin films as average grain sizes (estimated from SEM images) were found to be 10.8 ± 1.2, 11.0 ± 1.8, and 46.0 ± 8.5 nm for Si, MgO, and Al_2O_3, respectively. This matches well with the pattern of coercivity change. According to the theoretical analysis of Zou et al. [187], the coercivity of 1 Oe would occur for a grain size less than 2 nm, and it would be about 40–50 Oe for 25-nm grains. This estimation is consistent with our results.

The TEM image along with the SAD pattern of soft magnetic FeCo thin film deposited on the Si substrate using hydrogen is shown in Figure 4.46, which indicates the random polycrystalline texture of thin films. The XRD pattern of this sample (not shown) exhibited broad (110) peaks and hence the texture of these films was deduced to be that of randomly oriented (110) crystallites.

The coercivity of thin soft polycrystalline film can be estimated using Hoffmann's ripple theory [188], wherein the coercivity is expressed as

$$H_c = \frac{\beta}{L_{ex}} \cdot \frac{(2D)^{1/2}}{\pi\mu_o M_s} \cdot \left(\frac{K_\mu}{A_{ex}}\right)^{1/4} \cdot \delta$$

where β is a fitting factor, L_{ex} is the exchange length, K_μ is the magnetic anisotropy, A_{ex} is the exchange stiffness coefficient, D is the average grain diameter in plane, and δ is the structure factor. The structure factor depends on effective in-plane local anisotropy constant K_{local}, grain size D, and the number of particles in a sphere of radius of exchange length L_{ex}, and is given by

$$\delta = \frac{K_{local}D}{\sqrt{n}}$$

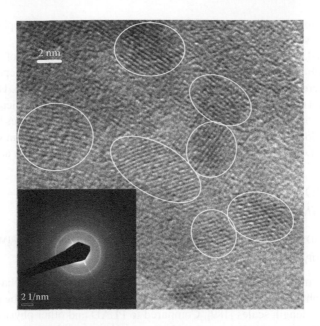

FIGURE 4.46 HRTEM image and inset SAD patterns of the sample deposited on Si(100) substrate using H_2 as filling gas. (Reprinted from *Phys. Lett. A*, 374, Z. Y. Pan et al., Miniature plasma focus as a novel device for synthesis of soft magnetic FeCo thin films, 1043–1048, Copyright 2010, with permission from Elsevier.)

The expression for K_{local} given by Zou et al. [187] is

$$K_{local} = ((AK_1 + BK_2 + C\Delta\lambda\sigma)^2 + (D\Delta\lambda\sigma)^2 + K_2(EK_1 + F\Delta\lambda\sigma))^{1/2}$$

where A–F are the texture-dependent coefficients, $\Delta\lambda = \lambda_{100} - \lambda_{111}$, with λ as the magnetostriction constant along different planes, and σ is the stress.

The FeCo sample has randomly oriented (110) textures and hence the magnetostriction constants $\lambda_{100} \approx \lambda_{111} \approx \lambda_s$; thus, $\Delta\lambda = \lambda_{100} - \lambda_{111} \approx 0$. Therefore, $K_{local} = ((AK_1 + BK_2)^2 + K_2EK_1)^{1/2}$. By taking the anisotropy constant $K_1 = 3.50 \times 10^4$ J/m^3 and $K_2 = 0.50 \times 10^4$ J/m^3 and a set of (110) oriented texture coefficients from A to F reported in Reference 187, the K_{local} can be estimated. For simplicity, it can be assumed that the exchange length L_{ex} is similar to the average grain size D. Hence, D, L_{ex}, and the number of particles n can be estimated using SEM images (not shown) to estimate the structure factor δ.

Finally, using $M_s = 9.61 \times 10^5$ A/m from the VSM results, the exchange stiffness coefficient of $A_{ex} = 2.015 \times 10^{-11}$ J/m, $\beta = 1$, and the structure factor δ into Hoffmann's expression for ripple coercivity, the ripple coercivity for FMPF-1 DPF-grown FeCo thin film was estimated to be about 4.8 Oe. Hence, the experimental coercivity value of about 6.3 Oe for the sample deposited on the Si substrate using the hydrogen-filled FMPF-1 DPF device agreed reasonably well with the theoretical ripple coercivity of 4.8 Oe, which also indicated that the FeCo granular thin films were a very soft thin-film system, since this ripple theory is only applicable to the soft thin-film system.

To conclude, this section presented results of the first-ever use of a low 120 J energy FMPF-1 device for any kind of thin-film deposition. The deposited FeCo thin films were found to be very soft with a coercivity of about 6 Oe. However, the magnetic data recording industry is looking for coercivities values of less than 1 Oe, which are theoretically possible if the FeCo nanoparticle size decreases below 2 nm. Growing FeCo nanoparticles with size less than 2 nm is a challenge, but with optimization of the DPF device operation parameters, it might be possible to achieve them.

4.5 NANOPHASE MAGNETIC MATERIAL SYNTHESIS BY ATMOSPHERIC MICROPLASMAS

4.5.1 ATMOSPHERIC MICROPLASMAS: ATMOSPHERIC PLASMAS, EXPERIMENTAL SETUP, PRINCIPLES OF OPERATION, AND FORMATION

Atmospheric microplasma (AMP) refers to the class of electrical discharges carried out at atmospheric pressure, where at least one of the dimensions of the plasma is of submillimeter length scales [189]. Compared to PLAD and other physical synthesis techniques, the biggest difference for AMP is that it can synthesize nanoscale materials in atmospheric ambience instead of a vacuum chamber. This saves considerable cost of building and maintaining a vacuum system, making AMP a strong candidate to be used in nanoscale material synthesis. The advantages of AMP techniques such as simplicity (no vacuum and heating system requirement), efficiency (short synthesis time of about a few minutes), and well-controlled particle size and self-arrangement also make it a strong candidate for nanostructured material synthesis. Different AMP systems and configurations have been studied [189–195]. They include hollow-cathode metal tubes (in Figure 4.47a) [191], gas jets with external electrodes (in Figure 4.47b) [192], gas jets with tube electrodes (in Figure 4.47c) [194], *in situ* deposition for inner microcapillaries (in Figure 4.47d) [195], and plasma–liquid interactions (in Figure 4.47e) [190], and so on. In this chapter, the AMP technique based on plasma–liquid interaction is studied. The plasma–liquid interaction has great potential for applications in water treatment [196], biomedicine [197], and chemical analysis [198]. AMP formed by combining atmospheric DC glow discharge with electrolyte cathode allows the use of liquids, making a variety of chemical reactions possible [199]. Nanoparticles synthesized by plasma–liquid interaction with an AC plasma jet formed with H_2 gas are found to be spherical, crystalline, uniform, and extremely small with mean diameters of 2 nm [200]. Moreover, metal nanoparticles can be synthesized by plasma–liquid interaction at atmospheric pressure and room temperature without the need for chemical reducing agents [201], making the plasma–liquid interaction approach safer and more biocompatible than conventional colloidal growth methods [202].

The experimental setup of AMP-induced liquid chemical method in this chapter is shown in Figure 4.48. A DC AMP discharge with a constant current of 2–5 mA (using TREK 615–10, ±10 kV AC/DC generator) was sustained between a hollow stainless-steel anode (of Ø 300 or 700 μm bore diameter) and the surface of the

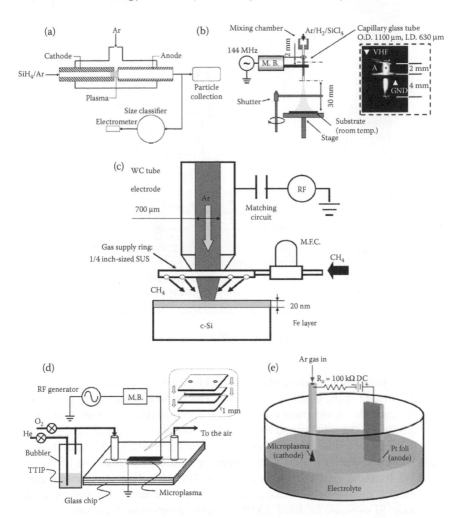

FIGURE 4.47 Various AMP systems and configurations; (a) hollow-cathode metal tubes; (b) gas jets with external electrodes; (c) gas jets with tube electrode; (d) *in situ* deposition for inner microcapillaries; and (e) plasma–liquid interactions. ((a) Reprinted with permission from R. M. Sankaran et al., Synthesis of blue luminescent Si nanoparticles using atmospheric-pressure microdischarges, *Nano Lett.* 5, 537–541. Copyright 2005 American Chemical Society; (b) Reprinted from T. Nozaki et al., Microplasma synthesis of tunable photoluminescent silicon nanocrystals, *Nanotechnology* 18, 235603, is acknowledged. Copyright 2007 © IOP Publishing. Reproduced with permission. All rights reserved. DOI: http://dx.doi.org/10.1088/0957-4484/18/23/235603; (c) Reprinted from *Thin Solid Films*, 515, Z. Yang et al., Synthesis of Si nanocones using rf microplasma at atmospheric pressure, 4153–4158, Copyright 2007, with permission from Elsevier; (d) Reprinted from *Surf. Coat. Tech.*, 202, H. Yoshiki and T. Mitsui, TiO2 thin film coating on a capillary inner surface using atmospheric-pressure microplasma, 5266–5270. Copyright 2008, with permission from Elsevier; (e) Reproduced with permission from F.-C. Chang, C. Richmonds, and R. M. Sankaran, Microplasma-assisted growth of colloidal Ag nanoparticles for point-of-use surface-enhanced Raman scattering applications, *J. Vac. Sci. Technol. A* 28, L5–L8. Copyright 2010, American Vacuum Society.)

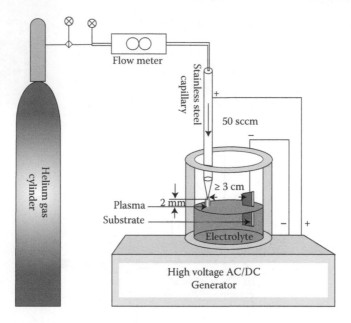

FIGURE 4.48   Experimental setup of the AMP-induced liquid chemical synthesis method.

aqueous electrolyte solution in a glass vessel. The tip of the anode was placed 2 mm above the surface of the electrolyte solution. The surface of the electrolyte solution acts as the cathode. Its electrical connection is completed with a Si substrate, which is half-inserted in the solution and connected to the cathode of the AC/DC generator. The helium gas, at a flow rate of 50 cm³ per minute at STP (SCCM) with the use of a mass flow controller, was used as the discharge medium because of its high-energy metastable state and excellent heat conductivity [189,203]. Nanostructured magnetic materials such as iron oxide nanoparticles and Pt_3Co nanoflowers were formed on the cathodic Si substrate for fixed AMP discharge duration of 2–20 min.

The mechanism of AMP operation and nanoparticle formation for the AMP-induced liquid chemical method has not been fully understood by researchers. Three reported types of AMP interactions with liquid [45] are (i) gas-phase species and plasma radiation-induced liquid reactions (Figure 4.49a) [45], and electron-initiated nonequilibrium reactions in liquid phase [45,204] depending on (ii) electron energy distribution (Figure 4.49b) and (iii) electron density (Figure 4.49c). If injected electrons have sufficient energy to induce molecular dissociation at the plasma–liquid interface, radicals can be formed and activate specific solution chemistry (Figure 4.49b) [204]. Otherwise, injected electrons "absorbed" in the solution become solvated electrons and directly contribute to reactions, for example, as reducing agents (Figure 4.49c) [204]. Taking the mechanism of surfactant-free Au nanoparticles synthesized in $HAuCl_4$ solutions, for example, Mariotti et al. [45] reported that the synthesis of Au nanoparticles may be initiated by reducing Au-salt via $\left[AuCl_4\right]^- + 3e_{aq}^- \rightarrow Au + 4Cl^-$. The Au-salt reduction may result from "cascaded" liquid chemistry by H^0 radicals, H^- anions, or H_2O_2.

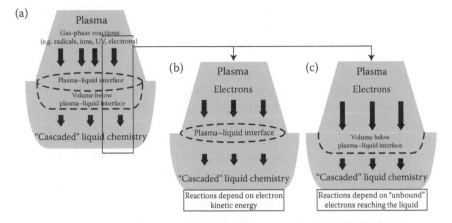

FIGURE 4.49 Schematic diagram of AMP interactions with liquids: (a) reactions from the plasma–liquid interface induced by gas-phase species and plasma radiations; (b and c) non-equilibrium reactions in liquid phase initiated by electrons from the plasma depend on (b) electron energy distribution and (c) electron density. (D. Mariotti et al.: Plasma–liquid interactions at atmospheric pressure for nanomaterials synthesis and surface engineering, *Plasma Process. Polym.* 2012, 9, 1074–1085. Copyright Wiley-VCH Verlag GmbH & Co. KGaA. Reproduced with permission.)

Dissociative electron attachment to water produces a range of radicals and ions via $H_2O + c_{gas}^- \rightarrow H^- + OH^0$ and $2OH^0 \rightarrow H_2O_2$ or $H_2O + 2e_{aq}^- \rightarrow OH^- + H^0$ and $2OH^- \rightarrow H_2O_2 + 2e_{aq}^-$. Moreover, water oxidation at the anode due to the ultra-high applied voltage (>800 V) produces quantities of positive ions such as H^+ [45], which can also participate in the overall liquid chemical reactions or recombine with other species in the solution or in the proximity of the plasma–liquid interface.

4.5.2 Near Superparamagnetic Fe_3O_4 Synthesis Using AMP

Superparamagnetic iron oxide nanoparticles are regarded as a useful tool in numerous medical applications such as MRI, enzyme immobilization, drug and gene targeting, cell separation, drug delivery, and magnetic hyperthermia [16]. The nature of their high magnetism and good biocompatibility is an asset specifically for *in vivo* biomedical applications such as controlled drug delivery [205] as it allows the movement through the blood to be controlled by the external magnetic field [16]. However, various iron oxide synthesis techniques such as hydrothermal and chemical methods take quite a long time and require high temperature for the synthesis of iron oxide nanoparticles. Liu et al. [206] reported that Fe_3O_4 nanoparticles specifically synthesized by thermal decomposition methods tend to agglomerate because of anisotropic dipolar attraction, which affects their dispersion and magnetic properties. An AMP-induced liquid chemical method has been used to synthesize various size-controlled metal nanoparticles at atmospheric ambience and room temperature with plasma discharging time as short as about 10 min [45]. Thus, in this chapter, the properties of iron oxide nanoparticles synthesized by thermal decomposition and the AMP-induced liquid chemical method are compared. The effect of plasma discharge time (3, 5, and 10 min) on the properties of AMP-synthesized iron oxide nanoparticles is also studied.

The experimental setup is shown in Figure 4.8. The electrolyte solutions are prepared by (i) dissolving 12.0 g of ferric chloride ($FeCl_3 \cdot 6H_2O$) and 4.9 g of ferrous chloride tetrahydrate ($FeCl_2 \cdot 4H_2O$) in 100 mL deionized water (stirring for 30 min at a temperature of 40°C), and (ii) subsequently slowly adding 25 mL of ammonium hydroxide (25% NH_4OH) into the solution (stirring for another 10 min at the same temperature of 40°C). Upon the addition of NH_4OH, a black suspension was instantly formed in the solution. The prepared final solution is distributed into three parts, one of which was labeled "as-prepared solution" (AS). Another part of the distributed solution was stirred at a temperature of 90°C for 90 min and labeled "thermally decomposed solution" (TD). The last part, labeled "AMP-treated solution" (AMP-3min), underwent the AMP-induced liquid chemical treatment with a DC constant current at 5 mA for the plasma discharging time of around 3 min. To study the effect of plasma discharging time on the properties of iron oxide nanoparticles, two sets of the same solutions have been prepared and treated with AMP at the same DC constant current but for a plasma discharging time of 5 and 10 min. The related samples here are addressed as "AMP-5min" and "AMP-10min," respectively. All the synthesized iron oxide nanoparticles are spin-coated on Si substrates after purification [207] for characterization.

The morphologies, particle sizes, and shapes of iron oxide nanoparticles obtained in three different treatment methods, that is, AS (as prepared), TD (thermally decomposed), and AMP-3min (3 min AMP treated), are shown by SEM images in Figure 4.50. It can be seen that the nanoparticles extracted from the AMP-treated solution (shown in Figure 4.50c) are well isolated and less agglomerated, while combined iron oxide nanoparticles are observed for as-prepared and thermally decomposed samples (shown in Figure 4.50a and b, respectively). It is also found that the shapes of nanoparticles are dominantly spherical for all the samples, while a small amount of nanotubes are observed for AS and AMP-3min samples. ImageJ (Wayne Rasband, 1.46r) software was used to calculate the average particle sizes for both the spherical nanoparticles and nanotubes. It was noticed that the AMP-3min sample shows smaller particle size of 11.7 ± 0.7 nm than the TD one of 30.7 ± 3.0 nm.

The EDX was performed to analyze the elemental composition of nanoparticles on AS, TD, and AMP-3min samples. Because the three samples show similar EDX spectra, only the EDX spectrum of the AMP-3min sample is shown in Figure 4.51a for reference. The presence of Fe and O characteristic x-ray peaks and the absence of other characteristic x-ray peaks, corresponding to other elements used in the preparation of the electrolyte solution, confirm the successful formation of iron oxide nanoparticles without any additional impurities in all AS, TD, and AMP-3min samples. The Si peak was contributed by the Si substrate on which these nanoparticles were dispersed. The average O at.% shown in Table 4.2 is higher than the standard Oat.% (57.10–58.02 at.%) of Fe_3O_4 as stated by Wriedt [208], resulting in a smaller average Fe:O ratio (average Fe at.% to average O at.% ratio) than that of Fe_3O_4 (about 0.75). This is normally because of the fact that the quantitative estimation for oxygen

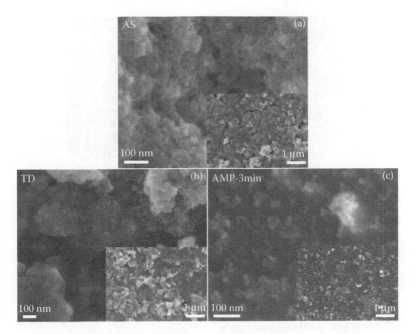

FIGURE 4.50 SEM images of (a) as-prepared (AS), (b) thermally decomposed (TD), and (c) AMP-treated (AMP-3min) samples; inset: low-magnification SEM images for the three samples.

in oxides cannot be reliably arrived at by direct measurement of the oxygen peak in the EDX results. Thus, XRD and UV characterizations are applied to further confirm the composition of the samples.

The XRD patterns of AS, TD, and AMP1 samples are shown in Figure 4.51b. All the diffraction peaks for the AS sample are in good agreement with the peaks of the JCPDS Pattern No. 01-082-1533 for Fe_3O_4, indicating that the formed iron oxide nanoparticles are Fe_3O_4. However, fewer and weaker Fe_3O_4 peaks and stronger Si peak (at around 56°) are observed in the XRD spectra of TD and AMP-3min samples. This might be due to the exposure of the surface of the Si substrate, resulting from the better isolated nanoparticles as seen in the SEM images in Figure 4.50.

The UV experiment was performed to further confirm the composition of nanoparticles for the AMP-3min sample. The absorption band within the region of 300–400 nm in the UV–vis spectrum (in Figure 4.51c) is reported to be due to the absorption and the scattering of UV radiation by the iron oxide nanoparticles, specifically Fe_3O_4 [209]. The broad absorption band indicates the formation of nanosized particles. The other small absorbance peak at about 580 nm might be due to the co-precipitation process in the initial stages of the Fe^{2+} and Fe^{3+} ions to form iron oxide nanoparticles as reported by Islam et al. [210]. The sharp dip peak at around 420 nm cannot match most of the reported UV radiation peaks [209,211] for Fe_3O_4 and we are unable to comment on its observation. However, the UV–vis spectrum still proves the formation of Fe_3O_4 in the AMP-3min sample.

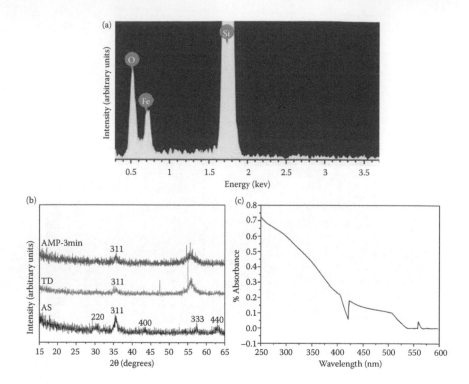

FIGURE 4.51   (a) EDX result of the AMP-3min sample; (b) XRD patterns of AS, TD, and AMP-3min samples; (c) UV–visible spectrum of the AMP-3min sample.

The magnetic properties of AS, TD, and AMP-3min samples were characterized by VSM at room temperature. The hysteresis loops (Figure 4.52) show that all three samples exhibit quite soft magnetic behaviors with coercivities less than 10 G. The similar magnetic behavior of TD and AMP-3min samples suggests the potential considerable interest in the AMP technique in the synthesis of iron oxide nanoparticles because of the ultra-shorter synthesis time (~3 min) used. However, the as-prepared nanoparticles exhibit the smallest coercivity of around 3 G among the three samples.

TABLE 4.2

Average Values of Fe:O Ratio and Individual Percentages from 12 EDX Spectra on Each Type of Sample

	Samples		
	AS	**TD**	**AMP1**
Average Fe:O ratio	0.33	0.18	0.27
Fe at.%	25	16	22
O at.%	75	84	78

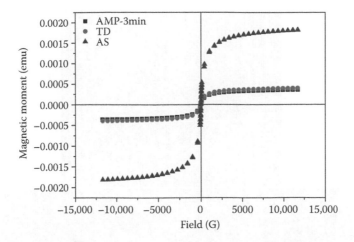

FIGURE 4.52   Vibrating sample magnetometer curve of AS, TD, and AMP-3min samples.

It seems that the AS sample shows the best properties among the three samples with the best crystallinity, the largest amount of nanoparticles, the smallest nanoparticle size, and the smallest coercivity. However, it may be noted that the AS sample is formed in the solution where the chemical reaction has not been completed, which means the properties of the AS sample are unstable and that is why the researcher used the thermal decomposition method to complete the reaction and obtain a stabilized product.

To further improve the properties of AMP-treated samples, two more samples are prepared with the AMP discharge time of 5 and 10 min (referred to as AMP-5min and AMP-10min, respectively). The effects of different AMP discharge durations on the samples' properties are studied. It can be seen from the SEM results shown in Figure 4.53 that with the increasing plasma discharge time, the amount of particles increases and the particles start to agglomerate, as clearly shown in Figure 4.53c, where the particles agglomerate to form small blocks for 10 min discharge as compared to the more scattered distribution in Figure 4.53a for 3 min discharge. Although nanoparticles agglomerate more in the AMP-10min sample, the particle size is still distinct. The calculated average particle size for AMP-3min, AMP-5min, and AMP-10min samples are 11.7 ± 0.7, 5.2 ± 0.2, and 6.3 ± 0.3 nm, respectively. Nanostructures in the samples also change with increasing plasma discharge time; both spherical nanoparticles and nanotube-like elongated nanoparticles are observed in the AMP-3min sample (Figure 4.53a), while only spherical nanoparticles with better even distribution are found in the AMP-5min and AMP-10min samples. The composition of nanoparticles seems to remain the same with increasing plasma discharge time as similar EDX results, which is not shown in this section, for the AMP-5min and AMP-10min samples compared to AMP-3min sample shown in Figure 4.51a.

The hysteresis curves in Figure 4.54 show increased magnetic moment with increasing plasma discharge time, which might be due to increased amount of nanoparticles that are being formed in samples with longer plasma discharge as shown in SEM results. The observed coercivities for sample AMP-3min, AMP-5min,

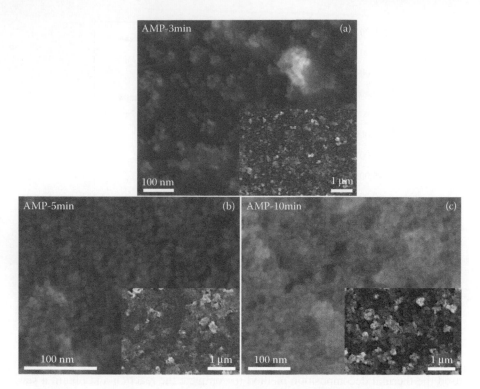

FIGURE 4.53 SEM images of AMP-treated samples with plasma discharge time of (a) 3, (b) 5, and (c) 10 min; inset: low-magnification SEM images for the three samples.

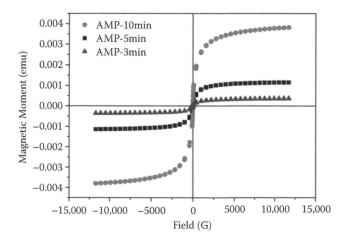

FIGURE 4.54 VSM curve of AMP-3min, AMP-5min, and AMP-10min samples.

and AMP-10min are 10, 6, and 3 G, respectively. This exhibits that with increasing plasma discharge time, Fe_3O_4 nanoparticles become increasingly more soft magnetic, approaching close to the superparamagnetic limit of zero coercivity. This indicates that superparamagnetic Fe_3O_4 nanoparticles may be achieved by tailoring plasma discharge time in the AMP-induced chemical synthesis method.

4.5.3 Nanophase Pt₃Co Synthesis Using AMP

Pt-based bimetallic nanoparticles have been reported to show higher catalytic activity compared with Pt nanoparticles for the oxidation of ethanol, methanol, and alcohol [19,212] for their application as a catalyst in fuel cells. For Pt-based bimetallic catalysts, He et al. [213] reported that PtM (M = Fe, Co, V, Mn) is highly corrosive and highly active compared to PtM composed of other second metals such as Zn, Cu, and Zr. They also found that PtCo alloy with 78 at.% Pt and 22 at.% Co (matching closely to the stoichiometry of Pt_3Co) is the best Pt–Co bimetallic system because of its largest mass-fraction-specific activity and positive half-wave potentials (HWP) as well as due to a relatively small composition change after stability testing [213]. To ensure high catalytic activity, the particles should be in the vicinity of each other instead of being separate islands [19]. Well-controlled three-dimensional (3D) Pt nanostructures (e.g., dendrites and nanoflowers) are reported to be particularly interesting in catalysis for their high net catalytic active surface area and high catalytic activity [214–217]. Meanwhile, the 3D nanostructure also means that less metal is needed, which could reduce the cost, especially for Pt, in commercial applications [215,216]. However, there are few reports about the synthesis of Pt-based bimetallic nanoflowers, one of which reported is Pt-loading Au bimetallic nanoflowers [218–220]. Catalysts prepared by plasma or catalysts modified by plasma exhibit higher activity and are reported to have improved selectivity and better stability than samples synthesized by conventional methods [221,222]. The AMP has also been used to synthesize 0D, 1D, and 2D nanostructured metal and metal oxide materials [189,193], but synthesis of 3D nanostructures using AMP has seldom been reported. In this work, the synthesis of 3D nanostructured Pt_3Co, in the form of nanoflowers (NFs), is explored and optimized using AMP.

The AMP setup used for Pt bimetallic nanostructures is similar to the one shown previously in Figure 4.8 except for the following special parameters: (i) a DC constant current of 2 mA, (ii) a hollow stainless-steel anode of Ø 700 µm bore diameter, (iii) the fixed plasma discharge time of 10 min, and (iv) an electrolyte solution composed of dissolved 0.002 M chloroplatinic acid hydrate ($H_2PtCl_6 \cdot xH_2O$), variable molar concentration (0.4, 0.5, and 0.6 M) of cobalt(II) sulfate heptahydrate ($CoSO_4 \cdot 7H_2O$) and 0.5 M sodium sulfate (Na_2SO_4) in distilled water. Pt_3Co NFs (Figure 4.55a) were found to be formed on the Si substrate for the sample synthesized with 0.6 M $CoSO_4 \cdot 7H_2O$.

Figure 4.55a presents SEM images of NFs with an average size of ~756 nm and an average petal thickness of a few nm. The cross-sectional SEM images (presented in Figure 4.55b) indicate hybrid formation of NFs all over on the substrate with an average height of ~416 nm. The EDX spectrum and mappings (Figure 4.55c and d) and XRD pattern (Figure 4.55e) indicate a Pt_3Co composition of NFs. Elements

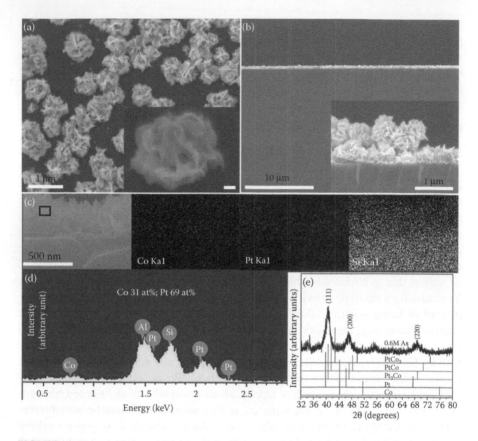

FIGURE 4.55 (a) Topographical SEM images of Pt₃Co NFs synthesized with 0.6 M $CoSO_4 \cdot 7H_2O$, inset: a single NF, scale bar is 100 nm; (b) cross-sectional SEM images of Pt₃Co NFs, inset: higher-resolution images of NFs; (c) EDX mapping of cross-sectional NFs; (d) EDX spectrum of cross-sectional NFs; (e) XRD pattern of the NF sample.

Co, Pt, Al, and Si are detected in the cross-sectional EDX results shown in Figure 4.55d. The element Si is mostly due to the Si substrate, while Al is due to the sample holder in the SEM system. The cross-sectional EDX mapping in Figure 4.55c shows main distribution of Co and Pt elements at the site of NFs, while the main distribution of Si is under NFs. The average atomic ratio of Co and Pt is found to be about 31 and 69, respectively, indicating the composition of NFs to be Pt-rich cobalt–platinum alloy. This is further proved by the XRD diffraction pattern shown in Figure 4.55e, which matches well with the standard diffraction pattern of cubic Pt₃Co (PDF #00-029-0499).

The SEM images in Figure 4.56a–c show the effect of increasing $CoSO_4 \cdot 7H_2O$ concentration, from 0.4 to 0.6 M, on the growth process of the topographical features. With increasing $CoSO_4 \cdot 7H_2O$ concentrations, the morphologies were observed to turn from ultrathin nanosheet structure (0.4 M, Figure 4.56a) to thicker anisotropic growing nanosheets (0.5 M, Figure 4.56b) to blooming nanoflowers (0.6 M, Figure 4.56c). Nanoparticles are observed among or under the nanosheets or NFs in

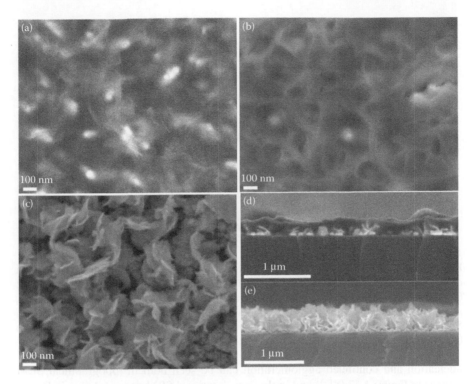

FIGURE 4.56 Topographical SEM images of samples synthesized with $CoSO_4 \cdot 7H_2O$ concentration of (a) 0.4, (b) 0.5, and (c) 0.6 M; cross-sectional SEM images of samples synthesized with $CoSO_4 \cdot 7H_2O$ concentration of (d) 0.4 and (e) 0.6 M.

all three samples, and nanoparticle agglomerates are only clearly found in the sample with the highest $CoSO_4 \cdot 7H_2O$ concentration of 0.6 M. Considering the cross-sectional SEM images (Figure 4.56d and e) for the samples with $CoSO_4 \cdot 7H_2O$ concentration of 0.4 and 0.6 M together, it seems that (i) both nanosheets and nanoflowers are grown on nanoparticles, (ii) nanoparticle quantities increase and nanoparticle agglomerates with increasing $CoSO_4 \cdot 7H_2O$ concentration, and (iii) nanosheets are more anisotropic with increasing $CoSO_4 \cdot 7H_2O$ concentration. It seems that NFs are formed by well-arranged nanosheets on nanoparticles.

The growth process NFs in the AMP discharging process is further outlined by carefully scanning through different SEM images at different parts of the deposited sample with a $CoSO_4 \cdot 7H_2O$ concentration of 0.6 M. The NF growth process can be classified into three main stages: stage 1 (Figure 4.57a), nucleation and agglomeration of nanoparticles; stage 2 (Figure 4.57b), formation of petals; and stage 3 (Figure 4.57c), petal growth and formation of NFs. At the first stage, the electrons injected from the AMP discharge into the electrolyte solution induce the reduction of the Co and Pt salt, forming metallic Co and Pt, which come in contact with each other, leading to Pt_3Co nanoparticle nucleation. With the increasing AMP discharge time, while new nuclear sites are being created, large agglomerates of Pt_3Co nanoparticles are also formed as the initial nucleation sites continue to grow in size. The new

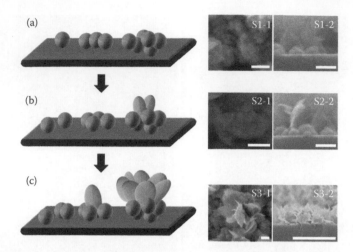

FIGURE 4.57 Three stages of NF growth process: (a) stage 1: nucleation and agglomeration of Pt₃Co; (b) stage 2: formation of petals; and (c) stage 3: petal growth and formation of NFs; (S1-1 and S1-2), (S2-1 and S2-2), and (S3-1 and S3-2) are SEM images for the three stages, respectively; S1-1, S2-1, and S3-1 are topographical SEM images, scale bar: 100 nm; S1-2, S2-2, and S3-2 are cross-sectional SEM images, scale bar for S1-2 and S2-2: 100 nm, scale bar for S3-2: 1 μm.

nucleation and continued agglomeration can be seen both in topographical SEM images in Figure 4.57 (S1-1) and in the cross-sectional SEM images shown in Figure 4.57 (S1-2).

In the second growth stage (Figure 4.57b), petals are observed to grow on the agglomerates. The observation of petal growth only on the bigger-sized agglomerates (Figure 4.57 (S2-2)) rather than on new nucleation sites (smaller nanoparticles) indicates the continued accumulation of material on agglomerates and their evolution into petals. In other words, the large agglomerates act as the nucleation sites for petals. Thus, petals start to grow on the surface of the agglomerates. In the last stage (Figure 4.57c), petals keep growing due to the anisotropic growth of Pt₃Co. The magnetic attraction of excessive Pt₃Co crystallites, hindrance between agglomerates, and other factors caused petals to curl and aggregate (Figure 4.57 (S3-1 and S3-2)). As a result, complete NFs are formed (Figure 4.55a).

The electrochemical properties of the Pt₃Co nanosheets and NFs were shown in their typical cyclic voltammograms (CV) curves, which were obtained from −0.25 to 1.20 V in N-saturated 0.5 M H_2SO_4 at a scan rate of 50 mV/s (as shown in Figure 4.58a). Pt₃Co NFs (solid line) show the formation/reduction of surface Pt oxides between 0.4 and 1.2 V [19,220,223], while Pt₃Co nanosheets (dash line) do not show any formation/reduction of Pt oxides. A similar phenomenon is also shown in the CV of Pt₃Co nanosheets and NFs in N-saturated 0.5 M H_2SO_4 aqueous solution containing 1 M CH_3OH at room temperature (in Figure 4.58b). Typical methanol oxidation peaks in the forward and backward scan are detected on the Pt₃Co NF sample (Figure 4.58b), while almost no electrocatalytic performance is shown by the Pt₃Co nanosheets. The anodic peak in the forward sweep between the potential of 0.6 and

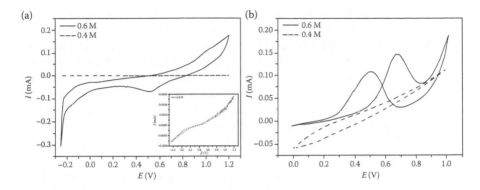

FIGURE 4.58 Cyclic voltammograms of nanosheets (dash line) and NFs (solid line) in (a) N-saturated 0.5 M H_2SO_4, inset: enlarged CV curve of nanosheets; and (b) 0.5 M H_2SO_4 + 1 M CH_3OH solutions at a scan rate of 50 mV/s.

0.8 V corresponds to the oxidation of methanol molecules, while the anodic peak in the backward sweep between the potential of 0.4 and 0.6 V is generally attributed to the oxidative removal of incompletely oxidized intermediates accumulated on the catalyst surface during the forward sweep [223,224]. It is well known that a noble metal catalyst such as Pt usually exhibits superior activity at the edges and corners [225]. The NF structure has more edges and corners than the nanosheet structure, which might be the reason for better electrocatalytic performance toward methanol for Pt_3Co NFs. The ratio of the forward anodic peak current (I_f) to the reverse anodic peak current (I_b) is an important factor for evaluating a catalyst [224,226]. A higher value of I_f/I_b indicates better oxidation of methanol to carbon dioxide and less poisoning of the catalyst by the CO-like intermediates [224,226]. The I_f/I_b of Pt_3Co NFs (~1.00) is comparable to the reported values of 0.98 by Ahamadi et al. [224] and 1.2 by Liu et al. [226]. However, the fabrication time (10 min) spent for Pt_3Co NFs in this work is much shorter than that (>21 h) reported by Ahamadi et al. [224] and that (>31 h) reported by Liu et al. [226]. This indicates strong potential applications of Pt_3Co NFs in catalysts and the AMP technique in catalyst fabrication.

4.6 SUMMARY AND KEY CONCLUSIONS

To summarize, the plasma-based and plasma-assisted processes are indeed one of the most versatile tools for fabrication of nanostructured magnetic materials. Plasmas provide a complex, reactive, and far-from-equilibrium chemical factory and are attractive alternatives for nanoscale magnetic material synthesis due to (i) higher throughput, (ii) greater versatility in synthesizing nanostructured material through both bottom-up and top-down synthesis approaches, (iii) better control of size distribution, (iv) less energy intensiveness as the desired material can be synthesized either at room or significantly lower temperature, and (vi) higher quality of nanomaterial synthesis. The plasmas used for nanomaterial synthesis exist over an enormous range of densities and temperatures, leading to the possibility of exotic ranges and combinations of novel nanomaterial synthesis with desired properties. In this chapter, we presented nanophase magnetic material synthesis using three different types of plasmas from (i) PLAD facility, (ii) high-energy-density DPF device, and (iii) AMP device. While PLAD and DPF plasmas are pulsed (transient) in nature, the AMP plasma is continuous. The plasma in the PLAD facility, at the substrate surface, has a high density of the order of 10^{15-16} cm^{-3} but the temperature is relatively low, of the order of a few electron volts only. In comparison, in the DPF device, both the plasma density and plasma temperature are relatively higher, to approximately about 10^{16-18} cm^{-3} and several tens to hundreds of electron volts. The AMP plasma on the other hand is relatively cold plasma with plasma density either lower or almost similar to that of PLAD plasmas.

It was demonstrated that FePt/FeCo nanoparticles and FePt nanocomposites can be synthesized by conventional PLAD and using special target–substrate geometry, known as BPD. In both cases, the as-deposited FePt nanoparticles were in low-K_u *fcc* phase and thermal annealing at the temperature of 600°C was required for the phase transition to high-K_u *fct* phase. FePt nanoparticle agglomerates deposited by BPD were found to have better crystallinity, after annealing, since backward-moving ablated species had higher kinetic energy due to shorter travel distance. They also had better uniformity in agglomerate size and size distribution. However, high-temperature annealing of BPD-grown FePt nanoparticles deposited by BPD resulted in multidomains instead of a single domain for single nanoparticles, which in turn changed the magnetostatic interaction to the exchange coupling.

FePt:Al$_2$O$_3$ composite thin films were successfully synthesized by conventional PLAD by the co-ablation of FePt and Al$_2$O$_3$ targets using two laser beams. A large number of aluminum oxide laser droplets were found on the deposited nanocomposite thin films due to the higher laser energy density and porous nature of aluminum oxide target. The as-deposited FePt:Al$_2$O$_3$ composite exhibited the magnetically soft *fcc* phase and thermal annealing was required for the phase transition to the magnetically hard *fct* phase. Compared with FePt nanoparticle agglomerates and nanoparticle networks, FePt:Al$_2$O$_3$ nanocomposite thin films achieved lower exchange coupling effects after thermal annealing for phase transition due to the introduction of nonmagnetic matrix material, Al$_2$O$_3$. FePt:Al$_2$O$_3$ nanocomposite thin films were also synthesized by magnetic trapping assisted PLAD. It was observed that magnetic trapping not only lowered the number of laser droplets but also lowered the annealing temperature from about 600°C to 300°C to induce high-K_u *fct* phase transition.

One of the main issues discussed with respect to the use of PLAD for FePt thin films is the formation of impurity phases and their effect on the film's magnetic properties. The XRD patterns of FePt thin films, synthesized using PLAD, showed several unidentified impurity peaks. A systematic investigation analyzing the effect of target composition, substrate material, annealing parameters such as temperature, duration, and ambience, and PLAD deposition parameters such as chamber ambience, laser energy fluence, and target–substrate distance on impurity phase formation was done. The impurity phase formation was observed only on Si substrates. It was found that the target composition, PLAD chamber ambience, and annealing ambience were not responsible for impurity phase formation. The annealing temperature and duration influenced the impurity phases, but were not the cause of their formation. A decrease in the laser energy fluence and increase of the target–substrate distance resulted in the elimination of the impurity phases and better magnetic and structural properties of FePt thin films. The energy of the ablated plasma species, controlled by the laser energy fluence and the target–substrate distance, was found to be the main factor responsible for the formation of the impurity phases. Finally, a modified PLAD setup with controlled condensation in a diffusion cloud chamber was used for the synthesis of FeCo magnetic nanoparticles with narrow size distribution. The ambient gas pressure played an important role in the morphology and particle size of FeCo nanoparticles.

The chapter also reviewed the successful application of the DPF device as an energetic ion irradiation facility for the processing of magnetic thin films grown by PLAD as well as a high deposition rate source for impurity-free magnetic thin films with very high coercivities. The DPF device offers a complex mix of high-energy ions (up to several MeV) of the filling gas species, immensely hot and dense decaying plasma from the post-pinch phase, a fast-moving ionization wavefront, and a strong shockwave that provides a unique physical/chemical environment that is completely unheard of in any other conventional plasma-based deposition or processing facility.

The DPF device, with hydrogen as the filling gas, was used as the processing facility of PLAD-grown FePt thin films of two different thicknesses of about 67 and 100 nm. Nanostructuring of PLAD-synthesized FePt thin films, particularly in nanoparticle form, was successfully achieved by energetic plasma focus H^+ ion irradiation. The annealing temperature for phase transition of FePt thin films from the low-K_u fcc phase to the high-K_u fct phase was successfully reduced to 400°C by plasma focus ion irradiation. It was postulated that the ion irradiation resulted in vacancy or interstitial point defects in irradiated thin films, which lowered the activation energy for diffusion and hence lowered the annealing temperature for phase transition. The DPF device reduced the phase transition temperature to the hard magnetic phase in a single shot ion exposure with ion pulse duration of the order of about a hundred to a few hundred nanoseconds, as opposed to that of a long exposure irradiation time of a few to several minutes from continuous ion sources. Hence, the DPF provides a fast and effective way to lower the phase transition temperature along with the nanostructurization of the irradiated thin films.

The NX2 plasma focus device, operating in the sub-kJ range, was successfully used to synthesize CoPt nanoparticle thin films of as small particle size as possible with narrow size distribution using different hydrogen-filling gas pressures (keeping the substrate–anode top distance and the number of plasma focus deposition shots fixed) and using different numbers of plasma focus deposition shots (at fixed substrate–anode distance at fixed filling hydrogen gas pressure of 6 mbar). The morphological features (shapes and sizes) showed a strong dependence on the filling gas pressure and the number of plasma focus deposition shots. The deposition rate at a deposition distance of 25 cm and at an optimized gas pressure of 6 mbar was estimated to be about 1.78 nm/shot, which is more than 30 times higher as compared to that of conventional PLAD. The DPF deposition has an added advantage of simultaneously processing the deposited material by a complex plasma/ion beam/shockwave mixture providing novel magnetic properties in the deposited material. The phase transition, from the low-K_u fcc structured $A1$ phase to high-K_u fct structured $L1_0$ phase, took place when the annealing temperature was about 600°C, resulting in significantly enhanced hard magnetic properties. The chapter also presented the first-ever use of a low 120 J energy FMPF-1 device for soft magnetic FeCo thin-film deposition. The deposited FeCo thin films were found to very soft with a coercivity of about 6 Oe.

The atmospheric microplasma (AMP), with at least one of the dimensions of the plasma in the submillimeter length scale, was used to synthesize nanoscale magnetic/catalytic materials in atmospheric ambience with resource-saving advantages of simplicity (no vacuum and heating system requirement) and efficiency (short synthesis time of about a few minutes). The iron oxide (Fe_3O_4) nanoparticles were synthesized by the conventional thermal decomposition method and the AMP-induced liquid chemical method. The effect of plasma discharge time (3, 5, and 10 min) on the properties of AMP-synthesized iron oxide nanoparticles was studied and it was demonstrated with the increasing plasma discharge time that Fe_3O_4 nanoparticles become increasingly more soft magnetic (with a coercivity of about 3 G for 10 min AMP discharge), approaching close to the superparamagnetic limit. The AMP has also been used to synthesize 0D, 1D, and 2D nanostructured metal and metal oxide materials, but the synthesis of 3D nanostructures using AMP has seldom been reported. We reported the synthesis of 3D nanostructured Pt_3Co in the form of NFs using AMP. Through the systematic analysis of SEM images of different parts of the deposited material, the nanoflower growth mechanism was understood and presented in the chapter. The nanoflower morphology of Pt_3Co was found to exhibit better electrochemical properties, indicating their potential application as catalysts.

ACKNOWLEDGMENTS

The authors are grateful to the National Institute of Education, Nanyang Technological University, Singapore, for providing AcRF Grants No. RI 17/03/RSR, RP 13/06/RSR, and RI 7/11/RSR, and various collaborators (in particular, Dr. Zhang Tao, Dr. Lin Jiaji, Dr. Pan Zhenying, Dr. Paul Lee, Dr. Augustine Tan, and Professor Lee Sing) for their help, support, and collaboration for various research work presented in this chapter.

REFERENCES

1. ITER Organization. The way to new energy, http://www.iter.org/.
2. National Ignition Facility & Photon Science, in: http://lasers.llnl.gov/, Lawrence Livermore National Laboratory.
3. Plasma Processing of Materials: Scientific Opportunities and Technological Challenges, 1991. National Academic Press, Washington, D.C.
4. Nanotechnology products @ National Nanotechololgy Infrastructure Network, in: http://www.nnin.org/news-events/spotlights/nanotechnology-products.
5. M. Roco, C. Mirkin, and M. Hersam, Nanotechnology research directions for societal needs in 2020: Summary of international study, *J. Nanoparticle Res.* 13, 2011, 897–919.

6. A. Clunan and K. Rodine-Hardy, *Nanotechnology in a Globalized World: Strategic Assessments of an Emerging Technology*, CCC PASCC Reports, Center on Contemporary Conflict (CCC), Monterey, USA, 2014.
7. K. Ostrikov and A. B. Murphy, Plasma-aided nanofabrication: Where is the cutting edge? *J. Phys. D: Appl. Phys.* 40, 2007, 2223–2241.
8. O. Kostya, C. Uros, and B. M. Anthony, Plasma nanoscience: Setting directions, tackling grand challenges, *J. Phys. D: Appl. Phys.* 44, 2011, 174001.
9. M. Meyyappan, Plasma nanotechnology: Past, present and future, *J. Phys. D: Appl. Phys.* 44, 2011, 174002.
10. M. Shiratani, K. Koga, S. Iwashita, G. Uchida, N. Itagaki, and K. Kamataki, Nanofactories in plasma: Present status and outlook, *J. Phys. D: Appl. Phys.* 44, 2011, 174038.
11. P. S. C. Committee on High Energy Density Plasma Physics, *National Research Council Frontiers in High Energy Density Physics: The X-Games of Contemporary Science*, National Academies Press, Washington, D.C., 2003.
12. L. Soto, New trends and future perspectives on plasma focus research, *Plasma Phys. Contr. Fusion* 47, 2005, A361–A381.
13. C. Ross, Patterned magnetic recording media, *Annu. Rev. Mater. Res.* 31, 2001, 203–235.
14. S. N. Piramanayagam, Perpendicular recording media for hard disk drives, *J. Appl. Phys.* 102, 2007, 011301.
15. R. Banerjee, Y. Katsenovich, L. Lagos, M. McIntosh, X. Zhang, and C. Z. Li, Nanomedicine: Magnetic nanoparticles and their biomedical applications, *Curr. Med. Chem.* 17, 2010, 3120–3141.
16. A. Akbarzadeh, M. Samiei, and S. Davaran, Magnetic nanoparticles: Preparation, physical properties, and applications in biomedicine, *Nanoscale Res. Lett.* 7, 2012, 1–13.
17. A. Lu, E. L. Salabas, and F. Schuth, Magnetic nanoparticles: Synthesis, protection, Functionalization, and Application, *Angew. Chem.* 46, 2007, 1222–1244.
18. Y. W. Jun, J. S. Choi, and J. Cheon, Heterostructured magnetic nanoparticles: Their versatility and high performance capabilities, *Chem. Commun.* 12, 2007, 1203–1214.
19. A. Santasalo-Aarnio, E. Sairanen, R. M. Aran-Ais, M. C. Figueiredo, J. Hua, J. M. Feliu, J. Lehtonen, R. Karinen, and T. Kallio, The activity of ALD-prepared PtCo catalysts for ethanol oxidation in alkaline media, *J. Catal.* 309, 2014, 38–48.
20. C. C. Berry, Progress in functionalization of magnetic nanoparticles for applications in biomedicine, *J. Phys. D: Appl. Phys.* 42, 2009, 224003.
21. Q. A. Pankhurst, N. T. K. Thanh, S. K. Jones, and J. Dobson, Progress in applications of magnetic nanoparticles in biomedicine, *J. Phys. D: Appl. Phys.* 42, 2009, 224001.
22. H. Zijlstra, Permanent magnets; theory, in: *Handbook of Magnetic Materials*, E. P. Wohlfarth (ed.), North-Holland Publishing Company, Amsterdam, the Netherlands, 1982, pp. 37–105.
23. E. Yang, Development of FePt/Oxide high anisotropy magnetic media, Materials Science and Engineering, Carnegie Mellon University, Pittsburgh, PA, 2010, p. 1.
24. T. Coughlin, *Invest in New Technologies or Divest in Market Share*, Coughlin Associates, Atascadero, CA, 2010.
25. E. Grochowski, J. Robert, and E. Fontana, Future technology challenges for NAND flash and HDD products, in: *Flash Memory Summit, Conference ConCepts, Inc.*, Santa Clara, CA, 2012.
26. A. C. Munce and J. U. Thiele, Hitachi's overseas research on hard disk drive, *Hitachi Rev.* 55, 2006, 5.
27. D. Weller and A. Moser, Thermal effect limits in ultrahigh-density magnetic recording, *IEEE Transactions on Magnetics* 35, 1999, 4423–4439.

28. S. Khizroev and D. Litvinov, Perpendicular magnetic recording: Writing process, *J. Appl. Phys.* 95, 2004, 4521–4537.

29. J. S. Chen, L. N. Zhang, J. F. Hu, and J. Ding, Highly textured $SmCo_5$ (001) thin film with high coercivity, *J. Appl. Phys.* 104, 2008, 093905.

30. Z. L. Zhao, *The Effects of Additional Nonmagnetic Layers on Structure and Magnetic Properties of $L1_0$ FePt Thin Films*, Department of Material Sience, National University of Singapore, Singapore, 2007, p. 159.

31. D. Weller, A. Moser, L. Folks, M. E. Best, W. Lee, M. F. Toney, M. Schwickert, J. U. Thiele, and M. F. Doerner, High K_u materials approach to 100 Gbits/in^2, *IEEE Trans. Magn.* 36, 2000, 10–15.

32. S. H. Charap, L. Pu-Ling, and H. Yanjun, Thermal stability of recorded information at high densities, *IEEE Trans. Magn.* 33, 1997, 978–983.

33. Binary Alloy Phase Diagrams, ASM International (OH), 1990.

34. S. H. Whang, Q. Feng, and Y. Q. Gao, Ordering, deformation and microstructure in L10 type FePt, *Acta Mater.* 46, 1998, 6485–6495.

35. H. S. Ko, A. Perumal, and S.-C. Shin, Fine control of L10 ordering and grain growth kinetics by C doping in FePt films, *Appl. Phys. Lett.* 82, 2003, 2311–2313.

36. Q. A. Pankhurst, J. Connolly, S. K. Jones, and J. Dobson, Applications of magnetic nanoparticles in biomedicine, *J. Phys. D: Appl. Phys.* 36, 2003, R167.

37. R. S. Molday and D. Mackenzie, Immunospecific ferromagnetic iron-dextran reagents for labeling and magnetic separation of cells, *J. Immunol. Methods* 52, 1982, 353–367.

38. H. Pardoe, W. Chua-anusorn, T. G. St Pierre, and J. Dobson, Structural and magnetic properties of nanoscale iron oxide particles synthesized in the presence of dextran or polyvinyl alcohol, *J. Magn. Magn. Mater.* 225, 2001, 41–46.

39. C. Sangregorio, J. K. Wiemann, C. J. O'Connor, and Z. Rosenzweig, A new method for the synthesis of magnetoliposomes, *J. Appl. Phys.* 85, 1999, 5699–5701.

40. B. Q. Lu, Y. J. Zhu, G. F. Cheng, and Y. J. Ruan, Synthesis and application in drug delivery of hollow-core-double-shell magnetic iron oxide/silica/calcium silicate nano-composites, *Mater. Lett.* 104, 2013, 53–56.

41. D. H. Kim, D. E. Nikles, D. T. Johnson, and C. S. Brazel, Heat generation of aque-ously dispersed $CoFe_2O_4$ nanoparticles as heating agents for magnetically activated drug delivery and hyperthermia, *J. Magn. Magn. Mater.* 320, 2008, 2390–2396.

42. W. J. Atkinson, I. A. Brezovich, and D. P. Chakraborty, Usable frequencies in hyper-thermia with thermal seeds, *IEEE Trans. Biomed. Eng.* 31, 1984, 70–75.

43. M. M. Yallapu, S. F. Othman, E. T. Curtis, B. K. Gupta, M. Jaggi, and S. C. Chauhan, Multi-functional magnetic nanoparticles for magnetic resonance imaging and cancer therapy, *Biomaterials* 32, 2011, 1890–1905.

44. Y. G. Wang, Q. Wei, Y. Zhang, D. Wu, H. M. Ma, A. P. Guo, and B. Du, A sandwich-type immunosensor using Pd-Pt nanocrystals as labels for sensitive detection of human tissue polypeptide antigen, *Nanotechnology* 25, 2014, 055102.

45. D. Mariotti, J. Patel, V. Svrcek, and P. Maguire, Plasma-liquid interactions at atmo-spheric pressure for nanomaterials synthesis and surface engineering, *Plasma Process. Polym.* 9, 2012, 1074–1085.

46. D. B. Chrisey and G. K. Hubler, *Pulsed Laser Deposition of Thin Films*, John Wiley & Sons, New York, 1994.

47. R. Eason, *Pulsed Laser Deposition of Thin Films*, John Wiley & Sons, Inc., Hoboken, New Jersey, 2007.

48. K. E. Youden, *Fabrication and Characterisation of Photorefractive Thin Films and Waveguides*, Physics Department, Faculty of Science, University of Southampton, Southampton, England, 1992.

49. P. R. Willmott and J. R. Huber, Pulsed laser vaporization and deposition, *Rev. Mod. Phys.* 72, 2000, 315–328.

50. J. J. Lin, S. Mahmood, T. L. Tan, S. V. Springham, P. Lee, and R. S. Rawat, Backward plume deposition as a novel technique for high deposition rate Fe nanoclusters synthesis, *Nanotechnology* 18, 2007, 115617.

51. J. J. Lin, S. Mahmood, T. Zhang, S. M. Hassan, T. White, R. V. Ramanujan, P. Lee, and R. S. Rawat, Synthesis of Fe_3O_4 nanostructures by backward plume deposition and influence of ambient gas pressure on their morphology, *J. Phys. D: Appl. Phys.* 40, 2007, 2548–2554.

52. Z. Y. Zhang and M. G. Lagally, Atomistic processes in the early stages of thin-film growth, *Science* 276, 1997, 377–383.

53. Happy, S. R. Mohanty, P. Lee, T. L. Tan, S. V. Springham, A. Patran, R. V. Ramanujan, and R. S. Rawat, Effect of deposition parameters on morphology and size of FeCo nanoparticles synthesized by pulsed laser ablation deposition, *Appl. Surf. Sci.* 252, 2006, 2806–2816.

54. J. J. Lin, Z. Y. Pan, S. Karamat, S. Mahmood, P. Lee, T. L. Tan, S. V. Springham, and R. S. Rawat, FePt: Al_2O_3 nanocomposite thin films synthesized by magnetic trapping assisted pulsed laser deposition with reduced intergranular exchange coupling, *J. Phys. D: Appl. Phys.* 41, 2008, 095001.

55. J. J. Lin, Pulsed laser deposition of nanostructured magnetic materials, PhD thesis, National Institute of Education, Nanyang Technological University, Singapore, 2009, p. 197.

56. K. Aimuta, K. Nishimura, H. Uchida, and M. Inoue, Change in element distributions in FePt films with Ag or Cu additives associated with their crystallographic phase transformation through thermal annealing, *Phys. Status Solidi (B)* 241, 2004, 1727–1730.

57. S. Kang, J. W. Harrell, and D. E. Nikles, Reduction of the fcc to $L1_0$ ordering temperature for self-assembled FePt nanoparticles containing Ag, *Nano Lett.* 2, 2002, 1033–1036.

58. B. Wang, K. Barmak, and T. J. Klemmer, A1 to $L1_0$ transformation in FePt films with ternary alloying additions of Ag and Au, *IEEE Trans. Magn.* 46, 2010, 1773–1776.

59. C. L. Platt, K. W. Wierman, E. B. Svedberg, R. van de Veerdonk, J. K. Howard, A. G. Roy, and D. E. Laughlin, $L1_0$ ordering and microstructure of FePt thin films with Cu, Ag, and Au additive, *J. Appl. Phys.* 92, 2002, 6104–6109.

60. T. Mahalingam, J. P. Chu, and S. F. Wang, Structure and magnetic properties of sputtered FePtM (M = Ti, Nb) thin films, *Mater. Lett.* 61, 2007, 4046–4049.

61. Q. Yan, T. Kim, A. Purkayastha, P. G. Ganesan, M. Shima, and G. Ramanath, Enhanced chemical ordering and coercivity in FePt alloy nanoparticles by Sb-doping, *Adv. Mater.* 17, 2005, 2233–2237.

62. S. S. Kang, D. E. Nikles, and J. W. Harrell, Synthesis, chemical ordering, and magnetic properties of self-assembled FePt–Ag nanoparticles, *J. Appl. Phys.* 93, 2003, 7178.

63. C. H. Lai, C. C. Chiang, and C. H. Yang, Low-temperature ordering of FePt by formation of silicides in underlayers, *J. Appl. Phys.* 97, 2005, 10H310.

64. J. S. Chen, B. C. Lim, Y. F. Ding, and G. M. Chow, Low-temperature deposition of $L1_0$ FePt films for ultra-high density magnetic recording, *J. Magn. Magn. Mater.* 303, 2006, 309–317.

65. T. Konagai, Y. Kitahara, T. Itoh, T. Kato, S. Iwata, and S. Tsunashima, Perpendicular anisotropy of MBE-grown FePt–Ag granular films, *J. Magn. Magn. Mater.* 310, 2007, 2662–2664.

66. C. Feng, Q. Zhan, B. H. Li, J. Teng, M. H. Li, Y. Jiang, and G. H. Yua, Magnetic properties and microstructure of FePt/Au multilayers with high perpendicular magnetocrystalline anisotropy, *Appl. Phys. Lett.* 93, 2008, 152513.

67. Y. Ogata, Y. Imai, and S. Nakagawa, Effect of multilayer configuration of [Fe/Pt] multilayer to attain (001) oriented FePt ordered alloy thin films, *J. Appl. Phys.* 107, 2010, 09A715.

68. J. J. Lin, M. V. Roshan, Z. Y. Pan, R. Verma, P. Lee, S. V. Springham, T. L. Tan, and R. S. Rawat, FePt nanoparticle formation with lower phase transition temperature by single shot plasma focus ion irradiation, *J. Phys. D: Appl. Phys.* 41, 2008, 135213.

69. Z. Y. Pan, R. S. Rawat, J. J. Lin, T. Zhang, P. Lee, T. L. Tan, and S. V. Springham, Nanostructuring of FePt thin films by plasma focus device: Pulsed ion irradiation dependent phase transition and magnetic properties, *Appl. Phys. A* 96, 2009, 1027–1033.

70. Z. Y. Pan, J. J. Lin, T. Zhang, S. Karamat, T. L. Tan, P. Lee, S. V. Springham, R. V. Ramanujan, and R. S. Rawat, Lowering of L10 phase transition temperature of FePt thin films by single shot H$^+$ ion exposure using plasma focus device, *Thin Solid Films* 517, 2009, 2753–2757.

71. C. H. Lai, C. H. Yang, and C. C. Chiang, Ion-irradiation-induced direct ordering of L1$_0$ FePt phase, *Appl. Phys. Lett.* 83, 2003, 4550–4552.

72. S. Kavita, V. R. Reddy, A. Gupta, S. Amirthapandian, and B. K. Panigrahi, Study of face-centred tetragonal FePt phase formation in as-deposited and heavy ion irradiated Fe/Pt multilayers, *Nucl. Instrum. Methods Phys. Res. Sect. B. Beam Interact. Mater. Atoms* 244, 2006, 206–208.

73. J. J. Lin, T. Zhang, P. Lee, S. V. Springham, T. L. Tan, R. S. Rawat, T. White, R. Ramanujan, and J. Guo, Magnetic trapping induced low temperature phase transition from fcc to fct in pulsed laser deposition of FePt:Al$_2$O$_3$ nanocomposite thin films, *Appl. Phys. Lett.* 91, 2007, 063120.

74. T. Kobayashi, H. Akiyoshi, and M. Tachiki, Development of prominent PLD (Aurora method) suitable for high-quality and low-temperature film growth, *Appl. Surf. Sci.* 197–198, 2002, 294–303.

75. S. H. Huh, A. Nakajima, and A. K. Kaya, Fabrication of ferromagnetic nanocluster rods by magnetic trapping, *J. Appl. Phys.* 95, 2004, 2732–2736.

76. U. Wiedwald, A. Klimmer, B. Kern, L. Han, H. G. Boyen, P. Ziemann, and K. Fauth, Lowering of the L1$_0$ ordering temperature of FePt nanoparticles by He$^+$ ion irradiation, *Appl. Phys. Lett.* 90, 2007, 062508.

77. J. P. Liu, Y. Liu, C. P. Luo, Z. S. Shan, and D. J. Sellmyer, Magnetic hardening in FePt nanostructured films, *J. Appl. Phys.* 81, 1997, 5644–5646.

78. K. L. Torres, R. R. Vanfleet, and G. B. Thompson, Comparison of simulated and experimental order parameters in FePt-II, *Microsc. Microanal.* 17, 2011, 403–409.

79. D. P. Chiang, S. Y. Chen, Y. D. Yao, H. Ouyang, C. C. Yu, Y. Y. Chen, and H. M. Lin, Microstructure and ordering parameter studies in multilayer [FePt(x)/Os]n films, *J. Appl. Phys.* 109, 2011, 07A732.

80. A. C. Sun, F. T. Yuan, and J. H. Hsu, Control of growth and ordering process in FePt(001) film at 300°C, in: *International Conference on Magnetism*, G. Goll, H. V. Lohneysen, A. Loidl, T. Pruschke, M. Richter, L. Schultz, C. Surgers, and J. Wosnitza (eds.), IOP Publishing Ltd, Bristol, 2010, p. 102009.

81. L. Shen, Z. M. Yuan, J. Q. Goh, T. J. Zhou, B. Liu, and Y. P. Feng, The effect of introduced defects on saturation magnetization and magnetic anisotropy field of L1(0) FePt, *IEEE Trans. Magn.* 47, 2011, 2422–2424.

82. T. Thomsona, B. D. Terris, M. F. Toney, S. Raoux, J. E. E. Baglin, S. L. Lee, and S. Sun, Silicide formation and particle size growth in high-temperature-annealed, self-assembled FePt nanoparticles, *J. Appl. Phys.* 95, 2004, 6738–6740.

83. X. Li, B. Liu, H. Sun, J. Guo, F. Wang, W. Li, and X. Zhang, L1$_0$ phase transition in FePt thin films via direct interface reaction, *J. Phys. D: Appl. Phys.* 41, 2008, 235001.

84. P. Jang, C.-S. Jung, K. Seomoon, and K.-H. Kim, Interfacial structure of ferromagnetic Fe-Pt thin films grown on a Si substrate, *Curr. Appl. Phys.* 11, 2011, S95–S97.

85. C. W. White, S. P. Withrow, J. D. Budai, L. A. Boatner, K. D. Sorge, J. R. Thompson, K. S. Beaty, and A. Meldrum, Formation of ferromagnetic FePt nanoparticles by ion implantation, *Mater. Res. Soc. Symp. Proc.* 704, 2002, W7.7.1–W7.7.6.

86. K. Leistner, J. Thomas, H. Schlörb, M. Weisheit, L. Schultz, and S. Fähler, Highly coercive electrodeposited FePt films by postannealing in hydrogen, *Appl. Phys. Lett.* 85, 2004, 3498–3500.

87. N. H. Luong, V. V. Hiep, D. M. Hong, N. Chau, N. D. Linh, M. Kurisu, D. T. K. Anh, and G. Nakamoto, High-coercivity FePt sputtered films, *J. Magn. Magn. Mater.* 290, 2005, 559–561.

88. Y. Wang, R. Medwal, N. Sehdev, B. Yadian, T. L. Tan, P. Lee et al. Elimination of impurity phase formation in FePt magnetic thin films prepared by pulsed laser deposition, *Appl. Surf. Sci.* 288, 2014, 381–391.

89. S. Mahmood, R. S. Rawat, M. Zakaullah, J. J. Lin, S. V. Springham, T. L. Tan, and P. Lee, Investigation of plume expansion dynamics and estimation of ablation parameters of laser ablated Fe plasma, *J. Phys. D: Appl. Phys.* 42, 2009, 135504.

90. P. Sharma, N. Kaushik, M. Esashi, M. Nishijima, and A. Makino, On the growth and magnetic properties of flower-like nanostructures formed on diffusion of FePt with Si substrate, *J. Magn. Magn. Mater.* 337–338, 2013, 38–45.

91. R. M. Bozorth, *Ferromagnetism*, IEEE, New York, 1993.

92. H. S. Jung, W. D. Doyle, J. E. Wittig, J. F. Al-Sharab, and J. Bentley, Soft anisotropic high magnetization Cu/FeCo films, *Appl. Phys. Lett.* 81, 2002, 2415–2417.

93. Y. Fu, Z. Yang, M. Mitsunori, X. X. Liu, and M. Akimitsu, Studies on high-moment soft magnetic FeCo/Co thin films, *Chinese Phys.* 15, 2006, 1351–1355.

94. H. Okamoto, Co-Fe (cobalt-iron), *J. Phys. Eqil. Diff.* 29, 2008, 383–384.

95. H. L. Su, N. J. Tang, R. L. Wang, B. Nie, S. L. Tang, L. Lv, and Y. W. Du, Chemical synthesis of face-centered-tetragonal FePt film using sol-gel method, *Chem. Lett.* 36, 2007, 180–181.

96. P. L. Ong, S. Mahmood, T. Zhang, J. J. Lin, R. V. Ramanujan, P. Lee, and R. S. Rawat, Synthesis of FeCo nanoparticles by pulsed laser deposition in a diffusion cloud chamber, *Appl. Surf. Sci.* 254, 2008, 1909–1914.

97. T. Zhang, K. S. Thomas. Gan, P. Lee, R. V. Ramanujan, and R. S. Rawat, Characteristics of FeCo nano-particles synthesized using plasma focus, *J. Phys. D: Appl. Phys.* 39, 2006, 2212–2219.

98. M. S. El-Shall, W. Slack, W. Vann, D. Kane, and D. Hanley, Synthesis of nanoscale metal oxide particles using laser vaporization/condensation in a diffusion cloud chamber, *J. Phys. Chem.* 98, 1994, 3067–3070.

99. Y. B. Pithawalla, M. S. El-Shall, S. C. Deevi, V. Ström, and K. V. Rao, Synthesis of magnetic intermetallic FeAl nanoparticles from a non-magnetic bulk alloy, *J. Phys. Chem. B* 105, 2001, 2085–2090.

100. N. X. Sun, A. M. Crawford, and S. X. Wang, Advanced soft magnetic materials for magnetic recording heads and integrated inductors, *MRS Online Proc. Library* 721, 2002, E6.3.1–E6.3.10.

101. T. Zhang, J. Lin, A. Patran, D. Wong, S. M. Hassan, S. Mahmood et al. Optimization of a plasma focus device as an electron beam source for thin film deposition, *Plasma Sources Sci. Technol.* 16, 2007, 250–256.

102. A. Patran, D. Stoenescu, R. S. Rawat, S. V. Springham, T. L. Tan, L. C. Tan, M. S. Rafique, P. Lee, and S. Lee, A magnetic electron analyzer for plasma focus electron energy distribution studies, *J. Fusion Energy* 25, 2006, 57–66.

103. M. Zakaullah, I. Ahmad, M. Shafique, G. Murtaza, M. Yasin, and M. M. Beg, Correlation study of ion, electron and X-ray emission from argon focus plasma, *Phys. Scripta* 57, 1998, 136–141.

104. V. Raspa, L. Sigaut, R. Llovera, R. Cobelli, P. Knoblauch, R. Vieytes, A. Clausse, and C. Moreno, Plasma focus as a powerful hard x-ray source for ultrafast imaging of moving metallic objects, *Brazilian J. Phys.* 34, 2004, 1696–1699.

105. M. Zakaullah, K. Alamgir, M. Shafiq, S. M. Hassan, M. Sharif, S. Hussain, and A. Waheed, Characteristics of x-rays from a plasma focus operated with neon gas, *Plasma Sources Sci. Technol.* 11, 2002, 377–382.

106. H. Bhuyan, S. R. Mohanty, N. K. Neog, S. Bujarbarua, and R. K. Rout, Comparative study of soft x-ray emission characteristics in a low energy dense plasma focus device, *J. Appl. Phys.* 95, 2004, 2975–2981.

107. M. Barbaglia, H. Bruzzone, H. Acuna, L. Soto, and A. Clausse, Experimental study of the hard x-ray emissions in a plasma focus of hundreds of Joules, *Plasma Phys. Contr. Fusion* 51, 2009, 045001.

108. T. Zhang, X. Lin, K. A. Chandra, T. L. Tan, S. V. Springham, A. Patran, P. Lee, S. Lee, and R. S. Rawat, Current sheath curvature correlation with the neon soft x-ray emission from plasma focus device, *Plasma Sources Sci. Technol.* 14, 2005, 368–374.

109. H. Bhuyan, S. R. Mohanty, T. K. Borthakur, and R. S. Rawat, Analysis of nitrogen ion beam produced in dense plasma focus device using Faraday cup, *Indian J. Pure Appl. Phys.* 39, 2001, 698–703.

110. H. Kelly, A. Lepone, A. Marquez, M. J. Sadowski, J. Baranowski, and E. Skladnik-Sadowska, Analysis of the nitrogen ion beam generated in a low-energy plasma focus device by a Faraday cup operating in the secondary electron emission mode, *IEEE Trans. Plasma Sci.* 26, 1998, 113–117.

111. H. Heo and D. K. Park, Measurement of argon ion beam and X-ray energies in a plasma focus discharge, *Phys. Scripta* 65, 2002, 350–355.

112. A. Szydlowski, A. Banaszak, B. Bienkowska, I. M. Ivanova-Stanik, M. Scholz, and M. J. Sadowski, Measurements of fast ions and neutrons emitted from PF-1000 plasma focus device, *Vacuum* 76, 2004, 357–360.

113. V. A. Gribkov, A. Banaszak, B. Bienkowska, A. V. Dubrovsky, I. Ivanova-Stanik, L. Jakubowski et al. Plasma dynamics in the PF-1000 device under full-scale energy storage: II. Fast electron and ion characteristics versus neutron emission parameters and gun optimization perspectives, *J. Phys. D: Appl. Phys.* 40, 2007, 3592–3607.

114. J. Moreno, C. Pavez, L. Soto, A. Tarifeño, P. Reymond, N. Verschueren, and P. Ariza, Preliminary studies of ions emission in a small plasma focus device of hundreds of Joules, *AIP Conference Proceedings* 1088, 2009, 215–218.

115. L. Rapezzi, M. Angelone, M. Pillon, M. Rapisarda, E. Rossi, M. Samuelli, and F. Mezzetti, Development of a mobile and repetitive plasma focus, *Plasma Sources Sci. Technol.* 13, 2004, 272–277.

116. A. Tartari, G. Verri, A. Da Re, F. Mezzetti, C. Bonifazzi, and L. Rapezzi, Improvement of calibration assessment for gold fast-neutron activation analysis using plasma focus device, *Measure. Sci. Technol.* 13, 2002, 939–945.

117. E. H. Beckner, Production and diagnostic measurements of kilovolt high-density deuterium helium and neon plasmas, *J. Appl. Phys.* 37, 1966, 4944–4952.

118. T. L. Tan, D. Wong, P. Lee, R. S. Rawat, S. Springham, and A. Patran, Characterization of chemically amplified resist for X-ray lithography by Fourier transform infrared spectroscopy, *Thin Solid Films* 504, 2006, 113–116.

119. Y. Kato and S. H. Be, Generation of soft x-ray using a rare gas-hydrogen plasma focus and its application to x-ray lithography, *Appl. Phys. Lett.* 48, 1986, 686–688.

120. V. A. Gribkov, A. Srivastava, P. L. C. Keat, V. Kudryashov, and S. Lee, Operation of NX2 dense plasma focus device with argon filling as a possible radiation source for micro-machining, *IEEE Trans. Plasma Sci.* 30, 2002, 1331–1338.

121. S. M. P. Kalaiselvi, T. L. Tan, A. Talebitaher, P. Lee, and R. S. Rawat, Optimization of neon soft X-rays emission from 200 J fast miniature dense plasma focus device: A potential source for soft X-ray lithography, *Phys. Lett. A* 377, 2013, 1290–1296.

122. R. Petr, A. Bykanov, J. Freshman, D. Reilly, J. Mangano, M. Roche, J. Dickenson, M. Burte, and J. Heaton, Performance summary on a high power dense plasma focus x-ray lithography point source producing 70 nm line features in AlGaAs microcircuits, *Rev. Sci. Instrum.* 75, 2004, 2551–2559.

123. Y. Kato, I. Ochiai, Y. Watanabe, and S. Murayama, Plasma focus x-ray source for lithography, *J. Vac. Sci. Technol. B* 6, 1988, 195–198.
124. E. P. Bogolyubov, V. D. Bochkov, V. A. Veretennikov, L. T. Vekhoreva, V. A. Gribkov, A. V. Dubrovskii et al. A powerful soft X-ray source for X-ray lithography based on plasma focusing, *Phys. Scripta* 57, 1998, 488–494.
125. R. S. Rawat, T. Zhang, G. J. Lim, W. H. Tan, S. J. Ng, A. Patran et al. Soft X-ray imaging using a neon filled plasma focus X-ray source, *J. Fusion Energy* 23, 2004, 49–53.
126. S. Hussain, A. Zakaullah, S. Ali, and A. Waheed, Low energy plasma focus as an intense X-ray source for radiography, *Plasma Sci. Technol.* 6, 2004, 2296–2300.
127. S. Hussain, M. Shafiq, R. Ahmad, A. Waheed, and M. Zakaullah, Plasma focus as a possible x-ray source for radiography, *Plasma Sources Sci. Technol.* 14, 2005, 61–69.
128. C. Moreno, V. Raspa, L. Sigaut, R. Vieytes, and A. Clausse, Plasma-focus-based tabletop hard x-ray source for 50 ns resolution introspective imaging of metallic objects through metallic walls, *Appl. Phys. Lett.* 89, 2006, 091502.
129. V. Raspa, C. Moreno, L. Sigaut, and A. Clausse, Effective hard x-ray spectrum of a tabletop Mather-type plasma focus optimized for flash radiography of metallic objects, *J. Appl. Phys.* 102, 2007, 123303.
130. F. Di Lorenzo, V. Raspa, P. Knoblauch, A. Lazarte, C. Moreno, and A. Clausse, Hard x-ray source for flash radiography based on a 2.5 kJ plasma focus, *J. Appl. Phys.* 102, 2007, 033304.
131. F. Castillo, Gamboa-deBuen, J. J. E. Herrera, J. Rangel, and S. Villalobos, High contrast radiography using a small dense plasma focus, *Appl. Phys. Lett.* 92, 2008, 051502.
132. V. A. Gribkov, Current and perspective applications of dense plasma focus devices, *AIP Conference Proceedings* 996, 2008, 51–64.
133. R. Verma, R. S. Rawat, P. Lee, M. Krishnan, S. V. Springham, and T. L. Tan, Miniature plasma focus device as a compact hard x-ray source for fast radiography applications, *IEEE Trans. Plasma Sci.* 38, 2010, 652–657.
134. M. V. Roshan, S. V. Springham, R. S. Rawat, and P. Lee, Short-lived PET radioisotope production in a small plasma focus device, *IEEE Trans. Plasma Sci.* 38, 2010, 3393–3397.
135. J. S. Brzosko, K. Melzacki, C. Powell, M. Gai, R. H. France, J. E. McDonald, G. D. Alton, F. E. Bertrand, and J. R. Beene, Breeding 10(10)/s radioactive nuclei in a compact plasma focus device, in: *Application of Accelerators in Research and Industry*, American Institute of Physics, Melville, J. L. Duggan and I. L. Morgan (eds.), 2001, pp. 277–280.
136. M. Sumini, D. Mostacci, F. Rocchi, M. Frignani, A. Tartari, E. Angeli, D. Galaverni, U. Coli, B. Ascione, and G. Cucchi, Preliminary design of a 150 kJ repetitive plasma focus for the production of 18-F, *Nucl. Instrum. Methods Phys. Res. Sect. A Accelerators Spectrometers Detectors Associated Equip.* 562, 2006, 1068–1071.
137. A. Talaei, S. M. S. Kiai, and A. A. Zaeem, Effects of admixture gas on the production of F-18 radioisotope in plasma focus devices, *Appl. Radiat. Isot.* 68, 2010, 2218–2222.
138. B. Shirani, F. Abbasi, and M. Nikbakht, Production of N-13 by C-12(d,n)N-13 reaction in a medium energy plasma focus, *Appl. Radiat. Isot.* 74, 2013, 86–90.
139. J. N. Feugeas, E. C. Lionch, C. O. de Gonzaez, and G. Galambos, Nitrogen implantation of AISI 304 stainless steel with a coaxial plasma gun, *J. Appl. Phys.* 64, 1988, 2648–2651.
140. R. S. Rawat, M. P. Srivastava, S. Tandon, and A. Mansingh, Crystallization of an amorphous lead zirconate titanate thin film with a dense plasma focus device, *Phys. Rev. B* 47, 1993, 4858–4862.
141. C. R. Kant, M. P. Srivastava, and R. S. Rawat, Thin carbon film deposition using energetic ions of a dense plasma focus, *Phys. Lett. A* 226, 1997, 212–216.

142. S. Lee, T. Y. Tou, S. P. Moo, M. A. Eissa, A. V. Gholap, K. H. Kwek, S. Mulyodrono, A. J. Smith, Suryadi, W. Usada, and M. Zakaullah, A simple facility for the teaching of plasma dynamics and plasma nuclear fusion, *Am. J. Phys.* 56, 1988, 62–68.

143. S. Lee, P. Lee, G. X. Zhang, Z. P. Feng, V. A. Gribkov, M. Liu, A. Serban, and T. K. S. Wong, High rep rate high performance plasma focus as a powerful radiation source, *IEEE Trans. Plasma Sci.* 26, 1998, 1119–1126.

144. M. Krishnan, The dense plasma focus: A versatile dense pinch for diverse applications, *IEEE Trans. Plasma Sci.* 40, 2012, 3189–3221.

145. S. M. Hassan, T. Zhang, A. Patran, R. S. Rawat, S. V. Springham, T. L. Tan et al. Pinching evidences in a miniature plasma focus with fast pseudospark switch, *Plasma Sources Sci. Technol.* 15, 2006, 614–619.

146. A. Bernard, A. Coudeville, A. Jolas, J. Launspach, and J. D. Mascureau, Experimental studies of plasma focus and evidence for non-thermal processes, *Phys. Fluids* 18, 1975, 180–194.

147. H. Krompholz, F. Ruhl, W. Schneider, K. Schonbach, and G. Herziger, A scaling law for plasma-focus devices, *Phys. Lett. A* 82, 1981, 82–84.

148. S. Lee, S. H. Saw, P. Lee, and R. S. Rawat, Numerical experiments on plasma focus neon soft x-ray scaling, *Plasma Phys. Contr. Fusion* 51, 2009, 105013.

149. A. J. Toepfer, D. R. Smith, and E. H. Beckner, Ion heating in dense plasma focus, *Bull. Am. Phys. Soc.* 14, 1969, 1013.

150. N. Qi, S. F. Fulghum, R. R. Prasad, and M. Krishnan, Space and time resolved electron density and current measurements in a dense plasma focus z-pinch, *IEEE Trans. Plasma Sci.* 26, 1998, 1127–1137.

151. P. Choi, C. Deeney, H. Herold, and C. S. Wong, Characterization of self-generated intense electron beams in a plasma focus, *Laser Particle Beams* 8, 1990, 469–476.

152. M. V. Roshan, S. V. Springham, A. Talebitaher, R. S. Rawat, and P. Lee, Magnetic spectrometry of high energy deuteron beams from pulsed plasma system, *Plasma Phys. Contr. Fusion* 52, 2010, 085007.

153. R. S. Rawat, T. Zhang, K. S. T. Gan, P. Lee, and R. V. Ramanujan, Nano-structured Fe thin film deposition using plasma focus device, *Appl. Surf. Sci.* 253, 2006, 1611–1615.

154. J. A. Christodoulides, Y. Huang, Y. Zhang, G. C. Hadjipanayis, I. Panagiotopoulos, and D. Niarchos, CoPt and FePt thin films for high density recording media, *J. Appl. Phys.* 87, 2000, 6938–6940.

155. Z. Y. Pan, R. S. Rawat, M. V. Roshan, J. J. Lin, R. Verma, P. Lee, S. V. Springham, and T. L. Tan, Nanostructured magnetic CoPt thin films synthesis using dense plasma focus device operating at sub-kilojoule range, *J. Phys. D: Appl. Phys.* 42, 2009, 175001.

156. R. S. Rawat, W. M. Chew, P. Lee, T. White, and S. Lee, Deposition of titanium nitride thin films on stainless steel–AISI 304 substrates using a plasma focus device, *Surf. Coat. Tech.* 173, 2003, 276–284.

157. M. Hassan, A. Qayyum, R. Ahmad, R. S. Rawat, P. Lee, S. M. Hassan, G. Murtaza, and M. Zakaullah, Dense plasma focus ion-based titanium nitride coating on titanium, *Nucl. Instrum. Methods Phys. Res. Sect. B Beam Interact. Mater. Atoms* 267, 2009, 1911–1917.

158. G. R. Etaati, M. T. Hosseinnejad, M. Ghoranneviss, M. Habibi, and M. Shirazi, Deposition of tungsten nitride on stainless steel substrates using plasma focus device, *Nucl. Instrum. Methods Phys. Res. Sect. B Beam Interact. Mater. Atoms* 269, 2011, 1058–1062.

159. M. T. Hosseinnejad, M. Ghoranneviss, G. R. Etaati, M. Shirazi, and Z. Ghorannevis, Deposition of tungsten nitride thin films by plasma focus device at different axial and angular positions, *Appl. Surf. Sci.* 257, 2011, 7653–7658.

160. R. S. Rawat, P. Lee, T. White, L. Ying, and S. Lee, Room temperature deposition of titanium carbide thin films using dense plasma focus device, *Surf. Coat. Tech.* 138, 2001, 159–165.

161. R. Gupta and M. P. Srivastava, Carbon ion implantation on titanium for TiC formation using a dense plasma focus device, *Plasma Sources Sci. Technol.* 13, 2004, 371–374.

162. Z. A. Umar, R. S. Rawat, K. S. Tan, A. K. Kumar, R. Ahmad, T. Hussain, C. Kloc, Z. Chen, L. Shen, and Z. Zhang, Hard TiCx/SiC/a-C:H nanocomposite thin films using pulsed high energy density plasma focus device, *Nucl. Instrum. Methods Phys. Res. Sect. B Beam Interact. Mater. Atoms* 301, 2013, 53–61.

163. H. Bhuyan, M. Favre, E. Valderrama, G. Avaria, H. Chuaqui, I. Mitchell, E. Wyndham, R. Saavedra, and M. Paulraj, Formation of hexagonal silicon carbide by high energy ion beam irradiation on Si(100) substrate, *J. Phys. D: Appl. Phys.* 40, 2007, 127–131.

164. Z. R. Wang, H. R. Yousefi, Y. Nishino, H. Ito, and K. Masugata, Preparation of silicon carbide film by a plasma focus device, *Phys. Lett. A* 372, 2008, 7179–7182.

165. R. S. Rawat, V. Aggarwal, M. Hassan, P. Lee, S. V. Springham, T. L. Tan, and S. Lee, Nano-phase titanium dioxide thin film deposited by repetitive plasma focus: Ion irradiation and annealing based phase transformation and agglomeration, *Appl. Surf. Sci.* 255, 2008, 2932–2941.

166. G. Macharaga, R. S. Rawat, G. R. Deen, P. Lee, T. L. Tan, and S. V. Springham, TiO$_2$ Nano-cluster thin films by dense plasma focus and ion implantation effect on its photocatalytic activity, *J. Adv. Oxidation Technol.* 14, 2011, 308–313.

167. Y. Malhotra, S. Roy, M. P. Srivastava, C. R. Kant, and K. Ostrikov, Extremely non-equilibrium synthesis of luminescent zinc oxide nanoparticles through energetic ion condensation in a dense plasma focus device, *J. Phys. D: Appl. Phys.* 42, 2009, 155202.

168. I. A. Khan, R. S. Rawat, R. Ahmad, and M. A. K. Shahid, Deposition of alumina stabilized zirconia at room temperature by plasma focus device, *Appl. Surf. Sci.* 288, 2014, 304–312.

169. I. A. Khan, S. Jabbar, T. Hussain, M. Hassan, R. Ahmad, M. Zakaullah, and R. S. Rawat, Deposition of zirconium carbonitride composite films using ion and electron beams emitted from plasma focus device, *Nucl. Instrum. Methods Phys. Res. Sect. B Beam Interact. Mater. Atoms* 268, 2010, 2228–2234.

170. I. A. Khan, M. Hassan, T. Hussain, R. Ahmad, M. Zakaullah, and R. S. Rawat, Synthesis of nano-crystalline zirconium aluminium oxynitride (ZrAlON) composite films by dense plasma Focus device, *Appl. Surf. Sci.* 255, 2009, 6132–6140.

171. I. A. Khan, M. Hassan, R. Ahmad, G. Murtaza, M. Zakaullah, R. S. Rawat, and P. Lee, Synthesis of zirconium oxynitride (ZrON) nanocomposite films on zirconium substrate by dense plasma focus device, *Int. J. Mod. Phys. B* 22, 2008, 3941–3955.

172. M. Hassan, R. S. Rawat, P. Lee, S. M. Hassan, A. Qayyum, R. Ahmad, G. Murtaza, and M. Zakaullah, Synthesis of nanocrystalline multiphase titanium oxycarbide (TiCxOy) thin films by UNU/ICTP and NX2 plasma focus devices, *Appl. Phys. A* 90, 2008, 669–677.

173. E. Ghareshabani, R. S. Rawat, S. Sobhanian, R. Verma, S. Karamat, and Z. Y. Pan, Synthesis of nanostructured multiphase Ti(C,N)/a-C films by a plasma focus device, *Nucl. Instrum. Methods Phys. Res. Sect. B Beam Interact. Mater. Atoms* 268, 2010, 2777–2784.

174. T. Devolder, H. Bernas, D. Ravelosona, C. Chappert, S. Pizzini, J. Vogel, J. Ferre, J. P. Jamet, Y. Chen, and V. Mathet, Beam-induced magnetic property modifications: Basics, nanostructure fabrication and potential applications, *Nucl. Instrum. Methods Phys. Res. Sect. B Beam Interact. Mater. Atoms* 175, 2001, 375–381.

175. D. Ravelosona, C. Chappert, V. Mathet, and H. Bernas, Chemical order induced by He$^+$ ion irradiation in FePt(001) films, *J. Appl. Phys.* 87, 2000, 5771–5773.

290   Advances in Magnetic Materials

176. D. Ravelosona, C. Chappert, H. Bernas, D. Halley, Y. Samson, and A. Marty, Chemical ordering at low temperatures in FePd films, *J. Appl. Phys.* 91, 2002, 8082–8084.
177. G. Sanchez and J. Feugeas, The themal evolution of tagets under plama focus pulsed ion implantation, *J. Phys. D: Appl. Phys.* 30, 1997, 927–936.
178. J. F. Ziegler, Interactions of ions with matter, in: http://www.srim.org/.
179. H. Trinkaus, Ion beam induced amorphization of crystalline solids: Mechanisms and modeling, *Mater. Sci. Forum* 248–249, 1997, 3–12.
180. R. Sagar and M. P. Srivastava, Amorphization of thin film of CdS due to ion irradiation by dense plasma focus, *Phys. Lett. A* 183, 1993, 209–213.
181. M. Sadiq, M. Shafiq, A. Waheed, R. Ahmad, and M. Zakaullah, Amorphization of silicon by ion irradiation in dense plasma focus, *Phys. Lett. A* 352, 2006, 150–154.
182. B. C. Lim, J. S. Chen, and J. H. Yin, Reduction of exchange coupling and enhancement of coercivity of $L1_0$ FePt(001) films by Cu top layer diffusion, *Thin Solid Films* 505, 2006, 81–84.
183. R. Verma, M. V. Roshan, F. Malik, P. Lee, S. Lee, S. V. Springham, T. L. Tan, M. Krishnan, and R. S. Rawat, Compact sub-kilojoule range fast miniature plasma focus as portable neutron source, *Plasma Sources Sci. Technol.* 17, 2008, 045020.
184. R. Verma, P. Lee, S. V. Springham, T. L. Tan, R. S. Rawat, and M. Krishnan, Order of magnitude enhancement in x-ray yield at low pressure deuterium-krypton admixture operation in miniature plasma focus device, *Appl.Phys. Lett.* 92, 2008, 011506.
185. R. Verma, P. Lee, S. Lee, S. V. Springham, T. L. Tan, R. S. Rawat, and M. Krishnan, Order of magnitude enhancement in neutron emission with deuterium-krypton admixture operation in miniature plasma focus device, *Appl. Phys. Lett.* 93, 2008, 101501.
186. Z. Y. Pan, R. S. Rawat, R. Verma, J. J. Lin, H. Yan, R. V. Ramanujan, P. Lee, S. V. Springham, and T. L. Tan, Miniature plasma focus as a novel device for synthesis of soft magnetic FeCo thin films, *Phys. Lett. A* 374, 2010, 1043–1048.
187. P. Zou, W. Yu, and J. A. Bain, Influence of stress and texture on soft magnetic properties of thin films, *IEEE Trans. Magn.* 38, 2002, 3501–3520.
188. H. Hoffmann and T. Fujii, The wall coercivity of soft magnetic films, *J. Magn. Magn. Mater.* 128, 1993, 395–400.
189. M. Davide and R. M. Sankaran, Microplasmas for nanomaterials synthesis, *J. Phys. D: Appl. Phys.* 43, 2010, 323001.
190. F.-C. Chang, C. Richmonds, and R. M. Sankaran, Microplasma-assisted growth of colloidal Ag nanoparticles for point-of-use surface-enhanced Raman scattering applications, *J. Vac. Sci. Technol. A* 28, 2010, L5–L8.
191. R. M. Sankaran, D. Holunga, R. C. Flagan, and K. P. Giapis, Synthesis of blue luminescent Si nanoparticles using atmospheric-pressure microdischarges, *Nano Lett.* 5, 2005, 537–541.
192. T. Nozaki, K. Sasaki, T. Ogino, D. Asahi, and K. Okazaki, Microplasma synthesis of tunable photoluminescent silicon nanocrystals, *Nanotechnology* 18, 2007, 235603.
193. D. Mariotti, A. C. Bose, and K. Ostrikov, Atmospheric-microplasma-assisted nanofabrication: Metal and metal-oxide nanostructures and nanoarchitectures, *IEEE Trans. Plasma Sci.* 37, 2009, 1027–1033.
194. Z. Yang, H. Shirai, T. Kobayashi, and Y. Hasegawa, Synthesis of Si nanocones using rf microplasma at atmospheric pressure, *Thin Solid Films* 515, 2007, 4153–4158.
195. H. Yoshiki and T. Mitsui, TiO_2 thin film coating on a capillary inner surface using atmospheric-pressure microplasma, *Surf. Coat. Tech.* 202, 2008, 5266–5270.
196. A. Yamatake, J. Fletcher, K. Yasuoka, S. Ishii, Water treatment by fast oxygen radical flow with DC-driven microhollow cathode discharge, *IEEE Trans. Plasma Sci.* 34, 2006, 1375–1381.

197. G. Fridman, A. D. Brooks, M. Balasubramanian, A. Fridman, A. Gutsol, V. N. Vasilets, H. Ayan, and G. Friedman, Comparison of direct and indirect effects of non-thermal atmospheric-pressure plasma on bacteria, *Plasma Process. Polym.* 4, 2007, 370–375.

198. K. W. Jo, M. G. Kim, S. M. Shin, and J. H. Lee, Microplasma generation in a sealed microfluidic glass chip using a water electrode, *Appl. Phys. Lett.* 92, 2008, 011503.

199. N. Shirai, M. Nakazawa, S. Ibuka, and S. Ishii, Atmospheric DC glow microplasmas using miniature gas flow and electrolyte cathode, *Jap. J. Appl. Phys.* 48, 2009, 036002.

200. I. G. Koo, M. S. Lee, J. H. Shim, J. H. Ahn, and W. M. Lee, Platinum nanoparticles prepared by a plasma-chemical reduction method, *J. Mater. Chem.* 15, 2005, 4125–4128.

201. H. Furusho, K. Kitano, S. Hamaguchi, and Y. Nagasaki, Preparation of stable water-dispersible PEGylated gold nanoparticles assisted by nonequilibrium atmospheric-pressure plasma jets, *Chem. Mater.* 21, 2009, 3526–3535.

202. G. Schmid, *Nanoparticles: From Theory to Applications*, Wiley-VCH Verlag GmbH & Co. KGaA, Weinheim, Germany, 2005.

203. K. H. Kale and A. N. Desai, Atmospheric pressure plasma treatment of textiles using non-polymerising gases, *Indian J. Fibre Textile Res.* 36, 2011, 289–299.

204. J. McKenna, J. Patel, S. Mitra, N. Soin, V. Svrcek, P. Maguire, and D. Mariotti, Synthesis and surface engineering of nanomaterials by atmospheric-pressure microplasmas, *Eur. Phys. J. Appl. Phys.* 56, 2011, 24020.

205. Y. B. Mou, Y. Y. Hou, B. A. Chen, Z. C. Hua, Y. Zhang, H. Xie et al. In vivo migration of dendritic cells labeled with synthetic superparamagnetic iron oxide, *Int. J. Nanomed.* 6, 2011, 2633–2640.

206. D. Liu, Y. Li, J. Deng, and W. Yang, Synthesis and characterization of magnetic Fe^3O^{4-} silica-poly(γ-benzyl-l-glutamate) composite microspheres, *Reactive Funct. Polym.* 71, 2011, 1040–1044.

207. Y. Wang, P. Kaur, A. T. L. Tan, R. Singh, P. C. K. Lee, S. V. Springham, R. V. Ramanujan, and R. S. Rawat, Iron oxide magnetic nanoparticles synthesized by atmospheric microplasmas, *Int. J. Mod. Phys.: Conf. Series* 32, 2014, 1460343.

208. H. A. Wriedt, The Fe-O (Iron-Oxygen) system, *JPE* 12, 1991, 170–200.

209. O. U. Rahman, S. C. Mohapatra, and S. Ahmad, Fe_3O_4 inverse spinal super paramagnetic nanoparticles, *Mater. Chem. Phys.* 132, 2012, 196–202.

210. M. D. S. Islam, Y. Kusumoto, J. Kurawaki, M. D. Abdulla-Al-Mamun, and H. Manaka, A comparative study on heat dissipation, morphological and magnetic properties of hyperthermia suitable nanoparticles prepared by co-precipitation and hydrothermal methods, *Bull. Mater. Sci.* 35, 2012, 1047–1053.

211. H. Wei, N. Insin, J. Lee, H. S. Han, J. M. Cordero, W. Liu, and M. G. Bawendi, Compact zwitterion-coated iron oxide nanoparticles for biological applications, *Nano Lett.* 12, 2011, 22–25.

212. M. N. Cao, D. S. Wu, and R. Cao, Recent advances in the stabilization of platinum electrocatalysts for fuel-cell reactions, *ChemCatChem* 6, 2014, 26–45.

213. T. He, E. Kreidler, L. Xiong, and E. Ding, Combinatorial screening and nano-synthesis of platinum binary alloys for oxygen electroreduction, *J. Power Sources* 165, 2007, 87–91.

214. Z. Q. Yao, M. S. Zhu, F. X. Jiang, Y. K. Du, C. Y. Wang, and P. Yang, Highly efficient electrocatalytic performance based on Pt nanoflowers modified reduced graphene oxide/carbon cloth electrode, *J. Mater. Chem.* 22, 2012, 13707–13713.

215. X. M. Chen, B. Y. Su, G. H. Wu, C. J. Yang, Z. X. Zhuang, X. R. Wang, and X. Chen, Platinum nanoflowers supported on graphene oxide nanosheets: Their green synthesis, growth mechanism, and advanced electrocatalytic properties for methanol oxidation, *J. Mater. Chem.* 22, 2012, 11284–11289.

216. J. N. Tiwari, R. N. Tiwari, K. L. Lin, Synthesis of Pt nanopetals on highly ordered silicon nanocones for enhanced methanol electrooxidation activity, *ACS Appl. Mater. Interfaces* 2, 2010, 2231–2237.

217. L. Wei, Y. J. Fan, H. H. Wang, N. Tian, Z. Y. Zhou, and S. G. Sun, Electrochemically shape-controlled synthesis in deep eutectic solvents of Pt nanoflowers with enhanced activity for ethanol oxidation, *Electrochim. Acta* 76, 2012, 468–474.

218. Y. X. Li, S. N. Wu, X. Cui, L. Wang, and X. M. Shi, Ultralow platinum-loading bimetallic nanoflowers: Fabrication and high-performance electrocatalytic activity towards the oxidation of formic acid, *Electrochem. Commun.* 25, 2012, 19–22.

219. L. Qian and X. R. Yang, Polyamidoamine dendrimers-assisted electrodeposition of gold-platinum bimetallic nanoflowers, *J. Phys. Chem. B* 110, 2006, 16672–16678.

220. Q. Q. Wu, Y. X. Li, H. Y. Xian, C. D. Xu, L. Wang, and Z. B. Chen, Ultralow Pt-loading bimetallic nanoflowers: Fabrication and sensing applications, *Nanotechnology* 24, 2013, 025501.

221. X. P. Yu, F. B. Zhang, N. Wang, S. X. Hao, and W. Chu, Plasma-treated bimetallic Ni-Pt catalysts derived from hydrotalcites for the carbon dioxide reforming of methane, *Catal. Lett.* 144, 2014, 293–300.

222. H. Zhang, W. Chu, H. Xu, and J. Zhou, Plasma-assisted preparation of Fe-Cu bimetal catalyst for higher alcohols synthesis from carbon monoxide hydrogenation, *Fuel* 89, 2010, 3127–3131.

223. C. Xu, Y. Liu, F. Su, A. Liu, and H. Qiu, Nanoporous PtAg and PtCu alloys with hollow ligaments for enhanced electrocatalysis and glucose biosensing, *Biosens. Bioelectron.* 27, 2011, 160–166.

224. R. Ahmadi, M. K. Amini, and J. C. Bennett, Pt–Co alloy nanoparticles synthesized on sulfur-modified carbon nanotubes as electrocatalysts for methanol electrooxidation reaction, *J. Catal.* 292, 2012, 81–89.

225. Z. Zhang, Y. Yang, F. Nosheen, P. Wang, J. Zhang, J. Zhuang, and X. Wang, Fine tuning of the structure of Pt-Cu alloy nanocrystals by glycine-mediated sequential reduction kinetics, *Small* 9, 2013, 3063–3069.

226. L. Liu, E. Pippel, R. Scholz, and U. Gösele, Nanoporous Pt–Co alloy nanowires: Fabrication, characterization, and electrocatalytic properties, *Nano Lett.* 9, 2009, 4352–4358.

5 Compositional Optimization and New Processes for Nanocrystalline NdFeB-Based Permanent Magnets

Zhongwu Liu and Lizhong Zhao

CONTENTS

ABSTRACT

Discovering the exchange coupling effect promotes the development of nanocrystalline NdFeB magnets. The magnetic properties for nanocrystalline and nanocomposite NdFeB-based alloys have been improved by both compositional modification and microstructural optimization. The effects of substitutions for Nd and Fe and other elemental doping have been investigated regarding to their roles in enhancing room temperature properties, thermal stability, and exchange interaction. To synthesize NdFeB magnets with energy product toward theoretically predicted value, new approaches have been proposed. Both top-down and bottom-up approaches, such as surfactant-assisted ball milling, chemical reduction process, and co-precipitation have been successful very recently. To assemble nanocrystalline NdFeB powders or nanoparticles into bulk magnets, various novel consolidation processes including spark plasma sintering (SPS) and high-velocity compaction (HVC) have been employed to obtain isotropic magnets with fine grain structure. To achieve anisotropy, hot deformation (HD) was selected as the follow-on process for SPS and HVC magnets. As for nanocrystalline bonded magnets, to enhance the thermal stability, melt-spun NdFeB ribbons and hard ferrite powders were mixed for achieving NdFeB/ferrite composite, which has potential to fill the market gap between the bonded NdFeB and ferrite hard magnets.

5.1   INTRODUCTION

Since the last century, our daily life has been significantly dependent on materials with outstanding magnetic properties. Although they emerged as permanent magnetic material almost 30 years ago, NdFeB-based alloys are still the most powerful magnets for industry. Their excellent hard magnetic properties result from the $Nd_2Fe_{14}B$-based phase with high saturation magnetization (M_S) and high anisotropy field (H_a). The maximum energy product $(BH)_{max}$ of sintered NdFeB magnets has pass 470 kJ/m³ [1]. Nevertheless, to improve the performance of electric and magnetic devices, the demand for stronger magnets has never ceased. Although not much progress has been made since the end of the last century in the field of permanent magnets, the effort toward improved properties and reduced production cost of NdFeB magnets has continued.

The production of NdFeB magnets starts from the alloy preparation and powdering. The subsequent processes include sintering, bonding or hot pressing, and hot deformation. The magnetic properties of NdFeB magnets are closely related to the processing parameters in each step. To obtain magnets with good properties, these parameters have to be carefully controlled. The most common preparation techniques for NdFeB permanent magnets at present are sintering and melt-spinning, leading to microcrystalline and nanocrystalline microstructures, respectively.

Sintered NdFeB magnets are fabricated by the traditional powder metallurgy route. The latest process involves ingot preparation by induction melting and strip casting, powdering by hydrogen decrepitation (HD) and jet milling, pressing and alignment under a magnetic field, sintering and heat treatment (HT) in vacuum, and machining and surface treatment. These processes produce the microcrystalline magnets containing at least $Nd_2Fe_{14}B$ type phase and some rare earth (RE)-rich phases. The presence of the nonmagnetic RE-rich phase enhances the coercivity but dilutes the magnetization of the magnets.

For the nanocrystalline NdFeB alloys, rapid solidification is the key process for powder preparation. The nanocrystalline ribbons are prepared by melt spinning followed by vacuum annealing. In the laboratory, the ribbons can also be prepared by direct quenching using melt spinning without HT by carefully controlling the wheel speed. The obtained ribbons are then crushed into powders with desired particle size. The standard nanocrystalline bulk magnets are generally prepared using these powders by three approaches, named bonding, hot pressing, and hot pressing followed by deformation. The magnets prepared by these methods are commercially named by MQ1, MQ2, and MQ3, which were initially established by the commercial company Magnequench [2].

Compared to sintered NdFeB magnets, nanocrystalline NdFeB magnets generally exhibit enhanced remanence in their isotropic state due to the so-called exchange coupling between the nanograins. In particular, nanocrystalline magnets are allowed to contain the soft magnetic phases of α-Fe or Fe_3B due to the existence of exchange coupling between the hard magnetic $Nd_2Fe_{14}B$ and soft phases. As one of the results, the nanocrystalline magnets normally have lower RE content, which results in a reduced cost of raw materials, with respect to sintered magnets. Except for the above, in principle, the NdFeB magnets with grain size close to their critical single domain size, which is about a few hundred nanometers, should display the highest coercivity [3]. This also attracts attention from researchers and engineers who aim to improve the magnetic properties of permanent magnets.

Remanence enhanced nanocrystalline NdFeB magnets were first discovered in the early 1990s [4]. After that, various processes like MQ1, MQ2, and MQ3 were introduced. Further developments led to improved compositions, new fabrication routes, enhanced properties, and novel applications [5–8]. During the last 20 years, compositional optimization and new processes for nanocrystalline NdFeB-based permanent magnets have emerged as the hot topics in the research and development of the permanent magnets. Substantial work is aiming at improving the properties and reducing the cost, and great progress has been made. In this chapter, we present an extensive introduction to these topics. The discussions are based on the compositional optimization of the nanocrystalline NdFeB magnets, new processes for nanocrystalline powders, densification processes for nanocrystalline alloys, anisotropy formation for nanocrystalline magnets, and bonded composite magnets.

It should be noted here that the so-called NdFeB magnets in this chapter embrace all NdFeB-based alloys containing various RE elements with Co substitution for Fe and a wide range of dopants, as long as the $Nd_2Fe_{14}B$ phases are presented as the main hard magnetic phase in the compositions. The so-called $Nd_2Fe_{14}B$ phases here include all 2:14:1 type hard magnetic phases whether they contains other elements or not.

5.2 STRUCTURE, MAGNETISM, AND PROPERTIES OF NANOCRYSTALLINE NdFeB ALLOYS

Nanocrystalline NdFeB alloys generally refer to NdFeB-based alloys with a grain size that varies from a few nanometers to a few tenths of a nanometer. They also include the isotropic or anisotropic magnets with grain size up to a few hundred nanometers, prepared from nanocrystalline powders, which are fabricated by melt spinning, HDDR, or wet chemistry methods. The reduced grain size of the nano-crystalline alloys leads to many novel properties and new magnetism.

5.2.1 REMANENCE ENHANCEMENT

Stoner and Wohlfarth [9] developed a theory for materials based on non-interacting uniaxial magnetic particles which are uniformly magnetized along their easy axis of magnetization. They predicted that remanent polarization J_r (or remanent mag-netization M_r) should have half the value of saturation polarization J_s (or saturation magnetization M_s) for a material with randomly oriented crystallites, that is,

$$J_r = \frac{1}{2}J_s \quad \text{or} \quad M_r = \frac{1}{2}M_s \qquad (5.1)$$

Therefore the maximum energy product of modern RE magnets with their high anisotropy and coercivity is usually limited by the remanence

$$(BH)_{\max} \leq \frac{J_R^2}{4\mu_0} = \frac{J_s^2}{16\mu_0} \qquad (5.2)$$

For conventional isotropic microcrystalline alloys having a stoichiometric $Nd_2Fe_{14}B$ or near stoichiometric composition, J_r is found to be $J_s/2$ or slightly smaller, which means that they obey the Stoner–Wohlfarth model, even though the crystallites in such a polycrystalline aggregate cannot be considered to be non-interacting [10].

However, many studies have shown that it is possible to enhance the J_r above the limit of $0.5J_s$ by means of decreasing grain size d_g of $Nd_2Fe_{14}B$ phase. This effect, called *remanence enhancement*, was clearly demonstrated by Manaf et al. in the early 1990s [4]. They observed in melt-spun $Fe_{70.6}Nd_{13.2}B_6Si_{1.2}$ alloy that as the mean grain size d_g is refined below ~40 nm, the J_r is progressively enhanced. The remanence enhancement due to exchange coupling in isotropic magnets means a great step in increasing J_r and $(BH)_{max}$.

The underlying physics of exchange coupling has been investigated by various researchers [11–13]. The enhanced J_r in isotropic hard magnets results from the ferromagnetic exchange interaction. This interaction is driven by a quantum mechanical coupling of the electron spins, leading to a reduction in the total magnetic energy when the magnetic moments are parallel. This coupling causes magnetization around the boundaries of the crystallites, whose easy axes are unfavorably aligned with respect to the original magnetizing field, to deviate from the easy axes in the two grains. In fact, this exchange interaction in a polycrystalline REFeB aggregate is always presented if the grains are not decoupled by the intergranular RE-rich phase but, it is not significant for $d_g \geq \sim 50$ nm, since the total exchange volume is too small compared with a fraction of the total grain volume.

The exchange coupling effect is exerted at a characteristic length scale called exchange length l_{ex}. l_{ex} is of the same order of magnitude as the domain wall width of the hard $Nd_2Fe_{14}B$ phase (~ 4 nm). Thereby, this exchange coupling is considered to occur in the outer 4 nm of each 2/14/1 grain and becomes progressively more significant as d_g is decreased into the nanocrystalline range.

5.2.2 THREE TYPES OF NANOCRYSTALLINE NdFeB ALLOYS BASED ON COMPOSITION

The commonly mentioned NdFeB-based alloys cover a number of compositions, namely RE-(Fe,M_1)-B-M_2. RE includes all RE elements. M_1 are Co or other transition metals which can be substituted into the Fe position in the $Nd_2Fe_{14}B$ lattice. M_2 are the dopings or additions. However, for nanocrystalline NdFeB-based alloys, stoichiometric $Nd_2Fe_{14}B$ composition is not required. The RE element content can be varied, which leads to various microstructures and magnetic properties. Figure 5.1 shows the three types of nanocrystalline $Nd_2Fe_{14}B$-based magnets with ideal microstructures, as described by Gutfleisch [14].

Type (I) is *decoupled nanocrystalline magnets (high-coercive RE-rich magnets)* with RE-rich composition and the individual crystallites are separated by a thin paramagnetic layer of so-called RE-rich phases. The composition of this type of magnet is similar to that of sintered magnets. The difference is that these magnets are isotropic and have nanocrystallites. The resulting microstructure obtained by melt spinning qualitatively agrees with the situation in sintered magnets where single crystallites of the hard magnetic phase are more or less magnetically decoupled by a paramagnetic RE-rich boundary phase. However, the scale is about a factor of 100 smaller now and the easy axes of the nanoscaled grains are isotropically distributed. The grain size ranges from 50 to 200 nm which guarantees that mainly single-domain particles exist. The isotropically distributed easy directions result in a remanence (J_r) of maximum one-half of the spontaneous polarization (J_S), although practically J_r is slightly higher than $J_S/2$ due to the unavoidable exchange coupling. This microstructure also results in high intrinsic coercivity $_jH_C$. The reason for the high coercivity is that each grain behaves like an elementary hard magnet. For demagnetizing, the reversed applied field first has to overcome the crystal anisotropy. As in the case of

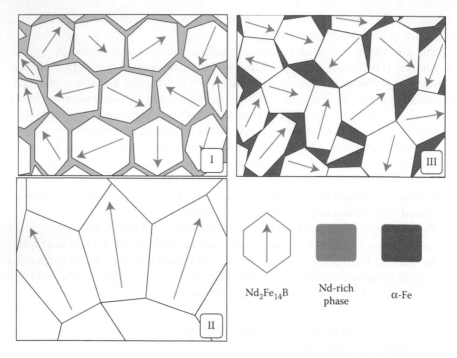

FIGURE 5.1   Three types of $Nd_2Fe_{14}B$ based magnets with idealized microstructures.

sintered magnets the decoupling, and therefore the coercivity, can be improved by adding small amounts of Ga and Nb and/or by subsequent annealing treatments.

Type (II) is *stoichiometric magnets (remanent enhanced single-phase magnets)*, which can be obtained using the stoichiometric $Nd_2Fe_{14}B$ composition [4,15,16] in which there is no intergranular phase. If the grains have mean diameters less than ~50 nm, they are significantly exchange coupled and, as a consequence, remanence enhancement is observed. Remanence enhancement in *single phase* nanostructured systems was first reported by McCallum et al. [17] and Clemente et al. [18] for melt-spun Nd–Fe–B alloys containing Si or Al, which, it was reported, help to promote nanoscale refinement. Subsequently Manaf et al. [19,20] showed that grain refinement down to 20 nm and consequent exchange enhancement could be achieved for melt-spun ternary NdFeB alloys without additions, providing the composition is such as to avoid formation of nonmagnetic Nd-rich intergranular layers, which weakens the exchange coupling between crystallites. It was reported that the addition of small amounts of Zr [17] and Nb [21] contributes to a further refinement of the hard magnetic grain structure. In the case of stoichiometric magnets, a significantly enhanced remanence and a slightly reduced coercivity—both quantities are in the order of 1 T now—are observed as a consequence of the exchange coupling between the grains. However, as the exchange interaction is a very short-range interaction its influence is largest at the grain boundary and disappears inside the grain. Thus, the polarization direction underlies a smooth transition from the easy axis of one grain to the easy axis of another grain. Since only this exchange coupled volume fraction can bring about the remanence enhancement, the grain size should be smaller than 50 nm [22]. With diminishing grain sizes the remanence J_r increases. The exchange coupling between the grains is also the reason for the smaller coercivity values compared to decoupled magnets.

Type (III) is *RE-deficient nanocomposite magnets (exchange coupled two-phase magnets)*. A further increase in remanence is found in the type (III) nanocomposite magnets, a two- or multi-phase exchange-coupled magnet, where a Nd-deficient composition is used and the exchange coupling occurs between both the $Nd_2Fe_{14}B$ grains and $Nd_2Fe_{14}B$ grains and the soft magnetic Fe-rich grains [23–26]. This class of remanence enhanced nanostructures was described as the *exchange spring* systems in 1991 by Kneller and Hawig [23]. The first experimental observation of the exchange spring effect was by Coehoorn et al. [27], in melt-spun Nd poor $Nd_4Fe_{78}B_{18}$ alloys, in which the hard phase $Nd_2Fe_{14}B$ contained only ~15% of the Fe atoms and in the soft phases, mainly Fe_3B with a small amount of α-Fe, were predominant. Similarly, remanence enhancement was reported, by Manaf et al. [28], in the melt-spun sub-stoichiometric NdFeB alloys consisting of $Nd_2Fe_{14}B$ with α-Fe phases. Further microstructural investigations on this system [24,29] confirmed that the nanoscale structure consisted of $Nd_2Fe_{14}B$ as the main phase, with a mean grain size of ~30 nm, and smaller (~15 nm) crystallites of α-Fe second phase located as isolated particles interspersed throughout the microstructure. The so-called composite magnets show a further significant increase of the remanence-enhancing effect compared to the single phase nanocrystalline alloys, accompanied by a further reduction of the coercivity. The remanence enhancement is due to the exchange coupling among the grains and also to the very large spontaneous polarization of the α-Fe grains ($J_S = 2.15$ T) which significantly intensifies the magnetic texturing effect. However, to obtain excellent hard magnetic properties in such composite magnets, it is imperative that the soft magnetic α-Fe grains are completely exchange coupled with the hard magnetic grains. Therefore, their size may not exceed a certain limit which is in the order of the Bloch wall width of the hard magnetic phase [30]. Otherwise, the hysteresis loop would show a two-step demagnetization behavior with drastically deteriorated hard magnetic properties. For this reason, sintered magnets with their mm-scaled grains cannot be realized with compositions in the Fe-rich region of the phase diagram. The simulation results [31] showed that with an increasing amount of α-Fe, the remanence becomes larger up to a maximum value of 1.42 T in $Pr_6Fe_{90}B_4$ containing 46.9 vol% α-Fe, a value which is quite comparable with well-oriented sintered magnets. This, however, comes at the expense of the coercivity, which is continually reduced with increasing α-Fe concentration. On the other hand, the $(BH)_{max}$ shows a maximum value of 180.7 kJ/m³ for 30.4 vol% α-Fe which means an increase by a factor of nearly two as compared with a conventional nanocrystalline decoupled NdFeB magnet.

Based on composition, there are two types of nanocomposite alloys: (1) those with a relatively low boron content composed of two phases ($Nd_2Fe_{14}B$ and α-Fe) and (2) those with high boron content containing three phases ($Nd_2Fe_{14}B$, Fe_3B, and α-Fe). For the latter case, an amorphous alloy is more readily formed by melt spinning even at a relatively low roll speed. The easy formation of fine Fe_3B and $Nd_2Fe_{14}B$ grains after subsequent annealing treatment was found to improve the exchange coupling effect and to lead to remanence enhancement [32]. For the $Nd_2Fe_{14}B/\alpha$-Fe system, higher J_r can be obtained due to higher J_s for α-Fe than for Fe_3B. However, it is a challenge to obtain exchanged coupled alloys by annealing over-quenched ribbons due to the increased grain size of α-Fe phase formed during devitrification [33].

5.2.3 MAGNETIC PROPERTIES OF THREE TYPES OF NANOCRYSTALLINE NdFeB ALLOYS

Nanocrystalline NdFeB hard magnets are well-suited for tailoring magnets with defined properties of the hysteresis loop. Three types of nanocrystalline NdFeB-based alloys can be obtained by simply modifying the RE content in the composition. Based on the phase diagram, these nanocrystalline permanent magnets with different microstructures and magnetic properties have been developed by means of the melt-spinning method. Figure 5.2a and b, and c show typical microstructures for stoichiometric magnets, magnets with RE-excess (decoupled magnets), and magnets with over-stoichiometric Fe (nanocomposite magnets) [34]. The typical hysteresis loops for three types of $RE_2(FeCo)_{94-z}B_6$ alloys are shown in Figure 5.2d, where the alloy with $z = 12$ is a single phase alloy, $z = 14$ for the RE-rich alloy, and $z = 8$ and 10 for the nanocomposite alloys [35].

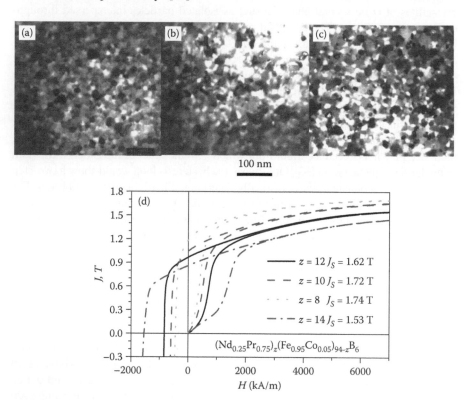

FIGURE 5.2 Typical microstructures (a–c) and hysteresis loops (d) for three types of $RE_2(FeCo)_{94-z}B_6$ alloys, where $z = 12$ indicates single phase alloy, $z = 14$ RE-rich alloy, and $z = 8$ and 10 nanocomposite alloys. (Adapted from Z. W. Liu and H. A. Davies, *J. Magn. Magn. Mater.* 313, 2007, 337–341.)

The effects of Nd content on the magnetic properties of nanoscale $Nd_xFe_{94-x}B_6$ were systematically studied by Davies et al. [36]. They varied the Nd content between 8 and 19 at.%, and found that the remanence decreases with increasing Nd content (x). The enhanced J_r up to 1.1 T was obtained at $x = 8$ at.%, which is due to the exchange coupling between the $Nd_2Fe_{14}B$ grains and the α-Fe grains which have larger J_s (2.2 T). Because the α-Fe grains are present on an ultrafine scale, it does not result in serious deterioration of the loop shape in the second quadrant and thus, $(BH)_{max}$ is also enhanced (up to 160 kJ/m³). The intrinsic coercivity $_jH_C$ on the other hand, decreases with the increasing volume fraction of soft α-Fe due to the reduced resistance to reverse magnetization; $_jH_C$ increases steeply owing to the appearance of a Nd-rich paramagnetic intergranular phase, which decouples the grains and hence reduces the remanence enhancement. This inverse correlation of J_r and $_jH_C$ was analytically predicted by Huo and Davies [37], considering the average intergrain interaction and the random anisotropy model.

The reduction of Nd concentration in nanocomposite NdFeB alloys provides less expensive magnets and also reduced corrosion rate, thus being of interest for technical use.

5.2.4 THERMAL STABILITY

Thermal stability, also called temperature stability, is a crucial indicator for evaluating whether a permanent magnet can be used at an elevated temperature, and is generally represented by temperature coefficients and irreversible/reversible magnetic losses.

The temperature coefficient of remanence (α) and temperature coefficient of coercivity (β), of a permanent magnet in the temperature range of T_0–T can be expressed as,

$$\alpha = \frac{J_r(T) - J_r(T_o)}{J_r(T_o)(T - T_o)} \times 100\% \qquad (5.3)$$

and

$$\beta = \frac{_jH_c(T) - _jH_c(T_o)}{_jH_c(T_o)(T - T_o)} \times 100\%, \qquad (5.4)$$

respectively.

Based on above definition, both coefficients are negative. The lower the absolute values of the coefficients, the higher the thermal stability.

The thermal stability of nanocrystalline NdFeB alloys with various phase constitutions has also been investigated by Liu et al. [38,39]. Figure 5.3a indicates that the nanocomposite has lower absolute values of the temperature coefficients α and β than single phase and RE-rich nanocrystalline alloys, indicating that nanocomposite alloys have higher thermal stability. Figure 5.3b clearly indicates that reduced grain size is beneficial to the temperature stability of remanence, thus the existence of exchange coupling between nanocrystalllites is an advantage for magnets used at elevated temperatures.

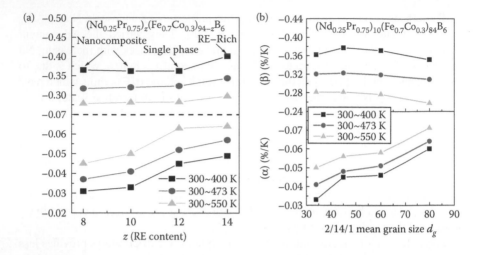

FIGURE 5.3 Temperature coefficients for nanocrystalline $(NdPr)_z(FeCo)_{94-z}B_6$ alloys with various RE contents (a) and for the $(NdPr)_{10}(FeCo)_{84}B_6$ alloys with various grain sizes (b). (Adapted from H. A. Davies and Z. W. Liu, *J. Magn. Magn. Mater.* 294, 2005, 213–225.)

Irreversible loss is another important criterion for evaluating thermal stability. For the ribbon sample, irreversible loss (IL) can be defined as the differences in J_r or $_jH_C$ at room temperature (300 K) before and immediately after being exposed to an elevated temperature [40], that is,

$$IL_{jHC} = \frac{_jH_C(300\,K) - _j H_C'(300\,K)}{_jH_C(300\,K)} \times 100\% \tag{5.5}$$

$$IL_{J_r} = \frac{J_r(300\,K) - J_r'(300\,K)}{J_r(300\,K)} \times 100\% \tag{5.6}$$

where $_jH_C$ and J_r, $_jH_C'$, and J_r' are the coercivity and remanence before and after exposure at an elevated temperature, respectively. Based on Equations 5.5 and 5.6, two types of irreversible loss can be defined. Long-term irreversible loss (LTIL) measures the irreversible losses (IL) in J_r or $_jH_C$ on the sample after exposure to an elevated temperature (such as 423 K) for various times, and short-time irreversible loss measures the IL in J_r or $_jH_C$ after exposure at various temperatures for a fixed time (such as 1 h).

Figure 5.4a and b shows the IL in coercivity for single phase ($z = 12$) and nanocomposite ($z = 8, 10$) $(NdPr)_z(FeCo)_{94-z}B_6$ alloys without and with Co addition, respectively, after exposure to 423 K for various times [40]. Both figures indicate that the nanocomposite alloys exhibit less irreversible loss than the single phase alloy. These results further demonstrated that nanostructure and exchange coupling are beneficial to the thermal stability of NdFeB magnets. Nanocrystalline alloys have advantages over microcrystalline alloys on the elevated temperature application.

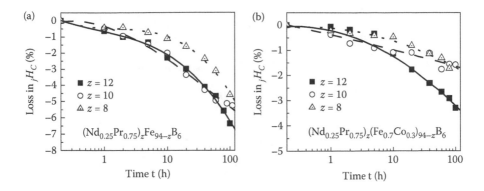

FIGURE 5.4 Irreversible losses in coercivity for single phase ($z = 12$) and nanocomposite ($z = 8, 10$) (NdPr)$_z$(FeCo)$_{94-z}$B$_6$ alloys. (Adapted from Z. W. Liu and H. A. Davies. *J. Phys. D: Appl. Phys.* 40, 2007, 315–319.)

5.2.5 EFFECTS OF EXCHANGE INTERACTION

The most common method of quantitatively analyzing the interactions between the grains is by constructing δM plots, known as Henkel plots [12], based on the relationship

$$\delta M(H) = M_d(H) - (1 - 2M_r(H)), \tag{5.7}$$

where $M_r(H)$ and $M_d(H)$ are defined as the remanent magnetizations after applying a field H on a thermally demagnetized sample and after applying a reverse field on a previously saturated sample, respectively. For an assembly of non-interacting single-domain particles, $\delta M = 0$; a deviation of δM from zero could be caused by interaction between grains. An increase in the positive deviation suggests that the exchange coupling be enhanced. A negative deviation of the plots is interpreted as being due to magnetostatic interactions dominating when the magnetization reversal occurs.

Figure 5.5 shows the typical δM points for three types of nanocrystalline RE(FeCo) B alloys [41]. For a single-phase alloy a large positive derivation of δM indicates that intergranular interactions are dominated by exchange until reversal occurs, when δM decreases abruptly to small negative values as a result of the cooperative switching of the exchange-coupled grains. For the nanocomposites δM is also initially positive though much weaker than that for the single phase alloy, as the hard phase prevents the demagnetization of the two-phase structure, but becomes much more strongly negative after reversal. The negative part is more pronounced in nanocomposites with larger α-(Fe,Co) volume fraction, showing that the importance of magnetostatic interactions increases with increasing soft phase content. For RE-rich alloys, the magnetostatic interaction becomes very small due to the existence of RE-rich phases at the grain boundaries.

For nanocrystalline NdFeB alloys, based on the nucleation mechanism of the magnetic reversal, the temperature dependence of coercivity $\mu_0 H_c$ can be analyzed using the modified Brown's equation [15]:

$$\mu_0 H_c(T) = \alpha_K \alpha_{ex} \mu H_N^{\min}(T) - N_{eff} J_S(T) \tag{5.8}$$

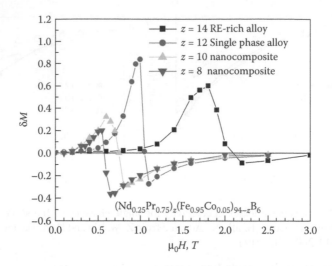

FIGURE 5.5 δM points for three types of nanocrystalline RE(FeCo)B alloys. (Adapted from Z. W. Liu and H. A. Davies, *J. Phys. D: Appl. Phys.* 42, 2009, 145006.)

The so-called microstructural parameters α_k, α_{ex}, and N_{eff} are related to the non-ideal microstructure of the real magnet. α_k describes the influence of the non-perfect grain surfaces on the crystal anisotropy. The effective demagnetization factor N_{eff} is due to enhanced stray fields at the edges and corners of the grains. α_{ex} takes into account the effect of exchange coupling. If the grains with the smallest nucleation field are assumed to govern the whole demagnetization process, the minimum nucleation field is $H_N^{min} = \alpha_\psi^{min} 2K_1/J_s$ [31]. For randomly oriented grains the parameter α_ψ^{min}, related to the grain orientation distribution, is equal to 1/2. For single crystal $RE_2Fe_{14}B$, the anisotropy field is $\mu_0 H_a = 2K_1/M_S$. Above equation can be rewritten as $\mu_0 H_c(T) = \alpha_K \alpha_{ex} \alpha_\psi^{min} \mu_0 H_a(T) - N_{eff} J_S(T)$. A plot of the experimental values of $\mu_0 H_C(T)/J_S(T)$ versus the theoretical values $\mu_0 H_a(T)/J_S(T)$ should yield a straight line, having a slope $1/2\alpha_k\alpha_{ex}$ and an ordinate intersection ($-N_{eff}$). Both $\alpha_k\alpha_{ex}$ and N_{eff} are temperature-independent. The value of α_k for the rapidly solidified de-coupled NdFeB magnet was found almost constant, that is, $\alpha_k = 0.8(\pm0.1)$, indicating a nearly perfect grain surface [42]. Any variation in $\alpha_k\alpha_{ex}$ mainly results from exchange coupling, with smaller values of $\alpha_k\alpha_{ex}$ indicating stronger exchange coupling.

Figure 5.6a shows the plots of $\mu_0 H_C(T)/J_S(T)$ versus $\mu_0 H_a(T)/J_S(T)$ and microstructural parameters for nanocomposites with various compositions [43]. A linear relationship is observed for all alloys. Figure 5.6b shows the derived microstructural parameters versus RE content for all alloys [41]. Decreasing RE content reduces both parameters, indicating that introducing and increasing the fraction of the α-(Fe,Co) phase leads to stronger exchange interaction between grains and also reduces stray fields by improving microstructure. N_{eff} values for exchange-coupled nanocomposites (<0.1) are much smaller than for RE-rich alloys. The reason, given by Kronmuller and Goll [31], is the more polyhedral shapes of the partly de-coupled grains in the RE-rich alloys.

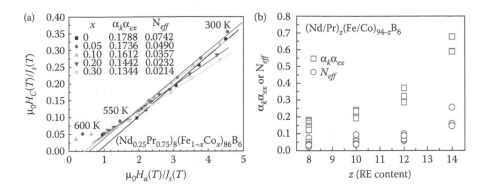

FIGURE 5.6 Plots of $\mu_0 H_C(T)/J_s(T)$ vs $\mu_0 H_a(T)/J_s(T)$ (a) and microstructural parameters (b) for a wide range of nanocrystalline alloys. (Adapted from Z. W. Liu and H. A. Davies, *J. Phys. D: Appl. Phys.* 42, 2009, 145006; Z. W. Liu and H. A. Davies, *J. Phys. D: Appl. Phys.* 39, 2006, 2647–2653.)

The rather small values of $\alpha_k \alpha_{ex}$ in remanence enhanced alloys (Figure 5.6) illustrate the drastic influence of exchange coupling on the coercivity. For nanocomposites with mean grain size of the soft phase of less than ~20 nm, soft magnetic grains are almost fully exchange-coupled to hard grains. Rotations within the soft grains are easier than in the hard grains, which induce, via exchange coupling, an enhanced rotation within the hard grains. Thus the hard grains approach their nucleation field at lower fields, and this leads to a decrease of H_C and a drastic decrease of $\alpha_k \alpha_{ex}$. The decrease of N_{eff} is also related to the collective behavior of the exchange-coupled grains. Due to the exchange coupling the reversal of magnetization in nanocomposites takes place as a collective nucleation process where a cluster of grains which extends over the domain wall width of α-Fe (~70 nm) is involved. The nucleation process takes place in more or less spherical ensembles of exchange-coupled grains which leads to a reduction of local stray fields. In particular, the smoothing effect of the exchange interaction at the grain boundaries reduces the stray field effect at the edges and corners of the grains.

5.2.6 Overview of the Magnetic Properties of Melt-Spun Nanocrystalline NdFeB Alloys

Compared to sintered NdFeB magnets, nanocrystalline NdFeB alloys prepared by melt spinning have relatively low magnetic properties since they are magnetically isotropic. Liu and Davies [35] summarized the magnetic properties of a variety of melt-spun nanocrystalline NdFeB-based alloys. The relationships between J_r and $_jH_C$, and between $(BH)_{max}$ and $_jH_C$ for the directly quenched nanocrystalline NdFeB alloys, including single phase, nanocomposite, and RE-rich alloys, are given in Figure 5.7. The compositions include Nd/Pr–(Dy)–Fe/Co–B [34], Nd–Fe–(B,V,Si,Ga,Zr,Cu,Ni) [44], Pr–(Fe,Co)–B–Cu [45], Pr–Fe–B [16,46], and Nd–Fe–B–(Si) [28,47]. J_r decreases approximately linearly with increasing $_jH_C$, whereas $(BH)_{max}$ has a maximum value of 160–170 kJ/m^3 for these isotropic bulk nanocrystalline alloys. The data indicate

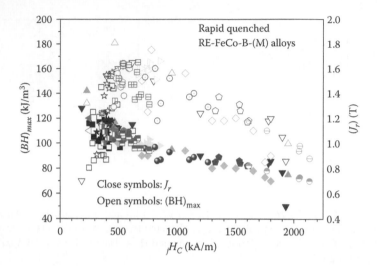

FIGURE 5.7   Relationships between J_r/(BH)$_{max}$ and $_jH_C$ for the directly quenched nanocrystalline NdFeB alloys, including single phase, nanocomposite and RE-rich alloys. The various types of symbols represent for different compositions.

that, in order to obtain the best $(BH)_{max}$, alloy compositions having $_jH_C$ generally in the range 400–800 kA/m should be chosen. The results consistently demonstrate that an approximately linear inverse relationship exists between J_r and $_jH_C$, but that, for the achievement of the best energy product, there is an optimum coercivity range. Although J_r and $_jH_C$ can be adjusted over wide ranges by substituting elements and changing RE:Fe ratios, the practical limits for $(BH)_{max}$ for the compositionally modified NdFeB-based nanocrystalline alloys are in the range ~160–180 kJ/m^3. Similar behavior was also reported earlier for nanocrystalline ternary NdFeB alloys by Pawlik et al. [48]. Manaf et al. [49] also predicted the relationship between $(BH)_{max}$ and $_jH_C$ for NdFeBSi alloys with various grain sizes, based on the standard expression $(BH) = (\mu_0 H + J) \times H$ and using the empirical relationship between J_r and $_jH_C$ obtained from experimental data and a hypothesis of $J = J_r + [(J_s - J_r)/H_a] \times H$ for an idealized rhomboidal second quadrant J–H loop. They predicted that a maximum in $(BH)_{max}$ would correspond to $_jH_C = $ ~400 kA/m, below which the second quadrant B–H loop would become nonlinear with a consequent rapid collapse of $(BH)_{max}$, and this prediction was in good agreement with the experimental observations. The extensive investigations indicate that this is also broadly true for a wide range of REFeB-based alloys, including various compositions and phase constitutions.

5.3   COMPOSITIONAL OPTIMIZATION FOR NANOCRYSTALLINE NdFeB MAGNETS

The magnetic properties of NdFeB magnets are closely related to both their intrinsic properties and their microstructure. Extensive efforts have been made via compositional adjustment to improve the room temperature properties and thermal stability of the nanocrystalline alloys.

5.3.1 Compositional Optimization with Respect to Intrinsic Properties

The hard magnetic properties of NdFeB magnets result from the $Nd_2Fe_{14}B$ phase. As is understood, in the $Nd_2Fe_{14}B$ (2:14:1) phase, Nd can be substituted by other RE elements, Fe can be substituted by Co or Ni, and B can be substituted by C. As a result, RE, Co, and C are generally employed to modify the intrinsic properties of the $Nd_2Fe_{14}B$ phase, thus the extrinsic properties of the magnets. The intrinsic properties of the $Nd_2Fe_{14}B$ phase include its saturation magnetization M_S, anisotropy field H_a, and Curie temperature T_C.

5.3.1.1 RE Elements Substitutions

The $Nd_2Fe_{14}B$ structure forms for all the RE elements, except Eu and Pm. The RE elements in the NdFeB alloys generally occupy the same positions as Nd atoms. For different $RE_2Fe_{14}B$ compounds, high values of J_S are exhibited for RE = Pr, Nd, and Sm, and these high values make the high J_r value feasible. The anisotropy field H_a has also large values for RE = Pr, Nd, Sm, Tb, and Dy [50], and the high H_a value also enables a large $_jH_C$ value. The intrinsic properties of $RE_2Fe_{14}B$ phases mentioned earlier are shown in Table 5.1. Although Tb has high T_C and the highest H_a value, the use of it has been restricted due to the high cost of Tb metal. In most cases, the effects of Pr, Sm, Dy, La, Ce, and Y on the magnetic properties of melt-spun nanocrystalline NdFeB alloys were investigated.

In NdFeB alloys, the most common used substitution is Pr. Substituting Pr for Nd has at least three advantages, including (1) reducing or eliminating the spin reorientation, (2) increasing coercivity because $Pr_2Fe_{14}B$ phase has higher H_a than $Nd_2Fe_{14}B$ as shown in Table 5.1, and (3) reducing the production cost because Nd and Pr coexist in the natural ores and the work of separating them can be avoided.

The beneficial effects on the coercivity and energy product of Pr substituting Nd have been demonstrated by Liu et al. [35,51]. Figure 5.8 shows that, in

TABLE 5.1
Intrinsic Magnetic Properties of $RE_2Fe_{14}B$ Compounds at 300 K

Compound	J_s (T)	H_a (T)	T_c (°C)
$Pr_2Fe_{14}B$	1.56	8.7	292
$Nd_2Fe_{14}B$	1.61	6.7	312
$Sm_2Fe_{14}B$	1.52	15 (planar)	339
$Tb_2Fe_{14}B$	0.66	22	347
$Dy_2Fe_{14}B$	0.71	15	319
$Y_2Fe_{14}B$	1.41	2.0	~305
$Ce_2Fe_{14}B$	1.17	3.0	~298
$La_2Fe_{14}B$	1.27	2.0	~285

Source: Adapted from E. Burzo, *Rep. Prog. Phys.* 61, 1998, 1099–1266.

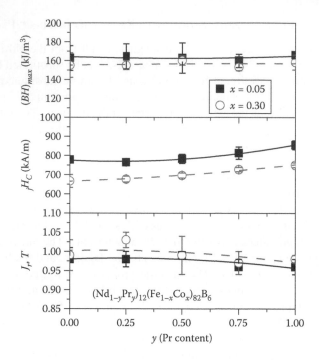

FIGURE 5.8 Effects of Pr substitution for Nd on room temperature magnetic properties of single phase $(Nd_{1-y}Pr_y)_{12}(Fe_{1-x}Co_x)_{82}B_6$ ($x = 0.05$ and 0.30) alloys. (Adapted from Z. W. Liu et al., *J. Magn. Magn. Mater.* 321, 2009, 2290–2295.)

nanocrystalline single phase $(Nd_{1-y}Pr_y)_{12}(Fe_{1-x}Co_x)_{82}B_6$ alloys, Pr substitution indeed enhances $_jH_C$ without reduction in J_r and $(BH)_{max}$. They also found that in nanocomposite $(Nd_{1-y}Pr_y)_{10}(Fe_{1-x}Co_x)_{84}B_6$ ($x = 0$–0.3; $y = 0$–1) alloys, increasing Pr substitution enhances both $_jH_C$ and $(BH)_{max}$. This is very important for nanocomposites and for Co substituted alloys since $_jH_C$ is reduced by the exchange coupling or Co addition in these alloys. For nanocrystalline RE-rich alloys, Pr substitution also enhances $_jH_C$, but it generally has no significant effect on J_r and $(BH)_{max}$.

The difference in $_jH_C$ for various Pr substituted alloys can be attributed to the large difference in H_a of the $Nd_2Fe_{14}B$ and $Pr_2Fe_{14}B$ phases. The H_a for $Pr_2Fe_{14}B$ (~8.7 T) is ~30% larger than that of $Nd_2Fe_{14}B$ (~6.7 T) at room temperature. On the other hand, the $Nd_2Fe_{14}B$ and $Pr_2Fe_{14}B$ phases have similar values of J_s, ~1.60 T and ~1.56 T, respectively. Therefore, similar values of J_r can also be expected in the alloys with various Pr substitutions. However, as far as exchange-coupled nanocrystalline alloys are concerned, on the basis of the larger H_a for the $Pr_2Fe_{14}B$ phase and no significant difference in exchange constant A, it would be expected that the exchange length l_{ex} ($l_{ex} = \pi(A/K)^{1/2}$) would be smaller than that for the $Nd_2Fe_{14}B$ phase, leading to a somewhat smaller exchange enhancement of J_r in the Pr substituted NdFeB alloys for a given d_g. However, this is not consistently observed in the experimental work [52].

It is worth noting that Pr substitution is more beneficial to $(BH)_{max}$ for nanocomposite alloys than for the single-phase alloys. This is because, although both $_jH_C$ and

J_r are important to the $(BH)_{max}$, the $_jH_C$ becomes more important for nanocomposite than for single phase alloys. In fact, when the second quadrant J–H curve has very good squareness and the B–H curve in the second quadrant is linear, increasing $_jH_C$ has no significant effect on $(BH)_{max}$. Thus enhancing $_jH_C$ further by Pr substitution does not contribute to higher $(BH)_{max}$ for alloys with high $_jH_C$. This explains why Pr substitution did not improve $(BH)_{max}$ for RE-rich alloys.

The heavy RE element Dy is generally employed in sintered NdFeB magnets for improving coercivity. The room temperature magnetic properties of melt-spun single phase, nanocomposite, and RE-rich NdFeB alloys with various Dy substitutions for RE have been investigated by Liu [34]. Generally, Dy substitution increases $_jH_C$ but slightly decreases J_r and $(BH)_{max}$, as shown in Figure 5.9a and b for single phase $[(Nd_{0.25}Pr_{0.75})_{1-w}Dy_w]_{12}(Fe_{0.95}Co_{0.05})_{82}B_6$ and nanocomposite $[(Nd_{0.25}Pr_{0.75})_{1-w}Dy_w]_{10}(Fe_{0.95}Co_{0.05})_{84}B_6$ alloys, respectively [34].

The effects of Dy substitution on the magnetic properties also result mainly from the changes in the intrinsic properties of the $RE_2Fe_{14}B$ phase. Dy addition reduces the saturation magnetization J_s because the Dy atomic magnetic moments align ferrimagnetically to both Nd and Fe moments in the 2/14/1 unit cell, but increases the anisotropy field H_a for the 2/14/1 phase, as clearly shown in Table 5.1, which leads to decreased J_r and increased $_jH_C$. To control the combined magnetic properties, the doping level of Dy should be controlled accordingly. For nanocomposite alloys, despite the reduced J_r, both the $_jH_C$ and the $(BH)_{max}$ can be improved by selecting the appropriate substitution level of Dy.

As shown later, for nanocrystalline exchange-coupled alloys and Co substituted alloys, $_jH_C$ tends to decrease. Due to the important effect of $_jH_C$ on the elevated temperature behavior, it is reasonable to expect improved temperature stabilities for Dy substituted alloys, especially for nanocomposite alloys. Due to this advantage, Dy is generally employed with Co substitution to compensate the possible reduction in $_jH_C$ caused by Co addition.

Another $Nd_2Fe_{14}B$ type compound with high H_a is $Sm_2Fe_{14}B$. However, the $Sm_2Fe_{14}B$ compound is not uniaxial and the magnetization is in the easy plane, which affects H_a. Figure 5.10 shows the nanocomposite $Nd_{10}Fe_{85}B_5$ alloys with Co, Dy, or

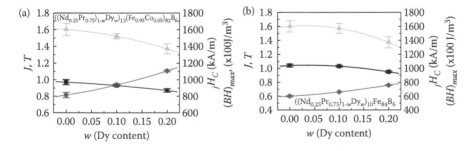

FIGURE 5.9 Effects of Dy substitution on the room temperature magnetic properties of single phase $[(Nd_{0.25}Pr_{0.75})_{1-w}Dy_w]_{12}(Fe_{0.95}Co_{0.05})_{82}B_6$ (a) and nanocomposite $[(Nd_{0.25}Pr_{0.75})_{1-w}Dy_w]_{10}(Fe_{0.95}Co_{0.05})_{84}B_6$ (b) alloys. (Adapted from Z. W. Liu, PhD thesis, University of Sheffield, 2004.)

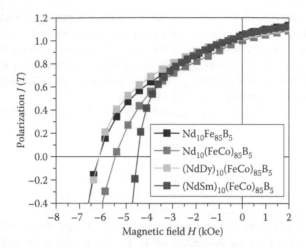

FIGURE 5.10 Demagnetization curves for Co, Dy, and Sm substituted $Nd_{10}Fe_{85}B_5$ nano-composites. (Adapted from Z. W. Liu et al., *J. Magn. Magn. Mater.* 321, 2009, 2290–2295.)

Sm substitutions [51]. The roles of various elements in adjusting the demagnetization curve are illustrated. The result indicates that Dy substitution can compensate $_jH_C$ loss due to Co addition. Although some researchers reported an improved $_jH_C$ with Sm substitution [53], this result indicates that Sm reduces the $_jH_C$ of the Nd–FeCo–B alloy. As pointed out by Chen et al. [54] and Zhang et al. [55], the decrease in H_c can be attributed to the decrease in the magnetocrystalline anisotropy of the material. On the other hand, despite the lower values of $Sm_2Fe_{14}B$ than $Nd_2Fe_{14}B$, 10% Sm addition did not greatly reduce J_r and, indeed, a very good squareness of second quadrant J–H curves is obtained for the Sm substituted alloy, which indicates excellent exchange coupling between the hard and soft phases. Thus the reduced $_jH_C$ may also be attributed to improved exchange coupling. Yang et al. [56] also noticed the enhanced exchange coupling interaction associated with a small amount of Sm substitution and obtained a reduced $_jH_C$ for Sm substituted $Nd_2Fe_{14}B/\alpha$-Fe nanocomposite. The reason can be explained as follows: compared to the $Nd_2Fe_{14}B$ phase, $Sm_2Fe_{14}B$ has a negative value of K [57]. If a small amount of Sm is substituted for Nd, the magnetocrystalline anisotropy constant K of the hard $RE_2Fe_{14}B$ phase is reduced. The exchange length L_{ex} and critical grain size d_{op} exchange-coupled soft phase (as described in the Introduction) would be increased. This would enhance the exchange coupling interaction between the hard and soft phases.

As is well known, the $Y_2Fe_{14}B$ compound has a higher M_s than $Dy_2Fe_{14}B$, though its H_a is lower. Most importantly, H_a of $Y_2Fe_{14}B$ exhibits a weaker temperature dependence compared to those of many other $RE_2Fe_{14}B$ compounds [58], which is beneficial for temperature stability. Liu et al. [59] employed Y to substitute Dy in nanocrystalline NdFeB alloys with the aim to reduce the material cost and achieved success. Figure 5.11a shows the hysteresis loops for as-spun $[Nd_{0.8}(Dy_{1-x}Y_x)_{0.2}]_{10}Fe_{84}B_6$ alloys at room temperature. For all the alloys with various Y substitutions, the hysteresis loops show a single hard magnetic phase behavior, which indicates a good exchange coupling between the magnetically hard and soft phases for these nanocomposite

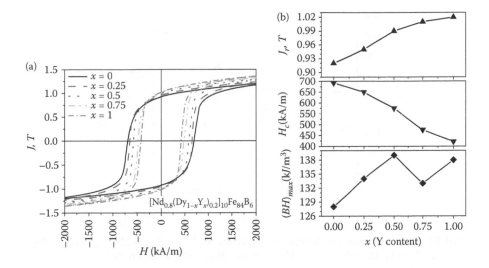

FIGURE 5.11 Hysteresis loops of as-spun $[Nd_{0.8}(Dy_{1-x}Y_x)_{0.2}]_{10}Fe_{84}B_6$ alloys (a) and the effects of Y substitution on the J_r, H_c and $(BH)_{max}$ (b). (Adapted from Z. W. Liu, D. Y. Qian, and D. C. Zeng, *IEEE Trans. Magn.* 48, 2012, 2797–2799.)

alloys. Consequently, good loop squareness has been obtained for all alloys. Also, as shown in Figure 5.11b, with increasing x value from 0 to 1, J_r increases from 0.92 to 1.02 T, but $_jH_C$ decreases from 692.0 to 421.3 kA/m. The reasons can be attributed to the higher M_S and lower H_a of $Y_2Fe_{14}B$ than $Dy_2Fe_{14}B$ [50]. As a result, a highest value of $(BH)_{max}$, 139 kJ/m³ (17.5 MGOe), is obtained at $x = 0.5$.

As we know, La and Ce are almost the least expensive RE metals in the world and coexist in ores in nature. To reduce the cost of RE permanent magnets, LaCe substitution for Nd and Dy in sintered NdFeB or NdDyFeB has been employed recently by various researchers [60]. Very recently, Hussain et al. [61] studied melt-spun La and Ce $[(La_{0.5}Ce_{0.5})]_{10}Fe_{84}B_6$-based nanocomposite alloy. The intrinsic coercivity $_jH_C$, remanence M_r and $(BH)_{max}$ for direct quenched alloy are 120 kA/m, 65 emu/g, and 20 kJ/m³, respectively. In order to improve the magnetic properties, especially the coercivity, Nd substitution for La and Ce is employed. Figure 5.12 shows the composition dependent magnetic properties for the directly quenched (DQ) and overquenched and annealed $[(La_{0.5}Ce_{0.5})_{1-x}Nd_x]_{10}Fe_{84}B_6$ alloys with various Nd substitutions ($x = 0$, 0.3, 0.5, and 0.7). Introducing and increasing Nd content generally improve $_jH_C$, J_r, and $(BH)_{max}$. For the optimally quenched alloys, 30% substitution of Nd of the total RE for (LaCe) gives a $(BH)_{max}$ value of 86 kJ/m³ (10.8 MGOe) and a M_r value of 104 emu/g. With 70% substitution of Nd, the optimal magnetic properties of $M_s = 150$ emu/g, $M_r = 105$ emu/g. $_jH_C = 365$ kA/m and $BH = 120$ kJ/m³ (15 MGOe) are recorded in the $(La_{1.5}Ce_{1.5})Nd_7Fe_{84}B_6$ alloys. The LaCe added alloys can be well explained by the intrinsic properties of various 2:14:1 phases, $La_2Fe_{14}B$, $Ce_2Fe_{14}B$, and $Nd_2Fe_{14}B$. These results indicate that La and Ce substitution are able to improve the performance/cost ratio of the nanocrystalline alloys.

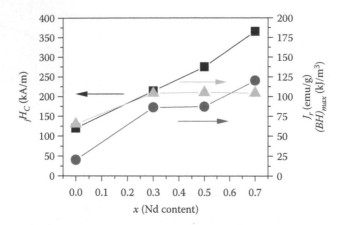

FIGURE 5.12 Composition dependent properties for $[(La_{0.5} Ce_{0.5})_{1-x}Nd_x]_{10} Fe_{84}B_6$ alloys. (Adapted from M. Hussain et al., *J. Magn. Magn. Mater.* 399, 2016, 26–31.)

5.3.1.2 Cobalt Substitution

In $Nd_2Fe_{14}B$ compound, the only element that can substitute Fe in a wide range of composition is Co. It is also the only element which leads to significant change in the magnetic ordering temperature. Co substitution for Fe increases the Curie temperature T_C monotonically with increasing x in the stoichiometric $RE_2Fe_{14-x}Co_xB$ compounds. Concerning the extrinsic properties, Co addition has been reported as enhancing the J_r and $(BH)_{max}$ values [62], but generally with decreasing $_jH_C$ [63–65]. The improvement of $(BH)_{max}$ by Co addition is attributed to the enhancement of J_r, presumably resulting from the increase in magnetization of both the 2/14/1 and α-(Fe,Co) phases and exchange coupling between the soft and hard magnetic phases. However, other results [66–68] indicated that a small amount of Co does not decrease $_jH_C$ and even improves both J_r and $_jH_C$, and the reason has been attributed to microstructural changes.

In Figure 5.13, the second quadrant demagnetization curves (J–H and B–H) are plotted for $(Nd_{0.25}Pr_{0.75})_{10}Fe_{84}B_6$ nanocomposites with various Co concentrations [51]. The results show that 5%–20% Co substitutions slightly increase J_r but that more than 5% Co substitutions decreases $_jH_C$. 5%–20% Co substitutions also enhance $(BH)_{max}$. As a result, when Co is employed to improve the elevated temperature properties, it is possible to maintain $(BH)_{max}$ at a high level by controlling its concentration. However, 30% Co substitution decreases J_r, $_jH_C$, and $(BH)_{max}$.

The effects of Co substitution are also relevant to the intrinsic properties. Cobalt enters into the Fe atomic position when substituting Fe. It was reported that Co substitution for Fe initially slightly increases then markedly decreases M_s [69]. Generally, Co substitution is thought to decrease $_jH_C$ due to the decreased H_a. This is the main drawback of Co-additions made to increase T_C [70]. However, the experimental results [34,67,71] show that a small amount of Co substitution does not decrease $_jH_C$ and 5% Co even slightly increases it. This has also been found by other researchers. The reason may be related to an improved microstructure. In addition,

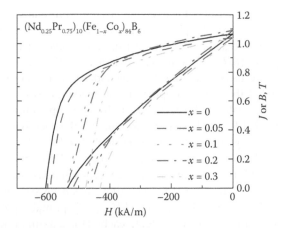

FIGURE 5.13   Demagnetization J–H and B–H curves for Co substituted $(Nd_{0.25}Pr_{0.75})_{10}Fe_{84}B_6$ nanocomposite alloys. (Adapted from Z. W. Liu et al., *J. Magn. Magn. Mater.* 321, 2009, 2290–2295.)

the slightly increased J_r for 5%–20% Co substituted nanocomposite alloys can be attributed to a combination of the higher magnetization of the α-(Fe,Co) phase and the enhanced exchange coupling due to increased exchange length L_{ex}, which results from decreased K in ($l_{ex} = \pi(A/K)^{1/2}$) with Co substitution.

Co-substituted exchange-coupled magnets have been produced by Melsheimer et al. [71]. The NdFeCoB alloys show the typical nanocomposite structure, and exhibit a significant remanence enhancement and, therefore, very high-energy products for the whole ferromagnetic temperature range. Due to the Co-substitution the T_C increases from 585 K for the unsubstituted material $(Nd_2Fe_{14}B)$ to 680 K for $Nd_{12}(Fe_{72}Co_{10})B_6$ and 780 K for $Nd_{12}(Fe_{62}Co_{20})B_6$.

In addition to the intrinsic properties, the effects of Co substitution on the magnetic properties have been attributed by some researchers to the changes in microstructure. The better combination of J_r, $_jH_C$, and $(BH)_{max}$ obtained for the alloys with small Co substitutions was thought to result from a more homogeneous and finer microstructure [67], but Chang et al. [72] reported a slightly coarser grain structure caused by Co substitution in nanocomposite $(Nd,La)_{9.5}Fe_{78-x}Co_xCr_2B_{10.5}$ alloys. The worse properties for the 30% Co substituted alloys have been attributed by some researchers [73] to the appearance of a new phase. However, no convincing evidence for this has been observed so far.

5.3.2   COMPOSITIONAL OPTIMIZATION WITH RESPECT TO THE MICROSTRUCTURE

The extrinsic magnetic properties of the NdFeB magnets can be enhanced by effective elemental doping. A proper concentration of Zr, Nb, Ti, Ta, or Ga addition were reported to improve the microstructure of nanocrystalline NdFeB alloys.

Doping elements such as Nb and Zr have been well studied, and the coercivity can be substantially increased by a small addition of these elements [74–76]. It was reported that Nb and Zr can retard the formation of α-Fe and enhance the formation of $Nd_2Fe_{14}B$. This is very useful for nanocomposites with relatively low coercivity and with large grains of the soft phase. Other transition metals such as Hf, Ti, and V have also been investigated in detail. An optimum level of these elements are beneficial for $_jH_C$ and $(BH)_{max}$ due to the microstructural changes [50]. Bao et al. [77] investigated the effects of adding Zr on the $Nd_{12.3}Fe_{81.7}B_{6.0}$ permanent magnetic alloys, and concluded that via Zr doping (1) the tendency to form an amorphous phase for this alloy is enhanced, (2) the crystallizing temperature is increased, (3) the grain size of the $Nd_2Fe_{14}B$ phase is refined, and (4) the coercivity and remanence are increased, so that high magnetic properties with $J_r = 1.041$ T, $_jH_C = 887.5$ kA/m, and $(BH)_{max} = 175.2$ kJ/m^3 are achieved for melt-spun NdFeB ribbons with $x = 1.5$.

The effect of Ti addition on the magnetic properties and microstructure of nano-crystalline $Nd_{12.3}Fe_{81.7-x}Ti_xB_{6.0}$ ($x = 0.5 - 3.0$) alloys has been systematically investigated by Wang et al. [78]. The J_r, $_jH_C$, and $(BH)_{max}$ of optimally processed ribbons increased first with an increase in Ti content, reached the maximum values at $x = 1.0$, then decreased with further increasing Ti content. The variation of magnetic properties with increasing Ti content can be attributed to both the variation of the exchange coupling interactions and the decrease in M_S in the ribbons.

Zhang et al. [79] has focused on the effects of Ti and C co-doping on the phase structure and magnetic properties of $(Nd_{9.5}Fe_{84.4}B_{6.1})_{100-2m}Ti_mC_m$ ($m = 0$, 1, 2, 3, 4, and 6) ribbons. Small additions of Ti and C promote noticeable increases in coercivity with only a very modest decrease in remanence. At a concentration of 2 at.% Ti and C, optimum properties of the bonded magnets are achieved with $B_r = 0.6868$ T, $_jH_C = 487$ kA/m, and $(BH)_{max} = 69$ kJ/m^3.

Another transition metal, Ta, has received relatively less attention in NdFeB-based permanent magnets. Although Ta has been frequently employed in soft magnetic materials such as Fe or FeCo-based thin films or bulk metallic glass for improving thermal stability, high frequency properties, and glass formability [80,81], there have been very few reports on the effect of Ta doping on the structure and magnetic properties of REFeB nanocrystalline alloys. Liu et al. [51] investigated the effects of Ta on the nanocomposite NdFeB alloys. Shown in Figure 5.14a are the hysteresis loops for nanocomposite $Nd_9Fe_{86-x}Ta_xB_5$ alloys with various Ta doping measured under a magnetic field of 1.5 T. The coercivity $_jH_C$ increases with the introduction of 1% Ta and with the increase of Ta concentration from 1% to 2%. Further increase of Ta to 3 at.% reduces the coercivity. J_S (at 1.5 T) and J_r decrease linearly with increasing Ta content. The best squareness of the second quadrant $J–H$ loop is obtained for 1% Ta doped alloy whereas 3 at.% Ta doping results in a deterioration in the squareness. The reason for the latter is attributed to the precipitated second phase, which decouples the nanograins and reduces the magnetization [82]. Figure 5.14b shows the dependence of magnetic properties on the Ta concentration. The maximum $(BH)_{max}$ of 139 kJ/m^3 is obtained in a Ta concentration of 1 at %. This $(BH)_{max}$ value is quite high if one considers that the maximum applied field used here is only 1.5 T, which did not saturate the material. Except for enhanced room temperature coercivity, an improved thermal stability was also found in 2 at.% Ta doped (NdDy)-(FeCo)-B alloys. The results indicate that Ta doping provide a possible compositional choice to adjust the combination of magnetic properties for nanocomposite NdFeB alloys.

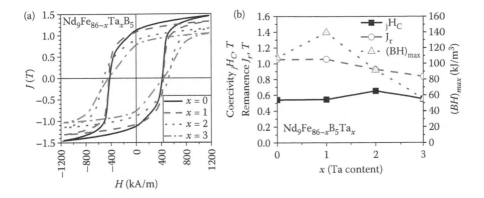

FIGURE 5.14 Hysteresis loops (a) and magnetic properties (b) for $Nd_9Fe_{86-x}Ta_xB_5$ ($x = 0$, 1, 2, 3, and 4) nanocomposites. (Adapted from Z. W. Liu et al., *J. Magn. Magn. Mater.* 321, 2009, 2290–2295.)

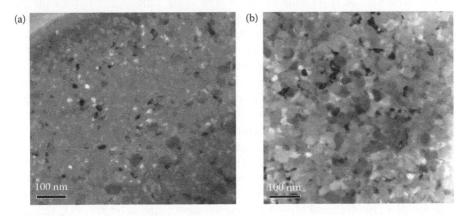

FIGURE 5.15 TEM micrographs for $Nd_9Fe_{86-x}Ta_xB_5$ nanocomposites without (a) and with 2% (b) Ta doping. (Adapted from Z. W. Liu et al., *J. Magn. Magn. Mater.* 321, 2009, 2290–2295.)

The microstructure of $Nd_9Fe_{86}B_5$ alloys without and with 2 at.% Ta doping are shown in Figure 5.15. The results, however, do not show reduced grain size associate with Ta doping. Thus, as initially observed by Chin et al. [82], there is no evidence of grain refinement due to alloying with Ta. Large Ta-containing inclusions were also not found.

5.3.3 ELEVATED TEMPERATURE MAGNETIC PROPERTIES FOR COMPOSITION MODIFIED MAGNETS

The main factor limiting the application of NdFeB magnets thus far has been their low Curie temperature (T_C) and poor thermal stability. The temperature coefficients of ternary NdFeB magnets are a factor of 3.0–3.6 greater than those of SmCo magnets so that the former cannot be used at temperatures >100–150°C depending on the precise composition. For the nanocrystalline alloys, these problems still exist, and can become even worse because the exchange coupling may decrease the $_jH_C$ at elevated temperatures to an unacceptable level. There is thus an increasing need to enhance the elevated temperature behavior of nanocrystalline alloys, and in particular, nanocomposites. However, lots of studies on the thermal stability on NdFeB-based magnets have concentrated almost exclusively on the sintered variants. The thermal stability of nanocrystalline materials has received relatively little attention, with only specific compositions having been investigated. Here the thermal stability of nanocrystalline materials can be defined in terms of IL and temperature coefficients of remanence a and coercivity β.

The IL of magnetization are attributed to the demagnetization of small volumes of the materials under the influence of the local demagnetizing field. Due to energy fluctuations, the relative balance between anisotropy and exchange energies changes, and thus affecting domain structures and reducing the magnetization. The range of IL of magnetization depends on several parameters, such as coercivity, working point, aging, and T_C. Therefore, irreversible loss is an important criterion for evaluating thermal stability.

The variations of IL of magnetization with aging time at 125°C for isotropic melt-spun and hot-pressed isotropic magnets $Nd_xFe_{94-x}B_6$ ($x = 11 - 14$) show a strong dependence on $_jH_C$ [83]. The magnets having $_jH_C = 1470$ kA/m exhibit only a few percent loss after 1000 h annealing, whereas those having $_jH_C = 300$ kA/m showed ~20% losses. The IL shows the same trend, even when different processing routes were used. The differences in the microstructure and composition are reflected in the value of $_jH_C$, which, irrespective of the material, appears to be a decisive parameter which determines the IL. Magnets with higher $_jH_C$ tend to have a higher resistance to demagnetization. All the microstructural elements which can stabilize the domain structure and retard domain wall propagation are effective in reducing the IL.

The heat resistance temperatures (HRT) defined as the temperature where the irreversible loss is 3% in $Nd_{13.6-x}Dy_xFe_{77.6-y}Co_{2.8}M_yB_6$ hot worked magnets, with $x = 0$, 1, 2 and $y = 0$, 0.5 and 1.0 after 2 h exposure at different temperatures, for base and $M =$ Dy-, Ga-, and Mo-added magnets were found to be 381, 453, 425, and 419 K, respectively [84]. HRT is dependent on initial $_jH_C$ and squareness ratio for the demagnetization curve.

A reduction in irreversible flux losses (FL) was reported in Cu-doped NdFeB magnets [85]. Besides, the IL is also reduced by Nb additions [86]. Fernow and Ervens [87] investigated the flux loss of bonded NdFeB magnets with various binders (nylon and epoxy). The results indicate that the binder also affects the FL. The IL was reduced by replacing the epoxy binder by nylon.

Kanai et al. [88] systematically evaluated the irreversible flux loss due to exposure to an elevated temperature for nanocomposite resin-bonded magnets with various coercivities. It was found that the short-term and long-term FL in some nanocomposite magnets are smaller than in a conventional isotropic NdFeB magnet despite their lower coercivity values. It was clarified that the observed small values for the short-term flux loss in the nanocomposite magnets can be attributed to the squareness of a demagnetization curve of the nanocomposite magnets not deteriorating at an elevated temperature.

Liu and Davies [38,40] investigated the LTIL and short-term irreversible loss (STIL) for nanocrystalline Nd/Pr–(Dy)–Fe/Co–B ribbon alloys. The measured values of T_C, $_jH_C$ at room temperature, and J_s, J_r/J_s, and LTIL $_{jH_C}$ are given in Table 5.2.

TABLE 5.2

Curie Temperature, Room Temperature Coercivity, Saturation Magnetization, Remanence Ratio, LTIL $_{jHc}$ for Tested Alloys

No.	Alloy	T_C (°C)	$_jH_C$ (kA/m)	J_s (T)	J_r/J_s	LTIL $_{jHc}$ (%)
1	$(Nd_{0.25}Pr_{0.75})_{12}Fe_{82}B_6$	296	821	1.63	0.59	6.4
2	$(Nd_{0.25}Pr_{0.75})_{12}(Fe_{0.7}Co_{0.3})_{82}B_6$	535	727	1.60	0.60	3.2
3	$(Nd_{0.25}Pr_{0.75})_{10}Fe_{84}B_6$	296	598	1.69	0.64	5.3
4	$(Nd_{0.25}Pr_{0.75})_{10}(Fe_{0.7}Co_{0.3})_{84}B_6$	535	503	1.68	0.62	1.8
5	$(Nd_{0.25}Pr_{0.75})_8Fe_{86}B_6$	296	422	1.73	0.67	4.5
6	$(Nd_{0.25}Pr_{0.75})_8(Fe_{0.7}Co_{0.3})_{86}B_6$	535	361	1.71	0.66	1.9
7	$(Nd_{0.75}Pr_{0.25})_{10}(Fe_{0.95}Co_{0.05})_{84}B_6$	348	548	1.68	0.64	–
8	$(Nd_{0.75}Pr_{0.25})_{10}(Fe_{0.7}Co_{0.3})_{84}B_6$	545	484	1.66	0.63	–
9	$[(Nd_{0.25}Pr_{0.75})_{0.8}Dy_{0.2}]_{12}Fe_{82}B_6$	298	1265	1.39	0.61	2.0
10	$[(Nd_{0.25}Pr_{0.75})_{0.9}Dy_{0.1}]_{10}Fe_{84}B_6$	297	659	1.63	0.64	2.1
11	$[(Nd_{0.25}Pr_{0.75})_{0.8}Dy_{0.2}]_{10}Fe_{84}B_6$	298	756	1.50	0.63	5.4

Source: Adapted from Z. W. Liu and H. A. Davies, *J. Phys. D: Appl. Phys.* 40, 2007, 315–319.

The result indicates that the magnitude of T_C is an important factor influencing the LTIL. Large values of the IL were observed in samples 1, 3, 5, and 11; these ribbons also had low T_C. Samples 2, 4, and 6 had high T_C and, correspondingly, relatively lower IL. This would largely explain why Co substitutions, which enhance T_C, significantly reduced the LTIL in J_r and in $_jH_C$, that is, Co additions improved the thermal stability. The values of LTIL $_{jHc}$ for samples 9 and 10, despite their low T_C, were lower than those for samples 1, 3, 5, and 11, which had similar T_C values. This can be attributed to the relatively higher room temperature $_jH_C$ for samples 9 and 10. Sample 9, a single phase alloy with 20% substitution of Nd + Pr by Dy, had a particularly high $_jH_C$ and, consistent with this, its IL value was even lower than for sample 10, a nanocomposite alloy with 10% Dy substitution. Thus, magnets with higher $_jH_C$ tend to have a higher resistance to demagnetization. The improvement in the irreversible loss for Dy addition can be attributed to the enhanced room temperature $_jH_C$. The importance of $_jH_C$ for IL is further emphasized by the data for samples 3 and 10. These two samples had similar T_C and J_r/J_s, but sample 10, having the higher coercivity, had much lower IL.

The temperature coefficients of remanence a and coercivity β are described by the percentage of change per unit temperature when temperature is raised from T_1 to T_2, as described earlier. The a and β values for the NdFeB magnets are dependent on the room temperature $_jH_C$. For a NdFeB magnet having $(BH)_{max} = 256$ kJm^{-3} and $_jH_C = 904$ kA/m, the irreversible coefficient of coercivity is $\beta = -1.09\%K^{-1}$ and of the $(BH)_{max}$ is $-0.398\%K^{-1}$ [89]. For a magnet having $(BH)_{max} = 31\ 0$ kJm^{-3}, it was found that $\alpha = -0.10\%K^{-1}$ and $\beta = -0.6\%K^{-1}$ [90]. In the case of magnets having higher $_jH_C$, values of α and β of $-0.089\%K^{-1}$ and $-0.54\%K^{-1}$, respectively, were obtained. The temperature coefficients α and β decrease in absolute magnitude were a result of Dy substitutions [91].

Although there are a number of elements, like Co, Ni, Si, and Cu, which increase the T_C when substituting for Fe, Co is the only element which leads to significant changes in the magnetic ordering temperature. A partial substitution of Fe by Co improves the temperature dependence of J_r, but in addition weakens the $_jH_C$. Co substitution also does not improve the temperature dependent $_jH_C$, because RE-Co coupling is also moderate and the Co moments are small. For this reason, when adding Co, additional substitutions on RE or Fe sites must be performed to compensate for their negative influence on the coercive fields.

Partial replacement of Nd by Dy increases the $_jH_C$ of the magnet through an enhanced H_a. Since the magnetic moment of Dy couples antiparallel to both the Nd and Fe moments, it partially compensates for the temperature dependence of the J_s, improving the linear behavior α with increasing Dy content. A combination of Dy and Co substitution leads to a further improvement in α [92]. Endoh et al. [93] investigated the region of high concentration of heavy RE to achieve α = 0%/K in the temperature range 25–150°C. However, such a system exhibits significantly reduced $(BH)_{max}$.

The magnets made by die-upsetting melt-spun ribbons show better temperature stability than sintered magnets [84]. The same behavior has been observed for magnets prepared by directly Joule heating from rapidly quenched powder [94]. Li et al. [95] studied the temperature characteristics of NdFeB magnets fabricated by single-stage hot deformation (HD). The temperature coefficient of J_r was found to be −0.12%K^{-1}, similar to those of isotropic MQPA powder, die-upset, and sintered NdFeB magnets. However, the value of α is −0.72%K^{-1} which is about twice that of the isotropic powder, and slightly higher than those of die-upset and sintered NdFeB magnets.

The thermal stability of nanocrystalline magnets has received less attention than conventional MQ magnets. For $Nd_{4.8}Fe_{77.5}B_{17.7}$, a value α = −(0.12 – 0.14) %K^{-1} is reported, which decreases to α = −0.10%K^{-1} in the case of $Nd_{3.5}Dy_1Fe_{73}Co_3Ga_1B_{18.5}$ in the temperature range 50–150°C. For the above alloys, values of α of −0.40%K^{-1} and −0.38%/K^{-1}, respectively, were obtained [96]. Jurczyk and Jakubowicz [97] studied the elevated temperature behavior of a range of remanence enhanced $Nd_{12.6}Fe_{69.8}Co_{11.6}M_xB_6$/α-Fe (M = Al–Cr, Cr, Zr) nanocomposites. The α and β for the hot pressed nanocomposite magnets with M = Zr and x = 0.5, containing 37.5 vol% α-Fe, are −0.07 and −0.35%K^{-1}, respectively, are much lower than those for the Nd–Fe–B sintered magnet. If the content of α-Fe phase increases, the thermal stability of the $_jH_C$ also increases. Partial replacement of Fe by Co, Zr or Al–Cr substantially reduces the values of α and β.

Chen et al. [98] reported that the thermal stability of nanocrystalline NdFeB magnets is significantly improved by a small amount of Nb substitution. Both the temperature coefficient β and the irreversible flux aging loss are significantly reduced. Enrichment of Nb along the grain boundaries is believed to be the main reason for the observed improved thermal stability in Nb-substituted powders.

The temperature dependence of $_jH_C$ in Nd–Fe–B-based magnets is related to that of the spontaneous magnetization and of the H_a of the main magnetic phase as well as to the microstructure of the sample. The part played by microstructure is described by the parameters $α_K$ and N_{eff}. Hirosawa et al. [99] computed the dependence of $_jH_C$ and β on $α_K$ and N_{eff} parameters according to Equations 5.3 and 5.4 and found that the theoretical limit of the temperature dependence of $_jH_C$ was about −0.36%K^{-1} for ternary NdFeB magnets in the temperature range 300–420 K. Practical limits were determined by actual values of $α_K$ and N_{eff} which must be obtained empirically. By using sintered magnets, as a model system, the temperature dependence of $_jH_C$ has been investigated [100]. The reduction of exchange coupling and magnetic interaction between $RE_2Fe_{14}B$ grains and local demagnetizing fields seem to be essential in increasing $_jH_C$ and improving the value of β. The magnetic interactions can be reduced when the grain size is small. For strongly exchange-coupled grains with no pinning present, if one grain can reverse at a lower field then this reversal can quickly propagate through the whole magnet, leading to a much reduced $_jH_C$. Besides the optimum HT, the magnetic separation of $RE_2Fe_{14}B$ grains by nonmagnetic boundary phases and/or suppression of grain growth are important factors which prevent degradation of not only $_jH_C$ but also of its temperature dependence. An increase in α_K, if possible, may be achieved by reducing the average grain size in the sintered magnets, and it will be most effective in improving β, while a further reduction of N_{eff} can be expected after the heat-treatment conditions are optimized. In magnets with additional solute elements changes in J_s and H_a must be considered in addition to boundary grain phase changes. Rapidly solidified Nd–Fe–B isotropic magnets have $\beta \approx -0.4\%K^{-1}$, which is smaller than the isotropic sintered magnet. When the samples are hot deformed, their value of β becomes identical to those observed in anisotropic sintered magnets. The N_{eff} value in rapidly solidified isotropic magnets is smaller than in anisotropic magnets, probably due to a decrease in the magnetic stray field as a result of the random orientation of neighboring grain magnetizations [99].

Liu et al. [38] systematically studied the elevated temperature magnetic properties of nanocrystalline $(Nd_{1-y}Pr_y)_z(Fe_{1-x}Co_x)_{94-z}B_6$ ($x = 0$–0.3, $y = 0$–1, $z = 8$–12) alloys, including single-phase alloys, nanocomposites, and RE-rich alloys. The results based on elevated temperature measurements, indicates that nanocrystalline alloys have much lower values of the temperature coefficients than conventional microcrystalline REFeB alloys. Co substitution substantially improves the thermal stability of all alloys through decreasing the values of both α and β, whereas Pr substitution has no advantage on the thermal stability and slightly increases the value of β. Nanocomposite alloys have lower values of β than single phase and RE-rich alloys. Introducing and increasing the fraction of soft phase (α-Fe,Co) decreases the value of α. RE-rich alloys have relatively worse thermal stability. Both T_C and $_jH_C$ are important to the temperature coefficients. Increased T_C improves both α and β. A high $_jH_C$ must be maintained at elevated temperatures in order to obtain a low temperature coefficient β. In addition, the enhanced exchange coupling also benefits the temperature stability.

Figure 5.16 shows the absolute values of temperature coefficients α and β for nanocomposite $[Nd_{0.8}(Dy_{1-x}Y_x)_{0.2}]_{10}Fe_{84}B_6$ alloys with various Y substituted for Dy in the temperature range of 293–393 K [59]. The value of $|\alpha|$ decreases from $0.097\%/°C$ to $0.088\%/°C$ with increasing x from 0 to 0.75, and a high value of $0.098\%/°C$ is obtained at $x = 1$. Also, $|\beta|$ decreases from $0.397\%/°C$ to $0.394\%/°C$ with increasing x from 0 to 0.5. When x increases from 0.5 to 1, the β increases up to $0.398\%/°C$. The results demonstrated that Y substitution for Dy has an obvious advantage for thermal stability although it leads to reduced coercivity. This improvement is mainly due to the higher M_s for $Y_2Fe_{14}B$ than for $Dy_2Fe_{14}B$ and the more weak dependence of H_a on the temperature for $Y_2Fe_{14}B$ than for $Dy_2Fe_{14}B$. In addition, our further research also showed that very low absolute values of α ($-0.072\%/°C$) and β ($-0.354\%/°C$) were obtained in a $[Nd_{0.5}Y_{0.5}]_{10}Fe_{84}B_6$ alloy without Dy addition.

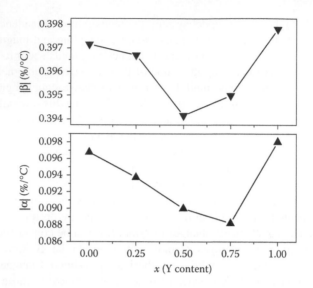

FIGURE 5.16 Temperature coefficients (293–393 K) for nanocomposite $[Nd_{0.8}(Dy_{1-x}Y_x)_{0.2}]_{10}Fe_{84}B_6$ ($x = 0$, 0.25, 0.5, 0.75, and 1) alloys. (Adapted from Z. W. Liu, D. Y. Qian, and D. C. Zeng, *IEEE Trans. Magn.* 48, 2012, 2797–2799.)

5.3.4 EXCHANGE COUPLING

Henkel plots [12,101] are suggested to be effective for analyzing the exchange interactions between the grains. The decrease of intergrain exchange coupling results in the drop of the maximum value of δM [102]. Figure 5.17 illustrates the Henkel plots for $[Nd_{0.8}(Dy_{1-x}Y_x)_{0.2}]_{10}Fe_{84}B_6$ alloys with $x = 0$ and $x = 5$. A positive δM is observed

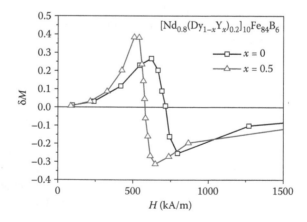

FIGURE 5.17 Henkel plots for as-spun $[Nd_{0.8}(Dy_{1-x}Y_x)_{0.2}]_{10}Fe_{84}B_6$ ($x = 0$ and 5) alloys. (Adapted from Z. W. Liu, D. Y. Qian, and D. C. Zeng, *IEEE Trans. Magn.* 48, 2012, 2797–2799.)

in both samples [59], confirming the existence of exchange coupling interaction between soft and hard phases. In addition, a higher value of positive δM is found in $[Nd_{0.8}(Dy_{0.5}Y_{0.5})_{0.2}]_{10}Fe_{84}B_6$ alloy, suggesting that Y substituted alloy has stronger exchange coupling. It can be explained by the increased exchange length L_{ex} ($L_{ex} = \sqrt{A/K}$), because Y substituting Dy decreases the anisotropy constant K. The exchange coupling is also believed to be partly responsible for the improved thermal stability for Y added alloys [41].

The effects of Co on exchange coupling and microstructure can be analyzed in the framework of the nucleation model by investigating the temperature dependence of $_jH_C$ based on the modified Brown's equation (Equation 5.8). According to the above equation, a plot of the experimental values of $\mu_0 H_C(T)/J_S(T)$ versus the theoretical values $\mu_0 H_a(T)/J_S(T)$ should yield a straight line, having a slope $1/2\alpha_k\alpha_{ex}$ and an ordinate intersection $(-N_{eff})$. Figure 5.6 earlier shows this plot for the nanocomposite $(Nd_{0.25}Pr_{0.75})_8(Fe_{1-x}Co_x)_{86}B_6$ alloys with various Co contents. The microstructural parameters calculated from the straight lines are shown in the figure. Increasing Co concentration decreases the values of both $\alpha_k\alpha_{ex}$ and N_{eff}, showing that Co substitution improves the exchange coupling, and alters the microstructure. The improved exchange coupling can be qualitatively explained by the increased exchange length due to the decreased anisotropy constant resulting from Co substitution. The improved microstructure resulting from Co substitution is demonstrated by the enhanced J_r and $(BH)_{max}$ at room temperature [66]. It is possible that Co addition leads to more equiaxed grains and smoother grain boundaries resulting in a reduction of local stray field. Co substitution decreases both $\alpha_k\alpha_{ex}$ and N_{eff}, indicating that Co improves not only the exchange coupling but also the microstructure. The improved exchange coupling can be quantitatively explained by the increased L_{ex} due to decreased anisotropy constant by Co substitution. The improved microstructure is demonstrated by the enhanced J_r and $(BH)_{max}$ at room temperature [43].

5.4 NEW PROCESSES FOR NANOCRYSTALLINE OR NANOCOMPOSITE NdFeB POWDERS

The most commonly used processes for producing NdFeB magnetic alloys are conventional powder metallurgy methods [103] and rapid quenching techniques [104,105]. These methods, however, are energy intensive and require high purity elements as starting materials. To prepare $Nd_2Fe_{14}B$-based bonded magnets, sintered magnets, or magnetic elastomers for engineering applications, magnetic powders are essential. Magnetic flakes from melt spinning or ingots need be crushed, ball milled, or hydrogen treated to obtain micron sized powders. Recently, nanostructured permanent alloys like SmCo and NdFeB have attracted considerable attention because of their excellent hard magnetic properties. Unfortunately, the fabrication of ultra-fine powders from these highly reactive materials has been found to be very challenging. In the last few years, new approaches, including bottom-up and top-down approaches, have been attempted to synthesize nanocrystalline NdFeB alloy powders.

5.4.1 Bottom-Up Approaches

5.4.1.1 Synthesis of NdFeB Nanopowders by Wet Chemistry

Early, Lin et al. [106] reported that, starting from superfine precursors, NdFeB-based magnetic materials can be prepared by reduction-diffusion reaction at relative low temperatures. The boride precursors of $Co_{2.0}B_{1.2}$ and $Fe_{1.0}B_{2.5}$ were obtained from the reduction of $CoCl_2$ and $FeSO_4$ with $NaBH_4$ in aqueous solution at room temperature. The other precursors of Nd_2O_3 and Fe_2O_3 were obtained by a polymer network gel process. The obtained $Nd_2Fe_{14}B$ and $Nd_2Fe_{12}Co_2B$ materials are fine crystalline powders exhibiting strong magnetic anisotropy. The particle sizes of these materials can be controlled to a few microns without using mechanical milling processes.

Afterward, synthesis of $Nd_2Fe_{14}B$ magnets by chemical routes has been frequently attempted; one such method involves the reduction of suitable salts of iron and neodymium by sodium borohydride [107,108]; another is the polyol reduction method [109], and the organometallic complex of iron and neodymium is reduced by a polyalcohol. In chemical methods, nanoparticle size can be controlled by adjusting reaction parameters such as time, temperature, and concentration of reagents. However, due to a high negative reduction potential of the RE element (−2.43 eV), it is quite difficult to co-reduce REs and transition metals simultaneously (−0.057 eV). $Nd_2Fe_{14}B$ alloys are reactive and oxidation prone, making the synthesis of nanoparticles challenging.

In 2001 Murray et al. experimentally produced $Nd_2Fe_{14}B$ nanoparticles by sol–gel followed by reduction-diffusion [110]. The sol–gel technique was used to synthesize a chemically homogenous oxide; reduction-diffusion of this oxide can produce $Nd_2Fe_{14}B$ magnetic nanoparticles. Particle growth can be minimized by controlling the reaction temperature and time. As a result, fine $Nd_2Fe_{14}B$ nanoparticles of ~25 nm size were successfully obtained by this method. In details, neodymium chloride hexahydrate ($NdCl_3.6H_2O$, 99.9%), iron chloride hexahydrate ($FeCl_3.6H_2O$, 97%–102%), boric acid (H_3BO_3, 99.8%), citric acid (99.5%), and ethylene glycol (99%) were used for synthesis of Nd–Fe–B oxide powder by a Pechini-type sol–gel process. $Nd_2Fe_{14}B$ powder was prepared by mixing Nd–Fe–B oxides powder with 1.5 wt.% of CaH_2 and annealing at 800°C for 2 h in vacuum. CaH_2 (90%–95%, Sigma aldrich) was used as the reducing agent instead of calcium metal due to the ease of mixing and pulverization caused by hydrogen gas released during heating; this results in greater surface area and accelerates reduction. In this process, the major byproduct, calcium oxide (CaO) was removed by water. This technique can be readily extended to the synthesis of exchange-coupled magnetic nanoparticles to obtain high-energy product magnets.

Using a similar process, Deheri et al. [111] synthesized the NdFeB powders with a coercivity of 6.5 kOe and a saturation magnetization of 21.1 emu/g (Figure 5.18). The saturation magnetization was low due to the presence of nonmagnetic CaO. After removal of CaO by washing, the saturation magnetization increased to 102 emu/g but the coercivity decreased to 3.9 kOe resulting in a $(BH)_{max}$ of 2.9 MGOe. The coercivity decrease may be due to the formation of the $Nd_2Fe_{14}BH_{4.7}$ phase with low coercivity during washing [112]. Interparticle interactions can also affect the coercive field and the reversal mechanism.

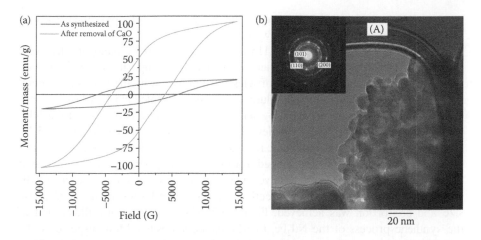

FIGURE 5.18 (a) Room temperature hysteresis loops and (b) TEM image of $Nd_2Fe_{14}B$ nanopowders synthesized by sol–gel followed by reduction-diffusion. (Adapted from P. K. Deheri et al., *Chem. Mater.* 22, 2010, 6509–6517.)

Swaminathan et al. [113] reported the novel synthesis of $Nd_2Fe_{14}B$ nanoparticles by a microwave assisted combustion process. The process consisted of NdFeB mixed oxide preparation by microwave assisted combustion, followed by the reduction of the mixed oxide by CaH_2. This combustion process is fast, energy efficient, and offers facile elemental substitution. The coercivity of the resulting powders was ~8.0 kOe and the saturation magnetization was ~40 emu/g. After removal of CaO by washing, saturation magnetization increased, and an energy product of 3.57 MGOe was obtained. A range of magnetic properties was obtained by varying the microwave power, reduction temperature, and Nd to Fe ratio. A transition from soft to exchange-coupled to hard magnetic properties was obtained by varying the composition of $Nd_xFe_{1-x}B_8$ (x varies from 7% to 40%). This synthesis procedure offers an inexpensive and facile platform to produce exchange-coupled hard magnets.

5.4.1.2 Preparation of Nanocomposite NdFeB Powders by Bottom-Up Approach

The theoretical energy product of exchange coupling nanocomposite magnets could be as high as 1 MJ/m^3, but this has never been achieved in the experiment so far. The main reason may be that conventional techniques have difficulty in controlling both hard and soft phases at the nanometer scale.

Early, Cha et al. [114] reported the synthesis of nanocomposite NdFeB powders. Soft magnetic α-Fe nanoparticles were prepared by a co-precipitation route and hard magnetic $Nd_{15}Fe_{77}B_8$ nanoparticles were prepared with ball milling for 20 h by using a shaker mill. A mechanical ball mill technique was applied to build up exchange-coupled nanoparticles. A mixture of $Nd_2Fe_{14}B$ and α-Fe nanoparticles in a stainless steel boat was milled for 2 h and annealed in a vacuum furnace under vacuum (10^{-5} Torr) at 650°C for 30 min. The crystal structure of the nanoparticles was confirmed by using X-ray powder diffraction and the magnetic hysteresis loops of the hard, soft, and nanocomposite particles are shown in Figure 5.19.

As reported by Kang and Lee [115], monodispersed α-Fe nanoparticles were synthesized under an argon atmosphere via thermal decomposition of Fe^{2+}-oleate$_2$. NdFeB ultrafine amorphous alloy particles were prepared by the reaction of metal ions with borohydride in aqueous solution. Exchange-coupled $Nd_2Fe_{14}B/\alpha$-Fe nanocomposite magnets were prepared by self-assembly of these two types of particles using a surfactant. The morphologies of the products are shown in Figure 5.20. However, unfortunately, no magnetic results were reported in their published work.

Recently, Yu et al. [116] developed a novel bottom-up strategy with a two-step thermal decomposition and reductive annealing process to synthesize $Nd_2Fe_{14}B/\alpha$-Fe nanocomposites, in which effective control of the hard/soft magnetic phase size and proportion was achieved. Figure 5.21 shows a schematic illustration of the synthetic process of the $Nd_2Fe_{14}B/\alpha$-Fe nanocomposites. Monodispersed α-Fe nanoparticles were firstly synthesized by thermal decomposition of $Fe(CO)_5$. Afterward, the obtained Fe nanoparticles were added into oleylamine(OAm)-contained organic precursors of Nd, Fe, and B. The mixture was subsequently heated to 300°C to generate Nd–Fe–B oxide/α-Fe composite. Finally, with a subsequent Ca-assisted high-temperature reductive annealing process, the as-synthesized Nd–Fe–B-oxide/α-Fe composites were further converted into $Nd_2Fe_{14}B/\alpha$-Fe exchange-coupled magnets in which the composition and magnetic properties are able to be tuned by changing the ratio between Nd–Fe–B-oxide and α-Fe. This work provides an effective approach to adjust the phase size and distribution for exchange-coupled, RE nanomagnets, which can be fundamental for high-energy magnets.

In Yu's work, high-resolution transmission electron microscopy (HRTEM) images demonstrated the formation of $Nd_2Fe_{14}B/\alpha$-Fe nanocomposites. The α-Fe

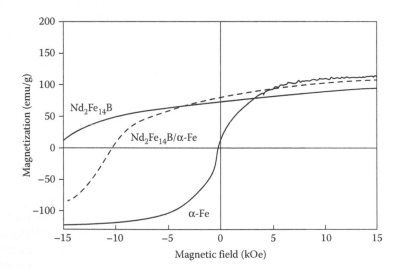

FIGURE 5.19 Demagnetization curves for NdFeB samples. (Adapted from H. G. Cha et al., *Curr. Appl. Phys.* 7, 2007, 400–403.)

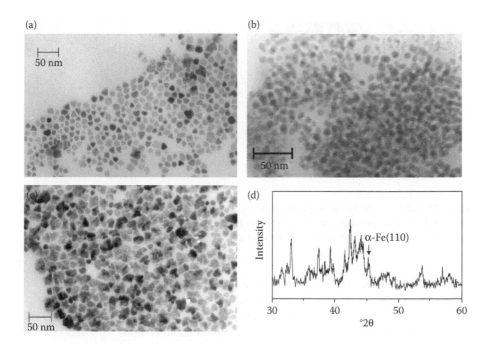

FIGURE 5.20 TEM images of α-Fe nanoparticles (a), NdFeB nanoparticles (b), and binary nanoparticle assemblies Nd–Fe–B alloy/α-Fe (c). (d) XRD pattern of $Nd_2Fe_{14}B$/α-Fe nanocomposite magnet. (Adapted from Y. S. Kang and D. K. Lee, *Int. J. Nanosci.* 5, 2006, 315–321.)

nanocrystals are well distributed in the $Nd_2Fe_{14}B$ matrix even after a high temperature reductive annealing process, which could be attributed to the fact that the Nd–Fe–B-oxide matrix could inhibit the crystal growth as well as α-Fe nanoparticles aggregation. Thus, in the nanocomposite powder, the strained size of the well-dispersed α-Fe (~8 nm) ensured the effective exchange coupling between α-Fe and hard magnetic phase. Hard magnetic properties have been successfully obtained in these

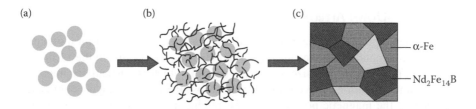

FIGURE 5.21 Schematic illustration of $Nd_2Fe_{14}B$/α-Fe nanocomposites synthesized by wet chemical method. (a) monodispersed α-Fe nanoparticles; (b) composites of α-Fe evenly embedded in net-like Nd–Fe–B-oxide matrix; (c) $Nd_2Fe_{14}B$/α-Fe nanocomposites obtained by thermal reductive diffusion process. (Adapted from L. Q. Yu, C. Yang, and Y. L. Hou, *Nanoscale* 6, 2014, 10638–10642.)

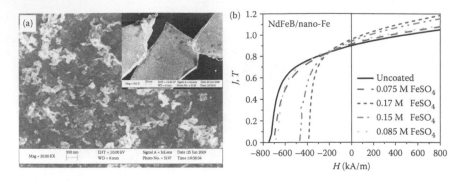

FIGURE 5.22 (a) SEM micrographs showing the surface of NdFeB/Fe composite powders (a) and (b) hysteresis loops for uncoated NdFeB and NdFeB/nano-Fe composite powders by chemical synthesis. (Adapted from K. P. Su et al., *J. Appl. Phys.* 109, 2011.)

magnets. For the $Nd_xFe_{10}B_1$ composites with tuned Nd/Fe ratio, an addition of Nd increases the coercivity to the maximum value of 12,000 Gs at $x = 2.6$.

Recently, Su et al. [117] reported a new wet chemistry approach to synthesize the nanocomposites of NdFeB/nano-Fe(Co) with controllable compositions and sizes of both phases. The synthesis was based on the precipitation of Fe and Co ions from their aqueous solutions on the melt-spun NdFeB powders. Figure 5.22a shows the micrograph of the surface layer of the nanocomposite powders. The coating layers are comprised of Fe nanoparticles with average particle size of ~61 nm. The hysteresis loops of NdFeB/α-Fe composite powders synthesized with various Fe^{2+} concentrations are shown in Figure 5.22b with uncoated powders. Smooth demagnetization curves for all nanocomposites suggest that the magnetic hard and soft phases are well exchange coupled. Coating of the Fe nanoparticles layer leads to an obvious increase of J_r. Using a high Fe^{2+} concentration solution (0.17 M $FeSO_4$), a high J_r was obtained, which gives a remanence enhancement of around 6.5%. A layer of $Fe_{65}Co_{35}$ alloy can further increase J_r of the nanocomposite since $Fe_{65}Co_{35}$ alloy has higher J_s than Fe or Co at room temperature. This work provides an easy approach to prepare exchange-coupled nanocomposites with controllable compositions and sizes of both hard and soft phases.

5.4.2 TOP-DOWN APPROACHES

While sintered permanent magnets are still the major type of magnet for most applications, the demand for bonded magnets is rapidly growing because of the advantages of bonded magnets in part size and shape control, which is particularly important for the miniaturization of electronic devices. Most RE bonded magnets are isotropic because the magnetic powders used for producing bonded magnets are isotropic. Although attempts have been made to produce anisotropic bonded RE magnets based on NdFeB micrometre-sized powder particles, the production of nanoscale anisotropic RE magnetic powder particles remains a significant challenge.

One of the approaches for the preparation of magnetic nanoparticles is by a ball milling process in the presence of an organic carrier liquid and surfactant, a technique used for making magnetic fluids [118]. Surfactants aid in achieving smaller particle sizes during milling and in dispersing the fine particles when appropriate solvents are used [119]. Recently, surfactant-assisted ball milling (SABM) has also been proven to be an effective technique to produce anisotropic hard magnetic nanoparticles including SmCo and NdFeB-based RE nanocrystalline materials [120–123]. The produced Sm-Co and Nd–Fe–B nanoparticles containing fine grains have a high aspect ratio with their thickness of tens of nanometers, and width and length of several hundred nanometers. The chip-like nanocrystals show strong magnetocrystalline anisotropy and can be aligned in a magnetic field.

SABM has become a hot topic for preparation of NdFeB nanocrystals by a top-down approach during the last few years [124–126]. It has been found that both nanoparticles and nanoflakes can be produced. Figure 5.23 shows the morphology and hysteresis loops for SABMed NdFeB powders. Ultrafine hard magnetic particles can be produced in this way and with increasing milling time the $_jH_C$ reduces. A $_jH_C$ value more than 300 kA/m can be still obtained in the nanoparticles with mean size of 20 nm [117]. A very promising characteristic of this process is that it can produce anisotropic nanostructured flakes with easy axis lying in the flake plane by controlling milling agents and parameters. These anisotropic powders are suitable for preparing anisotropic bonded magnets. The initial work was conducted by Chakka et al. [120] for SmCo magnets, and they demonstrated the synthesis of anisotropic nanostructured SmCo flakes. Afterward, much work [126,127] has been carried out on this topic of surfactant-assisted ball milling hard magnets and been quite successful.

The milling processes have an important effect on the magnetic properties of the alloy. The processing parameters include milling time, milling speed, ball to powder ratio, and additives. With increased milling time, the grain size of NdFeB powders will decrease, and the powders will become amorphous when the milling time is long enough. Then, with an optimal HT, the amorphous powders can be converted back to the isotropic crystalline tetragonal $Nd_2Fe_{14}B$ phase. Jurczyk's work [125] shows that when increasing the milling time from 10 min to 2700 min, the coercivity is found to

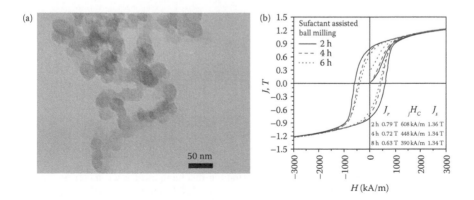

FIGURE 5.23 The morphology of NdFeB nanoparticles milled for 8 h and hysteresis loops for the nanoparticles after milling for 2, 4, and 8 h by SABM. (Adapted from K. P. Su et al., *J. Appl. Phys.* 109, 2011.)

increase with the milling time at first, passing through a maximum 493 kA/m (after about 90 min) and then decreasing to 4 kA/m after about 2700 min, when the powders have proved to be amorphous. After an optimal HT, a high $_jH_C$ of 943 kA/m has been achieved due to the formation of single domain size $Nd_2Fe_{14}B$ grains. However, the powders so produced are isotropic. To get the anisotropic powders directly, the milling time and the ball to powder ratio play an important role.

Su et al. [128] suggested that with increasing milling time the thickness of the nanoflakes decreases significantly, but the average width of the nanoflake has no obvious change. However, the coercivities decreased with the increasing milling time, which results from the reduced grain size. Thus, controlling the milling time is the key to getting the anisotropic $Nd_2Fe_{14}B$ nanoflakes. The schematic evolution and formation mechanism of micrometer and submicrometer flakes and textured nanocrystalline nanoflakes is shown in Figure 5.24. The process is (1) single-crystal irregular microparticles by the fragmentation of the isotropic polycrystalline ingot; (2) single-crystal micron and then submicron flakes by the cleavage of single-crystal $Nd_2Fe_{14}B$ microparticles/flakes along the easy glide (110) basal planes; (3) submicron flakes with small-angle grain boundaries via accommodation of localized deformation and dislocations; (4) textured nanocrystalline submicron flakes or nanoflakes with both small-angle and large-angle grain boundaries via recrystallization and severe plastic deformation because of the significant ductility exhibited by brittle materials in the nanocrystalline state [129].

According to Liu's work [130], the ball to powder ratio has a significant effect on the morphology of the powders. The larger the ball to powder ratio, the more strongly they collide, and the higher the broken efficiency, then more flake-like anisotropic powders can be obtained (shown in Figure 5.25). First, the raw materials are broken into blocks and flakes in the milling process. Afterward, most of the powders are refined and some are thinned continuously. Finally they are broken into homogeneous particles.

The additive during ball milling is very important for getting anisotropic flake-like particles with high magnetic magnetocrystalline anisotropy, which can be aligned in

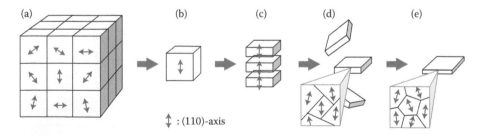

FIGURE 5.24 Schematic evolution and formation mechanism of micrometer and submicrometer flakes and textured nanocrystalline nanoflakes formed from melt-spun powders during ball milling. (a) Bulk ingot with polycrystalline structure; (b) single-crystal particles; (c) single-crystal micron and then submicron flakes; (d) submicron flakes with small-angle grain boundaries; and (e) textured poly-nanocrystalline nanoflakes. (Adapted from B. Z. Cui et al., *Acta Mater.* 60, 2012, 1721–1730.)

FIGURE 5.25 The morphology for of NdFeB nanoparticles with different ball to powder ratios. (a) 8:1; (b) 12:1; (c) 16:1; and (d) 20:1. (Adapted from J. Y. Liu et al., *J. Magn. Mater. Devices,* 45, 2014, 18–22.)

a magnetic field to make anisotropic magnets. The presence of solvent and surfactant in SABM is known to be very efficient at impeding cold welding and the agglomeration of particles during ball milling. The solvent and surfactant covering a particle or flake can lower the energies of freshly cleaved surfaces, enabling long-range capillary forces and lowering the energy required for crack propagation [131]. However, the use of a surfactant not only influences the size of the particle, but also plays an important role in determining the shape and texture of the particle. Generally, dry milling yields more or less equiaxed particles [132], whereas wet milling, particularly with a high-energy mill and/or at a high ball to powder ratio, results in high-aspect-ratio flakes, often with a submicron thickness [133,134]. Various contents of oleic acid (OA) or oleylamine (OY) were often chosen as the surfactant. It was reported that enhanced texture was observed in the samples with higher contents of surfactant; this is because a large amount of surfactant resulted in well-separated nanoflakes instead of the closely packed "kebab-like" flake morphology [123].

Meanwhile, ball milling or annealing within an applied magnetic field are also beneficial for the anisotropy of the powders [135]. Poudyal et al. [136] work indicates that field ball milling will induce anisotropy in magnetic particles. During the milling process, when introducing a magnetic field, the powder particles formed chains along the field direction. The particles inside the chains were aligned. While the chains were colliding with the balls, re-joined small particles preferentially aligned themselves along the chain's long axis direction, which is normally parallel to the magnetic field direction. Then the enhanced M_r/M_s ratio was observed in the field ball milling samples. The magnetic NdFeB powders are very easily agglomerated during the milling process. However, when annealing these powders in a magnetic field at an appropriate temperature, they will be separated and then magnetized in the easy direction. With these effects, the powders are easier to be aligned than the powders without magnetic annealing.

The coercivity of ball milling powders with ternary NdFeB composition is not high enough for application. To improve the coercivity $_jH_C$, the elemental additive is one of the most efficient ways. Dy, NdCu, and PrCu alloys were chosen as the additives [137]. The addition of heavy RE element Dy was accomplished before arc-melting, and the Dy added $Nd_{14}Dy_{1.5}Fe_{78.5}B_6$ as-milled nanoflakes have a coercivity H_c of 4.3 kOe while $Nd_{15.5}Fe_{78.5}B_6$ nanoflakes have an H_c of 3.7 kOe. The reason for enhanced coercivity was attributed to the higher magnetocrystalline anisotropy for Dy substituted $(DyNd)_2Fe_{14}B$ than that for $Nd_2Fe_{14}B$, whereas, the NdCu and PrCu were blended during the milling process. After milling for 5 h, part of the NdCu were diffused into the $Nd_2Fe_{14}B$ nanoflakes, however, lots of NdCu flakes were still mixed with $Nd_2Fe_{14}B$ nanoflakes. The coercivity of the NdCu added powders increased to 5.3 kOe. After post-annealing at 450°C for 0.5 h, the coercivity of the NdCu added powders enhanced to 7.0 kOe. It was found that the $Nd_{70}Cu_{30}$ phase was observed around the grain boundary and the $Nd_{70}Cu_{30}$ phase can act as the pinning center during the demagnetization process. The thickening of Nd-rich phase at grain boundaries in the $Nd_2Fe_{14}B$-based flakes was also observed, thus, the weakened exchange coupling was another reason for the enhanced coercivity.

5.5   DENSIFICATION PROCESSES FOR NANOCRYSTALLINE NdFeB ALLOYS

Nanostructured powders have to be assembled into bulk magnets for most applications. It is desirable to keep the nanostructure after consolidation. In this case, conventional densification techniques like hot press (HP) and hot isotropic press do not work since they will promote grain growth which lose the benefits of the nanocrystallites. To prepare isotropic or anisotropic nanocrystalline magnets, new processes have been recently developed, such as spark plasma sintering (SPS), HD, and high-velocity compaction (HVC).

5.5.1   SPARK PLASMA SINTERING

SPS has emerged as a new pressure sintering process to be performed at a relatively low temperature in a short period of time [138,139]. This process utilizes a high-frequency, momentary local high-temperature field generated by pulse energy, spark impact pressure, and Joule heating throughout sintering to heat the powders. One of the important advantages of the SPS is the short sintering time, which can effec-tively restrain the grain growth and allow preparation of high-density fine crystalline materials [140]. On the other hand, the surface of the powder can be activated by the intergranular discharge plasma in the process of SPS, which offers the possibility for uniform distribution of the low melting point Nd-rich phase in NdFeB alloys. It is also possible to compact NdFeB powders without the Nd-rich phase, such as a nanocomposite alloy, by SPS. These characteristics make the SPS method suitable for the production of bulk materials from nanocrystalline melt-spun ribbons [140]. In addition, the combination of SPS and the HD process has been suggested to fabricate anisotropic magnets with nanocrystallites and has been quite successful [141–143].

Liu et al. [144] investigated the microstructure and properties evaluation for SPSed nanocrystalline NdFeB alloys. Figure 5.26 shows the typical structure of the SPSed magnets made from melt-spun ribbons. The ribbons are aligned perpendicu-larly to the pressing direction.

The microstructure of SPSed magnets has significant change with the SPS tem-perature (T_{SPS}). Figure 5.27 show the SEM images for the Nd-rich magnets SPSed at 600–800°C for 5 min with SPS pressure $P_{SPS} = 50$ MPa. Two distinguished zones with different grain sizes were noticed, especially for the temperature higher than 700°C. Liu et al. [144] refer to these two zones as the coarse grain zone (CZ) and fine grain zone (FZ), as indicated in Figure 5.27d, and carried out detailed investi-gations on the variation of both zones. Since the SPS process utilizes a momentary local high-temperature field generated by pulse energy, and Joule heating to heat the specimen. The Joule heating is mainly located at the particle boundaries. This leads to abnormal grain growth at the particle boundaries, while the grains inside the particles have less tendency for growth. Therefore, the coarse grain zones in Figure 5.27b, c, and d are corresponding to the grain boundary area and the fine grain zones to the interior of the particles. It is found that the particle boundaries are not obvious and the overall grains are almost uniformly distributed for the magnets SPSed at 600°C (not shown here) and 650°C (Figure 5.27a). With the increase of

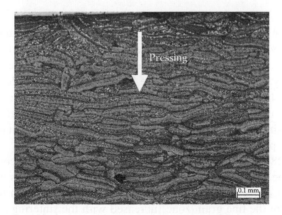

FIGURE 5.26 Optical images for the typical structure of SPSed magnet. (Adapted from Z. W. Liu et al., *J. Phys. D: Appl. Phys.* 44, 2011, 25003–25011.)

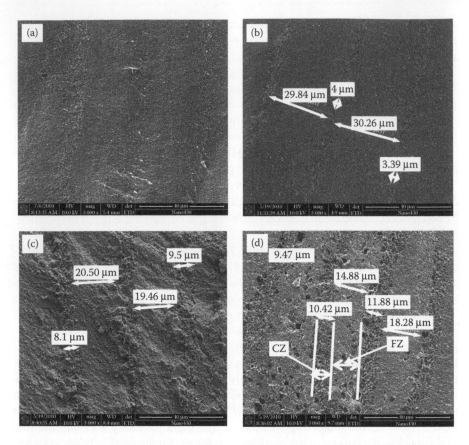

FIGURE 5.27 Microstructures of NdFeB magnets SPSed at various temperatures with $P_{SPS} = 50$ MPa and $t_{SPS} = 5$ min: (a) 650°C; (b) 700°C; (c) 750°C; (d) 800°C. CZ and FZ in (d) indicate coarse grain zone and fine grain zone. (Adapted from Z. W. Liu et al., *J. Phys. D: Appl. Phys.* 44, 2011, 25003–25011.)

temperature from 700 to 800°C, the average width of the coarse grain zone increases and the average width of the fine grain zone decreases. There is also a slight decrease of the widths of the coarse grain zone and fine grain zone (one particle) due to improved density and deformation. Figure 5.28a shows the dependences of the average width of the fine grain zone and the average ratio of fine grain zone area on the sintering temperature. It is very clear that the width of the fine grain zone increases with the decreasing T_{SPS}. This microstructure characteristic has an important effect on the magnetic properties.

In Figure 5.28b, both mean sizes in the fine grain zone and coarse grain zone increase with the SPS temperature, but the growth of the grain is more obvious in the coarse grain zones. The size difference between the grains in the coarse grain zone and those in the fine grain zone increased with the sintering temperature T_{SPS} (Figure 5.28b). For example, at $T_{SPS} = 600$ and 650°C, the grain sizes in the coarse grain zone and fine grain zone are almost similar or show no big difference, whereas

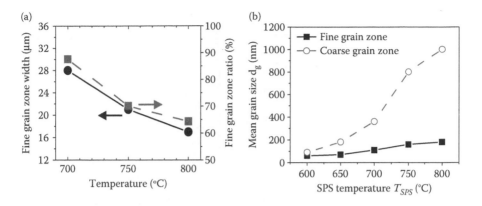

FIGURE 5.28 The fine grain zone width and ratio (a) and the mean grain sizes in coarse grain zone and in fine grain zone (b) for NdFeB magnets SPSed at various temperatures (P_{SPS} = 50 MPa; t_{SPS} = 5 min). (Adapted from Z. W. Liu et al., *J. Phys. D: Appl. Phys.* 44, 2011, 25003–25011.)

at T_{SPS} > 700°C, the difference in the grain sizes in the two zones is obvious. Hence, SPS at 600–650°C does not lead to significant grain growth and a large area of coarse grain zone in the sintered magnets and uniform grain structure can be obtained. It is also found that, for the temperatures of 750 and 800°C, some grains in the fine grain zonc also have abnormal growth due to the high temperature providing extra energy.

The current pulse ratio (CPR) also has a substantial influence on the width of the coarse grain zone. Wang et al. [145] reported that by tuning the CPR of power-on to power-off time during sintering from 12:2 to 2:2, the thickness of the surface diffusion layer of the Nd–Fe–B particles was reduced.

The SPS pressure P_{SPS} is also an important parameter. Hou et al. [146] found that with increasing pressure, the coarse grain zone was getting narrower. The reason can be attributed to the fact that high pressure improves density and reduces the surface area with high Joule heating by spark plasma, which inhabits the grain growth inside the particles. However, the SPS pressure does not have significant effect on the grain sizes.

The parts of magnetic hysteresis curves of the starting powders and NdFeB magnets and density and magnetic properties for NdFeB magnets SPSed at various temperatures (P_{SPS} = 50 MPa; t_{SPS} = 5 min) are plotted in Figures 5.29. The magnetic parameters for the starting powders before considering the demagnetizing factor are J_r = 0.76 T, $_jH_C$ = 1636 kA/m, and $(BH)_{max}$ = 99 kJ/m³. Not surprisingly, in the temperature range 600–800°C, the density of the magnet increases with the increases of SPS temperature, pressure, and holding time in a certain range. It was pointed out that the plasma discharges between the sample powders in the early stage of the sintering process are effective in promoting densification [147]. The density increases from 7.10 g/cm³ for the sample SPSed at 600°C to 7.57 g/cm³ for the sample prepared at 800°C. Low density at low T_{SPS} led to low magnetic properties at 600°C. The increase of density leads to a similar increase of M_S and J_r. With increase of T_{SPS}, the $_jH_C$ increases first and decreases after the temperature reaches 700°C. The $(BH)_{max}$ of the magnets reaches a maximum value of 116 kJ/m³ at 700 then gradually

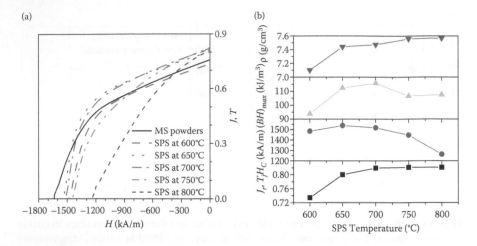

FIGURE 5.29 Magnetic hysteresis curves of the starting powders and NdFeB magnets (a) and density and magnetic properties for NdFeB magnets (b) SPSed at various temperatures (P_{SPS} = 50 MPa; t_{SPS} = 5 min). (Adapted from Z. W. Liu et al., *J. Phys. D: Appl. Phys.* 44, 2011, 25003–25011.)

decreases. The best combined hard magnetic properties are obtained at 700°C. It is also found that with increase of P_{SPS}, the higher density and smaller coarse grain zone lead to improved magnetic properties including J_r, $_jH_C$, and $(BH)_{max}$. On the other hand, increase of holding time also improves the magnetic properties, mainly due to the increase of density. The magnets SPSed at 650°C/50 MPa/5min also have good magnetic properties with J_r = 0.8 T, $_jH_C$ = 1537 kA/m, and $(BH)_{max}$ = 113 kJ/m³. The magnetic properties are also directly related to the microstructure. For P_{SPS} = 50 MPa and t_{SPS} = 5 min, the coercivity reduces from 1538 to 1263 kA/m when the T_{SPS} decreases from 650°C to 800°C due to the increase of crystallite size and the increase of coarse grain zone width. The results also show that, even SPSed at 800°C, the SPSed magnet still has good magnetic properties due to the high ratio of fine grain zone in the magnet (>60%), which indicates the obvious advantages of the SPS process for inhabiting grain growth.

For both the magnets with RE-rich and single-phase NdFeB alloys, the influence of starting powder sizes on the coercivity of SPSed magnets is very significant. Table 5.3 lists the optimal magnetic properties for magnets SPSed with four different particle size ranges under optimal sintering conditions. It is found that the magnetic properties for the SPSed single-phase magnets are much lower than those for RE-rich magnets. The main reason can be attributed to the low density and low coercivity. For single-phase (stoichiometric 2/14/1 composition) NdFeB ribbons, it is relatively difficult to consolidate micro-sized melt-spun powders into high density bulk magnets due to the deficiency of the liquid phase such as the rich-Nd phase. The latter also led to relatively low coercivity. However, it can be observed that a larger particle size is beneficial to achieve better magnetic properties, including higher coercivity $_jH_C$ and density ρ. The properties' variations are attributed to the sintering mechanism of SPS. Local high-temperature that existed between the particle

TABLE 5.3
Optimal Magnetic Properties for Single-Phase and Nd-Rich Magnets SPSed with Four Different Particle Size Ranges

Particle Size	Composition	$(BH)_{max}$ (kJ/m³)	$_jH_C$ (kA/m)	J_r (T)	ρ (g/cm³)
Less than 45 μm	Single-phase	65	435	0.70	6.68
	Nd-rich	117	1202	0.83	7.33
45–100 μm	Single-phase	69	538	0.67	6.41
	Nd-rich	115	1466	0.82	7.46
100–200 μm	Single-phase	79	504	0.73	6.69
	Nd-rich	116	1434	0.82	7.50
200–400 μm	Single-phase	81	604	0.73	6.42

Source: Adapted from Y. L. Huang et al., *Powder Metall.* 55, 2012, 124–129.

boundaries induced coarse grain zones in the vicinity of the particle boundaries. For the larger particle size ribbons, the volume fraction of the coarse grain zone was less than that with the smaller particle size ribbons due to less particle boundaries. Thus the magnetic properties of NdFeB magnets SPSed with smaller particle size ribbons are worse than those with larger particle size ribbons.

For Nd-rich phase NdFeB ribbons, with increasing initial particle size, the magnet density increases. The magnet prepared with powders 45 μm has the minimum value of intrinsic coercivity $_jH_C$. The remanence J_r and the maximum energy product $(BH)_{max}$ change slightly with varied particle size.

The SPSed magnets with nanocrystallites generally show exchange coupling between grains. Figure 5.30 shows the Henkel plots for spark plasma sintered magnets prepared from the powders with various sizes. The large positive values of δM

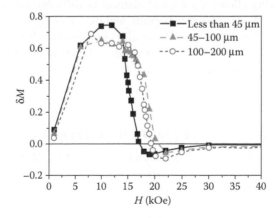

FIGURE 5.30 Henkel plots of Nd-rich NdFeB magnets prepared by SPS with different powder size ranges by $T_{SPS} = 700°C$, $t_{SPS} = 5$ min, and $P_{SPS} = 50$ MPa. (Adapted from Y. L. Huang et al., *Powder Metall.* 55, 2012, 124–129.)

verify the strong exchange coupling interaction due to the nanostructure in the magnets. For the magnets prepared from small powders (45 μm), the largest positive value of δM was obtained, which indicated that exchange coupling in the magnet spark plasma sintered from small powders is stronger than that from large powders. This can be manifested by the disappeared coarse grain zone in magnets prepared from smaller powders, as discussed earlier. Meanwhile, a relatively sharp peak obtained in the δM plot indicates a uniform grain size distribution [41]. For the spark plasma sintered magnet prepared from the particle size of less than 45 μm, a possible deficiency of spark plasma and stronger exchange coupling due to the uniform distribution of nanostructure grain size are the main reasons for the reduction in coercivity. In contrast, the high $_jH_C$ of 1466 and 1434 kA/m obtained in the spark plasma sintered magnets from large particles (Table 5.3) possibly resulted from the weak exchange coupling in the magnets. Meanwhile, the uniform distribution of the Nd-rich phase due to the activated powder surface by interparticle discharge plasma in the process of SPS also contributed to the high $_jH_C$.

For the conventional sintered magnets and hot deformed magnets, the RE-rich phase not only plays a significant role in the densification during the sintering process, but is also believed to have a strong influence on the magnetic, mechanical, and corrosion properties. Therefore, the RE-rich phase also plays the same role in SPSed magnets. The magnetic properties and density of SPSed magnets with different Nd contents are shown in Table 5.4. The magnetic properties increase with the increasing Nd content. These can be attributed to the role of the Nd-rich phase. First, the liquid Nd-rich phase plays a critical role in the densification of SPS. For SPS treated magnets with low net RE content, low density is attributed to the deficiency of the Nd-rich phase. Second, the pinning effect of the Nd-rich phase existing in the grain boundary phase leads to high coercivity for the Nd-rich composition. Therefore, the high RE content signified high coercivity and density in the same sintering conditions.

Since the SPS technique is a quick sintering process with high heating and cooling rates, it is expected that not all Nd-rich phases are in equilibrium state after sintering. Tempering allows a sufficient element of diffusion between the Nd-rich phase and epitaxial layer of $Nd_2Fe_{14}B$ grain, and the Nd-rich phase tends to be in an equilibrium state after tempering. Mo et al. [148] reported that the Nd-rich phase was found to distribute along the grain boundary after SPS. Besides, there are a few pores in the microstructure. Then, the sample was cooled to room temperature after

TABLE 5.4
Magnetic Properties and Density of SPS Treated Magnets

Content (Nd)	J_r (T)	$_jH_C$ (kA/m)	$(BH)_{max}$ (kJ/m³)	ρ (g/cm³)
11.40 at.%	0.73	448	73	6.64
11.86 at.%	0.75	585	81	6.81
13.11 at.%	0.81	1176	110	7.35
13.5 at.%	0.83	1516	118	7.56

Source: Adapted from Y. L. Huang et al., *Powder Metall.* 55, 2012, 124–129.

the SPS in vacuum and then subjected to tempering at 1000°C for 2 h. The number of pores and intergranular Nd-rich phase are obviously decreased and a continuous thin film of the Nd-rich phase is formed along the grain boundary after tempering. Then the well separated $Nd_2Fe_{14}B$ grain improved the magnetic properties.

During the SPS process, the locally formed liquid phase (here referred to as the Nd-rich phase) is rapidly homogenized via the capillary forces present between the fine precursor powders. As a result, a small amount of the Nd-rich phase in the magnet forms thin layers along the grain boundaries of the main phase, while a majority of the Nd rich phase agglomerates into the triple junctions as fine round particles. It is therefore believed that the microstructure restricted the pathways for corrosion propagation through the magnet material and effectively suppressed the intergranular corrosion process along the Nd-rich phase in the magnet, and therefore the corrosion resistance of the magnet rises remarkably. As a result, the SPS Nd–Fe–B magnet possesses excellent corrosion resistance [146].

To further improve the coercivity, the magnetic powders can be mixed with some metal powders (such as Zn and Cu powders) for SPS. During sintering, the metal powders can react with the magnetic matrix and presumably diffuse into the neodymium-rich phase surrounding the $Nd_2Fe_{14}B$ grains at the high temperature and pressure required for densification, which has been confirmed by electron microprobe analysis [149]. By this method, the enhancement of coercivity has been achieved by Cu and Zn additions. It was found that both elements increased the coercivity of hot-pressed magnets by ~37%. For die-upset magnets, coercivity was increased by ~91% for Zn addition, which is higher than ~75% by Cu addition [150]. Another effective way to enhance coercivity and maintain the low concentration of high cost RE element Dy for NdFeB magnets is the development of a core-shell microstructure, which is formed by segregation of Dy along the grain boundary area, by the addition of powdered Dy compounds (oxides, fluorides, or hydride) [151–153]. It was reported that adding 4 wt.% Dy_2O_3 enhanced the coercivity by 37.9% (from 11.6 to 16.0 kOe) for a sintered $Nd_{15}Fe_{77}B_8$ magnet [152].

Liu et al. [154] investigated the effects of separated and combined Dy_2O_3 and Zn additions on the magnetic properties, microstructure, and thermal stability of SPSed nanocrystalline NdFeB magnets. For Dy_2O_3 added magnets, with increasing Dy_2O_3 content, the porosity increases in the magnets, resulting in the decreasing density. As we know, the Dy_2O_3 compound has high thermal stability with high melting point and it is not easy to decompose. Also, Dy_2O_3 has poor electric and thermal conductivities, which are not beneficial for the densification during SPS when it stays at the particle interface. As a result, the relatively low densities are obtained for Dy_2O_3 added magnets. However, it is quite different for Zn added magnets. Figure 5.31 show the microstructure at the location of the Nd–Zn compounds in the particle boundary for 2 wt.% Zn added magnets. The Nd–Zn compound was confirmed by EDS. The relative atomic percentages of Zn at Position 1 and Position 2 in Figure 5.31 are 4.88 and 2.98 at.%, respectively. Unlike the Dy_2O_3 added magnets, no large pores existed in the Zn added magnets, as shown in Figure 5.31a. The density of the magnets decreases more slightly with Zn addition. Figure 5.31 also shows that in the existing Zn areas, either the coarse grains or porosity was observed. The former is attributed to the exothermic reaction between Zn and $Nd_2Fe_{14}B$, and the latter may

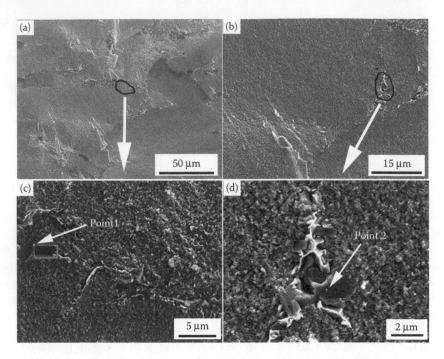

FIGURE 5.31 The SEM graphs for Nd–Zn compounds at the particle boundary in 2 wt.% Zn added SPSed magnets. (a) Boundary 1, (b) boundary 2, (c) enlarged graph for the marked area in (a), and (d) enlarged graph for the marked area in (b). Point 1 and 2 are for EDS. (Adapted from Z. W. Liu et al., *IEEE Trans. Magn.* 51, 2015, 1–4.)

result from a high vapor pressure formed by gasification of Zn due to a temperature of more than 3300 K [155] at the particle interface and high densification rate during the SPS process.

Figure 5.32 shows the grain structures of the fine grain zone for NdFeB magnets with different powder additions. Compared with additive free magnets, the grain size of the SPSed magnets with Zn or Dy_2O_3 addition is relatively smaller, roughly 100 nm for the fine grain zone. Compared with Dy_2O_3, Zn has a more significant effect on suppressing grain growth. It is, therefore, reasonable to consider that both additives have diffused into the grain boundaries and play an important role in suppressing grain coarsening of $Nd_2Fe_{14}B$ matrix phase during SPS [156,157].

Figure 5.33 shows the magnetic properties of the starting powders and the magnets sintered with different Dy_2O_3 additions. The result shows that the addition of Dy_2O_3 leads to a considerable increase of $_jH_C$. It was reported that the $(Nd,Dy)_2Fe_{14}B$ phase can be formed by a chemical reaction between Dy_2O_3 and NdFeB at sintering temperature (1343 K) [152]. Since a very high temperature [155] can be reached at particle interface during SPS, it is reasonable to consider that a small amount of $(Nd,Dy)_2Fe_{14}B$ have also formed at the particle boundaries in the magnets. The higher anisotropy field of the $(Nd,Dy)_2Fe_{14}B$ phase than that of the $Nd_2Fe_{14}B$ phase contributes to the enhancement of the coercivity. On the other hand, since the maximum value of J_r is determined by the saturation magnetization (J_s) and density of the

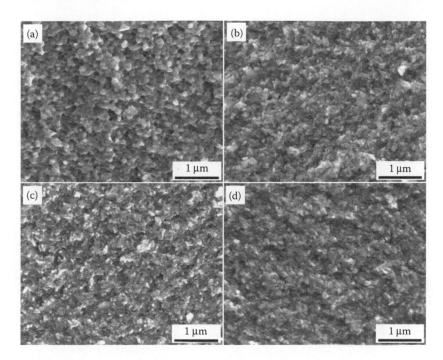

FIGURE 5.32 Grain structures in fine grain zone for NdFeB magnets SPSed with different powder additions: (a) additive free; (b) 0.6 wt.% Zn; (c) 2 wt.% Dy_2O_3; (d) 0.6 wt.% $Zn + 2$ wt.% Dy_2O_3. (Adapted from Z. W. Liu et al., *IEEE Trans. Magn.* 51, 2015, 1–4.)

alloy, the addition of Dy results in a reduction in J_r. The $(BH)_{max}$ slightly increases up to $x = 2$ wt.% and then decreases with the subsequent addition of the Dy_2O_3 content.

Figure 5.34 shows the magnetic properties of the starting powders and the magnets sintered with different Zn additions. The magnets with Zn content less than 2 wt.% show higher $_jH_C$ than the additive free magnet. The highest coercivity was obtained for 0.6 wt.% Zn addition, which can be explained by the reduced grain size, shown in Figure 5.32. The appearance of coarse grains and porosities in the vicinity of the Nd–Zn compound should be responsible for the decease of coercivity for the magnets with more than 0.6 wt.% Zn addition (shown in Figure 5.32b). J_r slightly decreases with the addition of Zn, which can be explained by the reduced volume fraction of $Nd_2Fe_{14}B$ phase and low density. $(BH)_{max}$ increased at first and then decreased, and a maximum value was obtained at 0.6 wt.% Zn addition.

To further improve the magnetic properties, the combination of 2 wt.% Dy_2O_3 and 0.6 wt.% Zn additions were investigated. The magnetic properties for the magnet with $Dy_2O_3 + Zn$ addition are also shown in Table 5.5. The combined addition considerably increases the magnetic properties of SPSed magnets, which produced a ~20% increase in coercivity. An optimal combination of the magnetic properties with $J_r = 0.73$ T, $_jH_C = 1142$ kA/m, and $(BH)_{max} = 91$ kJ/m^3 were obtained. These SPSed samples do not only work as isotropic permanent magnets but can also be employed as the ideal precursors for hot deformed anisotropic NdFeB magnets [158].

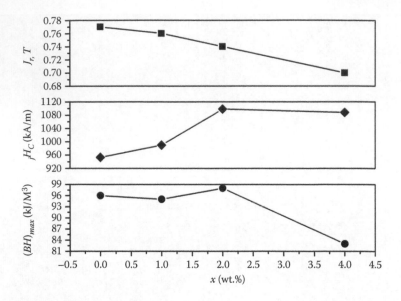

FIGURE 5.33 Magnetic properties of SPSed NdFeB magnets with different Dy_2O_3 additions. (Adapted from Z. W. Liu et al., *IEEE Trans. Magn.* 51, 2015, 1–4.)

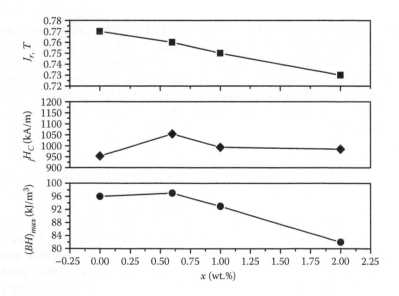

FIGURE 5.34 J_r, $_jH_C$, and $(BH)_{max}$ of $Nd_{10.15}Pr_{1.86}Fe_{80.41}Al_{1.67}B_{5.91} + xZn$ samples as a function of Zn addition. (Adapted from Z. W. Liu et al., *IEEE Trans. Magn.* 51, 2015, 1–4.)

TABLE 5.5

Magnetic Properties and Density for SPSed Nd–Fe–B Magnets with the Optimal Powder Addition at Room Temperature

SPSed Samples	J_r (T)	$_jH_C$ (kA/m)	$(BH)_{max}$ (kJ/m³)	Density (g/cm³)
Nd–Fe–B	0.77	953	96	7.43
Nd–Fe–B + 0.6 wt.% Zn	0.76	1054	97	7.38
Nd–Fe–B + 2 wt.% Dy_2O_3	0.74	1098	92	7.23
Nd–Fe–B + 0.6 wt.% Zn + 2 wt.% Dy_2O_3	0.73	1142	91	7.22

Source: Adapted from Z. W. Liu et al., *IEEE Trans. Magn.* 51, 2015, 1–4.

5.5.2 HIGH-VELOCITY COMPACTION

In the last few years, a new powder metallurgy technique, the explosive compaction technique, was suggested as a potential process to consolidate NdFeB powders [159,160]. However, the explosive compaction has to be carried out in an enclosed space and explosive materials such as TNT or hexogen are required to produce the shock wave. A Mach stem hole may form in the central part of the sample [161]. Therefore, it is difficult for the explosive compaction to be applied in large-scale production because of its devastation. Based on our opinion, another powder metallurgy technology, that is, HVC, is also a possible alternative method for the explosive compaction. HVC technology was put forward by the company Höganäs AB in 2001 and has had great development in recent years due to its high efficiency, low cost, and ability to produce bulk materials of high density and good mechanical property [162]. It is also an easy technology for large parts and large-scale production. The densification during HVC is achieved by intensive shock waves, created by a hammer, as in the schematic diagram shown in Figure 5.35. The compaction energy is transferred through the compaction tool to the powders. So far, HVC has been successfully employed for consolidating various kinds of powders, such as iron powders [163], stainless steel powders, titanium powders [164,165], copper powders, and polymers. However, as far as the authors know, the HVC process has never been reported for NdFeB magnets.

Deng et al. [166] used HVC as a consolidation technique for nanocrystalline NdFeB ribbons. The schematic illustration of the HVC process and self-designed facility are shown in Figure 5.35. The hammer is operated by a spring and the impact energy is provided by the elastic potential energy of the compressed spring. The mass of the hammer and its velocity at the moment of impact determine the compaction energy. The four key parameters in the process are the impact energy E, the diameter of the mold Φ, the mold temperature T, and the mass of the powders M. NdFeB (Nd: 26.7 wt.%, FeCoAl: 72.1 ~ 72.5 wt.%, B: 1 ± 0.2 wt.%) powders were consolidated into bulk magnets by HVC with impact energies E = 954 J, 1048 J, and 1284 J. Figure 5.36 shows the typical HVCed magnet and the magnetic hysteresis

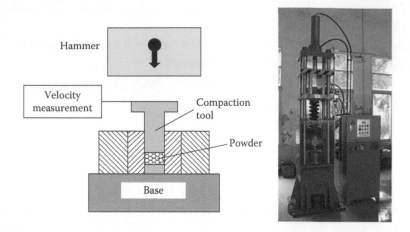

FIGURE 5.35 Schematic diagram of HVC. (Adapted from X. Deng et al., *J. Magn. Magn. Mater.* 390, 2015, 26–30.)

loops for the initial powders and compacted magnets prepared at various impact energies. Compacted green was obtained and a smooth surface was observed. No uncompacted powders were found on the surface. The result indicates that NdFeB powders have good forming ability.

The process parameters, the density, and the magnetic properties of the powders and HVCed magnets are shown in Table 5.6. The density increases slowly with increasing impact energy. The magnetic properties have no significant change for various HVC magnets. Based on powder metallurgy theory, the density variation during a conventional pressing process can be divided into three stages. In the first

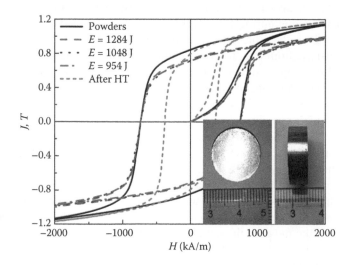

FIGURE 5.36 Hysteresis loops of starting power and HVC at room temperature. (Adapted from X. Deng et al., *J. Magn. Magn. Mater.* 390, 2015, 26–30.)

TABLE 5.6

Processing Parameters (the Impact Energy E and Velocity V), the Density and Magnetic Properties ($_jH_C$, J_r, $(BH)_{max}$ and Magnetization at 6T J_s) for the Initial Powders and HVCed Magnets

Sample	E (J)	V (m/s)	ρ (g/-cm³)	$_jH_C$ (kA/m)	J_r (T)	$(BH)_{max}$ (kJ/m³)	$J_{s,6T}$ (T)
Powders	0	0	7.55	755	0.830	109	1.37
Bulk A	954	5.26	6.51	748	0.730	82	1.19
Bulk B	1048	5.50	6.55	748	0.732	82	1.19
Bulk C	1284	6.07	6.59	748	0.735	82	1.20

Source: Adapted from X. Deng et al., *J. Magn. Magn. Mater.* 390, 2015, 26–30.

stage, the powders rearrange and fill the voids, and the density increases with the increasing compaction pressure linearly. In the second stage, the density has almost no change as compacting pressure increases. In the last stage the compaction pressure exceeds the critical stress and the density increases with the increase of the compacting pressure due to the plastic deformation of the powders. The second stage is manifested for brittle powders. The density remains the same under the impact energy from 954 to 1284 J in this work, indicating the compaction ceased in the second stage and the NdFeB powders were not deformed. Since the relative density has not changed much for three magnets, the saturation magnetization has no significant change for all compacted samples. The three compacted magnets have similar magnetic properties with remanence $J_r = 0.73$ T, intrinsic coercivity $_jH_C = 748$ kA/m, and maximum energy product $(BH)_{max} = 82$ kJ/m³. Compared to the starting powders, the compacted magnets have reduced J_r, resulting from the reduced density. The $_jH_C$ has almost no change, indicating the HVCed magnets maintain the microstructure of the nanocrystalline powders. The compacted magnets with relatively high density can inherit the coercivity of the starting powders.

The magnet prepared at $E = 1284$ J was subjected to HT at 700°C for 50 min. The phase structures for the magnets before and after HT were examined by X-ray diffraction. All XRD peaks (not shown here) are attributed to the tetragonal hard magnetic $Nd_2Fe_{14}B$ phase. Increased intensities and narrowed diffraction peaks for heat treated magnets indicate slight grain growth. The hysteresis loop of the magnet after HT is shown in Figure 5.36 and the magnetic properties of $J_r = 0.75$ T, $_jH_C = 379$ kA/m, and $(BH)_{max} = 76$ kJ/m³ have been obtained. The $_jH_C$ reduced up to 50% and the J_r increased slightly after HT. The reduced $_jH_C$ is believed to result from the growth of the nanocrystalline grains and the change of microstructure. The slightly increased J_r is due to the improved density, similar to the magnets fabricated by explosive compaction.

For the application, another concern about the compacted magnets is their mechanical strength. The compression properties and the density of the HVCed cylindrical magnets before and after HT have been investigated. The density increases from 6.59 to 6.69 g/cm³ after HT, which gives a relative density ρ_r of 88%. The as-HVCed

FIGURE 5.37   SEM images obtained from the directions parallel (a) and perpendicular (b) to the compaction direction for the heat treated magnet. (Adapted from X. Deng et al., *J. Magn. Magn. Mater.* 390, 2015, 26–30.)

magnet has a compression strength of 67.2 MPa. After HT, the strength increased to 311.6 MPa. A four-time increase in compression strength results from the fact that the HT can eliminate the internal stress caused by compaction, slightly improve the density, and reduce the defects in the HVCed magnet. It is worth noting that the strength of the HVCed magnet is comparable or higher than that of conventional bonded NdFeB magnets, normally below 250 MPa [167,168].

The morphologies obtained from the fractures along the directions parallel and perpendicular to the compaction direction for the heat treated magnet are shown in Figure 5.37a and b, respectively. Figure 5.37a shows that powders contact with each other closely due to the deformation caused by HVC. The edges of particles are broken into small pieces filling the voids between particles. Figure 5.37b shows some porosities in the sample, but dense consolidation can be realized. The flaky NdFeB powders stack layer by layer along the compaction direction after HVC.

HVC has been demonstrated as a feasible approach for preparing NdFeB magnets. Besides the simplicity of HVC, this process has some advantages over conventional fabrication methods. Compared with bonding techniques, HVC is a binder free process. Hence, the compacted magnets can be used at higher temperatures. Compared to the sintering process, HVC is able to maintain the nanostructure of the powders.

5.6   ACHIEVING ANISOTROPY FOR NANOCRYSTALLINE NdFeB MAGNETS

To obtain high magnetic properties, it is important to achieve anisotropy in nano-crystalline magnets. The methods for preparing anisotropic nanocrystalline NdFeB magnets are divided into field alignment and plastic deformation. The alignment of the nanocrystalline powders meets some difficulty in the practical process, since the polycrystalline powders are isotropic and anisotropic nanopowders are not easy to prepare. Alternatively, the anisotropic magnets can be prepared by HD, and good magnetic properties together with exceptional corrosion, thermal stability, and fracture toughness can be obtained. It is well known that the grains of HDed NdFeB magnets (by die-upsetting or extruding) are elongated normally to the press direction, and are platelet-shaped and oriented in such a way that the c-axis of the $Nd_2Fe_{14}B$ grain is normal to the platelet. The deformation mechanism, which is not capable of classic plastic flow through the dislocation mechanism, is believed to be a combination of stress-assisted grain growth via mass transport and grain boundary sliding. Up to now, the largest reported value of $(BH)_{max}$ for hot deformed magnets is 433 kJ/m^3, which was achieved in a $Nd_{13.5}(FeCo)_{80}Ga_{0.5}B_6$ composition [169].

5.6.1   Hot Press Followed by HD

HP + HD (die-upset) is a standard process for anisotropic nanocrystalline magnets, which has been well used for MQ3 magnets in the industry. Lin et al. [170] conducted HD on the hot pressed magnets, which was obtained at a temperature range of 580–650°C and a pressure of 200 MPa. The HD was performed at a temperature range of 750–950°C and pressure of 150 MPa. The results showed that HD at 800°C has the best wetting behavior of the Nd-rich phase among the grain boundaries, which results in a higher density and better magnetic properties of the magnet. When the HD temperature is higher than 850°C, the microstructure changes from small flat platelet-shaped grains to large spherical grains; the latter have bad grain alignment. The Nd-rich phase extrusion occurred at the same time, which will led to the reduction of density and magnetic properties, especially low remanence and low mechanical strength due to the reduction of the grain boundary.

The influences of the powder densification temperature on the microstructure and magnetic properties have been investigated by Lipiec and Davies [171]. It was found that, from the viewpoint of the efficiency of the orientation process, there is an optimum temperature for the densification. The results show that for the Magnequench MQP-A powder they used the optimum temperature is 650°C. However, the typical large grains and small grains area can be observed in all temperature range samples. The large grains prefer to grow at the particle boundaries, and the fraction of large grains increases with increasing temperature. The disadvantages of the HP + HD process include that two heating processes promote the extra grain growth and eventually increase production cost.

Lai et al. [172] prepared isotropic magnets from melt-spun powders at different hot pressing temperatures from 550°C to 700°C, then upset them into fully dense anisotropic magnets at the same die-upsetting temperature of 850°C. They found that die-upset magnets had the characteristics of inhomogeneous microstructure, including well-aligned grains structure and nonaligned grains layers transverse to press direction, which was a quasi-periodic layer structure with a total length of 5–15 µm. Nonaligned grains layers were mainly made of large grains and had a higher Nd content. They proposed a new interpretation for the formation of layer structure: the layer structure was correlated to the original ribbon interface which was divided into three types based on the contact forms. Because of the incomplete contact of neighboring ribbons, concentration of stress occurred in the contacted points and the Nd-rich phase was squeezed into interspaces at high temperature under stress.

Due to the release of interfacial energy and the fluidity of enough Nd-rich liquid phases, the nonaligned layers with large grains formed both in the hot compaction and subsequent HD process. The layer structure affected the magnetic properties of die-upset magnets. With increase of the hot pressing temperature, the nonaligned grains layers became thicker, and the magnetic performance of die-upset magnets decreased. They concluded that it would be necessary to reduce the thickness of large grains layers for the preparation of high-performance die-upset magnets.

5.6.2 SPS Followed by HD

SPSed magnets have the advantage of nanocrystalline structure, which is suitable for HD. Recently, various research groups have conducted studies on SPS + HDed nanocrystalline NdFeB magnets. Liu et al. [144] prepared the SPSed NdFeB magnets at $T_{SPS} = 700°C$, $P_{SPS} = 50$ MPa, and $t_{SPS} = 5$ min. The HD was executed at 750°C with a compression ratio of 54%. Figure 5.38a shows the XRD patterns before and after HD, a slight texture has been obtained in the HDed magnet, although the texture is

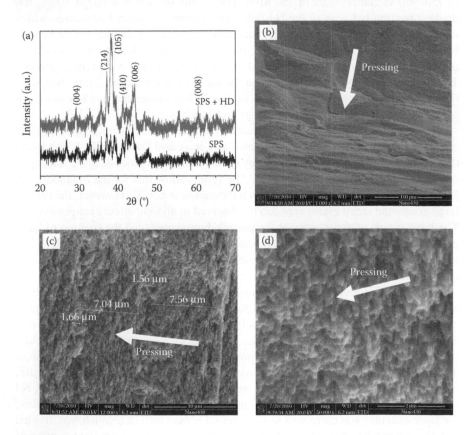

FIGURE 5.38 XRD patterns for SPSed magnet before and after HD (a) and microstructure of HDed magnets (b, c, d). (Adapted from Z. W. Liu et al., *J. Phys. D: Appl. Phys.* 44, 2011, 25003–25011.)

not as strong as the conventional anisotropic sintered NdFeB magnets [173]. Figure 5.38b is a low magnification SEM image of an HDed magnet, showing densified deformed particles and homogeneous structure. No pores or other macro-defects were observed. The high magnification image in Figure 5.38c shows that the two-zone structure still exists in the deformed magnet. Compared with the widths for the SPSed magnet, both the coarse grain zone and fine grain zone are deformed. The width ratio of the coarse grain zone and fine grain zone is almost no change. It can be concluded that the coarse zone area did not grow during HD. Figure 5.38d also confirms that there is no obvious grain growth during HD. In addition, Figure 5.38b–d clearly show that the grains are deformed into longitude shape by HD, which leads to magnetic anisotropy.

Figure 5.39 shows the hysteresis loops for SPS + HDed magnets parallel to the compression direction under various compression rates up to 73%, indicating an obvious magnetic anisotropy behavior [174]. Increasing deformation ratio enhanced the anisotropy, which increased the remanence but reduced the coercivity. As mentioned before, the grain of hot-deforming NdFeB magnets is platelet-shaped and has an orientation with the c-axis of the NdFeB grain (the easy axis) normal to the platelet. Using the isotropic magnetic powders, the isotropic SPSed magnet has magnetic properties of $J_r = 0.82$ T, $_jH_C = 1516$ kA/m, and $(BH)_{max} = 116$ kJ/m^3. After HD, excellent magnetic properties have been obtained at the temperature of 300 K in the magnets along the easy direction, including $J_r = 1.32$ T, $_jH_C = 847$ kA/m, and $(BH)_{max} = 303$ kJ/m^3 (~38 MGOe). Here the $(BH)_{max}$ is almost tripled after HD compared to the SPSed precursor, indicating that the current process provides a good approach to improve the magnetic properties of isotropic nanocrystalline NdFeB alloys.

According to the plastic deformation theory, the inhomogeneous chemical composition, microstructure, temperature distribution, and the friction between punch and sample can induce the nonuniform plastic deformation during HD. Hou et al. [146] investigated the nonuniform deformation in SPS + HDed magnets. According

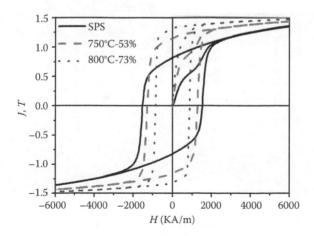

FIGURE 5.39 The hysteresis loops for SPS + HDed magnet parallel (ED) and perpendicular (HD) to the c-axis. (Adapted from Z. W. Liu, *Key Eng. Mater.* 510–511, 2012, 1–8.)

to the difference of stress distribution, the sample under HD can be divided into three zones, the conical contact zone, the central zone, and the vertical edge zone, respectively, indicated by I, II and III in Figure 5.40a. The conical contact zone is the most difficult to be deformed under the intense three-directional compressive stress. The central zone has the most serious deformation since it is far away from the punch in spite of the three-directional compressive stress. The plastic deformability of the vertical edge zone lies between the conical contact zone and the central zone due to two-directional tensile stress and one-directional compressive stress in the process of the deformation. Figure 5.40b–d show the microstructures of the conical contact zone, the central zone, and the vertical edge zone after deformation. For the conical contact zone, some undeformed grains are observed (Figure 5.40b inset), and the direction of oriented growth of $Nd_2Fe_{14}B$ grains is not consistent with each other. In Figure 5.40c, the central zone consists of platelet-shaped grains, and the $Nd_2Fe_{14}B$ grains are well oriented with the c-axis along the pressing direction due to the homogeneous distribution of stress. For the vertical edge zones (Figure 5.40d), although the microstructure also consists of platelet-shaped grains, the direction of

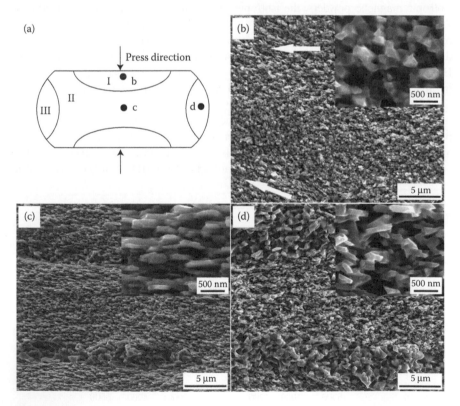

FIGURE 5.40 The nonuniform deformation map of magnet (a) and microstructure of various deformation zones for HDed NdFeB magnet with the compression ratio of 73%: (b) the conical contact zone, (c) the central zone, and (d) the vertical edge zone. (Adapted from Y. H. Hou et al., *Mater. Sci. Eng.* B 178, 2013, 990–997.)

oriented growth of $Nd_2Fe_{14}B$ grains is not perpendicular to the pressing direction. The magnetic properties for three zones are also different. The optimal magnetic properties with $J_r = 1.29$ T, $_jH_C = 995$ kA/m, and $(BH)_{max} = 293$ kJ/m^3 were obtained in the central zone due to the optimal c-axis crystallographic alignment. On the contrary, the relatively low magnetic properties with $J_r = 1.21$ T, $_jH_C = 978$ kA/m, and $(BH)_{max} = 251$ kJ/m^3 were obtained in the conical contact zone.

The magnetic properties at 293 K (20°C) and 393 K (120°C) for the magnets SPSed at 700°C and 800°C and for the SPS + HDed magnet are shown in Table 5.7. As expected, the remanence and coercivity decrease with increase of measured temperature due to the temperature dependences of magnetization and the anisotropy field. At 393 K, the excellent magnetic properties with $(BH)_{max}$ of 91 and 166 kJ/m^3 have been achieved in SPSed magnet and SPS + HDed magnet, respectively, which is very useful for high temperature applications.

The calculated temperature coefficients in the temperature range of 293–393 K for SPSed magnets and SPS + HDed magnet are also shown in Table 5.7. Comparing the data for 700°C and 800°C SPSed magnets, the results indicate that the grain size have important effect on the temperature stability, especially on the temperature coefficient. Since the magnet SPSed at 700°C has smaller and more uniform grain size, it indicates that refining grain size can improve the thermal stability. This conclusion is in good agreement with that obtained early in nanocomposite NdPrFeCoB alloys with various grain sizes [175]. In addition, the anisotropic magnet has similar temperature stability as their SPSed precursor before HD. For conventional anisotropic sintered magnets, based on much coarser grain sizes than in the present case, the published values of temperature coefficient of remanence (α) and temperature coefficient of remanence (β) are approximately −0.13%/K and −0.9%/K, respectively, over the range 0–150°C for a Neomax-35 magnet [176] and approximately −0.12%/K and −0.63%/K, respectively, over the range 20–140°C for another, unspecific commercial NdFeB magnet [177]. The obtained temperature coefficients have smaller absolute values than above and are also smaller than or comparable to those obtained on the NdFeB alloys without or with Dy substitution by other researchers [178]. The results hence indicate high temperature stability can be obtained for SPSed isotropic

TABLE 5.7

Temperature Coefficient of Remanence (α) and Temperature Coefficient of Remanence (β) for SPSed Isotropic and SPS + HDed Anisotropic NdFeB Magnets

Process of Sample	J_r (T)		$_jH_C$ (kA/m)		$(BH)_{max}$ (kJ/m^3)		α (%/K)	β (%/K)
	293 K	393 K	293 K	393 K	293 K	393 K		
800°C SPS	0.81	0.69	1315	650	107	62	−0.152	−0.50
700°C SPS	0.83	0.74	1573	813	121	91	−0.113	−0.48
700°C SPS + 750°C HD	1.15	1.03	1338	563	240	166	−0.112	−0.58

Source: Adapted from Y. H. Hou et al., *Mater. Sci. Eng.* B 178, 2013, 990–997.

or SPS + HDed anisotropic NdFeB magnets made from melt-spun ribbons. The HD process does not have a negative effect on the temperature stability of SPSed magnets.

Besides the magnetic properties, mechanical properties and corrosion resistance properties are also important for the practical applications of NdFeB magnets. Similar to structural materials, microstructure has important effects on the mechanical properties of magnetic materials. Figure 5.41a and b show the microhardness and potentiodynamic polarization curves, respectively, for the SPSed magnets and HDed magnets [146]. The SPSed magnet has the maximum value of microhardness, 740 Hv, due to the fine grain structure. The microhardness of the HDed magnets decreases from 660 to 599 Hv with increasing the compression ratio from 51% to 80%. The increase of mean grain dimensions for the HDed magnets should be responsible for the decrease of microhardness. The results of microhardness evolution for the SPSed and HDed magnets aroused by the microstructure changes are in accordance with the Hall–Petch relationship. Figure 5.41b shows that the corrosion potential of the HDed magnets is more positive than that of the SPSed and conventional sintered magnet. The corrosion potential of HDed magnets with various compression ratios are almost the same due to the same composition and similar microstructure. Conventional sintered magnets, being more prone to electrochemical corrosion, have the lowest corrosion potential. The corrosion current is related to the microstructure, chemical composition, and electric properties of the Nd-rich phase. The corrosion current densities results in Figure 5.41b indicated that the large grain size is beneficial in reducing corrosion current densities.

Recently, Liu et al. [179,180] investigated the relationship between the recoil loops and microstructure for hot deformed NdFeB magnets. Based on melt-spun ribbons, SPS was carried out at various temperatures with a pressure of 50 MPa for 5 min. HD was carried out at 750°C for 20 min under a uniaxial stress of 350 MPa. As expected, the density of SPSed magnets increased with increasing SPS temperature. The magnet SPSed at 700°C has a density of 7.33 g/cm^3, while the density of that SPSed at 650°C is only 7.03 g/cm^3. The low magnification images for these two samples are shown in Figure 5.42a and b. Some pores can be clearly observed for the

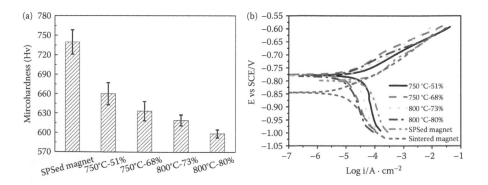

FIGURE 5.41 The microhardness (a) and potentiodynamic polarization curves (b) of the SPSed magnet and HDed magnets. (Adapted from Y. H. Hou et al., *Mater. Sci. Eng. B* 178, 2013, 990–997.)

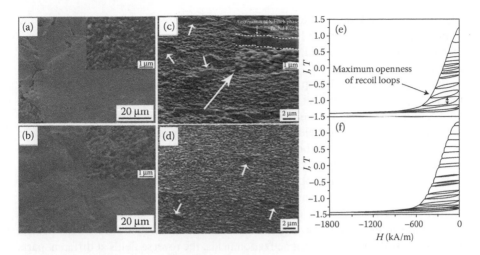

FIGURE 5.42 Microstructures for SPSed magnet prepared at 650°C (a) and 700°C (b), and corresponding HDed magnets (c, d) and recoil loops (e, f). (Adapted from Z. W. Liu et al., *IOP Conf. Series: Mater. Sci. Eng.* 60, 2014, 012013.)

low density magnet (Figure 5.42a). Reduced porosity is evident in Figure 5.42b. The microstructure of the SPSed magnet consists of uniform equiaxed grains (insets in Figures 5.42).

The aforementioned two samples were subjected to HD at 750°C, and the microstructures after HD are shown in Figure 5.42c and d. Platelet-shaped $Nd_2Fe_{14}B$ grains due to deformation are formed. Some strip-shaped grains can be observed in Figure 5.42c, indicated by the arrows. The Fe/Nd atomic ratios detected by EDS indicate that these grains are Nd-rich phase. The existence of porosity in the SPSed precursor is considered to be the main reason for Nd-rich phase aggregation in HDed magnets. It can be assumed that liquid Nd-rich phase aggregated in the pores at high temperature during the early stages of HD, and it was squeezed to form some strip-shaped grains under uniaxial stress during deformation. Comparing Figure 5.42c with Figure 5.42d, higher SPS temperature led to an improved density and less porosity, and therefore, it reduced Nd-rich phase aggregation in SPS + HDed magnets. To confirm the above assumption, a low porosity SPS magnet with high density close to the theoretical value of 7.50 g/cm³ was prepared by optimal process and raw powders. After HD, no Nd-rich phase aggregation was observed, as presented in Figure 5.43a, which is attributed to the elimination of the porosity in the SPSed precursor. Hence, the results demonstrated that the Nd-rich phase aggregation results from the existing porosity in the SPSed precursor. Reducing porosity in the SPS precursor is beneficial to eliminate Nd-rich phase aggregation.

The recoil loops for SPSed and SPS + HDed magnets were obtained. All SPSed magnets had closed loops, similar to the normal single phase NdFeB magnet and RE-rich sintered microcrystalline NdFeB magnets. Interestingly, the recoil loops of hot deformed magnets with the same composition as SPSed ones are open, as shown in Figure 5.42e and f, although there is no soft magnetic phase detected by XRD.

These open recoil loops for Nd-rich NdFeB magnets are interesting and are worthy of investigation. On the other hand, the maximum openness of recoil loops does not take place in the coercive field, which is different from that of nanocomposite NdFeB alloys [181].

To explain the recoil loop characteristics, considering the microstructure and magnetic properties, Liu et al. [180] suggested that the strip-shaped Nd-rich phase with dimension up to several micrometers (Figure 5.42c and d) generates non-negligible local demagnetizing fields. The Nd-rich phase aggregation with various sizes, larger than grains, leads to nonuniform distribution of local demagnetization fields. These demagnetization fields can reduce the local pinning field and lead to the inhomogeneity of magnetic anisotropy. This variation of local magnetic anisotropy in the magnets should be responsible for the formation of open recoil loops. In other words, the open recoil loops arise from the difference in local reversal fields due to nonuniform distribution of large local demagnetization fields. Since there exists nonuniformly distributed demagnetization fields, the reverse fields at different parts of the magnets are different, which eventually results in open recoil loops.

The above explanation has been verified by experiments. As shown in Figure 5.42c and d, with increasing SPS temperature from 650°C to 700°C, Nd-rich phase aggregation was reduced in SPS + HDed magnets. Correspondingly, the openness of recoil loops is also reduced, as presented in Figure 5.42e and f. Furthermore, for the HDed magnets without Nd-rich phase aggregation, the recoil loops were found to be fully closed (Figure 5.43b). These results confirmed that the openness of recoil loops arises from the aggregation of the Nd-rich phase. A more serious distribution of the local demagnetization field contributes to the more open recoil loops. The characteristics of recoil loops and magnetic properties for SPS and HDed magnets are summarized in Table 5.8. The clear relationships between Nd-rich phase aggregation, recoil loops openness, and coercivity have been demonstrated. The distribution of large local demagnetization fields not only leads to reduced coercivity, but also results in open recoil loops.

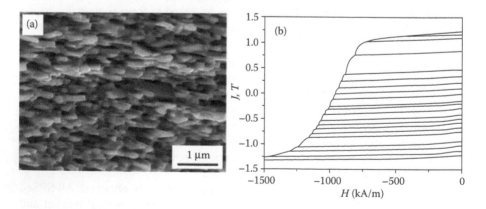

FIGURE 5.43 Microstructure (a) and recoil loops (b) for SPS + HDed magnets prepared with 68% deformation ratio using high density (7.50 g/cm³) SPS precursors. (Adapted from Z. W. Liu et al., *IOP Conf. Series: Mater. Sci. Eng.* 60, 2014, 012013.)

TABLE 5.8

A Summary of the Process, Density, Nd-Rich Aggregation, Recoil Loop Characteristics, and Coercivity for SPS + HDed Magnets

SPS Process	HD Process	SPS Precursor Density (g/cm³)	Nd-rich Phase Aggregation	Recoil Loops Openness	$_jH_C$ (kA/m)
650°C–50 MPa–5 min	750°C–20 min–350 MPa	7.03	Serious	Large	226
700°C–50 MPa–5 min	750°C–20 min–350 MPa	7.33	Medium	Medium	288
Optimal process and raw powders	750°C–20 min–350 MPa	7.50	Not observed	No	995

Source: Adapted from Z. W. Liu et al., *IOP Conf. Series: Mater. Sci. Eng.* 60, 2014, 012013.

The nanocrystallites alloys generally show exchange coupling between grains. Figure 5.44 shows the δM plots for SPSed and SPS + HDed magnets [144]. Positive values of δM were obtained in both magnets, indicating exchange coupling due to nanostructure, although the grain size in the fine grain zone is about 70 nm, as shown early. Interestingly, a higher positive δM peak is found in the HDed magnet, indicating that the deformed grains exchange coupled stronger than densified grains and improved densification is beneficial to the exchange coupling. The results also demonstrated that the HD did not lead to grain growth but enhanced the exchange coupling. The slight reduction of the coercive force of the HDed magnet from starting

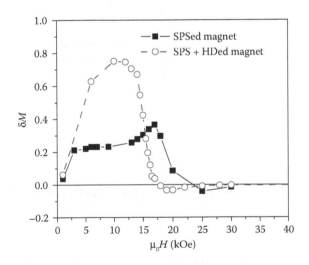

FIGURE 5.44 Henkle plots for SPSed and SPS + HDed NdFeB magnet. (Adapted from Z. W. Liu et al., *J. Phys. D: Appl. Phys.* 44, 2011, 25003–25011.)

powders and SPSed magnet to HDed magnet was, in our opinion, mainly attributed to the exchange coupling, not to the grain growth during the hot pressing process.

5.6.3 HVC FOLLOWED BY HD

Using HVCed NdFeB magnets as the precursors, anisotropic magnets can also be prepared by the follow-on process of HD. Preliminary work has been carried out by Deng et al. [166]. Figure 5.45 shows the XRD patterns for magnets before and after HD. For the HDed sample, the (00l) peaks and the peaks for the plane with direction close or parallel to (00l), such as (105), (314), (006), and (216) peaks, have enhanced intensity, indicating a slight c-axis crystallographic alignment of the magnet.

Figure 5.46 show the results on the anisotropic nanocrystalline NdFeB magnets prepared by the HVC + HD process. The nanocrystalline RE-rich composition with RE content of 29.2 wt.% was used as the starting powders. The as-HVCed sample has magnetic properties of $J_r = 0.64$ T and $_jH_C = 1151$ kA/m. After deformed at 750°C with deformation ratio of 40%, the magnet has demonstrated an obvious magnetic anisotropy, demonstrated by an enhanced remanence (Figure 5.46a). The remanence value along the pressing direction increased from 0.64 to 0.95 T. This anisotropy results from the textured microstructure with elongated grains caused by deformation (Figure 5.46b). The grains have even size distribution with their width of 50–80 nm and their length of 400–800 nm. The $(BH)_{max}$ before and after HD are 65 and 120 kJ/m³, respectively. After deformation, an almost twofold increase in $(BH)_{max}$ has been obtained. At the same time, the coercivity of the deformed magnets can still maintain a high value of 784 kA/m. No larger grains are observed here, because the HVC precursor was obtained at low temperature and is able to maintain

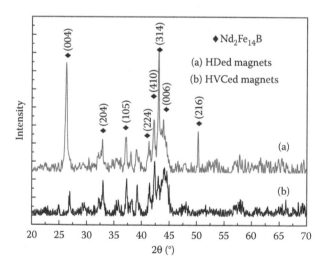

FIGURE 5.45 XRD patterns for HVCed magnets (a) and (b) HDed magnets. (Adapted from X. Deng et al., *J. Magn. Magn. Mater.* 390, 2015, 26–30.)

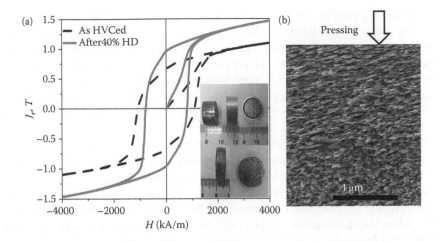

FIGURE 5.46 Magnetic hysteresis loops of HVCed (a) and HVC + HDed NdFeB magnets (inset) and the microstructure of HVC + HDed magnet (b). (Adapted from X. Deng et al., *J. Magn. Magn. Mater.* 390, 2015, 26–30.)

the nanostructure of the powders. It is believed that if the initial powders with better properties are used, much higher performance can be achieved in the HVCed and HVC + HDed magnets. Detailed work is now being undertaken in Liu's group. Nevertheless, this work has clearly demonstrated that HVC can be a good process for preparing an HD precursor. It is worth noting that the existing method for fabricating a nanocrystalline HD precursor is SPS or hot pressing, which requires a high temperature, vacuum environment, expensive facility, and can only produce magnets with limited dimension. On the contrary, HVC has shown promising advantages of low cost, low temperature (LT), and efficiency.

5.6.4 HOT EXTRUSION

Radially oriented NdFeB ring magnets find applications in electric motors where high flux densities are required, such as voice coil motors, brushless motors, alternating current servo motors, etc. These magnets can be produced successfully by two different methods, sintering and hot extrusion [8,182]. For the conventional powder sintering route, the crystal alignment is performed by an applied radially magnetic field during the pressing process of the powder. For the hot extrusion route, a radially oriented magnet is made from fine grained melt-spun powder by backward extrusion, and the crystal alignment along the easy c-axis is perpendicular to the direction of the plastic flow. The magnetic properties of the extruded magnets are independent of ring size unlike sintered magnets [183]. It means that the radially oriented magnets can be produced in a wide range of dimensions by hot extrusion. The hot extrusion method is especially advantageous in the case of the ring magnets with small wall thickness, small diameter, and high length-to-diameter ratio because the magnetic properties of these extruded magnets surpass those of sintered ones.

The starting powders for NdFeB ring magnets can be chosen from the melt-spun powders, mechanical alloying powders, and HDDR powders. The commercial MQ powders are used more frequently. Figure 5.47 shows the sketch for the deformation processes. Before hot extrusion, the powders are compacted into isotropic magnets at high temperature (500–700°C). Then, under the pressure of upper punch, the full-densed isotropic precursor is extruded along the opposite direction and forged into a ring magnet. At last, the back extrusion is employed to produce the anisotropic magnets [8].

During the hot extrusion process, the extrusion parameters (deformation ratio, velocity, and pressure) are critical for the magnetic and mechanical properties of the magnets. The higher deformation velocity needs larger pressure, and to reach a certain deformation ratio, certain deformation velocity and pressure are required. Gruenberger et al. [184] used the commercial MQP-A powder to make the ring magnets. When the deformation velocity was 10^{-3}/s, the required pressure was about 10 MPa, but for the 10^{-1}/s deformation velocity, the required pressure was up to about 100 MPa. However, the pressure also changes with different starting powders. To get the same deformation ratio, and using the same velocity, the pressure for MQP-B powders is six times as large as the pressure of MQP-B powders [183].

The grain alignment is determined by the deformation ratio. To improve the anisotropy of the magnets, it is quite essential to enhance the deformation ratio. On the contrary, the coercivity $_jH_C$ decreases with the increasing deformation ratio. Fuerst's work shows that the coercivity $_jH_C$ decreases from 1.5T for the raw powders to 1T after 0.8 deformation. Thus, it is hard to find the point with large remanence and coercivity [7].

The Nd-rich phase is also a critical effect for the deformation process. When the deformation temperature is high enough, the Nd-rich phase will melt into the liquid

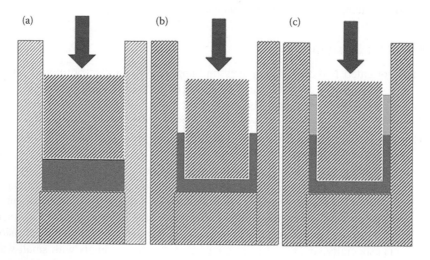

FIGURE 5.47 A sketch of the die (a) for hot pressing of ring-shaped preforms, (b) backward extrusion, and (c) the final stage after pressing out. The left-hand side represents the situation at the start and the right at the final stage of the deformation process.

phase. The liquid phase speeds up the deformation process. Due to the presence of the grain boundary phase with low melting point, the grain boundary sliding and grain rotation are conceivable under the compressive stress. The grains with the c-axis parallel to the press direction have the lowest strain energy. The applied pressure causes only grains with the c-axis parallel to the press direction to be allowed to grow. The deformation and the texture formation of melt-spun NdFeB alloys probably involve grain boundary sliding grain rotation, solution precipitation, and preferential growth of $Nd_2Fe_{14}B$ nanocrystalline grains along the easy growth a-axis [184].

5.7   BONDED NANOCRYSTALLINE NdFeB/ FERRITE COMPOSITE MAGNETS

One of the important types of nanocrystalline permanent magnets is bonded magnets, which have been widely used in the information, automotive, computer, and other fields because of their advantages over sintered magnets such as low cost, large freedom of shape, and simple process [185–187]. Bonded magnets therefore constitute the fastest growing segment of the permanent magnet market.

Generally, NdFeB-based materials have high room temperature magnetic properties but poor temperature stability and corrosion resistance, while hard ferrites, another important type of permanent magnet, have good corrosion resistance with low cost but relatively low magnetic properties. Most importantly, ferrites are one of the few ferromagnets whose anisotropy field increases with increasing temperature. As a result, the coercivity of the ferrite increases with the increasing temperature in a certain temperature range above the room temperature, which leads to its high temperature stability [188]. Therefore, it is expected by mixing the NdFeB and ferrite to produce a hybrid bonded magnet with moderate magnetic properties, high temperature stability, and low cost. The properties of this hybrid magnet can effectively bridge the gap between the sintered ferrite magnets and bonded NdFeB magnets, so as to meet the special magnetic and cost requirements.

Dospial and Plusa [189] prepared five hybrid-bonded magnets with different fractions of nanocrystalline NdFeB (MQP-B) and strontium ferrite powders. They found that the addition of the ferrite causes a decrease in the coercivity, remanence, and maximum energy product, however, these properties values are higher than those obtained by simply combining the hysteresis loops of the two kinds of magnets. Such enhancement of magnetic parameters, particularly the J_r, can be attributed to existence of magnetizing mean field interactions between the magnetic components. The shape of initial magnetization curve results from the magnetization processes of both hard magnetic components, like the domain walls displacement in multidomain NdFeB and ferrite grains at low field, the rotation of magnetization vectors in single-domain grains at higher fields, and the pinning of domain walls at the grain boundaries of both components. The Henkel plots (δM plots) for these hybrid magnets indicates the competition between the exchange coupling and dipolar interaction among the MQP-B grains.

Zhong et al. [190] also fabricated bonded composite magnets by mixing the commercial NdFeB powders (MQP-B) with nominal composition of $Nd_{11.5}Co_{1.9}Fe_{81.1}B_{5.5}$ and isotropic strontium ferrite ($SrFe_{12}O_{19}$) powders (PO2-S). The particle sizes of the

NdFeB powders and $SrFe_{12}O_{19}$ powders are in a range of 45–100 μm and 2–5 μm, respectively. The two types of powders were mixed with 3 wt.% epoxy resin as the binder. The hybrid bonded NdFeB/Sr-ferrite magnets were prepared by warm compaction process. The effects of the process on the magnet density and magnetic properties were investigated. The results showed that magnetic properties including B_r, $_jH_C$, $(BH)_{max}$, and density ρ of the composite bonded magnets first increase and then decrease with increasing warm compaction time and temperature. The optimized hybrid magnet samples were obtained by the optimal process with a compaction pressure of 1100 MPa, a molding temperature of 90°C, and a holding time of 10 min. The curing was carried out at 180°C for 2 h with maleic anhydride as the curing agent. The second quadrant demagnetization curves of the hybrid magnets with different NdFeB contents are displayed in Figure 5.48a. The demagnetization curves of the bonded strontium ferrite and the bonded lean-neo magnets have relatively good loop squareness. With increasing content of lean-neo added into strontium ferrite, the demagnetization curves of bonded hybrid magnets gradually become smooth and flat. In particular, the demagnetization curve of the bonded magnet with 40 wt. % lean-neo is almost a straight line.

Figure 5.48b shows the influence of the NdFeB powder content on the magnetic properties of the hybrid NdFeB/Sr-ferrite magnets (X. C. Zhong, Z. W. Liu, unpublished work). Since the samples were prepared by warm flow compaction without magnetic field, the particles of ferrite powder were not aligned during molding, resulting in isotropic magnets. The intrinsic coercivity ($_jH_C$), remanence (B_r), and maximum energy product ($(BH)_{max}$) significantly increase with increasing NdFeB content. However, the magnetic properties of hybrid magnets do not meet the "dilution law." The coercivity of NdFeB/Sr-ferrite hybrid magnets with 80–90 wt.% NdFeB is higher than that of pure bonded NdFeB magnet, indicating a coercivity enhancement effect. The bonded magnet with 20 wt.% Sr-ferrite shows high magnetic properties of B_r = 5.3 kG, $_jH_C$ = 8.1 kOe, and $(BH)_{max}$ = 5.5 MGOe. This phenomenon can be explained by the exchange effect between the two distinctly different magnetic powders [189]. The added strontium ferrite into bonded NdFeB magnet can effectively fill the porosity among the fish-scale flake NdFeB powders; thereby it can effectively

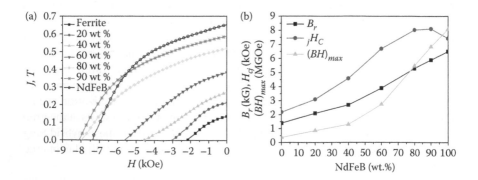

FIGURE 5.48 Second quadrant demagnetization curves (a) and magnetic properties (b) of bonded hybrid magnets with different NdFeB contents.

FIGURE 5.49 SEM images of the fracture for the bonded hybrid magnets with different NdFeB contents (a) 60 wt.%; (b) 100 wt.%.

transfer the pressure during the molding process. By this way, the density of the magnet can be improved and the magnetic powder fragmentation can be reduced. This will enhance the long-range magnetostatic exchange interaction between the two different magnetic powders in the hybrid magnets to some extent.

The fracture SEM images of the hybrid magnets (60 wt.% NdFeB + 40 wt.% Sr-ferrite) and the bonded NdFeB magnets are displayed in Figure 5.49. It is clear that the porosity in the hybrid magnet is lower than that in the bonded NdFeB magnet. The ribbon shaped NdFeB powders in the bonded magnet without ferrite addition were badly damaged. The experiment demonstrated that, by adding a small granular strontium ferrite, the relative density of bonded hybrid magnets can be increased. The bonded hybrid magnets with the content of 10 wt.% strontium ferrite have the highest density, and the relative density reaches 95.7%, as listed in Table 5.9.

As we know, the nonmagnetic binder and voids will increase the demagnetizing field and reduce the coercivity of the magnets. Hence, the magnet consisting of mixed ferrite and NdFeB powders is expected to benefit the coercivity. On the other hand, the strontium ferrite powders added in the magnet can separate the NdFeB powders. The effective contact between the NdFeB powders can reduce the long-range magnetostatic interaction in NdFeB powders. Theoretical calculations showed

TABLE 5.9

Experimental Density, Theoretical Density, and Relative Density of the Bonded Hybrid Magnets Prepared with Different Lean-Neo Contents

NdFeB Content (wt.%)	0	20	40	60	80	90	100
Experimental density (g/cm^3)	3.64	4.17	4.39	4.99	5.48	5.69	5.85
Theoretical density (g/cm^3)	4.49	4.75	5.04	5.37	5.74	5.95	6.17
Relative density (%)	80.9	87.7	87.14	92.9	95.4	95.7	94.7

that the long-range magnetostatic intergranular interactions make the coercivity of the ideal directional grain lower than that of the isolated grain by 20% [191]. Since the coercivity of the strontium ferrite magnetic powder is much lower than that of the NdFeB powder, the long-range magnetostatic exchange interaction between NdFeB and strontium ferrite magnetic powders is weaker than that between NdFeB powders. In other words, the addition of strontium ferrite can reduce the long-range magnetostatic interactions among magnetic powders, which is also beneficial for the coercivity. In addition, the crush of the NdFeB powders and the destruction of the magnetic grains may produce the nucleation centers of the reverse magnetization domain, which may also reduce the coercivity of the magnet. The above analysis explains why the coercivity of the bonded hybrid magnets with the contents of 10 and 20 wt.% Sr-ferrite (90 and 80 wt.% NdFeB) is slightly higher than that of the bond NdFeB magnet.

The thermal stability of the hybrid hard magnets was also investigated. Figure 5.50 shows the temperature coefficient of coercivity β, defined as

$$\beta(300\,K \sim T) = [{}_jH_C(T) - {}_jH_C(300\,K)]/[{}_jH_C(300\,K) \times (T - 300)] \times 100\%$$

for the bonded hybrid magnets with different NdFeB contents measured in the temperature range of 300–375 K. The value of β increases gradually from negative (–0.35%/°C) to positive (0.11%/°C) with the increase of strontium ferrite content. When the content of the ferrite is more than 40 wt.%, the trend of the compensation effect of the coercivity temperature coefficient is accelerated. As soon as the addition of the strontium ferrite reaches about 75 wt.%, the coercivity temperature coefficient becomes positive. This phenomenon can be well explained by the difference in the temperature dependences of anisotropy for ferrite and NdFeB alloy. In particular, when the strontium ferrite content reaches 40 wt.%, the volume fraction of the

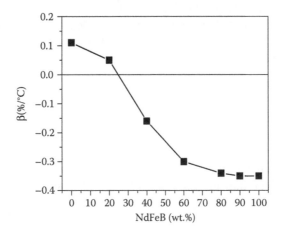

FIGURE 5.50 Temperature coefficient of coercivity (300–375 K) for the bonded hybrid magnets with different NdFeB contents.

low-density strontium ferrite in the bonded hybrid magnets is more than 50 vol.%. The strontium ferrite has been fully dispersed in the bonded hybrid magnets, which dominates the coercivity of the bonded hybrid magnets.

A bonded NdFeB/Sr-ferrite hybrid magnet contains a nonmagnetic binder and two different types of magnetic powders, which makes the magnetic domain structure of the magnet complicated. The microstructure of magnets, such as grain dimension, orientation, and structural defects, can affect magnetostatic interaction. Earlier, Schneider and Schmidt [192] reported the coercivity enhancement effect in the bonded NdFeB/strontium ferrite hybrid magnet, but no interpretation was given. However, Hua et al. [193] found that the tested coercivities of the bonded NdFeB/Sr-ferrite hybrid magnet are lower than the calculated values with increasing the NdFeB content, which showed a coercivity weakening phenomenon, but the effect was not explained, either. Wang et al. [188] investigated the magnetic properties of hybrid polymer bonded NdFeB/Sr-ferrite magnets and found that there existed the remanence enhanced effect when the NdFeB content was below 50 vol.%. This phenomenon was explained by exchange effect between the two distinctly different magnetic materials.

The magnetic interactions in the magnets can be evaluated by the Henkel plots [194]. The $M_d(H)$ plots, defined previously, of the bonded NdFeB/Sr-ferrite hybrid magnets with different lean-neo contents are displayed in Figure 5.51. The diagonal solid line corresponds to the Wohlfarth line. The $M_d(H)$ values above this line indicate that there is strong exchange coupling interaction between adjacent grains in the bonded magnets and the exchange coupling interaction is dominant. Otherwise, the $M_d(H)$ values below this line are indicative of magnetizing interactions, which suggests that the magnetostatic interaction mainly exists among the grains of the magnets. Figure 5.51 shows that the Henkel plots of bonded NdFeB/Sr-ferrite hybrid magnets have a typical S-shape curve, which can usually be observed in

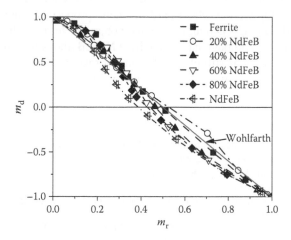

FIGURE 5.51 Henkel plots of bonded NdFeB/Sr-ferrite hybrid magnets with different NdFeB contents.

exchange-coupled magnets [195] and is also related to the dilution effect of strontium ferrite and the nonmagnetic binder. The bonded ferrite magnet showed strong magnetizing interactions at low fields. For bonded NdFeB magnets, the $M_d(H)$ values are nearly all located below the Wolhfarth line, indicating almost no exchange coupling between ferrite grains, while the intergranular exchange coupling exists in the NdFeB contained magnets. The destruction of magnetic powder and the increase of the demagnetizing field caused by the low relative density have a certain contribution to the enhancement of the magnetostatic exchange interaction. On the other hand, the exchange coupling is a short-range interaction, which mainly exists between adjacent NdFeB nanocrystals. The binder is only distributed among the magnetic powder of the hybrid magnet, apparently, which has little effect on the exchange coupling interaction.

The current investigations on the hybrid magnets have demonstrated the possibility of fabricating a permanent magnet with moderate magnetic properties and super thermal stability.

5.8   SUMMARY

Nanocrystalline NdFeB-based permanent magnets show significant advantages because of their reduced grain size, enhanced remanence, lowered RE content, and potential high coercivity. During the last 20 years, substantial research has aimed at improving their properties and reducing the cost based on the compositional optimization. To make full use of these magnets, process innovation is also undertaken. New processes for synthesizing nano powders, densifying nanocrystalline powders, producing magnetic anisotropy, and tailoring the properties combination are currently under development. Besides that described in this chapter, the progress in nanocrystalline NdFeB magnets embraces many other achievements and is frequently renewed.

Up to now, there are few prospects to discover a new compound as a substitution for the NdFeB alloy with excellent hard magnetic properties. Improvement of the well-known NdFeB magnets remains the main objective for both the scientific community and industry. Although nanocrystalline NdFeB alloys were discovered more than 20 years ago, there is still plenty of room for research and development of this type of material to meet various specific requirements for magnetic devices.

REFERENCES

1. Y. Matsuura, Recent development of Nd-Fe-B sintered magnets and their applications, *J. Magn. Magn. Mater.* 303, 2006, 344–347.
2. V. Panchanathan, Magnequench magnets status overview, *J. Mater. Eng. Perform.* 4, 1995, 423–434.
3. J. D. Livingston, Magnetic domains in sintered Fe-Nd-B magnets, *J. Appl. Phys.* 57, 1985, 4137–4139.
4. A. Manaf, R. A. Buckley, H. A. Davies, and M. Leonowicz, Enhanced magnetic properties in rapidly solidified Nd-Fe-B based alloys, *J. Mag. Mag. Mater.* 101, 1991, 360.
5. R. H. Richman, Permanent-magnet materials: Research directions and opportunities, *J. Electronic Mater.* 26, 1997, 415–422.

6. D. Lee, S. Bauer, A. Higgins, C. Chen, S. Liu, M. Q. Huang, Y. G. Peng, and D. E. Laughlin, Bulk anisotropic composite rare earth magnets, *J. Appl. Phys.* 99, 2006, 08B516.

7. C. Fuerst and E. Brewer, High remanence rapidly solidified NdFeB: Die upset magnets, *J. Appl. Phys.* 73, 1993, 5751.

8. D. Hinz, A. Kirchner, D. Brown, B. M. Ma, and O. Gutfleisch, Near net shape production of radially oriented NdFeB ring magnets by backward extrusion, *J. Mater. Proc. Technol.* 135, 2003, 358.

9. E. C. Stoner and E. P. Wohlfarth, A mechanism of magnetic hysteresis in heterogeneous alloys, *Phil. Trans. Roy. Soc.* 240, 1948, 599.

10. H. Sepehri-Amin, T. Ohkubo, M. Gruber, T. Schrefl, and K. Hono, Micromagnetic simulations on the grain size dependence of coercivity in anisotropic Nd–Fe–B sintered magnets, *Scripta Mater.* 89, 2014, 29.

11. T. Schrefl, J. Fidler, and H. Kronmuller, Remanence and coercivity in isotropic nanocrystalline permanent magnets, *Phys. Rev. B Condens. Matter.* 49, 1994, 6100–6110.

12. Q. Chen, B. M. Ma, B. Lu, M. Q. Huang, and D. E. Laughlin, A study on the exchange coupling of NdFeB-type nanocomposites using henkel plots, *J. Appl. Phys.* 85, 1999, 5917–5919.

13. H. Chiriac, M. Marinescu, P. Tiberto, and F. Vinai, Reversible magnetization behavior and exchange coupling in two-phase NdFeB melt spun ribbons, *Mater. Sci. Eng. A* 304, 2001, 957–960.

14. O. Gutfleisch, Controlling the properties of high energy density permanent magnetic materials by different processing routes, *J. Phys. D: Appl. Phys.* 33, 2000, R157–R172.

15. J. Bauer, M. Seeger, A. Zern, and H. Kronmuller, Nanocrystalline FeNdB permanent magnets with enhanced remanence, *J. Appl. Phys.* 80, 1996, 1667.

16. D. Goll, M. Seeger, and H. Kronmüller, Magnetic and microstructural properties of nanocrystalline exchange coupled PrFeB permanent magnets, *J. Magn. Magn. Mater.* 185, 1998, 49.

17. R. W. McCallum, A. M. Kadim, G. B. Clement, and J. E. Keem, High performance isotropic permanent magnet based on Nd-Fe-B, *J. Appl. Phys.* 61, 1987, 3577.

18. G. B. Clemente, J. E. Keem, and J. P. Bradley, The microstructural and compositional influence upon HIREM behavior in $Nd_2Fe_{14}B$, *J. Appl. Phys.* 64, 1988, 5299.

19. A. Manaf, M. Leonowicz, H. A. Davies, and R. A. Buckley, Magnequench magnets status overview, *J. Appl. Phys.* 70, 1991, 6366.

20. A. Manaf, M. Loeonowicz, H. A. Davies, and R. A. Buckley, Nanocrystalline Fe-Nd-B type permanent magnet materials with enhanced remanence, *Mater. Lett.* 13, 1992, 194.

21. G. C. Hadjipanayis, Nanophase hard magnets, *J. Magn. Magn. Mater.* 200, 1999, 373–391.

22. R. Fischer, T. Schrefl, H. Kronmüller, and J. Fidler, Phase distribution and computed magnetic properties of high-remanent composite magnets, *J. Magn. Magn. Mater.* 150, 1995, 329–344.

23. E. F. Kneller and R. Hawig, The exchange-spring magnet: A new material principle for permanent magnets, *IEEE Trans. Magn.* 27, 1991, 3588–3600.

24. A. Manaf, M. A. Al-Khafaji, P. Z. Zhang, H. A. Davies, R. A. Buckley, and W. M. Rainforth, Microstructure analysis of nanocrystalline NdFeB ribbons with enhanced hard magnetic properties, *J. Magn. Magn. Mater.* 128, 1993, 307–312.

25. W. Gong, G. C. Hadjipanayis, and R. I. Krause, Mechanically alloyed nanocomposite magnets, *J. Appl. Phys.* 75, 1994, 6649–6651.

26. L. Withanawasam, A. S. Murthy, and G. C. Hadjipanayis, Hysteresis behavior and microstructure of exchange coupled R2Fe14B/α-Fe magnets, *IEEE Trans. Magn.* 31, 1995, 3608–3610.

27. R. Coehoorn, D. B. de Mooij, and C. de Waard, Melt spun permanent magnet materials containing Fe_3B as the main phase, *J. Mag. Mag. Mater.* 80, 1989, 101.

28. A. Manaf, R. A. Buckley, and H. A. Davies, New nanocrystalline high-remanence Nd-Fe-B alloys by rapid solidification, *J. Mag. Mag. Mater.* 128, 1993, 302.

29. A. Zern, M. Seeger, J. Bauer, and H. Kronmuller, Microstructural investigations of exchange coupled and decoupled nanocrystalline NdFeB permanent magnets, *J. Mag. Mag. Mater.* 184, 1998, 89.

30. H. Kronmüller, R. Fischer, M. Seeger, and A. Zern, Micromagnetism and microstructure of hard magnetic materials, *J. Phys. D: Appl. Phys.* 29, 1996, 2274–2283.

31. H. Kronmuller and D. Goll, Micromagnetic analysis of nucleation-hardened nanocrystalline PrFeB magnets, *Scr. Mater.* 47, 2002, 551–556.

32. B. G. Shen, L. Y. Yang, H. G. Guo, and J. G. Zhao, Magnetic properties and crystallization of amorphous Fe-Nd-B alloys at constant Nd concentration, *J. Appl. Phys.* 75, 1994, 6312–6314.

33. H. Davies, *What Next in Nanocomposite Magnets? Permanent Magnets: Materials and Application*, One-day Seminar, Birmingham, UK, 2002.

34. Z. W. Liu, The sub-ambient, ambient and elevated temperature magnetic properties of nanophase Nd/Pr-Fe-B based alloys, PhD thesis, University of Sheffield, 2004.

35. Z. W. Liu and H. A. Davies, The practical limits for enhancing magnetic property combinations for bulk nanocrystalline NdFeB alloys through Pr, Co and Dy substitutions, *J. Magn. Magn. Mater.* 313, 2007, 337–341.

36. H. A. Davies, A. Manaf, M. Leonowicz, P. Z. Zhang, S. J. Dobson, and R. A. Buckley, Nanocrystalline structures and the enhancement of remanence and energy product in melt spun iron-rare earth-boron alloys for permanent magnets, *Nanostruct. Mater.* 2, 1993, 197.

37. S. Huo and H. A. Davies, *Proc. 8th Symp. on Magnetic Anisotropy and Coercivity in RE-TM Alloys*, C. A. F. Manwaring (ed.), University of Birmingham, UK, 1994, p. 155.

38. Z. W. Liu and H. A. Davies, Elevated temperature study of nanocrystalline (Nd/Pr)–Fe–B hard magnetic alloys with Co and Dy additions, *J. Magn. Magn. Mater.* 290–291, 2005, 1230–1233.

39. H. A. Davies and Z. W. Liu, The influence of processing, composition and temperature on the magnetic characteristics of nanophase RE–Fe–B alloys, *J. Magn. Magn. Mater.* 294, 2005, 213–225.

40. Z. W. Liu and H. A. Davies, Irreversible magnetic losses for melt-spun nanocrystalline Nd/Pr-(Dy)-Fe/Co-B ribbons, *J. Phys. D: Appl. Phys.* 40, 2007, 315–319.

41. Z. W. Liu and H. A. Davies, Intergranular exchange interaction in nanocrystalline hard magnetic rare earth-iron-boron-based melt-spun alloy ribbons, *J. Phys. D: Appl. Phys.* 42, 2009, 145006.

42. J. Bauer, M. Seeger, and H. Kronmuller, Magnetic properties and microstructural analysis of rapidly quenched FeNdBGaNb permanent magnets, *J. Magn. Magn. Mater.* 139, 1995, 323.

43. Z. W. Liu and H. A. Davies, Influence of Co substitution for Fe on the magnetic properties of nanocrystalline (Nd,Pr)-Fe-B based alloys, *J. Phys. D: Appl. Phys.* 39, 2006, 2647–2653.

44. X. Y. Xiong, Y. Q. Wu, and K. Hono, The effect of the addition of M element (M = Zr, Nb, V, Si and Ga) on the microstructural and magnetic properties of nanocomposite alpha-Fe/Nd2Fe14B and Fe3B/Nd2Fe14B magnets, in: *Proc. 17th Int. Workshop on Rare Earth Magnets and Their Applications*, G. C. Hadjipanayis and M. J. Bonder (eds.), Delaware, USA, 2002, p. 796.

45. Z. Wang, M. Zhang, S. Zhou, Y. Qiao, and R. Wang, Phase transformations and magnetic properties of melt-spun $Pr_7Fe_{88}B_5$ ribbons during annealing, *J. Alloy. Comp.* 309, 2000, 212.

46. G. Mendoza-Suarez, H. A. Davies, and J. I. Escalante-Garcia, jHc, Jr and (BH) max relationship in PrFeB melt spun alloys, *J. Magn. Magn. Mater.* 218, 2000, 97.

47. H. A. Davies, I. Ahmad, and R. A. Buckley, Processing and properties of nanocrystalline materials, in: *Processing and Properties of Nanocrystalline Materials*, C. Suryanarayana (ed.), TMS, Cleveland, Ohio, 1995, p. 441.

48. P. Pawlik, K. Pawlik, H. A. Davies, and O. Kaszuwara, Nanocrystalline (Pr,Dy)-(Fe,Co)-Zr-Ti-B magnets produced directly by rapid solidification, *J. Phys. Confer.* 1, 2009, 012060.

49. A. Manaf, P. Z. Zhang, I. Ahmad, H. A. Davies, and R. A. Buckley, Magnetic properties and microstructural characterization of isotropic nanocrystalline Fe-Nd-B based alloys, *IEEE Trans. Magn.* 29, 1993, 2866.

50. E. Burzo, Permanent magnets based on R-Fe-B and R-Fe-C alloys, *Rep. Prog. Phys.* 61, 1998, 1099–1266.

51. Z. W. Liu, Y. Liu, P. K. Deheri, R. V. Ramanujan, and H. A. Davies, Improving permanent magnetic properties of rapidly solidified nanophase RE-TM-B alloys by compositional modification, *J. Magn. Magn. Mater.* 321, 2009, 2290–2295.

52. G. Mendoza-Suárez and H. A. Davies, The coercivities of nanophase melt-spun PrFeB alloys, *J. Alloys Comp.* 281, 1998, 17–22.

53. Z. M. Chen, H. Okumura, and G. C. Hadjipanayis, Enhancement of magnetic properties of nanocomposite $Pr_2Fe_{14}B/\alpha$-Fe magnets by small substitution of Dy for Pr, *J. Appl. Phys.* 89, 2001, 2299.

54. W. Chen, R. W. Gao, L. M. Liu, M. G. Zhu, G. B. Han, H. Q. Liu, and W. Li, Effective anisotropy, exchange coupling length and coercivity in $Nd_{8-x}R_xFe_{87.5}B_{4.5}$ (R = Dy, Sm, x = 0–0.6) nanocomposite, *Mater. Sci. Eng. B* 110, 2004, 107.

55. M. Zhang, Z. D. Zhang, X. K. Sun, W. Liu, D. Y. Geng, X. M. Jin, C. Y. You, and X. G. Zhao, Remanent enhancement of nanocomposite $(Nd,Sm)_2Fe_{14}B/\alpha$-Fe magnets, *J. Alloys Comp.* 372, 2004, 267.

56. S. Yang, X. P. Song, B. X. Gu, and Y. W. Du, Enhancement of the exchange coupling interaction of nanocomposite $Nd_2Fe_{14}B/\alpha$-Fe magnets by a small amount of Sm substitution for Nd, *J. Alloys Comp.* 394, 2005, 1–4.

57. E. P. Wohlfarth and K. H. J. Buschow, *Ferromagnetic Materials: A Handbook on the Properties of Magnetically Ordered Substances*, Vol. 4, Elsevier, Amsterdam, 1988, p. 20.

58. S. Hirosawa, Y. Matsuura, H. Yamamoto, S. Fujimura, M. Sagawa, and H. Yamauchi, Magnetization and magnetic anisotropy of $R_2Fe_{14}B$ measured on single crystals, *J. Appl. Phys.* 59, 1986, 873–879.

59. Z. W. Liu, D. Y. Qian, and D. C. Zeng, Reducing Dy content by Y substitution in nanocomposite NdFeB alloys with enhanced magnetic properties and thermal stability, *IEEE Trans Magn.* 48, 2012, 2797–2799.

60. M. G. Zhu, W. Li, J. D. Wang, L. Y. Zheng, Y. F. Li, K. Zhang, H. B. Feng, and T. Liu, Influence of Ce content on the rectangularity of demagnetization curves and magnetic properties of Re-Fe-B magnets sintered by double main phase alloy method, *IEEE Trans. Magn.* 50, 2014, 1000104.

61. M. Hussain, J. Liu, L. Z. Zhao, X. C. Zhong, and Z. W. Liu, Composition related magnetic properties and coercivity mechanism for melt spun $[(La_{0.5}Ce_{0.5})_{1-x}RE_x]_{10}Fe_{84}B_6$ (RE = Nd or Dy) nanocomposite alloys, *J. Magn. Magn. Mater.* 399, 2016, 26–31.

62. R. Yapp and H. A. Davies, The effect of cobalt substitution on the magnetic properties of nanostructured NdFeB based melt spun ribbons, in: *Proc. 14th Int. Workshop on Rare-Earth Magnets and their Applications*, F. R. Missell et al. (eds.), World Scientific, Singapore, 1996, p. 556.

63. C. D. Fuerst, J. F. Herbst, and E. A. Alson, Magnetic properties of $Nd_2(Co_xFe_{1-x})_{14}B$ alloys, *J. Mag. Mag. Mater.* 54/57, 1986, 567.

64. J. Wecker and L. Schultz, Beneficial effect of Co substitution on the magnetic properties of rapidly quenched Nd-Fe-B, *Appl. Phys. Lett.* 51, 1987, 697.

65. C. D. Fuerst and J. F. Herbst, Hard magnetic properties of Nd-Co-B materials, *J. Appl. Phys.* 63, 1988, 3324.

66. A. Kojima, A. Makino, and A. Inoue, Effect of Co addition on the magnetic properties of nanocrystalline Fe-rich Fe-Nb-(Nd,Pr)-B alloys produced by crystallization of an amorphous phase, *Scr. Mater.* 44, 2001, 1383.

67. J. Wecker and L. Schultz, Magnetic hardening of Pr-Fe-Co-B alloys by rapid quenching, *Appl. Phys. Lett.* 54, 1989, 393.

68. H. A. Davies, C. L. Harland, R. J. I. Betancourt, and S. G. Mendoza, Praseodymium and neodymium-based nanocrystalline hard magnetic alloys, *Mat. Res. Soc. Symp. Proc.* 577, 1999, 27.

69. M. Q. Huang, E. B. Boltich, W. E. Wallace, and E. Oswald, Magnetic characteristics of $R_2(Fe,Co)_{14}B$ systems (R = Y, Nd, and Gd), *J. Magn. Magn. Mater.* 60, 1986, 270–274.

70. F. Bolzoni, J. M. D. Coey, J. Gavigan, D. Givord, O. Moze, L. Pareti, and T. Viadieu, Magnetic properties of $Pr_2(Fe_{1-x}Co_x)_{14}B$ compounds, *J. Magn. Magn. Mater.* 65, 1987, 123–127.

71. A. Melsheimer, M. Seeger, and H. Kronmüller, Influence of Co substitution in exchange coupled NdFeB nanocrystalline permanent magnets, *J. Magn. Magn. Mater.* 202, 1999, 458–464.

72. W. C. Chang, D. Y. Chiou, B. M. Ma, and C. O. Bounds, High performance α-Fe/R2Fe14B-type nanocomposites with nominal compositions of $(Nd,La)_{9.5}Fe_{78-x}Co_x$-$Cr_2B_{10.5}$ (x = 0–10), *J. Magn. Magn. Mater.* 189, 1998, 55–61.

73. Z. M. Chen, Y. Zhang, Y. Q. Ding, G. C. Hadjipanayis, Q. Chen, and B. M. Ma, Studies on magnetic properties and microstructure of melt-spun nanocomposite $R_8(Fe,Co,Nb)_{86}B_6$ (R = Nd, Pr) magnets, *J. Magn. Magn. Mater.* 195, 1999, 420–427.

74. R. Zhang, Y. Liu, J. W. Ye, W. F. Yang, Y. L. Ma, and S. J. Gao, Effect of Nb substitution on the temperature characteristics and microstructures of rapid-quenched NdFeB alloy, *J. Alloys Comp.* 427, 2007, 78.

75. R. J. I. Betancourt and H. A. Davies, High coercivity Zr and Co substituted (Nd-Pr)-Fe-B nanophase hard magnetic alloys, *IEEE Trans. Magn.* 37, 2001, 2480.

76. C. Wang, M. Yan, and W. Y. Zhang, Effects of Nb and Zr additions on crystallization behavior, microstructure and magnetic properties of melt-spun $(Nd,Pr)_2Fe_{14}B/\alpha$-Fe alloys, *J. Magn. Magn. Mater.* 306, 2006, 195.

77. X. Q. Bao, Y. Qiao, X. X. Gao, J. Zhu, and S. Z. Zhou, Microstructure refinement and magnetic properties enhancement of nanocrystalline $Nd_{12.3}Fe_{81.7}B_{6.0}$ ribbons by addition of Zr, *Sci. China Ser. E: Tech. Sci.* 52, 2009, 1891–1896.

78. C. Wang, Z. M. Guo, Y. L. Sui, X. Q. Bao, and Z. A. Zhen, Effect of titanium substitution on magnetic properties and microstructure of nanocrystalline monophase Nd-Fe-B Magnets, *J. Nanomater.* 2012, 2012, 425028.

79. R. Zhang, Y. Liu, J. Li, S. Gao, and M. Tu, Effect of Ti&C substitution on the magnetic properties and microstructures of rapidly-quenched NdFeB alloy, *Mater. Charact.* 59, 2008, 642–646.

80. V. Bekker, K. Seemann, and H. Leiste, Development and optimisation of thin soft ferromagnetic Fe-Co-Ta-N and Fe-Co-Al-N films with in-plane uniaxial anisotropy for HF applications, *J. Magn. Magn. Mater.* 296, 2006, 37.

81. H. Chiriac and N. Lupu, Bulk amorphous magnetic materials, *Phys. B* 299, 2001, 293.

82. T. S. Chin, S. H. Huang, and J. M. Yau, Enhanced thermal stability of sintered (Nd,Dy) (Fe,Co)B magnets by the addition of Ta or Ti, *IEEE Trans. Magn.* 29, 1993, 2791.

83. M. Leonowicz, W. Kaszuwara, S. Wojciechowski, and H. A. Davies, Irreversible losses of magnetisation in Fe-Nd-B type magnets, *J. Magn. Magn. Mater.* 157–158, 1996, 45–46.

84. N. Yoshikawa, Y. Kasai, T. Watanabe, S. Shibata, V. Panchanathan, and J. J. Croat, Effect of additive elements on magnetic properties and irreversible loss of hot-worked Nd-Fe-Co-B magnets, *J. Appl. Phys.* 69, 1991, 6049–6051.

85. H. Sakamoto, M. Fujikura, and T. Mukai, Cu-added Nd-Fe-B anisotropic powder for permanent magnet use, *J. Appl. Phys.* 69, 1991, 5832–5834.
86. S. Pandian, V. Chandrasekaran, K. J. L. Iyer, and K. V. S. Rama Rao, Microstructural and magnetic studies on p/m processed $(Nd_{14.9}Dy_{1.9})(Fe_{65.0}Co_{8.0}Cu_{1.0}Ga_{1.0}Nb_{0.7})$ $B_{7.5}$ alloy, *J. Mater. Sci.* 36, 2001, 5903–5907.
87. R. E. Fernow and W. Ervens, *Thermal properties of plastic-bonded NdFeB-magnets*, in: CEAM3 Topical Meeting Bonded Rare Earth Magnets, Barcelona, 1992, pp. 23–25.
88. Y. Kanai, S. Hayashida, H. Fukunaga, and F. Yamashita, *IEEE Trans. Magn.* 35, 1999, 3292–3294.
89. E. Potenziani, H. A. Leupold, J. P. Clarke, and A. Tauber, The temperature dependence of the magnetic properties of commercial magnets of greater than 25 MGOe energy product, *J. Appl. Phys.* 57, 1985, 4152–4154.
90. K. S. V. L. Narasimhan, Iron-based rare-earth magnets (invited), *J. Appl. Phys.* 57, 1985, 4081–4085.
91. W. Gong, Q. Li, and L. Yin, Temperature feature of NdDyFeB magnets with ultrahigh coercivity, *J. Appl. Phys.* 69, 1991, 5512–5514.
92. D. R. Gauder, M. H. Froning, R. J. White, and A. E. Ray, Elevated temperature study of Nd-Fe-B-based magnets with cobalt and dysprosium additions, *J. Appl. Phys.* 63, 1988, 3522–3524.
93. M. Endoh, M. Tokunga, E. B. Boltich, and W. E. Wallace, Magnetic properties of Ga-added die-upset Nd-Fe-B magnets. Magnetics, *IEEE Trans. Magn.* 25, 1989, 4114–4116.
94. H. Fukunaga, H. Tomita, H. Wada, M. Yamashita, and F. Toshimura, A systematic study on stability of flux in Nd-Fe-B magnets consolidated by direct joule heating, *J. Appl. Phys.* 76, 1994, 6846–6848.
95. Y. Li, Y. B. Kim, M. S. Song, T. S. Yoon, and C. O. Kim, The temperature dependence of the anisotropic Nd–Fe–B fabricated by single-stage hot deformation, *J. Magn. Magn. Mater.* 263, 2003, 11–14.
96. R. K. Mishra and V. Panchanathan, Microstructure of high-remanence Nd-Fe-B alloys with low-rare-earth content, *J. Appl. Phys.* 75, 1994, 6652–6654.
97. M. Jurczyk and J. Jakubowicz, Improved temperature and corrosion behaviour of nano-composite $Nd_2(Fe,Co,M)_{14}B/\alpha$-Fe magnets, *J. Alloys Compd.* 311, 2000, 292–298.
98. Z. Chen et al., A study on the role of Nb in melt-spun nanocrystalline Nd–Fe–B magnets, *J. Magn. Magn. Mater.* 268, 2004, 105–113.
99. S. Hirosawa, A. Hanaki, H. Tomizawa, and A. Hamamura, Current status of Nd-Fe-B permanent magnet materials, *Phys. B: Condensed Matt.* 164, 1990, 117–123.
100. K. D. Durst and H. Kronmüller, Determination of intrinsic magnetic material parameters of $Nd_2Fe_{14}B$ from magnetic measurements of sintered $Nd_{15}Fe_{77}B_8$ magnets, *J. Magn. Magn. Mater.* 59, 1986, 86–94.
101. K. O'Grady, M. El-Hilo, and R. W. Chantrell, The characterization of interaction effects in fine particle systems, *IEEE Trans. Magn.* 29, 1993, 2608–2613.
102. H. W. Zhang, C. B. Rong, X. B. Du, J. Zhang, S. Y. Zhang, and B. G. Shen, Investigation on intergrain exchange coupling of nanocrystalline permanent magnets by Henkel plot, *Appl. Phys. Lett.* 82, 2003, 4098–4100.
103. M. Sagawa, S. Fujimura, N. Togawa, and H. Yamamoto, New material for permanent magnets on a base of Nd and Fe (invited), *J. Appl. Phys.* 55, 1984, 2083–2087.
104. J. J. Croat, J. F. Herbst, R. W. Lee, and F. E. Pinkerton, High-energy product NdFeB permanent magnets, *Appl. Phys. Lett.* 1, 1984, 44.
105. N. C. Koon and B. N. Das, Crystallization of FeB alloys with rare earths to produce hard magnetic materials (invited), *J. Appl. Phys.* 55, 1984, 2063–2066.
106. J. H. Lin, S. F. Liu, Q. M. Cheng, X. L. Qian, L. Q. Yang, and M. Z. Su, Preparation of Nd-Fe-B based magnetic materials by soft chemistry and reduction-diffusion process, *J. Alloys Compd.* 249, 1997, 237–241.

107. Y. Haik, J. Chatterjee, and C. J. Chen, Synthesis and stabilization of Fe-Nd-B nanoparticles for biomedical applications, *J. Nanopart. Res.* 7, 2005, 675–679.

108. C. W. Kim, Y. H. Kim, H. G. Cha, and Y. S. Kang, Study on synthesis and magnetic properties of Nd-Fe-B alloy via reduction-diffusion process, *Phys. Scr.* T129, 2007, 321–325.

109. H. G. Cha, Y. H. Kim, W. K. Chang, and Y. S. Kang, Synthesis and charateristics of NdFeB magnetic nanoparticle, in: *IEEE NMDC 2006: IEEE Nanotechnology Materials and Devices Conference 2006, Proceedings.* 2006, IEEE, New York, NY, pp. 656–657.

110. C. B. Murray, S. Sun, W. Gaschler, H. Doyle, T. A. Betley, and C. R. Kagan, Colloidal synthesis of nanocrystals and nanocrystal superlattices, *IBM J. Res. Develop.* 45, 2001, 47–56.

111. P. K. Deheri, V. Swaminathan, S. D. Bhame, Z. W. Liu, and R. V. Ramanujan, Sol-gel based chemical synthesis of $Nd_2Fe_{14}B$ hard magnetic nanoparticles, *Chem Mater.* 22, 2010, 6509–6517.

112. S. Ram, E. Claude, and J. C. Joubert, Synthesis, stability against air and moisture corrosion, and magnetic properties of finely divided loose $Nd_2Fe_{14}B_x$ hydride powders, *IEEE Trans. Magn.* 31, 1995, 2200–2208.

113. V. Swaminathan, P. K. Deheri, S. D. Bhame, and R. V. Ramanujan, Novel microwave assisted chemical synthesis of $Nd_2Fe_{14}B$ hard magnetic nanoparticles, *Nanoscale* 5, 2011, 2718–2725.

114. H. G. Cha, Y. H. Kim, C. W. Kim, H. W. Kwon, and Y. S. Kang, Preparation for exchange-coupled permanent magnetic composite between α-Fe (soft) and $Nd_2Fe_{14}B$ (hard), *Curr. Appl. Phys.* 7, 2007, 400–403.

115. Y. S. Kang and D. K. Lee, Fabrication of exchange coupled hard/soft nanocomposite magnet and their characterization, *Int. J. Nanosci.* 5, 2006, 315–321.

116. L. Q. Yu, C. Yang, and Y. L. Hou, Controllable $Nd_2Fe_{14}B/\alpha$-Fe nanocomposites: Chemical synthesis and magnetic properties, *Nanoscale* 6, 2014, 10638–10642.

117. K. P. Su, Z. W. Liu, H. Y. Yu, X. C. Zhong, W. Q. Qiu, and D. C. Zeng, A feasible approach for preparing remanence enhanced NdFeB based permanent magnetic composites, *J. Appl. Phys.* 109, 2011, 07A710.

118. S. Odenbach, *Ferrofluids, Magnetically Controllable Fluids and Their Applications*, Springer, Berlin, 2002.

119. C. Suryanarayana, Mechanical alloying and milling, *Prog. Mater. Sci.* 46, 2001, 1.

120. V. M. Chakka, B. Altuncevahir, Z. Q. Jin, Y. Li, and J. P. Liu, Magnetic nanoparticles produced by surfactant-assisted ball milling, *J. Appl. Phys.* 99, 2006, 08E912.

121. Y. P. Wang, Y. Li, C. B. Rong, and J. P. Liu, Sm–Co hard magnetic nanoparticles prepared by surfactant-assisted ball milling, *Nanotechnology* 18, 2007, 465701.

122. N. Poudyal, C. B. Rong, and J. P. Liu, Effects of particle size and composition on coercivity of Sm–Co nanoparticles prepared by surfactant-assisted ball milling, *J. Appl. Phys.* 107, 2010, 09A703.

123. B. Z. Cui, A. M. Gabay, W. F. Li, M. Marinescu, J. F. Liu, and G. C. Hadjipanayis, Anisotropic $SmCo_5$ nanoflakes by surfactant-assisted high energy ball milling, *J. Appl. Phys.* 107, 2010, 09A721.

124. R. A. Varin and C. Chiu, Structural stability of sodium borohydride ($NaBH_4$) during controlled mechanical milling, *J. Alloys Compd.* 397, 2005, 276–281.

125. M. Jurczyk, J. Cook, and S. Collocott, Application of high energy ball milling to the production of magnetic powders from NdFeB-type alloys, *J. Alloys Comp.* 217, 1995, 65–68.

126. H. G. Cha, Y. H. Kim, W. K. Chang, H. W. Kwon, and Y. S. Kang, Characterization and magnetic behavior of Fe and Nd-Fe-B nanoparticles by surfactant-capped high-energy ball mill, *J. Phys. Chem. C* 111, 2006, 1219–1222.

127. N. G. Akdogan, G. C. Hadjipanayis, and D. J. Sellmyer, Anisotropic Sm-(Co,Fe) nanoparticles by surfactant-assisted ball milling, *J. Appl. Phys.* 105, 2009, 07A710.
128. K. P. Su, Z. W. Liu, D. C. Zeng, D. X. Huo, L. W. Li, and G. Q. Zhang, Structure and size-dependent properties of NdFeB nanoparticles and textured nano-flakes prepared from nanocrystalline ribbons, *J. Phys. D: Appl. Phys.* 46, 2013, 377–384.
129. B. Z. Cui, L. Y. Zheng, W. F. Li, J. F. Liu, and G. C. Hadjipanayis, Single-crystal and textured polycrystalline $Nd_2Fe_{14}B$ flakes with a submicron or nanosize thickness, *Acta Mater.* 60, 2012, 1721–1730.
130. J. Y. Liu, S. Guo, Z. W. Liu, and F. M. Xiao, Morphology and magnetic properties of anisotropic Nd-Fe-B powders by surfactant-assisted ball milling, *J. Magn. Mater. Devices*, 45, 2014, 18–22.
131. A. M. Gabay, N. G. Akdogan, M. Marinescu, J. F. Liu, G. C. Hadjipanayis, Rare earth-cobalt hard magnetic nanoparticles and nanoflakes by high-energy milling, *J. Phys.: Cond. Matter.* 22, 2010, 164213.
132. P. H. Zhou, L. J. Deng, J. L. Xie, and D. F. Liang, Effects of particle morphology and crystal structure on the microwave properties of flake-like nanocrystalline Fe_3Co_2 particles, *J. Alloys Compd.* 448, 2008, 303.
133. W. L. Zuo, R. M. Liu, X. Q. Zheng, R. R. Wu, F. X. Hu, J. R. Sun, and B. G. Shen, Textured $Pr_2Fe_{14}B$ flakes with submicron or nanosize thickness prepared by surfactant-assisted ball milling, *J. Appl. Phys.* 115, 2014, 17A734.
134. M. Yue, Y. P. Wang, N. Poudyal, C. B. Rong, and J. P. Liu, Preparation of $Nd_2Fe_{14}B$ nanoparticles by surfactant-assisted ball milling technique, *J. Appl. Phys.* 105, 2009, 07A708.
135. N. Poudyal, V. V. Nguyen, C. B. Rong, and J. P. Liu, Anisotropic bonded magnets fabricated via surfactant-assisted ball milling and magnetic-field processing, *J. Phys. D: Appl. Phys.* 44, 2011, 335002.
136. N. Poudyal, B. Altuncevahir, V. Chakka, K. Chen, T. D. Black, and J. P. Liu, Field-ball milling induced anisotropy in magnetic particles, *J. Phys. D: Appl. Phys.* 37, 2004, L45–L48.
137. B. Z. Cui, L. Y. Zheng, M. Marinescu, J. F. Liu, and G. C. Hadjipanayis, Textured $Nd_2Fe_{14}B$ flakes with enhanced coercivity, *J. Appl. Phys.* 111, 2012, 07A735.
138. G. Obara, H. Yamamoto, M. Tani, and M. Tokita, Magnetic properties of spark plasma sintering magnets using fine powders prepared by mechanical compounding method, *J. Magn. Magn. Mater.* 239, 2002, 464–467.
139. M. Yue, J. Zhang, Y. Xiao, G. Wang, and T. Li, New kind of NdFeB magnet prepared by spark plasma sintering, *IEEE Trans. Magn.* 39, 2003, 3551–3553.
140. H. Ono, N. Waki, M. Shimada, T. Sugiyama, A. Fujiki, F. Yamamoto, and M. Tani, Isotropic bulk exchange-spring magnets with 34 kJ/m³ prepared by spark plasma sintering method, *IEEE Trans. Magn.* 37, 2001, 2552–2554.
141. M. Yue, J. X. Zhang, H. Zeng, and K. J. Wang, Preparation, microstructure, and magnetic properties of bulk nanocrystalline Gd metal, *Appl. Phys. Lett.* 89, 2006, 232504.
142. T. Saito, T. Takeuchi, and H. Kageyama, Magnetic properties of Nd-Fe-Co-Ga-B magnets produced by spark plasma sintering method, *J. Appl. Phys.* 97, 2005, 10H103.
143. W. Q. Liu, Z. Z. Cui, X. F. Yi, M. Yue, Y. B. Jiang, D. T. Zhang, J. X. Zhang, and X. B. Liu, Structure and magnetic properties of magnetically isotropic and anisotropic Nd-Fe-B permanent magnets prepared by spark plasma sintering technology, *J. Appl. Phys.* 107, 2010, 09A719.
144. Z. W. Liu, H. Y. Huang, X. X. Gao, H. Y. Yu, X. C. Zhong, and J. Zhu, Microstructure and property evolution of isotropic and anisotropic ndfeb magnets fabricated from nanocrystalline ribbons by spark plasma sintering and hot deformation, *J. Phys. D: Appl. Phys.* 44, 2011, 25003–2501.

145. T. Wang, M. Yue, Y. Q. Li, M. Tokita, Q. Wu, D. T. Zhang, and J. X. Zhang, Tuning of microstructure and magnetic properties of nanocrystalline Nd–Fe–B permanent magnets prepared by spark plasma sintering, *IEEE Magn. Lett.* 6, 2015, 5500304.

146. Y. H. Hou, Y. L. Huang, Z. W. Liu, D. C. Zeng, S. C. Ma, and Z. C. Zhong, Hot deformed anisotropic nanocrystalline NdFeB based magnets prepared from spark plasma sintered melt-spun powders, *Mater. Sci. Eng. B* 178, 2013, 990–997.

147. N. Tamari, T. Tanaka, K. Tanaka, I. Kondoh, M. Kawahara, and M. Tokita, Effect of spark plasma sintering on densification and mechanical properties of silicon carbide, *J. Ceram. Soc. Jpn.* 103, 1995, 740–742.

148. W. Mo, L. Zhang, A. Shan, L. Cao, J. Wu, and M. Komuro, Microstructure and magnetic properties of NdFeB magnet prepared by spark plasma sintering, *Intermetallics* 15, 2007, 1483–1488.

149. C. D. Fuerst and E. G. Brewer, Enhanced coercivities in die-upset Nd-Fe-B magnets with diffusion-alloyed additives (Zn, Cu, and Ni), *Appl. Phys. Lett.* 56, 1990, 2252.

150. C. D. Fuerst and E. G. Brewer, Diffusion-alloyed additives in die-upset Nd-Fe-B magnets, *J. Appl. Phys.* 69, 1991, 5826.

151. M. H. Ghandehari, Reactivity of Dy_2O_3 and Tb_4O_7 with $Nd_{15}Fe_{77}B_8$ powder and the coercivity of the sintered magnets, *Appl. Phys. Lett.* 48, 1986, 548–550.

152. M. Doser and G. Keeler, Long-term stability of Fe-B-Nd-Dy alloys made by Dy_2O_3 additions, *J. Appl. Phys.* 64, 1988, 5311–5313.

153. F. Xu, L. Zhang, X. Dong, Q. Liu, and M. Komuro, Effect of DyF_3 additions on the coercivity and grain boundary structure in sintered Nd-Fe-B magnets, *Scripta Materialia* 64, 2011, 1137–1140.

154. Z. W. Liu, L. Z. Zhao, S. L. Hu, H. Y. Yu, X. C. Zhong, and X. X. Gao, Coercivity and thermal stability enhancement for spark plasma sintered nanaocrystalline Nd-Fe-B magnets with Dy_2O_3 and Zn additions, *IEEE Trans. Magn.* 51, 2015, 1–4.

155. S. Xiaoyan, L. Xuemei, and Z. Jiuxing, Neck formation and self-adjusting mechanism of neck growth of conducting powders in spark plasma sintering, *J. Am. Cera. Soc.* 89, 2006, 494–500.

156. H. T. Kim and Y. B. Kim, Additive blending effects on the magnetic properties of nanocrystalline NdFeB magnets, *Phys. Status Solidi (a)*, 201, 2004, 1938–1941.

157. L. Li, J. Yi, Y. Peng, and B. Huang, The effect of compound addition Dy_2O_3 and Sn on the structure and properties of NdFeNbB magnets, *J. Magn. Magn. Mater.* 308, 2007, 80–84.

158. Z. W. Liu, Y. L. Huang, H. Y. Huang, X. C. Zhong, H. Y. Yu, and D. C. Zeng, Isotropic and anisotropic nanocrystalline ndfeb-based magnets prepared by spark plasma sintering and hot deformation, *Key Eng. Mater.* 510–511, 2012, 307–314.

159. W. Liu, L. J. Cao, J. S. Wu, and T. C. Li, Characterization of melt-spun NdFeB magnets prepared by explosive compaction, *Materials Trans.* 44, 2003, 2094–2098.

160. S. Guruswamy, M. K. Mccarter, J. E. Shield, and V. Panchanathan, Explosive compaction of magnequench Nd-Fe-B magnetic powders, *J. Appl. Phys.* 79, 1996, 4851–4853.

161. S. Ando, Y. Mine, K. Takashima, S. Itoh, and H. Tonda, Explosive compaction of Nd-Fe-B powder, *J. Mater. Process. Technol.* 85, 1999, 142–147.

162. P. Skoglund, High density PM parts by high velocity compaction, *Powder Metall.* 44, 2001, 199–201.

163. B. Bos, C. Fors, and T. Larsson, Industrial implementation of high velocity compaction for improved properties, *Powder Metall.* 49, 2006, 107–109.

164. Z. Q. Yan, F. Chen, and Y. X. Cai, High-velocity compaction of titanium powder and process characterization, *Powder Technol.* 208, 2011, 596–599.

165. D. F. Khan, H. Q. Yin, H. Li, X. H. Qu, M. Khan, S. Ali, and M. Z. Iqbal, Compaction of Ti-6Al-4V powder using high velocity compaction technique, *Mater. Design.* 50, 2013, 479–483.

166. X. Deng, Z. W. Liu, H. Y. Yu, Z. Y. Xiao, and G. Q. Zhang. Isotropic and anisotropic nanocrystalline NdFeB bulk magnets prepared by binder-free high-velocity compaction technique, *J. Magn. Magn. Mater.* 390, 2015, 26–30.

167. J. Li, Y. Liu, S. J. Gao, M. Li, Y. Q. Wang, and M. J. Tu, Effect of process on the magnetic properties of bonded NdFeB magnet, *J. Magn. Magn. Mater.* 299, 2006, 195–204.

168. E. A. Périgo, M. F. D. Campos, R. N. Faria, and F. J. G. Landgraf, The effects of the pressing step on the microstructure and aging of NdFeB bonded magnets, *Powder Techn.* 224, 2012, 291–296.

169. T. Saito, M. Fujita, T. Kuji, K. Fukuoka, and Y. Syono, The development of high performance Nd-Fe-Co-Ga-B die upset magnets, *J. Appl. Phys.* 83, 1998, 6390–6392.

170. M. Lin, H. Wang, J. Zheng, and A. Yan, Effects of Fe fine powders doping on hot deformed NdFeB magnets, *J. Magn. Magn. Mater.* 379, 2015, 90–94.

171. W. Lipiec and H. A. Davies, The influence of the powder densification temperature on the microstructure and magnetic properties of anisotropic NdFeB magnets aligned by hot deformation, *J. Alloys Comp.* 491, 2010, 694–697.

172. B. Lai, Y. F. Li, H. J. Wang, A. H. Li, M. G. Zhu, and W. Li, Quasi-periodic layer structure of die-upset NdFeB magnets, *J. Rare Earths*, 31, 2013, 679.

173. W. Mo, L. Zhang, A. Shan, J. Wu, Study on microstructure of NdFeB magnet prepared by spark plasma sintering, *Rare Metal Mater. Eng.* 36, 2007, 2140–2143.

174. Z. W. Liu, New developments in NdFeB-based permanent magnets, *Key Eng. Mater.* 510–511, 2012, 1–8.

175. Z. W. Liu, R. V. Ramanujan, and H. A. Davies, Improved thermal stability of hard magnetic properties in rapidly solidified RE-TM-B alloys, *J. Mater. Res.* 23, 2008, 2733–2742.

176. Y. Xiao, S. Liu, H. Mildrum, K. J. Strnat, and A. E. Ray, The effects of various alloying elements on modifying the elevated temperature magnetic properties of sintered Nd-Fe-B magnets, *J. Appl. Phys.* 63, 1988, 3526.

177. J. F. Herbst, $R_2Fe_{14}B$ materials: Intrinsic properties and technological aspects. *Rev. Mod. Phys.* 63, 1991, 819–898.

178. W. Rodewald and B. Wall, Temperature stability and magnetizing behaviour of sintered Nd-Dy-Fe-Co-Mo-Al-B magnets, *J. Magn. Magn. Mater.* 101, 1991, 338–340.

179. Y. L. Huang, Z. W. Liu, X. C. Zhong, H. Y. Yu, and D. C. Zeng, NdFeB based magnets prepared from nanocrystalline powders with various compositions and particle sizes by spark plasma sintering, *Powder Metall.* 55, 2012, 124–129.

180. Z. W. Liu, Y. L. Huang, S. L. Hu, X. C. Zhong, H. Y. Yu, and X. X. Gao, Properties enhancement and recoil loop characteristics for hot deformed nanocrystalline NdFeB permanent magnets, *IOP Conf. Series: Mater. Sci. Eng.* 60, 2014, 012013.

181. B. Zheng, H. W. Zhang, S. F. Zhao, J. L. Chen, and G. H. Wu, The physical origin of open recoil loops in nanocrystalline permanent magnets, *Appl. Phys. Lett.* 93, 2008, 182503.

182. N. Yoshikawa, T. Iriyama, H. Yamada, Y. Kasai, and V. Panchanathan, Radially oriented high energy product Nd-Fe-B ring magnets. *IEEE Trans. Magn.* 35, 1999, 3268–3270.

183. O. Gutfleisch, A. Kirchner, W. Grünberger, D. Hinz, R. Schäfer, L. Schultz, I. R. Harris, and K. H. Müller, Backward extruded NdFeB HDDR ring magnets, *J. Magn. Magn. Mater.* 183, 1998, 359–364.

184. W. Gruenberger, D. Hinz, A. Kirchner, K. H. Mueller, and L. Schultz, Hot deformation of nanocrystalline Nd-Fe-B alloys, *IEEE Trans. Magn.* 33, 1997, 3889–3891.

185. D. Brown, B. M. Ma, and Z. M. Chen, Developments in the processing and properties of NdFeB-type permanent magnets, *J. Magn. Magn. Mater.* 248, 2002, 432.

186. S. Sugimoto, Current status and recent topics of rare-earth permanent magnets, *J. Phys. D: Appl. Phys.* 44, 2011, 1–11.

187. J. Ormerod and S. Constantinides, Bonded permanent magnets: Current status and future opportunities, *J. Appl. Phys.* 81, 1997, 4816.

188. X. F. Wang, D. Lee, and Z. L. Jiang, Magnetic properties of hybrid polymer bonded Nd-Fe-B/ferrite magnets, *J. Appl. Phys.* 99, 2006, 08B513.

189. M. Dospial and D. Plusa, Magnetization reversal processes in bonded magnets made from a mixture of Nd-(Fe,Co)-B and strontium ferrite powders, *J. Magn. Magn. Mater.* 330, 2013, 152–158.

190. X. C. Zhong, Y. L. Huang, and Y. H. Liu, Effects of NdFeB powder on the magnetic properties of bonded NdFeB/Sr-ferrite hybrid magnets, *Electron. Compon. Mater.* 31, 2012, 41–44.

191. M. Yan and X. L. Peng, *The Magnetism Basis and Magnetic Materials*, Zhejiang University Press, Hangzhou, 2006, pp. 1–254.

192. J. Schneider and R. K. Schmidt, Bonded hybrid magnets, *J. Magn. Magn. Mater.* 157/158, 1996, 27–28.

193. Z. H. Hua, S. D. Li, Z. D. Han, D. H. Wang, W. Zhong, B. X. Gu, M. Lu, J. R. Zhang, and Y. W. Du, Magnetic properties and intergranular action in bonded hybrid magnets, *J. Rare Earths* 25, 2007, 336–340.

194. Z. W. Liu, D. C. Zeng, R. V. Ramanujan, X. C. Zhong, and H. A. Davies, Exchange interaction in rapidly solidified nanocrystalline RE-(Fe/Co)-B hard magnetic alloys, *J. Appl. Phys.* 105, 2009, 07A736.

195. Y. Matsuyama, H. Kobayashi, and T. Mitamura, Positive temperature coefficient of magnetization characteristic for composite plastic-bonded magnet composed of $Nd_2Fe_{14}B$ and $SrO \cdot 6Fe_2O_3$, *J. Ceram. Soc. Jpn.* 112, 2004, 65–69.

6 Layered Two-Phase Magnetoelectric Materials

Zhaofu Du, Sam Zhang, Dongliang Zhao,
Tat Joo Teo, and Rajdeep Singh Rawat

CONTENTS

ABSTRACT

Multiferroic materials have attracted great interest for having more than one primary ferroic-order parameter in a single material. Magnetoelectric (ME) material is one of the multiferroic materials that produces a voltage under a stimulation of magnetic field and has stirred up much interest recently due to its potential applications in multifunctional devices. Natural single-phase multiferroic compounds are rare, and the ME effect is either weak or occurs at a too low temperature for practical applications. A two-phase ME material incorporates ferroelectric and ferri-/ferromagnetic phases into one to yield giant ME effect at room temperature, rendering it practically applicable. The elastic coupling interaction between the magnetostrictive phase and piezoelectric phase leads to giant ME response of these ME materials, such as ferrite and piezoelectric ceramics, magnetic metals/alloys, and piezoelectric ceramics. When a direct current (DC) bias field is superimposed with a small oscillating alternating field, an electric polarization is induced and/or a magnetization polarization appears. For application in microelectronic devices, nanostructured composite coatings of ferroelectric and magnetostrictive materials have been deposited on a substrate. In this chapter, multiferroism and magnetoelectricity are explained; experimental and theoretical aspects of laminate and multilayer structured ME composites are discussed; and transducers, sensors, microwave devices, etc. are introduced from the application point of view.

6.1   MULTIFERROISM AND MAGNETOELECTRICITY

Traditionally, ferroic properties include ferroelectricity, ferromagnetism, and ferroelasticity. However, current practice tends to exclude ferroelasticity but encompasses ferrotoroidicity (i.e., an ordered arrangement of magnetic vortices) and even antiferroicity [1]. Multiferroic materials are materials that exhibit more than one primary ferroic property in a single material. Figure 6.1 illustrates ferroic properties of multiferroic materials.

Magnetoelectricity is the most important effect of multiferroic materials that exhibits the change of electric polarity under an external magnetic field. The change of magnetization polarity in an external electric field is referred to as converse magnetoelectric effect (CME). Magnetoelectric (ME) effect and CME coefficients are defined as follows [2]:

$$\alpha_{ME} = \frac{\delta E}{\delta H} \quad \text{or} \quad \alpha_{ME} = \frac{dp}{dH} \tag{6.1}$$

$$\alpha_{CME} = \frac{\delta B}{\delta E} \quad \text{or} \quad \alpha_{CME} = \frac{\mu_0 dm}{dE} \tag{6.2}$$

where E, H, and B represent the electric field, the magnetic field, and the magnetic induction, respectively, and p and m represent the effective electrical and magnetic dipole moments of the entire system under equivalent conditions of electric and magnetic bias, while μ_0 is the permeability of free space.

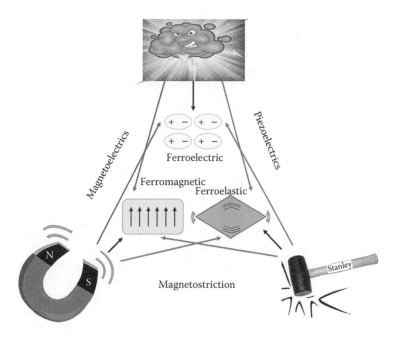

FIGURE 6.1 Ferroic properties of multiferroic materials.

In recent years, ME in engineered composites created huge impact by enabling highly sensitive yet low-noise ME magnetic field sensors and voltage-driven magnetic field generators, etc. [3,4]. Ideal ME devices would require no external power supply and external conditioning circuitry. They should also exhibit stable room temperature operation and should be inexpensive to fabricate. Currently, existing ME devices have exhibited several such characteristics and shown tremendous potential to compete with existing flux-gate, Hall-effect sensors, SQUID (superconducting quantum interference device), current measurement probes, and magnetoresistive magnetometers. Apart from magnetic field sensing, other promising applications such as transducers, filters, oscillators, phase shifters, and memory devices have also spurred numerous geometries and topologies of ME composites investigations, for example, the bulk heterostructural laminates, the thick- and thin-film devices, and more recently the quasi-one-dimensional tube topology [5].

ME effect was first discovered in 1888 when Rontgen found that a moving dielectric became magnetized when placed in an electric field [6], and the reverse effect, that is, electrical polarization of a dielectric moving in a magnetic field, was later reported by Wilson in 1905 [7]. In 1894, Pierre Curie pointed out the possibility of intrinsic ME behavior of crystals on the basis of symmetry considerations, in which the crystals can be electrically polarized in the presence of a magnetic field [8]. The term "magnetoelectric" first appeared in Debye's report after the initial (unsuccessful) attempts to demonstrate the static ME effect experimentally in 1926 [9]. By 1932, Wigner distinguished the magnetic quantities from the electric quantities in relation to space inversion and proposed the consideration of time inversion as a

symmetry criterion [10]. At the same time, Van Vleck clarified the role of external magnetic field on the appearance of the ME effect [11]. In 1937, Landau established the theory of phase transitions and the dependence of physical properties on
the symmetry of internal charge and current distribution [12]. Zeldovich initiated
the theoretical studies of toroidal current distributions as the sources of nondipo-
lar, nonquadrupolar magnetic fields in 1957 [13]. The considered systems, termed
as "anapoles," formed the embryo of the present-day concept of ferrotoroidic ME
effect. In 1959, Dzyaloshinskii predicted, in theory, on magnetic symmetry grounds
that ME effect should be observable in antiferromagnetic Cr_2O_3 [14], and the first
experimental observation was made by Astrov in 1960 [15]. The following year,
Rado and Folen also observed magnetically induced electric signals in Cr_2O_3 [16].
By 1964, Shubnikov and Belov had established all possible bicolor point groups [17],
while important numbers of ME boracites [18] and phosphates [19] were discovered
by the research groups led by Ascher and Schmid at the Battelle Institute in Geneva,
and of Newnham at the Pennsylvania State University.

Although the scientific findings about ME phases reached some form of saturation by mid-1970s, single-phase ME materials were less understood due to the con-
traindication between the conventional mechanism in ferroelectric materials, which
requires empty d orbitals, and formation of magnetic moments, which results from
partially filled d orbital, and the ME effect appears to be very weak in these materials. Hence, the single-phase ME materials were perceived as impractical because
of weak coupling and low temperature requirement, and the theoretical studies did
not indicate any feasibility of resolving those limitations. During the 1980s, some
research was continued, for example, by Tabares-Munoz et al. in 1985 [20], but it was
at a slower pace and scarce. By the late 1990s, the research efforts were stepped up,
and some factors that promoted the renaissance of magnetoelectricity are as follows:

- The "magnetoelectricity" concept found a conjunction with "multiferroicity"
  [21]. Such simultaneous occurrence of ferromagnetism and ferroelectricity
  was highly favorable for magnetoelectricity. The search for magnetoelec-
  trics evolved into a search for multiferroics with ME coupling.
- Development of theoretical tools, basically *ab initio* calculations [22,23],
  opened new and promising routes for designing potential ME materials.
- Solution–combustion [24] and other preparative techniques [25] made the
  synthesis of new families of compounds possible.
- Advances in thin-film growth methods [26] opened a new field of experi-
  mentation. This new physical dimensionality produced nanomaterials with
  surprising and exciting ME properties.

In the early days, magnetoelectricity was observed as an intrinsic effect in some
natural material systems at low temperatures [27–30]. Recently, there has been a
surge of activities in this area because of the better understanding of the multiferroic
properties and potential applications in information storage, spintronics, multiple-
state memories, etc. ME materials are also being pursued because of its ability to
couple the two order parameters (ferromagnetic and ferroelectric), allowing an additional degree of freedom in device design [31–34]. Currently, these materials are

prospected to control the charges by applied magnetic field or to spin polarize the charges by applied voltages so as to use these materials to construct new forms of multifunctional devices [35–37].

After the first single-phase ME material Cr_2O_3 with a linear ME effect was reported in 1960 [15], many other single-phase ME materials were subsequently discovered. The ME effect in a crystal is traditionally described by the Landau theory where the free energy F of the system is expressed in terms of an applied magnetic field $\bar{H}$ and an applied electric field $\bar{E}$. Using Einstein summation convention, F can be written as [38,39]

$$F(\bar{E},\bar{H}) = -P_i^s E_i - M_i^s H_i - \frac{1}{2}\varepsilon_0\varepsilon_{ij}E_iE_j - \frac{1}{2}\mu_0\mu_{ij}H_iH_j - \alpha_{ij}E_iH_j$$
$$- \frac{1}{2}\beta_{ijk}E_iH_jH_k - \frac{1}{2}\gamma_{ijk}H_iE_jE_k - \cdots \quad (6.3)$$

where P_i^s and M_i^s are the spontaneous polarization and magnetization, respectively. The third term on the right-hand side describes the contribution resulting from the electrical response to an electric field, where ε_{ij} is the relative permittivity. The fourth term is the magnetic equivalent of the third term, where μ_{ij} is the relative permeability. The fifth term describes linear ME coupling via α_{ij}. Other terms represent higher-order ME coupling via coefficients β_{ij}, γ_{ij}, etc.

The ME effect can be established in the form of $P_i(H_j)$ or $M_i(E_j)$ by differentiating F expressed as

$$P_i(H_j) = -\frac{\partial F}{\partial E_i} = P_i^s + \varepsilon_0\varepsilon_{ij}E_j + \alpha_{ij}H_j + \frac{1}{2}\beta_{ijk}H_jH_k + \gamma_{ijk}H_iE_j + \cdots \quad (6.4)$$

$$M_i(E_j) = \frac{\partial F}{\partial H_i} = M_i^s + \mu_0\mu_{ij}H_j + \alpha_{ij}E_j + \beta_{ijk}E_iH_j + \frac{1}{2}\gamma_{ijk}E_jE_k + \cdots \quad (6.5)$$

where α_{ij} is designated as the linear ME effect and corresponds either to the induction of electric polarization by a magnetic field or to a magnetization by an electric field. It was further shown that the ME response is limited by the following relationship:

$$\alpha_{ij}^2 < \varepsilon_{ii}\mu_{jj} \quad (6.6)$$

From Equation 6.5, a multiferroic material, which consists of ferromagnetic and ferroelectric materials, is responsible to display large ME effects because the ferromagnetic and ferroelectric materials often (but not always) possess large permeability and permittivity, respectively. Iniguez, who used the first-principle approach to compute the linear ME response of Cr_2O_3, found that the low-temperature response of such material has a significant lattice character [40].

The ME coupling of single-phase material is very weak. Most single-phase ME materials only exhibit the ME effect at very low temperatures (below 6 K) such as

R_2CuO_4 [41] and $Me_3B_7O_{13}X$ [42]. ME effects of some materials can only be detected at extremely high magnetic fields of >20 Tesla, for example, $BiFeO_3$ [43], while few be detected as a minute change in the dielectric constant at the Neel temperature or correspondingly a small change in the magnetic permeability at the Curie temperature such as $RMnO_3$ [44] and $Pb(Fe_{1/2}Nb_{1/2})O_3$ [45]. In fact, no single-phase material has ever been found to exhibit measurable ME coefficient at room temperature and in the presence of low magnetic field. As prior studies were primarily conducted out of scientific curiosity, the largest ME coefficient previously reported at either low temperature or high H is in the order of 1 mV/cmOe, which is too little to be of practical use in devices.

6.2   TWO-PHASE LAYERED LAMINATED ME MATERIALS

Composite materials consisting of ferroelectric and ferromagnetic phases have proved to exhibit higher ME effect than the single-phase materials. Ryu et al. [46] organized the resultant properties of the composites into three categories, that is, the sum properties, the product properties, and the combination properties. The schematic representation of the properties in a two-phase mixture is shown in Figure 6.2. A sum property of the composite material is defined as a weighted sum of the contributions of the individual component phases, proportional to the weight/volume fractions of the phases in a composite material. The sum properties can be observed in simple physical quantities such as density and resistivity. The dielectric property of a composite material can be a sum property if no reaction occurs for the

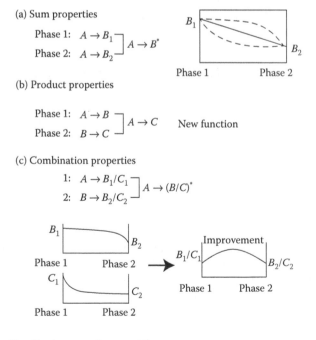

FIGURE 6.2   Resultant composite properties.

multiphase mixing, that is, simple mixing of multiphases. In this case, the resulted dielectric properties are restricted within Weiner's series and parallel rules. A product property is a resulted property reflected in a composite structure that is absent in the individual phases. As shown in Figure 6.2b, in a biphasic composite material, if one of the phases exhibits a property $A \rightarrow B$ (variable A results in B effect) with a relationship $dB/dA = X$, where X can be a constant (linear proportion) or a function (nonlinear), and the second phase exhibits $B \rightarrow C$ with a relationship $dC/dB = Y$ (Y can be a constant or a function), then the composite material will exhibit $A \rightarrow C$ that is absent in both initial phases. The proportionality tensor $dC/dA = (dC/dB)(dB/dA) = Y.X$ is the product of the proportionality functions of the two phases, termed as the product property. ME effect from a piezoelectric–magnetomechanical composite material is an example of such property. ME composite materials often consist of a piezoelectric phase and a ferromagnetic or magnetostrictive phase in which neither of the initial phases can obtain magnetic polarization (magnetization) resulting from the introduction of the electric field ($E \rightarrow M$) or electric polarization resulting from the presence of the magnetic field ($M \rightarrow E$). The electric/magnetic transformation phenomenon is a product property of a piezoelectric and a magnetic material through the intermediary of stress/strain. In some cases, the output property of the composite materials exceeded the initial phase. Such property enhancement is referred to as a combination effect, which is mainly due to the fact that the properties of interest are dependent on several parameters rather than just one factor. A simple demonstration of the combination effect is plotted in Figure 6.2c. Assuming that variables B and C follow the convex- and concave-type characteristics as illustrated on the left side of Figure 6.2c, the property of interest depends on B and C with the relationship B/C. Subsequently, the combination of five sets of B/C values will reach a maximum at an intermediate ratio of the two phases. Thus, the composite materials with desirable properties surpassing the natural materials can be obtained by using the combination effect.

In a composite material, the ME effect is a result of the magnetostrictive effect (the magnetic and mechanical effect in the magnetic phase) and the piezoelectric effect (the mechanical and electrical effect in the piezoelectric phase), in other words, a coupled electrical and magnetic phenomenon via mechanical interaction. The ME effect and the inverse ME effect in the two-phase composite material can be expressed as [47]

$$\mathrm{ME}_H = \frac{\text{magnetic}}{\text{mechanical}} \times \frac{\text{mechanical}}{\text{electrical}} \tag{6.7}$$

$$\mathrm{ME}_E = \frac{\text{electrical}}{\text{mechanical}} \times \frac{\text{mechanical}}{\text{magnetic}} \tag{6.8}$$

When a magnetic field is applied to the two-phase ME material, the magnetostrictive material changes its shape and the strain will be passed to the piezoelectric phase through the interface between the two phases resulting in an electric polarization in the piezoelectric phase causing voltage to appear along the polarization direction.

In order to fabricate composite materials with enhanced ME effect, one needs to make composites having ferromagnetic material with large magnetostrictive effect and ferroelectric material with large piezoelectric effect along with improved coupling between the two phases.

6.2.1 Theories

In 1972, van Suchtelen first proposed the ME effect in two-phase composite materials [48]. The ME effect in a composite material consisting of one magnetostrictive phase and one piezoelectric phase can be described by

$$\frac{\partial S}{\partial H} = e^m \text{ (magnetic phase)} \tag{6.9}$$

$$\frac{\partial P}{\partial S} = e \text{ (piezoelectric phase)} \tag{6.10}$$

where S represents the strain, e^m is the piezomagnetic coefficient, and e is the piezoelectric coefficient. As such, this two-phase ME material obeys

$$\frac{\partial P}{\partial H} = \alpha = k_c e^m e \tag{6.11}$$

where k_c is a coupling factor ($0 \leq |k_c| \leq 1$) between the two phases [49], and α is the ME coefficient of the composite material. This product response is due to the elastic coupling between the two phases and high magnetostriction and piezoelectric effect together with strong coupling between the two which will favor a large ME coefficient. Soon after, scientists found that a large ME effect could be produced at room temperature in two-phase ME materials [50,51].

6.2.1.1 Modeling of Laminated Composite Materials

Consider the piezoelectric/magnetostrictive laminated composite material in Figure 6.3 as an example, the bilayer in the (1,2) plane consists of the piezoelectric and magnetostrictive phases with free boundary but no epitaxial characteristics for the layers [52]. Assuming that only the (symmetric) extensional deformation will occur and ignoring any (asymmetric) flexural deformations of the layers, one would lead to a position-dependent elastic constant with the need for perturbation procedures [53].

An averaging method is used to derive the effective composite parameters and is carried out in two stages [54–56]. In the first stage, the sample is considered as a bilayer. For the polarized piezoelectric phase with the symmetry ∞m, the strain and electric displacement are given as

$$^P S_i = {}^P s_{ij} {}^P T_j + {}^P d_{ki} {}^P E_k \tag{6.12}$$

$$^P D_k = {}^P d_{ki} {}^P T_i + {}^P \varepsilon_{kn} {}^P E_n \tag{6.13}$$

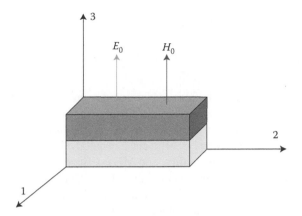

FIGURE 6.3   Sketch of a piezoelectric/magnetostrictive bilayer composite.

where $^{P}S_i$ and $^{P}T_j$ represent the strain and stress tensor components of the piezo-electric phase, respectively; $^{P}E_k$ and $^{P}D_k$ represent the vector components of electric field and electric displacement, respectively; $^{P}s_{ij}$ and $^{P}d_{ki}$ are the compliance and piezoelectric coefficients, respectively; and $^{P}\varepsilon_{kn}$ represents the permittivity matrix. The magnetostrictive phase is assumed to have a cubic symmetry and is described as

$$^{m}S_i = {}^{m}s_{ij}{}^{m}T_j + {}^{m}q_{ki}{}^{m}H_k \tag{6.14}$$

$$^{m}B_k = {}^{m}q_{ki}{}^{m}T_i + {}^{m}\mu_{kn}{}^{m}H_n \tag{6.15}$$

where $^{m}S_i$ and $^{m}T_j$ represent the strain and stress tensor components of the mag-netostrictive phase, respectively; $^{m}H_k$ and $^{m}B_k$ represent the vector components of magnetic field and magnetic induction, respectively; $^{m}s_{ij}$ and $^{m}q_{ki}$ are compliance and piezomagnetic coefficients, respectively; and $^{m}\mu_{kn}$ represents the permeability matrix. Equations 6.14 and 6.15 can be considered as a linearized equation describing the effect of magnetostriction. Assuming that the in-plane mechanical connectivity between the two phases has appropriate boundary conditions, ME voltage coef-ficients can be obtained by solving Equations 6.12 through 6.15. Previous models assumed ideal coupling at the interface [57]. Here, an interface coupling parameter $k = \left({}^{P}S_i - {}^{P}S_{i0} \right) / \left({}^{m}S_i - {}^{m}S_{i0} \right) (i = 1,2)$, where $^{P}S_{i0}$ represents the strain tensor compo-nents with no friction between layers, is introduced. The coupling parameter k depends on the interface quality and is a measure of differential deformation between piezo-electric and magnetostrictive layers. The coupling factor is $k = 1$ for an ideal interface and zero for the case with no friction. The significance of k and its relationship to structural, magnetic, and electrical parameters of the composite are discussed later.

In the second stage, the bilayer is considered as homogeneous [54–56], and the behavior is described as

$$S_i = s_{ij}T_j + d_{ki}E_k + q_{ki}H_k \tag{6.16}$$

$$D_k = d_{ki}T_i + \varepsilon_{kn}E_n + \alpha_{kn}H_n \tag{6.17}$$

$$B_k = q_{ki}T_i + \alpha_{kn}E_n + \mu_{kn}H_n \tag{6.18}$$

where S_i and T_j are strain and stress tensor components, respectively; E_k, D_k, H_k, and B_k are the vector components of electric field, electric displacement, magnetic field, and magnetic induction, respectively; s_{ij}, d_{ki}, and q_{ki} are effective compliance, piezoelectric, and piezomagnetic coefficients, respectively; and ε_{kn}, μ_{kn}, and α_{kn} are effective permittivity, permeability, and ME coefficient, respectively. Effective parameters of the composite are obtained by solving Equations 6.16 through 6.18, taking into account the solutions of Equations 6.12 through 6.15. The mechanical strain and stress for the bilayer and homogeneous material are assumed to be the same, and electric and magnetic vectors are determined using open- and closed-circuit conditions.

6.2.1.1.1 Longitudinal ME Effect

As shown in Figure 6.3, (1,2) is assumed as the film plane, and the direction 3 is perpendicular to the sample plane. Consider the polarization direction along with the axis 3. When the magnetic field is applied along the axis 3 and the resulting induced electric field δE is estimated across the sample thickness, such conditions give $\alpha'_{E,L} = \alpha'_{E,33} = \delta E_3/\delta H_3$. The nonzero components, that is, $^p s_{ij}$, $^p d_{ki}$, $^m s_{ij}$, $^m q_{ki}$, s_{ij}, d_{ki}, q_{ki}, and α_{kn} for this configuration are given in Table 6.1. Equations 6.12 through 6.18 are subsequently solved for the following boundary conditions:

$$^p S_i = k\,^m S_i + (1-k)\,^p S_{io} \quad (i=1,2) \tag{6.19}$$

$$^p T_i = -\frac{^m T_i(1-v)}{v} \quad (i=1,2) \tag{6.20}$$

$$^p T_3 = \,^m T_3 = T_3 \tag{6.21}$$

$$^m S_i = S_i (i=1,2) \tag{6.22}$$

$$S_3 = \frac{[^p S_3 + \,^m S_3(1-v)]}{v} \tag{6.23}$$

where $v = \,^p v/(^p v + \,^m v)$ with $^p v$ and $^m v$ denote the volume of piezoelectric and magnetostrictive phases, respectively, and $^p S_{10}$ and $^p S_{20}$ are the strain tensor components at $k=0$.

To obtain effective piezoelectric and piezomagnetic coefficients, it is necessary to consider the composite in an electric field $E_3 = V/t$ (V is the applied voltage, and t is the thickness of the composite) and a magnetic field H_3. The effective E

TABLE 6.1
Nonzero Coefficients of Piezoelectric and Magnetostrictive Phases and Homogeneous Material for Longitudinal Field Orientation

Piezoelectric Coefficients	Compliance Coefficients

Piezoelectric Phrase

Piezoelectric Coefficients	Compliance Coefficients
$^P d_{15} = {}^P d_{24}$	$^P s_{11} = {}^P s_{22}$
$^P d_{31} = {}^P d_{32}$	$^P s_{12} = {}^P s_{21}$
$^P d_{33}$	$^P s_{13} = {}^P s_{23} = {}^P s_{31} = {}^P s_{32}$
	$^P s_{33}$
	$^P s_{44} = {}^P s_{55}$
	$^P s_{66} = 2\left({}^P s_{11} + {}^P s_{12} \right)$

Magnetostrictive Phrase

$^m q_{15} = {}^m q_{24}$	$^m s_{11} = {}^m s_{22} = {}^m s_{33}$
$^m q_{31} = {}^m q_{32}$	$^m s_{12} = {}^m s_{21} = {}^m s_{13} = {}^m s_{23} = {}^m s_{31} = {}^m s_{32}$
$^m q_{33}$	$^m s_{44} = {}^m s_{55} = {}^m s_{66}$

Piezoelectric Coefficients	Piezomagnetic Coefficients	Compliance Coefficients

Homogeneous Material

Piezoelectric Coefficients	Piezomagnetic Coefficients	Compliance Coefficients
$d_{15} = d_{24}$	$q_{15} = q_{24}$	$^P s_{11} = {}^P s_{22}$
$d_{31} = d_{32}$	$q_{31} = q_{32}$	$^P s_{12} = {}^P s_{21}$
d_{33}	q_{33}	$^P s_{13} = {}^P s_{23} = {}^P s_{31} = {}^P s_{32}$
		$^P s_{33}$
		$^P s_{44} = {}^P s_{66}$
		$^P s_{66} = 2\left({}^P s_{11} + {}^P s_{12} \right)$

in the piezoelectric and H in the magnetostrictive phases are given by $^P E_3 = E_3/v$, $^m H_3 = (H_3 - v\,{}^m B_3/\mu_0)/(1-v)$. Applying continuity conditions for the magnetic field, the electric field, and the open- and closed-circuit conditions yields the expressions for effective permittivity, permeability, ME coefficient, and longitudinal ME voltage coefficient written as follows:

$$\varepsilon_{33} = \frac{\left\{ 2\left({}^P d_{31} \right)^2 (v-1) + {}^P \varepsilon_{33} \left[\left({}^P s_{11} + {}^P s_{12} \right)(1-v) + kv\left({}^m s_{11} + {}^m s_{12} \right) \right] \right\}}{\left\{ v\left[\left({}^P s_{11} + {}^P s_{12} \right)(1-v) + kv\left({}^m s_{11} + {}^m s_{12} \right) \right] \right\}} \tag{6.24}$$

$$\mu_{33} = \mu_0 \left\{ {}^m\mu_{33}\left[kv\left({}^ms_{11} + {}^ms_{12}\right) + (1-v)\left({}^ps_{11} + {}^ps_{12}\right)\right] - 2kv\left({}^mq_{31}\right)^2 \right\} \Big/$$
$$\left\{ \mu_0\left[v^2\left({}^ps_{11} + {}^ps_{12}\right) + (1-v)\left({}^ps_{11} + {}^ps_{12}\right) + kv(1-v)\left({}^ms_{11} + {}^ms_{12}\right)\right] \right.$$
$$\left. + {}^m\mu_{33}\{v(1-v)\left({}^ps_{11} + {}^ps_{12}\right) + kv^2\left({}^ms_{11} + {}^ms_{12}\right) - 2kv^2\left({}^mq_{31}\right)^2 \right\} \quad (6.25)$$

$$\alpha_{33} = \frac{-2\mu_0 kv(1-v)\,{}^pd_{31}\,{}^mq_{13}}{\left[\mu_0(v-1) - {}^m\mu_{33}v\right]\left[kv\left({}^ms_{11} + {}^ms_{12}\right) - \left({}^ps_{11} + {}^ps_{12}\right)(v-1)\right] + 2\left({}^mq_{31}\right)^2 kv^2} \quad (6.26)$$

$$\alpha'_{E,33} = \frac{E_3}{H_3} = 2\frac{\mu_0 kv(1-v)\,{}^pd_{31}\,{}^mq_{31}}{\left\{2\,{}^pd_{31}^2(1-v) + {}^p\varepsilon_{33}\left[\left({}^ps_{11} + {}^ps_{12}\right)(v-1) - v\left({}^ms_{11} + {}^ms_{12}\right)\right]\right\}}$$
$$\times \frac{\left[\left({}^ps_{11} + {}^ps_{12}\right)(v-1) - kv\left({}^ms_{11} + {}^ms_{12}\right)\right]}{\left\{\left[\mu_0(v-1) - {}^m\mu_{33}v\right]\left[kv\left({}^ms_{11} + {}^ms_{12}\right) - \left({}^ps_{11} + {}^ps_{12}\right)(v-1)\right] + 2\,{}^mq_{31}^2 kv^2\right\}} \quad (6.27)$$

Harshe et al. obtained an expression for the longitudinal ME voltage coefficient given as [57]

$$\alpha'_{E,33} = \frac{-2v(v-1)\,{}^pd_{13}\,{}^mq_{31}}{\left({}^ms_{11} + {}^ms_{12}\right)\,{}^p\varepsilon_{33}^T v + \left({}^ps_{11} + {}^ps_{12}\right)\,{}^p\varepsilon_{33}^T(1-v) - 2\left({}^pd_{13}\right)^2(1-v)} \quad (6.28)$$

and the longitudinal ME coefficient is given as [56]

$$\alpha_{E33} = \frac{E_3}{H_3} = \frac{2v(1-v)d_{31}q_{31}\mu_0\bar{s}}{\left(2vd_{31}^2 - {}^p\varepsilon_{33}\bar{s}\right)\left[\bar{\mu}\bar{s} + 2q_{31}^2(1-v)^2\right]} \quad (6.29)$$

where $\bar{s} = v\left({}^ps_{11} + {}^ps_{12}\right) + (1-v)\left({}^ms_{11} + {}^ms_{12}\right)$ and $\bar{\mu} = v\mu_0 + (1-v){}^m\mu_{33}$. S_{11} and S_{12} are compliances, d_{33} is the piezoelectric coefficient, q_{31} is the piezomagnetic coefficients, and ${}^m\mu_{33}$ is the permeability of the magnetic phase.

Equation 6.29 corresponds to a special case based on the assumption that ${}^m\mu_{33}/\mu_0 = 1$ and $k = 1$. Thus, it will lead to an expression for the longitudinal ME coupling and allows its estimation as a function of volume of the two phases, composite permeability and interface coupling. The ME voltage coefficients obtained in these works had similar trends as the experimental data. However, some deviations still remained in the overall magnitude of the ME voltage coefficients with the theoretical results being several times higher than the experimental values. Such deviations were probably due to the use of the electromagnetic boundary conditions when obtaining the theoretical model, which involved the quasi-static approximations and the open-circuit conditions of the piezoelectric phase [57,58]. To overcome this limitation, fundamental electromagnetic wave boundary conditions were applied on the fields within the composite structure. Using distinct theoretical models on the longitudinal ME effect in a piezoelectric/magnetostrictive bilayer, a more accurate model was derived to better approximate the experimental results [59]. The model was obtained by solving the constitutive equations of each layer for the all fields present, and subsequently by applying a field-averaging method [59] (along with the boundary conditions on the components of the fields at the composite interface) to obtain the homogenized material properties. The homogenized layer is characterized in terms of its effective permeability, effective permittivity, and the effective ME susceptibility tensor with constitutive equations expressed as

$$\vec{D} = \bar{\bar{\varepsilon}}\vec{E} + \bar{\bar{\xi}}\vec{H} \tag{6.30}$$

$$\vec{B} = \bar{\bar{s}}\vec{E} + \bar{\bar{\mu}}\vec{H} \tag{6.31}$$

where $\vec{D}$, $\vec{B}$, $\vec{E}$, and $\vec{H}$ represent the electric displacement field, the magnetic flux density, the electric field, and the magnetic field, respectively. $\bar{\bar{\varepsilon}}$ and $\bar{\bar{\mu}}$ represent the permittivity and permeability tensors, respectively, while $\bar{\bar{\xi}}$ and $\bar{\bar{s}}$ are the bianiso-tropic coupling tensors. In composite materials, the ME effect can be obtained using piezoelectric and magnetostrictive materials in a layered structure. Lead zirconium titanate (PZT) and barium titanate are examples of piezoelectric materials that have been used in ME composites [57,58]. The constitutive relationship for the piezoelectric phase is similar to Equations 6.12 and 6.13 but includes a relationship for the permeability of the piezoelectric phase $^{P}\mu_{kr}$ in its constitutive relationship expressed as

$$^{P}B_k = {}^{P}\mu_{kn}\,{}^{P}H_n \tag{6.32}$$

To model the media in terms of the fields, the relationship between all fields within each composite phase must be obtained. Hence, one must consider the magnetic flux/magnetic field relationship of the piezoelectric layer to obtain a homogeneous layer that combines the properties of both layers. Nickel ferrite $NiFe_2O_4$ and copper ferrite $CoFe_2O_4$ (CFO) are examples of magnetostrictive materials that have been used in ME composites. Magnetostrictive materials are represented by the constitutive Equations 6.14 and 6.15 that yields

$$^{m}D_k = {}^{m}\varepsilon_{kn}\,{}^{m}E_n \tag{6.33}$$

Here, the permittivity of the magnetostrictive phase has been accounted for as it is observed that Equation 6.33 relates the electric field and the electric displacement field. In this derivation, the composite is poled and biased perpendicular to the interface of the composite, which is named as "longitudinal ME configuration," as shown in Figure 6.3. Although higher ME voltage coefficients have theoretically been obtained for transverse and in-plane orientations [58], the longitudinal ME configuration model is used for there are experimental results in the literature with the same orientation which can test the theory. Harshe [57] obtained the experimental data to fabricate the composites in the longitudinal configuration. Hence, the published measured results can be served as a point of reference. In modeling the material, it is assumed that there is existing low-frequency fields in all directions, which may be time varying, on top of the poling and bias fields within the composite medium. For example, the fields in the axial direction can be expressed in terms of dc and ac components:

$$\bar{H}_3 = \hat{Z}H_0 + \hat{Z}H_{ac} \tag{6.34}$$

$$\bar{E}_3 = \hat{Z}E_0 + \hat{Z}E_{ac} \tag{6.35}$$

The total fields in each respective phase will be represented by a vector field expressed as

$$^{m,p}E = E_1\hat{1} + E_2\hat{2} + E_3\hat{3} \tag{6.36}$$

$$^{m,p}H = H_1\hat{1} + H_2\hat{2} + H_3\hat{3} \tag{6.37}$$

The ME effect in composite materials is known as a product property due to the resultant interaction between the piezoelectric and magnetostrictive phases. Modeling the ME effect requires good understanding of the strain and stress transfer relationship between the layers of the structure. Bichurin et al. introduced the use of coupling parameter, k, to model the mechanical interaction at the interface between the bilayers [21], which is defined as

$$k = \frac{(^{p}S_i - {}^{p}S_{i0})}{(^{m}S_i - {}^{p}S_{i0})} \quad (i = 1, 2) \tag{6.38}$$

where $^{p}S_{i0}$ are the strain tensor components with no friction between the phases. Here, a similar coupling parameter, k, is used as a damping factor to model the strain transfer relationship between the layers. It is assumed that the strain induced in one phase may not be completely transferred to the adjoining phase due to several factors such as the mechanical defects and losses, bonding methods, etc. These factors are contained and described by the interface coupling parameter.

In this derivation, only the symmetric or extensional deformation is considered. As such, the flexural deformations of the layers, which may lead to position-dependent elastic constants, are ignored [17,21]. The following assumptions are made [18]: (i) The shear stresses and strains are equal to zero such that $^{m,p}T_i = 0$, $^{m,p}S_i = 0$ for $i = 4$, 5, and 6. (ii) The thickness of each phase is much smaller than the width and length of the phase. Hence, the stress in the axial direction is approximated as zero, $^{m}T_3 = {}^{p}T_3 = 0$. (iii) The strain transfer between phases is related by an interface coupling parameter, k, such that $^{p}S_i = k \cdot {}^{m}S_i$. (iv) The summation of forces on the

1–2 plane boundaries are zero such that $^mT_i{}^mv + {}^pT_i{}^pv = 0$, for $i = 1, 2$, where ^mv and ^pv represent the magnetostrictive and piezoelectric volume fractions, respectively, and defined as

$$^mv = \frac{\text{volume}^m}{\text{volume}^{total}}, \quad ^pv = \frac{\text{volume}^p}{\text{volume}^{total}}$$

By solving the constitutive equations of each phase based on the given assumptions, the electric displacement field in the piezoelectric region is obtained using the nonzero components of the permittivity, permeability, compliance, and piezoelectric coefficient components as shown in Table 6.2.

The electric displacement field in the piezoelectric region is derived as

$$^pD_1 = {}^p\varepsilon_{11}{}^pE_1, \quad ^pD_2 = {}^p\varepsilon_{22}{}^pE_2, \quad ^pD_3 = K_1{}^mH_3 + K_2{}^pE_3 \qquad (6.39)$$

where

$$K_1 = \frac{2\,{}^pd_{31}\,{}^mq_{31}}{\left[k\left({}^ms_{11} + {}^ms_{12} \right)\left({}^pv/{}^mv \right) + \left({}^ps_{11} + {}^ps_{11} \right) \right]} \qquad (6.40)$$

TABLE 6.2

Nonzero Coefficients of Bulk Piezoelectric and Magnetostrictive Phases

Coefficient Type	Nonzero Components
	Piezoelectric Phase
Permittivity	$^p\varepsilon_{11} = {}^p\varepsilon_{22}, {}^p\varepsilon_{33}$
Permeability	$^p\mu_{11} = {}^p\mu_{22} = {}^p\mu_{33}$
Compliance	$^ps_{11} = {}^ps_{22}, {}^ps_{12} = {}^ps_{21}, {}^ps_{13} = {}^ps_{23} = {}^ps_{31} = {}^ps_{32} \;\; {}^ps_{11} = {}^ps_{22}, {}^ps_{33}$ $^ps_{44} = {}^ps_{55}, {}^ps_{66} = 2\left({}^ps_{11} + {}^ps_{12} \right)$
Piezoelectric	$^pd_{15} = {}^pd_{24}, {}^pd_{31} = {}^pd_{32}, {}^pd_{33}$
	Magnetostrictive Phase
Permittivity	$^m\varepsilon_{11}, {}^m\varepsilon_{22}, {}^m\varepsilon_{33}$
Permeability	$^m\mu_{11}, {}^m\mu_{22}, {}^m\mu_{33}$
Compliance	$^ms_{11} = {}^ms_{22} = {}^ms_{33}, {}^ms_{44} = {}^ms_{55} = {}^ms_{66}, {}^ms_{12} = {}^ms_{21} = {}^ms_{13} = {}^ms_{23} = {}^ms_{31} = {}^ms_{32}$
Piezoelectric	$^mq_{15} = {}^mq_{24}, {}^mq_{31} = {}^mq_{32}, {}^mq_{33}$

$$K_2 = \left\{ \frac{-2(^P d_{31})^2}{\left[k(^m s_{11} + {}^m s_{12})(^P v / {}^m v) + (^P s_{11} + {}^P s_{11}) \right]} \right\} + {}^P \varepsilon_{33} \qquad (6.41)$$

Here, the magnetic flux density in the piezoelectric region, which is required in the theoretical model, can be determined using Equation 6.32, while the form of the permeability tensor can be obtained using the components of the permeability piezoelectric phase shown in Table 6.2. When deriving the magnetic flux density in the magnetostrictive region, the effect of the dc magnetic field bias on the ferrite medium was considered. The ME effect in composite material exhibits a nonlinear behavior. A direct current (DC) magnetic field bias is usually applied so that the ME effect can be approximated as a linear over a short range [17]. However, the applied bias has a secondary effect when the magnetostrictive phase is a ferrite material. Ferrites have an intrinsic magnetic moment, and application of a DC magnetic field bias will lead to tensor permeability [60]. This implies a change to the nonzero permittivity values of the bulk magnetostrictive phase from what had been described in Table 6.1. For the case with the DC magnetic field bias in the axial direction as shown in Figure 6.3, the permeability tensor is defined as [60–62]

$$^m \mu = \begin{bmatrix} ^m \mu_{11}^* & ^m \mu_{12}^* & 0 \\ ^m \mu_{21}^* & ^m \mu_{22}^* & 0 \\ 0 & 0 & ^m \mu_{33}^* \end{bmatrix} \qquad (6.42)$$

The values of the nonzero components of the permeability as shown in Equation 6.42 depend on the intrinsic property of the magnetostrictive phase such as its magnetization saturation, Ms, and the magnitude of the applied DC magnetic field bias [60,61]. This change in permeability is a secondary effect from the DC magnetic field bias in the composite structure [63]. This secondary effect, obtained in the ferrite material magnetostrictive phase, is well known and has been used in several device applications based on the shape and values of the permeability tensor. The formulae for the components of the permeability tensor, in Equation 6.42, will not be shown since they are readily available in the past literature [60–62]. By solving the magnetostrictive constitutive equations, the magnetic flux density in the magnetostrictive phase is given as

$$^m B_1 = {}^m \mu_{11}^* \, {}^m H_1 + {}^m \mu_{12}^* \, {}^m H_2 \qquad (6.43)$$

$$^m B_2 = {}^m \mu_{21}^* \, {}^m H_1 + {}^m \mu_{22}^* \, {}^m H_2 \qquad (6.44)$$

$$^m B_3 = C_1 \, {}^P E_3 + C_2 \, {}^m H_3 \qquad (6.45)$$

where

$$C_1 = \frac{2 \, {}^P d_{31} \, {}^m q_{31}}{\left[k(^m s_{11} + {}^m s_{12}) + (^P s_{11} + {}^P s_{12})(^m v / {}^P v) \right]} \qquad (6.46)$$

$$C_2 = \left\{ -\frac{2k\left({}^{P}q_{31}\right)^2}{\left[k\left({}^{m}s_{11} + {}^{m}s_{12}\right) + \left({}^{P}s_{11} + {}^{P}s_{12}\right)\left({}^{m}\upsilon/{}^{P}\upsilon\right) \right]} \right\} + {}^{m}\mu_{33} \qquad (6.47)$$

The electric displacement field for the magnetostrictive phase, which is required for the theoretical model, can be determined using Equation 6.33, and the nonzero components of the permittivity magnetostrictive phase are shown in Table 6.2.

Based on Maxwell's equations, the conditions involving the normal and tangential fields at the interface are deduced. Assuming that there are no applied surface currents, the boundary conditions at the interface are as follows [64]: (i) the tangential components of the electric and magnetic fields are continuous across the interface, and (ii) the normal components of the electric displacement field and the magnetic flux density are continuous at the interface. From the first boundary condition, the tangential components of the electric and magnetic fields are continuous at the boundary. Thus, at the interface,

$$\bar{E}_x = {}^{m}E_1 = {}^{P}E_1, \quad \bar{E}_y = {}^{m}E_2 = {}^{P}E_2 \qquad (6.48)$$

$$\bar{H}_x = {}^{m}H_1 = {}^{P}H_1, \quad \bar{H}_y = {}^{m}H_2 = {}^{P}H_2 \qquad (6.49)$$

Here, $\bar{E}_x$, $\bar{E}_y$, $\bar{H}_x$, and $\bar{H}_y$ are the homogenized tangential components of the electric and magnetic fields in the ME layer. For case of understanding, all the homogenized fields are represented using the x, y, z coordinate system. The assumption made was that the homogenized layer is electrically thin with no field variation across the total thickness of the film. Using a field-averaging method [22], the tangential components of the electric field displacement and the magnetic flux density are expressed as

$$\left. \begin{aligned} \bar{D}_x &= {}^{m}D_1{}^{m}\upsilon + {}^{P}D_1{}^{P}\upsilon = \left[{}^{m}\varepsilon_{11}{}^{m}\upsilon + {}^{P}\varepsilon_{11}{}^{P}\upsilon \right]\bar{E}_x \\ \bar{D}_y &= {}^{m}D_2{}^{m}\upsilon + {}^{P}D_2{}^{P}\upsilon = \left[{}^{m}\varepsilon_{22}{}^{m}\upsilon + {}^{P}\varepsilon_{22}{}^{P}\upsilon \right]\bar{E}_y \end{aligned} \right\} \qquad (6.50)$$

$$\left. \begin{aligned} \bar{B}_x &= {}^{m}\bar{B}_1{}^{m}\upsilon + {}^{P}\bar{B}_1{}^{P}\upsilon = \left[{}^{m}\mu_{11}^{*}{}^{m}\upsilon + {}^{P}\mu_{11}{}^{P}\upsilon \right]\bar{H}_x + \left[{}^{m}\mu_{12}^{*}{}^{m}\upsilon \right]\bar{H}_y \\ \bar{B}_y &= {}^{m}\bar{B}_2{}^{m}\upsilon + {}^{P}\bar{B}_2{}^{P}\upsilon = \left[{}^{m}\mu_{12}^{*}{}^{m}\upsilon \right]\bar{H}_x + \left[{}^{m}\mu_{22}^{*}{}^{m}\upsilon + {}^{P}\mu_{22}{}^{P}\upsilon \right]\bar{H}_y \end{aligned} \right\} \qquad (6.51)$$

Here, $\bar{D}_x$, $\bar{D}_y$, $\bar{B}_x$, and $\bar{B}_y$ are the homogenized tangential components of the electric displacement field and magnetic flux density. Note that each component of the homogenized magnetic flux density in Equation 6.51 is related to magnetic fields in both the x and y directions. The homogenized tangential components can be expressed in matrix forms:

$$\begin{bmatrix} D_x \\ D_y \end{bmatrix} = \begin{bmatrix} {}^m\varepsilon_{11}^{*}\,{}^m\upsilon + {}^P\varepsilon_{11}\,{}^P\upsilon & 0 \\ 0 & {}^m\varepsilon_{22}^{*}\,{}^m\upsilon + {}^P\varepsilon_{22}\,{}^P\upsilon \end{bmatrix} \begin{bmatrix} E_x \\ E_y \end{bmatrix} \tag{6.52}$$

$$\begin{bmatrix} B_x \\ B_y \end{bmatrix} = \begin{bmatrix} {}^m\mu_{11}^{*}\,{}^m\upsilon + {}^P\mu_{11}\,{}^P\upsilon & {}^m\mu_{12}^{*}\,{}^m\upsilon \\ {}^m\mu_{21}^{*}\,{}^m\upsilon & {}^m\mu_{22}^{*}\,{}^m\upsilon + {}^P\mu_{22}\,{}^P\upsilon \end{bmatrix} \begin{bmatrix} H_x \\ H_y \end{bmatrix} \tag{6.53}$$

Note that the tangential components of the electric field displacement and the magnetic flux density do not depend on the ME susceptibility tensor. This indicates that the ME effect is only found in the direction of the biasing field as should be expected for the longitudinal configuration based on experimental results [57].

From the second boundary condition, the normal components of the electric displacement and the magnetic flux density are continuous across the interface. Thus,

$$\bar{D}_z = {}^mD_3 = {}^PD_3 \tag{6.54}$$

$$\bar{B}_z = {}^mB_3 = {}^PB_3 \tag{6.55}$$

Here, $\bar{D}_z$ and $\bar{B}_z$ are the homogenized normal components of the electric displacement field and magnetic flux density, respectively. Applying similar assumption of constant field variation along the thickness of the layer, the homogenized normal components of the electric and magnetic field can be expressed as

$$\bar{E}_z = {}^PE_3\,{}^P\upsilon + {}^mE_3\,{}^m\upsilon, \quad \bar{H}_z = {}^PH_3\,{}^P\upsilon + {}^mH_3\,{}^m\upsilon \tag{6.56}$$

Hence, solving the homogenized normal component of the electric displacement field yields

$$\bar{D}_z = \frac{(N_1N_3 - N_5)}{(N_1N_4 - N_7)}\bar{E}_z + \frac{(N_1N_2 - N_6)}{(N_1N_4 - N_7)}\bar{H}_z \tag{6.57}$$

where

$$N_1 = \left[1 + \left(\frac{C_2}{{}^m\upsilon}\frac{{}^P\upsilon}{{}^P\mu_{33}}\right)\right], \quad N_2 = \left[\frac{{}^P\mu_{33}}{{}^P\upsilon}\right], \quad N_3 = \left[\frac{K_2\,{}^m\upsilon^P\mu_{33}}{K_1({}^P\upsilon)^2}\right], \quad N_4 = \left[\frac{{}^m\upsilon^P\mu_{33}}{K_1\,{}^P\upsilon} + \left(\frac{K_2\mu_{33}({}^m\upsilon)^2}{K_1\,{}^m\varepsilon_{33}({}^P\upsilon)^2}\right)\right]$$

$$N_5 = \left(\frac{C_1}{{}^P\upsilon}\right), \quad N_6 = \left(\frac{C_2}{{}^m\upsilon}\right), \quad N_7 = \left(\frac{C_1}{{}^P\upsilon}\frac{{}^m\upsilon}{{}^m\varepsilon_{33}}\right) \tag{6.58}$$

Similarly, the normal component of the magnetic flux density is given as

$$B_z = \frac{(R_1R_3 - R_5)}{(R_1R_4 - R_7)}H_z + \frac{(R_1R_2 - R_6)}{(R_1R_4 - R_7)}E_z \tag{6.59}$$

where

$$R_1 = \left[1 + \left(\frac{K_2^m \upsilon}{{}^P \upsilon^m \varepsilon_{33}}\right)\right], R_2 = \left[\frac{{}^m \varepsilon_{33}}{{}^m \upsilon}\right], R_3 = \left[\frac{C_2 {}^P \upsilon^m \varepsilon_{33}}{C_1 ({}^m \upsilon)^2}\right], R_4 = \left[\frac{{}^P \upsilon^m \varepsilon_{33}}{C_1 {}^m \upsilon} + \left(\frac{C_2 {}^m \varepsilon_{33} ({}^P \upsilon)^2}{C_1 {}^P \mu_{33} ({}^m \upsilon)^2}\right)\right],$$

$$R_5 = \left(\frac{K_1}{{}^m \upsilon}\right), R_6 = \left(\frac{K_2}{{}^P \upsilon}\right), R_7 = \left(\frac{K_1 {}^P \upsilon}{{}^m \upsilon^P \mu_{33}}\right) \tag{6.60}$$

The results obtained show that the constitutive equations for the homogenized ME composite layer has the form of

$$\begin{bmatrix} D_x \\ D_y \\ D_z \end{bmatrix} = \begin{bmatrix} 0 & 0 & 0 \\ 0 & 0 & 0 \\ 0 & 0 & \alpha_{zz}^H \end{bmatrix} \begin{bmatrix} H_x \\ H_y \\ H_z \end{bmatrix} + \begin{bmatrix} \varepsilon_{xx} & 0 & 0 \\ 0 & \varepsilon_{yy} & 0 \\ 0 & 0 & \varepsilon_{zz} \end{bmatrix} \begin{bmatrix} E_x \\ E_y \\ E_z \end{bmatrix} \tag{6.61}$$

$$\begin{bmatrix} B_x \\ B_y \\ B_z \end{bmatrix} = \begin{bmatrix} 0 & 0 & 0 \\ 0 & 0 & 0 \\ 0 & 0 & \alpha_{zz}^E \end{bmatrix} \begin{bmatrix} E_x \\ E_y \\ E_z \end{bmatrix} + \begin{bmatrix} \mu_{xx} & \mu_{xY} & 0 \\ \mu_{YX} & \mu_{yy} & 0 \\ 0 & 0 & \mu_{zz} \end{bmatrix} \begin{bmatrix} H_x \\ H_y \\ H_z \end{bmatrix} \tag{6.62}$$

Table 6.3 is the theoretical model results comparison to experimental results obtained by Harshe et al. [57]. In computing the ME voltage coefficient, the same material characteristics of the media as given in literature are used. The material characteristics of the layers are shown in Table 6.4. Here, the theoretical model obtained allows for detailed analysis of the ME effect in the composite structure. Analysis can be done on the effects of the interface coupling parameter on the ME media. Increase in volume fraction of either the piezoelectric or magnetostrictive phase can also be analyzed.

TABLE 6.3

Experimental and Theoretical Magnetoelectric Voltage Coefficients

Material	Experimental α_{ME} Maximum [(V/m)/(kA/m)]	Theoretical α_{ME} [(V/m)/(kA/m)]
$CoFe_2O_4$: PZT-4	92.8	199–622
$CoFe_2O_4$: PZT-8	18.6	202–634
$CoFe_2O_4$: PZT-5H	74.4	143–475

TABLE 6.4

Material Parameters for Magnetostrictive and Piezoelectric Phases

Parameter (Units)	CFO	PZT-4	PZT-8	PZT-5H	PZT
s_{11} (10^{-12} m²/N)	6.5	12.3	11.5	16.5	15.3
s_{12} (10^{-12} m²/N)	−2.37	−4.05	−3.7	−4.78	−5
q_{31} (10^{-12} m/A)	566				
q_{33} (10^{-12} m/A)	1880				
d_{31} (10^{-12} C/N)		−123	−90	−274	−175
d_{33} (10^{-12} C/N)		289	225	593	400
μ_{33}/μ_0	2	1	1	1	1
$\varepsilon_{33}/\varepsilon_0$	10	1300	1000	3400	1750

6.2.1.1.2 Transverse ME Effect

This case corresponds to E and δE along the axis 3 and H and δH along the axis 1 (in the sample plane). Here, the ME coefficient is estimated by

$$\alpha'_{E,T} = \alpha'_{E,31} = \frac{\delta E_3}{\delta H_1} \tag{6.63}$$

For this case, the nonzero components, that is, $^P s_{ij}$, $^P d_{ki}$, $^m s_{ij}$, $^m q_{ki}$, s_{ij}, d_{ki}, q_{ki}, and α_{kn} are provided in Table 6.5. Equations 6.12 through 6.18 are used to solve for the boundary conditions in Equations 6.19 through 6.23, and subsequently the effective permittivity, the ME coefficient, and the transverse ME voltage coefficient are given as

$$\varepsilon_{33} = \frac{^P d_{31}^2 (1-v) + {}^P\varepsilon_{33} \left[\left({}^P s_{11} + {}^P s_{12} \right)(v-1) + kv \left({}^m s_{11} + {}^m s_{12} \right) \right]}{(v-1) \left[\left({}^P s_{11} + {}^P s_{12} \right)(v-1) + kv \left({}^m s_{11} + {}^m s_{12} \right) \right]} \tag{6.64}$$

$$\alpha_{31} = \frac{(v-1) \left({}^m q_{11} + {}^m q_{12} \right) {}^P d_{31} k}{(v-1) \left[\left({}^P s_{11} + {}^P s_{12} \right)(v-1) - kv \left({}^m s_{11} + {}^m s_{12} \right) \right]} \tag{6.65}$$

$$\alpha'_{E,31} = \frac{E_3}{H_1} = \frac{-kv(v-1) \left({}^m q_{11} + {}^m q_{12} \right) {}^P d_{31}}{{}^P\varepsilon_{33} \left({}^m s_{11} + {}^m s_{12} \right) kv + {}^P\varepsilon_{33} \left({}^P s_{11} + {}^P s_{12} \right)(1-v) - 2k {}^P d_{31}^2 (1-v)} \tag{6.66}$$

$$\alpha_{E31} = \frac{E_3}{H_1} = \frac{-v(1-v)d_{31}(q_{11} + q_{21})}{{}^P\varepsilon_{33}\bar{s} - 2vd_{31}^2} \tag{6.67}$$

Equation 6.66 describes the dependence of ME parameters on the volume fraction and is used to estimate the ME coupling for some representative systems.

TABLE 6.5
Nonzero Coefficients of Piezoelectric and Magnetostrictive Phases and Homogeneous Material for Transverse Field Orientation

Piezoelectric Coefficients	Compliance Coefficients
	Piezoelectric Phase
${}^{p}d_{15} = {}^{p}d_{24}$	${}^{p}s_{11} = {}^{p}s_{22}$
${}^{p}d_{31} = {}^{p}d_{32}$	${}^{p}s_{11} = {}^{p}s_{21}$
${}^{p}d_{33}$	${}^{p}s_{13} = {}^{p}s_{23} = {}^{p}s_{31} = {}^{p}s_{32}$
	${}^{p}s_{33}$
	${}^{p}s_{44} = {}^{p}s_{66}$
	${}^{p}s_{66} = 2\left({}^{p}s_{11} + {}^{p}s_{12}\right)$
	Magnetostrictive Phase
${}^{m}q_{35} = {}^{m}q_{26}$	${}^{m}s_{11} = {}^{m}s_{22} = {}^{m}s_{33}$
${}^{m}q_{12} = {}^{m}q_{13}$	${}^{m}s_{12} = {}^{m}s_{21} = {}^{m}s_{13} = {}^{m}s_{23} = {}^{m}s_{31} = {}^{m}s_{32}$
${}^{m}q_{11}$	${}^{m}s_{44} = {}^{m}s_{55} = {}^{m}s_{66}$

Piezoelectric Coefficients	Piezomagnetic Coefficients	Compliance Coefficients
	Homogeneous Material	
$d_{15} = d_{24}$	q_{35}, q_{36}	${}^{p}s_{11}, {}^{p}s_{22}, {}^{p}s_{33}$
$d_{31} = d_{32}$	q_{12}, q_{13}	${}^{p}s_{12} = {}^{p}s_{21}$
d_{33}	q_{11}	${}^{p}s_{13} = {}^{p}s_{31}, {}^{p}s_{23} = {}^{p}s_{32}$
		${}^{p}s_{44}, {}^{p}s_{55}, {}^{p}s_{66}$

For the 2–2 laminate $CoFe_2O_4$–$BaTiO_3$ ceramic composites, the variation in the ME coefficients with the volume fraction v of the magnetic phase is like Figure 6.4. The calculated results are similar. In the out-of-plane field mode, a maximum ME coupling appears around $v = 0.4$. In the transverse field mode, this maximum shifts to a slightly higher v and the transverse ME coefficient is larger than the longitudinal ME coefficient, which indicates a stronger transverse coupling than the longitudinal case in such 2–2 laminate composites.

When $k = 1$, it is perfect coupling at the interface between the two laminates. When $k = 0$, the ME effect disappears and the mechanical coupling is zero between the two laminates. Other values of k correspond to the actual interfaces, and the ME coefficients are between those values for the two extreme cases of $k = 0$ and 1. Figure 6.5 shows an example calculated from the laminate 2–2 composite of PZT and $CoFe_2O_4$ with different k values.

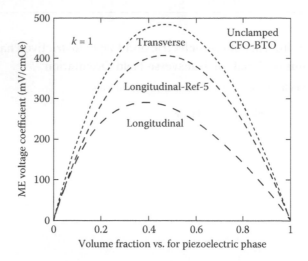

FIGURE 6.4 Transverse and longitudinal ME voltage coefficients for a perfectly bonded 2–2 structure of $CoFe_2O_4/BaTiO_3$.

Results obtained from Equations 6.63 and 6.67 fit well with the experimental results by choosing proper k values or facilitate characterization of the interface bonding status by determining k from the measured ME coefficients.

In these longitudinal and transverse modes, the laminate composite is poled along axis 3 and measured in the same direction (see Figure 6.3). If the composite can be poled with an in-plane electric field (i.e., along axis 1) and measured in the same axis 1 direction, an in-plane magnetic field would induce a large in-plane longitudinal ME coefficient [65] due to the absence of demagnetization fields and enhanced piezoelectric, and piezomagnetic coefficients as shown in Figure 6.6.

The maximum ME effect appears at a high v value, and the modeling for free boundary condition can be generalized to consider the clamped composites [65,66]. By considering that the composite will be clamped in the y direction, the in-plane stresses of the composite will be $T1 = T2 = 0$ (i.e., still free in plane), but the out-of-plane stress will be $T3 \neq 0$ ($T3 \to \infty$ for the rigidly clamped case). The longitudinal ME voltage coefficient is determined by

$$\alpha'_{E,33} = -\frac{\alpha_{33}(s_{33} + s_{c33}) - d_{33}q_{33}}{\varepsilon_{33}(s_{33} + s_{c33}) - d_{33}^2} \tag{6.68}$$

where

$$
\begin{aligned}
s_{33} = & \Big[\big(v(1-v)\big) \Big[2k \big({}^m s_{12} - {}^p s_{13}\big)^2 + {}^p s_{33}\big({}^p s_{11} + {}^p s_{12}\big) - k\,{}^m s_{11}\big({}^m s_{11} + {}^m s_{12}\big) \Big] \\
& + {}^m s_{11}\big({}^p s_{11} + {}^p s_{12}\big)(2v-1) - v^2 k\,{}^p s_{33}\big({}^m s_{11} + {}^m s_{12}\big) \\
& - v^2\,{}^m s_{11}\big({}^p s_{11} + {}^p s_{12}\big) \Big] \Big[\big({}^p s_{12} + {}^p s_{11}\big)(v-1) - kv\big({}^m s_{11} + {}^m s_{12}\big) \Big]
\end{aligned} \tag{6.69}
$$

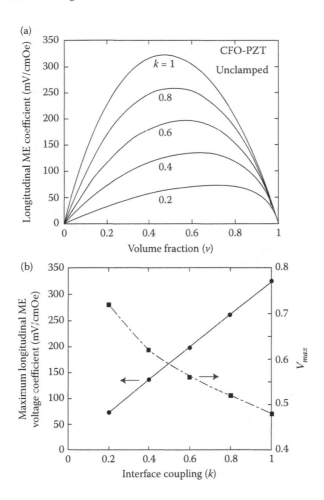

FIGURE 6.5 (a) Dependence of longitudinal ME voltage coefficient on interface coupling, k and volume fraction, v for CoFe$_2$O$_4$-PZT bilayer. (b) Variation with k of maximum longitudinal ME voltage coefficient and the corresponding V$_{max}$.

$$d_{33} = \frac{2\,{}^P d_{31} k (v-1)\left({}^m s_{12} - {}^P s_{13}\right) + {}^P d_{33}\left[\left({}^P s_{11} + {}^P s_{12}\right)(v-1) - kv\left({}^m s_{11} + {}^m s_{12}\right)\right]}{\left[\left({}^P s_{12} + {}^P s_{11}\right)(v-1) - kv\left({}^m s_{11} + {}^m s_{12}\right)^2\right]} \quad (6.70)$$

$$q_{33} = \frac{\mu_0(1-v)\left\{2kv\,{}^m q_{31}\left({}^P s_{13} - {}^m s_{12}\right) + {}^m q_{33}\left[(1-v)\left({}^P s_{11} + {}^P s_{12}\right) + kv\left({}^m s_{11} + {}^m s_{12}\right)\right]\right\}}{\left[(1-v)\mu_0 + {}^m\mu_{33}v\right]\left[kv\left({}^m s_{11} + {}^m s_{12}\right) + (1-v)\left({}^P s_{11} + {}^P s_{12}\right)\right] - 2k\,{}^m q_{31}^2 v^2} \quad (6.71)$$

A significant change in the ME coupling can be expected when the bilayer is subjected to a uniform out-of-plane stress as shown in Figure 6.6. In general, such axis 3 clamping leads to a large increase in α_{E33}. Yet in the transverse mode, such

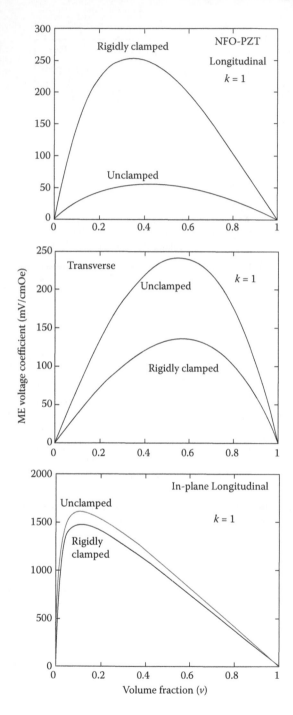

FIGURE 6.6 Comparison between f-dependence of the longitudinal, transverse, and in-plane longitudinal ME voltage coefficients for unclamped and rigidly clamped nickel ferrite _NiFe$_2$O$_4$_-PZT bilayer with $k = 1$.

clamping leads to a substantial reduction in α_{E31} as compared to the unclamped case. The clamping that causes changes to the in-plane longitudinal ME coupling is quite weak as compared to the longitudinal or transverse cases. For clamped sample conditions ($T1 = T2 = 0$ and $S3 = S_{c33}T3$), the transverse ME voltage coefficient is determined by

$$\alpha'_{E,31} = -\frac{\alpha_{31}(s_{33} + s_{c33}) - d_{33}q_{13}}{\varepsilon_{33}(s_{33} + s_{c33}) - d_{33}^2} \tag{6.72}$$

where S_{33} and d_{33} are defined by Equations 6.69 and 6.70, respectively, and

$$q_{13} = -\left\{ v^2\,{}^mq_{12}\left[\left({}^Ps_{11} + {}^Ps_{12}\right) - k\left({}^Ps_{13} + {}^ms_{11}\right)\right] + {}^mq_{12}(1 - 2v)\left({}^Ps_{11} + {}^Ps_{12}\right) + kv\,{}^mq_{12} \right.$$
$$\left. \left({}^ms_{13} + {}^ms_{11}\right) + kv(1 - v)\,{}^mq_{11}\left({}^Ps_{13} - {}^ms_{12}\right)\right\}\left[\left({}^Ps_{12} + {}^Ps_{11}\right)(v - 1) - kv\left({}^ms_{11} + {}^ms_{12}\right)\right] \tag{6.73}$$

6.2.1.1.3 Enhanced ME Effects at Resonance

Bichurin et al. first developed a theory for the ME effect at electromechanical, ferromagnetic, and magnetoacoustic resonances in the composite materials [56,67,68]. As the ME effect in the composite materials is due to the mechanical coupling between the piezoelectric and magnetic phases, it could be greatly enhanced when the piezoelectric or magnetic phase undergoes resonance [69–71], that is, electromechanical resonance (EMR) for the piezoelectric phase and the ferromagnetic resonance (FMR) for the magnetic phase. Mechanical oscillations of a medium are induced either by alternating the magnetic or the electric fields, and the wavelength is tens of meters and much larger than the composite sizes. Thus, it is possible to neglect space changing of the electric and magnetic fields within the sample volume. In order to describe frequency-dependent ME effect, the equations of elastodynamics are needed in addition to the constitutive equations, which describe the coupling of mechanical-electric-magnetic response in the ME composite materials. The linear approximation can be written by direct notation for tensors expressed as

$$\left. \begin{array}{l} \sigma = cS - e^T E - q^T H \\ D = eS + \varepsilon E + \alpha H \\ B = qS + \alpha^T E + \mu H \end{array} \right\} \tag{6.74}$$

The tensors c, e, q, ε, μ, and α are (6×6), (3×6), (3×6), (3×3), (3×3), and (3×3) matrices, respectively, by means of the compressive representation. For the piezoelectric phase in the composites, $q = 0$ and $\alpha = 0$; and for the magnetic phase in the composites, $e = 0$ and $\alpha = 0$. Yet for their composites, the effective ME coefficient $\alpha^* \neq 0$, which depends on details of the composite microstructures, that is, component phase properties, the volume fraction, the sample shape, the phase connectivity, etc.

The equations of elastic dynamics are

$$\frac{\partial^2 \mu_i}{\partial t^2} = \frac{\partial T_{ij}}{\partial x_j} \tag{6.75}$$

where u_i is the ith projection of the displacement vector, and T_{ij} still denotes the stress tensors, which are connected with the strain, electric, and magnetic fields via Equation 6.10. The joint solutions of Equations 6.74 and 6.75 under the boundary conditions yield the frequency-dependent ME coefficients.

The solution of Equation 6.75 depends on the shape of the sample and the orientations of the electric and magnetic fields. For example, consider a rectangular bilayer shown in Figure 6.3 with thickness t, width w, and length L. $L \gg t$ and $L \gg w$, and the polarization is still along axis 3. Under these boundary conditions, expressing the stress components through the strain components and substituting into Equation 6.75 yield the displacement μ_x [69]

$$\mu_x(x) = \frac{1}{k} \left\{ \left[\frac{\cos(kL) - 1}{\sin(kL)} \right] \cos(kx) + \sin(kx) \right\} (d_{31}E_3 + q_{31}H_3) \tag{6.76}$$

where $k = \omega(\rho s_{11})^{1/2}$, ω represents the angular frequency, while ρ, s_{11}, d_{31}, and q_{31} are the effective density, compliance, and piezoelectric and piezomagnetic coefficients of the composite material, respectively. Replacing q_{31} with q_{11} gives the expression for the case of the transverse orientation of fields. Furthermore, by using the open-circuit condition, that is, $\int_w dy \int_L D_3 dx = 0$, one can work out the induce voltage, E_3, in the composite material. Using the definition of ME coefficient as $E_3 = \alpha_{E,L}H_3 = \alpha_{E33}H_3$ at longitudinal and $E_3 = \alpha_{E,T}H_1 = \alpha_{E31}H_1$ at transverse orientation, the dynamic ME coefficients are obtained as [69]

$$\alpha_{E,L} = \frac{d_{31}q_{31}N - \alpha_{33}s_{11}}{\varepsilon_{33}s_{11} - d_{31}^2N} \tag{6.77}$$

$$\alpha_{E,T} = \frac{d_{31}q_{11}N - \alpha_{31}s_{11}}{\varepsilon_{33}s_{11} - d_{31}^2N} \tag{6.78}$$

where $N = 1 - (2/kL) \tan(kL/2)$. Here, all the property parameters denote the effective ones for the composite material. As seen from Equations 6.77 and 6.78 at a so-called frequency of an antiresonance where $\varepsilon_{33}s_{11} - Nd_{31}^2 = 0$, the ME coefficients sharply grow. This antiresonance frequency is dependent on the effective parameters of the composite material and its geometrical sizes. Taking a bilayer of spinel ferrites and PZT with L of about a centimeter as an example, its resonance frequency is around 300 kHz.

For example, using laminate PZT-CoFe$_2$O$_4$ composite rectangular plate with a length of 7.3 mm and a width of 2.15 mm, at the frequency of about 300 kHz, the

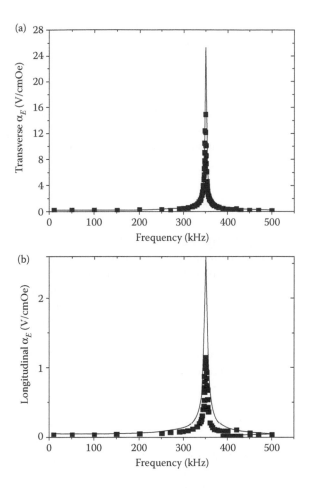

FIGURE 6.7 Electromechanical resonance frequency dependence of the (a) transverse and (b) longitudinal ME voltage coefficients of $CoFe_2O_4$-PZT bilayer.

resonance-induced enhancement in the ME effects is observed. The maximal value of the ME coefficient is observed at the transverse orientation of fields, and the resonant value of the ME coefficient is almost ten times higher than its longitudinal orientation value as shown in Figure 6.7.

6.2.1.2 Equivalent Circuit Method

For the laminate composites consisting of piezoelectric and magnetostrictive layers which named 2–2 type, Dong et al. proposed an equivalent circuit approach that is more convenient for modeling the ME coupling in the dynamic cases [72–74]. This approach is also based on the magnetostrictive and piezoelectric constitutive equations where the magnetostrictive and piezoelectric layers are mutually coupled through an elastic interaction via an equation of motion that is excited by the magnetic field. Based on magnetostrictive/piezoelectric laminates that have long-type configurations as shown in Figure 6.8, the piezoelectric constitutive equations for one-dimensional motion are given as

$$S_{1p} = s_{11}^E T_{1p} + d_{31,p} E_3, \quad D_3 = d_{31,p} T_{1p} + \varepsilon_{33}^T E_3 \tag{6.79}$$

Note: The long-type configuration typically describes the piezoelectric layer is either being polarized along its thickness or plane direction and also stressed by two magnetostrictive layers along its length, that is, principal strain direction. For the thickness poling case,

$$S_{3p} = s_{33}^D T_{3p} + g_{33p} D_3, \quad E_3 = -g_{33p} T_{3p} + \beta_{33}^T D_3 \tag{6.80}$$

In the case of length poling (where D_3 is the electric displacement), ε_{33}^T and β_{33}^T are the dielectric permittivity and impermeability under constant stress T, respectively; s_{11}^E and s_{33}^D are the elastic compliances of the piezoelectric material under constant electric field and displacement, respectively; $d_{31,p}$ and $g_{33,p}$ are the transverse piezoelectric constant and longitudinal piezoelectric voltage constant, respectively; and T_{1p}, T_{3p}, and S_{1p}, S_{3p} represent the stress and strain of the piezoelectric layer imposed by the magnetostrictive layers, respectively.

When H is applied parallel to the longitudinal axis of the laminate, a longitudinal strain is excited. The piezomagnetic constitutive equations for the longitudinal mode are given as

$$S_{3m} = s_{33}^H T_{3m} + d_{33,m} H_3, \quad B_3 = d_{33,m} T_{3m} + \mu_{33}^T H_3 \tag{6.81}$$

where B_3 is the magnetization along the length direction; μ_{33}^T is the permeability under constant stress; S_{33}^H is the elastic compliance of the magnetostrictive layer under constant H; and $d_{33,m}$ is the longitudinal piezomagnetic constant, while T_{3m} and S_{3m} are the stress and strain in the longitudinal direction of the magnetostrictive layers imposed on the piezoelectric layer, respectively. These constitutive equations

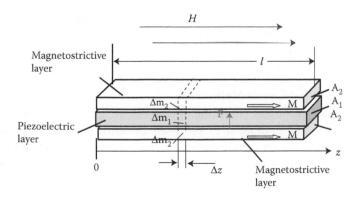

FIGURE 6.8 Long-type magnetostrictive/piezoelectric/magnetostrictive laminate.

are linear relationships, which do not account for loss components. Significant non-linearities in both piezoelectric and magnetostrictive materials are known to exist especially under resonance drive. A mechanical quality factor, Q_m, will be introduced later to include these losses.

In a harmonic motion, it will be assumed that three small mass units Δm_i in the laminate have the same displacement $\mu(z)$ along a given direction z. This follows from Figure 6.8, by assuming that the layers in the laminate act only in a coupled manner. Based on Newton's second law, an equation of motion to couple the piezoelectric and piezomagnetic constitutive Equations 6.79 and 6.80 is expressed as

$$\bar{\rho}\frac{\partial^2 u(z)}{\partial t^2} = n\frac{\partial T_{3,m}}{\partial z} + (1-n)\frac{\partial T_{i,p}}{\partial z}(0 < n < 1, \quad i = 1 \quad \text{or} \quad 3) \tag{6.82}$$

where $\bar{\rho} = (\rho_p A_p + \rho_m A_m)/A_{lam}$ is the average mass density of the laminate, $n = A_m/A_{lam} = t_m/t_{lam}$ is a geometric factor, and $t_m = t_{m1} + t_{m2}$ is the total thickness of the magnetic phase layers. For $A_m = A_{m1} + A_{m2}$, A_m and A_p are the cross-sectional areas of the magnetic phase and piezoelectric phase layers, respectively. ρ_p and ρ_m are the mass densities of the piezoelectric and magnetostrictive layers, respectively. For a given laminate width, w_{lam}, and thickness, t_{lam}, the total cross-sectional area of the laminate is $A_{lam} = t_{lam}w_{lam}$, and the total thickness is the sum of the layer thicknesses, $t_{lam} = t_p + t_m$, where t_p represents the thickness of the piezoelectric layer.

By combining Equations 6.79 and 6.80, the solution to Equation 6.82 can be obtained for longitudinally magnetized, M, and transversely or longitudinally poled, P, ME modes, respectively, named as L–T or L–L. Correspondingly, the two magneto–elasto–electric or ME equivalent circuits for both L–T and L–L modes under free boundary conditions can be derived as given in Figure 6.9 [72,74]. Here, an applied H acts as a magnetic-induced "mechanical voltage" $(\phi_m H_3)$ which then induces a "mechanical current" $(\dot{\mu}_1 \, \dot{u}$ and $\dot{\mu}_2)$ via the magnetoelastic effect with a coupling factor ϕ_m. In turn, $(\phi_m H_3)$ results in an electrical voltage, V, while $\dot{\mu}_1 \, \dot{u}$ and $\dot{\mu}_2$ result in a current, I_p, across the piezoelectric layer due to the electromechanical coupling. A transformer with a turn ratio of ϕ_p can be used to represent the electromechanical coupling. In the circuits shown in Figure 6.9, Z_1 and Z_2 are the characteristic mechanical impedances of the composite, and C_0 is the clamped capacitance of the piezoelectric plate.

As shown in Figure 6.9, under the open-circuit conditions, I_p from the piezoelectric layer is zero. Thus, the capacitive load, C_0 (and $-C_0$), can be moved to the main circuit loop. Applying Ohm's law to the mechanical loop, the following ME coefficients at low frequency for L–T and L–L modes can be directly derived as

$$\left.\frac{dE}{dH_3}\right|_{(L-T)} = \frac{nd_{33,m}g_{31,p}}{ns_{11}^E\left(1-k_{31}^2\right)+(1-n)s_{33}^H} \tag{6.83}$$

$$\left.\frac{dE}{dH_3}\right|_{(L-L)} = \frac{nd_{33,m}g_{33,p}}{ns_{33}^E\left(1-k_{33}^2\right)+(1-n)s_{33}^H} \tag{6.84}$$

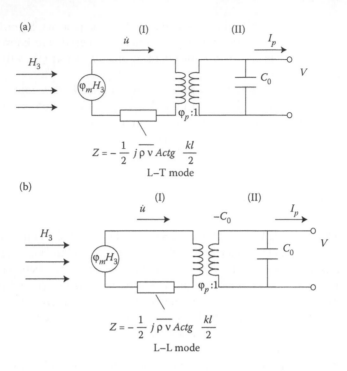

FIGURE 6.9 Magneto-elastic-electric bieffect equivalent circuits for (a) L–T mode and (b) L–L mode.

Thus, the ME coefficients are also proportional to the piezomagnetic constant, $d_{33,m}$, the piezoelectric voltage constants, $g_{31,p}$ or $g_{33,p}$, and the thickness ratio, n, of the Terfenol-D layers.

In order to derive the ME coefficients at the resonance frequency, the long-type ME laminate composites were assumed to be a L–L $\lambda/2$-resonator operating in a length extensional mode, the series angular resonance frequency is $\omega_s = \pi \bar{v}/l$ where l represents the length of the laminate and $\bar{v}$ represents the mean acoustic velocity. Under the resonant drive, the mechanical quality factor, Q_m, of the laminate is finite due to both mechanical and electric dissipations [74]. This limitation of the vibration amplitude must also be included in order to predict the resonant response. Finite values of Q_{mech} result in an effective motional mechanical resistance of $R_{mech} = \pi Z_0/8Q_{mech}$ accordingly, and the equivalent circuit of the laminates for the L–L mode under resonance drive is given in Figure 6.10. At EMR, $\omega = \omega_s$, dV/dH of the L–L mode reaches a maximum value of

$$\left(\frac{dV}{dH} \right)_{\omega s} = \frac{4Q_m \phi_m \phi_p}{\pi Z_0 \omega_s C_0} \tag{6.85}$$

where Q_m is the effective mechanical quality factor of the laminate composite including the contributions from the Terfenol-D and piezoelectric layers, and also from the

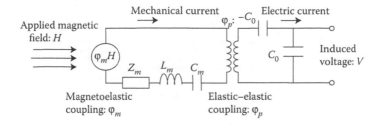

FIGURE 6.10   Magneto-elastic-electric equivalent circuits for L–L at resonance.

bonding between layers. Analysis has shown that dV/dH at the resonance frequency is Q_m times higher than that at subresonant frequencies. Using a similar approach, it is ready to obtain the resonance-equivalent circuit for the L–T mode. The 2–2 laminate structures can also be modeled numerically by using finite element method [75,76].

6.2.2   THE LAMINATES

After Ryu et al. reported first the giant magnetoelectric (GME) effect in the Terfenol-D/PZT laminate [77], a number of laminate composites combined by various magnetostrictive and piezoelectric materials are studied, FeCo, Ni, Fe–Ga, TbFe$_2$, Metglas, etc., as the magnetostrictive phases, and BTO, PZT, PMN-PT, etc., as the piezoelectric phases. Typically, the laminate composites are made of two magnetostrictive layers and a single piezoelectric layer, which is sandwiched between the two magnetostrictive layers. However, ME laminates can be made in many different configurations including disk, rectangular, and ring shapes. These various configurations can be operated in numerous working modes including T–T (transverse magnetization and transverse polarization), C–T (circumferential magnetization and transverse polarization), and L–L, L–T (which are defined earlier) as shown in Figure 6.11. The polarization can be longitudinal and circumferential also; hence, the modes become L–L, T–L, C–C, etc.

6.2.2.1   Transverse Magnetization and Transverse Polarization Mode Laminates

The first studies of Terfenol-D/PZT laminates were performed on three-layer, disk-shaped configurations that were operated in a transverse magnetization and transverse polarization (T–T) mode. Relatively large ME voltage coefficient (E) of 4.8 V/cmOe was reported by Ryu et al. [77] under a DC magnetic field bias of $H_{DC} = 4000$ Oe, though subsequent investigators reported the actual value to be 1.3 V/cmOe [78].

Although an ME coefficient of 7.5 V/cm Oe has been obtained in the shear-mode laminated PMN-PT/Terfenol-D composite [79], this composite structure encountered difficulties in practical applications because Terfenol-D is expensive and gets easily oxidized. In addition, PMN-PT is environmental unfriendly due to the lead-based content. Wang et al. used Fe–Ga as the magnetic phase, which has low hysteresis,

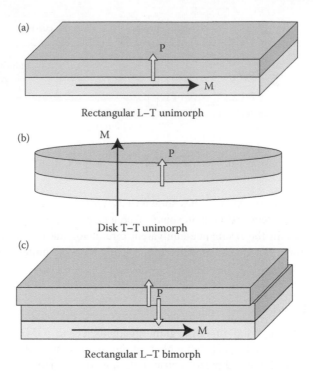

FIGURE 6.11 Different laminate working modes: (a) L–T, (b) T–T, and (c) L–T biomorph.

high tensile strength, good machinability, and low cost [80,81]. The piezoelectric BTO, with nothing injurious to the environment [82], is used as the piezoelectric core in the sandwich structure. (The following work is finished by Zhaofu Du and his group.) The Fe–Ga alloy was cut into disk-shaped plates of 10 mm in diameter and 1 mm in thickness. Using commercial $BaTiO_3$ disks of $\varphi 12 \times 1.5$ mm, three-layer, disk-shaped laminate of Fe–Ga/BTO/Fe–Ga is constructed using conductive epoxy resin where the magnetization and polarization were both oriented in its thickness directions, that is, the (T–T) mode. The ME voltage induced across the laminate composite was measured as a function of alternating current (AC) magnetic field frequency as shown in Figure 6.12. The results show that the Fe–Ga/BTO laminate has a much enhanced ME response when operated near the resonance frequency of ~95 kHz. The maximum ME voltage at resonance for the T–T mode was ~28.5 mV (or 190 mV/cmOe for the ME coefficient), which is ~7 times higher than that in the low-frequency range.

Figure 6.13 shows the ME voltage coefficient dV_{ME}/dH as a function of DC magnetic bias H_{DC}. These data were taken at a frequency of 1 kHz and a drive (H_{AC}) of 1 Oe. The value of dV_{ME}/dH was observed to be strongly dependent on H_{DC}. In the DC magnetic bias range, that is, $0 < H_{DC} < 750$ Oe, the ME voltage coefficient of the T–T mode of Fe–Ga/BTO laminate increased by increasing H_{DC}, reaching a maximum ME effect of $dV_{ME}/dH \sim 12.5$ mV/Oe at $H_{DC} = 750$ Oe (or, correspondingly, $dE_{ME}/dH = \sim 84$ mV/cmOe). For $H_{DC} > 750$ Oe, dV_{ME}/dH decreased dramatically by increasing H_{DC} because the Fe–Ga layers of the laminate approached saturation of its

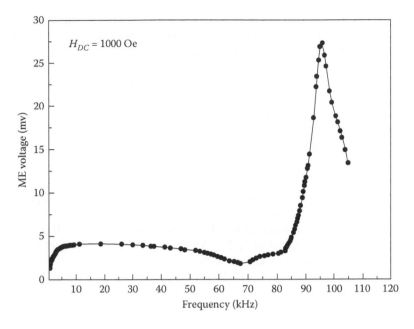

FIGURE 6.12 Induced ME voltage as a function of AC magnetic field frequency for Fe–Ga/BTO laminate.

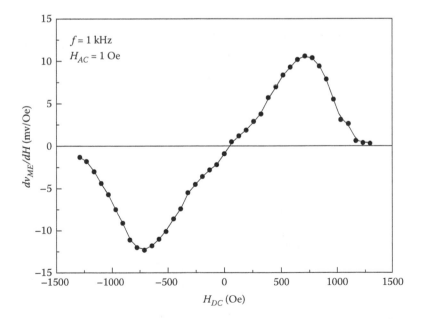

FIGURE 6.13 ME voltage coefficient dV_{ME}/dH as a function of DC magnetic bias H_{DC}.

406 Advances in Magnetic Materials

magnetostriction, which leads to the loss of piezomagnetic effects and ME coupling. In this report, the ME voltage coefficient dV_{ME}/dH was observed to be antisymmetric about H_{DC}, while previous work only reported the ME voltage amplitude as a function of H_{DC} [83].

The dependence of induced ME voltages on H_{AC} for the Fe–Ga/BTO laminate is shown in Figure 6.14. It shows that the induced ME voltages are almost linear functions of H_{AC}. At low frequency ($f = 1$ kHz) and with $H_{DC} = 750$ Oe, the value of ME voltage was ~12.5 mV under an AC magnetic field of 1.0 Oe. However, with $H_{DC} = 1000$ Oe, the ME voltage decreased to ~7.5 mV. As a comparison, Figure 6.14 also illustrates the induced ME voltage of the Fe–Ga/BTO laminate at a resonance frequency (f_r) of 95 kHz with $H_{DC} = 750$ Oe. It shows that a much higher ME voltage, that is, 28–30 mV, was obtained. Dong et al. [84] and Ryu et al. [77] previously studied the Terfenol-D/PZT laminates in T–T mode, which exhibited a large ME coupling. Although the low-frequency ME performance of the Fe–Ga/BTO laminate is not as good as the Terfenol-D/PZT laminate, its resonance ME performance is reasonably acceptable due to a high Q-factor.

The compressive stress in the BTO layer and the tensile stress in the Fe–Ga layers are derived from the simple beam theory and plane stress condition and expressed as [77]

$$\sigma_p^E = \frac{2E_f E_b t_f \Delta\varepsilon_0}{(1-v)(2E_f t_f + E_b t_b)} \tag{6.86}$$

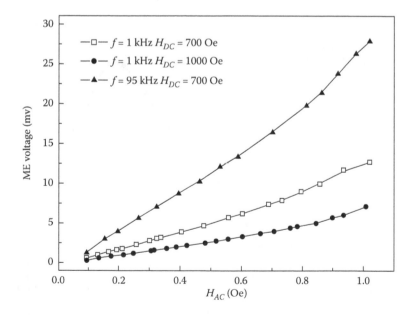

FIGURE 6.14 Induced ME voltages as a function of AC magnetic field H_{AC} for Fe–Ga/BTO laminate, taken at various DC magnetic biases.

$$\sigma_t^E = \frac{E_f E_b t_b \Delta\varepsilon_0}{(1-v)(2E_f t_f + E_b t_b)} \qquad (6.87)$$

where E, t, $\Delta\varepsilon_0$, and v represent the elastic modulus, thickness, the linear strain of the Fe–Ga layer, and Poisson's ratio, respectively. The subscript f or b represents the Fe–Ga or BTO, respectively. As shown in Equation 6.87, the compressive stress in the BTO layer increased with an increase in the linear strain of the Fe–Ga layer. At $H_{DC} = 750$ Oe, the slope of Fe–Ga magnetostriction reaches the maximum. Therefore, when an AC magnetic field H_{AC} is superimposed over the DC magnetic bias H_{DC}, the change in magnetostriction is the biggest. With higher DC magnetic bias, the slope of Fe–Ga magnetostriction decreased, and thus the compressive stress in the BTO layer also decreased. This shows that the induced ME voltage from the composite is proportional to the stress of BTO layer. Therefore, higher ME voltage may be obtained when the compressive stress in the BTO layer is higher. From Equation 6.86, it can be seen that this is achieved when the strain rate of the magnetostrictive layer is maximized, that is, largest $d\varepsilon/dH$. To apply this observation into practical applications, the electrical resistance load effect R_L on the ME coupling in a laminated composite material formed by Fe–Ga/BTO/Fe–Ga was studied.

For application, the ME material must connect with the external circuit, and study of the influence of the load on the ME materials is needed. Figure 6.15 illustrates the schematic diagram of the measurement principle under the load resistance. The composite material was made by sandwiching one thickness-polarized BaTiO$_3$ (BTO) piezoelectric disk between two thickness-magnetized Fe–Ga magnetostrictive alloy disks. The Fe–Ga rod was grown by a Bridgman method and was subsequently cut into disk-shaped plates, while the BTO disks are commercially available [85]. An electrical resistance load (R_L) with the resistance range of 1–200 kΩ was connected electrically in parallel with the ME voltage (V_{ME}) output of the composite material for the ME property measurement using a dynamic method. Details of the measurement system and procedure can be found in the previous literature [85].

The ME voltage coefficient dV_{ME}/dH induced across the ME laminated composite was measured as a function of DC magnetic field (H_{DC}) under various electrical resistance load (R_L) values as shown in Figure 6.16a. A constant AC magnetic field of $H_{AC} = 1$ Oe was applied, and the frequency of H_{AC} was 1 kHz. It is noted that dV_{ME}/dH has a strong dependence on H_{DC} for all cases due to the H_{DC}-dependent piezomagnetic coefficient of the Fe–Ga magnetostrictive alloy plate [86]. The maximum value of dV_{ME}/dH is found to be 10.8 mV/Oe under an optimal H_{DC} of

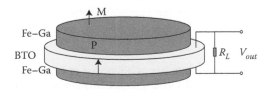

FIGURE 6.15 Schematic diagram of the effect of load resistance on ME properties measurement principle.

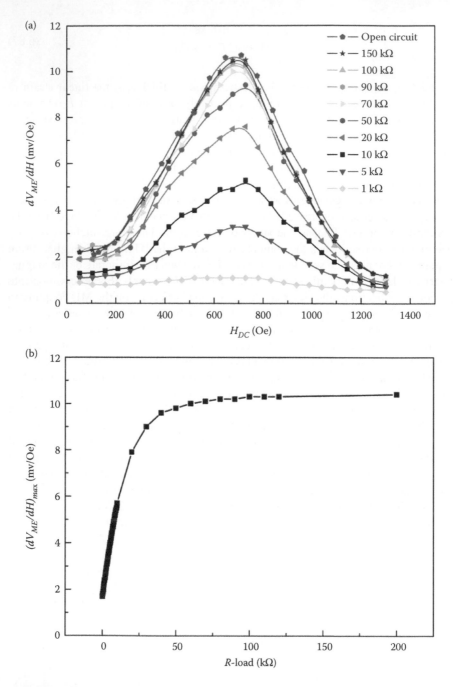

FIGURE 6.16 (a) ME voltage coefficient as a function of DC magnetic field (H_{DC}) under various electrical resistance load (R_L) values for the ME-laminated composite at an applied AC magnetic field (H_{AC}) of 1 Oe peak and a frequency of 1 kHz. (b) The $[dV_{ME}/dH]_{max}$ as a function of R_L.

700 Oe for the composite material in open-circuit condition (i.e., in the absence of R_L). Figure 6.16b plots $[dV_{ME}/dH]$ max as a function of R_L. It is clear that $[dV_{ME}/dH]$ max increased initially by increasing R_L and then saturated at $R_L > 50$ kΩ. This shows that the output power of the ME laminated composite material could not be increased further even by increasing R_L.

Figure 6.17a shows the frequency (f) dependence of ME voltage for the ME laminated composite material at an applied AC magnetic field (H_{AC}) of 1 Oe peak and under various electrical resistance load (R_L) values (including the open-circuit condition). At resonance, the ME response of the laminate is greatly enhanced. By increasing R_L, the resonance ME voltage increased gradually and approached the maximum value of ~65 mV under the open-circuit condition, while the resonance frequency (f_r) shifted to the higher-frequency side by decreasing R_L. In the low-frequency nonresonance range of 1–20 kHz in Figure 6.17b, there is a roll-off in ME voltage with a decrease in f for various R_L, and the cutoff frequency (f_{cut}) decreased by increasing R_L. Assuming that the measured magnetic field is sinusoidal, the cutoff frequency for detection is given as $f_{cut} = 1/2\pi\tau$ [87], where $\tau = NRC_0$ is the time constant, C_0 represents the capacitance of an individual piezoelectric layer, N is the number of piezoelectric layers, and R represents the parallel resistance of the composite's resistance (R_C), the electrometer's input resistance (R_E), and the electrical resistance load (R_L). R can be approximately set to R_L since R_L is generally smaller than R_C and R_E. In addition, an electrometer with high input resistance $(>10^9$ Ω) is important to obtain a large time constant. As a result, the value of f_{cut} is decreased by a factor of $1/R$ by increasing the electrical resistance load R_L.

Figure 6.18 shows the ME voltage coefficient dV_{ME}/dH and corresponding output power (P) as a function of electrical resistance load (R_L) under various frequencies. The ME output power was calculated by $P = V_{ME}^2/R_L$ at $H_{AC} = 1$ Oe peak. It was observed that the dV_{ME}/dH increased when the ME power increased initially reaching a maximum value and then decreased with an increasing electrical resistance load R_L. The similar load effect has also been observed in other systems with piezoelectric element [88]. As the ME output power is directly proportional to the square of ME voltage and the ME voltage is directly proportional to the AC magnetic field, the ME output power is directly proportional to the square of the AC magnetic field. In this manner, an enhanced P could be obtained if an increased H_{AC} could be used or composite fabrication technique could be improved. In Figure 6.18, the optimum load $R_{Load,opt}$ is labeled by red-colored line. When $R_{Load} = R_{Load,opt}$, the circuit has the maximum output power. At $f = 1$ kHz, the $R_{Load,opt}$ was 20 kΩ. However, when f elevated to 50 kHz, $R_{Load,opt}$ decreased to 0.3 kΩ. When close to the resonance frequency of ~92.5 kHz, the P reached the maximum value of ~3 µW (at $R_{Load,opt} = 0.6$ kΩ), compared to the value of ~2 × 10–3 µW at $f = 1$ kHz, an enhancement of ~1500 times was obtained.

Figure 6.19 shows the optimum electrical resistance load $(R_{Load,opt})$ as a function of frequency (f). It was observed that the $R_{Load,opt}$ first rapidly declined and then stabilized to a small value with increasing frequency. The inset of Figure 6.19 is the zoom-in view of the curve near the resonance frequency region. The $R_{Load,opt}$ ascended gradually with the frequency nearing the resonant frequency. In short, the

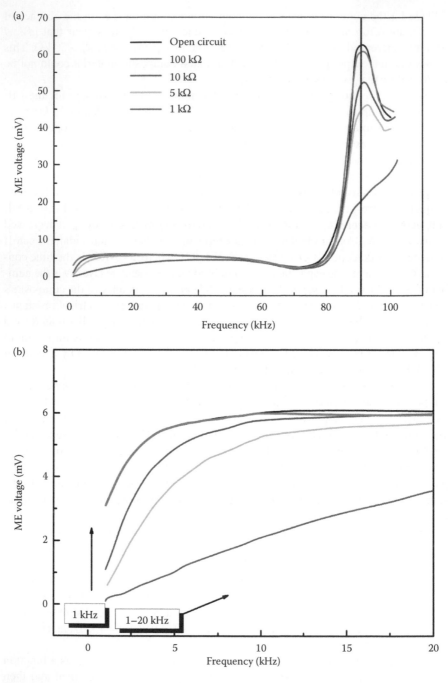

FIGURE 6.17 (a) Frequency (f) dependence of ME voltage for the ME-laminated composite at an applied AC magnetic field (H_{AC}) of 1 Oe peak and under various electrical resistance load (R_L) values (including the open-circuit condition). (b) Zoom-in view for the low-frequency nonresonance range of 1–20 kHz.

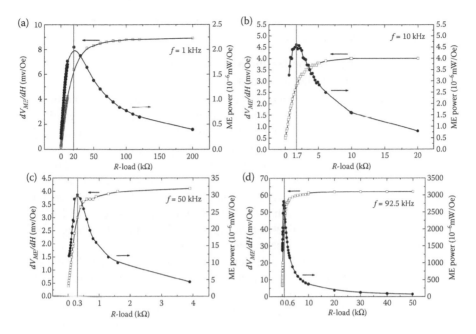

FIGURE 6.18 ME voltage coeflicient dV_{ME}/dH and corresponding output ME power (P) as a function of electrical resistance load (R-load) under the frequencies (a) 1 kHz; (b) 10 kHz; (c) 50 kHz; and (d) 92.5 kHz.

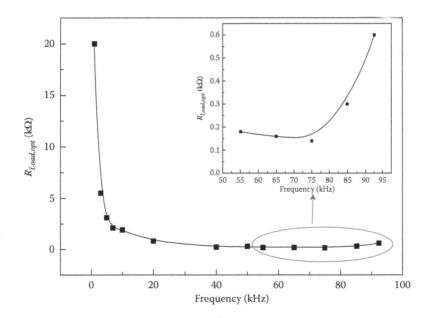

FIGURE 6.19 Optimum electrical resistance load ($R_{Load,opt}$) as a function of frequency. At $R_{Load} = R_{Load,opt}$, the circuit has a maximum output power.

P in the proposed device is adjustable by changing the attached electrical resistance load to achieve the best active status.

6.2.2.2 Longitudinal Magnetization and Transverse Polarization Mode Laminates

To reduce the demagnetization factor effect, a long-type configuration that uses a longitudinal magnetization was designed. This dramatic decrease in the demagnetization factor results in a large reduction in the H_{DC} required to achieve the maximum ME coefficient. Long, rectangular-shaped Terfenol-D/PZT/Terfenol-D and Terfenol-D/ PMN-PT/Terfenol-D three layer laminates with a longitudinal magnetization and transverse polarization L–T were then reported based on this consideration [89–91]. Experimental results confirmed that at low magnetic biases of $H_{DC} < 500$ Oe, much larger values of α_{ME} could be obtained for the L–T laminates relative to the T–T ones. Clearly, the long-type L–T laminates have significantly higher ME voltage coefficients than the T–T laminates under modest magnetic biases.

In order to obtain a high ME coefficient, the magnetic phase must have a large magnetostrictive coefficient at low magnetic field. As for soft magnetic materials, FeCoV is sensitive to the magnetic field and the strain over the magnetic field ($\Delta\lambda/\Delta H$) is large, it can be used as the magnetic phase in the composite. $Fe_{49.5}Co_{49}V_{1.5}$ alloy was cut into rectangular shape of 20×10 mm. Commercial single-crystal PMN-PT with (001) orientation was also cut into a rectangular shape of 20×10 mm using a diamond cut. The piezoelectric PMN-PT was poled along the thickness. Different thicknesses of PMN-PT (0.2, 0.4, 0.6, and 0.8 mm) and a FeCoV laminate, which is 0.8 mm thick, were assembled into a layered structure using conductive epoxy resin.

Figure 6.20 illustrates the magnetostrictive coefficient (λ) and the piezomagnetic coefficient versus magnetic field (H) of the FeCoV plate. By increasing H, the

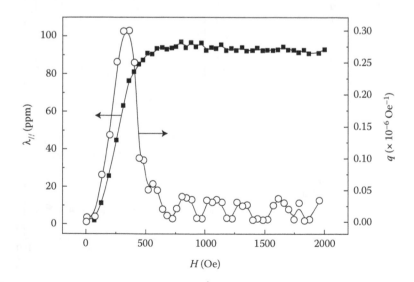

FIGURE 6.20 Magnetostrictive coefficient (λ) and piezomagnetic coefficient versus magnetic field (H) of the FeCoV plate.

magnetostrictive coefficient quickly increased to a saturation of ~90 ppm at 750 Oe. The piezomagnetic coefficient (q) increased sharply and peaked at ~430 Oe, and then it decreased to about 0 at ~750 Oe. The piezomagnetic coefficient is the slope of the magnetostriction line. At ~430 Oe, the magnetostriction line gets the highest slope, and at ~750 Oe, FeCoV is saturation magnetizing and the piezomagnetic is almost 0.

Figure 6.21 illustrates the ME coefficient as a function of static magnetic field strength of FeCoV/PMN-PT samples. Figure 6.21 shows that α_{ME} is strongly dependent on the static field strength. For all the samples, the α_{ME} versus H_{DC} shows the same trend: first increases sharply, peaks at 430 Oe, and then decreases quickly. According to the magneto–elasto–electric equivalent circuit method, the α_{ME} for L–T mode at resonance frequency can be expressed as [92]

$$\alpha_{ME} = \frac{8d_{33,m}d_{31,p}Q_{em}n_m}{\pi^2\varepsilon_{33}^T\left[(1-n_m)S_{33}^H+\left(1-k_{31}^2\right)n_mS_{11}^E\right]} \tag{6.88}$$

where α_{ME}, $d_{31,p}$, and $d_{33,m}$ are the resonant ME coefficient, the piezoelectric coefficient of piezoelectric material, and the piezomagnetic coefficient of magnetostrictive material, respectively. $Q_{em}, n_m, \varepsilon_{33}^T$, and k_{31} are the effective mechanical quality factor, the thickness ratio of magnetostrictive layers in the composites, the permittivity

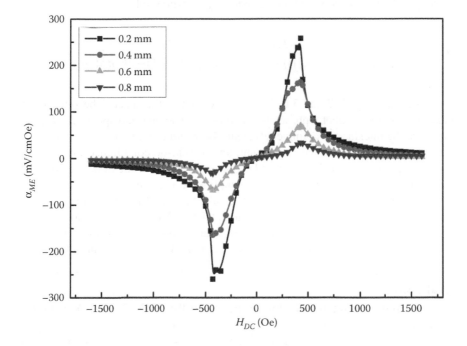

FIGURE 6.21 ME coefficient of different thickness samples (the thickness of the PMN-PT changed from 0.2 to 0.8 mm) as a function of static magnetic field strength.

under constant stress in the piezoelectric layer, and the electromechanical coupling coefficient of piezoelectric material, respectively. S_{33}^H and S_{11}^E are the longitudinal elastic compliance at constant H_{DC} and the elastic compliance at constant electric field, respectively. As illustrated in Figure 6.21, the ME coefficient first increased and then decreased gradually with the increment in H_{DC}; such a variation is primarily attributed to the piezomagnetic coefficient. According to Equation 6.88, the resonant ME field coefficient is directly proportional to the piezomagnetic coefficient ($\Delta\lambda/\Delta H$). The dependence of the piezomagnetic coefficient on the DC bias magnetic field (H_{DC}) results in the resonant α_{ME} on the DC bias magnetic field (H_{DC}). In addition, it can also be observed from Figure 6.21 that the optimum DC bias magnetic field ($H_{DC,0}$) increases with the increment in PMN-PT layers thickness.

According to Equation 6.86 and Figure 6.21, as the H_{DC} increased from zero, the magnetostriction of the FeCoV increased quickly and both the strain in the PMN-PT and the α_{ME} also increased. When the H_{DC} reached 750 Oe, the magnetic phase approached to saturation and the strain did not change anymore. Hence, the α_{ME} is almost zero. Referred to Figure 6.21, as the thickness of the PMN-PT increased (the volume fraction f of the piezoelectric phase increased), the experimental values of ME coefficient decreased. The main reason is that the stress created by the FeCoV is nearly the same under similar H_{AC}. Hence, as the thickness of the PMN-PT is increased, its strain decreased, which in turn decreased the α_{ME}.

According to Bichurin et al. [58], the surface combination is a parameter (k), and the longitudinal-transverse ME coefficient $\alpha_{E,31}$ is expressed as

$$\alpha'_{E,31} = \frac{E_3}{H_1} = \frac{-kv(1-f)(^mq_{11} + {}^mq_{21})^pd_{31}}{{}^p\varepsilon_{33}\left({}^ms_{12} + {}^ms_{11}\right)kf + {}^p\varepsilon_{33}\left({}^ps_{11} + {}^ps_{12}\right)(1-f) - 2k\,{}^pd_{31}^2(1-f)} \quad (6.89)$$

where the subscript 1, 2 and 3 denote the direction of the space. The superscript p and m denote the magnetic phase and the piezoelectric phase, respectively. $k, f, q, d,$ and s are the interface coupling coefficient, the volume fraction, the piezomagnetic coefficients, the piezoelectric coefficient, and the elastic compliances, respectively. Using the parameters in Table 6.1 and supposing $k = 0.1$, for the bonding strength of the conductive-epoxy resin is weak and simplified for calculate.

From Equation 6.89 and Table 6.6, the calculated and the experimental α_{ME} of the PMN-PT/FeCoV structure with different PMN-PT volume fraction is shown in Figure 6.22. The line is the calculated value, which is the volume fraction of PMN-PT from 0 to 1, and the circle is the calculated ME coefficient at the volume

TABLE 6.6

The Parameter Using in Calculation the α_{ME} of the PMN-PT/FeCoV Structure

Parameter	$^mq_{11}$ (10^{-12} m/A)	$^mq_{21}$ (10^{-12} m/A)	$^pd_{31}$ (10^{-12} m/V)	$^p\varepsilon_{33}$ (10^{-12} m/V)	$^ms_{11}$ (10^{-12} m²/N)	$^ms_{12}$ (10^{-12} m²/N)	$^ps_{11}$ (10^{-12} m²/N)	$^ps_{12}$ (10^{-12} m²/N)
Value	3750	260	160	−1800	13	−6	31.1	−28.2

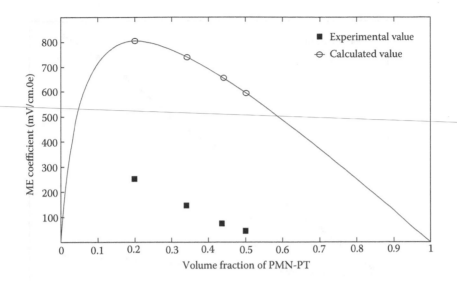

FIGURE 6.22 Calculated and experimental α_{ME} of the PMN-PT/FeCoV structure with different PMN-PT volume fraction.

fraction, which is the same as the experiment dots. The black dots are the experimental values. From Figure 6.22, the trend between the calculated value and the experimental value is similar within the experimental range. The experimental value being lower than the calculated value may be due to the stress loss from the magnetic to piezoelectric phase transition.

Figure 6.23 plots induced ME voltages over alternating field strength (H_{AC}). With an increase in the fraction of PMN-PT, the ME voltage decreased. For all the samples, a perfect linear relationship was observed under the static field of 400 Oe (with ME coefficient peaks at 430 Oe, 400 Oe was chosen to test the response to varying alternating field; see Figure 6.21).

According to Livingston's model of coherent rotation of magnetization, the strain generated from the magnetostriction under the small amplitudes of alternating magnetic field strength H_{AC}, $\Delta\varepsilon_0$ is expressed as [93]

$$\Delta\varepsilon_0 = \frac{3\lambda_s}{2}\left(\frac{H_{AC}^2}{H_0^2}\right) \tag{6.90}$$

where λ_s is the saturation magnetostriction constant and H_0 is the magnetic anisotropy field.

For PMN-PT, the induced strain is approximately proportional to the square of the electric field [94]. However, as the H_{AC} became very small (less than 2.5 Oe), the strain induced by H_{AC} was very small, and the resultant voltage of the PMN-PT is very small. Thus, the H_{AC} overlaps with the H_{DC} (about 400 Oe). Within this range, we assume that the ME voltage is proportional to the stress. As ME voltage and α_{ME} are directly proportional to the stress [77],

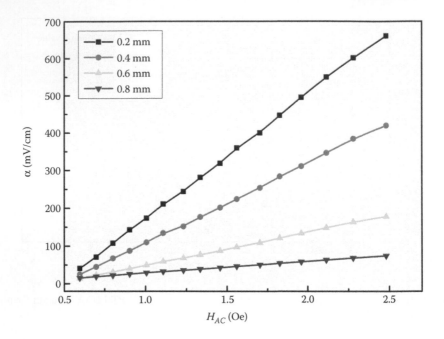

FIGURE 6.23 Induced ME voltages over alternating field strength (H_{AC}) at different volume fractions of PMN-PT.

$$V_i = 2g_{31} \times t \times \sigma_p^E \qquad (6.91)$$

where g_{31} is the piezoelectric voltage constant, σ_p^E is the stress in PMN-PT, and t is the thickness. Using the definition of $\alpha_{ME} = (\delta V_i / t)/\delta H$, combined Equations 6.86, 6.90, and 6.91 yield the ME voltage (α_V) of the PMN-PT/FeCoV structure

$$\alpha_V = 2g_{31} \times \frac{E_f E_p t_f}{(1-v)(2E_f t_f + E_p t_p)} \times \frac{3\lambda_s}{2} \left(\frac{H_{AC}}{H_0^2} \right)$$

$$= 3\lambda_s g_{31} \times \frac{E_f E_p t_f}{(1-v)(2E_f t_f + E_p t_p)} \times \left(\frac{1}{H_0^2} \right) \times H_{AC} \qquad (6.92)$$

From Equation 6.92, it is obvious that the ME voltage is directly proportional to H_{AC} when the other parameters are maintained. This makes PMN-PT/FeCoV laminate a perfect candidate in AC magnetic sensing.

The resonance frequency of the laminate ME composite materials is well studied. In the bilayer ME composite materials, there are two remarkable resonance peaks between 0 and 105 kHz (see Figure 6.24). The thicknesses of PMN-PT and Fe–Ga are 1 and 0.8 mm, respectively. Near the resonance peak, the ME voltage of the composite has a sharp rising. The bending resonance mode appears in the low-frequency region because of the nonsymmetrical stress distribution in the bilayer sample. With

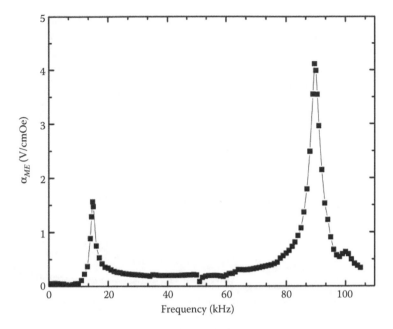

FIGURE 6.24 Magnetoelectric voltage coefficient α_{ME} as a function of AC magnetic field frequency for FeCoV(0.8 mm)/PMN-PT(1.0 mm) laminate composition.

further increase in frequency, the second resonance appeared, and it is generated at the planar acoustic resonance mode.

According to Bi et al. [95], the major resonance peak is planar acoustic resonance mode (88.6 kHz), and the secondary peak is the bending resonance mode (15.1 kHz), which are expressed by

$$f_{bending} = \frac{\pi d}{4\sqrt{3}l^2}\sqrt{\frac{1}{\bar{\rho}\bar{s}_{11}}}\beta_1^2 \tag{6.93}$$

$$f_{vabriction} = \frac{1}{2l}\sqrt{\frac{1}{\bar{\rho}\bar{s}_{11}}} \tag{6.94}$$

where d, l, $\bar{\rho}$, and $\bar{s}_{11}$ represent the total thickness, the length, the average density, and the equivalent elastic compliance of the sample, respectively. β_1 is the first mode order.

$$\bar{s}_{11} = \frac{s_{11}^F s_{11}^P}{V_F s_{11}^B + V_B s_{11}^F} \tag{6.95}$$

The effect of the PMN-PT thickness on the resonance peak is studied. Figure 6.25 shows the ME voltage coefficient α_{ME} as a function of AC magnetic field frequency for the samples having different PMN-PT thicknesses. The PMN-PT sample with

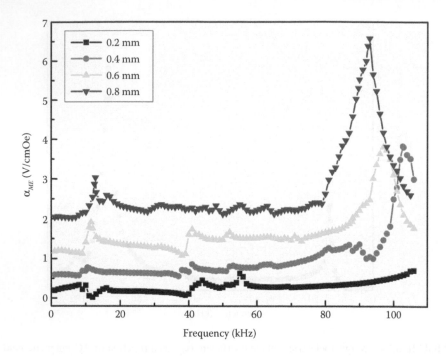

FIGURE 6.25 Magnetoelectric voltage coefficient α_{ME} as a function of AC magnetic field frequency for FeCoVPMN-PT bilayer sample, wherein the thickness of PMN-PT is 0.2, 0.4, 0.6, and 0.8 mm, respectively.

0.8 mm thickness has a major resonance peak at ~93 kHz. At the resonance, the α_{ME} is 5 V/cmOe higher than that at 1 kHz. With the thickness of PMN-PT decreased, the major resonance shifted to high frequency and resonance frequency of the 0.2 mm thickness sample is over 105 kHz (the maximum frequency of the measurement setup). There is a secondary peak at about 12 kHz, and it is the bending resonance mode. When the thickness of PMN-PT decreased, the bending resonance peak shifted to a lower frequency. The physical parameters are listed in Table 6.7, and the calculated results using Equations 6.93 and 6.94 are listed in Table 6.8. The calculated results agreed with the experiment results.

The density and elastic compliance of the PMN-PT are larger than that of the FeCoV. When the thickness of the PMN-PT decreased, the major resonance peak shifted to higher frequency. The total thickness of the sample is decreased when the thickness of the PMN-PT decreased, and subsequently the bending resonance shifted to a lower frequency.

TABLE 6.7
Physics Parameter of FeCoV and PMN-PT

Parameter	ρ_f	s_f	ρ_p	s_p	β_1
Value	8080 kg/m^2	5.85×10^{-12} m^2/N	8100 kg/m^2	28×10^{-12} m^2/N	1.5

TABLE 6.8

The Resonance Frequency of the Sample with the Different Thickness of PMN-PT

Thickness (PMN-PT)	0.2 mm	0.4 mm	0.6 mm	0.8 mm
$f_{bending}$ (kHz)	10.8	12.1	13.3	14.6
$f_{vibration}$ (kHz)	105.5	98.6	93.5	89.4

6.2.2.3 Circumferential Magnetization and Transverse Polarization Mode Laminates

In many situations, the magnetic fields are excited by the electric currents. In this case, the excited magnetic fields are vortexes. An ME ring-type laminate operated in circumferential magnetization and circumferential polarization has been designated as the C–C mode. Experimental investigations show that a three-layer Terfenol-D/PMN-PT/Terfenol-D laminate ring has a very high ME coefficient with maximum values of up to 5.5 V/cmOe at 1 kHz in response to a vortex magnetic field. High ME coupling in the C–C mode is due to the magnetic loop of the ring ME-type configuration, which is suitable for capturing a vortex-type field. Hence, such C–C ME laminate ring can be potential solution for electric current-sensing applications.

For the piezoelectric phase, it is easier to polarize along the thickness than along the circumstance. Using the ring-shaped sample, there is another advantage that the ME effect will be enhanced. To make the ME coefficient α_{ME} measureable, the following expression is used [96]:

$$\alpha_{ME} = \frac{V}{H_{AC} \cdot d} \tag{6.96}$$

where V is the voltage generated due to the ME effect, H_{AC} is the amplitude of the sinusoidal magnetic field, and d is the effective distance between the electrodes of the piezoelectric phase. From Equation 6.96, α_{ME} can be improved by decreasing H_{AC} at constant V. There are two ways to achieve that, first by using soft magnetic material to decrease H_{AC} [97]. Approximately 2.18 mV/cmOe was achieved in FeCoV film-based sandwich structure. The other way was to decrease the demagnetizing field (H_d). As the "magnetic poles" appear at the double-end surface when the material is noninfinite or magnetic circuit unclosed, H_d comes into existence in the material being magnetized working in the opposite direction of the applied field. H_{AC} is therefore decreased by the amount of H_d. High aspect ratio samples or closed magnetic circuit samples minimize or completely avoid H_d. Giang et al. analyzed the enhancement of the ME effects with a narrower width by taking into account the demagnetization contribution [98]. Leung et al. developed a ring-type electric current sensor for detection of vortex magnetic fields focusing on electric current sensitivity [99]. Dong et al. studied voltage effect in a ring-type ME laminate [100].

TABLE 6.9
Material Parameters of BaTiO$_3$ by Commercial Supplier

Material	Density (10^3 kg/m^3)	g_{31} (10^{-3} Vm/N)	Y_{11} (10^9 N/m^2)	S_{11} (10^{-12} m^2/N)	Poisson's Ratio
BaTiO$_3$	5.6	4.4	119	8.4	0.33

The effect of the demagnetizing field and the ME response of the closed magnetic circuit Fe–Ga/BTO/Fe–Ga sandwich structure with respect to the constant magnetic field, frequency, and the alternating field were also investigated. Fe-19at.% Ga alloy disks with a thickness of 0.8 mm were sliced from the rod and subsequently drilled in the center to achieve a "closed-loop circuit" with an outer diameter of 28 mm and an inner diameter of 10 mm. The magnetization and magnetostriction of the hollow disk along the circumference were measured by the hysteresis graph of soft magnetic materials [MATS-2010s] and standard strain gauge technique, respectively. Commercially available BTO disks (φ28 × 2 mm) were also drilled in the center to form a φ10 mm hole. Table 6.9 lists the basic parameters of the BTO disks.

The piezoelectric voltage constant, g_{31}, is 4.4 Vm/N. Two hollow Fe–Ga disks and one hollow BTO disk are assembled into a closed magnetic circuit Fe–Ga/BTO/ Fe–Ga sandwich structure using conductive epoxy resin as illustrated in Figure 6.26.

The ME effect of the closed magnetic circuit sandwich structure was measured under a constant magnetic field strength (H_{DC}) applied perpendicularly to the sample surface superimposed with a small alternating field strength (H_{AC}) along the circumferential direction (refer to Figure 6.26). The induced ME voltage was in the thickness direction. As the small alternating magnetic field was superimposed on the static field, the measurement was "dynamic" [101]. The dynamic magnetization was along circumference direction, and the polarization was oriented in transverse direction. The H_{DC} was controlled by a DC source variable from 0 to 800 Oe, and the alternating field H_{AC} ranging from 0.1 to 0.25 Oe was generated by a 10N coils, which was supplied by a signal generator, with the strength determined by [102]

$$H = \frac{N_1 I}{L_m} \tag{6.97}$$

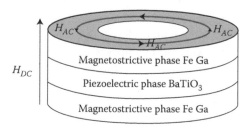

FIGURE 6.26	Schematic diagram of the closed magnetic circuit sandwich structure sample.

where N_1 represents the number of turns, I represents the current, and $\bar{L}_m$ represents the average magnetic circuit length.

Figure 6.27a plots X-ray powder diffraction (XRD) results of the Fe–Ga rod, and the observation confirmed that the A2 structure has three major peaks: (110), (200), and (211). The cross section along the radial direction (Figure 6.27b(i)) indicates equal-axial grains. Along the axial direction (Figure 6.27b(ii)), only a few grains are seen. These indicate that the Fe–Ga rod is heavily textured along the axial or (110) orientation.

In the closed magnetic circuit sandwich structure, the compressive stress in the BTO (σ) is expressed in Equation 6.86. It suggests that the stress in the BTO layer will enhance with an increase in the linear strain of the Fe–Ga layer. The voltage output from the piezoelectric layer (V_i) is determined by Equation 6.91, which suggests that the induced ME voltage from the composite is proportional to the stress of the BTO layer. As such, the ME voltage is proportional to the magnetostriction of Fe–Ga layer. Solving Equations 6.86, 6.91, and 6.96 simultaneously yields

$$\alpha_{ME} = \frac{V_i}{t \times H_{AC}} = 2g_{31} \frac{2E_f E_b t_f}{(1-v)(2E_f t_f + E_b t_b)} \times \frac{\Delta\varepsilon_0}{H_{AC}} \qquad (6.98)$$

Equation 6.98 suggests that higher ME coefficient (α_{ME}) may be obtained by increasing the ratio of the strain over the magnetic field ($\Delta\varepsilon_0/H_{AC}$).

When H_{AC} is small, $\Delta\varepsilon_0$ is less than 0.1 ppm, which is unmeasurable by standard strain gauges. Assuming that the magnetostriction is linear, the ratio of saturated magnetostriction over the saturated field is equal to $\Delta\varepsilon_0/H_{AC}$. Figure 6.28 shows the magnetization and magnetostriction property examined along circumference of the hollow Fe–Ga disk. Due to zero demagnetization, only ~45 Oe is needed to almost saturate the hollow Fe–Ga disk with a saturated magnetostriction of 25 ppm along the circumference. In comparison, ~1000 Oe is needed to saturate the Fe–Ga rod, and the saturated magnetostriction is ~100 ppm [103]. The orientation of the hollow Fe–Ga disk is perpendicular to the disk surface, which leads to the saturated magnetostriction along the circumference that is 75% less as compared to the Fe–Ga rod. The saturated field of the Fe–Ga solid disk along the thickness direction is larger than the Fe–Ga rod due to the demagnetization effect [103]. Thus, the ratio of saturated magnetostriction over saturated field for the solid disk is less as compared to the rod (100 ppm/1000 Oe [103]). From Figure 6.28, the ratio in the hollow Fe–Ga disk is 25 ppm/45 Oe or 5 folds larger than that of the rod (100 ppm/1000 Oe). According to Equation 6.98, higher ME coefficient (α_{ME}) can be obtained in the closed magnetic circuit sandwich structure.

The α_{ME} of the closed magnetic circuit sample has a sharp peak at the resonance frequency, f_r, of 68 kHz [104]. When approaching the f_r, α_{ME} rises sharply. The highest value reached is 1.88 V/cmOe, which is 8 times larger than the α_{ME} value obtained at a lower frequency. This phenomenon indicates that the device should work at the resonance frequency, and the sensitivity will be improved. The f_r has a relationship with the shape of the sample in vibration [105]. How to design the device and control the f_r at desired magnitude is very important.

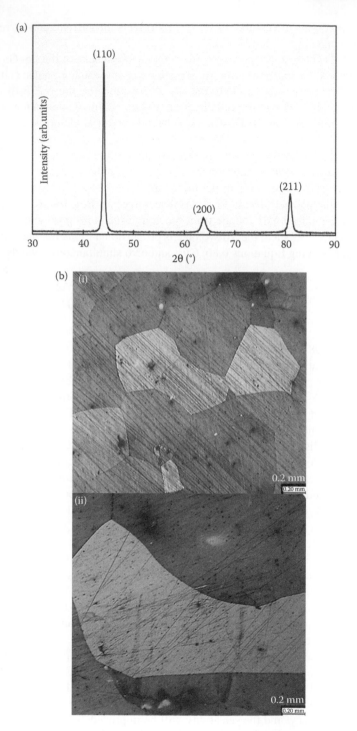

FIGURE 6.27 XRD results and the metallurgical microscope of the Fe–Ga rod.

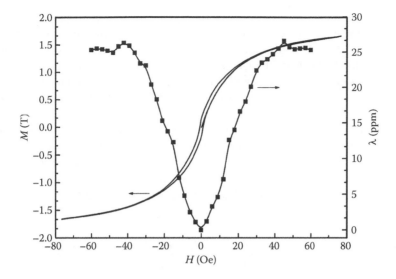

FIGURE 6.28 Magnetization and magnetostriction property of the hollow Fe–Ga disk examined along circumference.

From Figure 6.29, the sample only has one resonant peak. Being a sandwich structure, the sample has no bending resonant. The following assumptions were made: (1) the mode of the resonance is radial vibration. (2) Fe–Ga phase has a periodic force, F, and the mechanical property of Fe–Ga can be substituted by weighted average of the whole sample. For the vibration of the sample due to the periodic magnetic field

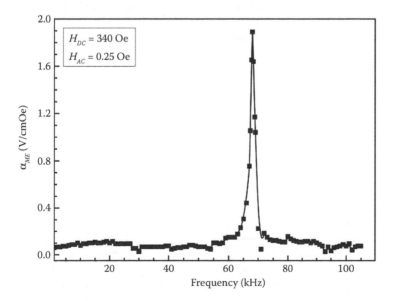

FIGURE 6.29 Magnetoelectric voltage coefficient α_{ME} as a function of AC magnetic field frequency for closed magnetic circuit Fe–Ga/BTO/Fe–Ga sandwich structure.

that caused the magnetomechanical vibration by stress coupled, the Fe–Ga phase applied the force. (3) There is no stress loss between the different phases (the interface is perfect). Based on these assumptions, the problem was simplified to study the vibration of the BTO phase, though its mechanical property is the weighted average of the whole sample. Figure 6.30a illustrates the draft of the closed magnetic circuit Fe–Ga/BTO/Fe–Ga sandwich structure. Figure 6.30b is the draft of Fe–Ga phase as a force, F.

Piezoelectric phase has a polarization along the thickness. The thickness of the sample is represented by h, the external diameter by a, and the internal diameter by b. v_{ra} and v_{rb} are the vibration velocity of the exterface and interface, respectively. Combining the structure of the sample and the direction of the magnetic field (along the circumstance), the direction of F is in-plane. The wave equations is expressed as [106]

$$\rho \frac{\partial^2 \xi_r}{\partial t^2} = \frac{\partial T_r}{\partial r} + \frac{1}{r}\frac{\partial T_{r\theta}}{\partial \theta} + \frac{\partial T_{rz}}{\partial z} + \frac{T_r - T_\theta}{r} \tag{6.99}$$

$$\rho \frac{\partial^2 \xi_\theta}{\partial t^2} = \frac{\partial T_{r\theta}}{\partial r} + \frac{1}{r}\frac{\partial T_\theta}{\partial \theta} + \frac{\partial T_{\theta z}}{\partial z} + \frac{T_{r\theta}}{r} \tag{6.100}$$

$$\rho \frac{\partial^2 \xi_z}{\partial t^2} = \frac{\partial T_{rz}}{\partial r} + \frac{1}{r}\frac{\partial T_{\theta z}}{\partial \theta} + \frac{\partial T_z}{\partial z} + \frac{T_{rz}}{r} \tag{6.101}$$

where ξ_r, ξ_θ, and ξ_z are the radial, the tangential, and the axial displacement components, respectively. T_r, T_θ, T_z, $T_{r\theta}$, T_{rz}, and $T_{\theta z}$ are the stresses in the ring. Using the polar coordinates, the relationships between the stresses and strains of the ring are expressed as

$$S_r = \frac{\partial \xi_r}{\partial r}, \quad S_\theta = \frac{1}{r}\frac{\partial \xi_\theta}{\partial \theta} + \frac{\xi_r}{r}, \quad S_z = \frac{\partial \xi_z}{\partial_z} \tag{6.102}$$

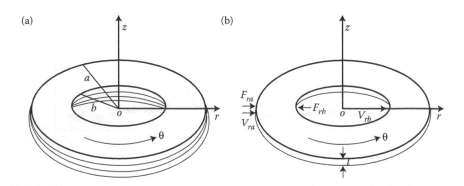

FIGURE 6.30 Schematic closed magnetic circuit Fe–Ga/BTO/Fe–Ga sandwich structure.

$$S_{r\theta} = \frac{1}{r}\frac{\partial \xi_r}{\partial \theta} + \frac{\partial \xi_\theta}{\partial r} - \frac{\xi_\theta}{r}$$

$$S_{\theta z} = \frac{1}{r}\frac{\partial \xi_z}{\partial \theta} + \frac{\partial \xi_\theta}{\partial z}$$

$$S_{rz} = \frac{\partial \xi_r}{\partial z} + \frac{\partial \xi_z}{\partial r} \qquad (6.103)$$

where S_r, S_θ, S_z, $S_{r\theta}$, $S_{\theta z}$, and S_{rz} are the strains in the ring. Here, the sample has a radial extensional vibration, and the thickness of the sample, h, is much smaller than a and b. Consider that $\xi_\theta = 0$, $(\partial T_\theta/\partial\theta) = 0$, $(\partial\xi_\theta/\partial\theta) = 0$, while T_z, $T_{r\theta}$, T_{rz}, $T_{\theta z}$, $S_{r\theta}$, $S_{\theta z}$, and S_{rz} can be ignored. The constitutive equations are expressed as

$$S_r = s_{11}^E T_r + s_{12}^E T_\theta + s_{13}^E T_z + d_{31}E_3 \qquad (6.104)$$

$$S_\theta = s_{12}^E T_r + s_{11}^E T_\theta + s_{13}^E T_z + d_{31}E_3 \qquad (6.105)$$

$$S_z = s_{13}^E(T_r + T_\theta) + s_{33}^E T_z + d_{33}E_3 \qquad (6.106)$$

$$D_3 = d_{31}(T_r + T_\theta) + d_{33}T_z + \varepsilon_{33}^T \qquad (6.107)$$

Here, s_{ij}^E is the elastic compliance constant at the static electric field. d_{31} and d_{33} are the piezoelectric strain constants, E_3 is the external exciting electric field, and ε_{33}^T is the dielectric constant measured at constant stress. Using the above equations and by combining the conditions of thin ring in radial vibration, the radial wave equation is obtained as

$$\frac{\rho(\partial^2\xi_r/\partial t^2)}{E_r} = \frac{\partial^2\xi_r}{\partial r^2} + \left(\frac{\partial\xi_r/\partial r}{r}\right) - \frac{\xi_r}{r^2} \qquad (6.108)$$

where ρ is the density, ξ_r is the radical displacement component, $\xi_r = \xi_{ra}\exp(jwt)$, and $E_r = s_{11}^E/\left((s_{11}^E - s_{12}^E)(s_{11}^E + s_{12}^E)\right)$.

The solution of Equation 6.108 is given as

$$\xi_{ra} = AJ_1(k_r r) + BY_1(k_r r) \qquad (6.109)$$

where $J_1(k_r r)$ and $Y_1(k_r r)$ are the Bessel functions of order one with A and B being the constants. From Figure 6.30b, the boundary conditions are $v_r|_{r=a} = -v_{ra}$ and $v_r|_{r=b} = v_{rb}$, and the constants are expressed as

$$A = -\frac{1}{jw}\frac{v_{ra}Y_1(k_r b) + v_{rb}Y_1(k_r a)}{J_1(k_r a)Y_1(k_r b) - J_1(k_r b)Y_1(k_r a)} \qquad (6.110)$$

$$B = -\frac{1}{j\omega}\frac{v_{ra}J_1(k_rb)+v_{rb}J_1(k_ra)}{J_1(k_ra)Y_1(k_rb)-J_1(k_rb)Y_1(k_ra)} \tag{6.111}$$

The radial stress, T_r, is determined by

$$T_r = v_{ra}\frac{J_1(k_rb)[k_rY_0(k_rr)-Y_1(k_rr)(1-v_{12})/r]-Y_1(k_rb)[k_rJ_0(k_rr)-J_1(k_rr)(1-v_{12})/r]}{j\omega[J_1(k_ra)Y_1(k_rb)-J_1(k_rb)Y_1(k_ra)]s_{11}^E\left(1-v_{12}^2\right)}$$

$$+v_{rb}\frac{J_1(k_ra)[k_rY_0(k_rr)-Y_1(k_rr)(1-v_{12})/r]-Y_1(k_ra)[k_rJ_0(k_rr)-J_1(k_rr)(1-v_{12})/r]}{j\omega[J_1(k_ra)Y_1(k_rb)-J_1(k_rb)Y_1(k_ra)]s_{11}^E\left(1-v_{12}^2\right)}$$

$$-\frac{d_{31}E_3}{S_{11}^E+S_{12}^E} \tag{6.112}$$

According to the boundary conditions of Figure 6.30, $F_{ra}=-T_r|_{r=a}S_a$ and $F_{rb}=-T_r|_{r=b}S_b$, whereby $S_a=2\pi al$ and $S_b=2\pi bl$. Combining with Equation 6.112 yields

$$F_{ra}' = \frac{Z_{ra}}{j}\frac{av_{ra}}{2\pi a^2}\left[\frac{Y_1(k_rb)J_0(k_ra)-J_1(k_rb)Y_0(k_ra)}{J_1(k_ra)Y_1(k_rb)-J_1(k_rb)Y_1(k_ra)}+\frac{1-v_{12}}{k_ra}\right]$$

$$+\frac{Z_{ra}}{j}\frac{bv_{rb}}{2\pi ab}\frac{Y_1(k_ra)J_0(k_ra)-J_1(k_ra)Y_0(k_ra)}{J_1(k_ra)Y_1(k_rb)-J_1(k_rb)Y_1(k_ra)}-\frac{d_{31}}{S_{11}^E+S_{12}^E}V_3 \tag{6.113}$$

$$F_{rb}' = \frac{Z_{rb}}{j}\frac{av_{rb}}{2\pi b^2}\left[\frac{Y_1(k_ra)J_0(k_rb)-J_1(k_ra)Y_0(k_rb)}{J_1(k_ra)Y_1(k_rb)-J_1(k_rb)Y_1(k_ra)}+\frac{1-v_{12}}{k_rb}\right]$$

$$+\frac{Z_{rb}}{j}\frac{av_{ra}}{2\pi ab}\frac{Y_1(k_rb)J_0(k_rb)-J_1(k_rb)Y_0(k_rb)}{J_1(k_ra)Y_1(k_rb)-J_1(k_rb)Y_1(k_ra)}-\frac{d_{31}}{S_{11}^E+S_{12}^E}V_3 \tag{6.114}$$

Let $v_{ra}'=-v_{ra}[J_0(k_rb)Y_1(k_rb)-Y_0(k_rb)J_1(k_rb)]$ and $v_{rb}'=-v_{rb}[J_0(k_ra)Y_1(k_ra)-Y_0(k_ra)J_1(k_ra)]$.

According to the spherical Bessel function, $J_{n+1}(x)Y_n(x)-Y_{n+1}(x)J_n(x)=2/(\pi x)$; thus,

$$\left.\begin{array}{l}v_{ra}'=\dfrac{2}{\pi k_r ab}v_{ra}a\\[2mm]v_{rb}'=\dfrac{2}{\pi k_r ab}v_{rb}b\end{array}\right\} \tag{6.115}$$

Substituting Equation 6.115 into Equations 6.113 and 6.114 yields

$$F_{ra}''=(Z_2+Z_3)v_{ra}'+Z_3v_{rb}'+N_{31}V_3 \tag{6.116}$$

$$F_{rb}''=(Z_1+Z_3)v_{rb}'+Z_3v_{ra}'+N_{31}V_3 \tag{6.117}$$

where Z_1, Z_2, and Z_3 are the impedances.

Letting the electric current into the sample as I, the harmonic vibration is given as $I = (dQ/dt) = J\omega Q$. The electrical charge, Q, is determined by

$$Q = 2\pi \int_a^b D_3 r\,dr \tag{6.118}$$

Here, D_3 represents the media electric displacement (the number of the electric charge in unit area). Using the above equations, Equation 6.118 is re-expressed as

$$Q = \frac{2\pi d_{31}}{s_{11}^E + s_{12}^E}[Ak_r C_1 + Bk_r C_2] + \pi(a^2 - b^2)\left[\varepsilon_{33}^T E_3 - \frac{2d_{31}E_3 d_{31}}{s_{11}^E + s_{12}^E}\right] \tag{6.119}$$

where

$$C_1 = \frac{1}{k_r^2}[k_r a J_1(k_r a) - k_r b J_1(k_r b)] \quad \text{and} \quad C_2 = \frac{1}{k_r^2}[k_r a Y_1(k_r a) - k_r b Y_1(k_r b)]$$

Substituting Equations 6.110 and 6.111 into Equation 6.119 yields

$$I = j\omega Q = j\omega\left(\frac{\varepsilon_{33}^T S}{l}\right)\left[1 - \left(\frac{2d_{31}^2}{\varepsilon_{33}^T\left(s_{11}^E + s_{12}^E\right)}\right)\right]V_3 - \frac{2\pi d_{31}}{s_{11} + s_{12}}(a v_{ra} + b v_{rb}) \tag{6.120}$$

where the area of the closed-loop sample is determined by $S = \pi(a^2 - b^2)$. With $C_{or} = \left(\varepsilon_{33}^T S/l\right)\left[1 - \left(2d_{31}^2/\left(\varepsilon_{33}^T\left(s_{11}^E + s_{12}^E\right)\right)\right)\right]$ representing the clamped electric capacitance of the sample in radial vibration, Equation 6.120 is re-written as

$$I = j\omega C_{or} V_3 - N_{31}(v'_{ra} + v'_{rb}) \tag{6.121}$$

From Equations 6.116, 6.117, and 6.121, the Mason equivalent circuit model of the sample, in which the force is substituted by voltage and the acoustic velocity is substituted by current, is shown in Figure 6.31.

When the ring sample is free from external forces, the input admittance of the sample in radial vibration can be determined by

$$Y = \frac{j\omega \varepsilon_{33}^T S}{l} \times \left\{1 - k_p^2 + k_p^2 \frac{2}{a^2 - b^2}\right\}$$

$$\times \frac{[a J_1(k_r a) - b J_1(k_r b)][Y(b) - Y(a)] + [a Y_1(k_r a) - b Y_1(k_r b)][J(a) - J(b)]}{(J(a)Y(b)) - (J(b)Y(a))} \tag{6.122}$$

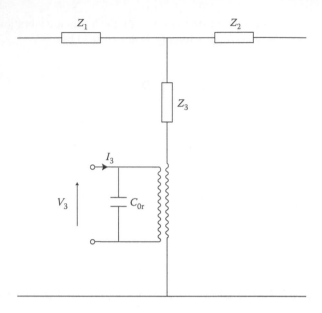

FIGURE 6.31 Mason equivalent circuit model of the ring.

where k_p is the plane electromechanical coupling coefficient, $k_p^2 = 2d_{31}^2/\left(\varepsilon_{33}^T(s_{11}+s_{12})\right)$, with $J(x)$ and $Y(x)$ expressions as

$$J(x) = \frac{-k_r J_2(k_r x)(1-v_{12})}{(1+v_{12})} + k_r J_0(k_r x) \tag{6.123}$$

$$Y(x) = \frac{-k_r Y_2(k_r x)(1-v_{12})}{(1+v_{12})} + k_r Y_0(k_r x) \tag{6.124}$$

Using the boundary conditions in Equation 6.122,

$$J(a) = J(x)\,|_{x=a},\, J(b) = J(x)\,|_{x=b},\, Y(a) = Y(x)\,|_{x=a},\, Y(b) = Y(x)\,|_{x=b}$$

When the input electric admittance is infinity, the resonance frequency is given as

$$f_r = \frac{k_r a J_0(k_r a)-(1-v_{12})J_1(k_r a)}{k_r b J_0(k_r b)-(1-v_{12})J_1(k_r b)} = \frac{k_r a Y_0(k_r a)-(1-v_{12})Y_1(k_r a)}{k_r b Y_0(k_r b)-(1-v_{12})Y_1(k_r b)} \tag{6.125}$$

This is the radical wave vibration equation, and the resonance frequency is closely related to the material parameters and the physical dimension of the sample. Equation 6.125 is the transcendental equation, and its closed-form solution is difficult to obtain. Thus, it needs to be solved via the numerical methods. Suppose that the ratio of the inner radius over the outer radius is $b/a = \gamma$, then $b = \gamma a$ and $k_r b = \gamma k_r a$. By substituting the relationships into the Equation 6.125, there are only three variables, that is, k, a, γ, and v_{12} with γ having a relation to the geometry parameter, v_{12} is related to the physical parameters. When the material and the geometry parameters are confirmed, k, a can be determined. If the value of Equation 6.125 is $R(n)$, the resonance of the frequency becomes

$$f_n = \frac{R(n)}{2\pi a}\sqrt{\frac{1}{\rho s_{11}\left(1-v_{12}^2\right)}} \tag{6.126}$$

where n is a positive integer. It represents the radial vibrational order of the circular ring sample. $R(n)$ has a relationship with the ratio of the outer and the inner diameters. With confirmed material parameters, it can be seen that the resonance frequency is related to the geometrical dimension and the radial vibrational order. ρ represents the density of the material, v_{12} is Poisson's ratio, and s_{11} represents the elastic compliance. At the beginning, the Fe–Ga phase was assumed as two forces, but in fact, the volume and the physical parameters of Fe–Ga phase should not be ignored. In Equation 6.126, Fe–Ga phase has effect on the density and the elastic compliance. According to the sum property of the composite, the average value $\bar{\rho}$ and $\bar{s}_{11}$ should be considered:

$$\bar{\rho} = \rho_F \times \frac{V_F}{V_F+V_B} + \rho_B \times \frac{V_B}{V_F+V_B} \tag{6.127}$$

$$\bar{s}_{11} = \frac{s_{11}^{F1} s_{11}^{B} s_{11}^{F2}}{V_{F1}s_{11}^{B}s_{11}^{F2} + V_B s_{11}^{F1}s_{11}^{F2} + V_{F2}s_{11}^{F1}s_{11}^{B}} \tag{6.128}$$

Therefore, the resonance frequency of the ring sample is given as

$$f_n = \frac{R(n)}{2\pi a}\sqrt{\frac{1}{\bar{\rho}\bar{s}_{11}\left(1-v_{12}^2\right)}} \tag{6.129}$$

For radial vibration in the closed magnetic circuit sample, this frequency is expressed as

$$f_r = \frac{R}{\pi D}\sqrt{\frac{1}{\bar{\rho}\bar{s}_{11}(1-v^2)}} \tag{6.130}$$

Here, D represents the outer diameter of the loop. $R = 1.5$ is calculated from Equation 6.125, $\bar{\rho}$ is the average density, and v is Poisson's ratio.

Using the following parameters (data of BTO from Table 2.9 and Fe–Ga from Datta et al. [107]), $R = 1.5$, $D = 28$ mm, $\bar{\rho} = 6.53 \times 10^3$ kg/m^3, $s_{11}^{F1} = s_{11}^{F2} = 16.95 \times 10^{-12}$ m^2/N, $s_{11}^{B} = 8.4 \times 10^{-12}$ m^2/N, and $v = 0.33$, Equation 6.130 determines the resonance frequency as 67.9 kHz for the closed magnetic circuit sandwich structure. Figure 6.29 plots α_{ME} as a function of H_{AC} frequency. The results show that the closed magnetic circuit sandwich structure has clear resonance peaks. α_{ME} peaks at 1.88 V/cmOe at a resonance frequency (f_r) of 68 kHz. This shows that Equation 6.130 can accurately predict the resonance frequency for such closed magnetic circuit sandwich structure.

Figure 6.32 shows the influence of the DC magnetic field H_{DC} on the ME voltage coefficient α_{ME}. The α_{ME} of the closed magnetic circuit Fe–Ga/BTO/Fe–Ga sandwich structure increased with an increasing H_{DC} and peaked at 1.88 V/cmOe with $H_{DC} = 340$ Oe under 68 kHz. When $H_{DC} > 340$ Oe, α_{ME} decreased with an increasing H_{DC}. From the phenomenological domain rotation model, the magnetostriction is caused by the domain rotation; thus, 180° rotation would result in 0, while 90° rotation would give rise to the maximum strain [108]. In equilibrium, the domain

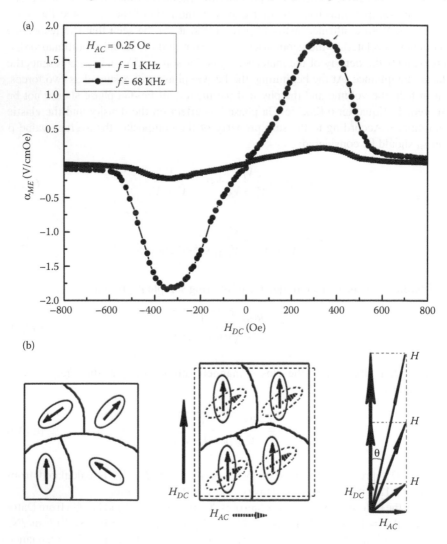

FIGURE 6.32 (a) Magnetoelectric voltage coefficient α_{ME} as a function of DC magnetic field H_{DC}. (b) Phenomenological domain rotation model of magnetostriction under DC magnetic field H_{DC} and alternate current magnetic field H_{AC}.

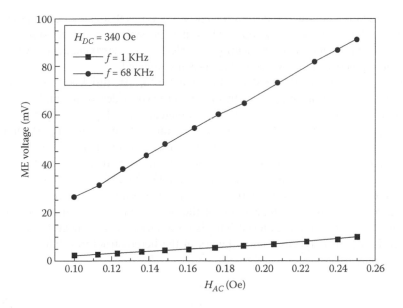

FIGURE 6.33 Magnetoelectric voltages as a function of alternate current magnetic field H_{AC} for closed magnetic circuit Fe–Ga/BTO/Fe–Ga sandwich structure.

orientations are completely random. Upon applying a magnetic field, the domains rotate to line up along the field direction. As the angle between H_{AC} and H_{DC} is perpendicular, the competition between the rotation in H_{AC} and H_{DC} direction will cause α_{ME} to peak at certain field strength (see Figure 6.32b).

Figure 6.33 exhibits the dependence of induced ME voltages on H_{AC} for the closed magnetic circuit sandwich structure. It shows that the induced ME voltage is a linear function of H_{AC}. When $H_{AC} = 340$ Oe and $H_{AC} = 0.25$ Oe, the ME voltage is ~10 mV at $f_r = 1$ kHz but 91 mV at resonance frequency ($f_r = 68$ kHz). Both values are eight times higher than that in the solid disk [103]. Compared with the solid disk that needs $H_{DC} = 750$ Oe [103], $H_{DC} = 340$ Oe can be easily obtained in practice by using a permanent magnet. Hence, the linear relationship between the ME voltage and H_{AC} is a favorable condition to be used in designing AC sensors.

6.3 TWO-PHASE LAYERED THIN-FILM MAGNETOELECTRIC MATERIALS

ME composite materials, which consist of magnetostrictive and piezoelectric phases, offer passive nature, high sensitivity, large effect enhancement at mechanical resonance, and large linear dynamic range. Thin-film ME 2–2 composite materials benefit from the perfect coupling between the piezoelectric and magnetostrictive phases and also from the reduction in size, which is essential for high spatial resolution. Such thin films are easy for on-chip integration, which is a prerequisite for incorporation into microelectronic devices [109]. Materials made in the form of super lattice structures offer unusual transportation properties that cannot be obtained by classical solid-state chemistry route. Thus, it is possible to construct superlattices whose structure consists of alternating ferroelectric and ferromagnetic layers, and the coupling between the two properties was investigated.

Qiu et al. [110] synthesized the $Fe_{73.5}Cu_1Nb_3Si_{13.5}B_9$ (thickness 30 μm)/PZT (20–50 μm) thick-film composites by electrostatic spray deposition and found that the appropriate thickness ratio between magnetostrictive layers and piezoelectric layers (t_m/t_p) will be favorable to raise the resonance ME field output performance. When the thickness of the PZT layers increases, the resonant ME field coefficient for $Fe_{73.5}Cu_1Nb_3Si_{13.5}B_9$/PZT thick-film composites increases first and then decreases, with a maximum value of 259.2 V/cmOe for a PZT thickness of 30 μm. Similar to the results obtained from the laminate samples, the resonance ME coefficient α_E is not only dependent on the performance of piezoelectric material and magnetostrictive material but also dramatically influenced by the thickness ratio, n_m, of magnetostrictive layers in the heterostructures. The n_m can be expressed as

$$n_m = \frac{t_m/t_p}{1 + t_m/t_p} \tag{6.131}$$

where t_m/t_p is the thickness ratio between the magnetostrictive layers and piezoelectric layers. In order to comprehend the increase in resonant ME field coefficient and the role of layer thickness ratio t_m/t_p in $Fe_{73.5}Cu_1Nb_3Si_{13.5}B_9$/PZT thick-film composites, the resonant ME field coefficient as a function of t_m/t_p has been measured. The maximum of resonance ME field coefficient versus the t_m/t_p in $Fe_{73.5}Cu_1Nb_3Si_{13.5}B_9$/PZT thick-film composites has been investigated.

As shown in Figure 6.34a, with the increase in the thickness ratio t_m/t_p, the resonance ME field coefficients α_E increased first and then stabilized. This is because the $Fe_{73.5}Cu_1Nb_3Si_{13.5}B_9$/PZT thick-film composites do not need adhesive layers. Compared to the bulk ME materials, the above-mentioned thick-film composites do not need adhesive layers, which resulted in lower mechanical losses and larger final ME response, and output power. Therefore, the optimum thickness ratio between the magnetostrictive and the piezoelectric layers will be favorable to raise the resonance ME field output performance.

At the resonance frequency, the ME effect experience a sharp increase. Many groups had focused their study on the resonance ME effect, but Palneedi et al. [111] found the enhanced off-resonance ME response in the laser-annealed PZT thick film. Approximately 4 μm PZT was deposited on 25 μm thick Metglas foil, and a 560 nm continuous-wave laser was used to anneal the PZT film. The experiment results show that the thermal influence from laser irradiation was localized only to PZT and did not cause any degradation of the quality of Metglas, and the inherited magnetic properties practically remained unchanged.

Figure 6.35b plots the change in the ME voltage coefficient (α_{ME}) as a function of DC magnetic field (H_{DC}) at an off-resonance frequency of 1 kHz. A maximum ME coefficient of 2.96 V/cmOe was recorded at a low bias field (H_{DC}) of 53 Oe. This ME output is quite comparable to the values for other thin-film ME composite systems,

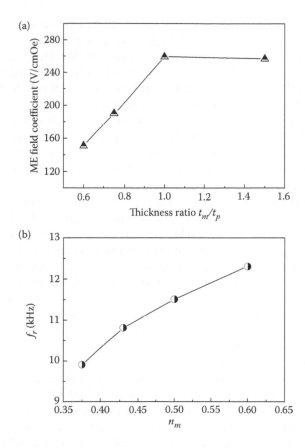

FIGURE 6.34 (a) Maximum of resonance ME field coefficient as a function of the thickness ratio t_m/t_p for $Fe_{73.5}Cu_1Nb_3Si_{13.5}B_9$/PZT thick-film composites. (b) The first longitudinal resonance frequency as a function of n_m.

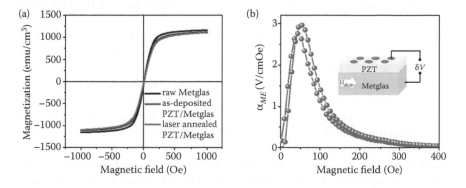

FIGURE 6.35 (a) Magnetization–magnetic field (M–H) loops of raw Metglas and as-deposited and laser-annealed PZT/Metglas; (b) ME voltage coefficient (α_{ME}) dependence of applied bias magnetic field (H_{dc}) for laser-annealed PZT/Metglas (inset shows a schematic illustration of α_{ME} measurement).

which employed sophisticated multilayer designs and cantilever structures to attain giant ME outputs [112,113]. Therefore, high intrinsic off-resonance magnetoelectricity can be obtained from a simple bilayer structure of PZT/Metglas by adopting a suitable fabrication approach with selectively chosen piezoelectric and magnetostrictive materials. Film deposition by granule spray in vacuum allowed highly favorable interfacial bonding with good electrical and mechanical connectivity between PZT and Metglas, which maximized the strain coupling and ME interactions. Localized heating through selective absorption of laser irradiation in PZT film not only induced significant improvement in its dielectric and ferroelectric properties but also contributed to the enhanced ME output by avoiding thermal damage to Metglas substrate and preserving its inherent magnetic properties.

An ME effect has been observed in the two-phase nanocomposite epitaxial films [114]. In the case of two-phase epitaxial films of piezoelectric BaTiO3 and magnetostrictive $CoFe_2O_4$ grown on $SrTiO_3$ substrates, one can either grow nano-thin layers of alternating phases in a laminate structure or a single layer consisting of both phases. For nano-thin laminate composites, the constraint imposed by the substrate restricts the in-plane deformation of both layers, resulting in weak ME properties compared with thicker-film or bulk laminate composites [115]. Two-phase nanocomposites offer similar opportunities and enhanced properties of the laminated bulk composites but on a miniaturized scale. They offer much promise with regard to integration with Si and other microelectronic materials [116]. Robert Jahns et al. used AlN and a plate capacitor or PZT with interdigital electrodes and magnetostrictive amorphous FeCoSiB single layers or exchanged biased multilayers and found that at mechanical resonance and depending on the geometry, extremely high ME coefficients of up to 9.7 kV/cmOe in air and up to 19 kV/cmOe under vacuum were obtained. To avoid external DC magnetic bias fields, composites consisting of exchanged biased multilayers serving as the magnetostrictive component with a maximum ME coefficient at zero magnetic bias field are employed. Furthermore, the anisotropic response of these exchanged biased composites can be utilized for three-dimensional vector field sensing. Sensitivity and noise of the sensors revealed limits of detection as good as to 2.3 $pT/Hz^{1/2}$ at mechanical resonance. Sensitivity between 0.1 and 1000 Hz outside resonance can be enhanced [117].

Bi et al. [95] studied the ME effect of negative magnetostrictive/piezoelectric and positive magnetostrictive laminate composite and found that the ME coefficient behavior of Ni/PZT/$TbFe_2$ with two resonance frequencies; one is the bending resonance mode, and the other is the planar acoustic resonance mode. Cheng et al. [118] studied the ME properties of Ni/PZT/FeCo laminated composites and also found two resonance frequencies. But Du et al. [119] used magnetron-sputtered Ni and FeCoV films to sandwich a piezoelectric PMN-PT foil and found that compared with the results of Bi and Cheng, the bending resonance mode of the Ni/PMN-PT/FeCoV sample is not obvious. The reason is that the thickness of the magnetostrictive phase is too small as compared to the thickness of PMN-PT (which is the substrate) the magnetostrictive phase is difficult to bending the sample. The Ni/PMN-Pt/Ni (NPN) and the FeCoV/PMN-PT/FeCoV (FPF) have a good ME effect.

Figure 6.36 plots the XRD patterns of the sputtered Ni and FeCoV thin films. Grazing-incidence diffraction at 1° was used to eliminate diffraction from the

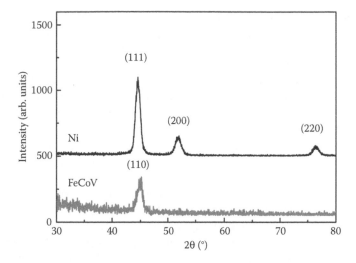

FIGURE 6.36 XRD patterns of sputtered Ni and FeCoV films after magnetoelectric measurement.

underneath PMN-PT foil. Crystalline face-centered cubic Ni film is confirmed. In FeCoV film, only one peak corresponding to (110) reflection is observed. The scanning electron microscope (SEM) images of the cross section and the surface of the magnetic thin films before and after the ME measurement are shown in Figure 6.37. Both films have columnar structures ranging from 50 to 100 nm in diameter. Before

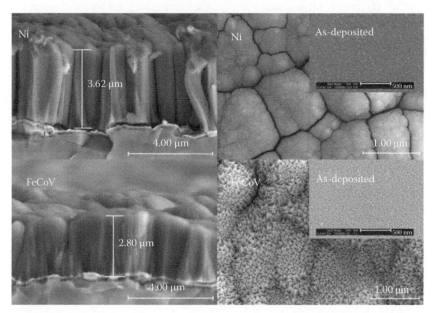

FIGURE 6.37 SEM cross-sectional image and the surface morphology of sputtered Ni and FeCoV thin films after ME measurement (the inset was as-deposited).

the ME measurement, both films have smooth surface with no cracks (see inset in Figure 6.37). Both surfaces developed cracks due to the magnetic straining. All ME responses were measured after the system was stabilized (i.e., the formation of surface cracks was completed).

Figure 6.38 illustrates the ME coefficient as a function of the static magnetic field strength for both NPN and FPF structures. As seen in Figure 6.38, α_{ME} is strongly dependent on the static field strength and peaked at 450 Oe in NPN with a peak value of 0.44 mV/cmOe and at 350 Oe in FPF with a peak value of 2.18 mV/cmOe. The saturation magnetic field strengths for both film-sandwiched structures (1800 Oe in FPF and 2000 Oe in NPN) are greater than that in the magnetic plate (750 and 1000 Oe, Figure 6.39). Such increase in the required magnetic field strength is believed to come from overcoming the pining effect from the underneath piezoelectric foil and the interfacial anisotropy [120]. Figure 6.40 plots the induced ME voltages over the alternating field strength. A perfect linear relationship is observed under a static field of 400 Oe (as ME coefficient peaks at 350 Oe in FeCoV and 450 Oe in Ni film-sandwiched structure, 400 Oe was chosen to test the response to varying alternating field). As seen in Figure 6.40, FPF has a maximum voltage response about five times more as compared to NPN, which corresponds to the same difference in the ME coefficients depicted in Figure 6.38. This makes FPF a perfect candidate for AC magnetic sensing.

The large ME voltage and α_{ME} in FPF come from the large magnetostriction of FeCoV as explained in Section 6.2. According to Livingston's model of coherent rotation of magnetization, the strain generated from magnetostriction λ under the small amplitudes of alternating magnetic field strength is expressed in Equation 6.104. As FeCoV has much larger saturation magnetostriction than Ni (cf. Figure 6.39), the corresponding strain $\Delta\varepsilon_0$ will be much larger, so will be the resultant stress,

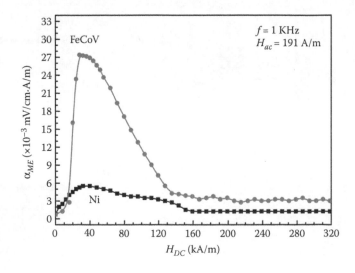

FIGURE 6.38 Magnetoelectric coefficient as a function of static field strength magnetic field H_{AC} for closed magnetic circuit.

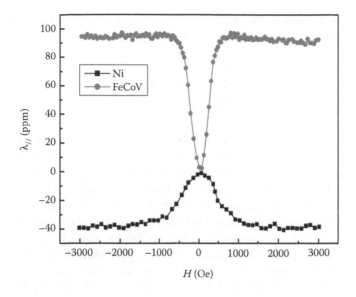

FIGURE 6.39 Magnetostrictive strains of Ni and FeCoV plate.

which can be seen clearly through Equation 6.100. For PMN-PT, the induced strain is approximately proportional to the square of the electric field [121]. However, as the H_{AC} is very small (less than 2.5 Oe), the strain induced by H_{AC} is also very small, the resultant voltage of the PMN-PT is very small and thus H_{AC} overlaps with the H_{DC} (about 400 Oe). Within this range, the ME voltage is assumed to be proportional to

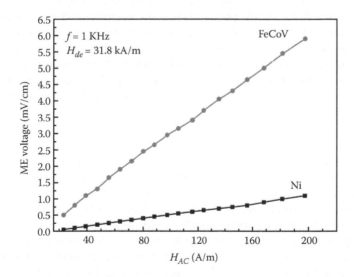

FIGURE 6.40 Dependence of the induced magnetoelectric voltages on alternating field strength field H_{AC} for closed magnetic circuit Fe–Ga/BTO/Fe–Ga sandwich.

the stress as seen in Figure 6.39. As ME voltage and α_{ME} are directly proportional to the stress, Equation 6.105 with $V_i = 2g_{31} \times t \times \sigma_p^E$ yields

$$\alpha_{ME} = \frac{V_i}{t \times H_{AC}} = \frac{2g_{31} \times \sigma_p^E}{H_{AC}} \quad \text{(V/cmOe)} \tag{6.132}$$

where g_{31} is the piezoelectric voltage constant. Greater stress (σ) in FPF therefore gives rise to greater ME voltage (V) and ME coefficient (α) in FPF structure.

The effect of the magnetic film thickness can also be seen from Equation 6.100 whereby the film thickness, t, appears in both the nominator and the denominator. However, in the denominator, since the piezoelectric foil (in this case, PMN-PT foil) thickness is few orders of magnitude larger than that of the magnetic film, the film thickness in the denominator is neglected (200 μm PMN-PT as compared to a few micrometer magnetic film). Hence, the stress σ_p^E is directly proportional to the magnetic film thickness, t_m. In this study, FeCoV film is 2.80 μm, while the Ni film is 3.62 μm; if the FeCoV film had the same thickness as Ni film, the ME response could have been even better.

6.4   MEASUREMENT OF ME EFFECT

There are three methods of measuring ME output, that is, the static method, the quasi-static method, and the dynamic method.

i. For the static method, the ME output (charge or voltage) is measured as a function of increasing magnetic field using an electrometer having high input impedance. While poling, there is a possibility of charges, which get accumulated at the grain boundaries, move toward the electrodes during the measurement. The discharge of the accumulated charges requires quite a lot of time before the final output is stabilized [122–125]. The microstructures within the materials have great effect on the results, and the accumulated charges in some materials may lead to erroneous conclusions, and therefore this method was soon discarded [123].

ii. For more precise data, the quasi-static method is required. It measures the electric charges induced by slowly increasing or decreasing the magnetic field at a constant temperature via an electromagnet with a linear sweep of H versus time. Using an example found in [126], the $(001)_{cub}$ platelets of $Cr_3B_7O_3Cl$ were fixed on a special rod in such a way that the magnetic field can be applied in the directions between the axes 3 and $\bar{3}$ by going through the axis 2, that is, the angle φ varying from 0° to 180°, as indicated in Figure 6.41. The samples with ferroelastic and ferroelectric single domain state underwent a cooling from 50 K down to 4.2 K in a magnetic field of 10 kOe parallel to the axis 3 or 2 in order to align the possible magnetic domains at low temperatures. The induced charges were measured on the planes perpendicular to the axis 3.

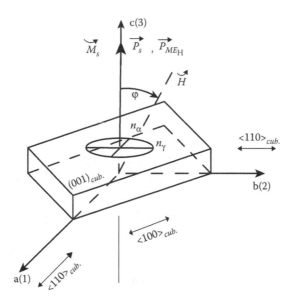

FIGURE 6.41 Correlation between crystallographic axes (mm21' phase), optical indicatrix, and orientation of the magnetic field H for the magnetoelectric measurements of Cr-CI.

Figure 6.42 gives the recorded variation in the magnetic field-induced charges (drift uncorrected) as a function of time at 4.2 K with different orientations of H. The magnetic field was programmed to increase linearly with time from 0 to 10 kOe within 2.5 min duration. It can be seen that the $(ME)_H$ signal is much more important with H parallel to the spontaneous polarization ($H\|3\|P_s$) than with H parallel to the axis 2, that is, perpendicular to P_s, in spite of a larger drift in the former case.

In order to calculate the ME coefficients, the recorded curves were corrected by taking into account the drift with time, which can be determined by joining the beginning and the end of the measurements at zero field (dashed lines in Figure 6.42) and subtracted from the measured values. At 4.2 K, the drift for various measurements shows a linear variation with time within the limits of a measuring cycle. After correction for the drift, the measured data can be fitted with a polynomial form, for example, $y_i = ax_i + bx_i^2$, where $x_i = 1, 2, \ldots, 10$ [kOe] and y_i represents for the related induced polarization. By matrix calculus [127], one can obtain the coefficients α and β of the linear and quadratic ME effects, respectively. In the case of Cr-Cl, the induced polarization is fitted according to the following forms:

$$H/\!/3 : P_3 = \alpha_{33}H_3 + \frac{1}{2}\beta_{333}H_3^2 \tag{6.133}$$

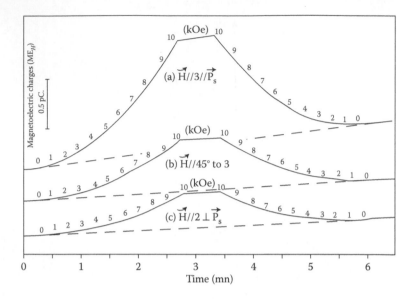

FIGURE 6.42 Variation (without correction of drift) of the magnetoelectric charges of Cr-Cl as a function of time, that is, of magnetic field strength (which increases linearly with time) at different orientations of H.

$$H//2 : P_3 = \alpha_{32}H_2 + \frac{1}{2}\beta_{322}H_2^2 \qquad (6.134)$$

Figure 6.43 shows the fitting results obtained from one of the measurements at 4.2 K for $H//3$. It can be seen that the coefficient, β_{333}, brings a large contribution to the ME effect in Cr-CI, as one could expect from the two-fold shape of the polarization curves (see Figure 6.42). Therefore, the ME effect is essentially of second order with a typical value of the ME tensor component $\beta_{333} = 1.5 \times 10^{-18}$ [s/A] at 4.2 K. However, a slight linear effect was observed, and the linear ME coefficient, α_{33}, shows very small and fluctuating values in the order of 5×10^{-14} [s/m] ($T = 4.2$ K). Figure 6.43 shows the angular dependence of the coefficients at 4.2 K, where β_{322} (4×10^{-19} [s/A]) is smaller than β_{333} and α_{32} smaller than 10^{-14} [s/m].

In order to study the temperature dependence of the ME coefficients, the variation in P_s was measured continuously down to 4.2 K. With such a variation in P_s, which lies parallel to the measured ME polarization, a slight instability in the temperature may give rise to the charges of the pyroelectric origin, which is much more important than the magnetic field-induced charges. Several measurements at the same temperature were therefore necessary in order to obtain recorded curves with linear drift and a mean value of the coefficients. Rivera used this technique in single-crystal boracites [128]. But this method cannot be employed for a polycrystalline material as the charge build up at the grain boundaries still affects the ME output.

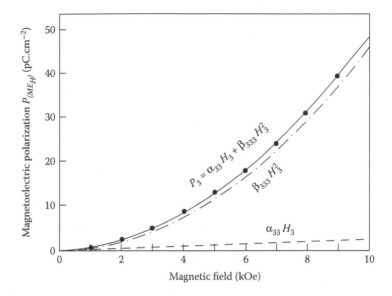

FIGURE 6.43 Dependence of the magnetoelectric polarization upon magnetic field (H//3), fitted according to the function P_3, showing a strong contribution of the second order ME effect and a weak contribution due to the linear term ($T = 4.2$ K).

iii. The dynamic method is conducted by a low-frequency AC magnetic field under a static magnetic bias field, which is superposed parallel to the AC field. This technique permits to measure the ME effect as a function of temperature continuously. In the dynamic method, the output ME voltage is measured with a bias AC magnetic field while keeping the quasi-static measurement in act, that is, measurement is carried out with a time-varying DC magnetic field in the presence of an AC field. The bias AC magnetic field employed will not allow the charges to move toward the electrodes since a suitable signal with an appropriate frequency is used where the polarity of the signal changes with time. This method avoids the leakage and accumulation of charges.

The details of dynamic method of measuring ME output are discussed in the following. The basic equations necessary to evaluate ME coefficients from the voltage can be found in Reference [128]. When a DC magnetic field is applied, the ME output voltage (V) in a polycrystalline material showing second-order effect can be given by

$$V = f(H) = C + \alpha H + \beta H^2 + \gamma H^3 + \cdots \qquad (6.135)$$

where C is constant; the α, β, and γ are the ME coefficients (α is the linear and β is the quadratic components of ME). Then,

$$\frac{dV}{dH} = \alpha + 2\beta H + 3\gamma H^2 + \cdots \qquad (6.136)$$

Suppose that a small AC field, that is, $H_{AC} = H_{AC0} \sin \omega t$, is superimposed over the DC field H_{DC}, the effective field is

$$H_{total} = H_{DC} + H_{AC0} \sin \omega t \tag{6.137}$$

From Equations 6.137 and 6.135, the ME output voltage is re-expressed as

$$
\begin{aligned}
V = f(H) &= C + \alpha(H_{DC} + H_{AC0} \sin \omega t) + \beta(H_{DC} + H_{AC0} \sin \omega t)^2 \\
&\quad + \gamma(H_{DC} + H_{AC0} \sin \omega t)^3 + \cdots \\
&= \frac{1}{8}\Big[\big(C + 4\beta H_{AC0}^2 + 3\delta H_{AC0}^4 + 8\alpha H_{DC} + 12\gamma H_{AC0}^2 H_{DC} + 8\beta H_{DC}^2 \\
&\quad + 24\delta H_{AC0}^2 H_{DC}^2 + 8\gamma H_{DC}^3 + 8\delta H_{DC}^4\big) + \big(8\alpha H_{AC0} + 6\gamma H_{AC0}^3 + 16\beta H_{AC0} H_{DC} \\
&\quad + 24\delta H_{AC0}^3 H_{DC} + 24\gamma H_{AC0} H_{DC}^2 + 32\delta H_{AC0} H_{DC}^3\big)\sin \omega t + \big(-4\beta H_{AC0}^2 \\
&\quad - 4\delta H_{AC0}^4 - 12\gamma H_{AC0}^2 H_{DC} - 24\delta H_{AC0}^2 H_{DC}^2\big)\cos 2\omega t + \cdots\Big]
\end{aligned}
\tag{6.138}
$$

The output ME signal, V_{out}, was measured by lock-in amplifier or other voltmeter and given as

$$
\begin{aligned}
V_{out} &= \frac{1}{8}\big(8\alpha H_{AC0} + 6\gamma H_{AC0}^3 + 16\beta H_{AC0} H_{DC} + 24\delta H_{AC0}^3 H_{DC} \\
&\quad + 24\gamma H_{AC0} H_{DC}^2 + 32\delta H_{AC0} H_{DC}^3\big) \\
&= \frac{H_{DC}^4}{8}\Bigg[\frac{8\alpha}{H_{DC}^3}\bigg(\frac{H_{AC0}}{H_{DC}}\bigg) + \frac{6\gamma}{H_{DC}}\bigg(\frac{H_{AC0}}{H_{DC}}\bigg)^3 + \frac{16\beta}{H_{DC}^2}\bigg(\frac{H_{AC0}}{H_{DC}}\bigg) \\
&\quad + 24\delta\bigg(\frac{H_{AC0}}{H_{DC}}\bigg)^3 + \frac{24\gamma}{H_{DC}}\bigg(\frac{H_{AC0}}{H_{DC}}\bigg) + 32\delta\bigg(\frac{H_{AC0}}{H_{DC}}\bigg)\Bigg]
\end{aligned}
\tag{6.139}
$$

For $(H_{AC0}/H_{DC}) \ll 1$ and neglecting high-order terms, the output ME signal is expressed as

$$
\begin{aligned}
V_{out} &= \frac{H_{DC}^4}{8}\bigg(\frac{8\alpha}{H_{DC}^3} + \frac{16\beta}{H_{DC}^2} + \frac{24\gamma}{H_{DC}} + 32\delta\bigg)\bigg(\frac{H_{AC0}}{H_{DC}}\bigg) \\
&= H_{AC0}\big(\alpha + 2\beta H_{DC} + 3\gamma H_{DC}^2 + 4\delta H_{DC}^3\big) \\
&= H_{AC0}\bigg(\frac{dV}{dH}\bigg)
\end{aligned}
\tag{6.140}
$$

Supposing that the effective thickness of piezoelectric phase is d, the ME coefficient, α_{ME}, is expressed as

$$\alpha_{ME} = \frac{dE}{dH} = \frac{dV}{dH} \cdot \frac{1}{d} = \frac{V_{out}}{d \cdot H_{AC}} \tag{6.141}$$

From Equation 6.141, one can recognize that the philosophy of the dynamic method is to measure the effective value the small AC ME voltage (V_{out}). The small AC ME voltage is appeared across the sample when applying a small AC magnetic field on it. However, in the static method, the ME charge or static voltage are measured. This method also allows one to measure the phase shift of the signal. However, this phase shift is approximately constant during varying the bias field measurements. By changing the DC magnetic bias field, one can explore the ME effect at different working points of the sample. Also by changing the frequency of the AC field, one can study the response of the sample under different time scale. Since the ME signal in this method has well-defined frequency (determined by the driving current) and is measured by a lock-in amplifier, the noise is dramatically reduced and the problem of charge accumulation is avoided. These are the advantages of this method. However, due to the assumption that the AC field is much smaller than the DC field (H_{AC0}/H_{DC}) ≪ 1, this method will give information about the ME effect at low-AC magnetic field, but at different working points of the magnetostrictive component.

Another weak point of this method is the discharging process, which may occur under the periodic condition during measurements. If low frequencies are employed, the discharge may happen through the resistance of the sample. At high frequencies, the discharge will occur through the capacitance formed by the two surfaces of the sample. Hence, this method may give smaller ME coefficients as compared to those measured by other methods. Since an AC magnetic field is employed, an induction voltage always exists and contributes as a zero signal. To avoid this, the lock-in amplifier has to work in the differential mode to subtract the common-mode induction contribution. Using this mode, not only the zero-signal problem is solved, but calibration is also not needed since the ME signal is zero without sample or with non-ME samples (based on well-shielded cables and good connection). When H_{DC} is zero, it is possible to find out the linear term α, and hence the second-order term β is evaluated in the presence of DC magnetic field. When H_{DC} is not zero, the above equation shows that α varies with β, giving rise to a pseudo-linear coefficient α (H_{DC}).

For the above measurement, a Helmholtz coil is mounted on the pole pieces of a DC electromagnet, whose magnetic field can be made time varying by interfacing it to a computer. The time-varying DC magnetic field can be achieved by the DC power supply [GP-HP-L/ ± 65A]. The block diagram is given in Figure 6.44.

The magnetic field strength was pre-calibrated from a Tesla meter [T-6 TYPE, ISAS]. The small solenoid coils produced a small AC magnetic field, which was supplied by a signal generator [CA1640–02, CALTEK] and detected by an AC induction magnetometer [CCG-1000]. If the signal current was very small to drive the coils, a power amplifier can be used to amplify the current. The AC magnetic field generated can be calculated since the current carried by the coils is known. The electric signal produced by the sample was input into a lock-in amplifier [LI 5640]. At the same time, the signal generator sent a signal that synchronized with the coil excitation signal to the lock-in amplifier as reference.

Keeping the DC magnetic field zero with an increasing AC field, the ME signal was recorded, and it provided the linear coefficient (α). By selecting a fixed value of AC magnetic field, the DC field was swept using the DC magnetic power

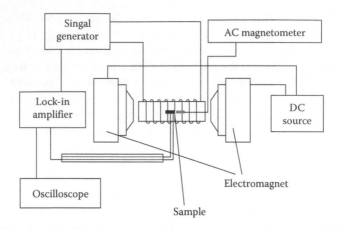

FIGURE 6.44 Block diagram of dynamic magnetoelectric experimental setup.

supply. The induced ME signal in the sample was measured with the lock-in amplifier (LI5640). With known values of α, H_{AC0}, H_{DC}, and V, the value of second-order coefficient (β) can be calculated. Since the total measuring time to complete the experiment was only a few minutes, no time was given to the charges to accumulate in the sample, as in the case of pure DC $(ME)_H$ measurements. The signal output in voltage represents the true value. The same experimental setup was used to take a temperature scan of the ME signal at fixed AC and DC bias fields by inserting the sample in a Dewar in between the pole pieces. The data on $(ME)_H$ at different temperatures would be useful to identify whether there are any magnetic anomalies in the material [129].

6.5 APPLICATIONS

6.5.1 MAGNETIC SENSORS

Making magnetic sensors from ME composites is simple and direct. When a magnetic field is applied, the magnetic phase in the ME composites strains and a proportional charge in the piezoelectric phase will produce a proportional voltage, which corresponds to the magnitude of the magnetic field. Hence, ME composites with high ME coefficients are suitable for developing highly sensitive magnetic field sensors, for example, a magnetic probe for detecting AC or DC fields.

6.5.1.1 AC Magnetic Field Sensors

The induced ME voltage has a linear response to H_{AC} over a wide range of fields from 10^{-11} T $< H_{AC} < 10^{-3}$ T [89,78]. When the laminates were operated under resonance drive, an enhancement in sensitivity to small magnetic field variations was observed. The sensitivity limit of the ME laminates at ambient conditions was 1.2×10^{-12} T [130]. These results unambiguously demonstrated that ME laminates have an ultrahigh sensitivity to small AC magnetic field variations.

6.5.1.2   DC Magnetic Field Sensors

ME laminates composites can detect the small DC magnetic field at the linear magnetostrictive area of the magnetostriction–magnetic field curve. In fact, small long-type ME laminates of Terfenol-D and PZT are quite sensitive to small H_{DC} variations when driven under a constant H_{AC} [131]. Based on the DC magnetic field sensor, Leung et al. designed a force-sensing device which is capable of sensing DC compressive forces as shown in Figure 6.45 [132].

6.5.1.3   ME Current Sensors

When electric current flows along a straight wire, there will be a vortex magnetic field. Hence, the ring-type ME laminates are ideal configurations for current

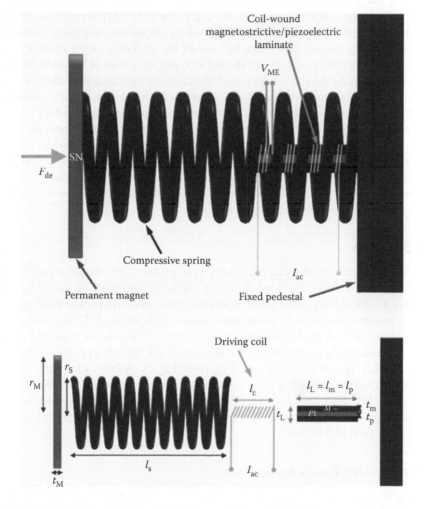

FIGURE 6.45   Schematic and exploded views of the proposed DC compressive force-sensing device.

detection. The ring type needs the wire to penetrate the ring, which is a disadvantage for practical application. Le et al. fabricated the ME bilayered structure, which consists of Cytop polymer and magnetic tape filled with magnetically soft particles, can attain significant ME properties from the laminate material of a transversely charging electret along with bias magnetic tape. This result demonstrated a possibility of making a low-cost flexible current sensor [133].

6.5.2 MICROWAVE DEVICES

Since the composite ME effect has pronounced linear ME response on an AC field oscillating in the presence of a stronger DC bias field, the composite ME effect is thus predetermined for microwave applications. Different frequency ranges are accessed using the enhancement from the EMR (~100 kHz), the FMR (~10 GHz), or the antiferromagnetic resonance (~100 GHz). The FeCoZr (100 nm) thin film was grown on PMN-PT ($10 \times 5 \times 0.5$ mm^3) by oblique deposition-sputtering technique. When a varying electric field is applied across the thickness of the PMN-PT substrate, the permeability spectra are shifted as typically shown in Figure 6.46a with different electric field values, which are indicatives of the changes in the resonance frequency with the applied electric field. Details of the variation in the resonance frequency f with the applied electric field E can be found in Figure 6.46b. It shows that when E reached 2 kV/cm, the resonance frequency drastically increased from 7.5 to 12 GHz and became almost saturated when $E > 3$ kV/cm. As the electric field (E) is decreased from 6 down to -6 kV/cm, first the resonance frequency is slightly reduced from 12.3 to 11.6 GHz as E is reduced from 6 to 0 kV/cm. When E further decreased negatively, the f versus E field curve exhibits a notch shape with a minimum f of 8.2 GHz at $E = -1.2$ kV/cm [134].

6.5.3 FILTERS

The lumped circuit model can effectively predict the center frequency and bandwidth of the resonator. The application of multi-ME laminates in filters not only broaden bandwidth but also allow controlling the work frequency band by tuning the external electrostatic and magnetostatic fields on the ME laminates [135]. Tatarenko et al. show the transmission characteristics [136]. An increase in E results in an increase in the magnitude of the downshift. An upshift in f can be observed when the direction of E was reversed by reversing the polarity of applied voltage and is attributed to a switch from compressive to tensile strain in YIG. Thus, a microwave ME filter based on YIG/PZT layered structure has been designed and characterized. This filter can be tuned by 2% of the central frequency with a nominal electric field of 3 kV/cm.

6.5.4 POSSIBLE FUTURE APPLICATIONS

Among the ME composites, Terfenol-D-based composites have shown the strongest ME coupling over a wide frequency range. However, it is expensive and gets easily oxidized. New material substitute for Terfenol-D should have some properties

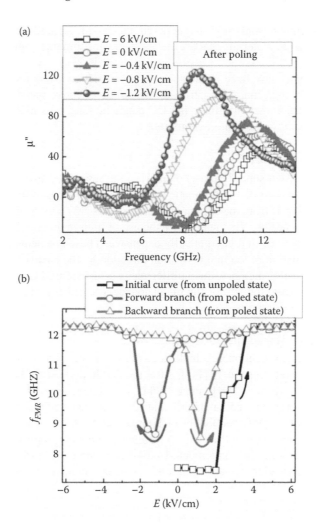

FIGURE 6.46 (a) Imaginary permeability spectra of the FeCoZr/PMN-PT(011) hetero-structure measured at room temperature with different applied electric fields. (b) Electric field dependence of FMR frequency for the FeCoZr/PMN-PT(011) heterostructure.

such as high piezomagnetic coefficient and relatively small demagnetization factor. Recently, using Metglas as the magnetic phase is widely studied as such a magnetic sensor is accurate and cheap. For the two-phase ME composites, the future directions include the following.

6.5.4.1 Optimization of the Material and Structure

Further optimization of ME laminate composites still remains a challenge which involves different configurations for structure optimization, such as C-type configurations; engineering-controlled connectivity of the magnetic–piezoelectric phases; and different magnetic circuits to improve the ME properties such as using the soft magnetic materials to make the magnetic circuit closed. Actually, there are many traditional magnetic materials that are cheap, stable, machinable, and have not been tested.

Metglas ribbons have been used as the magnetostrictive layers recently. Metglas has an extremely high magnetic permeability, so the effective piezomagnetic coefficient is high which may eliminate the DC magnetic bias, thus making it an ideal candidate for incorporation into ME composites.

6.5.4.2 State Memory

In the traditional two-state (0 and 1) memories, the resistance strongly depends on the relative orientation of the magnetic moments, which is used to determine the memory state (0 or 1) from the two magnetic electrodes [137]. In the writing process, the magnetic bits are usually encoded by the use of high magnetic fields which is a process relatively slow and energetically expensive. These problems can be solved with the manipulation of the magnetization direction by the electric field [138]. For this kind of multistate memory, the multiferroicity is an essential factor for the information storage, while the ME or the magnetodielectric effect is the mechanism for the reading and writing procedure [139].

6.5.4.3 Energy Harvesting

There has been significant interest in the area of the vibration energy based on piezoelectric and magnetic harvesters [140]. After the first hypothesis of ME materials as energy harvesting devices [141], some studies have been reported in this area. As described in the previous section, significant advances have been made in improving the magnitude of the ME coefficient of laminate composites, which will improve the ME energy harvesting efficiency. A combined magnetic and vibration energy harvesting device may be implemented on silicon using the thin-film deposition methods and fabrication process flow and combination with the micromachining technique [142].

REFERENCES

1. W. Erenstein, N. D. Mathur, and J. F. Scott, *Nature* 442, 2006, 759.
2. G. Wu, T. Nan, R. Zhang, N. Zhang, S. Li, and N. X. Sun, *Appl. Phys. Lett.* 103, 2013, 182905.
3. S. M. Gillette, T. Fitchorov, O. Obi, L. Jiang, H. Hao, S. Wu, Y. Chen, and V. G. Harris, *J. Appl. Phys.* 115, 2014, 17C734.
4. A. L. Geiler, S. M. Gillette, Y. Chen, J. Wang, Z. Chen, S. D. Yoon, P. He, J. Gao, C. Vittoria, V. G. Harris, *Appl. Phys. Lett.* 96, 2010, 053508.
5. J. Ma, J. Hu, Z. Li, C. W. Nan, *Adv. Mater.* 23, 2011, 1062.
6. W. C. Rontgen, *Ann. Phys.* 35, 1888, 264.
7. H. A. Wilson, *Philos. Trans. R. Soc. Lond., Ser. A* 204, 1905, 121.
8. P. Curie, *J. Phus. (Paris)* 3, 1894, 393.
9. P. Debye, *Z. Phys.* 36, 1926, 300.
10. E. P. Wigner, *Gott. Nachr. Math. Phys.* 1932, 545.
11. J. H. Van Vleck, *The Theory of Electric and Magnetic Susceptibilities*, Oxford University Press, London, 1932.

12. L. D. Landau, *Zh. Eksp. Teor. Fiz.* 7, 1937, 19.
13. Y. B. Zeldovich, *Eksp. Teor. Fiz.* 33, 1957, 1531.
14. I. E. Dzyaloshinskii, *J. Exp. Theor. Phys. (USSR)* 37, 1959, 881.
15. D. N. Astrov, *Phys.-JETP* 11, 1960, 708.
16. G. T. Rado and V. J. Folen, *Phys. Rev. Lett.* 7(8), 1961, 310.
17. A. V. Shubnikov and N. V. Belov, *Colored Symmetry*, Pergamon Press, Oxford, 1964.
18. E. Ascher, *Helv. Phys. Acta* 39, 1966, 40.
19. R. P. Santoro, D. J. Segal et al., *J. Phys. Chem. Solids* 27, 1966, 1192.
20. C. Tabares-Munoz, J. P. Rivera et al., *Jpn. J. Appl. Phys.* 242, 1985, 1051.
21. H. Schmid, *Ferroelectrics* 162(1), 1994, 317.
22. N. A. Hill, *J. Phys. Chem. B* 104(29), 2000, 6694.
23. M. Fiebig, *J. Phys. D* 38, 2005, R123.
24. E. Moran, M. Alario-Franco, M. Garcia-Guaderrama, and O. Blanco, *Materials Research Society Symposium Proceedings* 2008. doi: 10.1557/PROC-1148-PP01-09.
25. M. G. Kanatzidis, K. R. Poeppelmeier et al., *Prog. Solid State Chem.* 36(1–2), 2008, 1.
26. J. Wang, J. B. Neaton et al., *Science* 299, 2003, 1719.
27. I. E. Dzyaloshinshii, *Zh. Eksp. Teor. Fiz.* 37, 1959, 881.
28. G. T. Rado and V. J. Folen, *Phys. Rev. Lett.* 7, 1961, 310.
29. L. D. Landau and E. M. Lisfshitz, *Fluid Mechanics*, Addison-Wesley Co., Inc., Reading, MA, 1960.
30. A. M. J. G. Van Run, D. R. Terrell, and J. H. Scholing, *J. Mater. Sci.* 9, 1974, 1710.
31. Y. S. Koo, K. M. Song, N. Hur, J. H. Jung, T. H. Jang, H. J. Lee, T. Y. Koo, Y. H. Jeong, J. H. Cho, and Y. H. Jo, *Appl. Phys. Lett.* 94, 2009, 032903.
32. T. Lottermoser, T. Lonkai, U. Amann, D. Hohlwein, J. Ihringer, and M. Fiebig, *Nature (London)* 430, 2004, 541.
33. N. Hur, S. Park, P. A. Sharma, J. S. Ahn, S. Guha, and S. W. Cheong, *Nature (London)* 429, 2004, 392.
34. Y. Chen, J. Gao, T. Fitchorov, Z. Cai, K. S. Ziemer, C. Vittoria, and V. G. Harris, *Appl. Phys. Lett.* 94, 2009, 082504.
35. I. J. Busch-Vishniac, *Phys. Today* 56, 1998, 28.
36. W. Eerenstein, N. D. Mathur, and J. F. Scott, *Nature (London)* 442, 2006, 759.
37. N. Fujimura, T. Ishida, T. Yoshimura, and T. Ito, *Appl. Phys. Lett.* 69, 1996, 1011.
38. M. Fiebig, *J. Phys. D* 38, 2005, R123.
39. T. Kimura, S. Kawamoto, I. Yamada, M. Azuma, M. Takano, and Y. Tokura, *Phys. Rev. B* 67, 2003, R180401.
40. J. Iniguez, *Phys. Rev. Lett.* 101, 2008, 117201.
41. H. Wiegelmann, A. A. Stepanov, I. M. Vitebsky, A. G. M. Janseb, and P. Wyder, *Phys. Rev. B* 49, 1994, 10039.
42. J. P. Rivera and H. Schmid, *J. Appl. Phys.* 70, 1991, 6410.
43. V. A. Khomchenko, J. A. Paixao, V. V. Shvartsman, P. Borisov, and W. Kleemann, *Scr. Mater.* 62, 2010, 238.
44. H. J. Lewtas, T. Lancaster, P. J. Baker, S. J. Blundell, D. Prabhakaran, and F. L. Pratt, *Phys. Rev. B* 81, 2010, 014402.
45. Y. Yang, J. M. Liu, H. B. Huang, W. Q. Zou, P. Bao, and Z. G. Liu, *Phys. Rev. B* 70, 2004, 132101.
46. J. Ryu, S. Priya, K. Uchino, and H. Kim, *J. Electroceram.* 8, 2002, 107.
47. P. Martins and S. Lanceros-Méndez, *Adv. Funct. Mater.* 23, 2013, 3371.
48. J. van Suchtelen, *Philips Res. Rep.* 27, 1972, 28.
49. C. W. Nan, *Prog. Mater. Sci.* 37, 1993, 1.
50. A. M. J. G. Run, D. R. Terrell, and J. H. Scholing, *J. Mater. Sci.* 9, 1974, 1710.
51. J. Boomgard, A. M. J. G. Run, and J. Suchtelen, *Ferroelectrics* 10, 1976, 295.
52. I. A. Osaretin and R. G. Rojas, *Phys. Rev. B* 82, 2010, 174415.

53. B. K. Sinha, W. J. Tanski, T. Lukaszek, and A. Ballato, *J. Appl. Phys.* 57, 1985, 767.

54. M. I. Bichurin, V. M. Petrov, and G. Srinivasan, *J. Appl. Phys.* 92, 2002, 7681.

55. M. I. Bichurin, I. A. Kornev, V. M. Petrov, A. S. Tatarenko, Yu. V. Kiliba, and G. Srinivasan, *Phys. Rev. B* 64, 2001, 094409.

56. M. I. Bichurin, V. M. Petrov, Yu. V. Kiliba, and G. Srinivasan, *Phys. Rev. B* 66, 2002, 134404.

57. G. Harshe, J. O. Dougherty, and R. E. Newnham, *Int. J. Appl. Electromagn. Mater.* 4, 1993, 145.

58. M. I. Bichurin, V. M. Petrov, and G. Srinivasan, *Phys. Rev. B* 68, 2003, 054402.

59. D. R. Smith and J. B. Pendry, *J. Opt. Soc. Am. B* 23, 2006, 391.

60. D. M. Pozar, *Microwave Engineering*, Wiley, New York, 2005.

61. B. Lax and K. J. Button, *Microwave Ferrites and Ferrimagnetics*, McGraw-Hill, New York, 1962.

62. R. E. Collin, *Foundations for Microwave Engineering*, McGraw Hill, New York, 1992.

63. J. Ryu, S. Priya, K. Uchino, and H. E. Kim, *J. Electroceram.* 8, 2002, 107.

64. J. A. Kong, *Electromagnetic Wave Theory*, 2nd ed. Wiley, New York, 1990, 5.

65. M. I. Bichurin, V. M. Petrov, and G. Srinivasan, *Phys. Rev. B* 68, 2003, 054402.

66. C. L. Jia, T. L. Wei, C. J. Jiang, D. S. Xue, A. Sukhov, and J. Berakdar, *Phys. Rev. B* 90, 2014, 054423.

67. M. I. Bichurin, V. M. Petrov, O. V. Ryabkov, S. V. Averkin, and G. Srinivasan, *Phys. Rev. B* 72, 2005, 060408.

68. M. I. Bichurin and V. M. Petrov, *Sov. Phys. Tech. Phys.* 33, 1989, 1389.

69. M. I. Bichurin, D. A. Filippov, and V. M. Petrov, *Phys. Rev. B* 68, 2003, 132408.

70. D. A. Filippov, M. I. Bichurin, and V. M. Petrov, *Tech. Phys. Lett.* 30, 2004, 6.

71. M. I. Bichurin and V. M. Petrov, *Ferroelectrics* 162, 1994, 33.

72. S. X. Dong, J. F. Li, and D. Viehland, *IEEE Trans. Ultrason. Ferroelectr. Freq. Control* 50, 2003, 1253.

73. S. X. Dong, J. F. Li, and D. Viehland, *J. Mater. Sci.* 41, 2006, 97.

74. S. X. Dong, J. R. Cheng, J. F. Li, and D. Viehland, *Appl. Phys. Lett.* 83, 2003, 4812.

75. Y. X. Liu, J. G. Wan, J. M. Liu, and C. W. Nan, *J. Appl. Phys.* 94, 2003, 5111.

76. G. Liu, C. W. Nan, N. Cai, and Y. H. Lin, *J. Appl. Phys.* 95, 2004, 2660; *Int. J. Solids Struct.* 41, 2004, 4423.

77. J. Ryu, A. V. Carazo, K. Uchino, and H. E. Kim, *Jpn. J. Appl. Phys. Part 1* 40, 2001, 4948.

78. J. Y. Zhai, Z. Xing, S. X. Dong, J. F. Li, and D. Viehland, *Appl. Phys. Lett.* 88, 2006, 062510.

79. C. M. Kanamadi, J. S. Kim, H. K. Yang, B. K. Moon, B. C. Choi, and J. H. Jeong, *J. Alloys Compd.* 481, 2009, 781.

80. Y. J. Wang, X. Y. Zhao, J. Jiao, Q. H. Zhang, W. N. Di, H. S. Luo, C. M. Leung, and S. W. Or, *J. Alloys Compd.* 496, 2010, L4.

81. N. H. Duc and D. T. H. Giang, *J. Alloys Compd.* 449, 2008, 214.

82. Y. J. Wang, X. Y. Zhao, J. Jiao, L. H. Liu, W. N. Di, H. S. Luo, and S. W. Or, *J. Alloys Compd.* 500, 2010, 224.

83. A. A. Bush, K. E. Kamentsev, V. F. Meshcheryakov, Y. K. Fetisov, D. V. Chashin, and L. Y. Fetisov, *Tech. Phys.* 54, 2009, 1314.

84. S. X. Dong, J. F. Li, and D. Viehland, *J. Appl. Phys.* 95, 2004, 2625.

85. L. Wang, Z. F. Du, C. F. Fan, H. P. Zhang, and D. L. Zhao, *J. Alloys Compd.* 509, 2011, 508.

86. D. Seguin, M. Sunder, L. Krishna, A. Tatarenko, and P. D. Moran, *J. Cryst. Growth* 311, 2009, 3235.

87. S. X. Dong, J. Y. Zhai, Z. P. Xing, J. F. Li, and D. Viehland, *Appl. Phys. Lett.* 86, 2005, 102901.

88. Y. J. Wang, X. Y. Zhao, J. Jiao, L. H. Liu, W. N. Di, H. S. Luo, and S. W. Or, *J. Alloys Compd.* 500, 2010, 224.
89. S. X. Dong, J. F. Li, and D. Viehland, *Appl. Phys. Lett.* 83, 2003, 2265.
90. S. X. Dong, J. F. Li, and D. Viehland, *IEEE Trans. Ultrason. Ferroelectr. Freq. Control* 50, 2003, 1236.
91. S. X. Dong, J. F. Li, and D. Viehland, *J. Appl. Phys.* 95, 2004, 2625.
92. F. Yang, Y. M. Wen, P. Li et al., *Sens. Actuators A* 141, 2008, 129.
93. J. D. Livingston, *Phys. Status Solidi Appl. Res.* 70, 1982, 591.
94. C. L. Hom and N. Shankar, *Int. J. Solids Struct.* 33, 1996, 1757.
95. K. Bi, Y. G. Wang, D. A. Pan, and W. Wu, *Scr. Mater.* 63, 2010, 589.
96. M. Szklarska Lukasik, P. Guzdek, M. Dudek, A. Pawlaczyk, J. Chmist, W. Dorowski, and J. Pszczola, *J. Alloys Compd.* 549, 2013, 276.
97. Z. Du, S. Zhang, L. Wang, and D. Zhao, *Thin Solid Films* 544, 2013, 230.
98. D. T. Huong Giang, P. A. Duc, N. T. Ngoc, N. T. Hien, and N. H. Duc, *J. Magn.* 17, 2012, 308.
99. C. M. Leung, S. W. Or, S. Zhang, and S. L. Ho, *J. Appl. Phys.* 107, 2010, 09D918.
100. S. Dong, J. F. Li, and D. Viehland, *Appl. Phys. Lett.* 84, 2004, 4188.
101. J. Lu, D. Pan, B. Yang, and L. Qiao, *Meas. Sci. Technol.* 19, 2008, 045702.
102. Methods of measurement of a.c. magnetic properties of magnetically soft materials. GB/T 3658–2008, China.
103. L. Wang, Z. Du, C. Fan, L. Xu, H. Zhang, and D. Zhao, *J. Alloys Compd.* 509, 2011, 508.
104. M. Zeng, S. W. Or, and H. L. W. Chan, *Appl. Phys. Lett.* 96, 2010, 203502.
105. D. T. Huong Giang, P. A. Duc, N. T. Ngoc, and N. H. Duc, *Sensors Actuat. A Phys.* 179, 2012, 78.
106. S. Lin, *Sensors Actuat. A* 134, 2007, 505.
107. S. Datta, J. Atulasimha, C. Mudivarthi, and A. B. Flatau, *J. Magn. Magn. Mater.* 322, 2010, 2135.
108. D. C. Jiles and J. B. Thoelke, *J. Magn. Magn. Mater.* 134, 1994, 143.
109. C.-W. Nan, M. I. Bichurin, S. Dong, D. Iehland, and G. Srinivasan, *J. Appl. Phys.* 103, 2008, 031101.
110. J. Qiu, Y. Wen, P. Li, and H. Chen, *J. Appl. Phys.* 117, 2015, 17D701.
111. H. Palneedi, D. Maurya, G.-Y. Kim, S. Priya, S.-J. L. Kang, K.-H. Kim, S.-y. Choi, and J. Ryu, *Appl. Phys. Lett.* 107, 2015, 012904.
112. H. Greve, E. Woltermann, H.-J. Quenzer, B. Wagner, and E. Quandt, *Appl. Phys. Lett.* 96, 2010, 182501.
113. S. S. Nair, G. Pookat, V. Saravanan, and M. R. Anantharaman, *J. Appl. Phys.* 114, 2013, 064309.
114. H. M. Zheng, J. Wang, S. E. Lofland, Z. Ma, L. Mohaddes-Ardab, T. Zhao et al., *Science* 303, 2004, 661.
115. H. C. He, J. Wang, B. P. Zhou, and C. W. Nan, *Adv. Fun. Mater.* 17, 2007, 1333.
116. J. Zhai, Z. Xing, S. Dong, J. Li, and D. Viehland, *J. Am. Ceram. Soc.* 91, 2008, 351.
117. R. Jahns, A. Piorra, E. Lage, C. Kirchhor, D. Meyners, J. L. Gugat, M. Krantz, M. Gerken, R. Knochel, and E. Quandt, *J. Am. Ceram. Soc.* 96, 2013, 1673.
118. J. Cheng, Y.-G. Wang, and D. Xie, *Mater. Lett.* 143, 2015, 273.
119. Z. Du, S. Zhang, L. Wang, and D. Zhao, *Thin Solid Film* 544, 2013, 230.
120. F. Huang, G. J. Mankey, and R. F. Willis, *J. Appl. Phys.* 75, 1994, 6406.
121. C. L. Hom and N. Shankar, *Int. J. Solids Struct.* 33, 1996, 1757.
122. A. Hanumaiah, T. Bhimasankaram, S. V. Suryanarayana, and G. S. Kumar, *Bull. Mater. Sci.* 17, 1994, 405.
123. R. S. Singh, T. Bhimasankaram, G. S. Kumar, and S. V. Suryanarayana, *Solid State Commun.* 91, 1994, 567.
124. S. V. Suryanarayana, *Bull. Mater. Sci.* 17, 1994, 1259.

125. A. R. James, G. S. Kumar, M. Kumar, S. V. Suryanarayana, and T. Bhimasankaram, *Mod. Phys. Lett. B* 11, 1997, 633.

126. Z. G. Ye, J. P. Rivera, and H. Schmid, *Ferroelectrics* 161, 1994, 99.

127. P. Rivera and H. Schmid, *J. Physique*, Suppl. au no. 2, Tome 49, 1988, C8–849.

128. J. P. Rivera, *Ferroelectrics* 161, 1994, 147.

129. N. Ikeda, K. Saito, K. Kohn, H. Kita, J. Akimiku, and K. Siratori, *Ferroelectrics* 161, 1994, 111.

130. S. X. Dong, J. Y. Zhai, F. Bai, J. F. Li, and D. Viehland, *Appl. Phys. Lett.* 87, 2006, 062510.

131. M.-C. Lu, L. Mei, D.-Y. Jeong, J. Xiang, H. Xie, and Q. M. Zhang, *Appl. Phys. Lett.* 106, 2015, 112905.

132. C. Leung, S. W. Or, and S. L. Ho, *Rev. Sci. Instrum.* 84, 2013, 125003.

133. M.-Q. Le, F. Belhora, A. Cornogolub, P.-J. Cottinet, L. Lebrun, and A. Hajjaji, *J. Appl. Phys.* 115, 2014, 194103.

134. N. N. Phuoc and C. K. Ong, *Appl. Phys. Lett.* 105, 2014, 022905.

135. H. Zhou and J. Lian, *J. Appl. Phys.* 115, 2014, 193908.

136. A. S. Tatarenko, G. Srinivasan, and M. I. Bichurin, *Appl. Phys. Lett.* 88, 2006, 183507.

137. M. Julliere, *Phys. Lett. A* 54, 1975, 225.

138. J. M. Hu, Z. Li, J, Wang, and C. W. Nan, *J. Appl. Phys.* 107, 2010, 093912.

139. Y. Guo, Y. Liu, J. Wang, R. L. Withers, H. Chen, L. Jin, and P. Smith, *J. Phys. Chem. C* 114, 2010, 13861.

140. S. Adhikari, M. I. Friswell, and D. J. Inman, *Smart Mater. Struct.* 18, 2009, 115005.

141. F. Yang, Y. Wen, P. Li, and M. Zheng, *IEEE Trans. Magn.* 1–2, 2006, 1010.

142. S. Priya, J. Ryu, C. S. Park, J. Oliver, J. J. Choi, and D. S. Park, *Sensors* 9, 2009, 6362.

7 Rapidly Solidified Rare-Earth Permanent Magnets

Processing, Properties, and Applications

Shampa Aich, D. K. Satapathy, and J. E. Shield

CONTENTS

ABSTRACT

Rapidly solidified rare-earth-based permanent magnets are considered to have better potential as permanent magnets compared to the conventional bulk materials, which can be attributed to their improved microstructure and better magnetic properties compared to rare-earth magnets synthesized by the conventional (powder metallurgy) routes. The performance (quality) of these magnets depends on the thermodynamics and kinetics of the different processing routes, such as atomization, melt spinning, and melt extraction. Here, we review the various processing routes of rapidly solidified rare-earth permanent magnets and the related properties and applications. In the review, some specific alloy systems, such as Sm–Co-based alloys, Nd–Fe–B, and interstitially modified Fe-rich rare-earth magnets are discussed in detail mentioning their processing routes and subsequently achieved crystal structure, microstructure and magnetic properties, and the related scopes for various applications. Some newly developed nanocomposites and thin-film magnets are also included in the discussion.

7.1 INTRODUCTION

Rare-earth permanent magnets have revolutionized technology since their discovery in the 1970s and are ubiquitous in this information-technology-driven and energy-conscious world. Rare-earth magnets have allowed the miniaturization of countless devices and the development of highly efficient motors and generators. These magnets are stronger than the conventional magnets of ferrites or Alnico. Since the discovery of the naturally occurring mineral, magnetite (Fe_3O_4), magnetism and magnetic materials have been playing an important role in modern science and technology. In ancient times, the Chinese and the Greeks were using lodestones or "way-stones" in guiding mariners. In 1600, physicist William Gilbert experimented with lodestone, iron magnets, and the magnetic field of the earth. His experiments laid the foundation for current scientific applications and dispelled the folklore surrounding magnetism and magnetic material [1]. Research about magnetic materials expanded after the invention of electromagnets by physicist Hans Christian Oersted in 1820 [1]. Permanent magnets have brought much more attention to the field, because unlike powerful electromagnets, they can be used without any consumption of electricity or generation of heat.

Permanent magnets are used and extensively studied in academic and military research and energy laboratories. Another important area of application is in medical industries (MRI, hematology laboratories, and magnetic hyperthermia technique). About 160 magnets are used for different purposes in our daily lives. The applications range from refrigerator magnets, kitchen appliances, television, telephone, watches, computer, and audio systems to microelectronics. Another 100 magnets are used in the automobile industry. Permanent magnets are behind some of the most important inventions of our modern lives. They make our lives pleasant, comfortable, and easier. They have a promising future, because a number of new devices are waiting for them. Ultimately, there is a basic necessity to understand and improve their properties, as well as to look for new applications for them.

The first commonly used permanent magnets were made of carbon steel and were shaped like a horseshoe. Although this type of magnet is now obsolete, the horseshoe represents the symbol for magnetism [2]. In the past 60 years, the applications of permanent magnets have been diversified due to discoveries of new materials such as Alnicos (alloys of Al, Ni, Co, and Fe), ferrites (combination of iron oxide with another metal), Nd–Fe–B, and Sm–Co magnets. Although the Alnicos were extensively used in the mid-twentieth century as general-purpose permanent magnets, for their moderate magnetic properties achieved by relatively easy processing, they were replaced by much cheaper ferrites, which now occupy 55% of the permanent magnet world market.

The dawn of rare-earth permanent magnets was the discovery of the high anisotropy field of $SmCo_5$ in the late 1960s [3,4]. People were much more attentive to these Sm–Co magnets due to their high anisotropy field H_A, which was twice that of contemporary ferromagnetic Alnico alloys. Magnets made from rare-earth materials exhibit magnetic fields up to 1.4 T whereas ferrites and Alnico magnets exhibit magnetic fields in the range of 0.4–1 T. Since rare-earth magnets are extremely brittle and vulnerable to corrosive environment, they are coated with other materials to improve their corrosion resistance. Rare-earth elements are mostly alloyed with Co, Fe, and Ni since in pure form these elements have Curie temperatures below room temperature. This also leads to an increase in the magnetic anisotropy of the alloy. The high magnetic anisotropy of rare-earth magnets can be attributed to the unfilled f shells, which can contain up to seven unpaired electrons (as in gadolinium) with aligned electron spin. This anisotropy makes these alloys easy to magnetize in one direction while hard to magnetize in the other direction. These unpaired electrons behave as local paramagnets as they easily retain their magnetic moments. A higher anisotropy field increases the coercivity H_{ci}, which helps to increase the maximum energy product $(BH)_{max}$, the amount of energy stored inside the material. When forming compounds with magnetic transition metals Fe, Co, and Ni, the spin–orbit coupling results in extremely high magnetic anisotropy, which coupled with the relatively high magnetization of Fe, Co, and Ni results in the necessary recipe for high-energy densities. The important parameters that characterize the performance of a permanent magnet are as follows:

- High saturation magnetization M_s
- High remanence M_r
- Very high uniaxial magnetocrystalline anisotropy energy K_1: high coercivity H_c

- High maximum energy product $(BH)_{max}$
- High Curie temperature T_C

The other important factors are good temperature stability, mechanical strength, machinability, and low cost. A typical hysteresis curve (magnetization M vs. field H) for a permanent magnet has been shown in Figure 7.1 mentioning the important parameters discussed earlier. Another version of hysteresis curve is also available where the ordinate shows the magnetic induction (B) instead of the magnetization (M), and the curve is magnetic induction B versus field H type curve. Among all the aforementioned factors, the most important one is the maximum energy product $(BH)_{max}$ as this is the most representative quantity of a permanent magnet. The maximum energy product is the maximum value of the product of the magnetic induction B and the applied field H in the second quadrant of the B versus H hysteresis curve. Strnat reported the typical demagnetization curves for some important permanent magnets [5].

$(BH)_{max}$ is a quantity that measures the strength of a magnet of volume V, where V is inversely proportional to $(BH)_{max}$. So, a larger energy product means a stronger magnet, which implies that a smaller-sized magnet can be used according to the need for a specific application. Figure 7.2 explains the idea for equal energy output for different materials having different volumes.

The theoretical value of the maximum energy product $(BH)_{max}$ can be given as

$$(BH)_{max,theoretical} = \frac{(B_S)^2}{4} = \frac{(4\pi M_S)^2}{4} \tag{7.1}$$

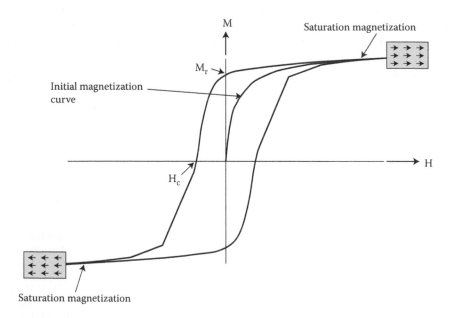

FIGURE 7.1   Typical hysteresis curve for a permanent magnet.

Ferrite V = 1   NdFeB V = 0.08   The future V = 0.03

FIGURE 7.2   Equal energy output for materials having different volumes.

Figure 7.3 illustrates the demagnetization quadrant and the variation of the BH product for a typical permanent magnetic material [6]. Here OP is the load line and P is the working point of the magnet where the load line intersects with the demagnetizing B–H curve.

Figure 7.4 shows the time evolution of the maximum energy product in a logarithmic scale for different permanent magnets over the last century [7]. The theoretical potential should be considered. The material having the highest saturation magnetization can limit the theoretical $(BH)_{max}$. The permanent magnet Nd–Fe–B has the highest theoretical energy product to be used in low-temperature applications and is already in large-quantity production in the United States [8]. The development of a hypothetical Fe–Co-based high-energy magnet is under progress. An excellent $(BH)_{max}$ value (500–1000 kJ/m³) could be achieved if the Fe–Co-based magnets with a large spontaneous polarization of 2.45 T could be realized by introducing strong planar pinning centers.

However, to avoid using relatively expensive and vulnerable sources of Co, the search for the Fe-based permanent magnets continued. This led to the discovery of $Nd_2Fe_{14}B$-based materials in 1983 [9–11], followed by interstitially modified Sm–Fe–N in 1992 [12–15]. But, both of these have some drawbacks compared to Sm–Co

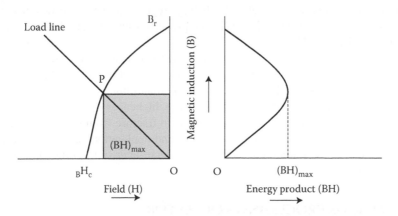

FIGURE 7.3   Demagnetization quadrant of a typical permanent magnet material and the variation of (BH) product with the demagnetizing field. (Adapted from R. A. McCurrie, *Ferromagnetic Materials—Structure and Properties*, Academic Press Limited, San Diego, CA, 1994, p. 193.)

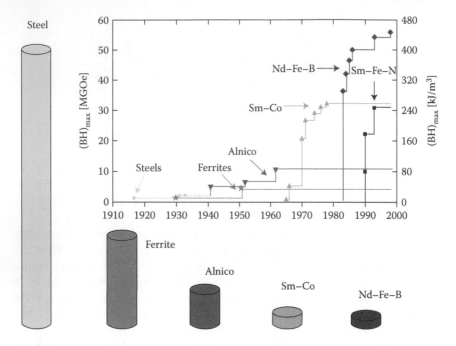

FIGURE 7.4 Development in the energy density $(BH)_{max}$ of hard magnetic materials in the twentieth century and presentation of different types of materials with comparable energy densities. (Reprinted with permission from O. Gutfleisch, *J. Phys. D Appl. Phys.* 33, 2000, R157–R172.)

magnets. The Sm–Fe compounds are not useful as permanent magnet materials (for their basal plane easy magnetization direction) unless nitrogen diffusion expands the crystal structure and produces $Sm_2Fe_{17}N_x$. The Nd–Fe–B magnets are not applicable at higher temperatures. Table 7.1 displays representative properties of various permanent magnet materials [7,16,17]. The properties have been tabulated with an increasing order of $(BH)_{max}$.

The Nd-based alloys have comparatively low Curie temperatures than Sm-based alloys. Although Nd-based magnets are the strongest and the cheapest and are most widely used, Sm-based alloys are better in maintaining high magnetic strength at higher temperatures. Sm-based alloys have higher oxidation resistance than Nd-based alloys, although they are more prone to fracture from thermal shock. The various processing routes, related properties, and applications of those rare-earth magnets are discussed and compared in more detail later in the following sections of this chapter.

7.2 VARIOUS PROCESSING ROUTES FOR RARE-EARTH PERMANENT MAGNETS

The microstructure and the magnetic properties obtained for a magnetic material produced by a specific processing route are always correlated and are strongly

TABLE 7.1
Crystal Structures and Magnetic Properties of Various Permanent Magnets

Magnet	Crystal Structure	B_r (T)	H_{ci} (kA/m)	$(BH)_{max}$ (kJ/m³)	T_C (°C)
Sr-ferrite	Hexagonal	0.2–0.4	100–300	10–40	450
Ba-ferrite	Hexagonal	0.38 MA/m	110–320	10–45	450
Alnico	Cubic	0.2–1.4	55	10–88	700–860
$SmCo_5$	Hexagonal	0.8–1.1	600–2000	120–220	720
Sm_2Co_{17}	Rhombohedral	0.9–1.15	450–1300	150–340	830
$Pr_2Fe_{14}B$	Tetragonal	1.1–1.3	8.7 T ($\mu_0 H_A$)	200–485	290–350
$Sm_2Fe_{17}N_x$	Rhombohedral	1.0–1.3	1050–2010	300–475	476
$Nd_2Fe_{14}B$	Tetragonal	1.0 -1.4	750–2000	250–520	310–400

Source: From O. Gutfleisch, *J. Phys. D Appl. Phys.* 33, 2000, R157–R172; J. M. D. Coey, *J. Magn. Magn. Mater.* 248, 2002, 441–456; S. Aich, Crystal structure, microstructure and magnetic properties of SmCo-based permanent magnets, PhD dissertation, University of Nebraska, Lincoln, NE, 2005.

dependent not only on the alloy composition but also on the processing parameters and heat treatments. The melt-spun ribbons of SmCo-based alloys produced by rapid solidification exhibited higher intrinsic properties, improved microstructures, and better magnetic properties (M_s ~8.5 kG, H_c ~4.1 kOe, $(BH)_{max}$ ~18.2 MGOe, and a high remanence ratio of 0.9) [18]. Zr and Cu substitution for Co helped to reduce the crystallographic texture, and $Sm(Co_{0.74}Fe_{0.1}Zr_{0.04}Cu_{0.12})_{8.5}$ ribbons were nearly isotropic [19]. In magnetically anisotropic $SmCo_5$ ribbons, well-crystallized grains with hexagonal structure (P6/mmm) were observed. Due to the addition of Fe in $SmCo_5Fe_x$ (x = 0, 1, and 2) melt-spun ribbons, produced by using a wheel speed of 25 m/s, the highest magnetic properties were observed for x = 2 ribbons due to their lowest content of Sm-rich phase and the smallest grain size [20]. Due to higher surface-to-volume ratio for x = 2, the intergrain exchange coupling enhanced the remanence. Improved magnetic properties (coercivity as high as 38.5 kOe) were reported for the melt-spun $Sm(Co_{0.74-x}Fe_{0.1}Cu_{0.12}Zr_{0.04}B_x)_{7.5}$ (x = 0.005–0.05) alloys [21]. Better magnetic properties were reported for the boron-containing samples than the carbon-containing samples in melt-spun $Sm(CoFeCuZr)_zM_x$ (M = B or C) nanocomposite magnets due to the finer grain size (30–50 nm) of the former [22].

Depending on the microstructure scale (grain size), the basic processing routes for the magnet production can be classified as either microcrystalline or nanocrystalline route [23]. The microcrystalline route follows the powder metallurgy technique and eventually provides anisotropic magnets having a maximum energy product as high as 50 MGOe, whereas the nanocrystalline route involves rapid solidification techniques (melt spinning or atomization) and other alternate routes (hydrogenation–disproportionation–desorption–recombination [HDDR] or mechanical alloying), which eventually results in isotropic ($(BH)_{max}$ ~10–15 MGOe) and anisotropic

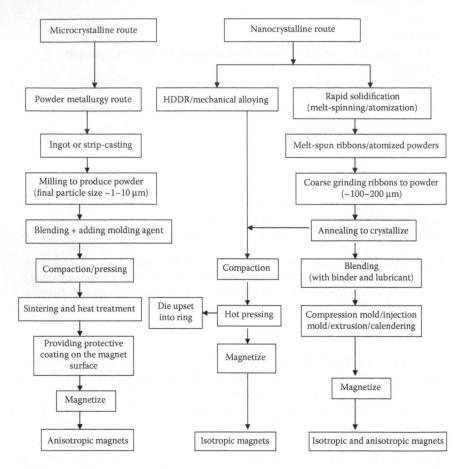

FIGURE 7.5 Schematic of the common processing routes (block diagram/flow chart). (From D. Brown, B.-M. Ma, and Z. Chen, *J. Magn. Magn. Mater.* 248, 2002, 432–440.)

magnets ($(BH)_{max}$ ~20–40 MGOe). Figure 7.5 shows the schematic of the basic processing routes for permanent magnet productions.

In the following sections, each of the common processing routes is described in brief.

7.2.1 POWDER METALLURGY ROUTE

For a long time, it has been a general trend to use powder metallurgy technique to produce anisotropic sintered magnets. Certain conditions are necessary to be fulfilled during the production of anisotropic sintered magnets using the powder metallurgy route [24–26]:

1. The oxygen content should be minimized.
2. The hard magnetic phase should be in a high-volume fraction.

3. The volume fraction of the nonmagnetic grain boundary material should be minimized.
4. A small crystallite size with narrow size distribution is required (for $Nd_2Fe_{14}B$-type magnet typically 2–6 μm).
5. Maximum alignment of the easy axis of magnetization of the crystallites should be maintained.

However, further discussion of the powder metallurgy route is out of the scope of this chapter.

7.2.2 HYDROGEN-ASSISTED PROCESSING

7.2.2.1 Hydrogenation–Disproportionation–Desorption–Recombination

The HDDR process is a well-known processing route for achieving the refined grain structure in the case of rare-earth transition-metal alloys (especially in Nd–Fe–B alloys). This process is very simple and mainly based on hydrogen-induced phase transformation, which can produce highly coercive $Nd_2Fe_{14}B$ powders that can be used to produce bonded magnets as well as fully dense hot-pressed magnets. The principal HDDR reaction of the $R_2Fe_{14}B$ phases can be mentioned as [27]

$$R_2Fe_{14}B + (2 \pm x)H_2 \Rightarrow 2RH_{2\pm x} + 11Fe + Fe_3B \Rightarrow 2RH_{2\pm x} + 12Fe + Fe_2B \quad (7.2)$$

where R = Nd or Pr. During HDDR of $Pr_{13.7}Fe_{63.5}Co_{16.7}Zr_{0.1}B_6$ alloy, an intermediate boride phase, $Pr(Fe,Co)_{12}B_6$ (R3m), has also been found after disproportionation [28]. Also, a high degree of texture has been reported for this type of alloy after conventional processing [29].

The HDDR reaction in the Nd–Fe–B system can be expressed as [30]

$$Nd_2Fe_{14}B + (2 \pm x)H_2 \Leftrightarrow 2NdH_{2\pm x} + 12Fe + Fe_2B \pm \Delta H \quad (7.3)$$

The whole reaction occurs in two stages:

1. *Stage I—Disproportionation*: $Nd_2Fe_{14}B$ phase decomposes into a finely divided mixture of neodymium hydride (NdH_2), iron (Fe), and ferro-boron (Fe_2B). The reaction occurs at ~800°C and at 1 bar hydrogen pressure.
2. *Stage II—Desorption and Recombination*: During desorption due to subsequent heat treatment under vacuum, hydrogen removal occurs from NdH_2 and the disproportionated NdH_2, Fe, and Fe_2B are recombined into $Nd_2Fe_{14}B$ phase with much finer grain structure. The HDDR process for Nd–Fe–B alloy is shown in Figure 7.6.

The application of HDDR-processed $Sm_2Fe_{17}N_3$ magnets is restricted to bonded magnets only because of their insufficient thermal stability (stable only up to 600°C) [7]. However, the thermal stability can be improved using $Sm_2Fe_{17-x}Ga_xC_y$

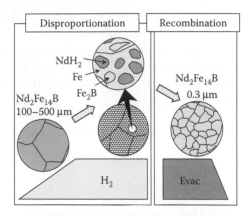

FIGURE 7.6 Schematic of HDDR process in NdFeB system. (Reprinted with permission from S. Sugimoto, *J. Phys. D Appl. Phys.* 44, 2011, 064001.)

alloy [31]. The reaction in HDDR-processed $Sm_2Fe_{17}N_3$ system can be expressed as [7]

$$Sm_2Fe_{17} + (2 \pm x)H_2 \Leftrightarrow 2SmH_{2\pm x} + 17Fe \pm \Delta H \tag{7.4}$$

The advantage of the HDDR process over melt spinning is production of anisotropic powders by aligning the c-axis of $Nd_2Fe_{14}B$ along one direction by changing the composition (addition of alloying elements) or by adjusting the process parameters (controlling the hydrogen pressure and temperature) of the HDDR process.

7.2.2.2 Reactive Milling in Hydrogen

In this special technique, mainly ball milling is done (under enhanced hydrogen pressure and temperature) for disproportionation, which is followed by vacuum annealing for desorption and recombination. Gutfleisch reported the effect of reactive milling on Sm_2Co_{17} alloy [7]. The following reaction occurs during reactive milling like the HDDR-processed Sm_2Fe_{17} alloy mentioned earlier:

$$Sm_2Co_{17} + (2 \pm \delta)H_2 \Leftrightarrow 2SmH_{2\pm\delta} + 17Co \pm \Delta H \tag{7.5}$$

The average grain size of rare-earth hydride phase (SmH) obtained in this process was much smaller (~9 nm) [32] compared to the same obtained during disproportionation of the conventional HDDR process [33,34]. Finally, the average grain size was dependent on the recombination temperature; the observed grain size was estimated as ~18, ~25, and ~29 nm at 600°C, 650°C, and 700°C recombination temperatures, respectively [35]. The remanence value ($J_r = 0.71$ T) was also significantly higher (for the sample recombined at 600°C) compared to the theoretical value ($J_s/2 = 0.65$ T) of single-domain Sm_2Co_{17} particles because of the strong exchange interaction between the nanosized grains in the former case. The higher

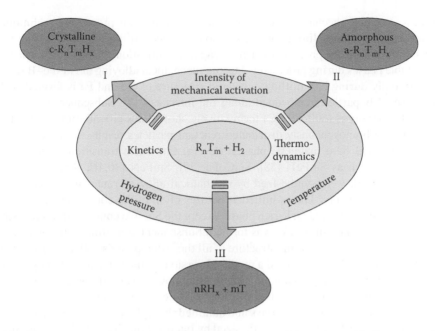

FIGURE 7.7 Schematic representation of hydrogen gas–solid reactions of R–T compounds. Depending on thermodynamics and kinetics, a ternary hydride (I, $c\text{-}R_nT_mH_x$), an amorphous hydride (II, $a\text{-}R_nT_mH_x$), or a binary R hydride and the T (III, $nRH_x + mT$) is formed. (Reprinted with permission from O. Gutfleisch, *J. Phys. D Appl. Phys.* 33, 2000, R157–R172.)

remanence enhancement was observed in the case of $Nd_2Co_{14}B$ alloys due to their smaller grain size obtained during reactive milling (disproportionation) of a series of $Nd_2(Fe,Co)_{14}B$ alloys [36].

Finally, the HDDR processing of R–T compounds (R_nT_m) is an alternative route to mechanical alloying, intensive milling, or rapid quenching for the synthesis of amorphous or nanocrystalline materials. The nature of the final product of the process depends on the thermodynamics and kinetics of the whole process [7,37]. Thermodynamics includes stabilities of the starting alloy and reaction products, and kinetics includes temperature, hydrogen pressure, and possibly mechanical activation. Depending on the thermodynamics and the kinetics, the final product can be an interstitial modified ternary hydride (crystalline $[c\text{-}R_nT_mH_x]$ or amorphous $[a\text{-}R_nT_mH_x]$) [7] or a binary R hydride and the T $(_nRH_x + mT)$ as a result of disproportionation (Figure 7.7).

7.2.3 Mechanical Alloying

Mechanical alloying uses a high-energy ball milling followed by a suitable annealing treatment. In this process, mixing of different elements occurs through an interdiffusional reaction resulting in the formation of ultrafine layered structure of composite particles. The feasibility of alloy formation depends on several factors, such as (i) thermodynamics of the alloy system, (ii) mechanical workability of the starting materials (powders), and (iii) the input energy used during the ball milling process. Because mechanical alloying is a nonequilibrium processing technique, it can overcome many limitations of the conventional milling processes. Like the other nonequilibrium processes (such as rapid solidification), it helps to form the metastable phases during processing. During mechanical alloying of Nd–Fe–B magnets, initially during ball milling, a layered structure of Nd and Fe is formed with undeformed B particles embedded along the interfaces [7]. Subsequent annealing at low temperatures (600–700°C) for relatively short times (10–30 minutes) for the ultrafine and homogeneously distributed reactant particles results in the formation of the $Nd_2Fe_{14}B$ hard magnetic phases. Several rare-earth transition-metal (R–T) compounds, such as SmFeTi [38], SmCoFe [39], SmFeN [40,41], SmFeGaC [42], and Sm–Co [43], were synthesized by mechanical alloying routes using elemental powders as precursors. The as-milled structures of all of the compounds consist of nanocrystalline α-Fe and an R-rich phase, except the Sm–Co compound where a single amorphous phase of $SmCo_5$ was formed. Subsequent annealing of the as-milled products forms nanocrystalline structure of all those compounds with crystallite size of 10–50 nm. Gutfleisch reported a modified version of mechanical alloying process called "intensive milling technique" where alloy powder is used instead of elemental powder during high-energy ball milling [7]. Coercivity of an intensively milled powder is relatively higher than the same obtained through mechanical alloying [44]. Nanostructured $PrCo_5$ powders synthesized by intensive milling for 4 hours and subsequent annealing at 800°C for 1 minute resulted in a coercivity of 16.3 kOe [45]. The nanocrystalline $Nd_{12}Fe_{82}B_6$ alloy powders prepared by HDDR and mechanical milling present high magnetic properties that can be attributed to the exchange coupling between the nanosized $Nd_2Fe_{14}B$ and α-Fe phases [46].

Among all the aforementioned processing routes, rapid solidification technique (rapid solidification processing [RSP]) is the most favorable route to produce permanent magnets because not only does it (RSP) produce the ultrafine grain size (nanostructure), but it also provides better chemical homogeneity and some desirable nonequilibrium metastable phases. The next section discusses various rapid solidification techniques as well as the principles and the advantages and disadvantages of those techniques.

7.2.4 Rapid Solidification Techniques

RSP is becoming a more important area in solidification and has significant potential in industrial use. It can be considered as nonequilibrium cooling as the cooling rate or solidification rate is very high here ($\sim 10^3$–10^9 K/s). In any RSP technique, the rate of advancement of the solidification front (solid/liquid interface) "V" is greater than 1 cm/s. During rapid solidification, as it is a nonequilibrium cooling, loss of local equilibrium occurs at the solid/liquid interface. Due to the interfacial nonequilibrium, the equilibrium phase diagram fails at the interface and the chemical potentials of liquid and solid are not equal anymore. The situation has been described schematically in Figure 7.8 along with related chemical potential gradients.

When the growth rate (V) is comparable or larger than the rate of diffusion over an interatomic distance (D_i/δ_i), that is, $V \geq D_i/\delta_i$, the crystal/atom will not

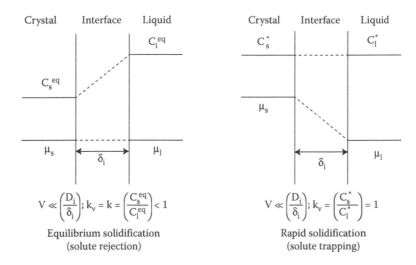

FIGURE 7.8 Loss of local equilibrium at the solid–liquid interface due to the increase in solidification rate/undercooling.

have time to change its composition (rearrange themselves) at the interface so as to equalize the chemical potential (μ) of both phases ($\mu_s \neq \mu_l$), which results in "solute trapping."

Rapidly solidified permanent magnets are getting much more attention since the last decade because of their significantly less complicated processing routes compared to the time-consuming and complicated heat treatment and solution treatment, which is normally required to achieve a remarkably high coercivity for bulk rare-earth permanent magnets. Among the several advanced rapid solidification techniques such as melt atomization, thermal spray coatings, melt spinning, laser melting and resolidification, and high-energy beam treatment of surfaces, melt spinning is the most commonly used rapid solidification technique for the processing of rare-earth permanent magnet alloy systems. The other techniques have specific advantage(s) with added disadvantage(s); for example, the atomization technique provides high production rate and uniform spherical particle morphology, but cannot provide the compositional changes required by the lower cooling rate involved [23]. Various techniques that can be used to produce rapidly solidified alloys can be categorized as the following:

1. Melt spinning, planar flow casting, or melt extraction, which produce thin (~25–100 μm) ribbon, tape, sheet, or fiber
2. Atomization, which produces powder (~10–200 μm)
3. Surface melting (by laser) and resolidification, which produce thin surface layers

Figure 7.9 shows the schematics of various RSP techniques. Each of the categories is discussed in brief in the following sections.

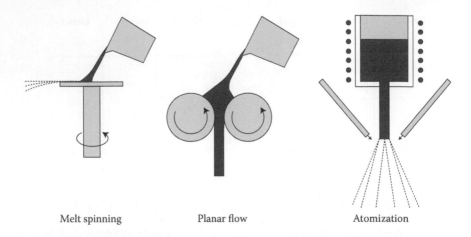

Melt spinning   Planar flow   Atomization

FIGURE 7.9 Schematics of some common RSP techniques.

7.2.4.1 Atomization

Atomization is a technique that uses high-pressure fluid jets to break up a molten metal stream into very fine droplets, which eventually solidify into fine particles. This is a versatile method for powder production. High-quality powders of different metals and alloys, such as aluminum, brass, iron, stainless steel, tool steel, and superalloys, are produced in this method.

The important *objectives* of atomization are as follows:

- Minimization of the average particle size
- Reduction of the particle size distribution width
- Technical production of complex melt systems for powder applications

7.2.4.1.1 *Various Atomization Processes*

Various atomization processes are available depending upon the atomizing medium to break up the liquids, requirements of powder characteristics, and related cost. Different types of atomization processes can be mentioned as follows:

- Water atomization
- Gas atomization
- Soluble gas or vacuum atomization
- Centrifugal atomization
- Rotating disk atomization
- Ultrarapid solidification process
- Ultrasonic atomization

Among all of the atomization techniques mentioned above, water atomization and gas atomization techniques are very popular and are mostly used (Figure 7.10) [47].

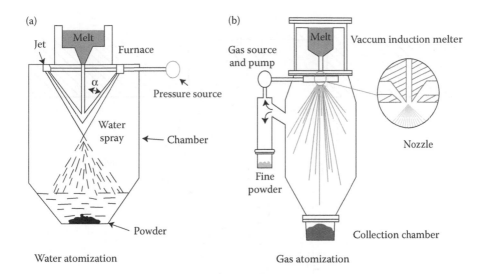

FIGURE 7.10 Schematics of atomization techniques: (a) water atomization and (b) gas atomization. (Adapted from R. M. German, *Powder Metallurgy Science*, 2nd ed., Metal Powder Industries Federation (MPIF), Princeton, NJ, 1994, ISBN-13: 978-1878954428.)

7.2.4.1.2 Mechanism of Atomization

In conventional (gas or water) atomization, a liquid metal is produced by pouring molten metal through a tundish with a nozzle at its base (a reservoir used to supply a constant, controlled flow of metal into the atomizing chamber). As the metal stream exits the tundish, it is struck by a high-velocity stream of the atomizing medium (water, air, or an inert gas). The molten metal stream is disintegrated into fine droplets, which solidify during their fall through the atomizing tank. Particles are collected at the bottom of the tank. Alternatively, centrifugal force can be used to break up the liquid as it is removed from the periphery of a rotating electrode or spinning disk/cup. The disintegration of liquid stream is shown in Figure 7.11. This has five stages: (i) formation of wavy surface of the liquid due to small disturbances (blobs); (ii) wave fragmentation and ligament formation (ligaments are nonspherical liquid sheets, sheared off the liquid jet column); (iii) disintegration of ligament into fine droplets; (iv) further breakdown of fragments into fine particles; and (v) collision and coalescence of particles.

Additional alloying can be performed in the liquid metal bath after the original charge has become molten. Also, the bath can be protected from oxidation by maintaining an inert gas atmosphere as a cover over the liquid metal. Alternatively, the top of the furnace can be enclosed in a vacuum chamber. The furnace type and degree of protection are determined by the chemical composition of the bath and the tendency of the metal to oxidize. Table 7.2 compares the shape and size of the particles obtained from different atomization techniques and cooling rates.

7.2.4.2 Surface Melting by Laser and Resolidification

Use of laser in material processing is attributed to the way it interacts with the materials (especially with the material surface). The laser–matter interaction within the

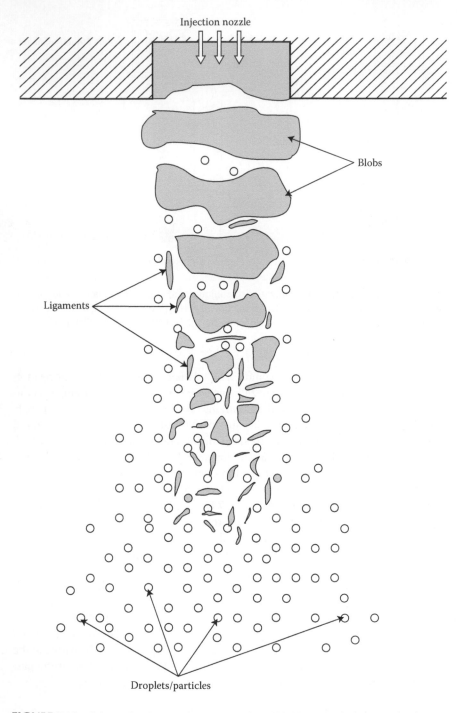

FIGURE 7.11   Schematic of the disintegration of the liquid stream during atomization.

TABLE 7.2
Particle Shape, Particle Size, and Cooling Rates for Various Atomization Techniques

Process/Technique	Particle Shape	Average Particle Size (μm)	Cooling Rate (K/s)
Water atomization	Irregular	75–200	10^2–10^4
Gas atomization (ultrasonic)	Spherical	10–50	$\geq 10^6$
Vacuum atomization (gas soluble)	Spherical	20–150	10^2–10^4
Centrifugal atomization (rotating electrode)	Spherical	150–250	10^4–10^6
Rotating disk atomization	Spherical	Variable (depending on disk speed)	

near-surface region achieves extreme heating and cooling rates (10^3–10^{10} K/s), while the total deposited energy (~0.1–10 J/cm^2) is insufficient to affect the temperature of the bulk material. This allows the near-surface region to be processed under extreme conditions with little effect on the bulk properties.

7.2.4.2.1 Laser Rapid Prototyping

One of the most recent applications of laser in material processing is development of rapid prototyping technologies, where lasers have been coupled with computer-controlled positioning stages and computer-aided engineering design to enable new capability [48–55]. This development implies that manufacturers are no longer constrained to shape metals by the removal of an unwanted material. Instead, components can now be shaped into near-net-shape parts by addition/building the object in lines or layers one after another. Rapid prototyping relies on "slicing" a three-dimensional computer model to get a series of cross sections that can then be made individually. The major techniques for making the slices are stereolithography, selective laser sintering (SLS), laminated object manufacturing, and fused deposition modeling. Laser can be a useful tool for *in situ* rapid prototyping fabrication of composite components such as cutting tools, shear blades, and so on [54].

7.2.4.2.2 Selective Laser Melting

Selective laser melting (SLM) is a powder-based additive manufacturing (AM) process that allows obtaining fully functional three-dimensional parts from a CAD model, able to produce functional components from materials having mechanical properties comparable to those of bulk materials. The competitive advantages of the AM process are geometrical freedom, shortened design to product time, reduction in process steps, mass customization, and material flexibility. SLM refers to the direct route of SLS when complete melting of powder occurs rather than sintering or partial melting. During the process, successive layers of metal powder are fully molten and consolidated on top of each other by the energy of a high-intensity laser beam (Figure 7.12) [56]. Consequently, almost fully dense parts with no need for

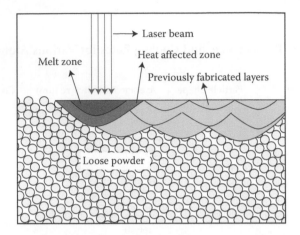

FIGURE 7.12   Schematic view of the transverse section showing different zones due to the process. (Adapted from J. P. Kruth et al., *Proceedings of the 16th International Symposium on Electro-Machining*, 2010.)

post-processing other than surface finishing are produced. The important applications in this area include customized medical parts, tooling inserts with conformal cooling channels, and functional components with high geometrical complexity. SLM is characterized by high-temperature gradients, causing no equilibrium to be maintained at the solid/liquid interface, thereby leading to rapid solidification as the melt pool undergoes transformation from liquid to solid. Formation of nonequilibrium phases and grain refinement are the basic characteristics of this process. Grain structure in SLM differs from the conventional manufacturing process not only because of the cooling rate but also due to the grain structure of the previously solidified layer and the SLM parameters resulting in different (improved) mechanical properties, such as yield strength, ductility, and hardness [56].

Although the objective in SLM is often to obtain 100% dense parts, the goal is difficult to achieve since there is no mechanical pressure, as in molding processes. SLM is characterized only by temperature effects, gravity, and capillary forces during SLM. Moreover, gas bubbles can become entrapped in the material during the solidification due to various causes, such as decrease in the solubility of the dissolved elements in the melt pool during solidification. Besides those melting and solidification phenomena, an insufficient surface quality can cause low density as well. Moreover, the laser energy may not be enough to melt the new layer completely since the depth of the powder in some regions will be thicker. Sometimes, a rough surface causes the entrapment of gas upon deposition of a new powder layer. When the new layer is being scanned, the gas is superheated and expands rapidly removing the liquid metal above it, thus creating a pore.

Building axis: So, the idea for remelting arrives. Laser remelting can improve the density when compared to parts made without remelting. The average porosity of parts without remelting is about 0.77% whereas the densest re-molten part obtained has a porosity of 0.032%. Higher remelting scan speed (200 mm/s) in combination with low laser power (85 W) resulted in better density values. Applying remelting once or multiple times after each layer does not significantly change the porosity for low laser energy inputs to the substrate.

7.2.4.3 Electrospinning

Electrospinning has been recognized as an efficient technique for the fabrication of polymer nanofibers [57]. Various polymers have been successfully electrospun into ultrafine fibers in recent years, mostly in solvent solution and some in melt form. Potential applications based on such fibers, specifically their use as reinforcement in nanocomposite development, have been realized. However, what makes electrospinning different from other nanofiber fabrication processes is its ability to form various fiber assemblies. This will certainly enhance the performance of products made from nanofibers and allow application-specific modifications. It is therefore vital for us to understand the various parameters and processes that allow us to fabricate the desired fiber assemblies. Fiber assemblies that can be fabricated include nonwoven fiber mesh, aligned fiber mesh, patterned fiber mesh, random three-dimensional structures, and submicron spring and convoluted fibers. Nevertheless, more studies are required to understand and precisely control the actual mechanics in the formation of various electrospun fibrous assemblies.

7.2.4.4 Melt Spinning

Melt spinning is one of the most commonly used rapid solidification techniques. People started to use this technique in 1872 with a simple version of melt spinning to produce wires of low-melting temperature alloys [58]. Later, some improved versions of the melt-spinning technique such as chill-block melt spinning (1908) (the precursor of modern single-roller melt spinning) and free-flight melt spinning (1961) (where a jet of molten alloy, coming out of a nozzle, is quenched by the surrounding gas, while it is still in flight) have been invented. The most recent improvement is the single-roller device (the modern version of the chill-block melt spinning), which has been described by Anantharaman and Suryanarayana [59,60]. This device can be used in the most sophisticated way where one or more melt streams are used to make wide or composite ribbons by impingement on single or twin chill roll surfaces [61]. Figure 7.13 represents an RSP unit.

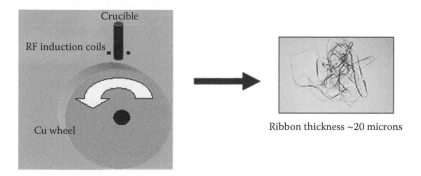

FIGURE 7.13 Schematic of rapid solidification unit/principle.

Here the molten metal is ejected through a small bore at the bottom of a quartz crucible on the surface of a rotating Cu wheel. When the red-hot molten metal touches the chilled surface of the rotating Cu wheel, the molten pool transforms to a thin ribbon due to very high cooling rate and the ribbons leave the wheel surface tangentially to a collecting chamber from where the final products (the melt-spun ribbons) are collected.

The melt-spun ribbons of SmCo alloys produced by rapid solidification exhibited higher intrinsic properties, improved microstructures, and better magnetic properties (M_r ~8.5 kG, H_c ~4.1 kOe, $(BH)_{max}$ ~18.2 MGOe, and a high remanence ratio of 0.9). Some of the other rapid solidification techniques mentioned above have been tried many times on Nd–Fe–B systems, but rarely on Sm–Co systems because of the high vapor pressure of Sm, making it difficult to control the composition. (Sm–Co was gas atomized a long time ago. A company called Crucible Industries worked on it.)

The advantages of the melt-spinning technique (as a rapid solidification technique) over the other solidification techniques on phase equilibria and microstructure of the materials can be mentioned as follows:

- The reduction of grain size as the cooling rate increases to achieve the typical scale (nanoscale) of microstructure
- Better chemical homogeneity with increasing cooling rate
- Production of nonequilibrium metastable crystalline phases
- Extension of solubility and homogeneity ranges of equilibrium phases as the cooling rate increases
- The formation of nonequilibrium glassy phases due to failure of the liquid to undergo complete crystallization

The kinetics of rapid solidification during melt-spinning technique is described in the following few lines. The fundamental feature of melt spinning related to the kinetics of rapid solidification is that the heat evolved during solidification must be transferred with sufficient rapidity to a heat sink, which involves the propagation of a solidification front at a high velocity. The typical cooling rate obtained in this method is ~10^5–10^6 K/s.

In Figure 7.14, three typical conditions have been considered: a molten sphere of radius "r" traveling in a cool gaseous medium (droplet rapid solidification process), a molten cylinder of radius r injected into a bath of liquid coolant (in production of rapidly solidified wire), or a parallel-sided slab of melt of thickness z in at least partial contact on one side with a chill substrate (chill-block melt spinning).

Assuming the interfacial heat-transfer coefficient h is sufficiently low to maintain an essentially uniform temperature throughout the sphere or slab or cylinder during cooling and solidification, the cooling rate ($\dot{T}$) can be expressed as the following

$$\dot{T} = -\frac{dT}{dt} = \frac{h(T - T_A)}{c\rho} \frac{A_0}{V_0} \tag{7.6}$$

where A_0 is the surface area losing heat, V_0 and ρ are the volume and the density, respectively, T_A is the final temperature after the heat lost to the gas, liquid coolant,

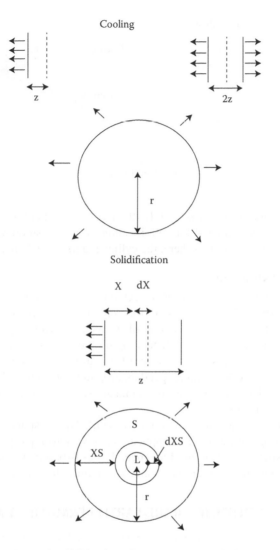

FIGURE 7.14 Cooling and solidification of a sphere or a cylinder or a slab. The arrows are indicating the direction of heat extraction. (Adapted from N. J. Grant, H. Jones, and E. J. Lavernia, *Elements of Rapid Solidification—Fundamentals and Applications*, M. A. Otooni (ed.), p. 35.)

or chill-block, and c is the specific heat released during a small increase in temperature dT to the heat removed in a corresponding time increment dt. Here A_0/V_0 is 3/r, 2/r, and 1/z for a sphere, a cylinder, and a slab, respectively. The average solidification front velocity ($\dot{X}$) can be given as

$$\dot{X} = -\frac{dX}{dt} = \frac{h(T_F - T_A)}{L\rho}\frac{A_0}{A_F} \tag{7.7}$$

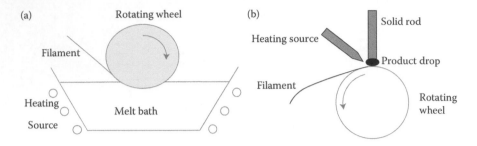

FIGURE 7.15   Schematic of (a) CME and (b) PDME.

where L is the latent heat released at T_F, the freezing temperature of the melt, and A_F is the instantaneous area of the solidification front at position X and A_0/A_F is $r^2/(r-x)^2$, $r/(r-x)$, and 1 for the sphere, the cylinder, and the slab, respectively.

7.2.4.5 Melt Extraction

The melt-extraction process is an RSP technique, which yields short metal fibers with equivalent diameters as low as 50 μm from almost arbitrary metals and alloys [62]. Smooth and uniform cross-section fibers with relatively high tensile strength can be obtained from the process depending on the process parameter. Generally, the melt-extraction process is divided into two subdivisions (Figure 7.15): (1) crucible melt extraction (CME) and (2) pendant drop melt extraction (PDME). Both the subdivisions individually as well as in combination are very beneficial for producing amorphous glassy ribbon.

In the following section and subsections, a detailed discussion is performed on the basis of processing, properties, and applications of various rapidly solidified permanent magnets, such as R–Co-based magnets, R–Fe–B-based systems, and R–Fe–T-based alloys, where R is the rare-earth and T is mainly the C and N.

7.3   RAPIDLY SOLIDIFIED RARE-EARTH PERMANENT MAGNETS

Rapidly solidified permanent magnets are getting much more attention since the last decade because of their significantly less complicated processing routes compared to the time-consuming and complicated heat treatment and solution treatment, which is normally required to achieve a remarkably high coercivity for bulk rare-earth permanent magnets. Among the several advanced rapid solidification techniques such as melt atomization, thermal spray coatings, melt spinning, laser melting and resolidification, and high-energy beam treatment of surfaces, melt spinning and melt atomization are the most commonly used rapid solidification techniques for the processing of rare-earth permanent magnet alloy systems. The melt-spun ribbons of Sm–Co alloys produced by rapid solidification exhibited higher intrinsic properties, improved microstructures, and better magnetic properties (M_r ~8.5 kG, H_c ~4.1 kOe, $(BH)_{max}$ ~18.2 MGOe, and a high remanence ratio of 0.9). Some of the other rapid solidification techniques mentioned earlier have been tried on Nd–Fe–B systems, but not on Sm–Co systems because of the high vapor pressure of Sm, making it difficult to control the composition. Several advantages of the melt-spinning technique as a rapid solidification technique over the other solidification techniques are reduction of grain size to achieve the typical scale (nanoscale) of microstructure, better chemical homogeneity, production of nonequilibrium metastable crystalline phases, and the formation of nonequilibrium glassy phase. The fundamental feature of the melt-spinning technique related to the kinetics of rapid solidification is that the heat evolved during solidification must be transferred with sufficient rapidity to a heat sink, which involves propagation of a solidification front at a high velocity. The typical cooling rate obtained in this method is ~10^5–10^6 K/s. In the following subsections, we discuss the processing, structure, properties, and applications of some important permanent magnets, such as R–Co-based (Sm–Co-based, Pr–Co-based) magnets, R–Fe–B (Nd–Fe–B, Pr–Fe–B) magnets, and SmFeN/SmFeC magnetic alloy systems.

7.3.1 RCo/Sm–Co-Based Magnets

As the second generation of rare-earth permanent magnets, Sm–Co-based magnets have been available since the early 1970s. The most interesting features of these magnets are high-energy products (14–30 MGOe), reliable coercive force, and the best temperature characteristics in the family of rare-earth materials.

Sm–Co-based magnets not only have better corrosion and oxidation resistance but also exhibit better temperature stability. This is the ideal material in applications such as pump couplings, sensors, and servomotors [63]. Two kinds of Sm–Co magnets are available in the market: (1) sintered magnets and (2) bonded magnets. The sintered magnets are formed through the powder metallurgy route as discussed earlier. In bonded magnets, thermoelastomer and thermoplastic resins are blended together with a variety of magnetic powders. The Sm–Co system forms two related equilibrium phases in Co-rich compositions [64]: (1) the $CaCu_5$-type $SmCo_5$ structure and (2) the Th_2Zn_{17}- or Th_2Ni_{17}-type Sm_2Co_{17} structure. The Sm_2Co_{17} structure is related to the $SmCo_5$ structure through the ordered substitution of one Sm by a pair of Co atoms (commonly referred to as Co dumbbells) (Equation 7.8).

$$3RCo_5 - R + 2Co = R_2Co_{17} \tag{7.8}$$

In addition to the ordered Sm_2Co_{17} dumbbell structures, the dumbbell arrangement can be randomized on the rare-earth sites as the disordered $TbCu_7$-type structure [64]. This metastable structure has the same unit cell as the $CaCu_5$ structure. The different crystal structures of the Sm–Co alloy system have been shown in Figure 7.16.

The suppression of the long-range order, leading to the formation of the $TbCu_7$-type $SmCo_7$ structure, has been accomplished by melt spinning [65], splat cooling [66], mechanical alloying [67], and some other special processes [68–70]. The formation of the disordered $SmCo_7$ structure has provided pathways to the development of materials with novel structures, as exemplified by recent advancements in the elevated-temperature performance of Sm–Co-based materials.

The microstructure and the magnetic properties obtained are strongly dependent on the alloy composition, processing parameters, and heat treatments. The melt-spun

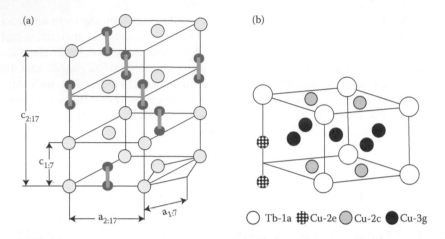

FIGURE 7.16 (a) Schematic showing Sm ($\bigcirc$) and Co ($\bullet$) dumbbell atoms. (Adapted from S. Aich, Crystal structure, microstructure and magnetic properties of SmCo-based permanent magnets, PhD dissertation, University of Nebraska, Lincoln, NE, 2005.) (b) Schematic diagram of hexagonal TbCu$_7$ crystal structure. (Reprinted with permission from J. Luo et al., *Intermetallics* 13, 2005, 710–716.)

ribbons produced by rapid solidification exhibited higher intrinsic properties, improved microstructures, and better magnetic properties (M_r ~8.5 kG, H_c ~4.1 kOe, (BH)$_{max}$ ~18.2 MGOe, and a high remanence ratio of 0.9) [18]. Zr and Cu substitution for Co helped to reduce the crystallographic texture, and Sm(Co$_{0.74}$Fe$_{0.1}$Zr$_{0.04}$Cu$_{0.12}$)$_{8.5}$ ribbons were nearly isotropic [19]. In magnetically anisotropic SmCo$_5$ ribbons, well-crystallized grains with hexagonal structure (P6/mmm) were observed. Due to the addition of Fe in SmCo$_5$Fe$_x$ (x = 0, 1, and 2) melt-spun ribbons, produced by using a wheel speed of 25 m/s, the highest magnetic properties were observed for x = 2 ribbons due to their lowest content of Sm-rich phase and the smallest grain size [20]. Due to higher surface-to-volume ratio for x = 2, the intergrain exchange coupling enhanced the remanence. Improved magnetic properties (coercivity as high as 38.5 kOe) were reported for the melt-spun Sm(Co$_{0.74-x}$Fe$_{0.1}$Cu$_{0.12}$Zr$_{0.04}$B$_x$)$_{7.5}$ (x = 0.005–0.05) alloys [21]. Better magnetic properties were reported for the boron-containing samples than the carbon-containing samples in melt-spun Sm(CoFeCuZr)$_z$M$_x$ (M = B or C) nanocomposite magnets due to the finer grain size (30–50 nm) of the former [22].

7.3.1.1 Addition of Alloying Elements/Stabilizing (1:7) Phase

Stabilizing the 1:7 phase with TbCu$_7$ structure, which was first discovered by Buschow and van der Goot [71], is an important and challenging aspect to consider during synthesizing Sm–Co binary compounds. The growing interest to stabilize the 1:7 phase can be attributed to the achievement of the combined merits of SmCo$_5$ (strong magnetic anisotropy) and Sm$_2$Co$_{17}$ (large saturation magnetization and high Curie temperature). The Sm–Co 1:7 compound is usually prepared by milling, melt spinning, and mechanical alloying followed by a suitable heat-treatment schedule to crystallize the amorphous 1:7 phase. But, annealing results in decomposition of the

Sm–Co 1:7 phase into 1:5 and 2:17 phases. The decomposition of SmCo$_7$ is attributed to the higher enthalpy of formation compared to its neighbor SmCo$_5$.

$$\text{SmCo}\,(1:7) \rightarrow \text{SmCo}\,(1:5) + \text{SmCo}\,(2:17) \tag{7.9}$$

Therefore, interest has been grown to focus on the addition of third element to synthesize the Sm(Co,M)$_7$ compound with TbCu$_7$ structure (space group P6/mmm) to use them as potential high-temperature rare-earth permanent magnets [67,69,72,73]. The feasibility of Sm(Co,M)$_7$ compound formation depends on several factors such as

- Enthalpy of formation of MCo$_7$
- Atomic radius ratio of Sm to (Co,M)
- Electronic configuration of M (the doping element)

While the magnetic properties of Sm(Co,M)$_7$ compounds depend on the site occupation of the stabilizing element M, which can be determined by the two factors:

- Enthalpies of solution of M in liquid Sm and Co
- Electronegativity difference in M, Sm, and Co

7.3.1.1.1 Electronegativity and Site Occupancy of the Doping Elements

The 1:7 phase can be regarded as a derivative of CaCu$_5$ structure (space group P6/mmm), a disordered structure where the Co–Co "dumbbell" pairs are randomly substituted for the Sm atoms occupying the 1a sites (see Figure 7.16) [71]. In Figure 7.16a, Sm$_2$Co$_{17}$ is a rhombohedral (R bar 3m) Th$_2$Zn$_{17}$-type structure. The relation between the lattice parameters for the two structures is

$$c_{2:17} = 3c_{1:7}$$
$$a_{2:17} = \sqrt{3}a_{1:7} \tag{7.10}$$

The magnetic anisotropy field depends on the content (amount) and the site occupancy of the third element M. The site occupancy of M depends on its electronegativity. When the electronegativity of M is less than that of Co, M prefers to occupy the 2e crystal position, whereas for higher electronegativity of M (higher than Co), M tends to be at the 3g site. In Table 7.3, the electronegativity values and corresponding site occupancy for a few doping elements have been mentioned. The stabilizing elements Si and Cu prefer to occupy the 3g site because of their higher electronegativity than that of Co, whereas Ti, Zr, and Hf have preference to occupy the 2e as they have lower electronegativity than that of Co. Considering the content of M, for M greater than 3 (as in SmCo$_3$Cu$_4$), M prefers to occupy the 2c site (Figure 7.16) [74,75]. The anisotropy field of the SmCo compound with 1:7 phase increases if M occupies the 2e and 3g sites, but decreases if it occupies the 2c site.

TABLE 7.3

Electronegativity and Corresponding Site Occupancy for Some Doping Elements

Element	Electronegativity	Site Occupancy
Cu	1.75	3g
Si	1.74	3g
Co	1.70	
Ti	1.32	2e
Hf	1.23	2e
Zr	1.22	2e

7.3.1.1.2 Enthalpy of Formation

Considering a binary alloy containing two kinds of atoms (A and B), the enthalpy of formation of binary transition-metal intermetallics can be estimated using Miedema's empirical formula [76–78]:

$$\Delta H^{form} = \frac{x_A V_A^{2/3} f_B^A \left[-P(\varphi_A - \varphi_B)^2 + Q\left\{ \left(n_{ws}^A\right)^{1/3} - \left(n_{ws}^B\right)^{1/3} \right\}^2 - R \right]}{\left\{ \left(n_{ws}^A\right)^{-(1/3)} + \left(n_{ws}^B\right)^{-(1/3)} \right\}} \tag{7.11}$$

where ΔH^{form} is enthalpy of formation; x_A is atomic concentration of element A in the binary alloy; V_A is atomic volume of atom A; f_A^B is the extent to which an A atom is in contact with its dissimilar atom B; φ_A and φ_B are the electronegativity of A and B atoms, respectively; n_{ws}^A and n_{ws}^B are electron density per Wanger–Seitz cell of A and B atoms, respectively; and P, Q, and R are constants for the given group of metals.

Theoretical calculations using Miedema's empirical formula mentioned above show that the enthalpy of formation of MCo_5 and MCo_7 (where M = Si, Ti, Zr, and Hf) is less than −16 and −12 kJ/mol, respectively. The experimental results match with the theoretical calculations and confirm that Si, Cu, Ti, Zr, and Hf can be used as effective stabilizing elements for $Sm(Co,M)_7$ compounds. According to theoretical calculations based on Miedema's formula, Al, Nb, and Ta can also be considered as the stabilizing elements as the enthalpy of formation of MCo_5 and MCo_7 satisfies the above requirement. However, Al, Nb, and Ta cannot be used (relatively difficult to use) to stabilize the Sm–Co-based 1:7 phase due to their electronic configurations [74]. In the case of Cu, theoretically calculated enthalpy of formation of MCo_5 and MCo_7 (M = Cu) does not satisfy the requirement for stabilizing the 1:7 phase. However, the $Sm(Co,Cu)_7$ compounds exhibit different stabilizing mechanisms, which may be related to the large mutual solubility between Co and Cu in the $Sm(Co,Cu)_7$. In the Co-rich part of the $Sm(Co,Cu)_7$ compound, the compound with the $TbCu_7$-type structure can be stabilized by Cu element, whereas in the Cu-rich part of the $Sm(Co,Cu)_7$ compound, the Sm–Cu-based 1:7 phase can be stabilized by Co element. Therefore, Co and Cu can have a large mutual solubility in the $Sm(Co,Cu)_7$ compound and the

$SmCo_{7-x}Cu_x$ compound has a large homogeneity region with $0.8 \leq x \leq 4.0$ [73], which results in the different stabilizing mechanism of Cu.

7.3.1.1.3 Atomic Radius of the Doping Element

To satisfy the geometrical requirement of forming the $Sm(Co,M)_7$ compound with the $TbCu_7$-type structure, the atomic radius of the doping element M must be larger than that of Co. The amount of the doping element M required to stabilize $SmCo_{7-x}M_x$ compound is inversely proportional to the atomic radius of the doping element. Table 7.4 shows how the amount required of different doping elements depends on their atomic radii [74].

The ratio of the atomic radius of the alloy $SmCo_{7-x}M_x$ can be expressed as [76–78]

$$r_p = \frac{r_{Sm}}{r_{Co+M}} = \frac{7r_{Sm}}{(7-x)r_{Co} + xr_M} \tag{7.12}$$

where r_{Sm} and r_{Co+M} are the atomic radius of Sm and the weighted average of the atomic radius of Co and M, respectively, and x is the amount of M. From Equation 7.12, it is clear that the ratio of the atomic radius r_p is a function of the content of doping element M. The structural stability of $SmCo_{7-x}M_x$ compounds depends on the effective $Sm/(Co,M)$ atomic radius ratio (r_p) and the difference in electronegativity (e_n) between Sm and (Co,M). Usually, the SmCo 1:7-type structure can be stabilized in the range from 1.08(5) to 1.12(9) for r_p and from $-1.04(4)$ to $-0.94(4)$ for e_n [75]. Another research group reported that the range of r_p should be considered from 1.421 to 1.436 for stabilizing the $SmCo_{7-x}M_x$ compounds, which is less than the atomic radius ratio of Sm to Co (1.44) [74].

The amount of doping (stabilizing) element also influences the magnetic properties of the $Sm(Co,M)_7$ compound. Both the saturation magnetization and Curie temperature of $Sm(Co,M)_7$ compounds decrease with increasing M content, while the magnetic anisotropy increases with increasing M content. The Curie temperature decreases almost linearly with the increase in M content since the doping of non-mangnetic stabilizing element weakens the exchange interaction between the Sm and Co sublattices. A remarkable large reduction in Curie temperature ($T_C = 445°C$) in the case of Si doping compared to other doping elements is attributed to its ability to be present in large amount ($x = 0.9$) in $Sm(Co,M)_7$ compounds. The doping element

---

TABLE 7.4

Dependence of Atomic Radius on the Amount Required of the Doping Elements

Doping Element	Atomic Radius (Å)	Amount Required (x)
Zr	1.60	0.19
Hf	1.58	0.21
Ti	1.46	0.30
Si	1.34	0.90

Cu has the least effect on the Curie temperature of $SmCo_{7-x}Cu_x$ compounds, and the Curie temperature decreases from 850.8°C for x = 0.8 to 810.8°C for x = 2.0 [73].

Using Hf and Zr as the stabilizing elements helps not only to stabilize the 1:7 phase but also to increase the magnetic anisotropy field of the $SmCo_{7-x}M_x$ alloy as they (Zr and Hf) as third metallic elements prefer to occupy the 2e site [67,74,75,79]. However, due to the poor formation ability of amorphous Sm–Co [80], it is difficult to achieve the fine grain size distribution, which is highly desirable to achieve high hard magnetic performance in ribbons [81]. One research group has reported that a small addition of carbon is helpful for grain refinement [82].

7.3.1.2 Addition of C/B as a Grain Refiner

Aich et al. reported the rapidly solidified melt-spun ribbons (isotropic) of binary Sm–Co alloys, which can achieve better microstructures and improved magnetic properties when modified with Nb/Hf and C/B addition [17,83–85]. The addition of Nb/Hf and C/B helps to decrease the size of the Co precipitate (~10 nm), which helps to improve exchange interactions between hard phase and soft phase resulting in improved remanence values. The addition of Nb/Hf stabilizes the 1:7 phase, also reduces size of (1:7) Phase, and helps to improve coercivity. Figure 7.17 represents some micrographs of SmCo(Nb/Hf)(B/C) alloys obtained using high-resolution transmission electron microscope (JEOL2010). Without any alloying addition (Figure 7.17a), the microstructure shows micron-sized big grain with larger-sized Co precipitates (~80 nm), whereas addition of Nb and C resulted in reduced size of (1:7) phase as well as smaller size of Co precipitates (~10 nm) (Figure 7.17b). The addition of Hf and C also helps to stabilize the 1:7 phase and results in grain refinement (Figure 7.17c). From Figure 7.17, it is clear that Hf and C/B addition results in more grain size reduction (nanograins) compared to the case of Nb and C addition.

Chang et al. [86,87] reported the microstructure, magnetic properties, and phase evolution of melt-spun $SmCo_{7-x}Hf_xC_y$ and $SmCo_{7-x}Zr_xC_y$ (x = 0–0.4; y = 0 and 0.1) ribbons. The phase transformation and microstructure of the ribbons with Hf substitution are similar to those with Zr. The ribbons with Hf substitution and a slight C addition exhibit a much higher coercivity and energy product than the Zr- and C-containing ribbons, because Hf substitution is more effective in increasing the anisotropy field of 1:7 phase than Zr substitution. The grain size distribution is almost unchanged with the element substitution. The Hf-substituted ribbons exhibited much higher coercivity and energy product compared to the Zr-substituted ribbons. For the alloys with x = 0.4, the maximum intrinsic coercivity ($_iH_c$) increases from 1.9 kOe for $SmCo_7$ ribbon to 17.2 kOe for M = Hf and 11.0 kOe for M = Zr. Due to the addition of a small amount of C in both Hf- and Zr-substituted ribbons, grain refinement occurs and the fcc-Co phase with a grain size of 5–10 nm appears, which results in a stronger exchange coupling effect between the grains leading to achievement of further improved magnetic properties. The optimal magnetic properties of $B_r = 6.8$ kG, $_iH_c = 11.7$ kOe, and $(BH)_{max} = 10.4$ MGOe have been achieved in $SmCo_{6.8}Hf_{0.2}C_{0.1}$ ribbons due to the larger volume fraction of Co phase and the stronger intergranular exchange coupling effect.

The addition of carbon resulted in not only the grain refinement (from micron scale to nanoscale) but also the morphological changes in the microstructures [17,88].

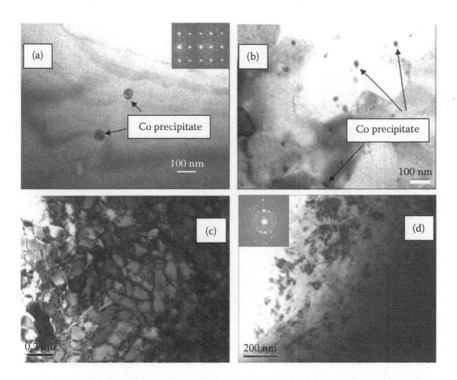

FIGURE 7.17 Bright-field TEM micrographs of $(Sm_{0.12}Co_{0.88})_{100-(x+y)}(Nb/Hf)_xC_y$ alloys melt spun at 40 m/s; (a) x = Nb = 0, y = C = 0; (b) x = Nb = 3, y = C = 0; (c) x = Hf = 3, y = B = 2; and (d) x = Hf = 3, y = C = 2.

At lower percentage of carbon addition (y = 1), equiaxed grains with a wide range of grain sizes (~100–700 nm) were observed (Figure 7.18a–c), with micron scale grains in a few regions. The electron beam diffraction pattern is shown in the inset of Figure 7.18a, indicating the formation of the nanocrystalline structure.

Figure 7.18d–f shows the microstructure for y = 3, which reveals a wide distribution of grain sizes. Figure 7.18d shows large elongated leaf-like structures embedded in a matrix of tiny substructures. In Figure 7.18e, small equiaxed grains, again embedded in a matrix of tiny substructures, were observed. The substructure contrast is caused by disorder (strain). Figure 7.18f represents a triple point grain boundary. At a higher percentage of carbon addition (y = 5), the microstructural morphology changes from equiaxed grains to dendrites, with the dendritic structure on the order of 150 nm long and 50 nm wide with a few coarse dendrites in some regions (Figure 7.18g–i). In Figure 7.18g–i, we can see dendritic microstructures exhibiting the presence of various sizes of dendrites. In Figure 7.18g, we see strangled fine dendritic structures. In Figure 7.18i, some relatively coarser dendrites are present and in Figure 7.18h, a mixture of coarse and fine dendrites are present with a few very coarse dendrites (~2.0 μm long and ~0.45 μm wide).

The coercivity was found to vary linear with x (%Co), ranging from 17.5 kOe at x = 0.67 to 2.75 kOe at x = 3 in $Sm_{(1/(1+x))}Co_{(5+x)/(6+x)}Nb_3C_3$ alloys (Figure 7.19) [17].

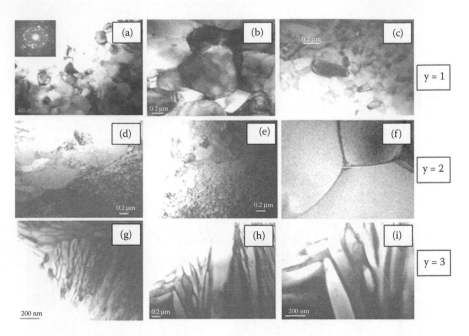

FIGURE 7.18 Transmission electron micrographs revealing the microstructure of the $(Sm_{12}Co_{88})_{100-y}C_y$ alloy melt spun at 40 m/s for $y = 1$, $y = 2$, and $y = 3$, showing morphological transition from equiaxed grains to dendritic structures; (a–c) for $y = 1$ showing equiaxed grains with a wide range of size distribution—(a) with an inset at the top left corner showing SAED pattern indicating the presence of nanocrystalline grains; (d–f) for $y = 3$ showing a mixture of smaller and larger equiaxed grains—(f) showing a triple point grain boundary; (g–i) for $y = 5$ showing dendritic microstructure—(g) very fine strangled dendritic structures, (h) a mixture of fine and coarse dendrites, and (i) relatively coarser dendritic structures.

The change in coercivity (H_c) was associated with a decrease in magnetocrystalline anisotropy (K_1) as x increased. Equation 7.13 describes the relation between the coercivity and the anisotropy

$$H_c = \frac{2\alpha K_1}{\mu_0 M_s} - NM_s \qquad (7.13)$$

where N is the demagnetization factor, M_s is the saturation magnetization, and μ_0 and α are the permeability and the microstructural parameter, respectively. Consequently, it appears that the intrinsic magnetism depends more on the concentration of transition-metal dumbbells than their ordering on the lattice.

7.3.1.3 Effect of Wheel Speed and Heat Treatment

Wheel speed also influences the microstructure and the magnetic properties of the rapidly solidified Sm–Co alloys (Figure 7.20). Higher wheel speed raises the chances of nonequillibrium cooling, which is directly related to the chances of formation of

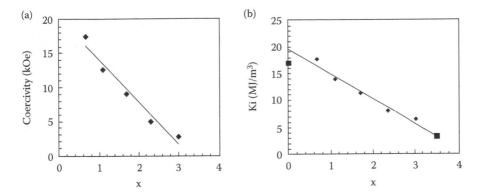

FIGURE 7.19 (a) Relationship between composition (x) and intrinsic coercivity for the $Sm_{(1/(1+x))}Co_{(5+x)/(6+x)}Nb_3C_3$ alloys. (From S. Aich, Crystal structure, microstructure and magnetic properties of SmCo-based permanent magnets, PhD dissertation, University of Nebraska, Lincoln, NE, 2005.) (b) Relationship between composition (x) and the magneto-crystalline anisotropy of the $Sm_{(1/(1+x))}Co_{(5+x)/(6+x)}Nb_3C_3$ alloys. The square end points denote the literature values for the $SmCo_5$ (x = 0) and Sm_2Co_{17} (x = 3.5) compounds.

Co precipitates. Also at higher wheel speed, reduced size of Co precipitates helps to improve the remanence and smaller 1:7 grain results in better coercivity.

Heat treatments were accomplished on the selected samples at temperatures ranging from 700°C to 900°C for 15 minutes. The ribbons were first wrapped in tantalum foil and then were sealed in quartz capsules in the presence of argon. The quartz capsules were then heat-treated in a tube furnace according to the heat-treatment schedule mentioned above followed by water quenching. The heat-treated samples

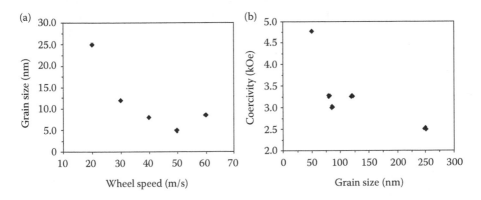

FIGURE 7.20 (a) Relationship between the grain size and the wheel speed of the $Sm_{11}Co_{89}$–NbC alloys. (From S. Aich, Crystal structure, microstructure and magnetic properties of SmCo-based permanent magnets, PhD dissertation, University of Nebraska, Lincoln, NE, 2005.) (b) Relationship between the grain size and the coercivity of the $Sm_{11}Co_{89}$–NbC alloys. (From S. Aich, Crystal structure, microstructure and magnetic properties of SmCo-based permanent magnets, PhD dissertation, University of Nebraska, Lincoln, NE, 2005.)

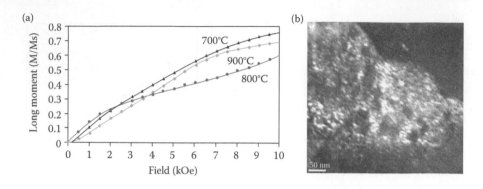

FIGURE 7.21 (a) Magnetization behavior of $(Sm_{12}Co_{88})NbC$ alloy annealed at 700°C, 800°C, and 900°C. (b) Dark-field image of transmission electron micrographs revealing the microstructures of $Sm_{12}Co_{88}$ alloy melt spun at 40 m/s and annealed at 800°C.

showed order–disorder transformations. The order–disorder transformations during heat treatment in Sm–Co alloys can be expressed as

$$R_2Co_{17} = R_{(1-r)}Co_{2r}Co_5 \qquad (7.14)$$

where R stands for Sm atoms and r = 1/3.

The transformation from the disordered $TbCu_7$-type $SmCo_7$ structure to the ordered Th_2Zn_{17}-type Sm_2Co_{17} structure was found to greatly influence the magnetic behavior of the Sm–Co-based alloys. Transition from nucleation-dominated magnetization to pinning-dominated magnetization was observed (Figure 7.21a). The magnetic behavior suggests that antiphase boundaries that developed during the ordering process acted as pinning centers. The bright regions in Figure 7.21b indicate presence of ordered regions correspond to the anti-phase domains (APDs) which provided significant interface area (anti-phase boundaries [APBs]) that acted as domain wall pinning sites.

7.3.1.4 Rare-Earth Magnets as High-Temperature Magnets

Sm–Co alloys are candidates for high-temperature applications (in the fields such as aeronautics, space, and electronic cars), as they exhibit excellent magnetic properties at ambient temperature, such as large magnetocrystalline anisotropy field (6–30 T), high Curie temperature (720–920°C), and large energy product (>200 kJ/m³) [89]. However, the highest service temperature of commercial 2:17-type Sm–Co magnets is only 300°C, and many efforts have been devoted to develop novel high-temperature permanent magnets, namely, development of high-temperature 2:17-type Sm–Co magnets, nanocrystalline Sm–Co magnets, and nanocomposite Sm–Co magnets.

The 2:17-type Sm–Co magnets are now available for application at 500°C or higher in the field of aeronautics and space limited by the cost. If the cost of the magnets can be lower, then the application will spread rapidly into the field of new energy, such as in electronic cars and wind turbines. The magnetic performance of the magnets is sensitive to the composition and heat treatment. Compared with traditional 2:17-type Sm–Co magnets, the high-temperature 2:17-type Sm–Co magnets have less Fe, higher Sm, a cellular structure with smaller size, and a much finer domain structure. The enhancement of coercivity of the magnets during the slow cooling process mainly correlates with the Cu concentration and gradient in the 1:5 phase or the increase of anisotropy of the 1:5 phase.

Both the nanocrystalline Sm–Co magnets and nanocomposite Sm–Co magnets have potential for high-temperature applications. The main problem of these two kinds of magnets is that the magnetic performance of the magnets is relatively low because it is hard to obtain high texture degree in the magnets. Many efforts have been devoted to prepare anisotropic magnets, and magnets with a certain degree texture have been produced by methods of hot compaction plus hot deformation, surfactant-assisted ball milling plus spark plasma sintering, directional annealing, and so on. However, the preparations of anisotropic bulk nanocrystalline Sm–Co magnets and nanocomposite Sm–Co magnets with high texture degree are still big challenges and need further research. If this problem can be solved, then the nanocrystalline Sm–Co magnets and nanocomposite Sm–Co magnets may surpass the 2:17-type Sm–Co magnets and become the new-generation high-temperature permanent magnets. And the improvement of the magnetic properties will speed the applications in the market.

7.3.1.5 Oxidation Protection for Sm–Co Magnets

During the operation of high-temperature permanent magnets, undesirable oxidation at high temperatures is a major issue for potential applications [89]. Two approaches—(1) alloying [90,91] and (2) surface modification [92,93]—were considered as the effective ways to protect rare-earth magnets at high temperature by increasing their oxidation resistance. Alloying nonmagnetic element Si can effectively improve the oxidation resistance of Sm–Co magnets. Using the Sm–Co magnets at high temperature (as high as 500°C) for long time (~500 hours) creates much thinner internal oxidation layer (IOL) ~3–4 µm if Si is added in the magnets compared to the magnets without any Si addition (IOL ~ 212 µm). Also the loss of energy product for Si addition is much less (~5%–6%) compared to the magnets without any Si addition (~52%–53%). Moreover, Liu et al. reported that formation of SiO_2 as the IOL reduces the oxidation rate and oxygen diffusion coefficient, which effectively enhances the oxidation resistance in $SmCo_{6.1}Si_{0.9}$ nanocrystalline magnets [90,91]. Although Si addition enhances the oxidation resistance in Sm–Co magnets, it deteriorates the magnetic properties of the magnets because of its nonmagnetic nature. Therefore, surface modification can be considered as the better way to protect the Sm–Co magnets from oxidation and has little effect on the magnetic properties at the same time. For instance, Ni-coated magnets show better stability (oxidation resistant) at 500°C than uncoated magnets [92,93]. For an uncoated magnet treated at 500°C for 500 hours, the $(BH)_{max}$ loss was ~40%, whereas for the Ni-coated magnet, the $(BH)_{max}$ loss was only ~4%. Also, the other magnetic properties (B_r and H_c) were much higher in the case of coated magnet than the uncoated one [92,93]. The reason behind the improvement of oxidation resistance and enhancement of the magnetic properties in the case of Ni-coated magnets can be attributed to the low oxygen invasion and less Sm vaporization compared to the uncoated magnet. Also, research reveals that at different operation temperatures, different types of coating have the best performance. For example, some coatings such as diffused Pt coating and paint-like overlay coating containing titanium and magnesium oxides can show the best performance at 450°C as well as at 550°C [94]. Other examples are that sputtered SiO_2 is more effective at 450°C but less effective at 550°C, whereas alumina-based overlay coating is more effective at 550°C but less effective at 450°C.

For the high-temperature permanent magnetic materials, a new research direction may be the rare-earth-free magnetic materials that do not rely on the limited supply of rare-earth metals and cost less. Further research is needed to bring the new rare-earth-free magnetic materials for high-temperature applications, and computation methods of combinational materials science may be helpful for the progress.

7.3.2 R–Fe–B-Based Magnets/Nd–Fe–B-Based Magnets

In the R–Fe–B magnets group, Nd–Fe–B magnets are considered as the most demanding and challenging material. In spite of the recent discovery, the Nd–Fe–B magnets are considered as very important magnetic materials in permanent magnet industries because of their enhanced coercivity and large energy product, sometimes significantly higher than the Sm–Co magnets (as mentioned in Table 7.1 and Figure 7.4). The excellent magnetic behavior of the Nd–Fe–B magnet can be attributed to the combined effect of the large spontaneous magnetization of 3d metals and the strong anisotropy fields of rare-earth transition-metal compounds, and at the same time, the magnets maintain a high value of the Curie temperature [95]. These attractive magnetic properties made the Nd–Fe–B magnets appropriate to be used as the powder products for the bonded magnet applications and the fully intermetallic magnets having energy product significantly higher than the best Sm–Co-based magnets. The attractive magnetic properties are attributed to the presence of $Nd_2Fe_{14}B$ phase. Also, Nd–Fe–B magnets attracted much more attention because of their lower cost due to greater availability of Nd and Fe compared to the same of Sm and Co. The following sections describe different categories of Nd–Fe–B magnets, various processing routes for those magnets, and their properties and applications.

The Nd–Fe–B magnets are mainly categorized as (1) sintered magnets and (2) bonded magnets. They are described in detail in the following.

7.3.2.1 Sintered Magnets

Sintered magnets or metallic magnets are produced by conventional casting and powder metallurgy route or hot deformation. Here, in this chapter, the discussion on metallic magnets has limited scope. So, our discussion is limited to bonded magnets.

7.3.2.2 Bonded Magnets

Bonded magnets are also produced by powder metallurgy routes, but here a binder is used to "glue" the powder particles together. The powder is produced by rapid solidification technique. $Nd_2Fe_{14}B$ stoichiometric ribbons are produced by melt spinning followed by milling to produce powders. The powders are then bonded using thermal-set or thermal-plastic polymers followed by compression. The powder magnetic properties, loading factor, and molding technique influence the BH_{max} of the magnets. Powder should have high BH_{max} to get a high BH_{max} compact. High loading factor within molding capability of the alloy is essential to achieve a high BH_{max} magnet. Thermal properties of the powder and the molding polymer should be good to produce a high BH_{max} magnet [23]. The thermal stability of a polymer-bonded magnet depends on several factors, such as particle–particle interaction, binder–particle interaction, amount of binder, and density of the magnet. These factors are needed to be optimized in a potential aggressive environment [23]. Bonded magnets have the several advantages over the sintered magnets, such as [27]

- Easily accomplished near-net-shape processing
- Avoidance of eddy currents
- Good mechanical properties

The main disadvantage of bonded magnet is the dilution of magnetic properties due to the polymer binder [27]. Typical values of B_r and H_{ci} of various metallic and bonded magnets are shown in Figure 7.22. The properties of bonded NdFeB magnets lie between metallic Nd–Fe–B and ferrites.

Table 7.5 compares the magnetic properties of metallic (sintered and hot pressed) and bonded Nd–Fe–B magnets. Four important magnetic properties—intrinsic coercivity ($_iH_c$), remanence (B_r), maximum energy product ($BH)_{max}$, and Curie temperature (T_C)—have been compared for those magnets.

For producing bonded magnets, injection molding is a more favorable technique than compression because of its low cost of processing [96]. Other techniques such as extrusion and calendaring are also used to produce bonded magnets, especially to produce flexible magnets. Magnets produced by calendaring are highly flexible and can be formed to any shape as required in the application. Also, calendaring

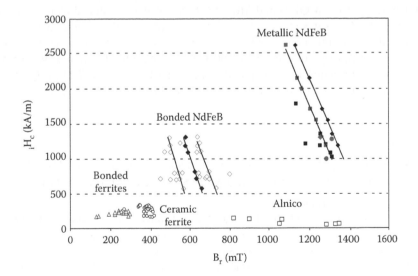

FIGURE 7.22 The B_r and $_iH_c$ plot of commercially available permanent magnets. (Reprinted with permission from B. M. Ma et al., *J. Magn. Magn. Mater.* 239, 2002, 418–423.)

TABLE 7.5

Magnetic Properties of Sintered and Bonded NdFeB Magnets

Magnetic Property	Bonded Magnets	Metallic Magnets (Hot Pressed)	Metallic Magnets (Sintered)
$_iH_c$ (kA/m)	>720	1280–1400	1035–2600
B_r (mT)	690	820–1310	1080–1370
$(BH)_{max}$ (kJ/m^3)	80	130–340	220–360
T_C (°C)	360	335–370	310

is the cheapest process compared to the other processes mentioned above. But, as the loading factor is less for calendaring, the B_r and $(BH)_{max}$ values obtained by this process is less compared to other processes (compression, injection, and extrusion). To design a permanent magnet, the three important factors—intrinsic coercivity $_iH_c$, remanence B_r, and maximum energy product $(BH)_{max}$ along with the temperature characteristics—should be considered other than the molding technique to achieve the desirable properties of the magnet. Powders produced by gas atomization have higher loading factor than those produced by melt spinning because of the spherical morphology of the achieved atomized powder. Spherical powder always shows better flow ability during injection molding compared to other melt-spun powder, as the shear viscosity is much less in the case of spherical powder. The magnetic powders are mainly used in the automotive sector that needs high B_r, high $_iH_c$ (>960 kA/m even at room temperature), and low flux aging loss when exposed to an elevated temperature. In Table 7.6, the loading factor, the flux aging loss, and the various magnetic properties are compared for a series of magnetic powder produced by Magnequench. Here MQP-B and MQP-13-9 are powders produced from melt-spun ribbons and MQP-S-9-8 is the spherical powders of NdFeB magnetic powder having bimodal distribution produced by inert gas atomization [96]. Among those three categories, when they are exposed to 180°C temperature for 100 hours, MQP-S-9-8 has the least flux aging loss compared to the other two (Figure 7.23). MQP-B has exhibited the maximum flux aging loss (as high as ~15%) in the same condition [96].

TABLE 7.6

Loading Factors, Flux Aging Loss, and Magnetic Properties of MQP-B, MQP-13-9 (Injection Molded), and MQP-S-9-8 (Gas Atomized) Powders

Magnet Powder	Loading Factor (vol.%)	Flux Aging Loss	B_r (mT)	$(BH)_{max}$ (kJ/m^3)	$_iH_c$ (kA/m)
MQP-B	62	Maximum	540	85	720
MQP-13-9	62	Intermediate	500	76	700
MQP-S-9-8	69	Minimum	500	73	700

Source: Reprinted with permission from B. M. Ma et al., *J. Magn. Magn. Mater.* 239, 2002, 418–423.

FIGURE 7.23 Morphology of MQP-S-9-8 powder. (Reprinted with permission from B. M. Ma et al., *J. Magn. Magn. Mater.* 239, 2002, 418–423.)

Spherical shape of the powder is important as production rate of powder is high, but cost of the powder is low, and mechanical strength of thin dimension magnet obtained is more in the case of spherical shaped powder compared to the same obtained from the other conventional processes [96].

Again, depending on the microstructure or grain size, the Nd–Fe–B magnets can be categorized as (1) nanocrystalline magnets and (2) microcrystalline magnets [23]. The microcrystalline magnets are produced by sintering the cast ingot (with relatively high rare-earth content ~15 at.%) using the conventional powder metallurgy route. At first, the cast ingot is pulverized using course grinder and then jet-miller to produce fine-powder particles (~5 μm). Then the powder particles are aligned in a magnetic field (10–20 kOe) followed by pressing into a dense compact. Finally, sintering is done at ~1100°C for several hours followed by a post-sintering heat treatment at 600°C to relieve the internal stresses [23]. It is beyond our scope to discuss this aspect here in more detail. So, we are focusing on nanocrystalline magnets.

7.3.2.3 Nanocrystalline Permanent Magnets

The nanocrystalline Nd–Fe–B magnets can be produced by melt spinning to achieve the spun ribbon composition close to 2:14:1 [23]. The ribbons then subsequently are crushed into powder and either bonded with polymer to get bonded magnet or hot deformed to get fully dense magnet.

In any rare-earth transition-metal system, the magnetic properties mainly depend on the grain size or the presence/absence of the intergranular phases that influence the surface/interface effects that differ from the conventional bulk or microcrystalline magnets. Below a critical grain size, large coercivity can be obtained when the crystallite size reaches the single-domain region (Figure 7.24). In the single-domain region, the coercivity decreases for the ultrafine grains due to the thermal effects and the coercivity eventually reaches to zero value in the superparamagnetism particle

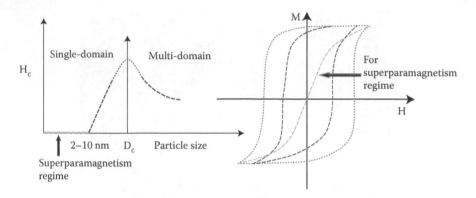

FIGURE 7.24 Magnetic properties of nanostructured materials.

size regime (2–10 nm). Usually, in rare-earth transition-metal compounds, the critical size for single-domain particle is a fraction of a micron (~10–10^3 nm) [97].

Nanocrystalline magnets can be obtained not only by rapid quenching (melt spinning) but also by mechanical alloying, high-energy ball milling, and hydrogen-assisted methods, such as reactive/intensive milling and HDDR process [27]. But, all the methods produce isotropic magnets with randomly oriented grain structure having poor remanence and magnetic energy density $(BH)_{max}$ compared to the ideal microstructure having some anisotropy with single-domain grain. Attempts have been made to maximize the energy density $(BH)_{max}$ by hot deformation or inducing texture/anisotropy by anisotropic HDDR and to improve the remanence by exchange coupling [27].

The rapidly solidified Nd–Fe–B magnets (melt-spun ribbons) are usually isotropic in nature because of their random orientation of the grains resulting in poor magnetic properties (remanence and energy product). Better magnetic properties can be achieved for this isotropic ribbon when the ribbons are crushed and hot pressed to full density and finally are crushed after hot deforming the compact [7]. Although this is a complicated process, it results in anisotropic magnet powder, which can provide aligned Nd–Fe–B-bonded magnets having energy product as high as 145 kJ/m^3 [7]. Similarly, Pr–Fe–B-based magnets can also be produced by rapid quenching (melt spinning) [98,99], which have significant potential for low-temperature applications as they show no spin reorientation down to 4.2 K (which occurs at 135 K for $Nd_2Fe_{14}B$). Also, the Pr–Fe–B magnets are used to produce hot-deformed (textured) magnets.

Consolidation (by conventional polymer bonding or hot pressing) of rapidly solidified Nd–Fe–B melt-spun ribbons were associated with a uniform and fine crystalline (<100 nm diameter) microstructure primarily based on $Nd_2Fe_{14}B$ phase [100]. Anisotropic magnets produced by thermomechanical deformation of the consolidated isotropic Nd–Fe–B magnets can achieve the energy product as high as 45 MGOe [101]. Rapidly solidified (melt spinning) rare earth–transition metal–boron (RE–TM–B) alloys achieved better performance by modifying the compositions and by optimizing the microstructure [102].

Excellent magnetic properties of highly die-upset Nd–Fe–B magnets can be lowered by reduced coercivity and nonuniform deformation. Those limitations can be overcome by heat treatment, substitution, addition, and multistage die upsetting. After multistage (75%) die upsetting, the Nd–Fe–Co–B–Ga magnet was able to perform superior magnetic properties, having remanence and energy product as high as 14.5 kG and 48.5 MGOe, respectively [103].

7.3.2.4 Nanocomposite Magnets

Nanocomposite magnets have been considered as high-performance magnets because of their superior energy products (higher than the conventional single-phase magnets), which is attributed to the exchange coupling between a hard magnetic phase with high coercivity and a soft magnetic phase with high saturation magnetization [104]. The $Fe_3B/Nd_2Fe_{14}B$ system is one of the good examples of nanocomposite magnets [105].

7.3.2.5 Protective Coating to Improve Corrosion Resistance

Nd–Fe–B magnets have poor corrosion resistance to humid environment [106]. The surface of the magnet should be coated with some protective coatings to protect the surface of the magnet from the aggressive environment.

Nd–Fe–B magnets can also be prepared using another well-known RSP technique named as high-pressure has atomization (HPGA) [107]. In HPGA technique, rapid solidification of mixed rare earth–iron–boron alloys, $MR_2Fe_{14}B$ (MR = Nd, Y, Dy) magnet alloys, produced almost similar properties and structures as closely related alloys produced by melt spinning at low wheel speeds. Additions of titanium carbide and zirconium to the permanent magnet alloy design in HPGA powder (using He atomization gas) have made it possible to achieve highly refined microstructures with magnetic properties approaching melt-spun particulate at cooling rates of 10^5–10^6 K/s. By producing HPGA powders with the desirable qualities of melt-spun ribbon, the need for crushing ribbon was eliminated in bonded magnet fabrication. The spherical geometry of HPGA powders is more ideal for processing of bonded permanent magnets since higher loading fractions can be obtained during compression and injection molding. This increased volume loading of spherical magnet powder can be predicted to yield a higher maximum energy product $(BH)_{max}$ for bonded magnets in high-performance applications.

Passivation of rare-earth-containing powder is warranted for the large-scale manufacturing of bonded magnets in applications with increased temperature and exposure to humidity.

Irreversible magnetic losses due to oxidation and corrosion of particulates is a known drawback of RE–Fe–B-based alloys during further processing, for example, injection molding, as well as during use as a bonded magnet. To counteract these effects, a modified gas atomization chamber allowed for a novel approach to *in situ* passivation of solidified particle surfaces through injection of a reactive gas, nitrogen trifluoride (NF_3). The ability to control surface chemistry during atomization processing of fine spherical RE–Fe–B powders produced advantages over current processing methodologies. In particular, the capability to coat particles while "in flight" may eliminate the need for post-atomization treatment, otherwise a necessary step for oxidation and corrosion resistance. Stability of these thin films was attributed to the reduction of each RE's respective oxide during processing; recognizing that fluoride compounds exhibit a slightly higher (negative) free-energy driving force for formation. Formation of RE-type fluorides on the surface was evidenced through x-ray photoelectron spectroscopy (XPS). Concurrent research with auger electron spectroscopy has been attempted to accurately quantify the depth of fluoride formation to grasp the extent of fluorination reactions with spherical and flake particulates. Gas fusion analysis on coated powders (dia. <45 μm) from an optimized experiment indicated an as-atomized oxygen concentration of 343 ppm, where typical, nonpassivated RE-atomized alloys exhibit an average of 1800 ppm oxygen. Thermogravimetric analysis (TGA) on the same powder revealed a decreased rate of oxidation at elevated temperatures up to 300°C, compared to similar uncoated powder.

University of Delware attempted to make isotropic nanocomposite $R_2Fe_{14}B$/α-Fe melt-spun ribbons to achieve the maximum energy product $(BH)_{max}$ greater than 20 MGOe [8]. Three alloy compositions—$Pr_9Fe_{85}B_6$, $Pr_5Nd_5Fe_{19}Co_5B_6$, and $Pr_9Co_5Fe_{80}B_6$—were melt spun at a wheel speed of 18 m/s in a temperature range of 1300–1500°C. The obtained $(BH)_{max}$ value was improved by optimizing the temperature of the melt used during the melt ejection onto the rotating wheel, which helped to result in a homogeneous nanophase structure. Use of an optimum ejection temperature of 1360°C was able to provide the highest value of the reduced remanence $M_r/M_s = 0.8$ and $(BH)_{max} = 21$ MGOe.

7.3.3 Interstitially Modified R_2Fe_{17}-Based Permanent Magnets

The Sm–Fe compounds are not useful as permanent magnet materials (for their basal plane easy magnetization direction) unless nitrogen or carbon diffusion expands the crystal structure and produces $Sm_2Fe_{17}M_x$ (M = N or C). Interstitial diffusion of nitrogen or carbon in Th_2Zn_{17}-type rhombohedral R_2Fe_{17} structure dramatically changes its intrinsic magnetic properties by increasing the Curie temperature (T_C) and by enhancing the magnetocrystalline anisotropy. The Curie temperature increases due to lattice expansion. In the case of interstitial diffusion of nitrogen in Sm_2Fe_{17} compound, due to 6% volume expansion, the Curie temperature was increased from 116°C to 476°C [7]. The increase in magnetocrystalline anisotropy of interstitially modified magnets is attributed to the change in crystalline electric field due to the presence of interstitial nitrogen or carbon atom in the large 9e sites [108]. Based on the crystal field theory, the anisotropy constant (K_1) can be expressed as [109]

$$K_1 = -\frac{3}{2}N_R\alpha_j\langle r^2\rangle\langle O_{20}\rangle A_{20} \tag{7.15}$$

where N_R is the rare-earth concentration, α_j is the Stevens factor, $\langle r^2\rangle$ is the expectation value of the square of 4f radius, A_{20} is the crystal field parameter, and $\langle O_{20}\rangle$ is the expectation value of $3J_z^2 - J(J + 1)$, where J is the total angular momentum of the rare-earth ions. Unlike the binary R_2Fe_{17} compound, the ternary $R_2Fe_{17}M_x$ $(0 \leq x \leq 3)$ nitrides or carbides are considered promising materials for permanent magnets. The high magnetic anisotropy of the $Sm_2Fe_{17}M_x$ magnet can persist even at high temperature, which made it challenging as a promising material for high-temperature applications replacing the Sm–Co-type expensive magnet. Relatively to a lower extent, the $RFe_{12-x}T_x$ (T = W, V, Ti, Mo, Si) compounds are also considered as the interstitially modified permanent magnets. In the case of 2:17 and 1:12 compounds, the potential to be used as permanent magnets is attributed to their enhanced magnetic anisotropy rather than the improved Curie temperature [109]. The magnetic anisotropy is a strong function of the crystal field parameter A_{20}, as mentioned in the above equation. The interstitial atoms occupy the positions very close to the rare-earth atom [110,111], which results in significant changes in A_{20} values. Some NMR experiments were carried out with the help of ^{89}Y spin-echo NMR on the compounds before and after charging with interstitial atom C, N, and H [112]. The experimental data were used to study the Y hyperfine field in Y_2Fe_{17} compounds as a function of increasing occupancy of the interstitial sites by the C, N, and H atoms. A strong decrease in the Y hyperfine field in the case of nitrides and carbides indicates the strong bonding effect (bonding between the on-site valence electrons of R atoms and the valence electrons of interstitial atoms). This bonding effect primarily determines the A_{20} value. In the case of interstitial hydrides, the change in the Y hyperfine field was negligible, which indicates a moderate bonding effect. The strong bonding effect of nitrides and carbides and related change in the A_{20} value explain the strong magnetic anisotropy associated with the nitride and carbide compounds. The changes in A_{20} value due to the interstitial bonding are different for 2:17 and 1:12 compounds, as they have different atomic arrangement for the interstitial atoms, as shown in Figure 7.25. To estimate the A_{20} value (the electric-field gradient at the 4f site), a rare-earth Mössbauer spectroscopy (^{155}Gd Mössbauer spectroscopy) was used [111], and from the measurement data it was derived that the electric-field gradient at the nuclear site (V_{zz}) of R_2Fe_{17} compound was changed from 4.4×10^{21} to 16.2×10^{21} V/m^2 due to nitrogenation. In the case of 1:12 compound, the V_{zz} value was changed from 1.6×10^{21} to -21.3×10^{21} V/m^2 (enhancement of field gradient is accompanied by a sign reversal) [112]. Usually, changes of similar magnitude and sign are expected for A_{20}. So, in the latter case, the enhancement of the field gradient accompanied by a sign reversal may be attributed to the different sign of α_j (Equation 7.15) of the R elements employed in 2:17 nitride compound (Sm) than in 1:12 nitride compound (Pr, Nd).

A simple gas-phase nitrogenation can be carried out according to the following reaction [7]:

$$2R_2Fe_{17} + 3N_2 \Rightarrow 2R_2Fe_{17}N_3 \qquad (7.16)$$

Temperature and pressure were maintained at ~1 bar and ~400–500°C, respectively. This type of gas–solid reaction takes place in three stages:

1. Adsorption of the gas molecules
2. Dissociation and subsequent chemisorption of the gas molecules
3. Long-range diffusion into the metal matrix

In the above nitrogenation process/reaction (Equation 7.16), the temperature should be high enough to overcome the activation barrier to initiate the adsorption of

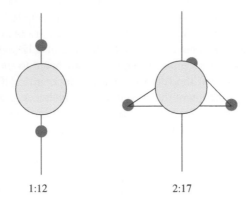

1:12   2:17

FIGURE 7.25 Different atomic arrangements for the interstitial atoms for 1:12 and 2:17 phases.

nitrogen atoms. However, a detailed investigation is needed to explore the efficiency of the irreversible nitrogenation process, whether, in a thermodynamic equilibrium, a solid solution of ternary nitrides with fully nitrided and nonnitrided layers are obtained as the diffusion of nitrogen atom is significantly slow into the matrix. To improve the rate of nitrogen diffusion into the matrix, people attempted to produce fine-powder particles of nitrogen using intensive milling [113] or HDDR technique [114], which eventually can be used during nitrogenation. It is well known that dislocations, grain boundaries, and any other free surfaces are considered as the high diffusivity paths as the mean jump frequency of atoms at those locations is much higher than that of the same atoms in the lattice. Microcracks and microcrystals provide faster and easier diffusion paths, leading to shorter annealing times and lower annealing temperatures in any processing. However, at lower temperatures, the following reaction (decomposition in RN and Fe) is thermodynamically preferable:

$$R_2Fe_{17} + N_2 \Rightarrow 2RN + 17Fe \qquad (7.17)$$

However, due to the difficulty to achieve the long-range diffusion of the metal atoms at the lower temperatures and insufficient kinetics, the formation of metastable ternary nitrides arc to some extent successful compared to the more stable binary nitride and metallic Fe. Due to this poor thermal stability, ternary nitrides have limited applications in metal- and polymer-bonded magnets. Addition of alloying elements, such as Al, Si, and Ga, can help to improve the thermal stability (or to increase the decomposition temperature) [115–117].

In general, carbides are considered to be more thermally stable than nitrides although the latter shows better magnetic properties. Usually, standard casting techniques are used to produce the ternary carbides $R_2Fe_{17}C_x$ ($x \le 1.5$). Better magnetic properties can be achieved by increasing the C concentrations either by partial substitution of Fe by Ga in the cast materials or by exposing fine particles of the binary compound to hydrocarbon gases at elevated temperatures (see reviews by Skomski et al. [118] or Fuji and Sun [15]. To achieve highly coercive powders, the grain size should be very small (grain refinement) and some special processing routes have been discussed by Muller et al. [119]. Figure 7.26 compares the demagnetization curves of $Sm_2Fe_{17}N_3$ and $Sm_2Fe_{15}Ga_2C_2$ compounds. The $Sm_2Fe_{17}N_3$ compounds were prepared by mechanical alloying of the binary compound and subsequent nitrogenation of the alloy powder annealed at 750°C, whereas $Sm_2Fe_{15}Ga_2C_2$ compounds were prepared by mechanical alloying of the melt-carburized compound followed by annealing at 800°C [120]. The alloy without any Ga addition shows higher remanence and higher coercivity compared to the alloy with Ga addition. Addition of nonmagnetic alloying element (Ga) results in the reduction in saturation magnetization and decrease in the anisotropy field of the $Sm_2Fe_{15}Ga_2C_2$ alloys, leading to lower remanence and lower coercivity in $Sm_2Fe_{15}Ga_2C_2$ compounds. However, the thermal stability of the $Sm_2Fe_{15}Ga_2C_2$ alloy was much improved compared to the $Sm_2Fe_{17}N_3$ compounds, which facilitated accessing/using the hot-pressing processing route to obtain fully dense magnets.

Discovery of Sm_2Fe_{17}-nitride has triggered a renewed interest in permanent magnets based on nitrides. Sm_2Fe_{17}-nitride-based permanent magnets were produced by RSP (melt spinning with a linear velocity of 55 m/s), followed by annealing at 650°C for 10 minutes and subsequent nitrogenation at 450°C for 4 hours [121]. These alloys exhibited attractive magnetic properties: intrinsic coercivity ~17.7 kOe, maximum energy product $(BH)_{max}$ ~12 MGOe and Curie temperature ~470°C. The secondary phases $SmFe_2$ and $SmFe_3$ were obtained during processing and Sm_2Fe_{17}-nitride was subsequently fully decomposed into SmN and α-Fe. The Sm-richer compositions are more recommended as Sm-poorer compositions are more prone to give α-Fe, an undesirable phase.

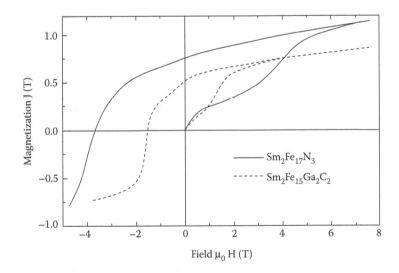

FIGURE 7.26 Comparison of demagnetization curves of $Sm_2Fe_{17}N_3$ prepared by mechanical alloying the binary compound and subsequent nitrogenation and annealing (750°C) and $Sm_2Fe_{15}Ga_2C_2$ prepared by mechanical alloying the melt-carburized compound followed by annealing at 800°C. (Reprinted with permission from O. Gutfleisch, *J. Phys. D Appl. Phys.* 33, 2000, R157–R172.)

Magnetically hard $SmFe_{7+x}M_x$ compounds were obtained by rapid solidification. Initially, the pure elements (99.9%) of Sm, Fe, and M (M = Mo, Ti, and V)
were melted by induction heating to obtain the precursor $SmFe_{7+x}M_x$, which was
subsequently melt spun at a linear velocity of 30 m/s followed by annealing at
650–750°C in an inert atmosphere [122]. The hardness was attributed to the crystalline $Sm(Fe,M)_7$ phase, which exhibited high Curie temperature ~355°C with a
high intrinsic coercivity of ~3–7 kOe, high remanence of 8 kG, and the maximum
energy product of 5 MGOe. The high Curie temperature value was attributed to the
extended solid solubility of the additives (transition elements) in the $Sm(Fe,M)_7$ cells
causing ~4.7%–6.7% volume expansion.

SmFeN magnets (isotropic as well as anisotropic) can be produced by three
routes: mechanical alloying, HDDR, and Zn-bonded technique. SmFeN magnets
obtained by mechanical alloying showed much higher coercivity compared to the
same obtained by Zn-bonded magnets. The higher coercivity obtained in the former
case can be attributed to the single-domain particles obtained through mechanical
alloying compared to the multi-domain structure obtained in the latter case [123].

The addition of tungsten in SmFeN–α-Fe nanocomposite magnets influences
both the microstructure and the magnetic properties (intrinsic properties) (of the
Sm–Fe–N phase). A series of Sm–Fe–W alloys were produced by high-energy ball
milling using elemental powders in Ar atmosphere (balls–to–powder ratio was 5).
After milling, the powders were annealed at a temperature of 620°C for 1 hour in a
vacuum of 10^{-3} Pa followed by nitriding carried out at 460°C for 1 hour [124].

Tungsten dissolves in the $Sm_2Fe_{17}N_3$ phase but does not dissolve in α-Fe. Some
reduction of α-Fe grain size is observed in samples containing W, whereas the grain
size of the $Sm_2(Fe,W)_{17}N_3$ phase remains constant. Tungsten exists in the microstructure in the form of small regularly dispersed grains. The remanence decreases
for higher W contents but up to 10 at.% W is still enhanced. As the effect of these
changes, the $(BH)_{max}$ value achieves its maximum of 120 kJ/m^3 for 2% W. The
decrease in the remanence for doped magnets results from the existence of inclusions of a nonferromagnetic phase (W) and for higher W contents, the formation of
$W/Sm_2Fe_{17}N_3$ boundaries, which lower the effect of remanence enhancement.

The addition of alloying elements changes the microstructure and magnetic properties of the SmFeN magnets significantly [125–128]. The main key to get success
in this process is to achieve α-Fe-free Sm_2Fe_{17} due to the stabilization of the TFe_2
Laves phases, where T is Ti, Nb, Zr, Ta, Hf, and V. A technique has been mentioned
to produce α-Fe-free Sm_2Fe_{17} by adding 4%–5% Nb in the melt and a subsequent
nitriding procedure of the material in the presence of N_2 to produce $Sm_2Fe_{17}N$ [125].
In this case, the soft α-Fe phase was replaced by the paramagnetic $NbFe_2$ phase. For
the unbounded powder, the coercivity was observed at 0.2 T for the alloy with 5%
Nb addition, and this material was highly comparable to the alloy obtained by long
annealing at 1000°C. And eventually, this process was able to develop the efficient
processing routes for the nitrogenated powder.

Since the discovery in 1990, SmFeN is considered as a potential candidate for
use as a high-energy permanent magnet due to its excellent magnetic properties. At
a higher temperature (~500°C), SmFeN dissociates into α-Fe and SmN, where α-Fe
is a soft phase and deteriorates the magnetic properties. As a result, the application

of this type of magnet is limited to as bonded magnets. Sm_2Fe_{17} phase is formed through a peritectic reaction where some α-Fe forms (~25% of the phases present). The formation of this detrimental α-Fe phase can be avoided/removed by extended vacuum or inert atmosphere heat treatment, which is not only expensive but also detrimental to the environment and difficult to maintain the precise composition due to Sm evaporation. Earlier research showed that the addition of third elements, such as Nb, Zr, and Ta, can reduce the formation of primary α-Fe to a large extent.

Ta addition in SmFeN improved the microstructure and magnetic properties (coercivity) significantly. Alloys of $Sm_{13.8}Fe_{82.2}Ta_{4.0}$ and binary SmFe were prepared using HDDR process and were subsequently nitrided/nitrogenated to obtain SmFeTaN and SmFeN alloys [128]. Partly, the alloys were milled using attritor milling prior to the HDDR process. In the case of HDDR-processed samples, higher coercivity was obtained in the Ta-containing sample due to the presence of soft magnetic phase α-Fe, which was replaced by $TaFe_2$ phase. Also, the coercivity of the pre-milled sample (sample attritor milled before HDDR) was higher because of the small particle size, produced by attritor milling prior to the HDDR process, which physically prevents the growth of large grains. Table 7.7 shows the coercivity values and grain size obtained in both the cases.

The same research group found some different phase formations during Ta addition in SmFe binary alloy using HDDR process [129]: Ta_3Fe_7 along with Sm_2Fe_{17}, $SmFe_2$, and $SmFe_3$ phases. In the case of Ta-containing alloy, dissolution of 2.0% of Ta into the 2:17 phase increased its stability with respect to decomposition in a hydrogen atmosphere. Initially, Ta was dissolved in the $Sm_2(Fe,Ta)_{17}$ phase in the cast structure; however, in the HDDR process, Ta-based precipitates were formed leaving the 2:17 phase with 1.2% dissolved Ta.

Zr addition (1 at.%) in the SmFe alloy can avoid the formation of soft magnetic phase α-Fe [127]. Without going for any time-consuming homogenization process, the nonhomogenized $Sm_{10.5}Fe_{88.5}Zr_{1.0}$ alloy was able to achieve the coercivity as high as 3.1 T and the maximum energy product $(BH)_{max}$ ~136 kJ/m³. The nonhomogenized $Sm_{10.5}Fe_{88.5}Zr_{1.0}$ was milled and annealed at vacuum and subsequently nitrogenated, resulting in anisotropic magnetic powders having coercivity of ~2.0 T and energy product $(BH)_{max}$ of ~136 kJ/m³, whereas the HDDR treatment and subsequent nitrogenation resulted in isotropic magnetic powder having coercivity as high as 3.1 T and energy product $(BH)_{max}$ ~103 kJ/m³.

TABLE 7.7
Coercivity and Grain Size Obtained in HDDR-Processed Material and Pre-Attritor Milled Material

Type of Processing	Coercivity (kA/m)		Grain Size (μm)
	SmFeN	SmFeTaN	
HDDR-processed material	360	680	~100
Pre-attritor milled material	1010	1280	~5

Injection-molded Sm–Fe–N magnets exhibited improved magnetic properties [130]. Yamamoto combined Nd–Fe–B powders with Sm–Fe–N powders to obtain an excellent energy product $(BH)_{max}$ over 160 kJ/m³.

Although several attempts have been made to develop the interstitially modified magnets, still they are thermodynamically unstable at high temperature, which prevents them to be used as high-temperature magnets.

7.3.4 NEW MATERIALS/NANOCOMPOSITES/THIN-FILM MAGNETS

Recently, in the nanotechnology era, developments in the microelectromechanical system (MEMS) technology and nanoelectromechanical system (NEMS) technology have stimulated the research activity on rare-earth-based permanent magnet thin films. Whether the film is rapidly solidified or not depends on the film deposition parameters (deposition temperature/substrate temperature, deposition time, rate of deposition, and substrate temperature), the substrate material (their heat-transfer coefficient), and the film deposition technology/method (sputtering, pulsed laser deposition, high-rate sputtering).

Jiang et al. reported an improved permanent magnet (hard ferromagnets) produced by deposition of multilayers of nanometer-thick Sm–Co and Fe, which behave as a nanocomposite [131]. Here, the energy product $(BH)_{max}$ of the nanocomposite hard magnets depends on the film thickness of the multilayered magnets. For a specific Sm–Co layer thickness, as the thickness of the Fe layer decreases in the nanocomposite magnet, the $(BH)_{max}$ of the film increases in the multilayered film.

Fabrication of $[Sm(Co,Cu)_5/Fe]_6$ multilayer film was reported with in-plane texture and a high maximum energy product $(BH)_{max}$ of 256 kJ/m³, which is larger than the theoretical limit of $(BH)_{max}$ for SmCo₅. The improved magnetic properties can be attributed to strong exchange coupling between the $Sm(Co,Cu)_5$ and Fe layers, which was clearly understood from the single-phase behavior and the irreversible rotation in the demagnetization process [132]. Uehara et al. reported a multilayered structure made of Co-free/Co-doped Nd–Fe–B film layer (~200 nm) with Ta layer (~10 nm), which showed better magnetic property (enhanced energy product) compared to the monolayered Nd–Fe–B film magnets [133]. The multilayered film was sputtered on a heated glass substrate. The higher energy product achieved in the multilayered film can be attributed to the formation of the highly textured (perpendicular alignment) $Nd_2Fe_{14}B$ phase having small sized grains (size was controlled by the thickness of the unit layer). This perpendicularly anisotropic multilayered magnetic film exhibited the magnetic properties of H_{cJ} = 979 kA/m, J_r = 1.44 T, and $(BH)_{max}$ = 364 kJ/m³. Recently, high-performance hard magnetic Nd–Fe–B-sputtered films were developed using high-rate sputtering onto 100 mm Si substrates [134]. The thickness of the film was ~5 μm and was deposited at 500°C. With a columnar grain structure, the film showed much improved magnetic properties of H_{cJ} = 1.28 MA/m and $(BH)_{max}$ = 400 kJ/m³. The coercivity was improved as high as 2.08 MA/m by varying (increasing) the Nd content in the films [135]. The increase in coercivity (enhanced coercivity up to H_{cJ} = 2.08 MA/m) was also reported by another group. Oka et al. [136] reported enhanced coercivity by the deposition of Cu layer on Nd–Fe–B thin-film layers. Sato et al. [137] reported a further enhancement of coercivity up to 2.50 MA/m by depositing of Nd–Cu layered structure. Thick-film magnets are ideal to use in magnetic microactuators. Nd–Fe–B thick-film magnets (thickness >10 µm) were fabricated by high-speed pulsed laser deposition and obtained good magnetic properties of $(BH)_{max} = 120$ kJ/m^3 [138].

7.4 APPLICATIONS OF RARE-EARTH PERMANENT MAGNETS

The rare-earth-based permanent magnets have a wide range of applications starting from the mobile phone and refrigerator in our daily life to modern science and technology such as MRI and other innovative systems. Some of those applications are discussed below. The important applications of the rare-earth-based permanent magnets can be mainly categorized as [23]

1. *Computer peripheral*: Disk drive spindle motors and voice coil motors, CD-ROM spindle motors, pick-up motors
2. *Office automation*: Printer and fax stepper motors, printer hammer, copy machine rollers
3. *Consumer electronics*: VCRs and camcorders, cameras, speakers (acoustic applications), headsets, microphones, pagers, DVD players, watches, cell phones
4. *Automotive*: Starter motors, electric steering, sensors, electric fuel pumps, motors in hybrid cars, instrumentation gauges, brushless DC motors, actuators, alternators
5. *Acoustic application*: Speaker (loud speakers)
6. *Appliances*: Portable power tools, household appliance motors, scales, air conditioners, water pumps, security systems
7. *Factory automation*: Magnetic couplings, pumps, motors, servo motors, generators, bearings
8. *Medical*: MRI, surgical tools, implants, "therapeutic"

Figure 7.27 shows the areas of the most important applications of the permanent magnets according to their sales distributions [139]. According to Figure 7.27, the largest application is in motors and generators. The second largest area of application is the acoustic devices (loudspeakers, headphones, microphones) and the measuring and control devices (NMR tomography). A substantial number of permanent magnets are used in telecommunications and data storage technology (computer peripheral, printers) and in magnetomechanical applications (couplings, bearings).

Figure 7.28 explains the extent of use of different permanent magnets. The most widely used ferrites occupy about 55% of the worldwide market for permanent magnets followed by the Nd–Fe–B and the Sm–Co magnets. The rare-earth magnets occupy about 37% of the world market and most of the remainder is Alnico [140]. Ferrites are used when the cost factor is very important, the Nd–Fe–B magnets are used when the size factor is important, and the Sm–Co magnets are used when the high-temperature stability is required [139].

Literature says that according to the recent trends of selling market, output of the bonded magnets has enormously increased in China showing the huge increase in

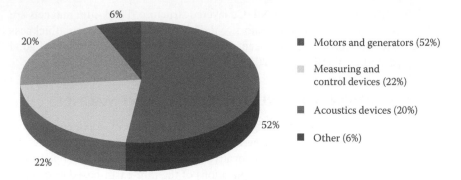

FIGURE 7.27   Use of rare-earth permanent magnets in different sectors.

share with the growth rate of over 110% each year [141,142], whereas Japan shows a significant decrease in its share in the worldwide share market of rare-earth-bonded magnets, although the output of ferrite-bonded magnets is much higher compared to the same of rare-earth-bonded magnets in Japan.

A major application area for Nd–Fe–B magnets is the data storage industry [143]. Spindle motors hard disk drives utilize bonded Nd–Fe–B magnets, while sintered $Nd_2Fe_{14}B$ magnets are used in CD-lens actuators and voice coil motors (VCM), which are the actuator of the arm of HDD read/write heads. Another growing area for Nd–Fe–B magnets is electric motor and related applications [23]. A huge expansion area for the Nd–Fe–B magnets is the mobile phone where Nd–Fe–B magnets are used in the isolators of the microwave stations [23]. About 4% of Nd–Fe–B magnets are used in loudspeakers (thinner loudspeakers in car doors and outside ring layout in TVs and monitors) [143]. They are used in MRIs and automotives as well. More widespread applications are considered in washing machines, refrigerators, and so on to improve energy efficiency and energy conservation [23]. The greatest potential markets are in the automobile industry where weight reduction, safe operation, and comfort improvements are required. To be used in HEVs (hybrid electric vehicles), the Nd–Fe–B magnets need high-temperature stability, which is difficult

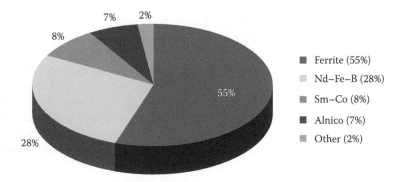

FIGURE 7.28   Estimated worldwide market for permanent magnet materials.

to achieve due to its low Curie temperature and high-temperature dependence of coercivity. To meet this requirement (high heat resistance) recently, the Nd–Fe–B magnets contains a large amount of Dy [30].

Sm–Co-based alloys are useful especially for high-temperature applications (sometimes as high as ~500°C) in the fields of space, aeronautics, and electric vehicles [89]. $SmCo_5$- and Sm_2Co_{17}-type magnets have high magnetocrystalline anisotropies and high Curie temperatures over a wide temperature ranging from 50°C to 250°C. Also, these magnets exhibit high corrosion resistance. This high-temperature property and anticorrosive nature made them ideal for those applications where high-energy density magnets are needed and the magnetic field is stable over various environmental conditions with a wide range of temperature. Magnetic bearings, sensors and actuators, microwave tubes, gyroscopes and accelerometers, and reaction and momentum wheels to control and stabilize satellites are examples for those types of applications [3,4,144,145]. However, costly ingredients and difficulty in magnetization limit the applications of these magnets.

Sm–Fe–N and other interstitial bonded magnets are used for small motors.

7.5 CONCLUSIONS AND FUTURE PERSPECTIVES

Rare-earth magnets play an important role as essential functional parts in many devices, such as motors, relays, sensors, actuators, encoders, and valves. Rare-earth permanent magnets have become the key materials for the development of a sustainable future because they are now used in many applications, contributing to energy saving and greenhouse-gas reduction. In this chapter, we discussed the various processing routes of rare-earth permanent magnets based on rapid solidification techniques. The highest cooling rate obtained in these processes was about 10^9–10^{10} K/s and are usually used for laser-surface melting/remelting. For the production of rare-earth permanent magnets, the most favorable rapid solidification techniques are atomization and melt spinning where the microstructure scale and morphology can be monitored by controlling the process parameters and processing media to achieve desired magnetic properties for specific applications. Recently, some newly developed nanocomposites and thin-film magnets are able to accept the challenge to be used as the high-performance magnets (because of their high-energy products) in the MEMS or NEMS industry.

However, future studies and investigations on finding and developing new permanent magnet materials are strongly in demand—rare-earth-free permanent magnets.

For the high-temperature permanent magnetic materials, a new research direction may be the rare-earth-free magnetic materials that do not rely on the limited supply of the rare-earth metals and cost less. Further research is needed to bring the new rare-earth-free magnetic materials for high-temperature applications, and computational methods of combinational materials science may be helpful for the progress. Another area to explore is the interstitially modified rare-earth magnets, which are still under development. A major problem is their thermodynamic instability at high temperatures, which hampers application of these materials in fully dense sintered magnets.

ACKNOWLEDGMENTS

The authors acknowledge the University of Nebraska-Lincoln, United States, and the Indian Institute of Technology, India, for their financial support throughout the study.

REFERENCES

1. B. D. Cullity, *Introduction to Magnetic Materials*, Addison-Wesley Publishing Company, Reading, MA, 1972, p. 1.
2. J. M. D. Coey, in *Proceedings of the 14th International Workshop on Rare-Earth Magnets and Their Applications*, F. P. Missell, V. Villas-Boas, H. R. Rechenberg, and F. J. G. Landgraf (eds.), Sao Paulo, Brazil, 1996, Vol. 1, p. 1.
3. K. Strnat, G. Hoffer, J. Olson, W. Ostertag, and J. J. Becker, A family of new cobalt-base permanent magnet materials, *J. Appl. Phys.* 38, 1967, 1001.
4. K. J. Strnat, Rare earth-cobalt permanent magnets, in *Ferromagnetic Materials*, E. P. Wohlfarth and K. H. J. Buschow (eds.), Elsevier, Amsterdam, 1988, Vol. 4, pp. 131–210, Chapter 2.
5. K. J. Strnat, Modern permanent magnets for applications in electro-technology, *Proc. IEEE* 78, 1990, 923.
6. R. A. McCurrie, *Ferromagnetic Materials—Structure and Properties*, Academic Press Limited, San Diego, CA, 1994, p. 193.
7. O. Gutfleisch, Controlling the properties of high energy density permanent magnetic materials by different processing routes, *J. Phys. D Appl. Phys.* 33, 2000, R157–R172.
8. G. C. Hadjipanayis, J.-F. Liu, A. Gabay, and M. Marinescu, Current status in rare-earth permanent magnet research in USA, *J. Iron Steel Res. Int.* 13, 2006, 12–22.
9. J. J. Croat, J. F. Herbst, R. W. Lee, and F. E. Pinkerton, Pr-Fe and Nd-Fe-based materials: A new class of high-performance permanent magnets, *J. Appl. Phys.* 55, 1984, 2078.
10. J. J. Croat, J. F. Herbst, R. W. Lee, and F. E. Pinkerton, High-energy product Nd-Fe-B permanent magnets, *Appl. Phys. Lett.* 44, 1984, 148.
11. M. Sagawa, S. Fujimura, N. Togawa, H. Yamamoto, and Y. Matsuura, New material for permanent magnets on a base of Nd and Fe, *J. Appl. Phys.* 55, 1984, 2083.
12. S. J. Collocott, R. K. Day, and J. B. Dunlop, in *Proceedings of 7th International Symposium on Magnetic Anisotropy and Coercivity in Rare-Earth Transition Metal Alloys (Canberra, Australia)*, Hi-Perm Laboratory, Research Center for Advanced Mineral and Materials Processing, The University of Western Australia (ed.), Scott Four Colour Print, Perth, 1992, pp. 437–444.
13. J. M. Cadogan, H. S. Li, A. Margarian, J. B. Dunlop, D. H. Ryan, S. J. Collocott, and R. L. Davis, New rare-earth intermetallic phases $R_3(Fe, M)_{29}X_n$: (R = Ce, Pr, Nd, Sm, Gd; M = Ti, V, Cr, Mn; and X = H, N, C), *J. Appl. Phys.* 76, 1994, 6138.
14. K. Kobayashi, Some aspects of the present status of Sm_2Fe_{17} nitrides, in *Proceedings of 13th International Workshop on REM and Their Applications (Birmingham, UK)*, C. A. F. Manwaring, D. G. R. Jones, A. J. Williams, and I. R. Harris (eds.), The University of Birmingham, Birmingham, 1994, pp. 717–732.
15. H. Fujii and H. Sun, Interstitially modified intermetallics of rare earth and 3D elements, in *Handbook of Magnetic Materials*, K. H. J. Buschow (ed.), Elsevier, Amsterdam, 1995, Vol. 9, pp. 303–404, Chapter 3.
16. J. M. D. Coey, Permanent magnet applications—Topical review. *J. Magn. Magn. Mater.* 248, 2002, 441–456.
17. S. Aich, Crystal structure, microstructure and magnetic properties of SmCo-based permanent magnets, PhD dissertation, University of Nebraska, Lincoln, NE, 2005.

18. R. A. McCurrie, *Ferromagnetic Materials—Structure and Properties*, Academic Press Limited, San Diego, CA, 1994, p. 255.
19. T. Ojima, S. Tomizawa, T. Yoniyama, and T. Hori, Magnetic properties of a new type of rare-earth cobalt magnets $Sm_2(Co, Cu, Fe, M)_{17}$, *IEEE Trans. Magn.* 13, 1977, 1317.
20. D. Li, H. F. Mildrum, and K. J. Strnat, Thermal stability of five sintered rare-earth–cobalt magnet types. *J. Appl. Phys.* 63, 1988, 3984.
21. T. Sun, A model on the coercivity of the hardened 2–17 rare earth-cobalt permanent magnets, *J. Appl. Phys.* 52, 1981, 2532.
22. J. F. Liu, Y. Zhang, D. Dimitar, and G. C. Hadjipanayis, Microstructure and high temperature magnetic properties of Sm(Co, Cu, Fe, Zr)z (z = 6.7–9.1) permanent magnets, *J. Appl. Phys.* 85, 1999, 2800.
23. D. Brown, B.-M. Ma, and Z. Chen, Developments in the processing and properties of NdFeb-type permanent magnets, *J. Magn. Magn. Mater.* 248, 2002, 432–440.
24. J. Fidler, S. Sasaki, and E. Estevez-Rams, High performance magnets—Microstructure and coercivity, in *Material Research Society Symposium Proceedings on Advanced Hard and Soft Magnets*, MRS Proceedings, Cambridge University Press, 1999, Vol. 577, p. 291.
25. M. Endoh and M. Shindo, Material design and fabrication of high energy Nd–Fe–B sintered magnets, in *Proceedings of 13th International Workshop on RE Magnets and Their Applications*, C. A. F. Manwaring et al. (eds.), Birmingham, UK, 1994, Vol. 1, p. 397.
26. W. Rodewald, B. Wall, and W. Fernengel, Grain growth kinetics in sintered Nd-Fe-B magnets, *IEEE Trans. Magn.* 33, 1997, 3841.
27. O. Gutfleisch, A. Bollero, A. Handstein, D. Hinz, A. Kirchner, A. Yan, K.-H. Müller, and L. Schultz, Nanocrystalline high performance permanent magnets, *J. Magn. Magn. Mater.* 242–245, 2002, 1277–1283.
28. O. Gutfleisch, A. Teresiak, B. Gebel, K.-H. Müller, N. B. Cannesan, D. N. Brown, and I. R. Harris, Metastable borides and the inducement of texture in $Pr_2Fe_{14}B$-type magnets produced by HDDR, *IEEE Trans. Magn.* 37, 2001, 2471.
29. R. N. Faria, A. J. Williams, and I. R. Harris, High anisotropy in Pr–Fe–Co–B–Zr HDDR powders, *J. Alloys Compd.* 287, 1999, L10.
30. S. Sugimoto, Current status and recent topics of rare-earth permanent magnets, *J. Phys. D Appl. Phys.* 44, 2011, 064001.
31. L. Cao, K. H. Muller, A. Handstein, W. Grunberger, and L. Neu Vand Schultz, High performance permanent magnets made by mechanical alloying and hot pressing, *J. Phys. D Appl. Phys.* 29, 1996, 271.
32. G. K. Williamson and W. H. Hall, X-ray line broadening from filed aluminium and wolfram, *Acta Metall.* 1, 1953, 22.
33. O. Gutfleisch, M. Matzinger, J. Fidler, and I. R. Harris, Characterisation of solid-HDDR processed $Nd_{16}Fe_{76}B_8$ alloys by means of electron microscopy, *J. Magn. Magn. Mater.* 147, 1995, 320.
34. M. Okada, K. Saito, H. Nakamura, S. Sugimoto, and M. Homma, Microstructural evolutions during HDDR phenomena in $Sm_2Fe_{17}N_x$ compounds, *J. Alloys Compd.* 231, 1995, 60.
35. O. Gutfleisch, M. Kubis, A. Handstein, K.-H. Muller, and L. Schultz, Hydrogenation disproportionation desorption recombination in Sm–Co alloys by means of reactive milling, *Appl. Phys. Lett.* 73, 1998, 3001.
36. A. Bollero, M. Kubis, O. Gutfleisch, K. H. Muller, and L. Schultz, Hydrogen disproportionation by reactive milling and recombination of $Nd_2(Fe_{1-x}Co_x)_{14}B$ alloys, *Acta Mat.* 48, 2000, 4929–4934.
37. X. L. Yeh, K. Samwer, and W. L. Johnson, Formation of an amorphous metallic hydride by reaction of hydrogen with crystalline intermetallic compounds—A new method of synthesizing metallic glasses, *Appl. Phys. Lett.* 42, 1983, 242.

38. L. Schultz, K. Schnitzke, J. Wecker, M. Katter, and C. Kuhrt, Permanent magnets by mechanical alloying (invited), *J. Appl. Phys.* 70, 1991, 6339.
39. J. Ding, P. G. McCormick, and R. Street, A study of $Sm_{13}(Co_{1-x}Fe_x)_{87}$ prepared by mechanical alloying, *J. Magn. Magn. Mater.* 135, 1994, 200.
40. J. Ding, P. G. McCormick, and R. Street, Remanence enhancement in mechanically alloyed isotropic Sm_7Fe_{93}-nitride, *J. Magn. Magn. Mater.* 124, 1993, 1.
41. K. Schnitzke, L. Schultz, J. Wecker, and M. Katter, High coercivity in $Sm_2Fe_{17}N_x$ magnets, *Appl. Phys. Lett.* 57, 1990, 2853.
42. J. Ding, B. Shen, P. G. McCormick, R. Street, and F. Wang, Magnetic hardening of mechanically milled $Sm_2Fe_{14}Ga_3C_2$, *J. Alloys Compd.* 209, 1994, 221.
43. Y. Yang, O. Gutfleisch, A. Handstein, D. Eckert, and K. H. Muller, High coercivity of Nd–Dy–Fe–(C, B) ribbons prepared by melt spinning, *Appl. Phys. Lett.* 76, 2000, 3627.
44. P. Crespo, V. Neu, and L. Schultz, Mechanically alloyed nanocomposite powders of $Nd_2Fe_{14}B/\alpha$-Fe with additional elements, *J. Phys. D Appl. Phys.* 30, 1997, 2298.
45. Z. Chen, X. Meng-Burany, and G. C. Hadjipanayis, High coercivity in nanostructured $PrCo_5$-based powders produced by mechanical milling and subsequent annealing, *Appl. Phys. Lett.* 75, 1999, 3165.
46. Shi Gang, Hu Lianxi, and Wang Erde, Preparation, microstructure, and magnetic properties of a nanocrystalline $Nd_{12}Fe_{82}B_6$ alloy by HDDR combined with mechanical milling, *J. Magn. Magn. Mat.* 301, 2006, 319–324.
47. R. M. German, *Powder Metallurgy Science*, 2nd ed., Metal Powder Industries Federation (MPIF), Princeton, NJ, 1994, ISBN-13: 978-1878954428.
48. W. M. Steen, *Laser Material Processing*, Springer Verlag, New York, 1991.
49. J. E. Gusic, H. W. Marcos, and L. G. von Uitert, Laser oscillations in Nd-doped yttrium aluminum, yttrium gallium and gadolinium garnets, *App. Phys. Lett.* 4, 1964, 182.
50. C. K. N. Patel, Continuous-wave laser action on vibrational-rotational transitions of CO_2, *Phys. Rev.* 136A, 1964, 1187.
51. J. Laeng, J. C. Stewart, and F. W. Liou, Laser metal forming processes for rapid prototyping—A review, *Int. J. Prod. Res.* 38, 2000, 3973–3996.
52. W. Wiehua-Wang, M. R. Holl, and D. T. Schwartz, Rapid prototyping of masks for through-mask electrodeposition of thick metallic components, *J. Electrochem. Soc.* 148, 2001, C363–C368.
53. A. Greco, A. Licciulli, and A. Maffezzoli, Stereolithography of ceramic suspensions, *J. Mater. Sci.* 36, 2001, 99–105.
54. L. Lu, J. Y. H. Fuh, Z. D. Chen, C. C. Leong, and Y. S. Wong, In situ formation of TiC composite using selective laser melting, *Mater. Res. Bull.* 35, 2000, 1555–1561.
55. K. Daneshvar, M. Raissi, and S. M. Bobbio, Laser rapid prototyping in nonlinear medium, *J. Appl. Phys.* 88, 2000, 2205–2210.
56. J. P. Kruth, M. Badrossamay, E. Yasa, J. Deckers, L. Thijs, and J. van Humbeeck, Part and material properties in selective laser melting of metals, in *Proceedings of the 16th International Symposium on Electro-Machining*, 2010, Wansheng Zhao et al. (eds.), Shanghai Jiao Tong University Press, Shanghai.
57. https://en.wikipedia.org/wiki/Electrospinning#/media/File:Electrospinning_Image_for_Wikipedia.tif.
58. R. W. Cahn, Background to rapid solidification processing, in *Rapidly Solidified Alloys*, H. H. Liebermann (ed.), Marcel Dekker, New York, 1993, p. 1.
59. T. R. Anantharaman and C. Suryanarayana, *Rapidly Solidified Metals*, Trans Tech Publications, Switzerland, 1987, p. 25.
60. C. Suryanarayana, Rapid solidification, in *Processing of Metals and Alloys*, R. W. Cahn (ed.), Trans Tech Publications, Switzerland, 1991, p. 57.

61. N. J. Grant, H. Jones, and E. J. Lavernia, Synthesis and processing, in *Elements of Rapid Solidification—Fundamentals and Applications*, M. A. Otooni (ed.), Springer, 1998, p. 35, ISBN 978-3-642-45755-5.

62. J. O. Strom-Olsen, Fine fibers by melt extraction, *Mater. Sci. Eng.* A 178, 1994, 239–243.

63. G. Hoffer and K. J. Strnat, Magnetocrystalline anisotropy of YCo_5 and Y_2Co_{17}, *IEEE Trans. Magn, Magn.* 2, 1966, 487.

64. A. E. Ray and K. J. Strnat, Research and development of rare earth-transition metal alloys as permanent-magnet materials, Technical Report AFML-TR-72-99, 1972.

65. H. Saito, M. Takahashi, T. Wakiyama, G. Kodo, and H. Nakagawa, Magnetocrystalline anisotropy for $Sm_2(Co-Mn)_{17}$ compound with $TbCu_7$ type disordered structure, *J. Magn. Magn. Mater.* 82, 1989, 322.

66. K. H. Buschow and F. J. A. den Breeder, The cobalt-rich regions of the samarium-cobalt and gadolinium-cobalt phase diagrams, *J. Less-Common Met.* 3, 1973, 191.

67. M. Q. Huang, W. E. Wallace, M. McHenry, Q. Chen, and B. M. Ma, Structure and magnetic properties of $SmCo_{7-x}Zr_x$ alloys (x = 0–0.8), *J. Appl. Phys.* 83, 1998, 6718.

68. M. Q. Huang, M. Drennan, W. E. Wallace, M. E. McHenry, Q. Chen, and B. M. Ma, Structure and magnetic properties of $RCo_{7-x}Zr_x$ (R = Pr or Er, x = 0–0.8), *J. Appl. Phys.* 85, 1999, 5663.

69. J. Zhou, I. A. Al-Omari, J. P. Liu, and D. J. Sellmyer, Structure and magnetic properties of $SmCo_{7-x}Ti_x$ with $TbCu_7$-type structure, *J. Appl. Phys.* 87, 2000, 5299.

70. S. Aich and J. E. Shield, Phase formation and magnetic properties of $SmCo_{5+x}$ alloys with $TbCu_7$-type structure, *J. Magn. Magn. Mater.* 279, 2004, 76–81.

71. K. H. J. Buschow and A. S. van der Goot, Composition and crystal structure of hexagonal Cu-rich rare earth–copper compounds, *Acta Cryst. Allogr. B* 27, 1971, 1085.

72. I. A. Al-Omari, Y. Yeshurun, J. Zhou, and D. J. Sellmyer, Magnetic and structural properties of $SmCo_{7-x}Cu_x$ alloys, *J. Appl. Phys.* 87, 2000, 6710.

73. J. Luo, J. K. Liang, Y. Q. Guo, Q. L. Liu, L. T. Yang, F. S. Liu, G. H. Rao, and W. Li, Effects of Cu on crystallographic and magnetic properties of $Sm(Co, Cu)_7$, *J. Phys. Condens. Matter* 15, 2003, 5621.

74. J. Luo, J. K. Liang, Y. Q. Guo, Q. L. Liu, F. S. Liu, Y. Zhang, L. T. Yang, and G. H. Rao, Effects of the doping element on crystal structure and magnetic properties of $Sm(Co,M)_7$ compounds (MZSi, Cu, Ti, Zr, and Hf), *Intermetallics* 13, 2005, 710–716.

75. Y. Q. Guo, W. Li, J. Luo, W. C. Feng, and J. K. Liang, Structure and magnetic characteristics of novel SmCo-based hard magnetic alloys, *J. Magn. Magn. Mater.* 303, 2006, 367–370.

76. A. R. Miedema, The heat of formation of alloys, *Philips Technol. Rev.* 36, 1976, 217.

77. A. R. Miedema, R. Boom, and F. R. De Boer, On the heat of formation of solid alloys, *J. Less-Common Met.* 41, 1975, 283.

78. A. K. Niessen, F. R. de Boer, R. Boom, P. F. De Chatel, M. C. M. Mattens, and A. R. Miedema, Model predictions for the enthalpy of formation of transition metal alloys II, *Calphad* 7, 1983, 51.

79. J. Luo, J. K. Liang, Y. Q. Guo, L. T. Yang, F. S. Liu, Y. Zhang, Q. L. Liu, and G. H. Rao, Crystal structure and magnetic properties of $SmCo_{7-x}Hf_x$ compounds, *Appl. Phys. Lett.* 85, 2004, 5299.

80. S. K. Chen, M. S. Chu, J. L. Tsai, and T. S. Chin, Magnetic properties of microcrystalline $SmCo_x$ alloys by melt spinning, *IEEE Trans. Magn.* 32, 1996, 4419.

81. G. C. Hadjipanayis, Nanophase hard magnets, *J. Magn. Magn. Mater.* 200, 1999, 373.

82. W. Gong and B. M. Ma, Comparison on the magnetic and structural properties of $Sm(Co_{0.67-x}Fe_{0.25}Cu_{0.06}Zr_{0.02}C_x)_{8.0}$, where x = 0–0.15, melt spun ribbons and cast alloys, *J. Appl. Phys.* 85, 1999, 4657.

83. J.-B. Sun, D. Han, C.-X. Cui, W. Yang, L. Li, and F. Yang, Effects of quenching speed on microstructure and magnetic properties of novel $SmCo_{6.9}Hf_{0.1}(CNTs)_{0.05}$ melt-spun ribbons, *Acta Mater.* 57, 2009, 2845–2850.

84. J.-B. Sun, D. Han, C.-X. Cui, W. Yang, H. Zhou, R.-T. Tian, and L.-G. Yang, Effect of Hf and CNTs on magnetic transformation and thermal stabilities of $TbCu_7$-type Sm-Co matrix ribbon magnets, *J. Alloys Compd.* 487, 2009, 626–630.

85. J.-B. Sun, D. Han, C.-X. Cui, W. Yang, L. Li, F. Yang, and L.-G. Yang, Effects of Hf and CNTs on structure and magnetic properties of $TbCu_7$-type Sm–Co magnets, *Intermetallics* 18, 2010, 599–605.

86. H. W. Chang, S. T. Huang, C. W. Chang, C. H. Chiu, W. C. Chang, A. C. Sun, and Y. D. Yao, Magnetic properties, phase evolution and microstructure of melt-spun $SmCo_{7-x}Hf_xC_y$ (x = 0–0.5; y = 0–0.14) ribbons, *J. Appl. Phys.* 101, 2007, 09K508.

87. H. W. Chang, S. T. Huang, C. W. Chang, C. H. Chiu, I. W. Chen, W. C. Chang, A. C. Sun, and Y. D. Yao, Comparison on the magnetic properties and phase evolution of melt-spun $SmCo_7$ ribbons with Zr and Hf substitution, *Scr. Mater.* 56, 2007, 1099–1102.

88. S. Aich and J. E. Shield, Effect of Nb and C additives on the microstructures and magnetic properties of rapidly solidified Sm-Co alloys, *J. Alloys Compd.* 425, 2006, 416–423.

89. C.-B. Jiang and S.-Z. An, Recent progress in high temperature permanent magnetic materials, *Rare Met.* 32, 2013, 431–440.

90. L. L. Liu and C. B. Jiang, The improved oxidation resistance of Si-doped $SmCo_7$ nanocrystalline magnet, *Appl. Phys. Lett.* 98, 2011, 252504.

91. L. L. Liu, T. Y. Jin, and C. B. Jiang, High-temperature oxidation resistance and magnetic properties of Si-doped Sm_2Co_{17}-type magnets at 500°C, *J. Magn. Magn. Mater.* 324, 2012, 2310.

92. Q. Y. Wang, L. Zheng, S. Z. An, T. Zhang, and C. Jiang, Thermal stability of surface modified Sm_2Co_{17}-type high temperature magnets, *J. Magn. Magn. Mater.* 331, 2013, 245.

93. C. H. Chen, M. Q. Huang, J. E. Foster, G. Monnette, J. Middleton, A. Higgins, and S. Liu, Effect of surface modification on mechanical properties and thermal stability of Sm–Co high temperature magnetic materials, *Surf. Coat. Technol.* 201, 2006, 3430.

94. W. M. Pragnell, H. E. Evans, and A. J. Williams, Oxidation protection of Sm_2Co_{17}-based alloys, *J. Alloys Compd.* 517, 2012, 92.

95. K.-H. Muller, G. Krabbes, J. Fink, S. Gruß, A. Kirchner, G. Fuchs, and L. Schultz, New permanent magnets, *J. Magn. Magn. Mater.* 226–230, 2001, 1370–1376.

96. B. M. Ma, J. W. Herchenroeder, B. Smith, M. Suda, D. N. Brown, and Z. Chen, Recent development in bonded NdFeB magnets, *J. Magn. Magn. Mater.* 239, 2002, 418–423.

97. A. P. Guimaraes, *Principles of Nanomagnetism*, Springer, Berlin, 2009, XII, 224pp, available in: http://www.springer.com/978-3-642-01481-9.

98. D. Goll, M. Seeger, and H. Kronmuller, Magnetic and microstructural properties of nanocrystalline exchange coupled PrFeB permanent magnets, *J. Magn. Magn. Mater.* 185, 1998, 49.

99. H. A. Davies, C. L. Harland, J. I. Betancourt, and G. Menoza, Document praseodymium and neodymium-based nanocrystalline hard magnetic alloys, in *Material Research Society Symposium Proceedings on Advanced Hard and Soft Magnets*, MRS Proceedings, Cambridge University Press, 1999, Vol. 577, p. 27.

100. J. F. Herbst, $R_2Fe_{14}B$ materials: Intrinsic properties and technological aspects, *Rev. Mod. Phys.* 63, 1991, 819.

101. J. J. Croat, Manufacture of $Nd_2Fe_{14}B$ permanent magnets by rapid solidification, *J. Less-Common Met.* 148, 1989, 7–15.

102. Z. W. Liu, Y. Liu, P. K. Deheri, R. V. Ramanujan, and H. A. Davies, Improving permanent magnetic properties of rapidly solidified nanophase RE–TM–B alloys by compositional modification, *J. Magn. Magn. Meter.* 321, 2009, 2290–2295.

103. C. D. Fuerst and E. G. Brewer, High-remanence rapidly solidified N&Fe-B: Die-upset magnets (invited), *J. Appl. Phys.* 73, 1993, 5751–5756.

104. E. F. Kneller and R. Hawig, The exchange-spring magnet: A new material principle for permanent magnets, *IEEE Trans. Magn.* 27, 1991, 3588.

105. S. Hirosawa, Nd-Fe-B-based nanocomposite permanent magnets suitable for strip casting, in *Proceedings of 18th International Workshop on High Performance Magnets and Their Applications (Annecy, France)*, N. M. Dempsey and P. de Rango (eds.), CNRS, Grenoble, 2004, Vol. 2, pp. 655–666.

106. O. M. Bovda, V. O. Bovda, V. V. Chebotarev, I. E. Garkusha, S. O. Leonov, L. V. Onischenko, V. I. Tereshin, and O. S. Tortika, Effect of pulsed plasma treatment on structure and corrosion behavior of Sm-Co and Nd-Fe-B magnets, *J. Iron Steel Res. Int.* 13, 2006, 337–342.

107. P. K. Sokolowski, Processing and protection of rare earth permanent magnet particulate for bonded magnet application, MSc thesis, Iowa State University, 2007.

108. J. M. D. Coey and H. Sun, Improved magnetic properties by treatment of iron-based rare earth intermetallic compounds in ammonia, *J. Magn. Magn. Mater.* 87, 1990, 251.

109. K. H. J. Buschow, F. H. Feijen, and K. de Kort, Rare earth permanent magnets, *J. Magn. Magn. Mater.* 140–144, 1995, 9–12.

110. R. Coehoorn, K. H. J. Buschow, M. W. Dirken, and R. C. Thiel, Valence-electron contributions to the electric-field gradient in hcp metals and at Gd nuclei in intermetallic compounds with the $ThCr_2Si_2$ structure, *Phys. Rev. B* 42, 1990, 4645.

111. M. W. Dirken, R. C. Thiel, R. Coehoorn, T. H. Jacobs, and K. H. J. Buschow, [155]Gd Mössbauer effect study of $Gd_2Fe_{17}N_x$, *J. Magn. Magn. Mater.* 94, 1991, L15.

112. D. P. Middleton, F. M. Mulder, R. C. Thiel, and K. H. J. Buschow, [155]Gd Mössbauer effect studies and magnetic properties of $GdFe_{12-x}Mo_x(N_y)$ compounds, *J. Magn. Magn. Mater.* 146, 1995, 123–128.

113. P. A. P. Wendhausen, B. Gebel, D. Eckert, and K. H. Muller, Effect of milling on the magnetic and microstructural properties of $Sm_2Fe_{17}N_x$ permanent magnets, *J. Appl. Phys.* 75, 1994, 6018.

114. N. M. Dempsey, P. A. P. Wendhausen, B. Gebel, K. H. Muller, and J. M. D. Coey, in *Proceedings of 14th International Workshop on Rare Earth Magnets and Their Applications*, F. P. Missell (ed.), Sao Paolo, Brazil, 1996, p. 349.

115. B. G. Shen, L. S. Kong, F. W. Wang, and L. Cao, A novel hard magnetic material for sintering permanent magnets, *J. Appl. Phys.* 75, 1994, 6253.

116. B. G. Shen, F. W. Wang, L. S. Kong, L. Cao, and W. S. Zhan, Structure and magnetic properties of $Sm_2Fe_{14}Ga_3C_x$ ($x = 0$–2.5) compounds prepared by arc melting, *Appl. Phys. Lett.* 63, 1993, 2288.

117. Z. H. Cheng, B. G. Shen, F. W. Wang et al., The formation and magnetic properties of $Sm_2Fe_{15}Al_2C_x$ ($x = 0$–2.0) compounds prepared by arc melting, *J. Phys. Condens. Matter* 6, 1994, 1185.

118. R. Skomski, S. Brennan, and S. Wirth, in *Interstitial Metallic Alloys*, F. Grandjean, G. J. Long, and K. H. J. Buschow (eds.), Kluwer, Dordrecht, 1995, NATO-ASI Series E, Vol. 281, Chapter 16.

119. K. H. Muller, L. Cao, N. M. Dempsey, and P. A. P. Wendhausen, Sm_2Fe_{17} interstitial magnets (invited), *J. Appl. Phys.* 79, 1996, 5045.

120. M. Kubis, D. Eckert, B. Gebel, K.-H. MuK ller, and L. Schultz, Intrinsic magnetic properties of $Sm_2Fe_{17-x}M_xN_y/C_y$ (M = Al, Ga or Si), *J. Magn. Magn. Mater.* 217, 2000, 14–18.

121. C. N. Christodoulou and T. Takeshita, Sm_2Fe_{17}-nitride-based permanent magnets produced by rapid solidification, *J. Alloys Compd.* 196, 1993, 161–164.

122. C. J. Yang, E. B. Park, and S. D. Choi, Magnetic hardening of rapidly solidified $SmFe_{7+x}M_x$ ($0.8 \leq x \leq 1.5$, M = Mo, V, Ti) compounds, *Mater. Lett.* 24, 1995, 347–354.

123. X. C. Kou, Coercivity of SmFeN permanent magnets produced by various techniques, *J. Alloys Compd.* 281, 1998, 41–45.

124. W. Kaszuwara, M. Leonowicz, and J. A. Kozubowski, The effect of tungsten addition on the magnetic properties and microstructure of SmFeN–α-Fe nanocomposite, *Mater. Lett.* 42, 2000, 383–386.

125. A. E. Platts, I. R. Harris, and J. M. D. Coey, Improvement in the cast structure of Sm(2) Fe(17) alloys by niobium additions, *J. Alloys Compd.* 185, 1992, 251.

126. B. Saje, A. E. Platts, S. KobeBesenicar, I. R. Harris, and D. Kolar, Microstructure and magnetic properties of Sm-Fe-Ta based alloys, *IEEE Trans. Magn.* 30, 1994, 690.

127. B. Gebel, M. Kubis, and K. H. Müller, Permanent magnets prepared from $Sm_{10.5}Fe_{88.5}Zr_{1.0}N_y$ without homogenization, *J. Magn. Magn. Mater.* 174, 1997, L1–L4.

128. K. Žužek, P. J. McGuiness, and S. Kobe, Bonded Sm–Fe–(Ta)–N materials produced via attritor milling and HDDR, *J. Alloys Compd.* 289, 1999, 265–269.

129. K. Zuzek, G. Drazíc, P. J. McGuiness, and S. Kobe, High coercivity powders based on Sm–Fe–Ta–N prepared by a novel technique, *Phys. Solid State* 44, 2002, 1540–1543.

130. M. Yamamoto, *BM News* (The Japan Association of Bonded Magnetic Materials [JABM]), 43, 2010, 72 (in Japanese).

131. J. Jiang, E. Fullerton, C. Sowers, I. Inomata, S. Bader, A. Shapiro, R. Shull, V. Gornakov, and V. Nikitenko, Spring magnet films, *IEEE Trans. Magn.* 35, 1999, 3229.

132. J. Zhang, Y. K. Takahashi, R. Gopalan, and K. Hono, $Sm(Co,Cu)_5$/Fe exchange spring multilayer films with high energy product $Sm(Co,Cu)_5$/Fe exchange spring multilayer films with high energy product, *Appl. Phys. Lett.* 86, 2005, 122509.

133. M. Uehara, N. Gennai, M. Fujiwara, and T. Tanaka, Improved perpendicular anisotropy and permanent magnet properties in Co-doped Nd-Fe-B films multilayered with Ta, *IEEE Trans. Magn.* 41, 2005, 3838.

134. N. M. Dempsey, A. Walther, F. May, D. Givord, K. Khlopkov, and O. Gutfleisch, High performance hard magnetic NdFeB thick films for integration into micro-electro-mechanical systems, *Appl. Phys. Lett.* 90, 2007, 092509.

135. N. M. Dempsey, Presentation in the *2nd International Symposium on Advanced Magnetic Materials and Applications (ISAMMA 2010)*, Sendai, Japan, 2010, AX-07

136. N. Oka, T. Sato, Y. Mishina, and T. Shima, Presentation in the *33rd Annual Conference on Magnetics in Japan*, Nagasaki, Japan, 2009, 12pE-4 (in Japanese).

137. T. Sato, T. Ohsuna, and Y. Kaneko, in *Digests of the 34th Annual Conference on Magnetics in Japan 2010 (Tsukuba, Japan)*, The Magnetic Society of Japan, Tokyo, 2010, p. 272.

138. M. Nakano, S. Sato, F. Yamashita, T. Honda, J. Yamasaki, K. Ishiyama, M. Itakura, J. Fidler, T. Yanai, and H. Fukunaga, Review of fabrication and characterization of Nd–Fe–B thick films for magnetic micromachines, *IEEE Trans. Magn.* 43, 2007, 2672.

139. K. H. J. Buschow, Permanent magnet materials based on 3d-rich ternary compounds, in *Ferromagnetic Materials*, E. P. Wohlfarth and K. H. J. Buschow (eds.), Elsevier, Amsterdam, 1988, Vol. 4, p. 1.

140. Z. Yao and C. B. Jiang, Structure and magnetic properties of $SmCo_xTi_{0.4}$-1:7 ribbons, *J. Magn. Magn. Mater.* 320, 2008, 1073–1077.

141. N. Ishigaki and H. Yamamoto, *Magnet. Japan* 3, 2008, 525.

142. H. Arakawa, *BM NEWS* (The Japan Association of Bonded Magnetic Materials [JABM]), 42, 2009, 9 (in Japanese).

143. R. H. J. Fastenau and E. J. Van Loenen, Applications of rare earth permanent magnets (invited paper), *J. Magn. Magn. Mater.* 157–158, 1996, 1–6.

144. K. Kumar, $RETM_5$ and RE_2TM_{17} permanent magnets development, *J. Appl. Phys.* 63, 1988, R13.

145. A. E. Ray and S. Liu, in *Proceedings of 12th International Workshop on RE Magnets and Their Applications*, F. P. Missel (ed.), Canberra, Australia, 1992, p. 552.

8 Magnetic Materials Prepared Using Polyacrylamide Gel Route

S. F. Wang, X. T. Zu, and Richard YongQing Fu

CONTENTS

ABSTRACT

This chapter summarizes recent developments in magnetic materials, including magnetic nanoparticles and magnetic nanocomposites, synthesized using polyacrylamide gel technology, for application purposes including photocatalysts, multiferroic materials, magnetocaloric materials, and solid oxide fuel cells. The polyacrylamide gel route is a fast, reproducible, cheap, and easy to scale-up method for obtaining highly dispersed nanopowders and nanocomposites. The reaction mechanisms and key influencing factors for the preparation of highly dispersed magnetic nanopowders and nanocomposites have been discussed based on experimental results. By appropriately selecting chelating agents/initiators and controlling the unique physicochemical processes, magnetic nanopowders or nanocomposites fabricated using polyacrylamide gel technology are promising for wide applications in catalysis, pigments, magnetocaloric materials, magneto-optical devices, and magneto-electronics.

8.1   INTRODUCTION

8.1.1   Definition and Classification of Magnetic Materials

Magnetic materials are classified into five categories, including ferrimagnetic materials, diamagnetic materials, paramagnetic materials, antiferromagnetic, and ferromagnetic materials according to their response to an externally applied magnetic field [1]. Ferrimagnetic materials are ferrites and magnetic garnets, and exhibit spontaneous magnetization because of the non-parallel arrangement of their atomic magnetic moments at a temperature below the Néel temperature [2]. Diamagnetic materials show repulsion to a magnetic field and are usually applied in magnetic levitation and sensing elements [3]. Paramagnetic materials show a weak attraction to a magnetic field and have a large positive value of susceptibility compared with diamagnetic materials [4]. In addition, some materials show exceptionally large Curie constants, and are known as superparamagnetic materials. Antiferromagnetic materials are transition metal compounds and exhibit an antiparallel arrangement of their atomic magnetic moments at a temperature below the Néel temperature.

Ferromagnetic materials show a strong attraction to a magnetic field. Because of the spontaneous magnetization, ferromagnetic materials are more suitable to be used in electrical and electromechanical devices including magnetic storage, electromagnets, and generators.

8.1.2 MAGNETIC NANOPARTICLES

Recently, magnetic nanoparticles have received significant attention because of their remarkable new phenomena such as extra anisotropy contributions, high saturation field, or superparamagnetism [5]. Recent advances in the development of magnetic nanoparticles show that a nonhydrolytic or non-condensation process in organic additives at an elevated sintering temperature is a highly effective and easy method for preparation of dispersed magnetic nanoparticles with well-adjusted particle sizes and particle size distributions [6–11].

8.1.2.1 Preparation of Magnetic Nanoparticles

Magnetic nanoparticles have been prepared with different compositions and structures, including iron oxides, pure metals, spinel-type ferrites, alloys, and so on [12]. The main preparation methods proposed for the production of magnetic nanoparticles include: (1) microbial methods, (2) physical methods, and (3) wet chemical preparation methods [13]. Among these methods, the wet chemical one is attractive because it has easy to control surface topography and particle growth using organic additives, surfactants, or chelating agents. Several commonly used wet chemical routes include the co-precipitation method [14], sol–gel method [15], hydrothermal method [16], oxidation method [17], electrochemical method [18], sonochemical method [19], polyacrylamide gel method [20], and micelle synthesis method [12].

8.1.2.1.1 Co-Precipitation Method

The co-precipitation method in an aqueous solution is the simplest and most efficient route to synthesize ferrite magnetic nanoparticles [21–25]. For instance, $NiFe_2O_4$ magnetic nanoparticles were synthesized by using nickel nitrate, ferric nitrate, and sodium hydroxide as starting materials and oleic acid as a surfactant [22]. Control of the particle size of $NiFe_2O_4$ magnetic nanoparticles can be achieved by changing the sintering temperature. The complex system substituted Sr-hexaferrite nanoparticles were also prepared by the co-precipitation method [21]. In this case, $SrCl_2 \cdot 6H_2O$, $FeCl_3 \cdot 6H_2O$, $MgCl_2 \cdot 6H_2O$, $CoCl_2 \cdot 6H_2O$, $TiCl_4$, and sodium hydroxide are used as starting materials. The SEM images of $SrFe_{12-2x}(Mg,Co)_{x/2}Ti_xO_{19}$ at different dopant concentrations are shown in Figure 8.1. As can be seen from Figure 8.1, the particle size decreases with the increase of dopant concentration. Gordani et al. [21] also discovered that with an increase in dopant concentration, the magnetic remanence (M_r) decreases.

8.1.2.1.2 Sol–Gel Method

The sol–gel method has been proposed to synthesize ferrite magnetic nanoparticles [26,27]. This method is based on the condensation and hydroxylation of a metal precursor solution to yield a sol of nanometer particles [13]. The synthesis of

FIGURE 8.1 SEM image of $SrFe_{12-2x}(Mg,Co)_{x/2}Ti_xO_{19}$ at different dopant concentrations: (a) x = 0, (b) x = 0.5, (c) x = 1, (d) x = 1.5, (e) x = 2, and (f) x = 2.5, prepared by co-precipitation method and sintered at 900°C for 1h. (Adapted from G. R. Gordani, A. Ghasemi, and A. Saidi, *Ceramics International* 40, 2014, 4945–4952. 2013 Elsevier Ltd and Techna Group S.r.l.)

$Ca_3Co_4O_9$ magnetic nanoparticles was achieved by a starch-assisted sol–gel combustion method [28]. This method has also been proposed to synthesize a complex system of manganese oxide magnetic nanoparticles including $La_{0.67}Sr_{0.33}MnO_3$ [29], $Bi_{1-x}Gd_xFe_{1-y}Mn_yO_3$ [30], $La_{0.6}Pr_{0.1}Ba_{0.3}Mn_{1-x}Ni_xO_3$ [31], etc. This sol–gel method was also proposed to synthesize Ni_3Fe nanoalloy nanoparticles [32]. The main advantage of this method is the good control of both the particle size and size distribution [13].

8.1.2.1.3 Hydrothermal Method

The hydrothermal method has been employed to synthesize iron oxide magnetic nanoparticles [33], ferrite magnetic nanoparticles [34], manganese oxide magnetic nanoparticles [35–37], and magnetic metal nanoparticles [38]. Using the hydrothermal method, it is possible to control the surface topography of the magnetic nanoparticles after optimization of the synthesis parameters such as the reactant concentration and stoichiometry. For example, Bouremana et al. [38] reported that the surface topography of Ni nanoparticles can be adjusted using different NaOH concentrations (Figure 8.2).

8.1.2.1.4 Oxidation Method

The oxidation method was used to synthesize Mn–Zn ferrite magnetic nanoparticles by the crystallization of ferrous hydroxide precipitate with a mild oxidant such as KNO_3 [39]. Bee et al. [40] reported that γ-Fe_2O_3 magnetic nanoparticles can be obtained by oxidizing Fe_3O_4 magnetic nanoparticles at 90°C in ferric nitrate solution.

8.1.2.1.5 Electrochemical Method

Preparation of ferrite magnetic nanoparticles employing the electrochemical method has recently been demonstrated [41–43]. $NiFe_2O_4$ nanoparticles with a size of 22 nm were synthesized in an electrochemical cell using three electrodes; with a sheet of iron as cathode and two sheets of iron and nickel as sacrificial anodes [43]. The optimization of the synthesis parameters was easily carried out by changing the current density [44].

8.1.2.1.6 Sonochemical Method

Development of a sonochemical method to prepare magnetic nanoparticles is currently under extensive investigation. Nabiyouni et al. [45] reported the sonochemical-assisted synthesis of hard magnetic $BaFe_{12}O_{19}$ nanoparticles by calcination of the precursor at 850°C. Figure 8.3 shows the schematic diagram for the experimental setup used for this sonochemical-assisted reaction to prepare $BaFe_{12}O_{19}$ nanoparticles [45]. Ghanbari et al. [46] synthesized Fe_3O_4 nanoparticles using a simple surfactant-free sonochemical reaction at room temperature. Kim et al. [47] prepared superparamagnetic iron oxide nanoparticles by the sonochemical method and the mean diameter of the iron oxide nanoparticles was about 15 nm. Saffari et al. [48] prepared $CoFe_2O_4$ nanoparticles via a facile surfactant-free sonochemical reaction and the process was done at a temperature of 80°C for 2 hours. The result showed that the $CoFe_2O_4$ nanoparticles dispersed inside the polymeric matrixes increased the coercivity [48]. The control of the particle size of $CoFe_2O_4$ nanoparticles could be successfully achieved by changing the solvents such as aqueous, alcoholic, and a mix of water/ethanol in 1:1 volume ratio [49].

8.1.2.1.7 Micelle Synthesis Method

Micelle synthesis methods include the normal micelle and reverse micelle methods [24]. $CoFe_2O_4$ magnetic nanoparticles of 5 nm were synthesized by the normal micelle method using cobalt (II) chloride and iron (II) chloride as starting materials. The reverse micelle synthesis method has been used to synthesize magnetic nanoparticles with uniform morphologies and excellent homogeneities [50]. For example, Thakur et al. [51] prepared 8.4 nm nickel-zinc ferrite magnetic nanoparticles using the reverse micelle technique and the particles were sintered at 500°C for 4 hours. Results showed that the nickel-zinc ferrite magnetic nanoparticles exhibited a reduced non-saturating magnetization compared to the nickel-zinc ferrite magnetic nanoparticles sintered at

FIGURE 8.2 SEM images of Ni nanoparticles prepared by hydrothermal method with different NaOH concentrations: (a) 5, (b) 10, (c) 15, (d) 20, (e) 25 mol/L. (Adapted from A. Bouremana et al., *Journal of Magnetism and Magnetic Materials* 358–359, 2014, 11–15. 2014 Elsevier B.V.)

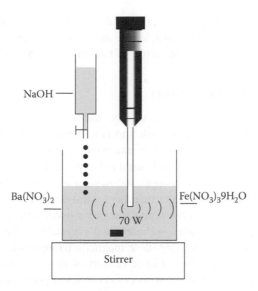

NaOH

Ba(NO$_3$)$_2$

Fe(NO$_3$)$_3$9H$_2$O

70 W

Stirrer

FIGURE 8.3 Schematic diagram of experimental setup used for the sonochemical reactions to preparation BaFe$_{12}$O$_{19}$ nanoparticles. (Adapted from G. Nabiyouni et al., *Journal of Industrial and Engineering Chemistry* 20, 2014, 3425–3429. 2013 The Korean Society of Industrial and Engineering Chemistry. Published by Elsevier B.V.)

1200°C [51]. In addition, cobalt ferrite nanoparticles of 4 nm were prepared by the reverse micelle method using an amphiphilic di iso-octyl sulfosuccinate surfactant with two different reactant salt (FeCl$_3$ · 6H$_2$O and CoCl$_2$ · 6H$_2$O) concentrations [52]. Another class of magnetic nanoparticles, namely, metal alloy magnetic nanoparticles (such as Fe$_3$Pt nanoparticles), have been synthesized by successively heating a mixture of platinum acetyl acetonate (Pt(acac)$_2$), iron penta carbonyl (Fe(CO)$_5$), triton X-100 (TX-100), cetyl trimethyl ammonium bromide (CTAB), and oleic acid at 500 K, followed by ethanol treatment to obtain Fe$_3$Pt magnetic nanoparticles [53].

8.1.2.1.8 Polyacrylamide Gel Method
The polyacrylamide gel method is a promising method for producing nanoparticles and nanocomposites such as magnetic nanoparticles, nanosheet structure, nanoporous structure, and core/shell structure, etc. This will be detailed in Section 8.2.

8.1.2.2 Factors Influencing Magnetic Properties
It is well known that the magnetic properties of magnetic materials are highly dependent on their morphologies, dimensions, particle size and shape, and chemical composition. Nanostructured magnetic materials are especially expected to exhibit enhanced properties or completely new properties which are usually absent in their bulk counterparts. By changing the particle size and shape, the composition, structure, and magnetism of magnetic materials can be adjusted. However, magnetic nanoparticles generally have a large surface free energy, thus they easily form aggregates to reach a stabilized state. The formation of aggregates decreases the magnetic

properties of magnetic materials. Factors which could result in the aggregation of magnetic nanoparticles include particle size and shape, solution composition, particle concentration, and surface chemistry of nanoparticles [54,55].

8.1.3 Magnetic Nanocomposites

As a result of advances in nanoscale science and technology, nanoscale magnetic particles can now be synthesized with superior ferromagnetism, magnetoelectric coupling, superparamagnetism, electron and phonon transport, etc. [56]. Magnetic nanoparticles have already been widely used in the magneto-electronics industry and are playing key roles in many other applications, such as optics and recording devices, purification of enzymes, waste-water treatment, catalysis, magnetic resonance imaging (MRI), drug delivery, magnetocooling, magnetic particle imaging, and biological materials, including disinfectants, sorbents, and recyclable catalysts [6–11,57–63]. However, there are drawbacks using these magnetic nanoparticles, for example, the severe aggregation and formation of large clusters of magnetic nanoparticles. These can be overcome by using a porous matrix of ceramics or polymers embedded with a small concentration of magnetic nanoparticles [57,58,64]. Therefore, novel designs of the interfacial structures of magnetic nanocomposites are crucial for their successful applications, for example, a core/shell structured Fe_3O_4/Au showed a great potential in nanoparticle-based therapeutic and diagnostic applications [65].

8.1.3.1 Classification of Nanocomposites

Nanocomposites generally are mixtures in which at least one of the components has one, two, or three dimensions less than 100 nm, and they often exhibit completely new properties or an improved property compared with those of bulk counterparts [66]. Nanocomposites can be classified into six categories (as illustrated in Figure 8.4), defined by the dimensions of the nanosized components or the matrix, that is, 0D nanoparticles coated in a nanoparticle to form core–shell or dumbbell structure (0–0 type); nanoparticle decorated nanowires (0–1 type); core–shell nanowire (1–1 type); 0D nanoparticles (NPs) dispersed in a bulk matrix (0–3 type); 1D nanorods/ nanowires/nanowhiskers in a bulk matrix (1–3 type); and 2D nanosheets in a bulk matrix (2–3 type) [57]. The matrix for the nanocomposites can be a porous network structure, a sheet structure, or dense bulk materials. By controlling the morphologies, dimensions, sizes, and defects of different components in the nanocomposite, it is possible to adjust the mechanical, optical, electrical, thermal, magnetic, and physicochemical properties of nanocomposites [67–70], leading to various applications from biomedicine to highly efficient phosphors [71,72].

Magnetic nanocomposites are based on magnetic nanoparticles or nanorods/ nanowires/nanowhiskers embedded inside a matrix of a porous network structure, a sheet structure, or dense bulk materials. They have attracted increasing attention due to their unique physicochemical properties for a variety of industrial applications [58]. Magnetic nanocomposites (such as $BiFeO_3$–$MgFe_2O_4$) with magnetic nanoparticles embedded in a bulk matrix (or in an epitaxial thin film) have good soft magnetic properties [73].

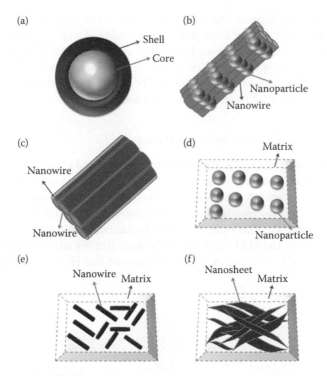

FIGURE 8.4 Different morphologies of nanocomposites: (a) core–shell structure (0-0 type), (b) nanoparticle decorated nanowires (0-1 type), (c) core–shell nanowire (1-1 type), (d) 0D nanoparticles (NPs) dispersed in a bulk matrix (0-3 type), (e) 1D nanorods in a bulk matrix (1–3 type), and (f) 2D nanosheets in a bulk matrix (2–3 type). (Adapted from S. F. Wang et al., *Nanoscience and Nanotechnology Letters* 6, 2014, 758–771.)

8.1.3.2 Preparation of Magnetic Nanocomposites

There are two common methods to make magnetic nanocomposites [57]. The first one is to uniformly disperse the magnetic nanoparticles into a solid or liquid matrix, and the second is to mix the magnetic nanoparticles with another nanoscale component. Up to now, many different types of magnetic nanocomposites have been synthesized, such as core/shell structure nanocomposites (e.g., SnO_2/Fe_3O_4 and $La_{0.7}Ca_{0.3}MnO_3/MgO$), metal/oxides nanocomposites (e.g., Fe/ZnO, Au-Co_3O_4, and Fe_3O_4/Se), and polymers/organics iron oxides (e.g., γ-Fe_2O_3/carbon and polyaniline/$BaFe_{12}O_{19}$). Several methods have been applied to prepare these magnetic nanocomposites, including hydrothermal synthesis [74–76], the solvothermal method [77,78], sol–gel and combustion method [79–83], one-step *in situ* polymerization method [84], sonochemical route [85,86], co-precipitation [87], nanocasting techniques [88,89], template approach [90], facile green synthesis route [91], polyacrylamide gel route [64,92], surfactant-mediated method [93], coaxial electrospinning [94], facile magnetosensitive catalysis process [95], etc. Among them, the polyacrylamide gel method has advantages of easy control of the topography by simply adjusting the pH value, organic additive or chelating agent, and elevating the calcination temperature. The polyacrylamide gel route allows preparation of highly pure nanopowders with a uniform and spherical shape [64,92,96], and has an ability to produce core–shell structured composites [64,92]. Successful applications of such magnetic nanocomposites are highly dependent on the stability of the magnetic nanoparticles and matrix under a range of different service conditions.

8.1.4 Application of Magnetic Materials

Magnetic materials, especially magnetic nanoparticles and nanocomposites, exhibit pronounced physicochemical properties, such as low toxicity, extra anisotropy contributions, high saturation field, and chemical stability that make them potential applications for waste-water treatment, drug delivery, information storage media, magnetic refrigeration materials, magnetic-levitation trains, microwave absorber materials, catalysis, and MRI (see Figure 8.5). Several applications of magnetic materials are highlighted including industrial applications and biomedical applications.

FIGURE 8.5 Different applications of magnetic materials including waste-water treatment, drug delivery, information storage medium, magnetic refrigeration materials, Maglev train, microwave absorber materials, catalysis, and magnetic resonance imaging.

8.1.4.1 Industrial Applications

Industrial applications of magnetic materials include magnetic refrigerators, colored pigments, waste-water treatment, information storage media, and magnetic-levitation trains [97–101]. It is well known that each industrial application requires the materials to have well-defined properties. For instance, magnetic nanoparticles with good chemical stability are required in catalysis and they can be used to assist an effective separation of catalysts [13]. In addition, in order to be used as an information storage medium, magnetic nanoparticles need to have a stable and switchable magnetic state to represent bits of information, a state which is not affected by temperature fluctuations [102,103].

8.1.4.2 Biomedical Applications

For biomedical applications, magnetic nanoparticles or magnetic nanocomposites need to exhibit superparamagnetism at room temperature [104–106]. Biomedical applications of magnetic nanoparticles or magnetic nanocomposites cover a wide scope such as therapy and diagnosis.

Applications using magnetic nanoparticles or magnetic nanocomposites in the therapy fields include drug delivery, hyperthermia/thermal ablation, radiotherapy combined with MRI, musculoskeletal system associated diseases and anemic chronic kidney disease as shown in Figure 8.6 [5].

Diagnosis applications of magnetic nanoparticles or magnetic nanocomposites are classified into two categories, *in vivo* applications and *in vitro* applications according to applications inside or outside the body [107]. *In vitro* applications could be further separated into sensing, cell sorting, bio-separation, enzyme immobilization, immunoassays, transfection, and purification applications. For *in vivo* applications, the main application is in MRI as shown in Figure 8.6.

8.2 POLYACRYLAMIDE GEL TECHNIQUE FOR MAGNETIC MATERIALS

The polyacrylamide gel technique is a technique whereby the gelation of the solution is achieved by polymerizing the acrylamide monomers or acrylamide monomers and bisacrylamide to form a polymer network, that is, polyacrylamide, which provides a structural framework to restrain the volume of metal precursor solution. Subsequently, the polyacrylamide gel is dried and sintered to obtain nanooxides. The technique is also called the acrylamide route [108], acrylamide polymerization [109,110], nitrate polyacrylamide gel process [110], polymer network method [111], and polyacrylamide gel method [112].

8.2.1 Recent Development of Polyacrylamide Gel Technique

Various wet chemical routes have been applied to produce nanopowder, such as the hydrothermal route [113–116], solvothermal technique [117], co-precipitation [118], bio-molecule assisted route [119], and sol–gel [120–122]. These methods are generally time-consuming if large quantities of nanopowders are required. Moreover, it is difficult to achieve homogeneous compositions because of different chemical

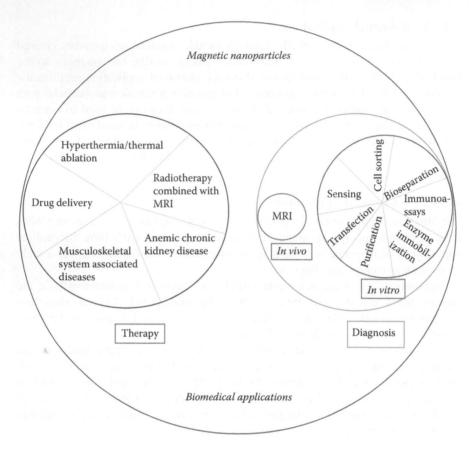

FIGURE 8.6 Magnetic nanoparticles or magnetic nanocomposites for biomedical applications. (Adapted from M. Arruebo et al., *Nano Today* 2, 2007, 22–32.)

behaviors of a large amount of cations [123]. The polyacrylamide gel technique, on the other hand, is reproducible, fast, cheap, and easy to scale-up, and can provide ultrafine powder at relatively low temperatures [124,125].

The polyacrylamide gel method was first reported by Douy and Odier [126] to prepare $YBa_2Cu_3O_{7-x}$ powders [127]. Recent advances in new techniques make it possible to synthesize not only high-temperature superconducting materials ($YBa_2Cu_3O_{7-x}$ [108,126], and $La_{1.85}Sr_{0.15}CuO_{4-x}$ [128], etc.), but also a variety of functional materials such as multiferroic materials (such as YMn_2O_5 [129], $YMnO_3$ [112], $LaA(A = Ca, Sr)MnO_3$ [130], $BiFeO_3$ [131], $SrTiO_3$ [132], $TbFeO_3$ [133], etc.), oxide thermoelectric materials ($Ca_3Co_4O_9$ [134], $Ca_{3-x}Y_xCo_4O_{9+\delta}$ [135], etc.), optical coating materials (such as YVO_4:Eu [136], Al_2O_3 [137,138], SnO_2:Eu [139], Ce^{3+}:YAG [127], $AlFeO_3$ [140], ZnO [141,142], $MgAl_2O_4$:Tb [143], etc.), wave-absorbing materials ($Co_{0.5}Zn_{0.5}Fe_2O_4$ [144], and $Ni_xZn_{1-x}Fe_2O_4$ [145]), photocatalytic materials ($CaTiO_3$ [146], $BiVO_4$ [147], etc.), nanocomposites (Al_2O_3/SiO_2 [109], $Li_2O/Al_2O_3/SiO_2$ [148], $La_xSr_{1-x}MnO_3/MgO$ [64,92], $(Ca_{0.95}Bi_{0.05})_3Co_4O_9/Ag$ [149], Mg_2SiO_4: Eu^{3+} [150], Ag/ZnO [151], Ag/CeO_2

[152], etc.), solid oxide fuel cell materials ($LaCoO_3$ [153], ($Zr_{0.84}Y_{0.16}O_{1.92}$, $Ce_{0.8}Gd_{0.2}O_{1.9}$, $La_{0.9}Sr_{0.1}Ga_{0.8}Mg_{0.2}$-$O_{2.85}$) [154], ZrO_2: Sc_2O_3 [155], ($La_{0.75}Sr_{0.25})_{1-x}Cr_{0.5}Mn_{0.5}O_3$ [156], ($La_{0.8}Sr_{0.2})_{0.9}MnO_3$ [157], etc.) and other oxide materials including Ca-α-SiAlON [158], $Ba_4YMn_{3-x}Cu_xO_{11.5\pm\delta}$ [159], $Sr_{1-x}La_{1+x}Al_{1-x}Mg_xO_4$ [160], $Ba_{0.99}Zr_{0.8}Y_{0.2}O_{3-\delta}$ [161], β-SiAlON [162], BeO [163], etc.

In 2001, Sin et al. [109] used the polyacrylamide gel method to prepare mullite composites, and showed that the addition of combustible species produced finer particle sizes and more easily separated powders. In 2003, Tarancón et al. [154] used this method to synthesize nanocrystalline materials of solid oxide fuel cells, including $Zr_{0.84}Y_{0.16}O_{1.92}$ (8YSZ), $Ce_{0.8}Gd_{0.2}O_{1.9}$ (CGO), $La_{0.9}Sr_{0.1}Ga_{0.8}Mg_{0.2}O_{2.85}$ (LSGM), $La_2Mo_2O_9$, $La_{0.8}Sr_{0.2}CoO_{3-\delta}$ (LSC), and $La_{0.8}Sr_{0.2}FeO_{3-\delta}$ (LSF), and demonstrated its efficiency in production and versatility compared to other synthetic routes (such as solid state reaction) [154]. In the late 2000s, a few researchers synthesized the magnetic nanocomposites with a view to improve the property of their magneto-transport properties by embedding 0D–3D nanoparticles into insulating phases (i.e., forming a core–shell structure) [64,92]. More recently, many researchers synthesized photocatalytic or multiferroic materials such as $BiFeO_3$, $LaCoO_3$, $YMnO_3$, YVO_3, etc. and magnetic composite materials including Ag/ZnO, Ag/CeO$_2$, etc. A statistical literature analysis of recent papers based on the Science Citation Index is shown in Figure 8.7, revealing a steady increase of the published scientific papers in this area,

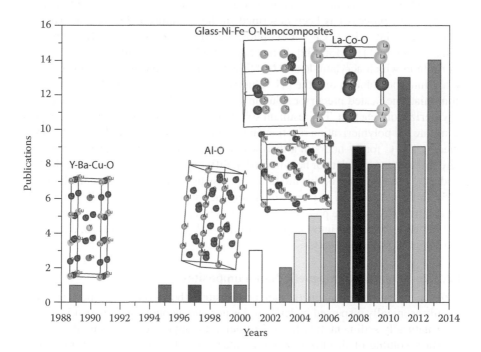

FIGURE 8.7 Publications per year for nanopowders or nanocomposites prepared by polyacrylamide gel route during the period 1989 to December 2014. (Adapted from S. F. Wang et al., *Nanoscience and Nanotechnology Letters* 6, 2014, 758–771.)

predominantly due to the enormous demands for multi-functional materials including multiferroic materials, solid oxide fuel cell materials, optical materials, etc. [164,165]. At the same time, advances in synthesis and characterization techniques have contributed to the novel designs of various multi-functional nanomaterials including nanoporous materials, nanopowders (including nanoparticles, nanorods), and nanocomposites [125].

8.2.2 Synthesis Techniques of Magnetic Nanomaterials

8.2.2.1 Synthesis of Magnetic Nanopowder

Figure 8.8 shows the chemical route for preparation of metal oxide nanoparticles and nanoporous ceramics using the polyacrylamide gel method. The starting materials (i.e., metal nitrates, metal sulfates, metal carbonates, metal acetate, or metal chlorates) are dissolved into an aqueous solution of nitric acid, hydrochloric acid, or sulfuric acid [125]. After the solution becomes transparent, a stoichiometric amount of chelating agent (i.e., citric acid, ethylenediamine-tetraacetic acid (EDTA), tartaric acid, oxalic acid, or acetic acid, etc.) is added to the solution with a given molar ratio with respect to the cations in order to complex the cations. Subsequently, an appropriate amount of initiator (e.g., persulfate aqueous solution $((NH_4)_2S_2O_8$ (APS) or α, α'- azoisobutyronitrile) is added [157,166]. In the final stage, acrylamide and N, N'-methylene-bisacrylamide monomers are added into the solution. The flow chart of a standard process has been described in the literature [125]. Nanopowders of YVO_4: Eu^{3+}, SnO_2: Eu^{3+}, Ce^{3+}: YAG phosphors have been successfully synthesized using this method [127,136,139]. Yang et al. [64,92] used a modified polyacrylamide gel route in which acrylamide and N, N'-methylene-bisacrylamide monomers were added into the solution and the pH value was adjusted to 3 by adding an aqueous ammonia. The heated precursor solution was then used as an initiator to trigger the polymerization reaction. The resultant solution was heated to 60–90°C on a hot plate to initiate the polymerization reaction and generate a polyacrylamide gel, which was dried at 120°C for 24 hours in a thermostat drier. The obtained xerogel precursors were ground into powders or heat-treated at different temperatures to make the final products. The flowchart of this modified process has been described in the literature [125]. In the gel process, an interesting phenomenon can be observed. When the Al^{3+} ion is introduced into the precursor, the gel is white. However, when the Fe^{3+} ion and W^{2+} ion are each added into the precursor, the gel becomes purple, red, and yellow (see Figure 8.8), respectively.

8.2.2.2 Synthesis of Magnetic Nanocomposites

In order to synthesize magnetic nanocomposites, the initial component is generally ultrasonically dispersed in deionized water and alcohol solution for 15–30 min. After naturally sedimentating for 5–24 hours, the upper liquid is poured away and a certain volume of deionized water is added into the solution again to make a uniform suspension. The second component is then dissolved into the above prepared suspension. Finally, magnetic nanocomposites can be prepared according to Section 8.2.2.1.

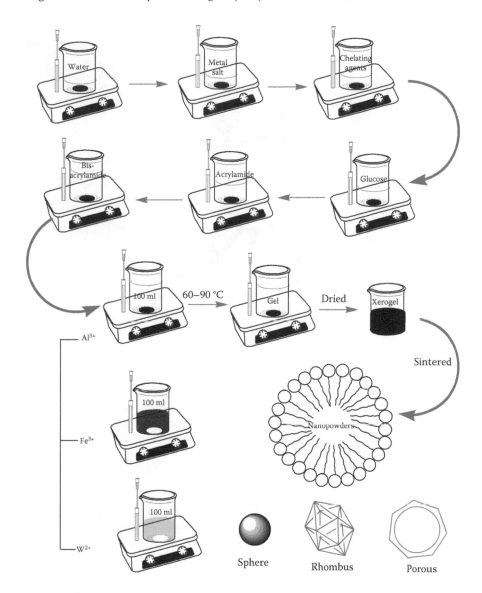

FIGURE 8.8 Flow-chart for preparation of metal oxide nanoparticles via polyacrylamide gel route.

In addition, the polyacrylamide gel method was also used with additional additives to prepare porous structure composites (such as $LaCoO_3$-B_2O_3-CuO composites [167] and $(Ca_{0.975}La_{0.025})_3Co_4O_9$-Ag [168]). EDTA was used as a chelating agent, and α, α'-azoisobutyronitrile was used as a thermo-chemical initiator. $LaCoO_3$ ceramics sintered at 1000°C showed a porous microstructure with a grain size around 1 μm (see Figure 8.9a). When a certain amount of B_2O_3-CuO was added into the $LaCoO_3$ precursor, the pores in the grain boundary were decreased (as shown in Figure 8.9b

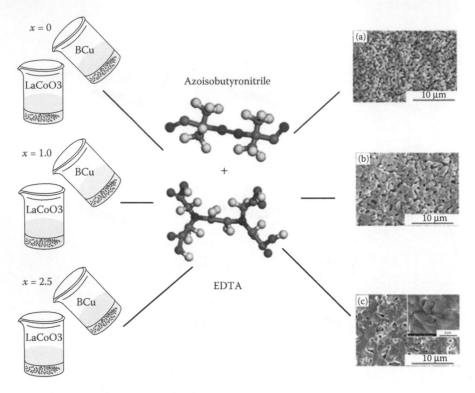

FIGURE 8.9 Scanning electron microscope (SEM) images of the fracture surface of the LaCoO₃ ceramics with different BCu additions ((a) $x = 0$, (b) $x = 1.0$, and (c) $x = 2.5$) sintered at 1000°C. The inset in (c) shows its magnified view. (Adapted from S. F. Wang et al., *Nanoscience and Nanotechnology Letters* 6, 2014, 758–771; Y. Song et al., *Journal of Alloys and Compounds* 536, 2012, 150–154.)

and c). Although the Seebeck coefficient S of the LaCoO₃ at a temperature below 200°C decreased with the increase of B₂O₃–CuO content x (wt.%), the electric conductivity and thermal resistivity were simultaneously enhanced with increasing values of x when $x \leq 2.5$ [167].

Wang et al. [138] utilized citric acid as a chelating agent based on the polyacrylamide gel method to prepare porous monolithic alumina. The chemical route for preparing α-Al₂O₃–carbon nanocomposites is shown in Figure 8.10, in which the formation process can be divided into three steps: complex reaction, crosslink reaction, and decompose process [169]. The fluorescent properties of the Al₂O₃/C were improved by using such a method. For the α-Al₂O₃–carbon composites, the C-O-Al bond gave rise to the formation of a defective structure, which resulted in the increase in the luminescence intensity. Of course, the increase of the C-O-Al bond and therefore the oxygen vacancies could also contribute to the increase in the luminescence intensity. Therefore, a certain content of carbon dopant could remarkably improve the morphology of α-Al₂O₃ and strongly increase its luminescent properties [169].

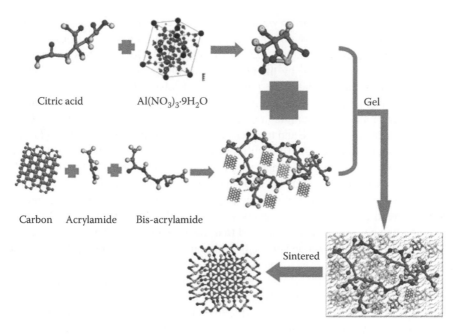

Citric acid   Al(NO₃)₃·9H₂O   Gel

Carbon   Acrylamide   Bis-acrylamide

Sintered

FIGURE 8.10 Chemical route for the preparation of α-Al₂O₃-carbon nanocomposites. (Adapted from S. Wang et al., *Journal of Luminescence* 153, 2014, 393–400. 2014 Elsevier B.V.)

8.2.3 Factors for Size and Distribution of Nanoparticles or Nanocomposites

Formation mechanisms of the magnetic nanopowder and nanocomposites based on the polyacrylamide gel method have been investigated [64,92,166,170,171]. During synthesis, many process parameters may influence the formation of nanopowder and nanocomposites, including amounts of monomer systems and organic additives, pH values, amounts of chelating agent and water, amounts of glucose and initiator, drying temperature and time, sintering temperature and aluminum precursor, crystallization temperature and time, as well as the heating rate in the crystallization region, etc. [125]. Among these influencing parameters, five key parameters are generally considered more important than the others, which will be discussed in detail in the following sections.

8.2.3.1 Effect of Monomer Systems

The polyacrylamide gel route is based on the polymer-based chemistry of combined metal oxides, semiconductor or traditional ceramics. In this route, to improve the forming process of metal oxides, semiconductor or ceramic, the key step of the gel-forming process is the addition of organic monomers, thus the selection and ratio of organic monomers play a crucial role.

When a small amount of acrylamide is introduced into the solution, the acrylamide monomer radicals are obtained because of the additional reactions between the monomer and primary radical [123–125,134,153,170]. Double bonds of the acrylamide monomer are opened by the primary radical and a completely new activation site is formed at the end of the polyacrylamide chain. In the chain propagation process, the primary radical in the first acrylamide unit attacks the carbon–carbon double bond of a new acrylamide monomer, leading to the linking-up of the second monomer unit to the first one and transfer of the radical site from the first acrylamide unit to the second one [125,170]. By this way, an increasing number of acrylamide units are linked resulting in an increase of the length of the polymer linear chains, resulting in fast growth. It has been observed that the acrylamide gel system only leads to a linear polymer chain (see Figure 8.11) [64]. However, the acrylamide monomers are neurotoxic chemical reagents verified by the toxicity research [125,172]. Laboratory experimental results showed that the acrylamide monomers could cause cancer for animals and human beings. In addition, the process is time consuming for preparing nanopowders or nanocomposites.

Recently, new types of less toxic organic monomers were investigated, and instead of using the acrylamide monomer, acrylamide and N, N′-methylene bisacrylamide were applied as new standard systems due to their rapid reaction processes and relatively low toxicity [96,125,173]. If a small amount of N,N′-methylene bisacrylamide is introduced into the acrylamide precursor solution, the polyacrylamide chains will be cross-linked through the bisacrylamide to grow into a complex web of interconnected loops and branches as shown in Figure 8.11 [137]. Generally speaking, the strength of the polyacrylamide gel increases with the concentration of the organic monomer or the ratio between the N,N′-methylene bisacrylamide and chain monomer, therefore, the obtained final products are denser and stiffer with the gel process [125,174]. When the above reactions are completed, the nitrates, sulfates, carbonates, and chlorates are trapped within the polymer network. Furthermore, the evaporation of water inside the polyacrylamide gel could cause the organic polymer framework to shrink, leading to trapping of a small amount of nitrates, sulfates, carbonates, or chlorates inside the organic polymer framework [64,125,132,137,166,170,175,176].

8.2.3.2　Effect of Initiator

The polymeric network formation of nanopowders generally has the following four steps:

1. Chain initiation
2. Chain propagation reaction
3. Chain termination reaction
4. Cross linking reaction [124,125]

In the polymerization reaction, the reaction rate of the chain initiation is generally the lowest, thus it is critical to control the polymerization reaction rate. The initiator plays a crucial role to promote the organic monomers including acrylamide and bisacrylamide. to react between each other and form a polymer network structure. The gel process commonly occurs at an elevated temperature of 40–90°C. The commonly

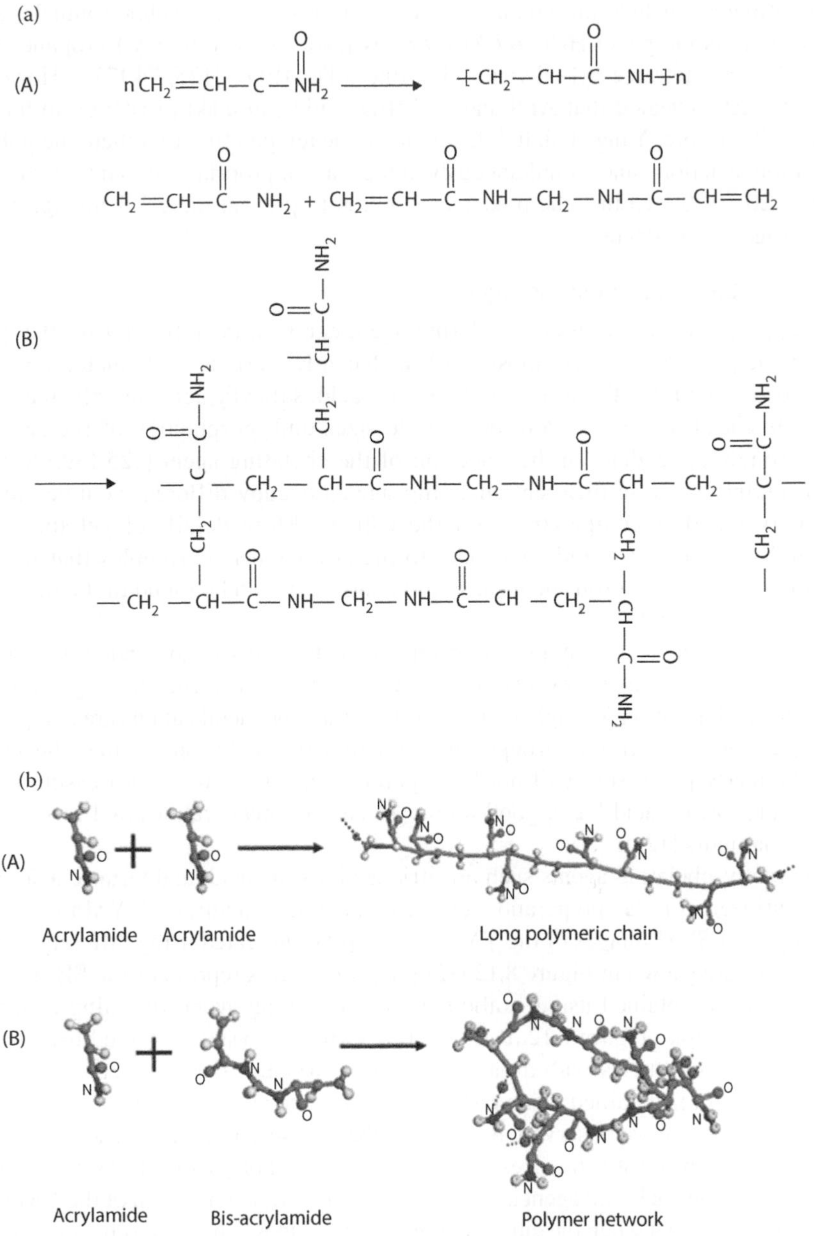

FIGURE 8.11 (a) Chemical structure and (b) chemical structural model of reaction mechanisms of acrylamide and bis-acrylamide. Carbon and hydrogen are indicated as black and white balls, respectively. (Adapted from S. F. Wang et al., *Nanoscience and Nanotechnology Letters* 6, 2014, 758–771; S. Wang et al., *The Journal of Physical Chemistry C* 117, 2013, 5067–5074. 2013 American Chemical Society.)

used initiators include ammonium persulfate (APS), a,a'-azoisobutyronitrile, azo-bis (2-amidinopropane) HCl (AZAP), azobis [2-(2-imidazolin-w-yl) propane] HCl (AZIP), and tetramethyl-ethylene diamine (TEMED) [125,174,177]. However, toxicity tests revealed that APS and TEMED could cause skin irritation in human beings. Therefore, Yang et al. [64,92] changed the temperature to initiate the polymerization reaction, and simultaneously added an appropriate amount of glucose (about 20 g/100 mL) into the precursor solution to prevent the gel from significant shrinkage during drying.

8.2.3.3 Effect of Chelating Agent

The appropriate selection of a chelating agent can significantly improve the quality of the prepared nanocomposite [125]. Recently, various acids including citric acid, oxalic acid, EDTA, acetic acid, tartaric acid, salicylic acid, and glycine, were used as chelating agents. Pore or particle sizes and morphology of the samples are strongly dependent on the selection of the chelating agent [125,139,140,178–181], mainly because different chelating agents occupy different volumes in the gel network after complexing with the cations. More details of gel formation through the polyacrylamide gel route to prepare nanosized samples that include nanoparticles, nanoporous materials, and nanosheets can be found in the literature [64,137,138,149,167].

Chelating agents usually prevent spontaneous hydrolysis and condensation reactions of different metal ions. Oxalic acid is a carboxyl type chelating agent, which permits the formation of single molecular units (i.e., the metal cations are completely occupied by the carboxyl groups) and suppress the diffusion, so that the initial stoichiometry precursors will not be apparently changed during processing [125]. Therefore, oxalic acid has a good ability to chelate metal cations and form stable molecular units [129].

Recently, chelating agents such as citric acid, oxalic acid, and tartaric acid have been utilized for the preparation of different nanostructures of YMn_2O_5 [129]. Examples of SEM images of the YMn_2O_5 samples fabricated using different chelating agents are shown in Figure 8.12 [129]. Figure 8.12aA represents the SEM image of the sample obtained using oxalic acid as a chelating agent, revealing nanorods inside the sample. Figure 8.12aB shows the YMn_2O_5 crystals obtained using citric acid, and nanoparticles with an average size of 100 nm. Figure 8.12aC shows the YMn_2O_5 crystals obtained using tartaric acid as a chelating agent with large porous structures. The above observation confirms that the selection of the chelating agent could have significant influences on the crystal morphology of a prepared sample. Meanwhile, the chelating agent also affects the phase transformation of the YMn_2O_5. It was reported that when the citric acid or tartaric acid was used as a chelating agent, the dominant orthorhombic YMn_2O_5 was decomposed into hexagonal $YMnO_3$ and Mn_3O_4 at 1000°C [125,129].

During the reaction process, oxalic acid was commonly used as a carboxyl chelating agent, whereas citric acid and tartaric acid were often used as carboxyl and hydroxyl chelating agents [125,129]. The corresponding reactions are shown in Figure 8.12b. Both these types of chelating agents could be employed to chelate inorganic precursors ($Y(NO_3)_3 \cdot 6H_2O$ and $Mn(CH_3COO)_2 \cdot 4H_2O$) to form a metal

(a)   (b)

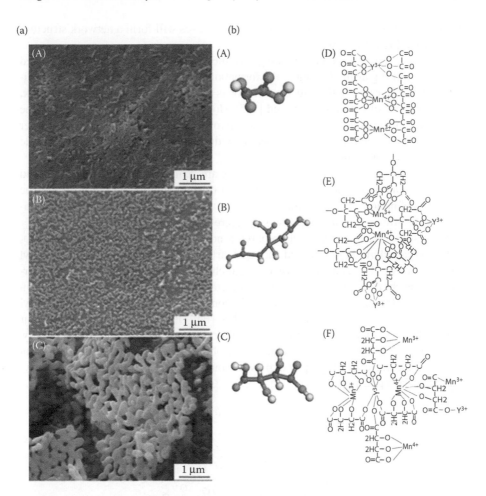

FIGURE 8.12 (a) SEM images of the YMn_2O_5 samples prepared using different chelating agents (oxalic acid (A), citric acid (B), and tartaric acid (C)) and calcined at 800°C. (b) The structure of chelating agents (oxalic acid (A), citric acid (B), and tartaric acid (C)) and metal ions-chelating agents (M-oxalic acid (D), M-citric acid (E), and M-tartaric acid (F)) complex. (Adapted from S. Wang et al., *Science China Chemistry* 57, 2014, 402–408. Science China Press and Springer-Verlag, Berlin, 2013.)

complex. They also prevent undesired spontaneous hydrolysis and condensation reactions for metal ions (i.e., Y and Mn). Oxalic acid permits the formation of individual molecular units (because the metal ions are occupied by carboxyl groups) which are free to diffuse, so that the initial stoichiometry remains [125]. At the same time, the molecular units form a shape of long rods from the complex reactions as shown in Figure 8.12bD [129]. When an appropriate amount of citric acid is used as the chelating agent, Mn^{3+} ions are coordinated by one hydroxyl group and five carboxyl groups, whereas Mn^{4+} ions are coordinated by one hydroxyl group and seven carboxyl groups to form different molecular units as shown in Figure 8.12bE [129]. If tartaric acid is used as chelating agent, the complexes will form a network structure as shown in Figure 8.12bF.

In the polyacrylamide gel process, N,N′-methylene-bisacrylamide is often used as a cross-linking agent. Owing to the formation of a 3D polymer network, the individual molecular unit in the aqueous solution is trapped within the polymer network, so the mobility of cation is restricted [125]. However, the network structures of the complexes in the aqueous solution are interwoven on the branches of the polymer network, and are subsequently wrapped up with the branches to form metal-organic polymer nanocomposites [125,129].

In the final stage, the organic compounds are removed during calcination, and the encapsulated complexes are decomposed during calcinations, resulting in the formation of nanoparticles or nanorods. The network structure decomposes simultaneously, forming a porous structure (see Figure 8.12) [129].

Interesting, Wang et al. [140] prepared a novel light emission material $AlFeO_3$ by the polyacrylamide gel method and used tartaric acid as a carboxyl and hydroxyl chelating agent. The chemical structure and chemical structure model of the coordination mechanisms of tartaric acid and metal ions (Al^{3+} and Fe^{3+}) are shown in Figure 8.13. In the complex reaction, Al^{3+} ions are coordinated by two hydroxyl groups and four carboxyl groups as well as Fe^{3+} ions are coordinated by four hydroxyl groups and two carboxyl groups to form a network structure [140]. Finally, the porous structures of $AlFeO_3$ ceramics are obtained. The result demonstrates that control of the surface topography of powder could be successfully achieved by varying the chelating agent.

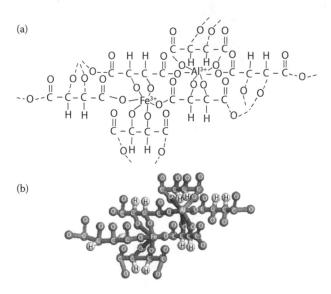

FIGURE 8.13 (a) Chemical structure and (b) chemical structural model of coordination mechanisms of chelating agent (tartaric acid) and metal ions (Al^{3+} and Fe^{3+}). Single ball (mark "O") represents oxygen atoms that will be coordinated by metal atoms (metal-O). (Adapted from S. Wang et al., *Optical Materials* 36, 2013, 482–488. 2013 Elsevier B.V.)

8.2.3.4 Effect of Sintering Temperature and Metal Precursor

Different processes result in the formation of intermediate metastable phases before a stable final product is formed. The formation sequence of the intermediate metastable phases is strongly dependent on the source materials, their microstructure and crystallinity, sintering temperature, pH value, impurities in the solution, etc. For example, in Reference [182], a temperature above 1200°C was reported necessary to obtain α-Al_2O_3 when using precursors such as bayerite or boehmite. However, in Reference [138], when citric acid was used as a chelating agent and aluminum nitrate was used as a precursor, pure α-Al_2O_3 could be prepared after sintering at a lower temperature of 1150°C. The sintering temperature affects not only the phase transition but also the particle sizes of the final product [125,127,139,148,171,183–185].

8.3 MAGNETIC MATERIALS DEVELOPMENT FROM THE POLYACRYLAMIDE GEL TECHNIQUE

Recently, the polyacrylamide gel technique has been developed to prepare magnetic nanoparticles and high-temperature superconductors. The synthesis of magnetic nanoparticles is more challenging as compared to the preparation of ceramic materials. The reported work has focused on the preparation of $BiFeO_3$, $LaFeO_3$, $AlFeO_3$, $MnFeO_3$, $TbFeO_3$, $CoFe_2O_4$, $NiFe_2O_4$, $MgFe_2O_4$, $CaFe_2O_4$, $BaFe_2O_4$, $SrFe_{12}O_{19}$, $BaFe_{12}O_{19}$, $Bi_2Fe_4O_9$, $YMnO_3$, $TbMnO_3$, $LaSrMnO_3$, $LaCaMnO_3$, and YMn_2O_5 nanostructures. In addition, many types of high-temperature superconductors prepared using the polyacrylamide gel technique have been reported, including $YBa_2Cu_3O_{7-x}$, $LaSrCuO_4$, and $Bi_2Sr_2Ca_{n-1}Cu_nO_{4+2n+d}$ [123,126,128,186].

8.3.1 POLYACRYLAMIDE GEL TECHNIQUE FOR MAGNETIC NANOPARTICLES

8.3.1.1 Ferrite Magnetic Nanoparticles

Ferrite magnetic nanoparticles are important magnetic functional materials and have extensive applications in various fields including multiple-state memory elements, microwave devices, spintronics, and magnetic–electric sensors [187,188]. Structurally, ferrites can be roughly divided into four categories: orthorhombic ferrites, cubic ferrites, tetragonal ferrites, and hexagonal ferrites. The orthorhombic M ferrites, such as $MFeO_3$ (M = Al, La, Tb, et al.), MFe_2O_4 (M = Ca^{2+}, Ba^{2+}, et al.), and $Bi_2Fe_4O_9$, are known to be superparamagnetic and exhibit weak ferromagnetism at room temperature [189,190]. The cubic M ferrites, such as MFe_2O_4(M = Mn^{2+}, Ni^{2+}, Fe^{2+}, Co^{2+}, Mg^{2+}, et al.) and $MnFeO_3$, are well-known spinel magnetic materials, exhibit magnetically soft behavior, and are easily magnetized and demagnetized [191,192]. The hexagonal M ferrites have the general formula $MFe_{12}O_{19}$, where M is a bivalent metal ion such as Ba, Sr, or La. Benefitting for the high coercivity, high magnetization, high Curie temperature, and very large magnetocrystalline anisotropy, hexagonal ferrites such as $BaFe_{12}O_{19}$ and $SrFe_{12}O_{19}$ are widely applied as magnetic recording materials [193]. In addition, $BiFeO_3$ is also a hexagonal M ferrite and exhibits a weak ferromagnetic behavior at room temperature [131]. The tetragonal ferrites $MnFe_2O_4$, a well-known magnetic material, has been widely applied in contrast-enhancement agents in MRI technology, electronic applications, and recording media [194]. Interestingly, all the aforementioned ferrites can be prepared by the polyacrylamide gel technique.

8.3.1.1.1 *M(M = Al, Bi, La, Mn, Tb, et al.) FeO₃ Ferrites Nanoparticles*

8.3.1.1.1.1 BiFeO₃ Magnetic Nanoparticles $BiFeO_3$ is an important multiferroic material and has wide applications in various fields. Yang et al. [131] reported the polyacrylamide gel technique synthesis of $BiFeO_3$ nanoparticles using an aqueous solution of nitric acid containing $Bi(NO_3)_3 \cdot 5H_2O$ and $Fe(NO_3)_3 \cdot 9H_2O$ at 600°C, and a variety of particle size distributions were obtained, including N,N'-methylene-bisacrylamide/acrylamide with mole ratios ranging from 0/25 to 5/25, and mean grain size of 110, 95, 70, and 52 nm. In the experiments, $BiFeO_3$ nanoparticles are generally prepared using EDTA as the chelating agent. However, when citric acid is used as a chelating agent to prepare $BiFeO_3$ xerogel, the XRD result of the xerogel sintered at 600°C shows no pure hexagonal $BiFeO_3$ phase obtained (see Figure 8.14). The results demonstrated that the proper choice of chelating agent can enhance $BiFeO_3$ phase purity [96].

Figure 8.15a shows an SEM image of the $BiFeO_3$ xerogel prepared using EDTA as the chelating agent and sintered at 600°C. The particles are nearly spherical in shape with a mean particle size of 55 nm. The room temperature hysteresis loops of the $BiFeO_3$ magnetic nanoparticles are shown in Figure 8.15b, and the saturation magnetization (M_s) is 1.56 emu/g.

8.3.1.1.1.2 LaFeO₃ Magnetic Nanoparticles $LaFeO_3$ is one important magnetic material and exhibits a very weak ferromagnetic behavior at room temperature [195]. To improve the magnetic properties of $LaFeO_3$ nanoparticles, $LaFeO_3$/

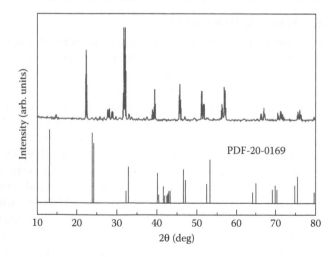

FIGURE 8.14 XRD patterns of the $BiFeO_3$ xerogel prepared using citric acid as chelating agent and sintered at 600°C.

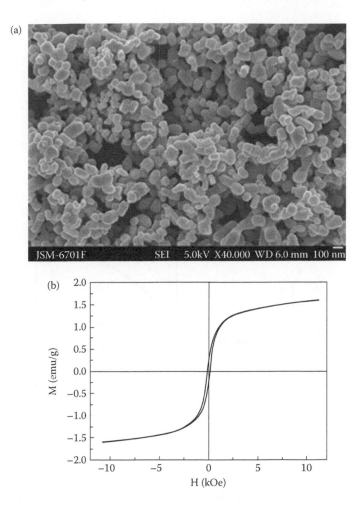

FIGURE 8.15 (a) SEM image and (b) room-temperature magnetic hysteresis curves of BiFeO$_3$ nanoparticles prepared using EDTA as chelating agent. (Adapted from H. Yang et al., *J Sol-Gel Sci Technol* 58, 2011, 238–243.)

Fe$_2$O$_3$ magnetic nanocomposites have been prepared using the polyacrylamide gel technique to enhance their magnetic properties. Wang et al. [196] prepared the gel of magnetic nanocomposites using La(NO$_3$)$_2$ · 9H$_2$O, Fe(NO$_3$)$_3$ · 9H$_2$O, EDTA, glucose, acrylamide, and N,N′-methylene-bisacrylamide in the aqueous solution of nitric acid by a heating process at 120°C for 10 min. The obtained xerogel was ground into powder and some powder was sintered at 800°C for 2 hours in air to prepare the LaFeO$_3$/Fe$_2$O$_3$ magnetic nanocomposites. Compared with the pure LaFeO$_3$ powder, the magnetic nanocomposites exhibit much stronger ferromagnetic behavior [196]. Figure 8.16a shows the SEM image of a LaFeO$_3$/Fe$_2$O$_3$ sample sintered at 800°C, revealing particles that are nearly spherical in shape and a mean particle size about 80 nm. Figure 8.16b shows the magnetic hysteresis loop

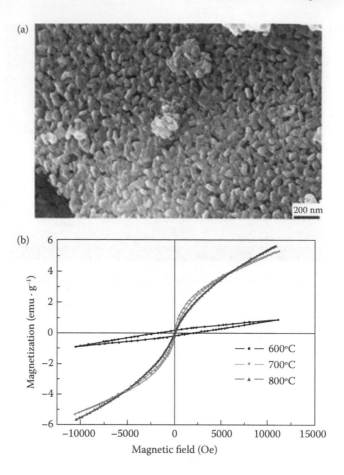

FIGURE 8.16 (a) SEM image of LaFeO$_3$/Fe$_2$O$_3$ sample sintered at 800°C. (b) Room-temperature magnetic hysteresis curves of LaFeO$_3$/Fe$_2$O$_3$ sample sintered at different temperatures. (Adapted from S. F. Wang, Q. P. Ding, and X. T. Zu, *Applied Mechanics and Materials* 563, 2014, 30–35.)

of LaFeO$_3$/Fe$_2$O$_3$ nanocomposites sintered at 600°C, 700°C, and 800°C. As can be seen in Figure 8.16b, the coercivity value decreases with the increase of sintering temperature. It is possible that the particle size increases with the increase of calcination temperature [196].

8.3.1.1.1.3 AlFeO$_3$ AlFeO$_3$ is a well-known perovskite-type oxide material, and exhibits piezoelectric, magnetoelectric, and ferromagnetic effects at low temperature [140]. Wang et al. [140] synthesized porous AlFeO$_3$ using tartaric acid as chelating agent using a modified polyacrylamide gel route. In addition, the nanostructure AlFeO$_3$ was fabricated without using tartaric acid as the chelating agent based on the same method. The crystallite growth procedure is shown in Figure 8.17a. Figure 8.17aA shows the growth procedure of the nanostructure AlFeO$_3$.

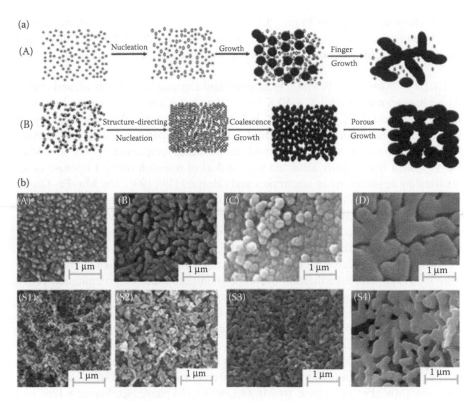

FIGURE 8.17 (a) Schematic illustration of growth phenomena of (A) AlFeO$_3$ particles and porous AlFeO$_3$ ceramics (B). (b) SEM images of the AlFeO$_3$ samples prepared without addition of tartaric acid ((A), (B), (C), and (D)) and with addition of tartaric acid ((S1), (S2), (S3), and (S4)) and sintered at 500, 600, 700, and 800°C, respectively. (Adapted from S. Wang et al., *Optical Materials* 36, 2013, 482–488. 2013 Elsevier B.V.)

The growth process has three steps:

1. Nucleation
2. Nucleation/growth
3. Finger growth

However, the growth processes of porous AlFeO$_3$ has three different steps (see Figure 8.17aB):

1. Nucleation
2. Nucleation/coalescence/growth
3. Porous structure growth

To confirm the crystal growth mechanism of AlFeO$_3$ samples, Figure 8.17b shows the SEM images of the AlFeO$_3$ samples fabricated without addition of tartaric acid as the chelating agent (see Figure 8.17bA–D and with addition of tartaric acid as the chelating agent (see Figure 8.17b(S1–S4)) sintered at 500°C, 600°C, 700°C, and 800°C. The results are consistent with the crystal growth mechanism.

8.3.1.1.1.4 MnFeO$_3$ FeMnO$_3$ is an important ferrimagnetic material, and exhibits a phase transition with $T_c \approx 40$ K [197]. Numerous studies are focused on multiferroic and catalytic properties of FeMnO$_3$ powders [197,198]. The stability of various compounds formed in the Mn–Fe–O system depends not only on Mn/Fe/O ratios but also on the synthesis method and sintering temperature [197]. The Mn–Fe–O system has recently attracted a great deal of research interest because of its outstanding applications in electronics and catalysts [198,199]. The Mn–Fe–O system includes rock-salt oxides Fe$_x$Mn$_{1-x}$O [200,201], mixed iron manganese oxides (Fe$_{1-x}$Mn$_x$)$_{1-y}$O($x \leq 0.3$) used as water-splitting reactions [202], cubic FeMnO$_3$ similar to the structure of Mn$_2$O$_3$ [197], γ-FeMnO$_3$ with the same structure as γ-Fe$_2$O$_3$ [203], and manganese-substituted magnetite Mn$_x$Fe$_{3-x}$O$_4$ [203,204].

Various methods have been adopted to synthesize FeMnO$_3$, such as mechanical alloying [197], oxidative thermal decomposition [205], and co-precipitation [203]. The polyacrylamide gel route has the advantage of easy control of the powder topography by using different chelating agents [178], organic additives [137], or sintering temperatures [206]. Wang et al. [207] reported a novel chemical route to synthesize cubic FeMnO$_3$ powders, and their phase and magnetic properties have been investigated. Figure 8.18a shows the XRD patterns of FeMnO$_3$ xerogel sintered at 500°C, 600°C, and 700°C. It is clear that the xerogel calcined at 500°C, 600°C, and 700°C are cubic (Mn^{+3},Fe^{+3})$_2$O$_3$ phase, (α-Mn$_2$O$_3$) · (α-Fe$_2$O$_3$) phase (see Figure 8.18b), and cubic structure of FeMnO$_3$ phase, respectively. The reaction can be described by the following equation [207]:

$$2\text{Mn(CH}_3\text{COO)}_2 \cdot 4\text{H}_2\text{O} + 2\text{Fe(NO}_3)_3 \cdot 9\text{H}_2\text{O} + 2\text{C}_4\text{H}_6\text{O}_6 + 2\text{NH}_3 \cdot \text{H}_2\text{O} \xrightarrow{500°C}$$
$$\text{cubic(Mn}^{+3},\text{Fe}^{+3})_2\text{O}_3 + 13\text{CO} + 43\text{H}_2\text{O} + 3\text{N}_2 + 3\text{CO}_2 + \text{NO} \tag{8.1}$$

$$(\text{Mn}^{+3},\text{Fe}^{+3})_2\text{O}_3 \xrightarrow{600°C} (\alpha - \text{Mn}_2\text{O}_3) \cdot (\alpha - \text{Fe}_2\text{O}_3) \tag{8.2}$$

$$(\alpha - \text{Mn}_2\text{O}_3) \cdot (\alpha - \text{Fe}_2\text{O}_3) \xrightarrow{700°C} 2(\text{FeMnO}_3) \tag{8.3}$$

To confirm if the FeMnO$_3$ xerogel sintered at 700°C contains iron ions, the XPS spectra of Fe$_{2p}$ electron levels was obtained and is shown in Figure 8.18c. The Fe 2p core level spectra are split because of the spin–orbit coupling into 2$p_{3/2}$ (711.19 eV) and 2$p_{1/2}$ (724.75 eV) components, which can be assigned to Fe^{3+} of FeMnO$_3$ [207]. Figure 8.18d shows the temperature dependence of magnetization of FeMnO$_3$ prepared at pH = 7 and sintered at 700°C under field cooling (FC) and zero field cooling (ZFC) conditions. As shown in Figure 8.18d, there is no sharp ferrimagnetic transition. The result is inconsistent with the results observed by Seifu et al. [197]. In addition, the thermal expansion behavior of FeMnO$_3$ was examined by a thermal

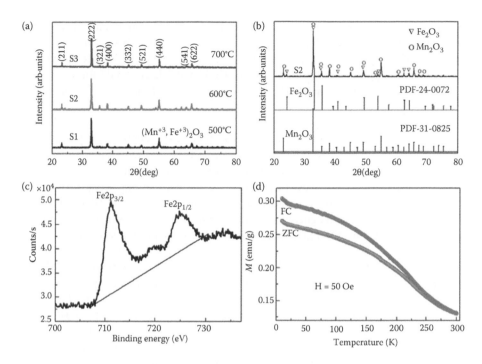

FIGURE 8.18 (a) XRD patterns of the FeMnO$_3$ ceramics prepared at pH = 7 and sintered at different temperatures. (b) XRD patterns of the FeMnO$_3$ xerogel sintered at 600°C. (c) XPS spectra of Fe$_{2p}$ electron levels and (d) M-T curves for FeMnO$_3$ prepared at pH = 7 and sintered at 700°C. (Adapted from Z. J. Li et al., *Journal of Nano Research* 37, 2016, 122–131.)

dilatometer. The result shows that FeMnO$_3$ has an average thermal expansion coefficient of 9.0139×10^{-6}/K [207].

8.3.1.1.1.5 TbFeO$_3$ TbFeO$_3$ is a typical orthoferrite oxide material and exhibits an orthorhombically distorted perovskite structure [208]. Sivakumar et al. [209] prepared TbFeO$_3$ nanocrystals with an average diameter of about 60 nm using a Fe(CO)$_5$ precursor in a simple sonochemical method. The maximum value of coercivity (*Hc*) was 500 Oe for the TbFeO$_3$ measured at 20 K. TbFeO$_3$ nanoparticles were also prepared using high-purity Tb$_4$O$_7$, Fe(NO$_3$)$_3 \cdot$ 9H$_2$O, tartaric acid, glucose, and acrylamide at pH = 3 in dilute nitric acid solution at 80°C, and then submitted to calcination at 650°C for 10 hours [133]. The particles have a narrow size distribution without any adhesive phenomenon. The mean particle size was about 50 nm. TbFeO$_3$ nanoparticles have been applied for the removal of organic dyes from the environment [133].

8.3.1.1.2 M(M = Mg, Ca, Ba, Co, Ni et al.)-Fe$_2$O$_4$ Ferrites Nanoparticles

8.3.1.1.2.1 CoFe$_2$O$_4$ To study the effect of crosslinking agent bis-acrylamide on the particle size, powder morphology, and magnetic properties of CoFe$_2$O$_4$ nanoparticles, the CoFe$_2$O$_4$ nanoparticles were prepared by Wang et al. [210] using a polyacrylamide gel route. In their work, two samples were prepared at pH = 2 without N,N'-methylene-bisacrylamide monomers and acrylamide: bis-acrylamide = 1:5 (molar ratio), respectively. The crosslinking agent bis-acrylamide was introduced into the $CoFe_2O_4$ precursor solution to decrease the particle size and saturation magnetization of $CoFe_2O_4$ nanoparticles. However, the $CoFe_2O_4$ nanoparticles appeared severely aggregated.

8.3.1.1.2.2   NiFe₂O₄

8.3.1.1.2.2   NiFe$_2$O$_4$ Zhao et al. [211] reported $NiFe_2O_4$ nanoparticles prepared by the polyacrylamide gel route using acetic acid or EDTA as the chelating agent. The oxalic acid, citric acid, and tartaric acid were used as chelating agents to prepare $NiFe_2O_4$ nanoparticles. However, the above three samples contain minor impurity phases except for the spinel $NiFe_2O_4$ phase. Similarly, the obtained magnetic nanoparticles appeared severely aggregated. The magnetic properties of the $NiFe_2O_4$ nanoparticles have been remarkably affected by different chelating agents due to the particle size effect.

8.3.1.1.2.3   (Mg, Ca, Ba)-Fe$_2$O$_4$ Figure 8.19 shows the XRD patterns of (a) $Ba_2Fe_2O_5$, (b) $Ba_2Fe_6O_{11}$, (c) $BaFe_4O_7$, (d) $BaFeO_3$, (e) $BaFe_2O_4$, and (f) $BaFe_{12}O_{19}$ prepared using a modified polyacrylamide gel route. As can be seen in Figure 8.19, the $Ba_2Fe_2O_5$, $Ba_2Fe_6O_{11}$, $BaFe_4O_7$, and $BaFeO_3$ xerogel sintered at 800°C are mixed phases. However, the $BaFe_2O_4$ and $BaFe_{12}O_{19}$ xerogel sintered at 800°C are a pure orthorhombic structure of $BaFe_2O_4$ and hexagonal structure of $BaFe_{12}O_{19}$, respectively. The results indicate that the polyacrylamide gel route is suitable for preparing the stable oxides at ambient temperature. Figure 8.19g and h shows the typical hysteresis curves for $BaFe_2O_4$ and $BaFe_{12}O_{19}$ magnetic field-dependent magnetization at room temperature, respectively. The VSM result confirmed that hexagonal $BaFe_{12}O_{19}$ was a hard magnet.

Spinel ferrite $MgFe_2O_4$ is a well-known magnetic material and has been extensively applied in electronic applications and recording media more than a half century [212]. Recently, $MgFe_2O_4$ magnetic nanoparticles have been prepared by the combustion synthesis method, high-energy ball milling method, hydrothermal synthesis, mechanochemistry synthesis, and microwave-assisted ball milling [213].

However, many unavoidable factors are associated with spinel $MgFe_2O_4$ magnetic nanoparticles. For instance, magnetic nanoparticles easily form agglomerates because of their intrinsic instability over long periods. In fact, these aggregate $MgFe_2O_4$ particles can obviously reduce their interfacial area, and then lead to the loss of the magnetism of the $MgFe_2O_4$ particles [214]. Wang et al. [214] present a method to synthesize Mg-ferrites magnetic nanoparticles at a low sintering temperature using a modified polyacrylamide gel route. This chemical route for the preparation of highly dispersed Mg-ferrites magnetic nanoparticles is shown in Figure 8.20. The $MgFe_2O_4$ nanoparticles were successfully prepared by different polysaccharides: glucose (S1), maltose (S1-1), and sucrose (S1-2). The added carbon (S4) in the precursor solution decreases the crystallite size and suppresses precursor powders from forming the spinel $MgFe_2O_4$ phase.

Figure 8.21a shows the hysteresis loops of the $MgFe_2O_4$ magnetic nanoparticles prepared by adding glucose at different sintering temperatures under an applied magnetic field of 1.25 T. As shown in Figure 8.21a, as sintering temperature increases, the remanent magnetization (M_r) and saturation magnetization (M_s)

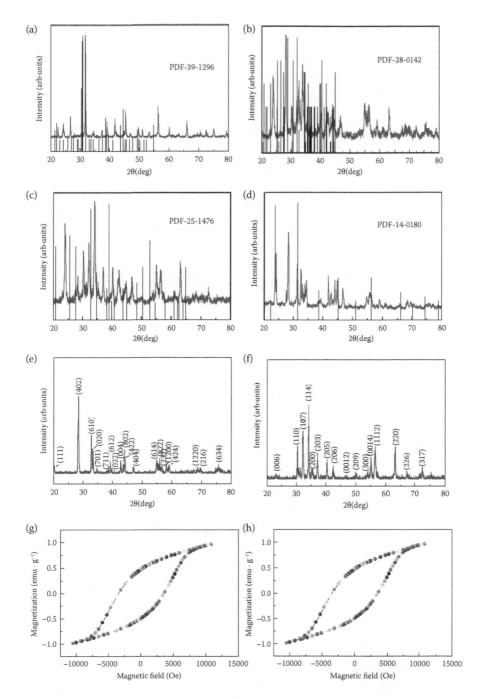

FIGURE 8.19 XRD patterns of (a) $Ba_2Fe_2O_5$, (b) $Ba_2Fe_6O_{11}$, (c) $BaFe_4O_7$, (d) $BaFeO_3$, (e) $BaFe_2O_4$, and (f) $BaFe_{12}O_{19}$ prepared by a modified polyacrylamide gel route. (g) and (h) Represent the VSM curves of $BaFe_2O_4$ and $BaFe_{12}O_{19}$, respectively.

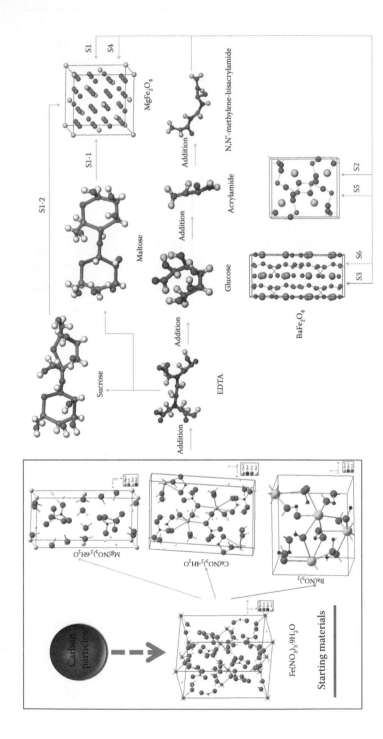

FIGURE 8.20 Chemical route for the preparation of highly dispersed (Mg, Ca, Ba)-ferrites magnetic nanoparticles.

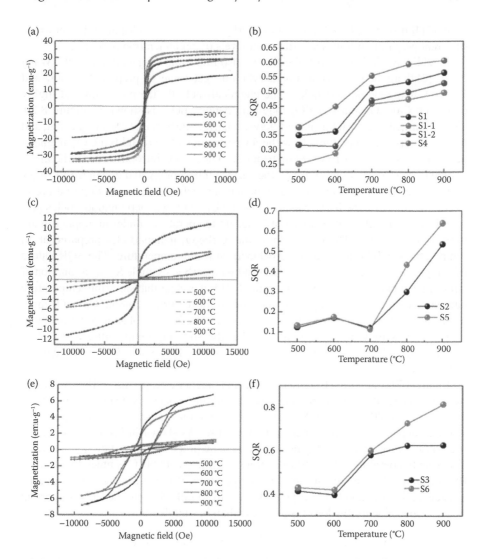

FIGURE 8.21 Room-temperature magnetic hysteresis curves of (a) $MgFe_2O_4$, (c) $CaFe_2O_4$, (e) $BaFe_2O_4$ samples prepared with glucose addition and sintered at different temperatures. Relationships between SQR and sintering temperature of (b) $MgFe_2O_4$, (d) $CaFe_2O_4$, (f) $BaFe_2O_4$ xerogel powders. (Adapted from S. F. Wang et al., *Ceramics International* http://dx.doi.org/10.1016/j.ceramint.2016.09.075.)

increase for the $MgFe_2O_4$ samples. It can be inferred that the M_r and M_s of $MgFe_2O_4$ magnetic nanoparticles increase with the increase of $MgFe_2O_4$ crystallinity [214]. Figure 8.21b shows the squareness ratio (SQR) of the different $MgFe_2O_4$ samples. As can be seen in Figure 8.21b, the SQR value of $MgFe_2O_4$ magnetic nanoparticles increases with the sintering temperature. Results indicate that sintering temperature can enhance the crystallinity and phase purity of the $MgFe_2O_4$ nanoparticles. Wang et al. [214] discovered the SQR values of $MgFe_2O_4$ nanoparticles are relevant to the different preparation routes and different levels of agglomeration particles.

In addition, $CaFe_2O_4$ and $BaFe_2O_4$ are also important magnetic materials. Wang et al. [214] applied the same polyacrylamide gel route to prepare $CaFe_2O_4$ and $BaFe_2O_4$ magnetic nanoparticles. The chemical route for the preparation of highly dispersed $CaFe_2O_4$ and $BaFe_2O_4$ magnetic nanoparticles is shown in Figure 8.20. The optimized chemical formula of polysaccharide, carbon, and chelating agent, and proper sintering temperature allow the formation of $CaFe_2O_4$ and $BaFe_2O_4$ magnetic nanoparticles with a narrow size distribution. The carbon introduced from the metal source precursor solution improved the surface morphology and enhanced the SQR value of $CaFe_2O_4$ and $BaFe_2O_4$ magnetic nanoparticles as shown in Figure 8.21c–f. Figure 8.21c shows the hysteresis loops of the $CaFe_2O_4$ magnetic nanoparticles prepared by adding glucose at different sintering temperatures under an applied magnetic field of 1.25 T. The results show that $CaFe_2O_4$ nanoparticles prepared at all temperatures exhibit paramagnetic behavior at room temperature. The SQR values of samples without carbon (Sample S2) and with carbon (Sample S5) as a function of sintering temperature are shown in Figure 8.21d. It is clear that sintering temperature can enhance the crystallinity and phase purity of the orthorhombic $CaFe_2O_4$ magnetic nanoparticles. Also from Figure 8.21(d), the SQR value of sample S5 is higher than that of sample S2. This confirms that the sintering temperature and carbon enhance the SQR value and phase purity of the orthorhombic $CaFe_2O_4$ magnetic nanoparticles.

Figure 8.21e shows the hysteresis loops and magnetic properties of $BaFe_2O_4$ magnetic nanoparticles (sample S3) prepared by adding glucose ND sintered at 500°C, 600°C, 700°C, 800°C, and 900°C. It can be found that values of M_r, M_s, and H_c are strongly dependent on the sintering temperature. As shown in Figure 8.21e and f, the phase transformation influences the magnetic properties of $BaFe_2O_4$ at a certain sintering temperature. The SQR of sample S6 sintered at 900°C is much higher than that of sample S3 sintered at the same temperature. Wang et al. discovered that the existence of some additional crystal defects in sample S3 was the most important factor responsible for the lower SQR value than in sample S6.

8.3.1.1.3   M(M = Sr, Ba, et al.)-Fe₁₂O₁₉ Ferrites Nanoparticles

M-type strontium hexaferrite ($SrFe_{12}O_{19}$) is an important magnetic material which exhibits pronounced magnetic properties and chemical stability. Therefore, the material has attracted wide interest because of its potential applications in recording media, telecommunications, and magneto-optical devices [215]. Generally, the large SQR value of magnetic materials is preferred in the above applications [216]. Zi et al. [215] found that the SQR value of the M-type strontium hexaferrite $SrFe_{12}O_{19}$ are closely correlated to the grain size and preparation method. Therefore, a new synthesis route is expected to enhance the SQR value of $SrFe_{12}O_{19}$.

Wang et al. [20] reported a modified polyacrylamide gel method to prepare $SrFe_{12}O_{19}$ magnetic nanoparticles with a hexagonal magnetoplumbite structure by adding glucose to restrain the precursor gel from drastically shrinking during the gel drying process. In addition, the carbon introduced by the metal source precursor solution improved the surface morphology of $SrFe_{12}O_{19}$ magnetic nanoparticles.

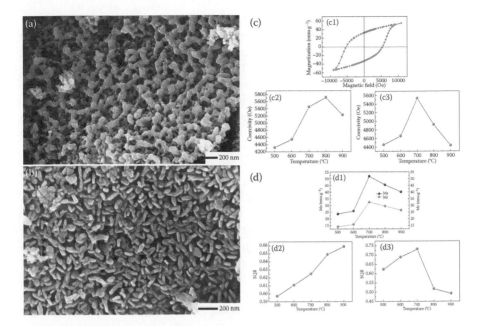

FIGURE 8.22 SEM images of (a) SrFe$_{12}$O$_{19}$ and (b) C- SrFe$_{12}$O$_{19}$ xerogel sintered at 700°C. (c) (c1) Room-temperature magnetic hysteresis curves of SrFe$_{12}$O$_{19}$ xerogel sintered at 700°C. Relationships between (c2) Hc, (c3) SQR and sintering temperature of SrFe$_{12}$O$_{19}$ xerogel powders. (d) Relationships between (d1) M$_r$ and M$_s$, (d2) H$_c$, (d3) SQR and sintering temperature of C-SrFe$_{12}$O$_{19}$ xerogel powders. (Adapted from S. Wang et al., *Journal of Sol-Gel Science and Technology* 73, 2015, 371–378.)

XRD results showed that the carbon and sintering temperature accelerated SrFe$_{12}$O$_{19}$ phase formation and did not change the hexagonal structure [215]. Metastable hexaferrite (SrFe$_{12}$O$_{19}$) was found to form at 700°C. The metastable hexaferrite decomposed into stable hexagonal SrFe$_{12}$O$_{19}$ phase and minor Fe$_2$O$_3$ above 800°C [215].

Figure 8.22a shows an SEM image of SrFe$_{12}$O$_{19}$ xerogel sintered at 700°C. The nanoparticles are bonded to fine particles to form agglomerates as shown in the SEM micrographs in Figure 8.10a, and thus present a compact structure. Figure 8.22b shows an SEM image of C-SrFe$_{12}$O$_{19}$ xerogel sintered at 700°C. The results indicate that the highly dispersed SrFe$_{12}$O$_{19}$ nanocrystals were obtained from the precursor solution of carbon. The prepared rod-like SrFe$_{12}$O$_{19}$ particles are uniform and the average length is 70 nm as shown in Figure 8.22b. The SEM results clearly demonstrates the effect of carbon on the modification of the sample surface morphologies of SrFe$_{12}$O$_{19}$ samples.

Figure 8.22cc1 shows the room temperature hysteresis loops of the SrFe$_{12}$O$_{19}$ xerogel sintered at 700°C under an applied magnetic field of 1.25 T. The results show that SrFe$_{12}$O$_{19}$ nanoparticles prepared at 700°C exhibit a hard magnetic behavior at room temperature. Figure 8.22cc2 illustrates the coercivity of the SrFe$_{12}$O$_{19}$ nanoparticles as a function of sintering temperature form 500°C to 900°C. The coercivity of the SrFe$_{12}$O$_{19}$ nanoparticles increased when the sintering temperature was increased from 500°C to 800°C. When the sintering temperature was further increased from 800°C to 900°C, the coercivity decreased. This could explain that the evolution of particle size with the increase of sintering temperature of xerogel may lead to the decrease of coercivity. The results from the $SrFe_{12}O_{19}$ sample are consistent with those in the previous reports [217]. However, the decrease of coercivity appeared in the C-$SrFe_{12}O_{19}$ sample sintered at 800°C, which is ascribed to the existence of a new Fe_2O_3 phase in the C-$SrFe_{12}O_{19}$ xerogel precursor sintered at or above 800°C [20].

8.3.1.1.4 $Bi_2Fe_4O_9$ Ferrites Nanoparticles

Orthorhombic $Bi_2Fe_4O_9$ is a well-known multiferroic material, which shows special magnetic and electric properties [181]. Singh et al. [218] reported the antiferromagnetic transition temperature (T_N) was 260 K for $Bi_2Fe_4O_9$ bulk materials. However, Tian et al. [219] reported that $Bi_2Fe_4O_9$ nanoparticles prepared by the Pechini method using nitrates as metal precursors exhibited a weak ferromagnetism at room temperature. In order to control the sample morphology, Zhang et al. [181] employed a polyacrylamide gel route to prepare $Bi_2Fe_4O_9$ nanoparticles. To obtain different sizes of $Bi_2Fe_4O_9$ nanoparticles, a chelating agent, EDTA or citric acid, was added in the metal salt precursor solution. The use of EDTA as the chelating agent (see Figure 8.23a) allowed the preparation of $Bi_2Fe_4O_9$ nanoparticles with a relatively narrow particle size distribution [181]. For the EDTA and citric acid processed samples (see Figure 8.23c), the remanent magnetization (M_r) was ~0.0084 and ~0.0054 emu/g, respectively. The coercivity value (H_c) of above two samples is ~200 and ~130 Oe, respectively. In addition, enhanced magnetic properties have also been reported in Sc ions doped $Bi_2Fe_4O_9$ ceramics [220] and Ti ions $Bi_2Fe_4O_9$ thin films [221].

8.3.1.2 Manganese Oxide Magnetic Materials

$RMnO_3$ (R = Y, Bi, La, Ce, Pr, Nd, et al.) manganites form an outstanding family exhibiting a wide variety of physicochemical properties and industrial applications [222]. $RMnO_3$ manganites can be divided into two kinds: orthorhombic structures and hexagonal structures. The R ionic radius of orthorhombic structures (R = La, Pr, Nd, Sm, Eu, Gd, Dy, and Tb) is larger than that of hexagonal structures (R = Ho, Er, Tm, Yb, Lu, Sc, and Y) [223]. Recently, $RMnO_3$ manganites have been prepared by the chemical solution method [224], co-precipitation method [225], hydrothermal treatment [226], epitaxial thin film growth techniques [227], sol–gel synthesis [228,229], etc. It is noted that $YMnO_3$, $TbMnO_3$, $La_{1-x}Sr_xMnO_3$, $La_{1-x}Ca_xMnO_3$, $La_{1-x}Na_xMnO_3$, and YMn_2O_5 nanostructures have been prepared by the polyacrylamide gel route.

8.3.1.2.1 $YMnO_3$ Magnetic Nanoparticles

The perovskite-type oxide $YMnO_3$ with both ferroelectric and anti-ferromagnetic orders is one type of multiferroic material [230], which has potential applications in magnetoelectric sensors, nonvolatile ferroelectric random access memories, and ferroelectric-gate field-effect transistors [231–234]. $YMnO_3$ multiferroics material was first reported by Yakel et al. in 1963 [235]. The hexagonal $YMnO_3$ structure

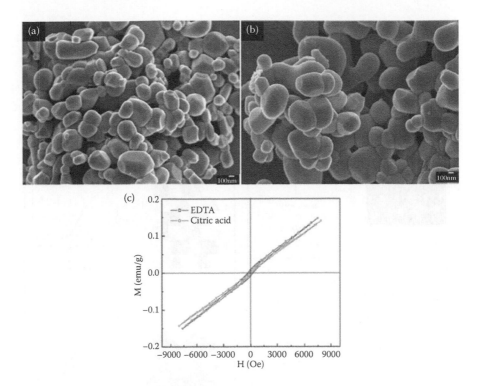

FIGURE 8.23 SEM images of $Bi_2Fe_4O_9$ samples synthesized using (a) EDTA and (b) citric acid as the chelating agent. (c) Room-temperature magnetic hysteresis loops for $Bi_2Fe_4O_9$ particles. (Adapted from M. Zhang et al., *Journal of Alloys and Compounds* 509, 2011, 809–812. 2010 Elsevier B.V.)

is characterized by MnO_5 bipyramids which form layers in the a–b plane [236] as shown in Figure 8.24a. In the a–b plane, the pyramids are linked at the base corners to form a triangular lattice [237]. The Y^{3+} ions are located among these MnO_5 bipyramid layers and are linked with O atoms [237].

Singh et al. [238] reported that $YMnO_3$ exhibited ferroelectric properties with the Curie temperature $T_C = \sim 920$ K and antiferromagnetic properties with the Néel temperature $T_N = \sim 74$ K. In 2010, Zheng et al. [239] reported the exchange bias effect of hexagonal $YMnO_3$ magnetic nanoparticles prepared by a facile hydrothermal method. In the same year, Wang et al. [240] reported that $YMnO_3$ nanoparticles exhibited a weak ferromagnetism behavior at room temperature as shown in Figure 8.24b. It was demonstrated that high-purity $YMnO_3$ nanoparticles can be obtained by a polyacrylamide gel route and the xerogel sintered at 800°C, which is nearly 300°C lower than the synthesis temperature based on the solid-state reaction method.

8.3.1.2.2 TbMnO₃ Magnetic Nanoparticles

The orthorhombic $TbMnO_3$ is an important multiferroic material and exhibits a giant magnetoelectric effect [241]. Walker et al. [242] reported that $TbMnO_3$ underwent two important magnetic phase transitions includes paramagnetic to

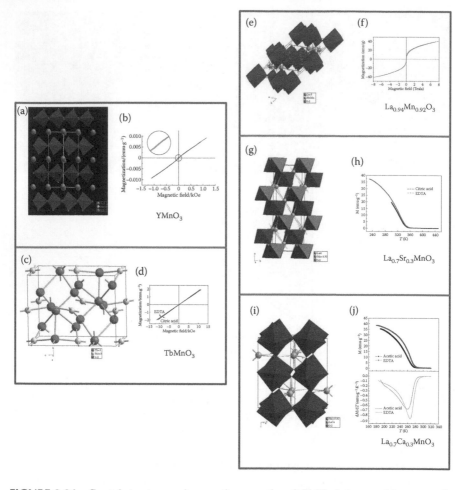

FIGURE 8.24 Crystal structure and magnetic properties of (Y, Tb, La) perovskite manganite and doped (Y, Tb, La) perovskite manganite magnetic materials, respectively. (a), (c), (e), (h), and (j) are crystal structure of YMnO$_3$, TbMnO$_3$, La$_{0.94}$Mn$_{0.92}$O$_3$, La$_{0.7}$Sr$_{0.3}$MnO$_3$ and La$_{0.7}$Ca$_{0.3}$MnO$_3$, respectively. (b), (d), (f), (i), and (k) are related to magnetic properties of YMnO$_3$, TbMnO$_3$, La$_{0.94}$Mn$_{0.92}$O$_3$, La$_{0.7}$Sr$_{0.3}$MnO$_3$, and La$_{0.7}$Ca$_{0.3}$MnO$_3$, respectively. (Adapted from Y. F. Wang and H. Yang, *Journal of Superconductivity and Novel Magnetism* 26, 2013, 3463–3467; H. Yang et al., *Journal of Alloys and Compounds* 555, 2013, 150–155; S. F. Wang et al., *Journal of the Chinese Ceramic Society* 38, 2010, 2303–2307; H. Yang et al., *Nanotechnology and Precision Engineering* 10, 2012, 46–51; G. Dezanneau et al., *Solid State Communications* 121, 2002, 133–137.)

antiferromagnetic phase transitions (T_{N1} = 41 K) and spiral antiferromagnetic phase transitions (T_{N2} = 28 K). Figure 8.24c shows the atomic model of the orthorhombic TbMnO$_3$ phase. The space group is *Pbnm* and the lattice parameters are a = 5.2931 Å, b = 5.8384 Å, and c = 7.4025 Å [243].

Generally, nanoscale materials display enhanced properties or a completely new set of properties from their bulk forms. Therefore, recently much effort has been devoted to the fabrication of highly dispersed $TbMnO_3$ nanoparticles to enhance their magnetic performance [244–246]. To obtain high-quality nanoparticles with a uniform spherical shape, a polyacrylamide gel route was used to prepare the $TbMnO_3$ nanoparticles [180].

In the sol process, the chelating agent citric acid, EDTA, or a combination of both in a 1:1 M ratio can be added to the precursor solution. It should be noted that all chelating agents do not affect the phase purity of $TbMnO_3$ nanoparticles. However, the particle size of the $TbMnO_3$ nanoparticles is strongly dependent on the choice of chelating agent [180,247]. $TbMnO_3$ magnetic nanoparticles show a typical hysteresis for their magnetic field-dependent magnetization at room temperature. Figure 8.24d shows the hysteresis loops of the $TbMnO_3$ magnetic nanoparticles prepared using citric acid and EDTA as chelating agent. Results showed that $TbMnO_3$ nanoparticles prepared using the citric acid or EDTA as chelating agent exhibited a paramagnetic behavior at room temperature [247].

8.3.1.2.3 LaMnO₃ and Ca, Sr Doped-LaMnO₃ Magnetic Nanoparticles

8.3.1.2.3.1 LaMnO₃

Orthorhombic $LaMnO_3$ is a p-type semiconductor with ABO_3 perovskite structure [248]. Orthorhombic $LaMnO_3$ exhibits oxygen mobility, high thermal stability, and unique catalytic, electrical, and magnetic properties [249,250]. The material has wide applications in solid oxide fuel cells, ferromagnets, electrocatalysts, giant magneto resistance and photocatalysts for hydrogen production [248,250]. Figure 8.24e shows the crystal structure of the orthorhombic $LaMnO_3$. The orthorhombic structure shows a Jahn-Teller distortion, as well as the tilting and rotation of the MnO_6 octahedra [251].

In order to obtain pure orthorhombic $LaMnO_3$ powders, various methods have been tested to prepare $LaMnO_3$ powders such as the conventional solid state reaction, hydrothermal treatment, sol–gel method, polymeric gel, molten salt reaction, and co-precipitation method [250,252].

In addition, in order to obtain $La_{1-x}Mn_xO_{3\pm\delta}$ nanocrystalline powder, Dezanneau et al. [253] reported a new sol–gel method (polyacrylamide gel route). In this method, the acrylamide is used to form a 3D tangled network. The AIBN and EDTA were used as polymerization initiator and chelating agent, respectively. In the $La_{1-x}Mn_xO_{3\pm\delta}$ nanocrystalline powder, a spin-glass behavior has been observed by ZFC-FC curves [253]. The as-prepared powders exhibited a ferromagnetic behavior at 10 K as shown in Figure 8.24f.

8.3.1.2.3.2 La₁₋ₓCaₓMnO₃

Development of metal-doped $LaMnO_3$ with enhanced magnetic and magnetocaloric properties are currently under extensive investigation. Various routes have been proposed for the preparation of (Ca, Sr, or Na)-$LaMnO_3$ magnetic materials [254–256]. For instance, the $La_{0.8}Sr_{0.2}MnO_3$ powders were prepared using La_2O_3, $MnCO_3$, and $SrCO_3$ powder as starting materials, with 24 hours ball milling followed by 2 hours sintering at 1000°C [257].

Yang et al. [179] synthesized $La_{0.7}Ca_{0.3}MnO_3$ nanoparticles with different sizes using a polyacrylamide gel route. The chelating agents, such as acetic acid, oxalic acid, citric acid, EDTA, or tartaric acid are frequently used to tailor the particle size of $La_{0.7}Ca_{0.3}MnO_3$ nanoparticles. Small amounts of impurities including $CaCO_3$ and

$La_2O_2CO_3$ are often obtained using tartaric acid as the chelating agent. It can be explained that tartaric acid has a relatively weak coordination capacity toward the three metal ions in the precursor solution. In the complex structure, some metal ion active sites which are not coordinated by tartaric acid anions will be coordinated by $-OH$ of water molecules since the non-coordinated metal ion active sites are very unstable. Figure 8.24j shows the crystal structure of $La_{0.7}Ca_{0.3}MnO_3$. Figure 8.24k shows the M-T and dM/dT-T curves of $La_{0.7}Ca_{0.3}MnO_3$ nanoparticles prepared using acetic acid and EDTA as the chelating agents, respectively. Yang et al. [179] discovered that the magnetization of samples prepared using acetic acid as the chelating agent is slightly weaker than that of samples prepared using EDTA as the chelating agent, which can be ascribed to its relatively smaller particle size and enhanced surface effects.

8.3.1.2.3.3 $La_{0.7}Sr_{0.3}MnO_3$ Wang et al. [130] synthesized $La_{0.7}Sr_{0.3}MnO_3$ nanoparticles with a spherical morphology and a particle size range of 19–60 nm. The chelating agents of acetic acid, oxalic acid, citric acid, EDTA, or tartaric acid were used to tailor the particle size of $La_{0.7}Sr_{0.3}MnO_3$ nanoparticles. Results demonstrated that $La_{0.7}Sr_{0.3}MnO_3$ nanoparticles were successfully prepared through adding different chelating agents. From references [130] and [179], it can be inferred that chelating agent has a high selectivity in the aspect of complexing different cations during the chelating reaction process [214]. Figure 8.24h shows the crystal structure of $La_{0.7}Sr_{0.3}MnO_3$. Figure 8.24i shows the M–T curves of $La_{0.7}Sr_{0.3}MnO_3$ nanoparticles prepared using citric acid and EDTA as the chelating agents, respectively. It is noted that the magnetization of samples prepared by citric acid as the chelating agent is slightly weaker than that of samples prepared by EDTA as the chelating agent because of the relatively smaller particle size of the samples prepared by citric acid as the chelating agent [130].

8.3.1.2.3.4 $La_{1-x}Na_xMnO_{3+\delta}$ Malavasi et al. [258] reported a polyacrylamide gel route or polyacrylamide-based sol gel route to prepare the $La_{1-x}Na_xMnO_{3+\delta}$ nanopowders. Basically, $La_{1-x}Na_xMnO_{3+\delta}$ particles with an average size of 35 nm were synthesized using a double distilled water with metal salt starting materials, including manganese nitrate, lanthanum nitrate, and sodium nitrate. In this case, the EDTA and AIBN were used as the chelating agent and polymerization initiators, respectively. EDTA was added into the solution in the molar ratio 1.1:1 with respect to the cations to complex the cations. The acrylamide: $N,N,$-methylene bisacrylamide molar ratio of 6:1 were introduced into the solution [258]. In the sintering process, La ions can be substituted for the Na ion in $LaMnO_3$ and the crystal is present as $La_{1-x}Na_xMnO_{3+\delta}$. In addition, the ionic radius of the Na ion is close to that of the La ion; hence, Na ion easily enters into the $LaMnO_3$ structure to form $La_{1-x}Na_xMnO_{3+\delta}$. It is demonstrated that superparamagnetic $La_{1-x}Na_xMnO_{3+\delta}$ nanoparticles can be synthesized at a sintering temperature of 973 K. The particle size of $La_{1-x}Na_xMnO_{3+\delta}$ nanoparticles is 35 nm, which is nearly 20 nm smaller than those obtained from the propellant synthesis route under identical sintering temperatures. Compared with the Ca-doped $LaMnO_3$, the Na doped $LaMnO_3$ has an obvious enhancement of the value of magnetoresistivity (MR) [258].

8.3.2   Polyacrylamide Gel Technique for Magnetic Nanocomposites

8.3.2.1   Nanocomposites with Porous Spheres Structure

Magnetic materials coated with porous glass (or silica) spheres using a polyacryl-amide gel route were first reported by Zhao et al. [177]. A model of magnetic materials coated with hollow glass microspheres is shown in Figure 8.25a. In those nanocomposite structures, some magnetic nanoparticles were immersed within the hollow glass microspheres, whereas the other nanoparticles were coated on the surfaces of the hollow glass microspheres. The magnetic-glass nanocomposites were formed by magnetic nanoparticles loaded on the surfaces of the hollow glass microspheres [177]. The size of these magnetic nanoparticles is less than 80 nm. Figure 8.25b shows the SEM images of the hollow glass microspheres and $NiFe_2O_4$-hollow glass microspheres (NFGMs) magnetic nanocomposites with different weight percentages (15%, 40%, and 65%) of the hollow glass microspheres after being sintered at 800°C. The hollow glass microspheres with the magnetic nanoparticles show many eroded speckles on their surfaces (see Figure 8.25bA). By using

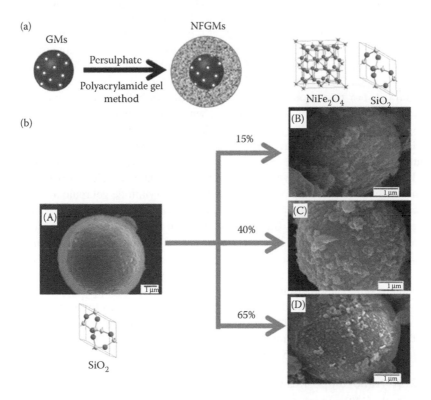

FIGURE 8.25   (a) A model of hollow glass microspheres with insulator cores and disordered nanoparticles shells, (b) SEM images of (A) GMs, NFGMs with different weight percent of glass microspheres (B) 15%, (C) 40%, and (D) 65%. (Adapted from S. F. Wang et al., *Nanoscience and Nanotechnology Letters* 6, 2014, 758–771; H. Zhao et al., *Chinese Journal of Chemical Physics* 20, 2007, 801–805.)

N,N′-methylene-bisacrylamide as a cross-linking agent, the magnetic nanoparticles can be easily coated on the hollow glass microsphere surfaces [177]. From Figures 8.25bB–D, the surfaces of the hollow glass microspheres are covered with a layer of magnetic nanoparticles with different contents of ferrite. The microstructures of the samples were found to be remarkably affected by the increase of the glass micro-sphere contents in the sample [177]. The magnetic nanocomposite with 40% hollow glass microspheres shows the best uniformity and continuity for the coverage of the nanoparticles as shown in Figure 8.25bC.

8.3.2.2 Nanocomposites with Core–Shell Structure

Pristine magnetic nanoparticles (such as $La_{0.67}Ca_{0.33}MnO_3$ [LCMO] nanoparticles) easily form agglomerates and large clusters, thus losing their magnetic, optical, and electrical properties associated with the presence of individual nanoparticles. Therefore, the magnetic nanoparticles have to be encapsulated with the insulating materials to suppress them from irreversible aggregation. Yao et al. [259] reported $La_{2/3}Ca_{1/3}MnO_3/CeO_2$ nanocrystalline composites prepared using a dispersing par-ticle polymer-network gel method, that is, polyacrylamide gel route. The CeO_2 was chosen as a nonmagnetic insulating material. However, Yao et al. did not observe the core–shell structure of $La_{2/3}Ca_{1/3}MnO_3/CeO_2$ nanocrystalline composites. A modi-fied polyacrylamide gel route has been developed to generate a structure with a mag-netic core and an insulator shell [64,92]. Different molar contents of MgO was added in the $(1 - x)$ $La_{0.67}Ca_{0.33}MnO_3$ (LCMO)/x MgO composites, that is, $x = 0$, 0.1, 0.2, 0.3, and 0.4. For these core (magnetic nanoparticles)/shell (insulator matrix) com-posites, the most promising connectivity scheme is core/shell type composites with nano-size magnetic particles uniformly enclosed in an insulator matrix.

Magnetic nanoparticles are normally spherical in shape and have a uniform size about 10s of nm, depending on the decomposition temperature and coverage ratio (see Figure 8.26aA) [64]. Coverage ratio is an important parameter for the formation of uniformly distributed nanoparticles via the polyacrylamide gel route method, and the obtained magnetic nanoparticles are optimal in a typical $(1 - x)$LCMO/xMgO composite with $x = 0.3$ [92]. Transmission electron microscopy (TEM) observation revealed clearly the core–shell structure as shown in Figure 8.26aB. Composition analysis confirmed the formation of 0–3 type core–shell structure of LCMO coated with MgO as shown in Figures 8.26aB–D. The magnetic nanocomposites exhibited a pronounced magnetoresistance (MR) sensitivity after being applied with the mag-netic fields, and the low-field sensitivity of MR was enhanced by increasing MgO content (see Figure 8.26b) [92]. This method has also been proposed to synthesize 0–3 type magnetoelectric composite materials $xCoFe_2O_4/(1 - x)BaTiO_3$ composites, that is, $x = 0$, 0.3, 0.4, and 0.5 [260].

8.3.2.3 Nanocomposites with Sheet Structure

Song et al. [149] reported the synthesis of $(Ca_{0.95}Bi_{0.05})_3Co_4O_9/Ag$ nanocompos-ites, in which Ag particles were dispersed inside the $(Ca_{0.95}Bi_{0.05})_3Co_4O_9$ matrix to form a sheet structure. The $(Ca_{0.95}Bi_{0.05})_3Co_4O_9/Ag$ nanopowders were synthe-sized using a polyacrylamide gel route and then the dense ceramic samples were prepared using a spark plasma sintering method [149]. Acetate solutions (including

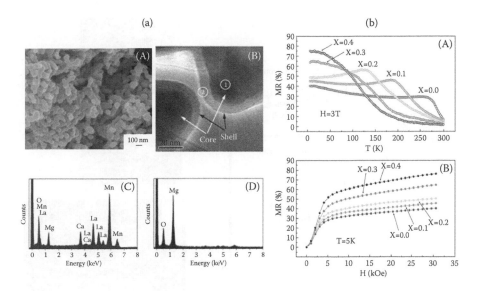

FIGURE 8.26 (a) (A) SEM and (B) transmission electron microscopy (TEM) image of a typical $(1 - x)$ LCMO/xMgO composite powder with $x = 0.3$. (C) and (D) energy dispersive X-ray (EDX) spectra obtained from the core region (spot 1) and the shell layer (spot 2) of a composite particle in (B). (b) (A) Temperature-dependent MR (at 3 T) and (B) magnetic field dependence MR (at 5 K) for $(1 - x)$ LCMO/x MgO composite powder. (Adapted from H. Yang et al., *Materials Letters* 63, 2009, 655–657; H. Yang et al., *Journal of Applied Physics* 106, 2009, 104317; S. F. Wang et al., *Nanoscience and Nanotechnology Letters* 6, 2014, 758–771.)

Ca(CH$_3$COO)$_2$ · H$_2$O and Co(CH$_3$COO)$_2$ · 4H$_2$O), Bi(NO$_3$)$_3$ · 5H$_2$O and AgNO$_3$) were used as the starting materials. The polymerization was promoted by adding a small amount of azoisobutyronitrile as the thermo-chemical initiator. Microstructure analysis showed that Ag was precipitated as the second phase on the grain boundaries of the (Ca$_{0.95}$Bi$_{0.05}$)$_3$Co$_4$O$_9$. SEM images of the fractured surface of the (Ca$_{1-x}$Bi$_x$)$_3$Co$_4$O$_9$ with different Bi contents are shown in Figure 8.27. The Ca$_3$Co$_4$O$_9$ sample prepared using the polyacrylamide gel technique showed sheet-like grains of about 1–3 µm (see Figure 8.27a). The texture of the plate-like Ca$_3$Co$_4$O$_9$ grains has been enhanced with the Bi ion substitution (see Figure 8.27b) [149]. Densification and texture of the sheet nanocomposites deteriorated with increase of the Ag addition. Ag particles (light gray region in the inset of Figure 8.27) are observed to disperse uniformly inside the Ca$_3$Co$_4$O$_9$ matrix (dark gray region in the inset of Figure 8.27), forming 2–3 type (2D nanosheets in a bulk matrix) nanocomposites.

8.3.2.4 Nanocomposites with Porous Networks Structure

Nanocomposites of α-Al$_2$O$_3$/LaMnO$_3$ were prepared using a modified polyacrylamide gel method by Wang et al. [261]. Porous α-Al$_2$O$_3$ was used as a hard template. The synthesis process starts with preparing a monolithic template (α-Al$_2$O$_3$) with various shapes. α-Al$_2$O$_3$ was found to be a promising candidate for the enhanced mass transportation of chemicals in liquid or biological processes in macroporous materials [138,262]. In the polyacrylamide gel route, citric acid was utilized as a

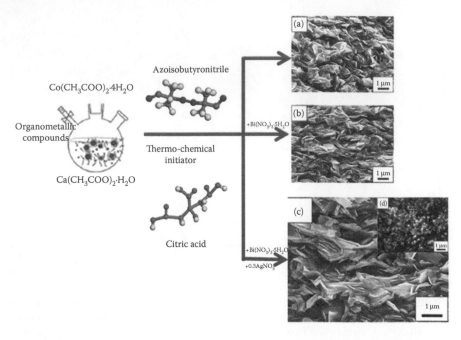

Co(CH$_3$COO)$_2$·4H$_2$O

Organometallic compounds

Ca(CH$_3$COO)$_2$·H$_2$O

Azoisobutyronitrile

Thermo-chemical initiator

Citric acid

+Bi(NO$_3$)$_3$·5H$_2$O

+Bi(NO$_3$)$_3$·5H$_2$O

+0.3AgNO$_3$

FIGURE 8.27 A schematic diagram for the preparation and typical SEM images of (a) Ca$_3$Co$_4$O$_9$, (b) (Ca$_{0.95}$Bi$_{0.05}$)$_3$Co$_4$O$_9$/Ag, (c) (Ca$_{0.95}$Bi$_{0.05}$)$_3$Co$_4$O$_9$/Ag – 0.3Ag, and (d) the back-scattered electron mode of (Ca$_{0.95}$Bi$_{0.05}$)$_3$Co$_4$O$_9$/Ag – 0.3Ag, respectively. Azoisobutyronitrile was applied as a thermo-chemical initiator of polymerization reaction. (Adapted from S. F. Wang et al., *Nanoscience and Nanotechnology Letters* 6, 2014, 758–771; Y. Song et al., *Materials Chemistry and Physics* 113, 2009, 645–649.)

chelating agent to complex with the cations in order to stabilize the solution against hydrolysis or condensation. The gelation of the solution was achieved by the formation of a polymer network, that is, polyacrylamide, which provides a structural framework to restrain the volume of precursor solution [125]. The detailed formation mechanisms of the porous α-alumina have been reported in Reference [138].

The procedures for the preparation of the α-Al$_2$O$_3$/LaMnO$_3$ nanocomposites are listed as follows [125,261]:

1. The precursor sol of the target product (LaMnO$_3$ precursor) is prepared using a polyacrylamide gel method.
2. The LaMnO$_3$ precursor fills the void space of the monolithic template using a polyacrylamide gel method.
3. The organics in the nanocomposites are removed by a high temperature sintering method.
4. The organics were decomposed during calcinations, thus forming the frame-structure of the α-Al$_2$O$_3$/LaMnO$_3$ nanocomposites.

Some of the LaMnO$_3$ nanoparticles adhere onto walls or surfaces of the α-Al$_2$O$_3$ to form agglomerates or particles, whereas other LaMnO$_3$ nanoparticles are embedded inside the porous α-Al_2O_3 to form a complex structure [125,261]. Owing to the formation of the porous structures, the metal ions (La and Mn) in the aqueous solution were trapped within the porous α-Al_2O_3, so the mobility of metal ions was limited [125,261]. When a small amount of acrylamide monomers and bisacrylamide was introduced into the solution, the growing polyacrylamide chains were cross-linked through the bisacrylamide to grow into a complex web of interconnected loops and branches, which further restricted the mobility of metal ions [64,112,125,129,137,138]. When the polymer was removed after calcination, the porous α-Al_2O_3 inside the citrate was degraded during the calcination, resulting in the formation of fine particles [125,261].

An SEM image of the obtained α-Al_2O_3 porous network is shown in Figure 8.28(S1) [261]. The average pore size of the α-Al_2O_3 sample is about 100 nm. When the La and Mn metal ions were added to the precursor, the α-Al_2O_3 porous network surface only showed a small amount of nanoparticles as shown in Figure 8.28(S2). When a large amount of the La and Mn metal ions were used, the agglomerates or small particles were observed, with one example shown in Figure 8.28(S3). The porous structures were gradually filled with $LaMnO_3$ nanoparticles as the concentration of the $LaMnO_3$ was increased. The α-Al_2O_3 porous network was almost invisible when the molar ratio of α-Al_2O_3 and $LaMnO_3$ was 1:1. In this case, the particles were observed to be nearly spherical in shape with a narrow diameter distribution and an average particle size of about 100 nm. However, aggregation became severe

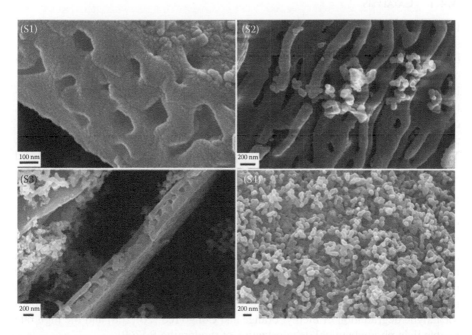

FIGURE 8.28 SEM images of x α-Al_2O_3/(1 − x) $LaMnO_3$ powders with $x = 1$(S1), $x = 0.9$(S2), $x = 0.7$(S3) and $x = 0.5$(S4). (Adapted from S. F. Wang et al., *Nanoscience and Nanotechnology Letters* 6, 2014, 758–771; S. F. Wang et al., *Advanced Materials Research* 1004–1005, 2014, 103–109.)

when the concentration of the LaMnO$_3$ was increased as shown in Figure 8.28(S4), in which the formed α-Al$_2$O$_3$/LaMnO$_3$ composites exhibited a weak ferromagnetism at room temperature [125].

8.4 APPLICATIONS

The polyacrylamide gel method is a promising route of preparing magnetic nanoparticles, and the key reason for using various chelating agents, organic additives, or sintering temperatures is that the corresponding crystal size may be adjusted under certain reaction conditions and through the selection of chelating agents or organic additives. The as-prepared magnetic nanoparticles have attracted extensive attention because of their uniform particle sizes and novel physicochemical properties. One attractive feature is their environment friendly nature as compared to bulk oxide materials. For example, the perovskite-type photocatalysts such as BiFeO$_3$, TbFeO$_3$, Bi$_2$Fe$_4$O$_9$, YMnO$_3$, YMn$_2$O$_5$, and LaSrMnO$_3$ have been prepared by the polyacrylamide gel method and showed outstanding photocatalytic properties. Another attractive application is its application to refrigeration and colored pigments. For instance, lanthanum manganese perovskite oxides La$_{1-x}$Ca$_x$MnO$_3$ and La$_{1-x}$Sr$_x$MnO$_3$ and blue pigment CoAl$_2$O$_4$ have been prepared by the polyacrylamide gel method and have exhibited magnetocaloric properties and coloring properties, respectively.

8.4.1 CATALYSIS

Photocatalysts are outstanding materials that provide a simple way for the conversion of light energy in oxidation and reduction processes [263]. Perovskite-type photocatalysis is a very promising recent green technique for photocatalytic degradation of organic pollutants. Perovskite-type photocatalysts for the degradation of contaminants in water are largely related to the photocatalysts' physicochemical properties, including particle size, morphology, and other surface properties. Therefore, developing a novel preparation method for perovskite-type photocatalysts is critical. Perovskite-type photocatalysts including BiFeO$_3$, TbFeO$_3$, Bi$_2$Fe$_4$O$_9$, YMnO$_3$, YMn$_2$O$_5$, and LaSrMnO$_3$ prepared by the polyacrylamide gel route will be discussed in detail below.

8.4.1.1 BiFeO$_3$, BiFeO$_3$-Graphene, TbFeO$_3$, Bi$_2$Fe$_4$O$_9$ Photocatalysts

8.4.1.1.1 BiFeO$_3$

Bismuth ferrite (BiFeO$_3$) is an important perovskite-type photocatalytic material and is simultaneously a chemically and thermally stable multiferroic material. Its multiferroic properties make it useful in the emerging field of spintronics, magnetic recording media, sensors, and high-density data storage [264–266]. BiFeO$_3$ possesses special photocatalytic properties for many applications, including the photocatalytic degradation of methyl orange and Congo red [267,268] and the high rate visible light photochemical decolorization of rhodamine B [269].

Methyl orange is a well-known azo dye and its structure is shown in Figure 8.29a. It consists of two benzene rings connected by an azo group, in which one of the rings contains a dimethyl amine and the other contains a sulfonic acid group [263].

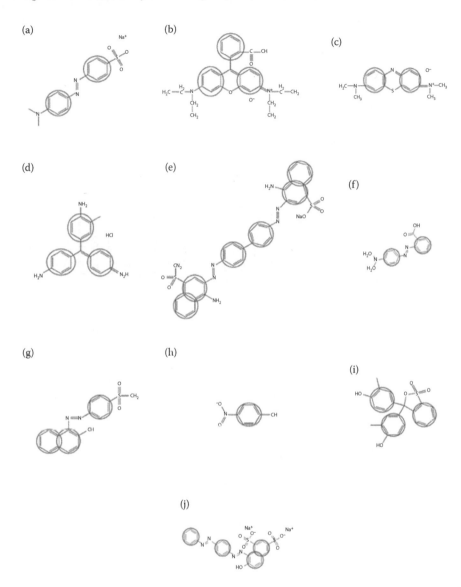

FIGURE 8.29 Structure of (a) methyl orange, (b) rhodamine B, (c) methyl blue, (d) acid fuchsine, (e) Congo red, (f) methyl red, (g) acid orange, (h) 4-nitrophenol, (i) cresol red, and (j) acid brilliant scarlet GR dye.

In addition, the structures of other dyes such as rhodamine B, methyl blue, acid fuchsine, Congo red, methyl red, acid orange, 4-nitrophenol, cresol red, and acid brilliant scarlet GR are also shown in Figure 8.29b–j. The degradation processes of methyl orange, rhodamine B, methyl blue, acid fuchsine, Congo red, methyl red, acid orange, 4-nitrophenol, cresol red, or acid brilliant scarlet GR dyes have been widely studied to determine photocatalytic activity of perovskite-type photocatalytic materials, and the results are summarized in Table 8.1.

TABLE 8.1

Photocatalyst Reacted with Methyl Orange, Rhodamine B, Methyl Blue, Acid Fuchsine, Congo Red, Methyl Red, Acid Orange, 4-Nitrophenol, Cresol Red, and Acid Brilliant Scarlet GR Dye

Photocatalyst	Dye	Dye (mg/L)	Catalyst (g/L)	Irradiation Time (h)	Irradiation Source	Degradation (%)	References
BiFeO$_3$	Methyl orange	10	2.5	6	UV irradiation	71	[270]
	Methyl orange	10	2.5	14	Visible irradiation	39	
BiFeO$_3$–graphene	Methyl orange	10	2.5	6	Visible irradiation	About 55	[271]
TbFeO$_3$	Methyl orange	10	1	6	Visible irradiation	36	[133,308]
	Rhodamine B methyl blue	10	1	6	Visible irradiation	41	
		10	1	6	Visible irradiation	58	
	Acid fuchsine Congo red	10	1	6	Visible irradiation	69	
		10	1	6	Visible irradiation	72	
Bi$_2$Fe$_4$O$_9$	Methyl red	25	1.25	6	UV irradiation	77	[181,273]
	Methyl red	25	1.25	6	Visible irradiation	40	[274]
	Methyl blue	5	0.5	6	Visible irradiation	68	
	Methyl orange	10	1.5	6	Visible irradiation	50	
Bi$_2$Fe$_4$O$_9$–graphene	Methyl orange	10	1.5	6	Visible irradiation	52	[274]
YMnO$_3$ 85 nm	Methyl red	15	1.2	6	UV irradiation	56	[112,309]
85 nm	Methyl red	15	1.2	16	Visible irradiation	51	
57 nm	Methyl red	15	1.2	6	UV irradiation	60	
57 nm	Methyl red	15	1.2	16	Visible irradiation	56	
45 nm	Methyl red	15	1.2	6	UV irradiation	62	
45 nm	Methyl red	15	1.2	16	Visible irradiation	57	

(Continued)

TABLE 8.1 (*Continued*)

Photocatalyst Reacted with Methyl Orange, Rhodamine B, Methyl Blue, Acid Fuchsine, Congo Red, Methyl Red, Acid Orange, 4-Nitrophenol, Cresol Red, and Acid Brilliant Scarlet GR Dye

Photocatalyst	Dye	Dye (mg/L)	Catalyst (g/L)	Irradiation Time (h)	Irradiation Source	Degradation (%)	References
$La_{0.7}Sr_{0.3}MnO_3$	Methyl red	5	0.5	6	UV irradiation	42	[275]
$La_{0.7}Sr_{0.3}MnO_3$-graphene	Methyl red	5	0.5	6	UV irradiation	53	[276]
YMn_2O_5	Methyl red	15	1.2	6	UV irradiation	57	[277,278]
	Methyl red	15	1.2	10	Visible irradiation	51	
$SrTiO_3$	Congo red	10	1	10	UV irradiation	90	[132]
	Rhodamine B	10	1	10	UV irradiation	37	
	Methyl orange	10	1	10	UV irradiation	30	
	Methyl blue	10	1	10	UV irradiation	50	
$SrTiO_3$-graphene	Acid orange	5	0.5	6	UV irradiation	88	[279]
$BaTiO_3$	Methyl red	10	1	2	UV irradiation	90	[280,281]
ZnO	Methylene blue	1.2	3 g	2	250 W high-pressure mercury lamp 125	71	[310]
	Methylene blue	10	0.15	3		85	[282]
Ag/ZnO Ag:Zn = 0.1:100	4-Nitrophenol	50 mL/150 mL	0.1 g	3	315 nm	About 98	[151]
TiO_2	Methyl orange	10	5	1	UV irradiation	38.5	[283–285]
	Cresol red	100 mL	0.2 g	2	UV irradiation	94.7	
	Acid orange 7	5	0.5	2	UV irradiation	91.5	
SnO_2/TiO_2	Acid brilliant scarlet GR	40	5	1	UV irradiation	95	[286,287]
$BiVO_4$	Methyl blue	0.5 g/L	0.1	5	200 W xenon light source	80	[288]

As can be seen in Table 8.1, methyl orange can be degraded by $BiFeO_3$ nanoparticles using both UV and visible light irradiations. In most cases, $BiFeO_3$ nanoparticles alone showed a lower degradation percentage, but could be remarkably enhanced when combined with other materials, including graphene. For example, $BiFeO_3$ nanoparticles (52 nm) showed a low effective degradation efficiency of MO dye (39%) [270], but when irradiated with visible light and combined with 5% graphene, degradation efficiency as high as 55% could be achieved [271].

8.4.1.1.2 $TbFeO_3$

$TbFeO_3$ is also an important perovskite-type photocatalytic material. Methyl orange, rhodamine B, methyl blue, acid fuchsine, and Congo red can be degraded by $TbFeO_3$ nanoparticles using visible light irradiation. Table 8.1 summarizes the results of studies of the degradation of above five dyes using $TbFeO_3$ photocatalysts. For example, $TbFeO_3$ prepared as nanospheres has been shown to degrade 37% of the methyl orange under visible light after 6 hours [133]. Similarly, nearly 72% of the rhodamine B was degraded by using $TbFeO_3$ nanoparticles with a light wavelength of 450 nm [133], proving that rhodamine B can be highly degraded under visible light by using $TbFeO_3$ nanoparticles.

8.4.1.1.3 $Bi_2Fe_4O_9$

$Bi_2Fe_4O_9$ is an important multiferroic material and exhibits excellent multiferroic properties, high catalytic oxidation activity, and high gas sensitivity [218,219,272]. It is noted that nanostructured $Bi_2Fe_4O_9$ exhibits high photocatalytic efficiency for the photodegradation of methyl red under visible light irradiation for 6 hours [181]. The degradation of methylene blue has also been investigated by photo-reducing the $Bi_2Fe_4O_9$ photocatalyst prior to adding it into the dye [273]. Xian et al. [273] explained the photocatalytic mechanism of the $Bi_2Fe_4O_9$ photocatalyst using the conventional semiconductor energy band theory. They noticed that hydroxyl radicals and photogenerated holes played important roles in the degradation of the dye.

$Bi_2Fe_4O_9$ photocatalyst or its composites could efficiently degrade methyl orange [274]. For example, the nanostructured $Bi_2Fe_4O_9$, which is of approximately 200 nm particle size, degraded 50% of the methyl orange [274]. However, when 2% (wt.) graphene was added into the dye, the degradation rate was increased to 59%. Adding 6% graphene increased the methyl orange degradation efficiency up to 72%, however, on further adding graphene up to 8% and 10%, the degradation efficiency dropped to 60% and 64%, respectively [274].

8.4.1.2 $YMnO_3$, YMn_2O_5, $LaSrMnO_3$ and $LaSrMnO_3$-Graphene Photocatalysts

Methyl red is one of the common azo dyes and its structure is shown in Figure 8.29f. Methyl red can be highly degraded by hexagonal $YMnO_3$ when in combination with minor orthorhombic $YMnO_3$ [112]. The particle size of the $YMnO_3$ nanoparticles is an important factor for the photodegradation of methyl red. Generally, smaller particles have a larger surface area and thus provide more reaction active sites, and then result in a higher photocatalytic activity. For $YMnO_3$ nanoparticles with particle sizes of about 85 nm, only 51% of the methyl red dye was found to be degraded.

However, by decreasing particle size to an average value of 57 nm, the degradation rate was increased to 56%. The photocatalytic activity can be further enhanced with a degradation of 57% if the particle size is furthered decreased up to 45 nm.

Similarly to $YMnO_3$ nanoparticles, $LaSrMnO_3$ nanoparticles can also efficiently degrade methyl red dye [275]. For instance, $LaSrMnO_3$ nanoparticles prepared by citric acid as a chelating agent has been shown to degrade 37% of the methyl red dye under visible light after 6 hours. $LaSrMnO_3$-graphene nanocomposites also played an important role in the degradation of methyl red dye [276]. When 10% of graphene was added to the dye, the degradation reached 53% [276].

Combining orthorhombic YMn_2O_5 photocatalysts with hexagonal YMn_2O_5 is commonly used to degrade azo dyes [277,278]. As can also be seen in Table 8.1, methyl red can be degraded by YMn_2O_5 photocatalysts using both UV and visible light irradiation sources. YMn_2O_5 photocatalysts prepared using the citric acid and EDTA as chelating agents could degrade 51% and 48% of the methyl red dye under visible light after 10 hours [278].

8.4.1.3 Other Photocatalysts

Table 8.1 also lists other photocatalysts that have been prepared using the polyacrylamide gel route. Congo red, rhodamine B, methyl orange, acid orange, and methyl blue can be degraded using $SrTiO_3$ nanoparticles using UV irradiation sources [132]. The combination of $SrTiO_3$ photocatalysts with graphene is commonly applied to degrade acid orange [279]. When 7.5% graphene was added to the acid orange dye solution using the $SrTiO_3$ photocatalysts, the degradation reached to 88% [279]. In addition, the photocatalytic activity of $BaTiO_3$ [280,281], ZnO [282], Ag/ZnO [151], TiO_2 [283–285], SnO_2/TiO_2 [286,287], and $BiVO_4$ [288] have all been studied. Similarly to $BiFeO_3$, $Bi_2Fe_4O_9$, $LaSrMnO_3$, and $SrTiO_3$ photocatalysts, 4-nitrophenol and acid brilliant scarlet GR dyes were degraded efficiently using ZnO and TiO_2 nanoparticles when in combination with other photocatalysts, respectively [151,287].

8.4.2 Magnetic Refrigeration Applications

Refrigeration technology was developed based on observations by Peltier over 100 years ago [289]. Newly developed refrigeration technology has attracted wide attention because of its potential applications in energy savings, space cooling, food refrigeration, and the environmental field. Refrigeration applications could be classified into ten categories according to the fundamental physicochemical properties of materials, that is, thermoacoustic refrigeration, thermotunneling, thermoelectric refrigeration, Malone cycle refrigeration, Stirling cycle refrigeration, magnetic refrigeration, pulse tube refrigeration, compressor driven metal hydride heat pumps, absorption refrigeration, and adsorption refrigeration [289,290].

Among these applications, magnetic refrigeration is a predominant technology for the refrigerator, and exhibits a magnetocaloric effect at extremely low temperature. The magnetocaloric effect was first reported by Weiss and Piccard in 1917 [291]. In 1933, sustained efforts of several research groups were directed toward the design of magnetic refrigerators.

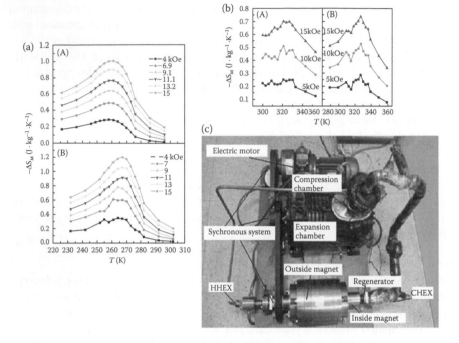

FIGURE 8.30 (a) Temperature dependence of $-\Delta_{SM}$ at different applied magnetic fields for samples prepared by using (A) acetic acid and (B) EDTA as chelating agent; (b) Temperature dependence of magnetic entropy change $-\Delta_{SM}$ at fields of 5, 10 and 15 kOe for samples by using (A) citric acid and (B) EDTA as chelating agent S1; (c) Photograph of the refrigerator. (Adapted from Y. F. Wang and H. Yang, *Journal of Superconductivity and Novel Magnetism* 26, 2013, 3463–3467; H. Yang et al., *Journal of Alloys and Compounds* 555, 2013, 150–155; X. N. He et al., *International Journal of Refrigeration* 36, 2013, 1465–1471.)

Recently, many materials such as $Gd_5Si_2Ge_2$ [292], FeRh, $MnP_{1-x}As_x$, Cr_3Te_4, $La(Fe,Co)_{11.83}Al_{1.17}$ [293], $LaFe_{11.4}Si_{1.6}$ [294], and Heusler alloys Ni–Mn–Ga [295,296] have been studied with respect to their magnetocaloric properties. In addition, the lanthanum manganese perovskite oxides $La_{1-x}Ca_xMnO_3$ and $La_{1-x}Sr_xMnO_3$ have also been investigated [130,179]. The $La_{1-x}Ca_xMnO_3$ and $La_{1-x}Sr_xMnO_3$ synthesized using the polyacrylamide gel route showed significant magnetic entropy changes with increasing the applied magnetic field [130,179], as shown in Figure 8.30a and b. Figure 8.30c shows a photograph of the hybrid refrigerator with several major parts such as a concentric Halbach cylinder permanent magnet, a stainless steel tube, one cold heat exchanger and one hot heat exchanger, two pistons and cylinders, and an electric motor [297].

8.4.3 Other Applications

8.4.3.1 Colored Pigments

Multifunction ceramic pigments known as thermoresistant materials are colored inorganic multiple compounds, which can keep their coloring properties after calcination at high temperatures up to 1500°C. $CoAl_2O_4$ is an important blue pigment. It is widely used for paint, decorating porcelain, coloration of plastics, and other ceramic products because of its excellent thermal and chemical stability [298]. Torkian et al. [299] synthesized $CoAl_2O_4$ nanoparticles using the combustion solution method with different ratios of β-alanine as a new mixed fuel. Figure 8.31a shows the photograph of blue pigment $CoAl_2O_4$ prepared using different β-alanine: (A) 0, (B) 5.5, (C) 35.6, and (D) 54.7 wt.% and the presence of the blue color (the reader is referred to reference [299] of this color figure) can be observed in the four samples,

FIGURE 8.31 (a) Photograph of pigment $CoAl_2O_4$ prepared with different β-alanine: (A) 0, (B) 5.5, (C) 35.6 and (D) 54.7 wt.%. (b) An application of pigment. (c) (A) UV–vis absorption spectra and (B) NIR reflectance spectra of $YMnO_3$ samples sintered at different temperatures. (d) Images of $CaZr_{1-x}Cr_xO_3$ (x = 0, 0.01, 0.03, 0.05 and 0.07) pigment. Gray version is black, the reader is referred to reference [299] and [305] of this figure. (Adapted from A. Han et al., *Dyes and Pigments* 99, 2013, 527–530; L. Torkian and M. Daghighi, *Advanced Powder Technology* 25, 2014, 739–744; A. Han et al., *Solar Energy* 91, 2013, 32–36; R. Ianoş and R. Lazău, *Dyes and Pigments* 105, 2014, 152–156; L. F. Vieira Ferreira et al., *Spectrochimica Acta Part A: Molecular and Biomolecular Spectroscopy* 104, 2013, 437–444.)

particularly for samples (B) and (C). Nanostructured $CoAl_2O_4$ was synthesized by Jafari et al. [300] using a polyacrylamide gel method. They noticed that the color and size of the samples were highly dependent on the sintering temperature. The blue pigments have been widely used in the pottery field as shown in Figure 8.31b. In addition, nanocrystalline Ba^{2+} added blue pigment $CoAl_2O_4$ was prepared using the polyacrylamide gel method [301]. The results showed that crystal sizes increased with increase of the Ba^{2+} content, whereas the absorption band at 400–500 nm was decreased and the color of the powder turned into light blue [301].

Besides $CoAl_2O_4$ blue pigment, the novel pale blue pigment $YMnO_3$ has also been prepared using the polyacrylamide gel route [302]. Figure 8.31c shows the (A) UV–vis absorption spectra and (B) NIR reflectance spectra of $YMnO_3$ samples sintered at 700°C, 800°C, 900°C, and 1000 °C. As shown in Figure 8.31cA, the UV–vis absorption spectra is consistent with the literature of $YMnO_3$ nanoparticles [112]. The NIR reflectance spectra confirmed the coloring properties of $YMnO_3$. Han et al. [101,303,304] reported the thermal stability of colored cool pigments of Fe doped $YMnO_3$, Fe doped $Y_2Ce_2O_7$, and Fe doped $La_2Mo_2O_7$ compounds prepared using the polyacrylamide gel route. A similar method has been used to prepare $CaZrO_3$-based electrolytes [305]. When a certain amount of Cr^{4+} ions were introduced into $CaZrO_3$ crystal to substitution Zr^{4+} ions, a potential red shade pigment was obtained [306]. The images of $CaZr_{1-x}Cr_xO_3$ ($x = 0$, 0.01, 0.03, 0.05, and 0.07) red shade pigment are shown in Figure 8.31d.

8.5   SUMMARY AND FUTURE WORK

The polyacrylamide gel technique is a fast, reproducible, cheap, and easy to scale-up method for obtaining highly disperse nanopowders and nanocomposites. This chapter covers the development, influencing factors, reaction mechanisms, and preparation techniques for the magnetic materials prepared using polyacrylamide gel technology. A few influencing factors for the preparation of nanopowders and nanocomposites have been discussed in detail, including monomer systems, organic additives, pH values, chelating agents and water, glucose, initiator, drying temperatures and duration, sintering temperatures and metal source precursors, crystallization temperatures and duration, and heating rate in the crystallization region. Based on the appropriate selection of chelating agents, monomer systems, and initiators, and a good understanding of the unique physicochemical properties of magnetic nanoparticles such as $BiFeO_3$, $TbFeO_3$, $Bi_2Fe_4O_9$, $YMnO_3$, $TbMnO_3$, $La_{1-x}Sr_xMnO_3$, $La_{1-x}Ca_xMnO_3$, $La_{1-x}Na_xMnO_3$, and YMn_2O_5 and magnetic nanocomposites, core–shell structures such as $La_{0.67}Ca_{0.33}MnO_3/MgO$ and $CoFe_2O_4/BaTiO_3$ or matrix–nanoparticle/nanorods/nanowires/nanowhiskers nanocomposites can be assembled and synthesized, which are promising for catalysis, colored pigments, magnetic refrigeration, magneto-optical and magneto-electronic applications. Although magnetic materials prepared using the polyacrylamide gel route have been widely investigated recently, many fundamental and critical issues, such as optical, electrical, thermal, and magnetic properties, as well as the formation mechanisms of magnetic nanocomposites, are still unsolved, and further advances in this field include synthesizing high-purity multiple complex oxides and improving their physicochemical properties, as well as exploration of their industrial applications.

ACKNOWLEDGMENTS

The authors are highly indebted to all their coworkers, who over the years have contributed to the field and whose names are found in the relevant references below. This work was financially supported by the Joint Fund of the National Natural Science Foundation of China and the China Academy of Engineering Physics (Grant No. 11076008) and by the Outstanding Doctoral Student Support Plan (A1098524023901001074), as well as the Royal Academy of Engineering Research Exchange with China and India.

REFERENCES

1. M. Faraji, Y. Yamini, and M. Rezaee, Magnetic nanoparticles: Synthesis, stabilization, functionalization, characterization, and applications, *Journal of the Iranian Chemical Society* 7, 2010, 1–37.
2. W. P. Wolf, Ferrimagnetism, *Reports on Progress in Physics* 24, 1961, 212.
3. D. M. Pozar and S. D. Targonski, Improved coupling for aperture coupled microstrip antennas, *Electronics Letters* 27, 1991, 1129–1131.
4. Y. Wang, Y. Yu, D. Li, K. T. Bae, J. J. Brown, W. Lin, and E. M. Haacke, Artery and vein separation using susceptibility-dependent phase in contrast-enhanced MRA, *Journal of Magnetic Resonance Imaging* 12, 2000, 661–670.
5. T. Pedro, M. María del Puerto, V. V. Sabino, G. C. Teresita, and J. S. Carlos, The preparation of magnetic nanoparticles for applications in biomedicine, *Journal of Physics D: Applied Physics* 36, 2003, R182–R197.
6. S. Gai, P. Yang, C. Li, W. Wang, Y. Dai, N. Niu, and J. Lin, Synthesis of magnetic, up-conversion luminescent, and mesoporous core–shell-structured nanocomposites as drug carriers, *Advanced Functional Materials* 20, 2010, 1166–1172.
7. H. Kim, J. Ahn, S. Haam, Y. Shul, S. Song, and T. Tatsumi, Synthesis and characterization of mesoporous Fe/SiO_2 for magnetic drug targeting, *Journal of Materials Chemistry* 16, 2006, 1617–1621.
8. P. Wang, Q. Shi, Y. Shi, K. K. Clark, G. D. Stucky, and A. A. Keller, Magnetic permanently confined micelle arrays for treating hydrophobic organic compound contamination, *Journal of the American Chemical Society* 131, 2008, 182–188.
9. J. Kim, J. Lee, H. B. Na, B. C. Kim, J. K. Youn, J. H. Kwak et al. A magnetically separable, highly stable enzyme system based on nanocomposites of enzymes and magnetic nanoparticles shipped in hierarchically ordered, mesocellular, mesoporous silica, *Small* 1, 2005, 1203–1207.
10. Z. Beji, A. Hanini, L. S. Smiri, J. Gavard, K. Kacem, F. Villain, J. M. Grenèche, F. Chau, and S. Ammar, Magnetic properties of Zn-substituted $MnFe_2O_4$ nanoparticles synthesized in polyol as potential heating agents for hyperthermia. Evaluation of their toxicity on endothelial cells, *Chemistry of Materials* 22, 2010, 5420–5429.
11. F. Q. Hu, L. Wei, Z. Zhou, Y. L. Ran, Z. Li, and M. Y. Gao, Preparation of biocompatible magnetite nanocrystals for in vivo magnetic resonance detection of cancer, *Advanced Materials* 18, 2006, 2553–2556.
12. A. Lu, E. L. Salabas, and F. Schüth, Magnetic nanoparticles: Synthesis, protection, functionalization, and application, *Angewandte Chemie International Edition* 46, 2007, 1222–1244.

13. L. H. Reddy, J. L. Arias, J. Nicolas, and P. Couvreur, Magnetic nanoparticles: Design and characterization, toxicity and biocompatibility, pharmaceutical and biomedical applications, *Chemical Reviews* 112, 2012, 5818–5878.

14. R. Massart, Preparation of aqueous magnetic liquids in alkaline and acidic media, *IEEE Trans. Magn.* 17, 1981, 1247–1248.

15. G. M. da Costa, E. De Grave, P. M. A. de Bakker, and R. E. Vandenberghe, Synthesis and characterization of some iron oxides by sol-gel method, *Journal of Solid State Chemistry* 113, 1994, 405–412.

16. F. Chen, Q. Gao, G. Hong, and J. Ni, Synthesis and characterization of magnetite dodecahedron nanostructure by hydrothermal method, *Journal of Magnetism and Magnetic Materials* 320, 2008, 1775–1780.

17. Y. Amemiya, A. Arakaki, S. S. Staniland, T. Tanaka, and T. Matsunaga, Controlled formation of magnetite crystal by partial oxidation of ferrous hydroxide in the presence of recombinant magnetotactic bacterial protein Mms6, *Biomaterials* 28, 2007, 5381–5389.

18. L. Cabrera, S. Gutierrez, N. Menendes, M. P. Morales, and P. Herrasti, Magnetite nanoparticles: Electrochemical synthesis and characterization, *Electrochemistry Acta* 53, 2008, 3436.

19. N. Enomoto, J. Akagi, and Z. Nakagawa, Sonochemical powder processing of iron hydroxides, *Ultrasonics Sonochemistry* 3, 1996, S97–S103.

20. S. Wang, C. Zhang, G. Sun, B. Chen, W. Liu, X. Xiang et al. Effect of carbon and sintering temperature on the structural and magnetic properties of $SrFe_{12}O_{19}$ nanoparticles, *Journal of Sol-Gel Science and Technology*, 73, 2015, 371–378.

21. G. R. Gordani, A. Ghasemi, and A. Saidi, Enhanced magnetic properties of substituted Sr-hexaferrite nanoparticles synthesized by co-precipitation method, *Ceramics International* 40, 2014, 4945–4952.

22. S. Joshi, M. Kumar, S. Chhoker, G. Srivastava, M. Jewariya, and V. N. Singh, Structural, magnetic, dielectric and optical properties of nickel ferrite nanoparticles synthesized by co-precipitation method, *Journal of Molecular Structure* 1076, 2014, 55–62.

23. M. Vadivel, R. Ramesh Babu, K. Sethuraman, K. Ramamurthi, and M. Arivanandhan, Synthesis, structural, dielectric, magnetic and optical properties of Cr substituted $CoFe_2O_4$ nanoparticles by co-precipitation method, *Journal of Magnetism and Magnetic Materials* 362, 2014, 122–129.

24. I. Sharifi, H. Shokrollahi, M. M. Doroodmand, and R. Safi, Magnetic and structural studies on $CoFe_2O_4$ nanoparticles synthesized by co-precipitation, normal micelles and reverse micelles methods, *Journal of Magnetism and Magnetic Materials* 324, 2012, 1854–1861.

25. H. Shokrollahi, Magnetic, electrical and structural characterization of $BiFeO_3$ nanoparticles synthesized by co-precipitation, *Powder Technology* 235, 2013, 953–958.

26. P. Sivakumar, R. Ramesh, A. Ramanand, S. Ponnusamy, and C. Muthamizhchelvan, Preparation and properties of nickel ferrite ($NiFe_2O_4$) nanoparticles via sol–gel auto-combustion method, *Materials Research Bulletin* 46, 2011, 2204–2207.

27. C. Yang, C. Z. Liu, C. M. Wang, W. G. Zhang, and J. S. Jiang, Magnetic and dielectric properties of alkaline earth Ca^{2+} and Ba^{2+} ions co-doped $BiFeO_3$ nanoparticles, *Journal of Magnetism and Magnetic Materials* 324, 2012, 1483–1487.

28. K. Agilandeswari and A. Ruban Kumar, Synthesis, characterization, temperature dependent electrical and magnetic properties of $Ca_3Co_4O_9$ by a starch assisted sol–gel combustion method, *Journal of Magnetism and Magnetic Materials* 364, 2014, 117–124.

29. A. Rostamnejadi, H. Salamati, P. Kameli, and H. Ahmadvand, Superparamagnetic behavior of $La_{0.67}Sr_{0.33}MnO_3$ nanoparticles prepared via sol–gel method, *Journal of Magnetism and Magnetic Materials* 321, 2009, 3126–3131.

30. P. Tang, D. Kuang, S. Yang, and Y. Zhang, The structural, optical and enhanced magnetic properties of $Bi_{1-x}Gd_xFe_{1-y}Mn_yO_3$ nanoparticles synthesized by sol–gel, *Journal of Alloys and Compounds* 622, 2015, 194–199.

31. E. Oumezzine, S. Hcini, E.-K. Hlil, E. Dhahri, and M. Oumezzine, Effect of Ni-doping on structural, magnetic and magnetocaloric properties of $La_{0.6}Pr_{0.1}Ba_{0.3}Mn_{1-x}Ni_xO_3$ nanocrystalline manganites synthesized by Pechini sol–gel method, *Journal of Alloys and Compounds* 615, 2014, 553–560.

32. L. Q. Xu, L. Y. Chen, H. F. Huang, R. Xie, W. B. Xia, J. Wei, W. Zhong, S. L. Tang, and Y. W. Du, A novel sol–gel process to facilely synthesize Ni_3Fe nanoalloy nanoparticles supported with carbon and silica, *Journal of Alloys and Compounds* 593, 2014, 93–96.

33. F. Ozel and H. Kockar, Growth and characterizations of magnetic nanoparticles under hydrothermal conditions: Reaction time and temperature, *Journal of Magnetism and Magnetic Materials* 373, 2015, 213–216.

34. N. Li, M. Zheng, X. Chang, G. Ji, H. Lu, L. Xue, L. Pan, and J. Cao, Preparation of magnetic $CoFe_2O_4$-functionalized graphene sheets via a facile hydrothermal method and their adsorption properties, *Journal of Solid State Chemistry* 184, 2011, 953–958.

35. R. Dhinesh Kumar and R. Jayavel, Low temperature hydrothermal synthesis and magnetic studies of $YMnO_3$ nanorods, *Materials Letters* 113, 2013, 210–213.

36. W. Zhang, J. Chen, Y. Wang, L. Fan, J. Deng, R. Yu, and X. Xing, Morphology evolution and physical properties of $Bi_2Mn_4O_{10}$ synthesized by hydrothermal method, *Journal of Crystal Growth* 380, 2013, 1–4.

37. S. Wu and Y. Mei, Hydrothermal synthesis of $DyMn_2O_5$ nanorods and their magnetic properties, *Journal of Alloys and Compounds* 583, 2014, 309–312.

38. A. Bouremana, A. Guittoum, M. Hemmous, B. Rahal, J. J. Sunol, D. Martínez-Blanco, J. A. Blanco, P. Gorria, and N. Benrekaa, Crystal structure, microstructure and magnetic properties of Ni nanoparticles elaborated by hydrothermal route, *Journal of Magnetism and Magnetic Materials* 358–359, 2014, 11–15.

39. R. Justin Joseyphus, A. Narayanasamy, K. Shinoda, B. Jeyadevan, and K. Tohji, Synthesis and magnetic properties of the size-controlled Mn–Zn ferrite nanoparticles by oxidation method, *Journal of Physics and Chemistry of Solids* 67, 2006, 1510–1517.

40. A. Bee, R. Massart, and S. Neveu, Synthesis of very fine maghemite particles, *Journal of Magnetism and Magnetic Materials* 149, 1995, 6–9.

41. H. Yin, Y. Zhou, X. Meng, T. Tang, S. Ai, and L. Zhu, Electrochemical behaviour of Sudan I at Fe_3O_4 nanoparticles modified glassy carbon electrode and its determination in food samples, *Food Chemistry* 127, 2011, 1348–1353.

42. D. Ramimoghadam, S. Bagheri, and S. B. A. Hamid, Progress in electrochemical synthesis of magnetic iron oxide nanoparticles, *Journal of Magnetism and Magnetic Materials* 368, 2014, 207–229.

43. R. Galindo, E. Mazario, S. Gutiérrez, M. P. Morales, and P. Herrasti, Electrochemical synthesis of $NiFe_2O_4$ nanoparticles: Characterization and their catalytic applications, *Journal of Alloys and Compounds* 536, 2012, S241–S244.

44. C. Pascal, J. L. Pascal, F. Favier, M. L. Elidrissi Moubtassim, and C. Payen, Electrochemical synthesis for the control of γ-Fe_2O_3 nanoparticle size. *Morphology, Microstructure, and Magnetic Behavior, Chemistry of Materials* 11, 1998, 141–147.

45. G. Nabiyouni, D. Ghanbari, A. Yousofnejad, and M. Seraj, A sonochemical-assisted method for synthesis of $BaFe_{12}O_{19}$ nanoparticles and hard magnetic nanocomposites, *Journal of Industrial and Engineering Chemistry* 20, 2014, 3425–3429.

46. D. Ghanbari, M. Salavati-Niasari, and M. Ghasemi-Kooch, A sonochemical method for synthesis of Fe_3O_4 nanoparticles and thermal stable PVA-based magnetic nanocomposite, *Journal of Industrial and Engineering Chemistry* 20, 2014, 3970–3974.

47. E. Hee Kim, H. Sook Lee, B. Kook Kwak, and B.-K. Kim, Synthesis of ferrofluid with magnetic nanoparticles by sonochemical method for MRI contrast agent, *Journal of Magnetism and Magnetic Materials* 289, 2005, 328–330.

48. J. Saffari, D. Ghanbari, N. Mir, and K. Khandan-Barani, Sonochemical synthesis of $CoFe_2O_4$ nanoparticles and their application in magnetic polystyrene nanocomposites, *Journal of Industrial and Engineering Chemistry* 20, 2014, 4119–4123.

49. M. Abbas, B. Parvatheeswara Rao, M. Nazrul Islam, K. W. Kim, S. M. Naga, M. Takahashi, and C. Kim, Size-controlled high magnetization $CoFe_2O_4$ nanospheres and nanocubes using rapid one-pot sonochemical technique, *Ceramics International* 40, 2014, 3269–3276.

50. P. Xu, X. Han, H. Zhao, Z. Liang, and J. Wang, Effect of stoichiometry on the phase formation and magnetic properties of $BaFe_{12}O_{19}$ nanoparticles by reverse micelle technique, *Materials Letters* 62, 2008, 1305–1308.

51. S. Thakur, S. C. Katyal, A. Gupta, V. R. Reddy, S. K. Sharma, M. Knobel, and M. Singh, Nickel–zinc ferrite from reverse micelle process: Structural and magnetic properties, mössbauer spectroscopy characterization, *The Journal of Physical Chemistry C* 113, 2009, 20785–20794.

52. S. Rana, J. Philip, and B. Raj, Micelle based synthesis of cobalt ferrite nanoparticles and its characterization using Fourier transform infrared transmission spectrometry and thermogravimetry, *Materials Chemistry and Physics* 124, 2010, 264–269.

53. M. Mandal, M. Shamsuzzoha, and D. E. Nikles, Micelles-mediated synthesis of magnetic Fe_3Pt nanoparticles of cubic morphology and their characterisation, *Journal of Magnetism and Magnetic Materials* 320, 2008, 630–633.

54. J. D. Hu, Y. Zevi, X. M. Kou, J. Xiao, X. J. Wang, and Y. Jin, Effect of dissolved organic matter on the stability of magnetite nanoparticles under different pH and ionic strength conditions, *Science of the Total Environment* 408, 2010, 3477–3489.

55. A. R. Petosa, D. P. Jaisi, I. R. Quevedo, M. Elimelech, and N. Tufenkji, Aggregation and deposition of engineered nanomaterials in aquatic environments: Role of physicochemical interactions, *Environmental Science & Technology* 44, 2010, 6532–6549.

56. S. H. Li, M. M. Lin, M. S. Toprak, D. K. Kim, and M. Muhammed, Nanocomposites of polymer and inorganic nanoparticles for optical and magnetic applications, *Nano Reviews* 1, 2010, 5214 (5211–5219).

57. W. Tianlong and M. K. Kannan, Cobalt-based magnetic nanocomposites: Fabrication, fundamentals and applications, *Journal of Physics D: Applied Physics* 44, 2011, 393001.

58. J. Liu, S. Z. Qiao, Q. H. Hu, and G. Q. Lu, Magnetic nanocomposites with mesoporous structures: Synthesis and applications, *Small* 7, 2011, 425–443.

59. J. Dong, Z. Xu, and S. M. Kuznicki, Magnetic multi-functional nano composites for environmental applications, *Advanced Functional Materials* 19, 2009, 1268–1275.

60. L. Su, J. Feng, X. Zhou, C. Ren, H. Li, and X. Chen, Colorimetric detection of urine glucose based $ZnFe_2O_4$ magnetic nanoparticles, *Analytical Chemistry* 84, 2012, 5753–5758.

61. S. Huang, P. Yang, Z. Cheng, C. Li, Y. Fan, D. Kong, and J. Lin, Synthesis and characterization of magnetic Fe_xO_y@SBA-15 composites with different morphologies for controlled drug release and targeting, *The Journal of Physical Chemistry C* 112, 2008, 7130–7137.

62. S. Huang, Y. Fan, Z. Cheng, D. Kong, P. Yang, Z. Quan, C. Zhang, and J. Lin, Magnetic mesoporous silica spheres for drug targeting and controlled release, *The Journal of Physical Chemistry C* 113, 2009, 1775–1784.

63. M. M. A. Rodrigues, A. R. Simioni, F. L. Primo, M. P. Siqueira-Moura, P. C. Morais, and A. C. Tedesco, Preparation, characterization and *in vitro* cytotoxicity of BSA-based nanospheres containing nanosized magnetic particles and/or photosensitizer, *Journal of Magnetism and Magnetic Materials* 321, 2009, 1600–1603.

64. H. Yang, Z. E. Cao, X. Shen, J. L. Jiang, Z. Q. Wei, J. F. Dai, and W. J. Feng, A polymer-network gel route to oxide composite nanoparticles with core/shell structure, *Materials Letters* 63, 2009, 655–657.

65. Z. Xu, Y. Hou, and S. Sun, Magnetic core/shell Fe_3O_4/Au and Fe_3O_4/Au/Ag nanoparticles with tunable plasmonic properties, *Journal of the American Chemical Society* 129, 2007, 8698–8699.

66. P. M. Ajayan, L. S. Schadler, and E. P. V. Braun, Polymer-based and polymer-filled nanocomposites, *Nanocomposite Science and Technology*, Wiley, New York, 2003.

67. C. Sanchez, B. Lebeau, F. Chaput, and J. P. Boilot, Optical properties of functional hybrid organic–inorganic nanocomposites, *Advanced Materials* 15, 2003, 1969–1994.

68. L. L. Beecroft and C. K. Ober, Nanocomposite materials for optical applications, *Chemistry of Materials* 9, 1997, 1302–1317.

69. G. Zhan and A. K. Mukherjee, Carbon nanotube reinforced alumina-based ceramics with novel mechanical, electrical, and thermal properties, *International Journal of Applied Ceramic Technology* 1, 2004, 161–171.

70. W. Zheng and S. Wong, Electrical conductivity and dielectric properties of PMMA/expanded graphite composites, *Composites Science and Technology* 63, 2003, 225–235.

71. K. M. Krishnan, Biomedical nanomagnetics: A spin through possibilities in imaging, diagnostics, and therapy, magnetics, *IEEE Transactions on Magnetics* 46, 2010, 2523–2558.

72. C. Chiang and C. M. Ma, Synthesis, characterization and thermal properties of novel epoxy containing silicon and phosphorus nanocomposites by sol–gel method, *European Polymer Journal* 38, 2002, 2219–2224.

73. D. H. Kim, N. M. Aimon, and C. A. Ross, Self-assembled growth and magnetic properties of a $BiFeO_3$-$MgFe_2O_4$ nanocomposite prepared by pulsed laser deposition, *Journal of Applied Physics* 113, 2013, 17B510.

74. W. Wang and J. Yao, Hydrothermal synthesis of SnO_2/Fe_3O_4 nanocomposites and their magnetic property, *The Journal of Physical Chemistry C* 113, 2009, 3070–3075.

75. Z. X. Yang, W. Zhong, C. T. Au, X. Du, H. A. Song, X. S. Qi, X. J. Ye, M. H. Xu, and Y. W. Du, Novel photoluminescence properties of magnetic Fe/ZnO composites: Self-assembled ZnO nanospikes on Fe nanoparticles fabricated by hydrothermal method, *The Journal of Physical Chemistry C* 113, 2009, 21269–21273.

76. J. Su, M. Cao, L. Ren, and C. Hu, Fe_3O_4–Graphene nanocomposites with improved lithium storage and magnetism properties, *The Journal of Physical Chemistry C* 115, 2011, 14469–14477.

77. J. Wang, H. Wang, J. Jiang, W. Gong, D. Li, Q. Zhang, X. Zhao, S. Ma, and Z. Zhang, Nonpolar solvothermal fabrication and electromagnetic properties of magnetic Fe_3O_4 encapsulated semimetal Bi nanocomposites, *Crystal Growth & Design* 12, 2012, 3499–3504.

78. Y. Wang, B. Li, L. Zhang, and H. Song, Multifunctional mesoporous nanocomposites with magnetic, optical, and sensing features: Synthesis, characterization, and their oxygen-sensing performance, *Langmuir* 29, 2013, 1273–1279.

79. Y. Xu, A. Karmakar, D. Wang, M. W. Mahmood, F. Watanabe, Y. Zhang et al. Multifunctional Fe_3O_4 cored magnetic-quantum dot fluorescent nanocomposites for RF nanohyperthermia of cancer cells, *The Journal of Physical Chemistry C* 114, 2010, 5020–5026.

80. R. K. Selvan, V. Krishnan, C. O. Augustin, H. Bertagnolli, C. S. Kim, and A. Gedanken, Investigations on the structural, morphological, electrical, and magnetic properties of $CuFe_2O_4$-NiO nanocomposites, *Chemistry of Materials* 20, 2008, 429–439.

81. J. Cao, W. Fu, H. Yang, Q. Yu, Y. Zhang, S. Liu et al. Large-scale synthesis and microwave absorption enhancement of actinomorphic tubular ZnO/$CoFe_2O_4$ nanocomposites, *The Journal of Physical Chemistry B* 113, 2009, 4642–4647.

82. C. M. Liu, L. M. Fang, X. T. Zu, and W. L. Zhou, The magnetism and photoluminescence of nickel-doped SnO_2 nano-powders, *Physica Scripta* 80, 2009, 065703.

83. C. M. Liu, X. T. Zu, and W. L. Zhou, Magnetic interaction in Co-doped SnO_2 nanocrystal powders, *Journal of Physics: Condensed Matter* 18, 2006, 6001.

84. P. Xu, X. Han, J. Jiang, X. Wang, X. Li, and A. Wen, Synthesis and characterization of novel coralloid polyaniline/$BaFe_{12}O_{19}$ nanocomposites, *The Journal of Physical
Chemistry C* 111, 2007, 12603–12608.

85. X. Liu, Z. Fang, X. Zhang, W. Zhang, X. Wei, and B. Geng, Preparation and characterization of Fe_3O_4/CdS nanocomposites and their use as recyclable photocatalysts,
Crystal Growth & Design 9, 2008, 197–202.

86. X. Liu, Q. Hu, X. Zhang, Z. Fang, and Q. Wang, Generalized and facile synthesis of
Fe_3O_4/MS (M = Zn, Cd, Hg, Pb, Co, and Ni) nanocomposites, *The Journal of Physical
Chemistry C* 112, 2008, 12728–12735.

87. M. Feyen, C. Weidenthaler, F. Schüth, and A.-H. Lu, Regioselectively controlled synthesis of colloidal mushroom nanostructures and their hollow derivatives, *Journal of
the American Chemical Society* 132, 2010, 6791–6799.

88. T. Valdés-Solís, A. F. Rebolledo, M. Sevilla, P. Valle-Vigón, O. Bomatí-Miguel,
A. B. Fuertes, and P. Tartaj, Preparation, characterization, and enzyme immobilization
capacities of superparamagnetic silica/iron oxide nanocomposites with mesostructured
porosity, *Chemistry of Materials* 21, 2009, 1806–1814.

89. X. Dong, H. Chen, W. Zhao, X. Li, and J. Shi, Synthesis and magnetic properties of
mesostructured γ-Fe_2O_3/carbon composites by a Co-casting method, *Chemistry of
Materials* 19, 2007, 3484–3490.

90. K. R. Lee, S. Kim, D. H. Kang, J. I. Lee, Y. J. Lee, W. S. Kim, D. Cho, H. B. Lim,
J. Kim, and N. H. Hur, Highly uniform superparamagnetic mesoporous spheres with
submicrometer scale and their uptake into cells, *Chemistry of Materials* 20, 2008,
6738–6742.

91. W. Lu, Y. Shen, A. Xie, X. Zhang, and W. Chang, Novel bifuncitonal one-dimensional
Fe_3O_4/Se nanocomposites via facile green synthesis, *The Journal of Physical Chemistry
C* 114, 2010, 4846–4851.

92. H. Yang, Z. E. Cao, X. Shen, T. Xian, W. J. Feng, J. L. Jiang, Y. C. Feng, Z. Q. Wei,
and J. F. Dai, Fabrication of 0–3 type manganite/insulator composites and manipulation of their magnetotransport properties, *Journal of Applied Physics* 106, 2009,
104317.

93. L. G. Teoh, K.-D. Li, and Y. H. Liu, Synthesis and characterization of Ni/NiO core/
shell nanoparticles prepared by surfactant-mediated method, *Nanoscience and
Nanotechnology Letters* 3, 2011, 798–804.

94. Y. Xie, Y. Ou, F. Ma, Q. Yang, X. Tan, and S. Xie, Synthesis of multiferroic $Pb(Zr_{0.52}Ti_{0.48})$
O_3/$CoFe_2O_4$ core/shell nanofibers by coaxial electrospinning, *Nanoscience and
Nanotechnology Letters* 5, 2013, 546–551.

95. Q. He, J. Liu, and R. Hu, Facile magnetosensitive catalyst fabrication of palladium/
platinum coated maghemite nanocomposites and characterization, *Nanoscience and
Nanotechnology Letters* 5, 2013, 995–1001.

96. T. Xian, H. Yang, X. Shen, J. L. Jiang, Z. Q. Wei, and W. J. Feng, Preparation of high-
quality $BiFeO_3$ nanopowders via a polyacrylamide gel route, *Journal of Alloys and
Compounds* 480, 2009, 889–892.

97. B. M. Berkovsky, V. F. Medvedev, and M. S. Krokov, *Magnetic Fluids: Engineering
Applications*, Oxford University Press, Oxford, 1993.

98. S. W. Charles and J. Popplewell, Properties and applications of magnetic liquids, in:
Hand Book of Magnetic Materials, K. H. J. Buschow (ed.), 2, 1986, p. 153, New York.

99. A. E. Merbach and E. Tóth, *The Chemistry of Contrast Agents in Medical Magnetic
Resonance Imaging*, Wiley, Chichester, UK, 2001.

100. I. Hilbert, W. Andra, R. Bahring, A. Daum, R. Hergt, and W. A. Kaiser, Evaluation of temperature increase with different amounts of magnetite in liver tissue samples, *Investigative Radiology* 32, 1997, 705–712.

101. A. Han, M. Ye, M. Zhao, J. Liao, and T. Wu, Crystal structure, chromatic and near-infrared reflective properties of iron doped $YMnO_3$ compounds as colored cool pigments, *Dyes and Pigments* 99, 2013, 527–530.

102. M. A. G. Soler, S. W. da Silva, V. K. Garg, A. C. Oliveira, R. B. Azevedo, A. C. M. Pimenta, E. C. D. Lima, and P. C. Morais, Surface passivation and characterization of cobalt–ferrite nanoparticles, *Surface Science* 575, 2005, 12–16.

103. H. M. Lee, Y. R. Uhm, and C. K. Rhee, Phase control and characterization of Fe and Fe-oxide nanocrystals synthesized by pulsed wire evaporation method, *Journal of Alloys and Compounds* 461, 2008, 604–607.

104. L. B. Bangs, New developments in particle-based immunoassays: Introduction, *Pure and Applied Chemistry*, 68, 1996, 1873–1879.

105. J. C. Joubert, *Magnetic Microcomposites as Vectors for Bioactive Agents: The State of Art*, Springer, Heidelberg, Allemagne, 1997.

106. P. D. Rye, Sweet and sticky: Carbohydrate-coated magnetic beads, *Bio/Technology* 14, 1996, 155.

107. M. Arruebo, R. Fernández-Pacheco, M. R. Ibarra, and J. Santamaría, Magnetic nanoparticles for drug delivery, *Nano Today* 2, 2007, 22–32.

108. G. V. Rama Rao, D. S. Surya Narayana, U. V. Varadaraju, G. V. N. Rao, and S. Venkadesan, Synthesis of $YBa_2Cu_3O_7$ through different gel routes, *Journal of Alloys and Compounds* 217, 1995, 200–208.

109. A. Sin, J. J. Picciolo, R. H. Lee, F. Gutierrez-Mora, and K. C. Goretta, Synthesis of mullite powders by acrylamide polymerization, *Journal of Materials Science Letters* 20, 2001, 1639–1641.

110. A. Sin, B. El Montaser, and P. Odier, Nanopowders by organic polymerisation, *Journal of Sol-Gel Science and Technology* 26, 2003, 541–545.

111. N. Zhang, T. Fu, F. Yang, H. Kan, X. Wang, H. Long, and L. Wang, Preparation of nano aluminum nitride powders by polymer network method, *Journal of Ceramic Processing Research* 152, 2014, 93–96.

112. S. F. Wang, H. Yang, T. Xian, and X. Q. Liu, Size-controlled synthesis and photocatalytic properties of $YMnO_3$ nanoparticles, *Catalysis Communications* 12, 2011, 625–628.

113. Z. Li, W. Shen, Z. Wang, X. Xiang, X. Zu, Q. Wei, and L. Wang, Direct formation of SiO_2/SnO_2 composite nanoparticles with high surface area and high thermal stability by sol–gel-hydrothermal process, *Journal of Sol-Gel Science and Technology* 49, 2009, 196–201.

114. W. Zeng, Y. Li, B. Miao, and Z. Gou, Facile synthesis of highly VOCs-response MoO_3 nanosheets using hydrothermal process, *Nanoscience and Nanotechnology Letters* 5, 2013, 986–989.

115. L. M. Fang, X. T. Zu, and C. M. Liu, Microstructure and magnetic properties in $Sn_{1-x}Fe_xO_2$ ($x = 0.01, 0.05, 0.10$) nanoparticles synthesized by hydrothermal method, *Journal of Alloys and Compounds* 491, 2010, 679–683.

116. X. Li, F. Zhang, C. Ma, Y. Deng, L. Zhang, Z. Lu, E. Sauli, Z. Li, and N. He, Controlling the morphology of $ZnS:Mn^{2+}$ nanostructure in hydrothermal process using different solvents and surfactants, *Nanoscience and Nanotechnology Letters* 5, 2013, 271–276.

117. S. S. Arbuj, S. R. Bhalerao, S. B. Rane, N. Y. Hebalkar, U. P. Mulik, and D. P. Amalnerkar, Influence of triethanolamine on physico-chemical properties of cadmium sulphide, *Nanoscience and Nanotechnology Letters* 5, 2013, 1245–1250.

118. S. Shetty, V. R. Palkar, and R. Pinto, Size effect study in magnetoelectric $BiFeO_3$ system, *Pramana-Journal of Physics* 58, 2002, 1027–1030.

119. S. Ravi and V. S. Prabvin, Nanostructured copper oxide synthesized by a simple bio-molecule assisted route with wide bandgap, *Nanoscience and Nanotechnology Letters* 5, 2013, 879–882.

120. S. Ghosh, S. Dasgupta, A. Sen, and H. Sekhar Maiti, Low-temperature synthesis of nanosized bismuth ferrite by soft chemical route, *Journal of the American Ceramic Society* 88, 2005, 1349–1352.

121. P. Yadav and M. C. Bhatnagar, Synthesis of NASICON nanoparticles and their optical properties, *Nanoscience and Nanotechnology Letters* 5, 2013, 530–535.

122. N. R. Panchal and R. B. Jotania, Physical properties of strontium hexaferrite nano magnetic particles synthesized by a sol-gel auto-combustion process in presence of non ionic surfactant, *Nanoscience and Nanotechnology Letters* 4, 2012, 623–627.

123. A. Sin and P. Odier, Gelation by acrylamide, a quasi-universal medium for the synthesis of fine oxide powders for electroceramic applications, *Advanced Materials* 12, 2000, 649–652.

124. R. T. Wang, X. P. Liang, Y. Peng, X. W. Fan, and J. X. Li, A novel method for the synthesis of nano-sized $MgAl_2O_4$ spinel ceramic powders, *Journal of Ceramic Processing Research* 11, 2010, 173–175.

125. S. F. Wang, H. B. Lv, X. S. Zhou, Y. Q. Fu, and X. T. Zu, Magnetic nanocomposites through polyacrylamide gel route, *Nanoscience and Nanotechnology Letters* 6, 2014, 758–771.

126. A. Douy and P. Odier, The polyacrylamide gel: A novel route to ceramic and glassy oxide powders, *Materials Research Bulletin* 24, 1989, 1119–1126.

127. C. Liu, R. Yu, Z. Xu, J. Cai, X. Yan, and X. Luo, Crystallization, morphology and luminescent properties of $YAG:Ce^{3+}$ phosphor powder prepared by polyacrylamide gel method, *Transactions of Nonferrous Metals Society of China* 17, 2007, 1093–1099.

128. B. Buffeteau, T. Hargreaves, B. Grevin, and C. Marin, Oxygen dependence on superconducting properties of $La_{1.85}Sr_{0.15}CuO_{4-\delta}$ ceramic and crystal samples, *Physica C: Superconductivity* 294, 1998, 55–70.

129. S. Wang, C. Zhang, G. Sun, B. Chen, X. Xiang, Q. Ding, and X. Zu, Chelating agents role on phase formation and surface morphology of single orthorhombic YMn_2O_5 nanorods via modified polyacrylamide gel route, *Science China Chemistry* 57, 2014, 402–408.

130. Y. F. Wang and H. Yang, Synthesis of different-sized $La_{0.7}Sr_{0.3}MnO_3$ nanoparticles via a polyacrylamide gel route and their magnetocaloric properties, *Journal of Superconductivity and Novel Magnetism* 26, 2013, 3463–3467.

131. H. Yang, T. Xian, Z. Q. Wei, J. F. Dai, J. L. Jiang, and W. J. Feng, Size-controlled synthesis of $BiFeO_3$ nanoparticles by a soft-chemistry route, *Journal of Sol-Gel Science and Technology* 58, 2011, 238–243.

132. T. Xian, H. Yang, J. F. Dai, Z. Q. Wei, J. Y. Ma, and W. J. Feng, Photocatalytic properties of $SrTiO_3$ nanoparticles prepared by a polyacrylamide gel route, *Materials Letters* 65, 2011, 3254–3257.

133. H. Yang, J. X. Zhang, G. J. Lin, T. Xian, and J. L. Jiang, Preparation, characterization and photocatalytic properties of terbium orthoferrite nanopowder, *Advanced Powder Technology* 24, 2013, 242–245.

134. Y. Song and C. Nan, Preparation of $Ca_3Co_4O_9$ by polyacrylamide gel processing and its thermoelectric properties, *Journal of Sol-Gel Science and Technology* 44, 2007, 139–144.

135. H. Q. Liu, Y. Song, S. N. Zhang, X. B. Zhao, and F. P. Wang, Thermoelectric properties of $Ca_{3-x}Y_xCo_4O_{9+\delta}$ ceramics, *Journal of Physics and Chemistry of Solids* 70, 2009, 600–603.

136. H. Zhang, X. Fu, S. Niu, G. Sun, and Q. Xin, Low temperature synthesis of nanocrystalline YVO$_4$:Eu via polyacrylamide gel method, *Journal of Solid State Chemistry* 177, 2004, 2649–2654.

137. S. Wang, X. Xiang, G. Sun, X. Gao, B. Chen, Q. Ding, Z. Li, C. Zhang, and X. Zu, Role of pH, organic additive, and chelating agent in gel synthesis and fluorescent properties of porous monolithic alumina, *The Journal of Physical Chemistry C* 117, 2013, 5067–5074.

138. S. Wang, X. Xiang, Q. Ding, X. Gao, C. Liu, Z. Li, and X. Zu, Size-controlled synthesis and photoluminescence of porous monolithic α-alumina, *Ceramics International* 39, 2013, 2943–2948.

139. X. Fu, H. Zhang, S. Niu, and Q. Xin, Synthesis and luminescent properties of SnO$_2$:Eu nanopowder via polyacrylamide gel method, *Journal of Solid State Chemistry* 178, 2005, 603–607.

140. S. Wang, C. Zhang, G. Sun, B. Chen, X. Xiang, H. Wang, L. Fang, Q. Tian, Q. Ding, and X. Zu, Fabrication of a novel light emission material AlFeO$_3$ by a modified polyacrylamide gel route and characterization of the material, *Optical Materials* 36, 2013, 482–488.

141. S. Sakohara and K. Mori, Preparation of ZnO nanoparticles in amphiphilic gel network, *Journal of Nanoparticle Research* 10, 2008, 297–305.

142. N. O. Dantas, A. F. G. Monte, W. A. Cardoso, A. G. Brito-Madurro, J. M. Madurro, and P. C. Morais, Growth and characterisation of ZnO quantum dots in polyacrylamide, *Microelectronics Journal* 36, 2005, 234–236.

143. S. A. Hassanzadeh-Tabrizi, Polymer-assisted synthesis and luminescence properties of MgAl$_2$O$_4$: Tb nanopowder, *Optical Materials* 33, 2011, 1607–1609.

144. R. T. Ma, Y. W. Tian, H. T. Zhao, G. Zhang, and H. Zhao, Synthesis, Characterization and electromagnetic studies on nanocrystalline Co$_{0.5}$Zn$_{0.5}$Fe$_2$O$_4$ synthesized by polyacrylamide gel, *Journal of Materials Science &Technology* 24, 2008, 628–632.

145. R. T. Ma, Y. Wang, Y. W. Tian, C. L. Zhang, and X. Li, Synthesis, characterization and electromagnetic studies on nanocrystalline nickel zinc ferrite by polyacrylamide gel, *Journal of Materials Science &Technology* 24, 2008, 419–422.

146. Y. S. Huo, H. Yang, T. Xian, J. L. Jiang, Z. Q. Wei, R. S. Li, and W. J. Feng, A polyacrylamide gel route to different-sized CaTiO$_3$ nanoparticles and their photocatalytic activity for dye degradation, *Journal of Sol-Gel Science and Technology* 71, 2014, 254–259.

147. J. H. Sun and H. Yang, A polyacrylamide gel route to photocatalytically active BiVO$_4$ particles with monoclinic scheelite structure, *Ceramics International* 40, 2014, 6399–6404.

148. S. Wu, Y. Liu, L. He, and F. Wang, Preparation of β-spodumene-based glass–ceramic powders by polyacrylamide gel process, *Materials Letters* 58, 2004, 2772–2775.

149. Y. Song, Q. Sun, L. Zhao, F. Wang, and Z. Jiang, Synthesis and thermoelectric power factor of (Ca$_{0.95}$Bi$_{0.05}$)$_3$Co$_4$O$_9$/Ag composites, *Materials Chemistry and Physics* 113, 2009, 645–649.

150. S. A. Hassanzadeh-Tabrizi and E. Taheri-Nassaj, Polyacrylamide gel synthesis and sintering of Mg$_2$SiO$_4$: Eu^{3+} nanopowder, *Ceramics International* 39, 2013, 6313–6317.

151. B. Divband, M. Khatamian, G. R. K. Eslamian, and M. Darbandi, Synthesis of Ag/ZnO nanostructures by different methods and investigation of their photocatalytic efficiency for 4-nitrophenol degradation, *Applied Surface Science* 284, 2013, 80–86.

152. S. Ahmadi, M. Manteghian, H. Kazemian, S. Rohani, and J. Towfighi Darian, Synthesis of silver nano catalyst by gel-casting using response surface methodology, *Powder Technology* 228, 2012, 163–170.

153. C. S. Cheng, L. Zhang, Y. J. Zhang, and S. P. Jiang, Synthesis of LaCoO$_3$ nano-powders by aqueous gel-casting for intermediate temperature solid oxide fuel cells, *Solid State Ionics* 179, 2008, 282–289.

154. A. Tarancón, G. Dezanneau, J. Arbiol, F. Peiró, and J. R. Morante, Synthesis of nano-crystalline materials for SOFC applications by acrylamide polymerisation, *Journal of Power Sources* 118, 2003, 256–264.

155. G. C. C. Costa and R. Muccillo, Comparative studies on properties of scandia-stabilized zirconia synthesized by the polymeric precursor and the polyacrylamide techniques, *Journal of Alloys and Compounds* 503, 2010, 474–479.

156. L. Zhang, S. Ping Jiang, C. Siang Cheng, and Y. Zhang, Synthesis and performance of $(La_{0.75}Sr_{0.25})_{1-x}$ $(Cr_{0.5}Mn_{0.5})O_3$ cathode powders of solid oxide fuel cells by gel-casting technique, *Journal of the Electrochemical Society* 154, 2007, B577–B582.

157. L. Zhang, Y. Zhang, Y. D. Zhen, and S. P. Jiang, Lanthanum strontium manganite powders synthesized by gel-casting for solid oxide fuel cell cathode materials, *Journal of the American Ceramic Society* 90, 2007, 1406–1411.

158. I. Najafi Hajivar and M. Kokabi, Polymer-network hydrogel facilitated synthesis of Ca-α-SiAlON balls composed of nanoparticles, *Ceramics International* 39, 2013, 3321–3327.

159. T. Barbier, C. Autret-Lambert, P. Andreazza, A. Ruyter, C. Honstettre, S. Lambert, F. Gervais, and M. Lethiecq, Cu-doping effect on dielectric properties of organic gel synthesized $Ba_{4y}Mn_{3-x}Cu_xO_{11.5\pm\delta}$, *Journal of Solid State Chemistry* 206, 2013, 217–225.

160. A. Magrez, M. Caldes, O. Joubert, and M. Ganne, A new "Chimie Douce" approach to the synthesis of $Sr_{1-x}La_{1+x}Al_{1-x}Mg_xO_4$ with K_2NiF_4 structure type, *Solid State Ionics* 151, 2002, 365–370.

161. A. Magrez and T. Schober, Preparation, sintering, and water incorporation of proton conducting $Ba_{0.99}Zr_{0.8}Y_{0.2}O_{3-\delta}$: Comparison between three different synthesis techniques, *Solid State Ionics* 175, 2004, 585–588.

162. A. R. Bahramian and M. Kokabi, Carbonitriding synthesis of β-SiAlON nanopowder from kaolinite–polyacrylamide precursor, *Applied Clay Science* 52, 2011, 407–413.

163. X. F. Wang, R. C. Wang, C. Q. Peng, T. T. Li, and B. Liu, Growth of BeO nanograins synthesized by polyacrylamide gel route, *Journal of Materials Science & Technology* 27, 2011, 147–152.

164. http://apps.webofknowledge.com/summary.do?product=UA&parentProduct=UA&search_mode=Refine&parentQid=1&qid=2&SID=3A8pnV6svaWM9K6FRoI&&page=2.

165. http://apps.webofknowledge.com/summary.do?product=WOS&parentProduct=WOS&search_mode=GeneralSearch&qid=2&SID=1FAzWy7UlZMQ4cxlvnJ&page=1&action=changePageSize&pageSize=50.

166. A. Amirshaghaghi and M. Kokabi, Tailoring size of α-Al_2O_3 nanopowders via polymeric gel-net method, *Iranian Polymer Journal* 19, 2010, 615–624.

167. Y. Song, Q. Sun, Y. Lu, X. Liu, and F. Wang, Low-temperature sintering and enhanced thermoelectric properties of $LaCoO_3$ ceramics with B_2O_3–CuO addition, *Journal of Alloys and Compounds* 536, 2012, 150–154.

168. Y. Song and C. W. Nan, High temperature transport properties of Ag-added $(Ca_{0.975}La_{0.025})_3Co_4O_9$ ceramics, *Physica B: Condensed Matter* 406, 2011, 2919–2923.

169. S. Wang, C. Zhang, G. Sun, Y. Yuan, L. Chen, X. Xiang, Q. Ding, B. Chen, Z. Li, and X. Zu, Self-assembling synthesis of α-Al_2O_3-carbon composites and a method to increase their photoluminescence, *Journal of Luminescence* 153, 2014, 393–400.

170. H. Liu, S. Gong, Y. Hu, J. Zhao, J. Liu, Z. Zheng, and D. Zhou, Tin oxide nanoparticles synthesized by gel combustion and their potential for gas detection, *Ceramics International* 35, 2009, 961–966.

171. A. Calleja, X. Casas, I. G. Serradilla, M. Segarra, A. Sin, P. Odier, and F. Espiell, Up-scaling of superconductor powders by the acrylamide polymerization method, *Physica C: Superconductivity* 372–376 (Part 2), 2002, 1115–1118.

172. O. O. Omatete, M. A. Janney, and S. D. Nunn, Gelcasting: From laboratory development toward industrial production, *Journal of the European Ceramic Society* 17, 1997, 407–413.
173. K. B. McAuley, The chemistry and physics of polyacrylamide gel dosimeters: Why they do and don't work, *Journal of Physics: Conference Series* 3, 2004, 29–33.
174. M. A. Janney, S. D. Nunn, C. A. Walls, O. O. Omatete, R. B. Ogle, G. H. Kirby, and A. D. McMillan, Gelcasting, in *Handbook of Ceramic Engineering*, M. N. Rahaman (ed.), Marcel Dekker, New York, 1998.
175. M. Tahmasebpour, A. A. Babaluo, S. Shafiei, and E. Pipelzadeh, Studies on the synthesis of α-Al_2O_3 nanopowders by the polyacrylamide gel method, *Powder Technology* 191, 2009, 91–97.
176. Y. Song, Q. Sun, L. R. Zhao, and F. P. Wang, Rapid synthesis of Bi substituted $Ca_3Co_4O_9$ by a polyacrylamide gel method and its high-temperature thermoelectric power factor, *Key Engineering Materials* 434–435, 2010, 393–396.
177. H. Zhao, R. Ma, G. Zhang, and X. Xiao, Preparation and electromagnetic properties of composites of hollow glass microspheres coated with $NiFe_2O_4$ nanoparticles, *Chinese Journal of Chemical Physics* 20, 2007, 801–805.
178. S. He, S. Wang, Q. Ding, X. Yuan, W. Zheng, X. Xiang, Z. Li, and X. Zu, Role of chelating agent in chemical and fluorescent properties of SnO_2 nanoparticles, *Chinese Physics B* 22, 2013, 058102.
179. H. Yang, Y. H. Zhu, T. Xian, and J. L. Jiang, Synthesis and magnetocaloric properties of $La_{0.7}Ca_{0.3}MnO_3$ nanoparticles with different sizes, *Journal of Alloys and Compounds* 555, 2013, 150–155.
180. G. J. Lin, H. Yang, T. Xian, Z. Q. Wei, J. L. Jiang, and W. J. Feng, Synthesis of $TbMnO_3$ nanoparticles via a polyacrylamide gel route, *Advanced Powder Technology* 23, 2012, 35–39.
181. M. Zhang, H. Yang, T. Xian, Z. Q. Wei, J. L. Jiang, Y. C. Feng, and X. Q. Liu, Polyacrylamide gel synthesis and photocatalytic performance of $Bi_2Fe_4O_9$ nanoparticles, *Journal of Alloys and Compounds* 509, 2011, 809–812.
182. W. H. Gitzen, *Alumina as a Ceramic Material*, American Ceramic Society, Ohio, 1970.
183. G. Ramanathan, R. John Xavier, and K. R. Murali, Dye sensitized solar cells with ITO films prepared by the acrylamide sol gel route, *IOSR Journal of Applied Physics* 2, 2013, 47–50.
184. X. F. Wang, R. C. Wang, C. Q. Peng, T. T. Li, and B. Liu, Synthesis and sintering of beryllium oxide nanoparticles, *Progress in Natural Science: Materials International* 20, 2010, 81–86.
185. D. X. Zhou, H. Liu, S. P. Gong, L. H. Huang, and L. Gan, Preparation of Nanocrystalline tin dioxide particles via the acrylamide polymerization process, *Rare Metal Materials and Engineering* 35, 2006, 581–582.
186. S. Menassel, M.-F. Mosbah, S. P. Altintas, A. Varilci, and F. Bouaicha, Synthesis of BiSrCa(Y)CuO superconductor from the sol-gel method and the effect of Y substitution, *AIP Conference Proceedings* 1476, 2012, 374–377.
187. N. A. Hill, Why are there so few magnetic ferroelectrics? *The Journal of Physical Chemistry B* 104, 2000, 6694–6709.
188. M. Fiebig, T. Lottermoser, D. Frohlich, A. V. Goltsev, and R. V. Pisarev, Observation of coupled magnetic and electric domains, *Nature* 419, 2002, 818–820.
189. A. Samariya, S. N. Dolia, A. S. Prasad, P. K. Sharma, S. P. Pareek, M. S. Dhawan, and S. Kumar, Size dependent structural and magnetic behaviour of $CaFe_2O_4$, *Current Applied Physics* 13, 2013, 830–835.
190. Q. He, H. Z. Wang, G. H. Wen, Y. Sun, and B. Yao, Formation and properties of $Ba_xFe_{3-x}O_4$ with spinel structure by mechanochemical reaction of α-Fe_2O_3 and $BaCO_3$, *Journal of Alloys and Compounds* 486, 2009, 246–249.

191. V. Šepelák, A. Feldhoff, P. Heitjans, F. Krumeich, D. Menzel, F. J. Litterst, I. Bergmann, and K. D. Becker, Nonequilibrium cation distribution, canted spin arrangement, and enhanced magnetization in nanosized $MgFe_2O_4$ prepared by a one-step mechanochemical route, *Chemistry of Materials* 18, 2006, 3057–3067.

192. T. Bala, C. R. Sankar, M. Baidakova, V. Osipov, T. Enoki, P. A. Joy, B. L. V. Prasad, and M. Sastry, Cobalt and magnesium ferrite nanoparticles: Preparation using liquid foams as templates and their magnetic characteristics, *Langmuir* 21, 2005, 10638–10643.

193. A. S. Teja and P. Y. Koh, Synthesis, properties, and applications of magnetic iron oxide nanoparticles, *Progress in Crystal Growth and Characterization of Materials* 55, 2009, 22–45.

194. J. Wang, Q. Chen, B. Hou, and Z. Peng, Synthesis and magnetic properties of single-crystals of $MnFe_2O_4$ nanorods, *European Journal of Inorganic Chemistry* 2004, 2004, 1165–1168.

195. D. Treves, Studies on orthoferrites at the weizmann institute of science, *Journal of Applied Physics* 36, 1965, 1033–1039.

196. S. F. Wang, Q. P. Ding, and X. T. Zu, Influence of Fe_2O_3 on structural and magnetic properties of $LaFeO_3/Fe_2O_3$ and Mn_2O_3/Fe_2O_3 magnetic nanocomposites, *Applied Mechanics and Materials* 563, 2014, 30–35.

197. D. Seifu, A. Kebede, F. W. Oliver, E. Hoffman, E. Hammond, C. Wynter et al. Evidence of ferrimagnetic ordering in $FeMnO_3$ produced by mechanical alloying, *Journal of Magnetism and Magnetic Materials* 212, 2000, 178–182.

198. I. R. Leith and M. G. Howden, Temperature-programmed reduction of mixed iron-manganese oxide catalysts in hydrogen and carbon monoxide, *Applied Catalysis* 37, 1988, 75–92.

199. P. A. Montano, Technological applications of mössbauer spectroscopy, *Hyperfine Interactions* 27, 1986, 147–159.

200. C. A. Goodwin, H. K. Bowen, and W. D. Kingery, Phase separation in the system (Fe,Mn)O, *Journal of the American Ceramic Society* 58, 1975, 317–320.

201. P. Franke and R. Dieckmann, Defect structure and transport properties of mixed iron-manganese oxides, *Solid State Ionics* 32–33 (Part 2), 1989, 817–823.

202. K. Ehrensberger, A. Frei, P. Kuhn, H. R. Oswald, and P. Hug, Comparative experimental investigations of the water-splitting reaction with iron oxide $Fe_{1-y}O$ and iron manganese oxides $(Fe_{1-x}Mn_x)_{1-y}O$, *Solid State Ionics* 78, 1995, 151–160.

203. H. L. Roux, A mossbauer study of paramagnetic and magnetic components in an uncalcined iron manganese oxide powder, *Journal of Physics: Condensed Matter* 2, 1990, 3391–3398.

204. B. Kolk, A. Albers, G. R. Hearue, and H. L. Roux, Evidence of a new structural phase of manganese-iron oxide, *Hyperfine Interactions* 42, 1988, 1051–1054.

205. I. D. Lick and D. B. Soria, Synthesis of $MnFeO_3$ from the oxidative thermal decomposition of $Mn[Fe(CN)_5NO]\cdot 2H_2O$, *Journal of the Argentine Chemical Society* 97, 2009, 102–108.

206. M. F. Zhang, J. M. Liu, and Z. G. Liu, Microstructural characterization of nanosized $YMnO_3$ powders: The size effect, *Applied Physics A* 79, 2004, 1753–1756.

207. Z. J. Li, S. F. Wang, B. Li, and X. Xiang, A new method for synthesis of $FeMnO_3$ ceramics and its phase transformation, *Journal of Nano Research* 37, 2016, 122–131.

208. Y. H. Zou, W. L. Li, S. L. Wang, H. W. Zhu, P. G. Li, and W. H. Tang, Spin dependent electrical abnormal in $TbFeO_3$, *Journal of Alloys and Compounds* 519, 2012, 82–84.

209. M. Sivakumar, A. Gedanken, D. Bhattacharya, I. Brukental, Y. Yeshurun, W. Zhong, Y. W. Du, I. Felner, and I. Nowik, Sonochemical synthesis of nanocrystalline rare earth orthoferrites using $Fe(CO)_5$ precursor, *Chemistry of Materials* 16, 2004, 3623–3632.

210. W. P. Wang, H. Yang, T. Xian, and J. L. Jiang, XPS and magnetic properties of $CoFe_2O_4$ nanoparticles synthesized by a polyacrylamide gel route, *Materials Transactions* 53, 2012, 1586–1589.

211. D. Zhao, H.Yang, T. Xian, W. Wang, Z. Wei, R. Li, W. Feng, and J. Jiang, Preparation and properties of $NiFe_2O_4$ nanoparticles by a polyacrylamide gel route, *Journal of Synthetic Crystals* 42, 2013, 316–321.

212. S. Yu and M. Yoshimura, Direct fabrication of ferrite MFe_2O_4 (M = Zn, Mg)/Fe composite thin films by soft solution processing, *Chemistry of Materials* 12, 2000, 3805–3810.

213. D. Chen, Y. Zhang, and C. Tu, Preparation of high saturation magnetic $MgFe_2O_4$ nanoparticles by microwave-assisted ball milling, *Materials Letters* 82, 2012, 10–12.

214. S. F. Wang, X. T. Zu, G. Z. Sun, D. M. Li, C. D. He, X. Xiang, W. Liu, S. B. Han, and S. Li, Highly dispersed spinel (Mg, Ca, Ba)-ferrite nanoparticles: Tuning the particle size and magnetic properties through a modified polyacrylamide gel route, *Ceramics International* http://dx.doi.org/10.1016/j.ceramint.2016.09.075.

215. Z. F. Zi, Y. P. Sun, X. B. Zhu, Z. R. Yang, J. M. Dai, and W. H. Song, Structural and magnetic properties of $SrFe_{12}O_{19}$ hexaferrite synthesized by a modified chemical co-precipitation method, *Journal of Magnetism and Magnetic Materials* 320, 2008, 2746–2751.

216. P. Shepherd, K. K. Mallick, and R. J. Green, Magnetic and structural properties of M-type barium hexaferrite prepared by co-precipitation, *Journal of Magnetism and Magnetic Materials* 311, 2007, 683–692.

217. L. A. García-Cerda, O. S. Rodríguez-Fernández, and P. J. Reséndiz-Hernández, Study of $SrFe_{12}O_{19}$ synthesized by the sol–gel method, *Journal of Alloys and Compounds* 369, 2004, 182–184.

218. A. K. Singh, S. D. Kaushik, B. Kumar, P. K. Mishra, A. Venimadhav, V. Siruguri, and S. Patnaik, Substantial magnetoelectric coupling near room temperature in $Bi_2Fe_4O_9$, *Applied Physics Letters* 92, 2008, 132910–132913.

219. Z. M. Tian, S. L. Yuan, X. L. Wang, X. F. Zheng, S. Y. Yin, C. H. Wang, and L. Liu, Size effect on magnetic and ferroelectric properties in $Bi_2Fe_4O_9$ multiferroic ceramics, *Journal of Applied Physics* 106, 2009, 103912–103914.

220. D. P. Dutta, C. Sudakar, P. S. V. Mocherla, B. P. Mandal, O. D. Jayakumar, and A. K. Tyagi, Enhanced magnetic and ferroelectric properties in scandium doped nano $Bi_2Fe_4O_9$, *Materials Chemistry and Physics* 135, 2012, 998–1004.

221. C. Murugesan Raghavan, J. Won Kim, J. Ya Choi, J. Kim, and S. Su Kim, Effects of Ti-doping on the structural, electrical and multiferroic properties of $Bi_2Fe_4O_9$ thin films, *Ceramics International* 40, 2014, 14165–14170.

222. A. Muñoz, J. A. Alonso, M. J. Martínez-Lope, M. T. Casáis, J. L. Martínez, and M. T. Fernández-Díaz, Magnetic structure of hexagonal $RMnO_3$ (R = Y, Sc): Thermal evolution from neutron powder diffraction data, *Physics Review B* 62, 2000, 9498.

223. A. K. Singh, S. Patnaik, S. D. Kaushik, and V. Siruguri, Dominance of magnetoelastic coupling in multiferroic hexagonal $YMnO_3$, *Physics Review B* 81, 2010, 184406.

224. H. W. Brinks, H. Fjellvåg, and A. Kjekshus, Synthesis of metastable perovskite-type $YMnO_3$ and $HoMnO_3$, *Journal of Solid State Chemistry* 129, 1997, 334–340.

225. M. Uehara, K. Takahashi, T. Asaka, and S. Tsutsumi, Preparation of $(La_{1-x}Sr_x)MnO_3$ by coprecipitation method, *Journal of the Ceramic Society of Japan* 106, 1998, 1248–1251.

226. V. R. Choudhary, S. Banerjee, and B. S. Uphade, Activation by hydrothermal treatment of low surface area ABO_3-type perovskite oxide catalysts, *Applied Catalysis A-General* 197, 2000, 183–186.

227. P. A. Salvador, T. Doan, B. Mercey, and B. Raveau, Stabilization of $YMnO_3$ in a perovskite structure as a thin film, *Chemistry of Materials* 10, 1998, 2592–2595.

228. H. J. Hwang and M. Awano, Preparation of $LaCoO_3$ catalytic thin film by the sol–gel process and its NO decomposition characteristics, *Journal of the European Ceramic Society* 21, 2001, 2103–2107.

229. Y. Y. Li, L. H. Xue, L. F. Fan, and Y. W. Yan, The effect of citric acid to metal nitrates molar ratio on sol–gel combustion synthesis of nanocrystalline $LaMnO_3$ powders, *Journal of Alloys and Compounds* 478, 2009, 493–497.

230. T. Katsufuji, S. Mori, M. Masaki, Y. Moritomo, N. Yamamoto, and H. Takagi, Dielectric and magnetic anomalies and spin frustration in hexagonal $RMnO_3$ (R = Y, Yb, and Lu), *Physical Review B* 64, 2001, 104419.

231. N. Fujimura, S. I. Azuma, N. Aoki, T. Yoshimura, and T. Ito, Growth mechanism of $YMnO_3$ film as a new candidate for nonvolatile memory devices, *Journal of Applied Physics* 80, 1996, 7084–7088.

232. N. Fujimura, T. Ishida, T. Yoshimura, and T. Ito, Epitaxially grown $YMnO_3$ film: New candidate for nonvolatile memory devices, *Applied Physics Letters* 69, 1996, 1011–1013.

233. K. Choi, W. Shin, and S. Yoon, Ferroelectric $YMnO_3$ thin films grown by metal-organic chemical vapor deposition for metal/ferroelectric/semiconductor field-effect transistors, *Thin Solid Films* 384, 2001, 146–150.

234. D. Ito, N. Fujimura, T. Yoshimura, and T. Ito, Ferroelectric properties of $YMnO_3$ epitaxial films for ferroelectric-gate field-effect transistors, *Journal of Applied Physics* 93, 2003, 5563–5567.

235. H. L. Yakel, W. C. Koehler, E. F. Bertaut, and E. F. Forrant, On structure of the crystal manganese troxides of the lanthanides and heavy yttrium, *Acta Crystallographica.* 16, 1963, 957–962.

236. V. Goian, S. Kamba, C. Kadlec, D. Nuzhnyy, P. Kužel, J. Agostinho Moreira, A. Almeida, and P. B. Tavares, THz and infrared studies of multiferroic hexagonal $Y_{1-x}Eu_xMnO_3$ (x = 0–0.2) ceramics, *Phase Transitions: A Multinational Journal* 83, 2010, 931–941.

237. P. Gao, Z. Chen, T. A. Tyson, T. Wu, K. H. Ahn, Z. Liu, R. Tappero, S. B. Kim, and S. W. Cheong, High pressure structural stability of multiferroic hexagonal $REMnO_3$, *Physics Review B* 83, 2011, 224113/224111–224113/224119.

238. K. Singh, N. Bellido, C. Simon, J. Varignon, M. B. Lepetit, A. De Muer, and S. Pailhès, Role of the ferromagnetic component in the ferroelectricity of $YMnO_3$, arXiv preprint arXiv:1112.1011, 2011.

239. H. W. Zheng, Y. F. Liu, W. Y. Zhang, S. J. Liu, H. R. Zhang, and K. F. Wang, Spin-glassy behavior and exchange bias effect of hexagonal $YMnO_3$ nanoparticles fabricated by hydrothermal process, *Journal of Applied Physics* 107, 2010, 053901–053904.

240. S. F. Wang, H. Yang, T. Xian, J. L. Jiang, Z. Q. Wei, Y. C. Feng, R. S. Li, and W. J. Feng, Preparation and characterization of $YMnO_3$ nanoparticles, *Journal of the Chinese Ceramic Society* 38, 2010, 2303–2307.

241. T. Kimura, T. Goto, H. Shintani, K. Ishizaka, T. Arima, and Y. Tokura, Magnetic control of ferroelectric polarization, *Nature* 426, 2003, 55–58.

242. H. C. Walker, R. A. Ewings, F. Fabrizi, D. Mannix, C. Mazzoli, S. B. Wilkins, L. Paolasini, D. Prabhakaran, A. T. Boothroyd, and D. F. McMorrow, X-ray resonant scattering study of the magnetic phase diagram of multiferroic $TbMnO_3$, *Physica B: Condensed Matter* 404, 2009, 3264–3266.

243. S. Venkatesan, C. Daumont, B. J. Kooi, B. Noheda, and J. T. M. D. Hosson, Nanoscale domain evolution in thin films of multiferroic $TbMnO_3$, *Physical Review B* 80, 2009, 214111.

244. R. Das, A. Jaiswal, S. Adyanthaya, and P. Poddar, Effect of particle size and annealing on spin and phonon behavior in $TbMnO_3$, *Journal of Applied Physics* 109, 2011, 064309.

245. S. Kharrazi, D. C. Kundaliya, S. W. Gosavi, S. K. Kulkarni, T. Venkatesan, S. B. Ogale, J. Urban, S. Park, and S. W. Cheong, Multiferroic $TbMnO_3$ nanoparticles, *Solid State Communications* 138, 2006, 395–398.

246. V. Dyakonov, A. Szytuła, R. Szymczak, E. Zubov, A. Szewczyk, Z. Kravchenko et al., Phase transitions in $TbMnO_3$ manganites, *Low Temperature Physics* 38, 2012, 216–220.

247. H. Yang, G. J. Lin, T. Xian, Z. Q. Wei, and W. J. Feng, Influence of chelating agents and crosslinking on $TbMnO_3$ nanoparticles prepared by a gel route, *Nanotechnology and Precision Engineering* 10, 2012, 46–51.

248. J. Hu, J. Ma, L. Wang, and H. Huang, Synthesis and photocatalytic properties of $LaMnO_3$–graphene nanocomposites, *Journal of Alloys and Compounds* 583, 2014, 539–545.

249. Y. Lu, Q. Dai, and X. Wang, Catalytic combustion of chlorobenzene on modified $LaMnO_3$ catalysts, *Catalysis Communications* 54, 2014, 114–117.

250. Y. Li, L. Xue, L. Fan, and Y. Yan, The effect of citric acid to metal nitrates molar ratio on sol–gel combustion synthesis of nanocrystalline $LaMnO_3$ powders, *Journal of Alloys and Compounds* 478, 2009, 493–497.

251. D. Bhattacharya,. P. Das, A. Pandey, A. K. Raychaudhuri, A.Chakraborty, and V. N. Ojha, On the factors affecting the high temperature insulator-metal transition in rare-earth manganites, *Journal of Physics: Condensed Matter* 13, 2001, L431–L439.

252. N. Das, D. Bhattacharya, A. Sen, and H. S. Maiti, Sonochemical synthesis of $LaMnO_3$ nano-powder, *Ceramics International* 35, 2009, 21–24.

253. G. Dezanneau, A. Sin, H. Roussel, H. Vincent, and M. Audier, Synthesis and characterisation of $La_{1-x}MnO_{3\pm\delta}$ nanopowders prepared by acrylamide polymerisation, *Solid State Communications* 121, 2002, 133–137.

254. M. Wang, K. Woo, and C. Lee, Preparing $La_{0.8}Sr_{0.2}MnO_3$ conductive perovskite via optimal processes: High-energy ball milling and calcinations, *Energy Conversion and Management* 52, 2011, 1589–1592.

255. K. P. Shinde, N. G. Deshpande, T. Eom, Y. P. Lee, and S. H. Pawar, Solution-combustion synthesis of $La_{0.65}Sr_{0.35}MnO_3$ and the magnetocaloric properties, *Materials Science and Engineering: B* 167, 2010, 202–205.

256. W. P. Stege, L. E. Cadús, and B. P. Barbero, $La_{1-x}Ca_xMnO_3$ perovskites as catalysts for total oxidation of volatile organic compounds, *Catalysis Today* 172, 2011, 53–57.

257. D. P. Lim, D. S. Lim, J. S. Oh, and I. W. Lyo, Influence of post-treatments on the contact resistance of plasma-sprayed $La_{0.8}Sr_{0.2}MnO_3$ coating on SOFC metallic interconnector, *Surface and Coatings Technology* 200, 2005, 1248–1251.

258. L. Malavasi, M. C. Mozzati, S. Polizzi, C. B. Azzoni, and G. Flor, Nanosized sodium-doped lanthanum manganites: Role of the synthetic route on their physical properties, *Chemistry of Materials* 15, 2003, 5036–5043.

259. L. D. Yao, W. Zhang, J. S. Zhang, H. Yang, F. Y. Li, Z. X. Liu, C. Q. Jin, and R. C. Yu, Enhanced magnetoresistance of $La_{2/3}Ca_{1/3}MnO_3$ /CeO_2 nanocrystalline composites synthesized by polymer-network gel method, *Journal of Applied Physics* 101, 2007, 063905.

260. W. P. Wang, H. Yang, T. Xian, and R. C. Yu, Observation of abnormal magnetoelectric behavior in 0–3 type $CoFe_2O_4$–$BaTiO_3$ nanocomposites, *Chemical Physics Letters* 618, 2015, 72–77.

261. S. F. Wang, G. A. Sun, Q. P. Ding, and X. T. Zu, Synthesis and magnetic properties of $LaMnO_3/\alpha$-Al_2O_3 magnetic nanocomposites, *Advanced Materials Research* 1004–1005, 2014, 103–109.

262. S. W. Bian, Y. L. Zhang, H. L. Li, Y. Yu, Y. L. Song, and W. G. Song, γ-Alumina with hierarchically ordered mesopore/macropore from dual templates, *Microporous and Mesoporous Materials* 131, 2010, 289–293.

263. E. Casbeer, V. K. Sharma, and X. Z. Li, Synthesis and photocatalytic activity of ferrites under visible light: A review, *Separation and Purification Technology* 87, 2012, 1–14.

264. S. J. Clark and J. Robertson, Band gap and schottky barrier heights of multiferroic $BiFeO_3$, *Applied Physics Letters* 90, 2007, 132903.

265. J. Wang, J. B. Neaton, H. Zheng, V. Nagarajan, S. B. Ogale, B. Liu et al., Epitaxial $BiFeO_3$ multiferroic thin film heterostructures, *Science* 299, 2003, 1719–1722.

266. T. J. Park, G. C. Papaefthymiou, A. J. Viescas, A. R. Moodenbaugh, and S. S. Wong, Size-dependent magnetic properties of single-crystalline multiferroic $BiFeO_3$ nanoparticles, *Nano Letters* 7, 2007, 766–772.

267. F. Gao, X. Y. Chen, K. B. Yin, S. Dong, Z. F. Ren, F. Yuan, T. Yu, Z. G. Zou, and J. M. Liu, Visible-light photocatalytic properties of weak magnetic $BiFeO_3$ nanoparticles, *Advanced Materials* 19, 2007, 2889–2892.

268. S. Li, Y. H. Lin, B. P. Zhang, Y. Wang, and C. W. Nan, Controlled fabrication of $BiFeO_3$ uniform microcrystals and their magnetic and photocatalytic behaviors, *The Journal of Physical Chemistry C* 114, 2010, 2903–2908.

269. C. Hengky, X. Moya, N. D. Mathur, and S. Dunn, Evidence of high rate visible light photochemical decolourisation of Rhodamine B with $BiFeO_3$ nanoparticles associated with $BiFeO_3$ photocorrosion, *RSC Advances* 2, 2012, 11843–11849.

270. T. Xian, H. Yang, J. F. Dai, Z. Q. Wei, J. Y. Ma, and W. J. Feng, Photocatalytic properties of $BiFeO_3$ nanoparticles with different sizes, *Materials Letters* 65, 2011, 1573–1575.

271. J. F. Dai, T. Xian, L. J. Di, and H. Yang, Preparation of graphene nanocomposites and their enhanced photocatalytic activities, *Journal of Nanomaterials* 2013, 2013, 5.

272. A. S. Poghossian, H. V. Abovian, P. B. Avakian, S. H. Mkrtchian, and V. M. Haroutunian, Bismuth ferrites: New materials for semiconductor gas sensors, *Sensors and Actuators B: Chemical* 4, 1991, 545–549.

273. T. Xian, H. Yang, W. Xian, X. F. Chen, and J. F. Dai, Photocatalytic mechanism of $Bi_2Fe_4O_9$ nanoparticles in the degradation of methylene blue, *Progress in Reaction Kinetics and Mechanism* 38, 2013, 417–424.

274. T. Xian, H. Yang, L. J. Di, and J. F. Dai, Graphene-assisted enhancement of photocatalytic activity of bismuth ferrite nanoparticles, *Research on Chemical Intermediates* 41, 2015, 433–441.

275. Y. F. Wang, H. Yang, T. Xian, H. M. Zhang, and J. Y. Su, Preparation and photocatalytic properties of $La_{0.7}Sr_{0.3}MnO_3$ nanoparticles, *Chinese Journal of Materials* 26, 2012, 476–482.

276. T. Xian, H. Yang, Y. F. Wang, L. J. Di, J. L. Jiang, R. S. Li, and W. J. Feng, Visible-light photocatalysis of LSMO-graphene nanocomposites toward degradation of methyl red, *Materials Transactions* 55, 2014, 245–248.

277. S. F. Wang, H. Yang, T. Xian, Z. Q. Wei, J. Y. Ma, and W. J. Feng, Nano YMn_2O_5 visible-light-driven semiconductor photocatalyst, *Journal of Inorganic Materials* 26, 2011, 1164–1168.

278. H. Yang, S. F. Wang, T. Xian, Z. Q. Wei, and W. J. Feng, Fabrication and photocatalytic activity of YMn_2O_5 nanoparticles, *Materials Letters* 65, 2011, 884–886.

279. T. Xian, H. Yang, L. Di, J. Ma, H. Zhang, and J. Dai, Photocatalytic reduction synthesis of $SrTiO_3$-graphene nanocomposites and their enhanced photocatalytic activity, *Nanoscale Research Letters* 9, 2014, 1–9.

280. W. P. Wang, H. Yang, T. Xian, Z. Q. Wei, J. Y. Ma, R. S. Li, and W. J. Feng, Polyacrylamide gel synthesis of $BaTiO_3$ nanoparticles and its photocatalytic properties for methyl red degradation, *Chinese Journal of Catalysis* 33, 2012, 354–359.

281. W. P. Wang, H. Yang, T. Xian, R. S. Li, J. Y. Ma, and J. L. Jiang, Photocatalytic degradation of methyl red by $BaTiO_3$ nanoparticles via a direct hole oxidation mechanism, *Advanced Science, Engineering and Medicine* 4, 2012, 479–483.

282. C. Y. Wang, Z. B. Shao, and S. J. Yang, Photocatalysis and preparation of nanometer - sizde ZnO by polyacrylamide gel method, *Non-Ferrous Mining and Metallurgy* 21, 2005, 29–32.

283. T. Xian, H. Yang, J. F. Dai, R. S. Li, Y. C. Feng, and J. L. Jiang, Preparation, photocatalytic properties and photogenerated hydroxyl radicals of TiO_2 nanoparticles, *Nanotechnology and Precision Engineering* 11, 2013, 111–117.

284. T. Xian, H. Yang, L. J. Di, X. F. Chen, and J. F. Dai, Polyacrylamide gel synthesis and photocatalytic properties of TiO_2 nanoparticles, *Journal of Sol-Gel Science and Technology* 66, 2013, 324–329.

285. M. Saket-Oskoui, M. Khatamian, K. Nofouzi, and A. Yavari, Study on crystallinity and morphology controlling of titania using acrylamide gel method and their photocatalytic properties, *Advanced Powder Technology* 25, 2014, 1634–1642.

286. X. R. Guan, Z. C. Shao, Y. W. Tian, and J. L. Gao, Preparation of nanometer TiO_2 composite powders by polyacrylamide and their photo-catalytic properites, *Non-Ferrous Mining and Metallurgy* 22, 2006, 41–44.

287. X. R. Guan, Z. C. Shao, Y. W. Tian, and S. Yang, Preparation of TiO_2 composite powder by polymer network gel method and their photocatalysis performance, *The Chinese Journal of Nonferrous Metals* 18, 2008, S321–S325.

288. J. H. Sun, H. Yang, T. Xian, W. P. Wang, and W. J. Feng, Polyacrylamide gel preparation, photocatalytic properties, and mechanism of $BiVO_4$ particles, *Chinese Journal of Catalysis* 33, 2012, 1982–1987.

289. P. Bansal, E. Vineyard, and O. Abdelaziz, Status of not-in-kind refrigeration technologies for household space conditioning, water heating and food refrigeration, *International Journal of Sustainable Built Environment* 1, 2012, 85–101.

290. S. A. Tassou, J. S. Lewis, Y. T. Ge, A. Hadawey, and I. Chaer, A review of emerging technologies for food refrigeration applications, *Applied Thermal Engineering* 30, 2010, 263–276.

291. P. Weiss and A. Piccard, Le phénomène magnétocalorique, *Journal of Physics (Paris)* 5th Ser, 1917, 103–109.

292. V. K. Pecharsky and K. A. G. Jr. Giant, Magnetocaloric effect in $Gd_5(Si_2Ge_2)$, *Physical Review Letters* 78, 1997, 4494–4497.

293. F. X. Hu, B. G. Shen, J. R. Sun, and Z. H. Cheng, Large magnetic entropy change in $La(Fe,Co)_{11.83}Al_{1.17}$, *Physics Review B* 64, 2001, 012409.

294. F. Hu, B. Shen, J. Sun, Z. Cheng, G. Rao, and X. Zhang, Influence of negative lattice expansion and metamagnetic transition on magnetic entropy change in the compound $LaFe_{11.4}Si_{1.6}$, *Applied Physics Letters* 78, 2001, 3675–3677.

295. F. Hu, B. Shen, and J. Sun, Magnetic entropy change in $Ni_{51.5}Mn_{22.7}Ga_{25.8}$ alloy, *Applied Physics Letters* 76, 2000, 3460–3462.

296. F. X. Hu, B. G. Shen, J. R. Sun, and G. H. Wu, Large magnetic entropy change in a Heusler alloy $Ni_{52.6}Mn_{23.1}Ga_{24.3}$ single crystal, *Physics Review B* 64, 2001, 132412.

297. X. N. He, M. Q. Gong, H. Zhang, W. Dai, J. Shen, and J. F. Wu, Design and performance of a room-temperature hybrid magnetic refrigerator combined with Stirling gas refrigeration effect, *International Journal of Refrigeration* 36, 2013, 1465–1471.

298. S. Akdemir, E. Ozel, and E. Suvaci, Solubility of blue $CoAl_2O_4$ ceramic pigments in water and diethylene glycol media, *Ceramics International* 37, 2011, 863–870.

299. L. Torkian and M. Daghighi, Effects of β-alanine on morphology and optical properties of $CoAl_2O_4$ nanopowders as a blue pigment, *Advanced Powder Technology* 25, 2014, 739–744.

300. M. Jafari and S. A. Hassanzadeh-Tabrizi, Preparation of $CoAl_2O_4$ nanoblue pigment via polyacrylamide gel method, *Powder Technology* 266, 2014, 236–239.

301. M. Jafari, S. A. Hassanzadeh-Tabrizi, M. Ghashang, and R. Pournajaf, Characterization of Ba^{2+}-added alumina/cobalt nanoceramic pigment prepared by polyacrylamide gel method, *Ceramics International* 40, 2014, 11877–11881.
302. A. Han, M. Zhao, M. Ye, J. Liao, Z. Zhang, and N. Li, Crystal structure and optical properties of YMnO$_3$ compound with high near-infrared reflectance, *Solar Energy* 91, 2013, 32–36.
303. A. Han, M. Ye, L. Liu, W. Feng, and M. Zhao, Estimating thermal performance of cool coatings colored with high near-infrared reflective inorganic pigments: Iron doped La$_2$Mo$_2$O$_7$ compounds, *Energy and Buildings* 84, 2014, 698–703.
304. M. Zhao, A. Han, M. Ye, and T. Wu, Preparation and characterization of Fe^{3+} doped Y$_2$Ce$_2$O$_7$ pigments with high near-infrared reflectance, *Solar Energy* 97, 2013, 350–355.
305. M. Dudek, Usefulness of gel-casting method in the fabrication of nonstoichiometric CaZrO$_3$-based electrolytes for high temperature application, *Materials Research Bulletin* 44, 2009, 1879–1888.
306. R. Ianoş and R. Lazău, Chromium-doped calcium zirconate – A potential red shade pigment: Preparation, characterization and testing, *Dyes and Pigments* 105, 2014, 152–156.
307. L. F. Vieira Ferreira, T. M. Casimiro, and P. Colomban, Portuguese tin-glazed earthenware from the 17th century. Part 1: Pigments and glazes characterization, *Spectrochimica Acta Part A: Molecular and Biomolecular Spectroscopy* 104, 2013, 437–444.
308. G. J. Lin, H. Yang, and T. Xian, Polyacrylamide gel synthesis and visible-light photocatalytic activity of TbFeO$_3$ nanoparticles, *Chemical Journal of Chinese Universities* 33, 2012, 1565–1571.
309. S. F. Wang, H. Yang, and T. Xian, A novel visible-light-driven semiconductor photocatalyst: Nano-yttrium manganite, *Chinese Journal of Catalysis* 32, 2011, 1199–1203.
310. Z. B. Shao, C. Y. Wang, X. B. Chen, C. M. Han, and Y. Wang, Photocatalysis and preparation of nanometer-sized ZnO/Ag by polyacrylamide gel method, *Chinese Journal of Materials Research* 19, 2005, 59–63.

9 Perpendicular Magnetic Anisotropy in Magnetic Thin Films

Prabhanjan D. Kulkarni, Somnath Bhattacharyya, and Prasanta Chowdhury

CONTENTS

ABSTRACT

The physical basis that underlies a preferred magnetic moment orientation in magnetic thin films can be quite different from that observed in bulk materials. The presence of an interfacial structure in a multilayered stack is the basic ingredient for this changeover. By varying the thickness of the individual layer and choosing appropriate materials, it is possible to tailor the magnetic anisotropy in thin films. The most dramatic manifestation in this regard is the change of the preferential direction or the easy axis of the magnetization from the commonly observed in-plane orientation to the perpendicular direction. This phenomenon is usually referred to as perpendicular magnetic anisotropy (PMA). In this chapter we have reviewed PMA in thin films; the theory of magnetic anisotropy and crystallographic, structural (strain), and interfacial contributions to PMA. It is important that the films retain the PMA on thermal annealing if they are to be used in device applications. The main reason for the reduction in PMA on annealing is intermixing at the interface, and therefore, the ways to avoid or reduce this intermixing upon annealing are discussed. We have also incorporated the applications of thin films with PMA in magnetic recording and magnetic tunnel junctions with PMA (*p*-MTJ) at the end of the chapter.

9.1 INTRODUCTION

Magnetic anisotropy in a magnetic specimen arises due the directional dependance of the magnetic properties. When a magnetic specimen is subjected to an external magnetic field, the specimen properties depend on both the magnitude and the direction of the applied field. Moreover, depending on the shape and the crystallographic structure, the required field to magnetize a specimen becomes less in a particular direction, known as the *easy axis*, and more in other direction, known as the *hard axis*. In other words, the easy axis is the energetically favorable direction to spontaneous magnetization. Magnetic anisotropy is defined by the difference between the energy required to magnetize in the hard and easy axes directions.

Magnetic anisotropy has a great influence in the field of spintronics. Especially, perpendicular magnetic anisotropy (PMA) observed in magnetic thin films has led to many peculiar properties and initiated technological advancement in devices like magnetoresistive random access memories (MRAMs), spin-transfer-torque random access memories (STT-RAMs), and read heads in magnetic memories.

In this chapter, we will discuss "perpendicular magnetic anisotropy in thin films" in detail. In the first section, we will see the two phenomenological contributions of anisotropy: surface and volume. The origin of magnetic anisotropy and its various contributors are discussed in Section 9.2. Section 9.3 consists of some observations of PMA in magnetic thin films. The effects of thermal annealing on the PMA and film structure is discussed in detail in Section 9.4. Section 9.5 comprises the phenomenon of exchange bias in thin films with PMA, which is essential in various device applications. Then in Sections 9.6 and 9.7, we will briefly see applications of thin films with PMA: in magnetic recording and in magnetic tunnel junctions (MTJs) with a PMA, which have a high potential toward miniaturization of future devices beyond Moore's law. The chapter will be summarized in Section 9.8.

9.1.1 COEFFICIENT OF MAGNETIC ANISOTROPY

The retentivity of a magnetic specimen is strongly correlated with its magnetic anisotropy value. In a PMA system with thin film structure, the PMA should be high enough to overcome the demagnetization field. While fabricating a device with such a material, the stability toward its magnetization is characterized by the Néel–Arrhenius equation [1],

$$\tau = \tau_0 \exp\left(\frac{|K_{eff}|V}{k_B T}\right), \tag{9.1}$$

where τ is the magnetic reversal time, τ_0 is a time constant with the typical value of 10^{-9}–10^{-12} s, k_B is the Boltzmann's constant, T is the temperature in Kelvin; and K_{eff} is magnetic anisotropy energy per volume V, also called the coefficient of magnetic anisotropy. To keep the magnetic states stabilized for 10 years' duration, the ratio $|K_{eff}|V/k_B T$ has to be very high (>60). Hence, to get better thermal stability at smaller volume, the coefficient of magnetic anisotropy (K_{eff}) has to be larger.

9.1.2 SURFACE AND VOLUME CONTRIBUTIONS

The coefficient of magnetic anisotropy, K_{eff}, can be expressed in terms of the phenomenological volume contribution, K_V, and surface contribution, K_S [2], and defined as

$$K_{eff} = K_V + 2K_S/t. \tag{9.2}$$

Equation 9.2 represents a weighted average of the magnetic anisotropy energy of the interface atoms and the inner atoms of a magnetic multilayer stack of thickness t. The equation is presented under the convention that $2K_S/t$ represents the difference between the anisotropy of the interface atoms with respect to the inner or bulk atoms. The factor 2 appears as there are two interfaces per magnetic layer and the interfaces are considered to be identical. In many experimental studies, Equation 9.2 in the form of $t \cdot K_{eff} = t \cdot K_V + 2K_S$ is considered and therefore, the values of the individual contributors K_V and K_S were calculated by plotting $t \cdot K_{eff}$ versus t. In the plot, the slope of the line and Y-intercept relates with the magnitude of K_V and $2K_S$, respectively.

9.2 THEORY

Though magnetic anisotropy can be expressed in its phenomenological interface and volume contributions as given in Equation 9.2, magnetic anisotropy has different characteristic contributors. We will view them in detail in this section.

9.2.1 ORIGIN OF MAGNETIC ANISOTROPY

Magnetic anisotropy originates from two main sources—magnetic dipolar interaction and spin–orbit interaction. The dipolar interaction is a long range interaction and generates a magnetic anisotropy that depends on the shape. Anisotropy of this nature is also commonly known as *shape anisotropy*. Apart from this, the spin–orbit interaction among the atoms in the lattice structure also contributes to magnetic anisotropy. When an external magnetic field is applied, it tries to align the electron spins in the same direction. Because of spin–orbit coupling, this also forces the orbits to adjust their orientation accordingly. But since orbits are strongly coupled to the lattice structure, the attempt to reorient the spin and orbits is resisted. Hence, this results in a total energy which depends on the orientation of the magnetic moments relative to the axes of the lattice. This is known as *magnetocrystalline anisotropy*. This magnetocrystalline anisotropy reflects the crystal symmetry of the specimen.

The above two anisotropies are the main contributors to net anisotropy in bulk materials and contribute to the volume anisotropy, K_V. In 1954, for the first time, Louis Néel predicted [3] a new type of magnetic anisotropy which arises at the interfaces due to the symmetry breaking. He argued that the symmetry breaking at interfaces may develop due to the quenching of orbitals and produce uncompensated magnetic moments. This type of anisotropy is called *Néel's anisotropy* and contributes in surface or interface anisotropy, K_S. Apart from the above, the lattice strain, if present in the structure, may induce an anisotropy contribution which is considered as *magnetoelastic anisotropy*. Similar to magnetocrystalline anisotropy, it originates from spin–orbit coupling. We will see each of the above anisotropies in detail in the following subsections.

9.2.2 SHAPE ANISOTROPY

Shape anisotropy in a magnetic specimen is purely a magnetostatic self energy which originates from the interaction among magnetic dipoles. To understand the effect of the shape of a specimen on developing this anisotropy, let us define the demagnetizing field.

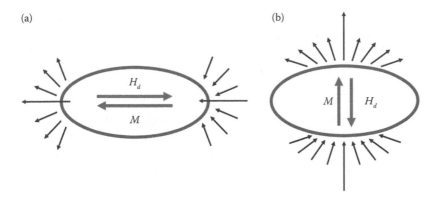

FIGURE 9.1 Magnetization (M) and demagnetizing field (H_d) in an ellipsoid when the external field is applied in the direction of the long axis (a) and short axis (b). H_d is higher for (b) because of smaller distance between dipoles.

Consider an external field H_{ext} applied to an ellipsoid of ferromagnetic material in the direction parallel to the long axis, as shown in Figure 9.1a. This causes the ellipsoid to get magnetized with the north and south poles as shown in Figure 9.1a. As the magnetic lines of force emerge out of the **N** pole and end at the **S** pole, the field line inside the ellipsoid goes in the direction opposite to that of the external field H_{ext}. This develops an intrinsic magnetic field which opposes the external magnetic field H_{ext} [4]. This field is called the demagnetizing field H_d, which originates from the magnetization of the specimen (M) and depends on the shape of the specimen. It can be written as:

$$H_d = N_d \cdot M, \tag{9.3}$$

where N_d is a demagnetization factor and depends solely on the shape of the specimen. Theoretical calculations have shown that N_d is smaller for the longer axis [5]. This means, higher N_d values are expected for field H_{ext} applied as shown in Figure 9.1b and hence higher H_d values. In other words, it is hard to magnetize in the direction as shown in Figure 9.1b and easier in the direction as shown in Figure 9.1a. In the case of magnetic thin films, the in-plane demagnetization field, $N_d^{in-plane}$ is very much lower in comparison to the out-of-plane value, $N_d^{out-of-plane}$. Therefore, the magnetic easy axis always lies within the plane of the film and results in in-plane magnetic anisotropy. The detailed discussion on the demagnetization factor calculation is out of the scope of this chapter, but readers may refer to Reference [5]. For higher aspect ratio, the developed shape anisotropy is higher.

9.2.3 Shape Anisotropy in Thin Films

The shape anisotropy in thin films always favors in-plane magnetic anisotropy and is one of the major contributors to net effective magnetic anisotropy. Therefore, to achieve net PMA, it is essential to overcome the shape anisotropy contribution.

The magnetostatic energy per unit volume for a specimen with its saturation magnetization M_s can be written as

$$E_d = \frac{\mu_o}{2} M_S^2 \cos^2\theta, \tag{9.4}$$

where μ_o is permeability of free space and θ is the angle with the film normal. Lets define $K_{sh} = (\mu_o/2)M_S^2$ such that, $E_d = K_{sh}\cos^2\theta$. We will call this constant K_{sh} the "shape anisotropy constant," which has a unit of *energy/volume*. According to the equation above the shape anisotropy favors in-plane anisotropy in thin films.

9.2.4 SPIN–ORBIT COUPLING

Spin–orbit coupling is the origin of several types of magnetic anisotropies: magnetocrystalline, magnetoelastic, and Néel's anisotropy. The Dirac equation, which defines the electron relativistically, reduces to the Pauli equation in limit of low velocities as

$$H_{Pauli} = \frac{p^2}{2m} - e\Phi - \frac{p^4}{8m^3c^2} + \frac{e\hbar^2}{8m^2c^2} divE + \frac{e\hbar}{4m^2c^2} \sigma \cdot (E \times p), \tag{9.5}$$

where all the notations have the usual meaning. The first four terms of the above equation (9.5) do not consider electron spin ($s = \sigma/2$) and form a scalar-relativistic Hamiltonian. The last term is spin–orbit coupling. It takes into account the coupling between the spin of an electron and the magnetic field generated due to its orbital motion around the nucleus. As the orbital is directly coupled with the crystal structure, this contribution gives rise to magnetocrystalline anisotropy. The spin–orbit interaction is also responsible for the anisotropy contributions for lattice strain and lattice symmetry breaking at interfaces which are magnetoelastic and Néel's magnetic anisotropy, respectively.

9.2.5 MAGNETOCRYSTALLINE ANISOTROPY

As mentioned earlier in this section, spin–orbit interactions are the origin of magnetocrystalline anisotropy. In a well-defined crystal lattice, the electron orbits are linked to the crystallographic structure and they are also coupled with the spins through spin–orbit interaction. These interactions lead to a situation where the spins get aligned to specific crystallographic directions. Therefore, it is energetically favorable to magnetize the crystal in some well-defined directions. These are often referred to as an easy axes of magnetization. There are also some other directions which are energetically less favorable, known as hard axes.

Though there are reports to theoretically derive the magnetocrystalline anisotropy and the spin–orbit interactions from the basic principles [6], the accuracy is inadequate. Therefore, more commonly, the anisotropies are written in the form of phenomenological terms, which is a power series expansions taking into consideration the crystal symmetry, that is, uniaxial and cubic.

9.2.5.1 Uniaxial

For a hexagonal structure, as in hcp Cobalt, the equation for magnetocrystalline uni-axial anisotropy energy density can be written in its simplest form as

$$E_u = -K_1\cos^2\theta + K_2\cos^4\theta, \tag{9.6}$$

where θ is the angle between the magnetization direction and the crystallographic c-axis. The constants K_1 and K_2 are anisotropy constants. In expression (9.6), higher order terms are neglected as the anisotropy constant values are very small. Depending on the values of K_1 and K_2, there are different kinds of anisotropy (i) with $K_1 = K_2 = 0$, the isotropic magnetic system, (ii) with $K_1 > 0$, the magnetic system with an easy axis along the c-axis and in rare cases, and (iii) with $K_1 < 0$, the magnetic system with an easy *plane* perpendicular to the c-axis.

9.2.5.2 Cubic

For the cubic crystals like Fe and Ni, the magnetocrystalline anisotropy shows cubic symmetry and has more than one easy and hard axis. For cubic crystals, the anisotropy energy density can be written as

$$E_c = K_1(m_x^2 m_y^2 + m_y^2 m_z^2 + m_z^2 m_x^2) + K_2(m_x^2 m_y^2 m_z^2). \tag{9.7}$$

Here $\hat{m} = \vec{M}/(|\vec{M}|) = m_x\hat{x} + m_y\hat{y} + m_z\hat{z}$ is a unit vector parallel to the magnetization direction. As we can see from the above equation, if $K_1 > 0$ then easy axes will be in (100) directions, and if $K_1 < 0$ then easy axes will be in body diagonal directions, that is, (111).

9.2.6 NÉEL'S INTERFACE ANISOTROPY

Néel predicted that at the surface/interface, the discontinuity can generate a magnetic anisotropy due to an uncompensated orbital moments, which is now commonly referred to as Néel's anisotropy. Though Néel predicted this effect in 1954 [3], it was confirmed experimentally in 1968 by Gradmann and Müller in ultrathin NiFe films grown on Cu(111) [7].

In ultrathin magnetic films, Néel's magnetic anisotropy plays a crucial role in developing PMA. Especially, if the thickness of a magnetic layer is small enough, then it can overcome the other bulk anisotropy contributors. The three couplings spin–spin, spin–orbit, and orbit–crystal-field act simultaneously, along with uncompensated orbital moments at the interface due to chemical discontinuity, is the source of Néel's anisotropy, and therefore the interface characteristics play an important role in defining it.

9.2.6.1 Néel's Anisotropy and Intermixing at Interfaces

Consider a multilayer structure of ferromagnetic (black) and non-ferromagnetic (white) materials as shown in Figure 9.2. Now the ferromagnetic atoms can have two moments: spin and orbital. A spin moment does not depend on the direction

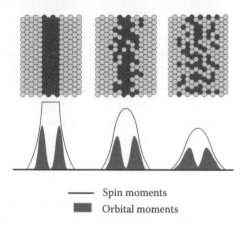

——— Spin moments

■ Orbital moments

FIGURE 9.2 The effect of intermixing at the interface on the spin–orbit interaction in a PM–FM interface. Black circles indicate FM atoms and white circles indicate PM atoms. As shown, the moments, hence the anisotropy is maximum for sharper interfaces and decreases with intermixing.

and contributes only to the magnitude of the anisotropy. On the other hand, though orbital moments have magnitude much smaller than spin moments, they define the direction of magnetic anisotropy. In case of the ideal interface with absence of intermixing, there exists a high spin moment value in the ferromagnetic (FM) layer and at the interface, and it decreases rapidly in an exponential manner while entering into the non-FM layer. Such an exponential decrease in magnetic moment value is due to the induced ferromagnetism in paramagnetic atoms. In other words, the paramagnetic atoms in the vicinity of the ferromagnetic atoms at the interface gain ferromagnetism. Further, due to the presence of chemical discontinuity at the interface, a difference in the chemical bonding arises among the ferromagnetic atoms. This results in uncompensated orbital moments at the interface. These uncompensated orbitals generate an orbital moment in a direction which depends on the bonding strength between ferromagnetic–ferromagnetic (FM–FM) and paramagnetic–ferromagnetic (PM–FM) atoms. If PM–FM bonds are stronger than FM–FM, we expect that the orbital moments are in a direction perpendicular to the interface, that is, out-of-plane direction. The orbital moments will be in in-plane direction when the FM–FM bonds are stronger. In Co/Pt, Co/Pd, Co/Au, etc., the PM–FM bond is stronger and hence the anisotropy at the interface contributes to PMA. Here we will assume this case while understanding the intermixing in Figure 9.2 and will not discuss the other case as that is out of the scope of this chapter.

Therefore in case of an ideal interface we get maximum orbital and spin moments and hence maximum interface anisotropy. Now when intermixing takes place at the interface, the spin moments get reduced (see Figure 9.2). This is because, similar to induced ferromagnetism in paramagnetic atoms, paramagnetism gets induced in ferromagnetic atoms in the vicinity of paramagnetic atoms. Hence, we expect a decrease in spin moment but in the broader area. Similarly, since the abruptness of symmetry breaking is less than that of an ideal case, the orbital moments are also expected to decrease. In short, we get decrease in Néel's anisotropy in an intermixed interface.

In case of a very broad interface, both the spin and orbital moments become very low and Néel's anisotropy ceases to exist.

9.2.7 Magnetoelastic Anisotropy

From the anisotropies explained earlier, it is now clear that the magnetic anisotropy depends on the crystal structure and any changes occurring in the crystal structure will be reflected in the magnetic anisotropy. In fact, the presence of a lattice strain introduces an extra energy term in the total magnetic anisotropy and the direction of this anisotropy depends on the features of the lattice strain (compressive or tensile strain), and its crystallographic direction. The anisotropy contribution due to lattice strain is called magnetoelastic anisotropy energy and can be written as

$$E_{me} = -K_{me}\cos^2\theta, \tag{9.8}$$

Here, θ is the angle between the magnetization and direction of stress σ ($=E\epsilon$, E being elastic modulus and ϵ as strain). The anisotropy constant, K_{me}, can be further calculated as

$$K_{me} = -\frac{3}{2}\lambda\sigma = -\frac{3}{2}\lambda E\epsilon, \tag{9.9}$$

where λ is a magnetostriction constant and can be positive or negative. The magnetoelastic contribution can be of a volume or surface type (refer to Equation 9.2) and mainly depends on how strain is present in the structure. Strain can be induced in a structure in various ways such as thermal strain, growth imperfection during deposition, and lattice mismatch between successive layers. Strain induced due to lattice mismatch plays an important role in defining a magnetic anisotropy in multilayer thin films. Let us review the lattice mismatch strain and its impact on anisotropy in detail.

9.2.7.1 Lattice Mismatch Strain and K_{me}

Let us consider a layer A with the lattice constant d_A being deposited on a layer B of the lattice constant d_B. In this case, two different growth mechanisms are possible (see Figure 9.3): coherent and incoherent.

Then if the thickness of layer A, t_A, is very small, the lattice of layer A gets matched with the lattice of layer B, though they have different lattice constants. This develops a strain in the multilayer structure, and the growth is termed as "coherent growth." Due to compressive and tensile strains arising within the lattice structure, the spin–orbit coupling is modified, especially for the magnetic layer and in turn the

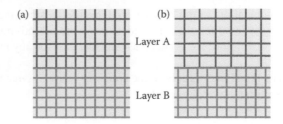

FIGURE 9.3 Two growth mechanisms: (a) presence of lattice strain throughout the layer—*coherent*; and (b) without lattice strain in the layer—*incoherent*.

magnetic anisotropy is changed. In Equation 9.8, the developed anisotropy due to lattice strain can be written in the coherent region as

$$K_{me}^V = \frac{3}{2}\lambda E_A \eta, \tag{9.10}$$

where $\eta = (d_A - d_B)/d_A$ is a lattice misfit. The superscript "V" denotes that it contributes to volume anisotropy (most common unit is erg/cm³). This equation holds good only for *coherent* growth with lattice matching.

In general, when the thickness t_A is restricted to few nanometers, a coherent structure with a large lattice strain exists. With further increasing of the value of t_A, it becomes energetically favorable to induce a lattice misfit and simultaneously this will release the strain. Hence, the coherent growth persists only till a critical thickness, say t_C, and above t_C we get *incoherent* growth with a lattice misfit at the interface. Den Broeder et al. [8] had formulated an equation for this critical thickness t_C as

$$t_C = \frac{2Gb}{|\eta| E_A}, \tag{9.11}$$

where G the shear modulus and b is the Burgers vector of the dislocation. This equation is developed for the case of single interface—layer A on layer B. But if layer A is sandwiched between the two layers of B then t_C gets enhanced four times.

The strain in the incoherent region is comparatively complex in nature. Theoretically, the strain depends on the thickness of a layer t_A, and decreases with it. In this case the energy minimization gives the strain in layer A as [9,10]

$$\varepsilon_A = -\frac{\eta}{2}\left(\frac{t_C}{t_A}\right). \tag{9.12}$$

Replacing ϵ in Equation 9.9, it becomes

$$K_{me} = \frac{3}{4}\lambda E_A \eta \frac{t_C}{t_A}. \tag{9.13}$$

It also can be expressed as

$$t_A \cdot K_{me} = \frac{3}{4} \lambda E_A \eta t_C. \tag{9.14}$$

Let $t_A \cdot K_{me} = 2K_{me}^s$ (the constant 2 is added to take into account the two interfaces of a sandwiched layer), then we get

$$K_{me}^s = \frac{3}{8} \lambda E_A \eta t_C. \tag{9.15}$$

This is the magnetoelastic anisotropy constant in the incoherent region and contributes to a surface anisotropy (the common unit is erg/cm^2). Remember, that K_{me}^s does not depend on η and E_A, as these terms cancel against t_C. Simplifying Equation 9.14 by putting t_C from Equation 9.10 we get for $\eta < 0$,

$$K_{me}^s = -\frac{3}{4} Gb\lambda. \tag{9.16}$$

Therefore, the strain in a magnetic layer develops magnetic anisotropy and this anisotropy contributes to K_V in coherent growth and K_S if the growth is incoherent. It should be noted that the above explanation holds for $\eta < 0$ as it develops the magnetoelastic anisotropy in the out-of-plane direction contributing PMA and it is also expected that for $\eta > 0$ the contribution is toward the in-plane direction. Also, in this chapter the anisotropy constant $K > 0$ denotes the contribution is toward PMA. It is to be mentioned that the magnitude of K_{me} is found to be one order higher than K_{mc} in most cases [11]. This suggests that even a little amount of strain contributes considerably to the net magnetic anisotropy.

Also, we get a phenomenon which is opposite of the strain-induced anisotropy. The phenomenon is called *magnetostriction*. When a specimen is magnetized in a particular direction, strain gets induced in ferromagnetic material. These effects are governed by two energies: elastic energy and magnetoelastic energy. It may be thermodynamically favorable, in a particular system, to reduce the magnetoelastic energy by inducing strain. This trend is counter balanced by the elastic energy, which increases with strain.

9.2.8 OTHER ANISOTROPY CONTRIBUTIONS FROM MAGNETIC DIPOLAR INTERACTIONS

As discussed in Section 9.2.2 before, the dipolar interaction leads to a magnetic anisotropy which depends on the shape of the specimen. The dipolar interaction also contributes to the magnetocrystalline, magnetoelastic, and surface anisotropies. In this section, these contributions are discussed.

In cubic crystals (Fe and Ni), it can be shown from symmetry arguments that the net magnetocrystalline anisotropy from magnetic dipolar interactions is zero. On the other hand, for hcp crystals (as the case for bulk Co) generally we get a net non-zero dipolar magnetocrystalline anisotropy. The contribution depends on the c/a ratio of the hcp structure. It vanishes for $c/a = \sqrt{8/3}$ (≈ 1.63299) and for the ratio not equal to $\sqrt{8/3}$, we get a nonzero contribution from the dipolar magnetocrystalline anisotropy. For Co hcp $c/a = 1.622$, which generates the dipolar magnetocrystalline anisotropy with $K_{mc}^{dip} = 5.7 \times 10^4$ erg/cm^3 (=4.7×10^{-7} eV/atom) [11]. But the point to be noted is that this anisotropy is much smaller than the conventional magnetocrystalline anisotropy (for Co hcp $K_{mc} \approx 5 \times 10^6$ erg/cm^3) and can be considered negligible.

As the symmetry is lowered in strained cubic structures, we get the magneto-elastic contribution to total magnetic anisotropy originated due to dipolar inter-actions. Becker [12] has formulated the equation for this dipolar magnetoelastic contribution as

$$K_{me}^{dip} = -3\alpha M_S^2, \tag{9.17}$$

where α is 0.6 for fcc and 0.8 bcc structures. The calculated K_{me}^{dip} in Fe and Ni by R. Becker is $\approx -7.3 \times 10^6$ erg/cm^3 and -5.0×10^5 erg/cm^3, respectively. These values are much smaller in magnitude than other contributions (in general $K_{me}^{dip} < 10\%$ of K_{me}^{V}) and hence in practical cases other contributions dominate.

Similarly, as discussed earlier, the symmetry gets lowered also at surfaces/inter-faces. Due to this, a dipolar magnetic anisotropy gets developed at surface/interface. This dipolar surface contribution can be written as

$$K_S^{dip} = -2\pi M_S^2 dk_s. \tag{9.18}$$

In this equation, d is the distance between neighboring atomic planes and k_s is the surface parameter and depends on the surface plane. In Reference [13] the calculated values for different surfaces are mentioned.

It is to be noted that, the above-mentioned anisotropy contributions due to dipolar interactions (viz., dipolar magnetocrystalline, dipolar magnetoelastic, and dipolar surface) have comparatively a negligible effect on the net magnetic anisotropy. Therefore, in general, these effects are not considered while quantifying the anisotropy of a system.

9.2.9 FINDING MAGNETIC ANISOTROPY CONSTANTS FROM EXPERIMENTAL OBSERVATIONS

The magnetic anisotropy constants can be derived experimentally from magnetic hysteresis curves. To derive those parameters, the magnetic hysteresis curves are to be measured along both magnetic easy and hard axes using instruments like the vibrating sample magnetometer (VSM), SQUID, etc. When the field is applied along the *easy axis*, the external field required to get magnetic saturation, let us define as H_{sat}^{easy}, is *minimum*. Likewise, the value of H_{sat}^{hard} is *maximum* when the field is applied in the direction of one of the *hard axes*. Figure 9.4 is an example of the *M–H* hysteresis curves measured in both easy (gray) and hard (black) axes.

A magnetic hysteresis curve is defined by different physical parameters, such as coercive field (H_C), remnant magnetization (M_R), and saturation magnetization (M_S). These parameters are shown in Figure 9.4. The value of M_S is the same irrespective of the direction. However, this is not true for a material, in which the anisotropy energy is larger than the exchange energy. For such materials, if exist, M_S is different in different directions. The term H_K is generally called the *anisotropy field* and there is a difference between fields required for saturation along the hard and easy axes:

$$H_K = H_{sat}^{hard} - H_{sat}^{easy}. \tag{9.19}$$

Then the overall or effective anisotropy constant can be calculated from experimental observations using the equation

$$K_{eff} = \frac{H_K \cdot M_S}{2}. \tag{9.20}$$

Here, if H_K is in Oersted (or Gauss) and M_S is in emu/cm³ then we get K_{eff} in erg/cm³. Note that the value we get from the above equation comprises all the anisotropies we have mentioned in this section. In thin films, $K_{eff} > 0$ gives PMA, and $K_{eff} < 0$ corresponds to in-plane anisotropy.

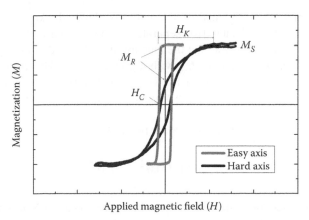

FIGURE 9.4 Typical *M–H* hysteresis curves. The curves are measured with the applied field in easy axis (gray) and hard axis (black) directions. The corresponding parameters like saturation magnetization (M_S), coercive field (H_C), remnant magnetization (M_R), and anisotropy field (H_K) are also shown in the figure.

9.2.10 INDIVIDUAL ANISOTROPY CONSTANTS

Recalling Equation 9.2, the same can be written as

$$t \cdot K_{eff} = 2K_S + t \cdot K_V. \qquad (9.21)$$

This can be considered as an equation of straight line with $2K_S$ as y-interception and K_V as a slope. Further, these phenomenological anisotropies can be subdivided into individual anisotropies as

In coherent region:

$$K_S = K_N, \qquad (9.22)$$

and

$$K_V = -K_{sh} + K_{mc} + K_{me}^V. \qquad (9.23)$$

In incoherent region:

$$K_S = K_N + K_{me}^S, \qquad (9.24)$$

and

$$K_V = -K_{sh} + K_{mc}. \qquad (9.25)$$

A typical plot of $t \cdot K_{eff}$ against the magnetic layer thickness t is shown in Figure 9.5. This plot indicates two separate regions having different slopes with crossover at $t = t_C$. For $t < t_C$, we get the increase in $t \cdot K_{eff}$ with t, giving $K_V > 0$; and for $t > t_C$, we get a decrease in $t \cdot K_{eff}$ with t, giving $K_V < 0$. These two regions correspond to different growth mechanisms: I-*coherent* and II-*incoherent*. Applying the equations mentioned above in these two regions we get the following:

1. If M_S is the saturation magnetization per unit volume then,

$$K_{sh} = (\mu_o/2)M_S^2 = 2\pi M_S^2.$$

2. The y-intersection of the line in the coherent region gives Néel's anisotropy constant. The y-intersection $\Rightarrow 2K_S$ and in the coherent region $K_S^c = K_N$.
3. The slope of line $= K_V$, and in the coherent region,

$$K_V^c = -K_{sh} + K_{mc} + K_{me}^V.$$

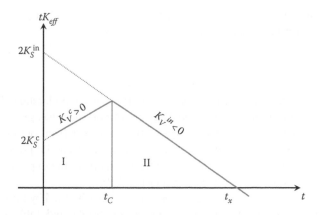

FIGURE 9.5 Expected theoretical graph of $t \cdot K_{eff}$ against the thickness of the magnetic layer, t. The slope of the straight line gives volume contribution K_V and the y-intercept gives the surface contribution K_S. Region I with positive slope corresponds to coherent growth whereas region II with negative slope corresponds to incoherent growth. The changeover from region I to region II happens at $t = t_C$. The magnetic anisotropy changes from PMA to in-plane anisotropy at $t = t_x$ with dominating shape anisotropy.

Therefore, $K_{me}^V = K_V^c + K_{sh} - K_{mc}$.

4. In the incoherent region, the value of the y-intersection gets increased, which is as expected, because,

$$K_S^{in} = K_N + K_{me}^S.$$

Therefore, we get, $K_{me}^S = K_S^{in} - K_N$.

5. Similarly, we have $K_V = -K_{sh} + K_{mc}$ in the incoherent region, therefore,

$$K_{mc} = K_V^{in} + K_{sh}.$$

9.3 EXPERIMENTAL OBSERVATIONS OF PMA IN MAGNETIC FILMS

As noted in theory section, there are different types of magnetic anisotropies present in a system. The net magnetic anisotropy in this system is nothing but the sum of all individual contributions. Because of the dipolar interactions (otherwise referred to as the shape anisotropy), the magnetic thin films show in-plane magnetic anisotropy. Therefore, to reorient the anisotropy in the direction perpendicular to the plane, we need to develop the anisotropies based on the spin–orbit interactions which can dominate over the shape anisotropy. In this section, we will look at some of the thin film systems which show a net PMA, instead of conventional in-plane anisotropy, and the sources for this changeover.

9.3.1 Cu/Ni

Extensive studies have been carried out by several groups to understand the magnetic anisotropy with different FM materials [2]. The importance was given to highlight the critical thickness of the FM layer where the crossover from out-of-plane to in-plane anisotropy occurred. We will look into some experimental reported results and will calculate the individual anisotropy contributions from the experimental results by following the steps as mentioned in Section 9.2.10.

Normally, magnetic anisotropy contributions can be obtained by fitting the plot of $t \cdot K_{eff}$ versus t to Equation 9.2. For example, see Figure 9.6 and Table 9.1 which are reproduced from Reference [14]. The plots in Figure 9.6 show $t \cdot K_{eff}$ versus t curves for Cu/Ni/Cu stacks with two different crystalline orientations and are qualitatively similar to the plot shown in Figure 9.5. Both the curves maintained a positive slope for $t_{Ni} < t_c$ and at $t = t_c$ the slope becomes negative.

Table 9.1 comprises the critical thickness (t_C) along with different anisotropy contributions above and below t_C extracted from plots of Figure 9.6. The critical thicknesses (t_C) reported were 13 Å (32 Å) and 15 Å (42 Å) for a trilayer (sandwich) in (111) and (100) crystal orientations, respectively. It is to be noted that this study had reported negative Néel's anisotropy. This means that, in a given case, Néel's anisotropy is favoring the in-plane direction. This can be understood by considering a simple ligand field model developed by J. Stöhr [15]. The model developed by J. Stöhr restates that the orbital moment of an atom becomes anisotropic at the interface due to quenching effects and further suggests that the orbital moment will be larger in the direction of stronger atomic bonding. And considering that the Ni–Ni bond strength is more than that of Cu–Ni [16], we get a situation with larger orbital moment in the in-plane direction resulting in $K_N < 0$.

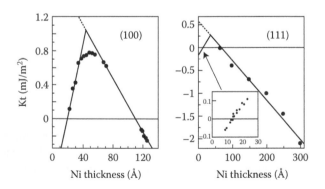

FIGURE 9.6 The PMA per area, $t \cdot K_{eff}$, as a function of the magnetic layer thickness (t). Also it shows that in Cu/Ni/Cu, PMA is more in (100) orientation. (Reprinted from R. Jungblut et al., *J. Appl. Phys.* 75, 1994, 6424–6426.)

TABLE 9.1

Magnetic Anisotropy Data of Cu/Ni Multilayers

			t_C (Å)		
Orientation	Layer	Sandwich	K below t_C	K above t_C	Stress Induced Anisotropy
(111)	13	32	$K_N = -0.08$ mJ/m^2	$K_S = +0.27$ mJ/m^2	$K_{me}^S = 0.35$ mJ/m^2 (0.64)
	(11)	(35)	$K_V = +0.13$ MJ/m^3	$K_V = -0.09$ MJ/m^3	$K_{me}^V = 0.22$ MJ/m^3 (0.36)
(100)	15	42	$K_N = -0.4$ mJ/m^2	$K_S = +0.9$ mJ/m^2	$K_{me}^S = 1.3$ mJ/m^2 (0.8)
	(12.5)	(40)	$K_V = +0.39$ MJ/m^3	$K_V = -0.16$ MJ/m^3	$K_{me}^V = 0.55$ MJ/m^3 (0.39)

Source: Reprinted from R. Jungblut et al. *J. Appl. Phys.*, 75, 1994, 6424–6426.
Note: The theoretical values are mentioned in parentheses.

9.3.2 Co/Pt

The lattice mismatch in Co/Pt is about 10.66% (considering Co/Pt (111) lattice), higher than most of the systems. Therefore it is possible to have higher PMA at lower thicknesses due to more strain in Co. At the same time it is expected to have smaller t_C as the mismatch, η, is more.

In 1991, Lin et al. [17] reported that the lattice orientation plays an important role in defining the magnetic anisotropy. They have observed M-H hysteresis curves in [Co/Pt] multilayer stack with (001), (110), and (111) orientations and reported that PMA is more in the (111) orientation (the observations from Reference [17] are shown in Figure 9.7). The study shows different surface and volume magnetic anisotropies for different orientation.

We have carried out magnetic anisotropy studies on [Co/Pt] multilayer films with varying Co layer thicknesses (t_{Co}) ranging from as low as 0.2–2 nm. The nonmagnetic Pt layer thickness (t_{Pt}) was kept constant at 1.2 nm.

The magnetic hysteresis curves of the multilayer stack with varying t_{Co} are shown in Figure 9.8. The stacks showed PMA with positive slope for $t_{Co} < t_c$ and the slope becomes negative at $t_c = 0.6$ nm. The cross-over from out-of-plane (PMA) to in-plane anisotropy is observed at $t_{Co} \approx 1.35$ nm, where K_{eff} became zero. For $t_{Co} > 1.35$ nm, K_{eff} became negative. It is to be noted here that the saturation magnetization (M_S) is observed to be more than that of bulk Co, especially at lower Co thicknesses. This is due to the contribution from Pt atoms [18]. When a Pt atom is surrounded by Co atoms, it gains ferromagnetism due to its high Stoner factor and acts as a ferromagnetic atom. This, in turn, gives extra contribution to magnetic saturation per Co volume, M_S [19].

More interesting is the plot of $t_{Co} \cdot K_{eff}$ versus t_{Co}. As shown in Figure 9.9, we can distinguish three regions based on slope variation. The straight lines are guidelines to the eye. The first region has a positive slope (i.e., $K_V > 0$) and is a region of coherent growth (readers may recall Figure 9.5). In this region, Co layers have a lattice strain which develops a magnetoelastic contribution to volume anisotropy, termed as K_{me}^V in this chapter. As explained in theory section, for higher t_{Co} the strain relaxation takes place, and thus the shape anisotropy becomes the dominating part in

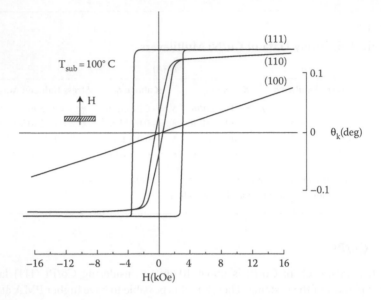

FIGURE 9.7 Dependence of structural orientation on the magnetic anisotropy in [Co/Pt] multilayers. (111) orientation gives better PMA. (Reprinted from C. J. Lin et al., *J. Magn. Magn. Mater.* 93, 1991, 194–206.)

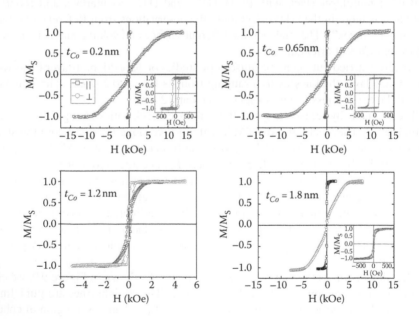

FIGURE 9.8 The magnetic hysteresis curves of Co/Pt multilayer stack with different Co layer thicknesses with an external field applied in direction parallel (square) and perpendicular (circle) to thin film. It can be observed that the easy axis of magnetic anisotropy gradually changes from out-of-plane to in-plane with increase in the FM thickness (t_{Co}).

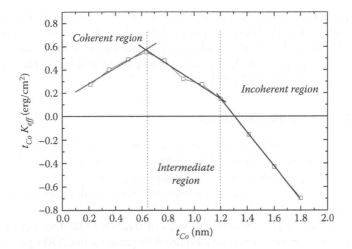

FIGURE 9.9   Plot of $t_{Co} \cdot K_{eff}$ against t_{Co}. The trend is similar to what is shown in Figure 9.5. At lower t_{Co}, the positive slope indicates a coherent growth. The growth becomes incoherent at higher t_{Co}, giving negative slope.

K_V. This makes the slope of a graph K_V negative (see Figure 9.5 for details). This region is mentioned here as the incoherent region and is region III in the graph. The experimental data indicates an intermediate region between transition from coherent to incoherent growth. The magnetic anisotropy constants of different types are given in Table 9.2.

Using the equations given in Section 9.2.10 for coherent and incoherent regions, we derive the following parameters for the Co/Pt multilayer stack:

(For the calculations, the magnetic saturation of Co is considered as $M_S = 1440$ emu/cm³.)

$$K_N = K_S^c = 0.15/2 = 0.075 \text{ erg/cm}^3.$$
$$K_{sh} = 2\pi M_S^2 = 1.302 \times 10^7 \text{ erg/cm}^3.$$
$$K_{mc} = K_V^{in} + K_{sh} = 0.183 \times 10^7 \text{ erg/cm}^3.$$
$$K_{me}^S = K_S^{in} - K_N = 1.85/2 - 0.075 = 0.85 \text{ erg/cm}^3.$$
$$K_{me}^V = K_V^c + K_{sh} - K_{mc} = 1.729 \times 10^7 \text{ erg/cm}^3.$$

TABLE 9.2

The Surface (K_S) and Volume (K_V) Contributions to Magnetic Anisotropy, Calculated from Figure 9.9

	$2K_S$ (erg cm⁻²)	K_V (×10⁷erg cm⁻³)
Region I	0.15	0.61
Region II	1.02	−0.70
Region III	1.85	−1.12

Note that here $K_N > 0$, which is as expected in Co/Pt multilayers. This is because the out-of-plane Co–Pt bond is approximately 1.6 times stronger than the in-plane Co–Co bond [15]. And according to a simple ligand field model proposed in Reference [15], this gives the larger orbital moment of Co atoms in the out-of-plane direction and in turn makes $K_N > 0$.

9.3.3 Structural Investigations

As mentioned before, in multilayers with ultrathin individual layer thicknesses, the lattice strain plays an important role in defining magnetic anisotropy. The presence of considerably high magnetoelastic anisotropy ($K_{me}^V = 1.729 \times 10^7$ erg/cm³) at lower FM thicknesses indicates a coherent growth of Co/Pt multilayer forming a superlattice structure.

To understand the growth in detail, x-ray diffraction (XRD) studies on the multilayer stack have been carried out. Figure 9.10 presents the XRD patterns of the multilayer stack with varying Co layer thickness. Interestingly, for $t_{Co} = 2.1$ Å, only one peak at 39.88° was observed. This peak is very near to bulk Pt (111) (39.80°) and corresponds to the Co/Pt(111) coherent structure.

This indicates that the Co layer, which is sandwiched between two Pt layers, gets strained to attain the lattice parameter the same as that of the Pt layer. Nevertheless, with increase in Co layer thickness, this Co/Pt (111) peak shifts gradually away from bulk Pt (111) and toward bulk Co (111) position (44.1°). This suggests the decrease in strain in the Co layer.

High-resolution lattice image in transmission electron microscopy (TEM) mode (parallel static illumination) designated as the HRTEM image of Co/Pt multilayers on an Si (100) substrate is presented in Figure 9.11a where the visibility of the multilayer structures is almost absent. This image is sensitive to the

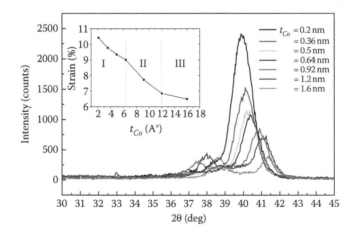

FIGURE 9.10 XRD pattern of samples with different t_{Co} showing qualitatively the decrease in strain in Co layer with increase in thickness. Insert shows a qualitative strain in the Co layer. (Adapted from P. Chowdhury et al., *J. Appl. Phys.* 112, 2012, 023912.)

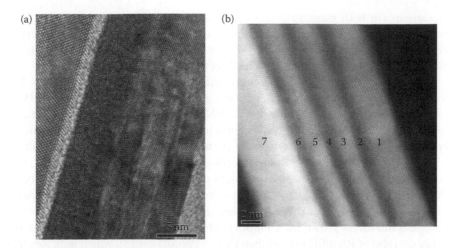

FIGURE 9.11 (a) HRTEM image is unable to reveal the multilayer structure due to coherent lattice growth. In contrast, (b) HAADF-STEM image shows a Co/Pt multilayer structure (dark layers represent Co which are sandwiched between brighter Pt layers).

variation in lattice parameters and XRD studies confirmed that the multilayer stack consists of similar lattice parameters throughout the structure (indicating a coherent growth of Co/Pt stack). The differences in lattice parameters within layers are much smaller than the spatial resolution of the image presented in Figure 9.11a which hinders the clear representation of the multilayer structure. In contrast, even the medium magnification image taken on a high angle annular dark field (HAADF) detector in scanning transmission electron microscopy (STEM) mode (convergent rastering illumination) designated as a HAADF-STEM image represents clear visibility of the multilayer structure. The HAADF-STEM image intensity is very sensitive to the square of the atomic number (Z) of the constituent elements within the imaged region. The large difference in atomic number of Co (27) and Pt (78) is the reason for the clearly distinguishable multilayer structure in Figure 9.11b with bright and dark layers representing Pt- and Co-enriched regions, respectively.

The results, hence, establish that the sample contains a coherent multilayer Co/Pt stack. It also suggests, from the magnitudes of different anisotropy contributions that the coherency strain developed in Co is in fact the main contributor to PMA along with Néel's anisotropy. Readers may refer to References [18] and [20] for a more detailed description.

9.3.4 THIN ALLOY FILMS

Apart from the multilayer films discussed in the preceding subsections, one more way to induce PMA is in fabricating the thin films of magnetic alloys. As we know, the lattice of alloys can be different from the materials involved in it. This, along with the fact that the nodes of lattice contain different atoms, makes the spin–orbit interactions (in both magnitude and direction) different from that of the bulk of contributing magnetic elements. Therefore, the alloying of appropriate materials with appropriate parameters can give the magnetocrystalline and/or magnetoelastic anisotropy contributions an out-of-plane direction and lead to net PMA.

In alloys, magnetic anisotropy strongly depends on the lattice characteristics. For example, the study [21] based on alloys of $Fe_{1-x}Co_x$ grown on buffer layers of Pd(001), Ir(001), and Rh(001) indicated that the net PMA can be induced due to the tetragonal distortions in these otherwise cubic structures. Figure 9.12 shows the schematic representation of magnetic anisotropy in these films at different Co concentrations "x" [21]. The induced uniaxial perpendicular anisotropy (PMA), K_U at 60 K and room temperature are represented by solid curves and dotted curves, respectively. The uniaxial PMA K_U (and hence PMA) is maximum in $Fe_{0.5}Co_{0.5}$ alloys. Further, it has been observed that the PMA is more in films grown on a Rh(001) substrate than the same films grown on a Pd(001) substrate. The reason behind this is the higher c/a ratio of Rh (001) (=1.24) than that of Pd(001) (=1.13), which indicates that Rh(001) can induce more distortion. This shows the importance of the substrate lattice constant and the distortion in thin films in deciding the net magnetic anisotropy. Further, since the distortions are likely to get removed with thicker thin films, the PMA gets reduced with the thickness of a film and vanishes at a thickness higher than 15 ML.

One more way to generate the PMA in thin alloy films is by enhancing the Néel's interface anisotropy contribution. If the thickness of the alloy film is small enough, the Néel's anisotropy dominates the dipolar (shape) anisotropy and gives net PMA. This effect is well demonstrated by Ikeda et al. [22] in MTJs of CoFeB/MgO/CoFeB with PMA. This study is very crucial from the technology point of view and is discussed separately in Section 9.7.2.

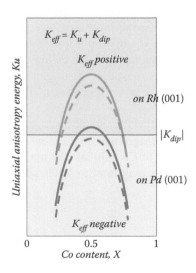

FIGURE 9.12 Schematic representation of effective anisotropy with respect to *Co* concentration "*x*." (Adapted from F. Yildiz et al., *Phys. Rev. B* 80, 2009, 064415.)

Alloys with PMA are gaining more interest nowadays in STT systems especially because of their low damping constant α comparative to most common multilayer thin films of Co, Fe, with Pt, Pd, etc. Though the multilayer films give a very high PMA, the use of heavy elements like Pt gives a higher damping constant α (e.g., typical α of Co/Pt is more than 0.1[23]). This makes them less attractive for devices like MTJs, as a higher α leads to higher switching currents for current induced magnetization switching [22]. On the other hand, the magnetic alloys like CoFe, CoFeB, and NiFe provide a smaller damping constant. In fact, Mizukami et al. [24] have demonstrated $\alpha = 0.008$ for $\delta = 1.46$ in $Mn_{3-\delta}Ga$ alloy films with PMA of about 15×10^6 erg/cm³. Recently [25], low α (up to 0.007) with good PMA (14 ± 1 M erg/cm³) is reported in $L1_0$ ordered FePd alloys films grown on $SrTiO_3$ substrates when annealed at 500°C. These studies for thin films with large PMA and low damping ratio are gaining in importance as these films are promising for use in devices based on STT mechanism.

9.3.5 Dependence of PMA on the Buffer Layer

The characteristics of structural growth is crucial in defining the magnetic anisotropy of thin magnetic films. In this regard, the quality of substrate on which the thin film is grown plays an important role, mainly because it is one of the factors affecting the crystal orientation (i.e., preferred texture growth) and interface quality, especially when the thickness of the film is a few nanometers or less. In particular, to have a good PMA in most magnetic thin films, sharper interfaces are required and it is difficult to have a PMA on rougher interfaces. Therefore, to reduce the substrate roughness and to provide a smooth surface for magnetic material, buffer layers can be deposited over the substrate. Further, it is always favorable to have a magnetocrystalline anisotropy (which depends on crystal structure, as mentioned in Section 9.2.5) with an easy axis out-of-plane, since this ensures a better PMA. The buffer layer can also be used to ensure the growth with a particular crystal orientation such that it provides magnetocrystalline anisotropy in an out-of-plane direction.

In general, materials like Ta [18,26–28] and Ru [21,22] are considered for buffer layers mainly because these layers have a tendency to reduce roughness and provide a better crystal structure. For example, the study on Co/Pt films showed an enhancement in PMA in multilayer films when grown on 5 nm Ta/10 nm Cu, instead directly on 10 nm Cu [29]. The same study has reported that the Co/Pt (111) texture, which is known to favor PMA, is considerably more on a 5 nm Ta/10 nm Cu buffer. The degree of (111) alignment is proportional to $1/2\sigma$, where 2σ is the standard deviation of Gaussian fit to XRD rocking curves. An enhancement in $1/2\sigma$ from 0.047 to 0.250 is reported in Reference [29] with the addition of a Ta layer in the buffer, inferring a better Co/Pt (111) texture, which gives enhancement in PMA from 446 to 921 kJ/m³ [4].

9.4 THERMAL STABILITY OF PMA

The thermal stability is critically important for device application as most of the device fabrication techniques require a post annealing process. It is reported that post annealing causes different structural changes in multilayer films based on the annealing temperature and duration, materials used, and the layer thicknesses. Further, magnetic anisotropy gets affected by lattice strain and grain boundaries which are highly sensitive to post annealing treatment. But it is the intermixing through interface between the layers which plays an important role in defining magnetic anisotropy and in fact is one of the main reason of the disappearance of PMA upon thermal annealing.

We will look into some interesting mechanisms in magnetic multilayers which gives stable PMA even on annealing. First is the transient interface sharpening which is reported in miscible systems. In this process, the intermixing takes place, but it is the uneven rate of intermixing which makes the interface sharper. Then we will look into ultrathin superlattice multilayer thin films, which has a so-called ordered-alloy-like structure. This ordered-alloy-like structure suppresses intermixing and keeps the structure unchanged on annealing keeping the PMA intact. Then we will see the interface sharpening in multilayers due to "uphill" or "back" diffusion which, in fact, has reportedly enhanced PMA upon annealing in multilayers like Co/Au and Co/Pt [20,30].

9.4.1 TRANSIENT INTERFACE SHARPENING IN MISCIBLE SYSTEMS

Materials are said to be miscible when they are mutually soluble. While annealing the multilayer stack of miscible materials, at a certain point, the multilayer structure ceases to exist and the intermixed state gets formed. But, depending on the coefficient of diffusion of the two materials, the intermixing takes place through an intermediate interface sharpening. And in magnetic multilayers if this stage is achievable in the process of post annealing, the PMA remains intact and often may get enhanced, due to chemically sharper interfaces.

In 2002, by using computer simulations based on deterministic kinetic equations and the Monte Carlo technique, Erdélyi et al. [31] showed that during a diffusion process in two mutually soluble materials, the initial broader interface can become chemically sharper. This transient interface sharpening is possible if the diffusion coefficient (D) strongly depends on the localized concentration. This can be understood by considering Fick's first law which states the formula for atomic flux as

$$j = -D \ grad \ C, \qquad (9.26)$$

with C and D being the concentration and diffusion coefficients, respectively. In a miscible system $D > 0$ and in immiscible systems $D < 0$. Therefore in miscible systems the direction of the atomic flux is opposite to the direction of the concentration gradient, causing intermixing. In the original Fick's first law, D was considered as a constant which is independent of concentration and is the same throughout the system. However, at the interfaces of the multilayers structure where the localized concentration itself is changing, Erdélyi et al. [31] suggested the concentration dependance of D and modified Fick's first law. In their assumption, D varies as defined by the formula

$$D(C) = D(0) \ exp(mC). \qquad (9.27)$$

In Equation 9.27, the compositional parameter "m" characterizes the degree of asymmetry of the diffusion coefficient. With concentration dependent D, it is possible to get a interface sharpening, especially if there is an asymmetry in partial mobilities.

Now let us consider Figure 9.13, which is reprinted from Reference [31]. This figure depicts the thermally driven diffusion at the Ni/Cu interface. In this figure, the x and y axes represent the layered structure of two elements Ni and Cu and the atomic fraction of Ni in the corresponding atomic layer, respectively. Therefore, the values of $y = 1$ and 0 represent the Ni and Cu layers, respectively. The initial concentration gradient throughout the interface is considered constant. Nevertheless, the localized concentration of Ni is considered to be changing linearly, which gives different values of $D(C)$ at different points of interface. Hence, we get a situation with parabolic increase in the atomic flux j at different points of the interface as illustrated in Figure 9.13. This asymmetric atomic flux distribution then leads to a situation with sharper interface (unfilled circles in Figure 9.13). Note that if the annealing time is sufficiently large then the system eventually becomes homogeneous.

This type of transient sharpening has been observed experimentally [32] in Mo/V multilayers by XRD technique. The study reported that the interface thicknesses decreased by about a factor of 2 from 1.4 to 0.78 nm after annealing the Mo/V multilayers at 973 K. The authors of [32] reported that as per this mechanism, as the interface becomes sharper, the diffusion rate (or j) becomes smaller. In fact, at a fixed temperature, they reported no observable change at the interface after a certain time. This suggests that practical stability can be achieved in these multilayers after annealing.

Recently, similar interface sharpening, has been experimentally reported in a miscible Cu/Ni multilayer system, by using atomic probe tomography measurements [33]. Figure 9.14 represents the experimental results of the as-prepared and annealed samples (reprinted from Reference [33]). The authors of this study have divided the multilayers into three regions. The first region consists of almost pure Ni, the thickness of which is decreased on annealing. The second is the Cu layer, the Ni content in which is increased considerably after annealing (up to 20%). The observations

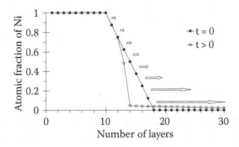

FIGURE 9.13 Schematic drawing of the flux distribution in the initial state in the case of concentration-dependent diffusion coefficients. The arrows represent the atomic flux and their lengths are proportional to the absolute value of the flux. (Adapted from Z. Erdélyi, I. A. Szabó, and D. L. Beke, *Phys. Rev. Lett.* 89, 2002, 165901.)

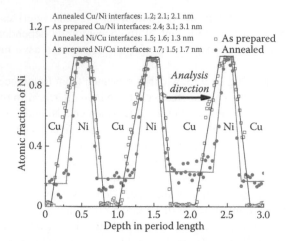

FIGURE 9.14 Transient interface sharpening in Cu/Ni multilayer results of atomic probe tomography. Ni concentration profile of as-prepared (open symbol) and annealed specimen (full symbol) is shown. Results also indicate the mobility difference, as Ni layers remain almost free of Cu. Cu layers grow but absorb a large amount of Ni. (Reprinted from Z. Balogh et al., *Appl. Phys. Lett.* 99, 2011, 181902.)

also suggest a decrease in the Ni layer thickness (first region) and increase in Cu layer thickness (second region). The third region consists of the interfaces. In the as-prepared multilayer the two interfaces—Cu/Ni and Ni/Cu—were not symmetrical. In fact the Cu/Ni interface has an average width of 1.4 nm and the Ni/Cu interface has the average width of 3.1 nm. On the other hand, the interfaces seem symmetrical and sharper in the annealed sample. The average interface thicknesses reported were 1.5 nm and 1.8 nm for Cu/Ni and Ni/Cu interfaces, respectively, after annealing at 773 K for 15 min.

This mechanism of having transient interface sharpening in miscible multilayer systems can give an enhanced PMA in the magnetic multilayers upon annealing, due to better Néel's interface anisotropy. But, as this is a transient response, the annealing time plays a crucial role. On the other hand, the detailed theoretical or experimental study to understand the effect of the transient interface sharpening on magnetic properties such as magnetic anisotropy of multilayer films is still lacking.

9.4.2 Ultrathin Ordered-Alloy-Like Structures with PMA

The thin film with PMA are used directly to develop a perpendicular magnetic tunnel junctions (p-MTJs) device. The specific application of p-MTJs is as a memory cell in STT devices like STT-RAMs and MRAMS. To achieve long-term stability for a pillar with a diameter of 20–30 nm, MTJ should have the PMA $K_{eff} > 10^6$ erg/cm^3. At the same time, low STT switching currents demand that the thickness of the free layer should be as thin as possible.

In 2010, Yakushiji et al. [34] studied ultrathin Co/Pt and Co/Pd multilayer films that showed K_{eff}: 3–9 × 10^6 erg/cm³ with a total layer thickness in 1–3 nm. The films possessed a very good thermal stability and, as shown in Figure 9.15a and b, the PMA remained unchanged even after annealing at 370°C for 1 h. Figure 9.15c and d show the TEM image of (Co 0.2 nm /Pt 0.2 nm)_6 sample, which revealed the fcc (111) ordered-alloy-like structure with atomically flat interfaces. In fact, this ordered-alloy-like structure suppresses the tendency of intermixing and in turn enhances thermal stability.

We have studied the magnetic anisotropy of the ultrathin ordered-alloy-like Co/Pt (111) films. The thin film Ta(4 nm)/[Pt/Co]_3/Pt(2 nm) has been deposited with individual layer thicknesses $t_{Co} = t_{Pt} = 0.2$ nm. The magnetic M–H hysteresis curves show a clear PMA with easy axis of magnetization normal to the film surface (see Figure 9.16). The M–H curve measured in the out-of-plane direction exhibits a rectangle behavior (remnant magnetization, $M_r/M_S = 1$) with rather sharp magnetization reversal. On the other hand, the M–H curve measured in the in-plane direction has a linear reversible behavior with almost no hysteresis with remanent magnetization (M_r/M_S) and coercive field (H_C) tending to zero. The measurements show a magnetic anisotropy field, $H_K = 4.5$ kG, and the magnetic anisotropy constant, as calculated by using Equation 9.20, is $K_{eff} = 5.5 × 10^6$ erg/cm³.

The sample was then post annealed at 350°C in vacuum for duration of 1 h. The M–H curves of the post annealed sample evidently showed no change (red circles in

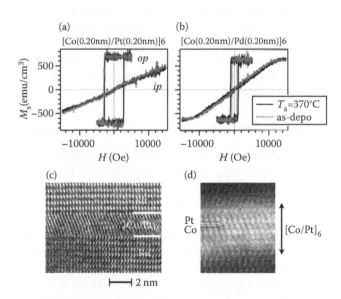

FIGURE 9.15 M–H hysteresis curves of (a) Co/Pt and (b) Co/Pd superlattice films in the as-deposited state and after post annealing for 370°C. "op" and "ip" denote M–H loops with out-of-plane and in-plane magnetic fields, respectively. Cross-sectional (c) HRTEM and (d) HAADF-STEM images of [(Co 0.20 nm)/(Pt 0.20 nm)]_6 superlattice film. (Adapted from K. Yakushiji et al., *App. Phys. Lett.* 97, 2010, 232508.)

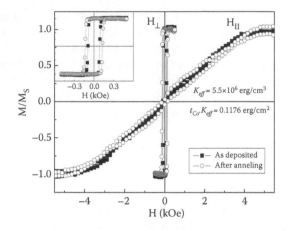

FIGURE 9.16   M–H hysteresis curves of [Pt 0.2 nm/Co 0.2 nm]$_3$ stack. Clear out-of-plane anisotropy (PMA) is seen in the stack. The magnetic anisotropy is K_{eff} = 5.5 × 10^6 erg/cm^3 and remained unchanged after post annealing at 350°C.

Figure 9.16) apart from a minor enhancement in the coercive field (H_C) in the out-of-plane direction, may be due to the enhancement in grain size. These results are similar to those reported in Reference [34]. The thin film can be considered to have an artificial superlattice structure, which as mentioned before, suppresses atomic interdiffusion.

These ordered-alloy-like structures, sometimes referred to as artificial superlattices, have emerged as a good candidate in applications requiring PMA and stability toward thermal annealing. However, the thin film structure, and therefore the magnetic properties it possesses, strongly depends on the deposition conditions. Also the substrate properties, especially roughness, also play a critical role in defining the magnetic anisotropy of these ultrathin films.

9.4.3   Enhancement in PMA Due to "Uphill" Diffusion at Interface

Another method which enhances PMA in thin films on thermal annealing is by undergoing interface sharpening due to uphill diffusion. If the multilayer stack consists of mutually immiscible materials, then by suitable thermal annealing it is possible to have sharper interfaces owing to "uphill" diffusion at interfaces. In this process, since the materials are insoluble, the atoms that were intermixed at the interface due to some practical reasons such as ballistic mixing or dynamical segregation during production, get separated making the interface chemically sharper.

In 1988, F.J.A. den Broeder et al. [30] reported an interesting phenomenon of enhancement in the PMA of Co/Au multilayer thin films after annealing the samples at 250–300°C. In addition to this, they reported an enhancement in the saturation magnetization per unit Co volume (M_S) in samples after thermal annealing. Since Co and Au are mutually insoluble, it has been inferred that the enhancement in PMA and M_S is due to the sharpening of interfaces which is caused by "uphill" diffusion upon annealing. The PMA observed in these multilayers after annealing is about 5 × 10^5 erg/cm^3.

Similar results were confirmed in another study by Spörl and Weller [35]. These authors studied the multilayer stacks of Co/Au, Co/Pt, and Co/Pd. In agreement with Reference [30], the magnetic hysteresis curves showed a clear enhancement in PMA only in Co/Au on annealing at 300°C for 1 h. XPS measurements carried on these multilayers confirmed the origin of this enhanced PMA in annealed Co/Au is due to the chemical sharpening of Co and Au indicating the "uphill" diffusion.

The Co/Pt multilayer structure is one of the most favorable stacks for PMA. The value of $K_{eff} \sim 10^7$ erg/cm^3 has been reported in these stacks. However, if these stacks are not ultrathin as discussed in Section 9.4.2, the thermal stability is a considered a real issue. Many studies have reported the reduction, even in some cases removal, of PMA upon annealing. For example, Bertero et al. [36] have reported the reduction in PMA in Co/Pt stacks and correlated it to intermixing at interfaces. They have used different characterization techniques like high-resolution transmission electron microscopy (HRTEM), XRD, X-ray magnetic circular dichroism (XMCD), Kerr rotation, magnetooptic hysteresis, etc. for structural and magnetic studies. They observed a good PMA ($K_{eff} > 10^6$ erg/cm^3) in as-deposited samples which reduced and vanished upon annealing. The reduction in PMA was attributed due to intermixing at the interfaces of Co/Pt layers. References [35,37–39] include several other studies which reported a similar kind of decrease in PMA upon thermal annealing of Co/Pt. Therefore, the thermal budget is one of the main concerns in using Co/Pt multilayers.

Contradictory to these above-mentioned studies on Co/Pt, recently we have observed an unexpected enhancement in PMA after annealing at 350°C [20]. The behavior is anomalous as it suggests the interface sharpening due to uphill diffusion at Co/Pt interfaces.

9.4.3.1 Enhanced PMA in Co/Pt Layers on Annealing Due to Uphill Diffusion

Stacks of Co/Pt have been deposited in ultrahigh vacuum DC magnetron sputtering system in base pressure of 7×10^{-9} torr Ar pressure while deposition has been kept to 3.5×10^{-3} torr. The thin films with structure Ta(4 nm)/[Pt(1.2 nm)/Co(t_{Co})]$_3$/Pt(2 nm) were fabricated with varying t_{Co}. In as-deposited state, samples showed PMA till $t_{Co} = 1.3$ nm, and for $t_{Co} > 1.3$ nm easy axis shifted to the in-plane direction. The different magnetic properties of the films in as-deposited state and after annealing are discussed below.

9.4.3.1.1 Saturation Magnetization (M_S) (emu/cm^3)

The saturation magnetization per Co volume is calculated from M–H curves measured by the vibrating sample magnetometery system. In films with lower t_{Co}, the value of M_S is observed to be more than bulk Co, even in as-deposited samples. The value, nevertheless, decreased exponentially with t_{Co}, reaching toward a value of bulk Co (1440 emu/cm^3) for larger t_{Co}. This enhancement can be attributed to induced magnetization in Pt atoms generated by nearby Co atoms. The Pt atom has a higher Stoner factor and though it behaves as a paramagnetic at room temperature, ferromagnetism can get induced in it in the vicinity of ferromagnetic atoms like Co.

When measuring the same samples after annealing at 350°C, a further enhancement in M_S is reported especially in stacks with smaller t_{Co} (see Figure 9.17a). This observation suggests a possible clustering of Co atoms upon annealing. However, in the stacks with little larger t_{Co} (=0.9 nm), $M_S = 1440$ erg/cm³ ± 5% is observed in measurements carried over before and after annealing. At larger t_{Co}, the contribution to M_S due to Pt atoms becomes negligible and hence we expect to get constant M_S.

Here, it is to be noted that similar values of M_s for as grown and post annealed samples indicate contradictory observations of intermixing as reported by several authors.

9.4.3.1.2 The Anisotropy Field (H_K) (Oe)

The anisotropy field (H_K) was calculated using Equation 9.19. This H_K was found to decrease with magnetic layer thickness t_{Co} in samples both before and after annealing. Nevertheless, in any individual sample, H_K is increased after thermal annealing (Figure 9.17b).

This observation suggests two possibilities, enhancement in strain in Co or a possible chemical sharpening at the interface. The enhancement in strain, which is unlikely, can enhance the orbital moments in the out-of-plane direction enhancing

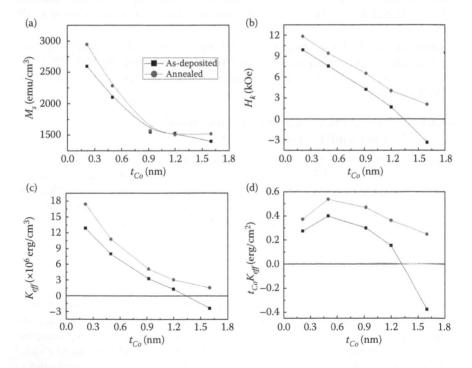

FIGURE 9.17 (a) M_S and (b) H_K as a function of t_{Co} in stacks calculated before and after annealing. (c) gives effective magnetic anisotropy constant, K_{eff} (=$H_KM_S/2$), versus t_{Co}, and shows decrease in K_{eff} with t_{Co}. Note that the anisotropy constant K_{eff}, hence PMA, increased after annealing in all samples. Graph in (d) is $t_{Co} \cdot K_{eff}$ versus t_{Co} and shows the enhancement in surface anisotropy constant K_S on annealing.

magnetoelastic anisotropy. On the other hand, with enhancement in chemical sharpening at the interface, the orbital moments in the out-of-plane direction become larger, as explained in Section 9.2.6.1 (readers may refer to [15,20] and [36]). This enhanced orbital moment makes it harder to rotate the moments near the interface in the in-plane direction and requires more external field to do so. Therefore, in both cases, it needs a larger external magnetic field for saturation in the in-plane direction. This, in turn enhances the anisotropy field H_K.

9.4.3.1.3 Magnetic Anisotropy Constant (K_{eff}) (erg/cm^3)

The value of K_{eff}, which can be calculated from Equation 9.20, is given in Figure 9.17c. As expected from observed M_S and H_K, an enhancement in K_{eff} is observed after annealing in the individual sample. This increase is quite anomalous, as discussed in the previous section.

9.4.3.1.4 Effective PMA Energy Density ($t_{Co} \cdot K_{eff}$) (erg/cm^2)

To understand the change in magnetic anisotropy after annealing and find out individual contributions to the anisotropy, $t_{Co} \cdot K_{eff}$ is plotted against t_{Co}. From Figure 9.17d, K_S in as-deposited multilayers is estimated as 0.8 erg/cm^2 (recalling Equation 9.21 from which, $2K_S = Y$-intercept = 1.6 crg/cm^2, in this case) which is enhanced by: 50%–0.12 erg/cm^2 after annealing. The thermal stability factor ($t_{Co} \cdot K_{eff}A/k_BT$) is reported highest for $t_{Co} \approx 0.5$–0.6 nm. Therefore, the multilayers will be most stable with $t_{Co} \approx 0.5$–0.6 nm.

9.4.3.2 Structural Studies

The effect of annealing on strain can be found out, at least qualitatively, by XRD studies. Figure 9.18 compares XRD patterns measured on same samples before and after annealing. A shift, though very small, is observed toward Co (111) and away from Pt (111) position. This indicates that the lattice strain has not increased, but has rather decreased a little, in the Co layer. Therefore, the observed enhancement in the PMA cannot be due to lattice strain.

Further, the full-width-half-maximum (FWHM) in XRD is decreased after annealing suggesting enhanced grain size. This, along with the shift observed in the XRD plot and enhanced value of M_S leads to a hypothesis that there is a some kind of clustering or separation of Co atoms from Pt, caused by post annealing. To confirm this, a quantitative composition determination of each layer using energy-dispersive X-ray (EDX) spectroscopy in STEM mode and a newly developed method which determines the compositional variation using intensity variation of a HAADF-STEM image within a wedge shape sample of multilayer thin films containing two elements were used [20].

The results are exciting. Average values of the compositions with standard deviations for a stack Ta(4 nm)/[Pt(1.2 nm)/Co(1.2 nm)]$_3$/Pt(2 nm) are plotted in Figure 9.19a and b. The structure is divided into seven layers consisting of four layers of Pt (including 2 nm Pt cap layer) and three layers of Co. The layers 2, 4, and 6 contained relatively higher Co than the others and the other 4 layers are considered Pt layers.

Comparing Figures 9.19a and b it can be stated that after annealing the brighter layers (layers 3 and 5) became more Pt enriched and the darker layers (layers 2, 4,

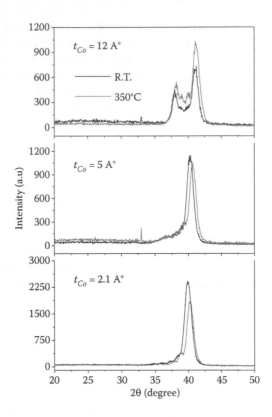

FIGURE 9.18 The XRD patterns measured before (black) and after (gray) annealing for the samples with structure Ta(4 nm)/[Pt(1.2 nm)/Co(t_{Co})]$_3$/Pt(2 nm) with different t_{Co} values. Little shift in peak toward Co (111) is observed in XRD patterns.

and 6) became more Co enriched. Since Co has a lower melting point than Pt this observation infers to only one possibility that annealing has indeed resulted in the uphill diffusion of Co causing the stack to become chemically sharper.

Using the intensity distribution of HAADF-HRSTEM images (images not added here, readers may refer to Reference [20]), the one-dimensional distribution of both elements, Co and Pt, across the layers has been determined. Intensity profiles (averaged over few nm along the length of the layers) taken across the layers, were used to determine the elemental distribution by assuming that layers 1 (i.e., Pt cap layer) and 7 (i.e., Pt layer nearest to Ta buffer) contain Pt only. One compositional profile from each as-deposited as well as annealed stack are shown in Figure 9.19c and d, respectively. The profiles indicated a periodic distribution of Co and Pt within the stack. Comparing these profiles it can be stated that after annealing, the chemical gradient across interfaces of the layers gets steeper which further confirms the uphill diffusion upon annealing. The Co concentration in the Co-rich layer enhanced by 20% after annealing.

Though the observed uphill diffusion in the Co/Pt multilayer stack is unexpected while considering many of the studies on these films, there are some reasons which

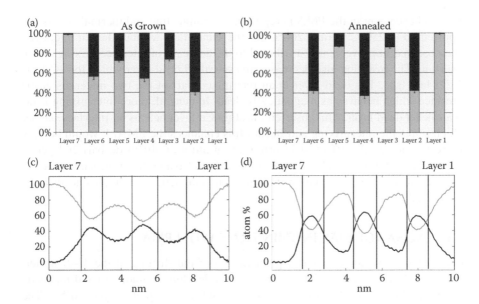

FIGURE 9.19 EDX spectroscopic quantifications within each layers of as grown and annealed sample are plotted respectively in (a) and (b). Average concentration of each layer (Pt in gray and Co in black) is plotted with standard deviation. The atomic distribution (averaged over 2 nm along the length of the layers) of Pt and Co across the layers in as grown and annealed samples are plotted in (c) and (d), respectively. (Adapted from T. Das et al., *Sci. Rep.* 4, 2014, 5328.)

can interpret these observations. First, the Co–Pt binary bulk phase diagram may be incomplete and there exists a miscibility gap. Rather, it may be necessary to consider magnetic ordering while studying a phase diagram [40]. The other possibility, perhaps more likely, is that the coherent lattice strain in ferromagnetic material along with magnetic exchange interaction can alter the diffusion mechanism. It has been suggested that in the Co–Pt alloy system, the magnetic interaction energy may introduce a positive energy of mixing which can cause the clustering of ferromagnetic atoms. This phenomenon is reported by Rooney et al. [41] in Co–Pt alloys when deposited at higher substrate temperatures. Therefore, the magnetic interactions along with the observed coherency strain in multilayer stacks can drive Co-atoms toward higher concentration resulting in uphill diffusion and leading to enhancement in the PMA.

9.4.4 Reduced Intermixing Due to Insertion of Intermediate Layer

The one more way to enhance the thermal stability toward the PMA in multilayer films is to introduce an intermediate layer at the interface. The inserted interface layer acts as a diffusion barrier and reduces the intermixing at the interface.

Recently [37], it has been reported that the insertion of ultrathin Cu in between the Co and Pt layers has enhanced the PMA of the stack at room temperature.

Also, enhancement in the PMA is reported after annealing and thermal stability is reported up to 250°C against 200°C for the same stack without the Cu interlayer. The observed improvement is attributed to the decrease in interface mixing on Cu insertion. The insertion also decreased the so-called dead layer which forms at the Co/Pt interface. The dead layer is an intermixed part of the two materials at the interface which behaves as magnetically dead. In general, more intermixing leads to a thicker dead layer. This inserted ultrathin Cu layer acts as an interface barrier which limits the intermixing and in turn this enhances the PMA. It would be very interesting to see the effect of insertion of an ultrathin diffusion barrier layer of some other materials which are immiscible in both Co and Pt. At the same time the effect of the inserted material on the interfacial magnetic anisotropies needs to be considered.

9.5   EXCHANGE BIAS IN THIN FILMS WITH PMA

The magnetic exchange bias is a phenomenon of the shifting of the $M–H$ hysteresis curve from the origin. The exchange bias develops a unidirectional magnetic anisotropy in the FM when coupled with an antiferromagnetic layer (AFM). The exchange bias was first observed in 1956 by Meiklejohn and Bean [42] in Co dots coated with a CoO layer. Since then the exchange bias has gained a lot of interest mainly because of its potential in various applications such as magnetic read heads, MRAMs, and STT-RAMs, etc. The magnetic exchange biasing is used, rather needed, in the above-mentioned applications to fix or pin the magnetization of a FM to a particular direction. Though the phenomenon is studied experimentally widely, it can be said that theoretical understanding is still lacking, especially in thin films with PMA, which is the subject of this chapter. In this section, we will focus on the exchange bias in films with PMA.

The exchange bias can be observed at the interface between the FM and AFM layers. In general, when the FM/AFM stack is heated at a temperature above Néel's temperature of AFM and below the Curie temperature of the FM layer and field cooled in the presence of an external magnetic field, then an exchange bias gets generated. In systems with PMA, the shift in the measured out-of-plane $M–H$ hysteresis loop is observed generally in the direction opposite to the external field, giving a negative exchange bias. This exchange bias can be characterized by exchange bias field H_{ex} and denotes the shift of a curve from the origin. The theoretical models presented for exchange bias indicate that this exchange bias in FM/AFM is an interfacial phenomenon and very strongly depends on interface properties; both structural and magnetic. In fact, the experimental values of exchange bias (H_{ex}) were reportedly smaller than that of the ideal interface. The grain boundaries, interfacial roughness, interfacial diffusion, and crystallographic orientation are among the factors affecting the exchange bias.

Though many studies have been performed on the exchange bias since 1956, in 2001, S. Maat et al. [43] experimentally measured the exchange bias between perpendicularly anisotropic Co/Pt multilayer stacks and CoO. In this section, the effects of certain factors on the exchange bias in thin films with PMA is explained.

9.5.1 EFFECT OF FM THICKNESS ON EXCHANGE BIAS

The studies carried on [FM–NM]/AFM stacks with PMA show initial increment in exchange bias followed by decrement when the FM layer thickness increases. Figure 9.20 shows the observations of exchange bias in [Pt/Co]-IrMn (IrMn as AFM) with varying the thickness of Co layer (t_{Co}) [44].

The vanishing of the exchange bias at higher thicknesses of ferromagnetic material is expected, since with the higher thicknesses of FM layer the shape anisotropy dominates over other out-of-plane anisotropies (K_N and K_{me}) and changes the magnetic anisotropy from the out-of-plane to in-plane direction. On the other hand, since the AFM is field cooled with the external field in out-of-plane direction, the magnetic moments remain in the out-of-plane direction (with antiparallel alignment). Therefore, the exchange coupling vanishes as the spins at the FM/AFM interface get aligned normal to each other.

Ideally, the exchange bias should be enhanced with lowering the FM layer thicknesses. Moreover, authors of Reference [44] reported decrease in H_C, K_{eff}, and squareness of M–H curve (M_r/M_S; M_r = remnant magnetization) along with lower H_{ex}. These observations suggest few reasons for the reduction of exchange bias. The most probable among them is the intermixing in Co and Pt, resulting in reduction in ferromagnetism in Co atoms, raising the possibility of a formation of magnetically dead layer. Also, there is fair chance of having a Pt atom at, interface, making it paramagnetic/AFM which cannot induce, exchange bias. One more possibility is of Co island formation [45], instead of a thin film which can lead to superparamagnetic Co. The possible mechanisms at different thicknesses of FM layer are given in Figure 9.21.

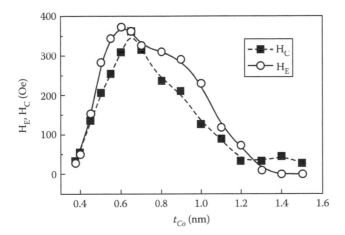

FIGURE 9.20 Dependence of the exchange bias field (H_{ex}) and the magnetic coercivity H_C on the Co thickness, t_{Co}, for perpendicular exchange-biased multilayers. (Reprinted from J. Sort et al., *Phys. Rev. B* 71, 2005, 054411.)

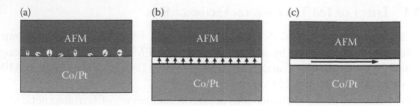

FIGURE 9.21 A schematic representation showing (a) superparamagnetic, (b) out-of-plane, and (c) in-plane magnetization of the FM layer, which is in contact with the AFM layer.

9.5.2 EXCHANGE BIAS ACROSS NON-MAGNETIC SPACER LAYER

For many years, it has been theoretically believed that the exchange bias is a pure interfacial phenomenon, that is, it solely depends on an interface and requires a interface between FM and AFM materials.

In 1997, experimental observations contradicting this assumption were reported by Gökemeijer et al. [46]. They showed the existence of an exchange bias across a non-magnetic layer in Py/spacer/CoO layers with spacer layer of materials like Ag, Au, and Cu. The measurements show a monotonic exponential decay in H_{ex} with increase in spacer layer thicknesses. In addition, this decay was observed to be material dependent. The decay rate reported was comparatively less for Ag reaching to zero at thickness of Ag $\approx$ 5 nm. The results show dependence on spacer material, and suggest that this long range phenomenon is probably electronic in nature. Recently [47], a quantum model for this exchange bias coupling across a non-magnetic metallic layer was proposed which fairly matched the analytical results with the experimental results of Gökemeijer et al. [46]. The model suggests that the long range dipole field interaction, which was induced due to quantum fluctuations by symmetry breaking at the AFM interface, is the reason for this exchange bias across the spacer layer. In another study, Meng et al. [48] reported the presence of a non-monotonous exchange bias (and also coercive field) in CoO/Ag/Fe and observed the changeover back to monotonous behavior in samples with a rougher interface. Nevertheless, they reported the highest exchange bias when the spacer layer was absent.

In multilayer thin films with PMA the effect of spacer layer thickness gives some interesting features. The studies reported initial enhancement in the exchange bias H_{ex} measured in the out-of-plane direction followed by exponential decay. It has been observed by Sort et al. [44] in [Pt/Co]-IrMn multilayers that, at lower thicknesses of Co (0.38 nm) an exchange bias in the out-of-plane direction was enhanced by the insertion of a Pt spacer between the [Pt/Co] and IrMn layers. But the same insertion caused reduction in exchange bias when t_{Co} = 1.1 nm.

In another study by Garcia et al. [49], the exchange bias was measured on a film of [Pt/Co]/spacer/FeMn with a spacer of different materials like Pt, Cu, Al, and Ru. The results reported by them are very interesting and reprinted in Figure 9.22. Note, all the measurements were carried out by applying a field perpendicular to thin films. As shown in Figure 9.22a, the exchange bias (H_{ex}) is enhanced with insertion of a Pt spacer. Interestingly, this initial enhancement is not observed in thin films with spacer layers of other metallic layers: Cu, Al, and Ru (see Figure 9.22b).

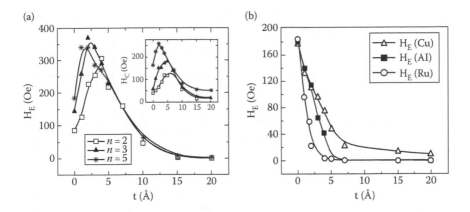

FIGURE 9.22 (a) Exchange field H_{ex} increasing with insertion of ultrathin Pt layer in [Pt/Co] and FeMn layers. (b) Exponential decrease of H_{ex} with thickness of nonmagnetic (NM) metallic layers of Cu, Al, and Ru in [Pt(20 Å)/Co(4 Å)]5/NM(t)/FeMn(130 Å). (Adapted from F. Garcia et al. *Appl. Phys. Lett.*, 83, 2003, 3537.)

Figure 9.23 shows our observations on perpendicular exchange bias in sputtered Pt(13 nm)/[Pt(2.5 nm)/Co(0.4 nm)]$_3$/Pt(t_{Pt} nm)/FeMn(13 nm)/Ta(6.5 nm) films. The observations are in agreement with the results reported by Garcia et al. [49]. Evidently, the sample without Pt spacer layer ($t_{Pt} = 0$) showed no exchange bias. Nevertheless, with a little addition of Pt spacer (<2 Å) in between the Co and FeMn layers, the stacks started showing a clear negative exchange bias. This exchange bias, H_{ex}, increased with t_{Pt}, reaching to the maximum of 236 Oe for $t_{Pt} = 0.45$ nm. However, with further increase in t_{Pt}, the exchange bias started decreasing and vanished at t_{Pt}: 1.6 nm.

The result can be explained considering three cases which are schematically explained in Figure 9.24:

1. *In absence of Pt spacer layer*, with Co/FeMn interface, Co tends to regain in-plane anisotropy. This causes a situation wherein Co atoms at the interface with magnetic moments are aligned in the in-plane direction and FeMn with magnetic moments in the out-of-plane direction. Thus, in this situation, the magnetic moments of Co and FeMn are orthogonal to each other and, therefore, give no exchange coupling.

2. Nevertheless, when *Pt spacer layer is added in between Co and FeMn*, it changes the magnetic anisotropy in this top Co layer back to the out-of-plane direction. However, a better exchange bias is possible at the FM/AFM interface. Contradictorily, the exchange bias enhanced initially in Co/Pt/FeMn. Also, other spacer materials have not shown any enhancement in H_{ex} [49]. This means that this initial enhancement is spacer material specific.

As discussed in earlier sections, although bulk Pt shows paramagnetism, Pt atoms gain ferromagnetism in the vicinity of Co atoms due to their high Stoner

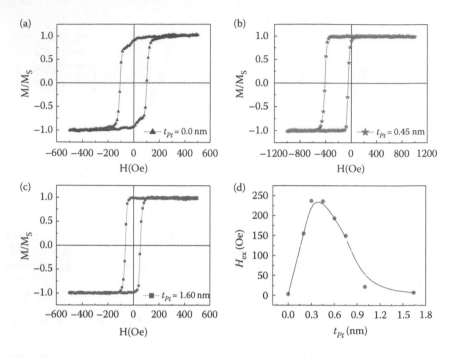

FIGURE 9.23 (a, b, and c) Show M–H curves of structure [Pt(2.5 nm)/Co(0.4 nm)]3/Pt(t_{Pt} nm)/ FeMn(13 nm) with $t_{Pt} = 0$ nm, 0.45 nm, and 1.60 nm, respectively. (d) Shows graph of H_{ex} against t_{Pt}. Maximum $H_{ex} = 236$ Oe is reported for $t_{Pt} = 0.45$ nm.

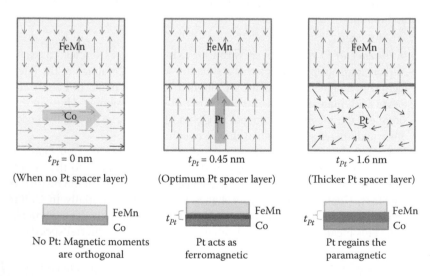

FIGURE 9.24 Different spin projection at different Pt spacer layer thicknesses (t_{Pt}).

factor. In the current case, with the insertion of the appropriate amount of Pt, Co atoms get aligned to the out-of-plane direction and Pt atoms get FM with magnetic anisotropy the same as Co. Therefore, the Pt/FeMn interface behaves as the FM/AFM interface with easy axis in the out-of-plane direction; and can produce an exchange bias.

3. But, *for Pt layer thickness high enough*, Pt atoms away from the Co/Pt interface regain the paramagnetism resulting in weaker FM/AFM spin projection at the Pt/FeMn interface. And eventually the Pt/FeMn interface becomes paramagnetic/AFM, which cannot generate exchange bias. As a result, the exchange bias starts decreasing in accordance to the previously reported long range exchange bias in the FM/paramagnetic/AFM stack [46,47], and almost vanished as $t_{Pt} = 1.0$ nm. This also indirectly suggests that Pt atoms get magnetized to a depth of 0.6 nm adjacent to the Co/Pt interface and then gradually change to paramagnetic behavior at 1.0 nm.

The results indicate that the insertion of an ultrathin layer—even of a paramagnetic material—gives enhancement to the exchange bias in thin films with PMA, due to increased perpendicular effective anisotropy at interface of the AFM layer.

9.6 PMA IN MAGNETIC RECORDING

The most common use of magnetic thin films with PMA nowadays is in magnetic hard disks. The perpendicular recording media has allowed the manufacture of hard disk drives (HDD) with higher data storage and is the reason for the enhanced computer storage and portable HDDs in the terabyte range in the last few years.

Figure 9.25 shows the basic concept of both longitudinal and perpendicular recording techniques. In longitudinal recording, the recording medium has in-plane magnetization and the magnetic direction defines logical "1" and "0" bits. The head consists of a GMR/TMR-read element which detects the field direction of bits and

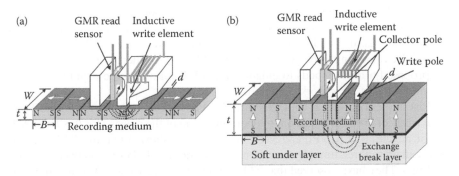

FIGURE 9.25   (a) Longitudinal and (b) perpendicular recording techniques. (Adapted from A. Moser et al., *J. Phys. D: Appl. Phys.* 35, 2002, R157–R167.)

hence reads the data. The other part consists of a write element which generates the external field and aligns the magnetization of a corresponding bit to a particular direction, and hence writes a data bit to the recording medium. Although, the longitudinal recording has been in use for many years, the poor thermal stability at smaller size puts limits on the bit density.

Nowadays, almost every magnetic hard disk manufacturer has adopted perpendicular recording over longitudinal (i.e., in-plane) recording due to its advantages over the latter. One of the main reasons for adopting thin films with PMA as a recording media is the higher thermal stability over longitudinal recording media, especially at higher density [51,52]. Further, the size and shape limitations of longitudinal magnetic medium with respect to magnetization curling are eliminated in perpendicular recording.

In the perpendicular recording technique, a magnetic media is deposited over a soft magnetic layer. This soft layer mirrors the magnetic field of the write head. In perpendicular recording, as shown in Figure 9.25b, the head has a write pole with smaller area, which concentrates the magnetic field to a single bit under consideration. The other part consists of a thicker collector pole with a larger area which completes the loop of magnetic lines. Since the field gets concentrated over a small area of a single bit, we can expect smaller crosstalk and the effect of stray field in perpendicular recording.

9.7 MAGNETIC TUNNEL JUNCTIONS WITH PMA

Though people have achieved a very high-tunnel magnetoresistance (TMR) ratio (several hundreds at room temperature) in MTJs, getting TMR in MTJs with PMA or so-called p-MTJs is difficult and has certain issues which do not arise in conventional in-plane MTJ structures. The origin of these issues lies in the restrictions over the ferromagnetic materials that can be used, as the FM should possess a stable PMA along with other properties needed to get better TMR.

9.7.1 p-MTJs with Multilayers as Electrodes

Since multilayer stacks like Co/Pd and Co/Pt show good PMA, they can be used as FM electrodes in p-MTJs. There are several studies reported on p-MTJs with these multilayers as electrodes, but unlike conventional in-plane MTJs which show TMR in few hundreds, the studies on p-MTJs with multilayers like Co/Pt as electrodes show TMR values in the range of 10%–30% [51,53–57]. There are studies reported that proposed a few methods to enhance the TMR ratio in these p-MTJs. We will look into some of these works in detail.

Most studies on p-MTJs with structure [Co/Pt]/ MgO/[Co/Pt] have reported TMR in the range of 10%–15% at room temperature. One of the reasons for this low TMR is poor matching of [Co/Pt] and MgO. Ye et al. [57] showed that this TMR can be enhanced up to three times by insertion of an ultrathin Mg layer in between [Co/Pt] and MgO. They have observed that TMR in the structure without Mg insertion was 10.4% and had increased to 32% with the insertion of 0.4 nm of Mg on both sides of the MgO barrier layer.

Yakushiji et al. [58] fabricated MTJs consisting of Co/Pt multilayers as a free layer with CoFe/CoFeB/TbFeCo as a fixed layer which showed a TMR ratio up to 85% at room temperature along with low resistance-area (RA) product (4.4 $\Omega\mu m^2$). The results of this study are shown in Figure 9.26.

The structure consists of a CoFeB/CoFe (0.3 nm) intermediate layer between the free [Co/Pt] electrode and MgO barrier layer. The structure show the highest TMR when annealed at 225°C and TMR started decreasing for higher annealing temperatures. It has been observed that the insertion of CoFeB/CoFe is necessary in getting a better TMR ratio. As can be seen in Figure 9.26c the TMR is almost negligible without the CoFeB/CoFe layers and increases with the thickness of the CoFeB/CoFe layer reaching to 97% for 1.4 nm of CoFeB/CoFe. The reason for this can be the fcc crystal structure of Co/Pt which does not match well with the bcc structure of MgO. Further, studies indicate that in MgO-based MTJ structures, the giant TMR originates from the fully-spin-polarized Δ_1 band in the bcc (001) alloys of Co and Fe. Therefore, the MTJ developed by Yakushiji et al. consists of two parts. The first is [Co/Pt] multilayer which generates the PMA and the second is the CoFeB/CoFe layer which is required for better spin dependent tunneling across the barrier and has a better TMR ratio.

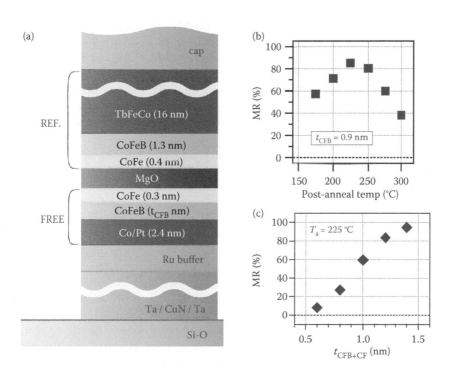

FIGURE 9.26 (a) Structure of TMR stack with [Co/Pt] as a free layer. (b) Effect of annealing temperature and (c) an intermediate CoFeB/CoFe layer on TMR. (Adapted from K. Yakushiji et al., *Appl. Phys. Express* 3, 2010, 053003.)

9.7.2   CoFeB/MgO/CoFeB: MTJs with PMA

A MTJ is a structure in which two magnetic layers are separated by a thin insulating layer. This MTJ structure shows a high TMR effect and this TMR can be controlled by an external magnetic field. Also, the resistance can be altered simply by passing an electric current through MTJ by the concept of the STT. This has made MTJs of standard use in nonvolatile magnetic random access memories.

With advancement in technology, the need for making devices smaller and smaller rises. In this context, the MTJ with PMA—lets call it p-MTJ—is of great interest. This is for the fact that PMA shows certain advantages over conventional in-plane anisotropy. For example, PMA promises a better thermal stability, lower threshold currents for spin transfer switching and current induced domain wall motions. The size and shape limitations of planar magnetic electrodes with respect to magnetization curling are eliminated by using PMA. Moreover, in magnetic memories, the use of p-MTJs enhances the recording density by reducing the area needed for storage of per unit data bit.

Nowadays, the CoFeB/MgO/CoFeB structure has emerged as the most common structure for MTJs because of the higher TMR ratios exhibited by them (>500). But most of the studies carried on these MTJs consist of an in-plane anisotropy. On the other hand, there are other structures consisting of thin films of Co/Pt, Co/Pd, etc. as an FM layer in the MTJ structure. But the structures were not able to reproduce the high TMR ratio shown by CoFeB/MgO/CoFeB. This mainly because the crystal structure of these multilayers do not match well with MgO which has a body-centered-cubic (bcc) structure. On the other hand, CoFeB matches well with the bcc MgO structure. In fact, experimental observations apparently show that the structure gains better crystallinity after a thermal treatment. Moreover, theoretically more spin–orbit scattering is expected in the multilayers if they have heavy elements like Pt, which gives poor spin-transport and hence poor device performance [59].

The breakthrough occurred in MTJs with PMA when Ikada et al. [22] fabricated for the first time a p-MTJ with a CoFeB/MgO/CoFeB system without using any extra magnetic layer for PMA induction. In this study they observed a PMA originating from CoFeB and MgO interfaces (see Section 9.2.6 for details). Though p-MTJ showed no TMR in as-deposited MTJs, TMR ratio ~121% in a stack of $Co_{20}Fe_{60}B_{20}$/$MgO/Co_{20}Fe_{60}B_{20}$ on annealing at 350°C is observed, due to enhanced structural matching and better crystallinity.

Figure 9.27 consists of few observations of Reference [22]. The schematic of the p-MTJ structure under study is represented in Figure 9.27a. The M–H hysteresis curves, which are given in Figure 9.27b, indicate a clear PMA with easy axis in the out-of-plane direction for the structure with a CoFeB thickness of 1.3 nm. The PMA coefficient in this structure is reported to be $K_{eff} = 2.1 \times 10^6$ erg/cm^3. This value is high enough to have a good thermal stability factor even at dimensions of several tens of nanometers. The inset of the figure shows the graph of $t \cdot K_{eff}$ versus magnetic layer thickness t, and indicates the change-over from perpendicular to in-plane magnetic anisotropy around CoFeB thickness $t = 1.5$ nm.

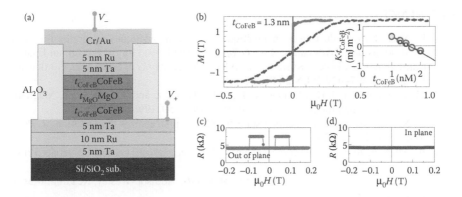

FIGURE 9.27   Studies on p-MTJ. (Reprinted from S. Ikeda et al., A perpendicular-anisotrpy CoFeB-MgO magnetic tunnel junction., *Nat. Mat.* 9, 2010, 721.) (a) Schematic of MTJ. (b) In-plane (dashes) and out-of-plane (lines) *M–H* hysteresis curves. Inset of (b) gives plot of $t \cdot K$ versus magnetic layer thickness t. Graphs (c) and (d) give *R–H* curves in out-of-plane and in-plane direction, respectively.

The electrical resistance of a junction (R) against an external magnetic field (H), otherwise called *R–H* curves, measured by applying H in the out-of-plane and in-plane of MTJ are shown in Figure 9.27c and d, respectively. The junction has not shown any change in resistance in in-plane measurement, which is expected as it is a hard axis. Nevertheless, the out-of-plane measurements show the resistance change of about 100%.

This system promises the advantages of being a CoFeB/Mgo/CoFeB system, which is now an established system to provide higher TMR ratios, along with the advantages of having PMA.

9.8   SUMMARY

In this chapter, the theory behind PMA in thin magnetic films has been explained along with experimental studies. Magnetic anisotropy originates from dipolar and spin–orbit interactions and the interfacial effects are the reason for developing a PMA in thin films. Especially in multilayer films, the lattice strain and the symmetry breaking at interface are the main sources of this PMA. In thin films, the sharper interfaces give better PMA. Moreover, for device applications, the PMA needs to be stable on thermal annealing at about 300°C. The stability can be achieved by selecting appropriate materials, materials which do not intermix on annealing. Along with this, certain other mechanisms also can produce stability on annealing for a certain finite time. These consist of a transient interface sharpening and ultrathin ordered-alloy-like structures. The phenomenon of exchange bias is crucial in devices to make pinning of magnetic anisotropy in a particular direction. In thin films with PMA, the exchange bias can be enhanced by adding certain paramagnetic material in between the FM and AFM materials, but the phenomenon cannot be observed in films with in-plane magnetic anisotropy. We have also seen one of the most interesting developments in MTJs, viz. *p*-MTJs.

REFERENCES

1. L. Néel, Théorie du tranage magnétique des ferromagnétiques en grains fins avec applications aux terres cuites. *Ann. Géophys* 5, 1949, 99–136. (in French; an English translation is available in Kurti, N., ed. 1988. *Selected Works of Louis Néel*. Gordon and Breach. pp. 407–427. ISBN 2-88124-300-2.).
2. M. T. Johnson, P. J. H. Bloemen, F. J. A. den Broeder, and J. J. de Vries, Magnetic anisotropy in metallic multilayers. *Rep. Prog. Phys.* 59, 1996, 1409.
3. L. Néel, Anisotropie magnétique superficielle et surstructures d'orientation. *J. Phys. Rad.* 15, 1954, 225–239.
4. A. Aharoni, *Introduction to the Theory of Ferromagnetism*. Oxford University Press, New York, 1996.
5. B. D. Cullity, *Introduction to Magnetic Materials*. Addison-Wesley, Reading, Massachusetts, 1972.
6. N. Mori, Calculation of ferromagnetic anisotroy energies for Ni and Fe metals. *J. Phys. Soc. Jpn.* 27. 1969, 307–312.
7. U. Gradmann and J. Müller, Flat ferromagnetic, epitaxial 48Ni/52Fe (111) films of few atomic layers. *Phys. Status Solidi* 27, 1968, 313–324.
8. F. J. A. den Broeder, W. Hoving, and P. J. H. Bloemen, Magnetic anisotropy of multilayers. *J. Magn. Magn. Mater.* 93, 1991, 562–570.
9. J. W. Matthews, Misfit dislocations, in *Dislocations in Solids*. F. R. N. Nabarro (ed.), Amsterdam–North-Holland, vol. 2, 1979, pp 461–545.
10. C. Chappert and P. Bruno, Magnetic anisotropy in metallic ultrathin films and related experiments on cobalt films. *J. Appl.* Phys. 64, 1988, 5736–5741.
11. P. Bruno, *Physical Origins and Theoretical Models of Magnetic Anisotorpy*. Ferienkurse des Forschungszentrums Jülich, Jülich, 1993, pp 24.1–24.28.
12. R. Becker, Theory of the magnetization curve. *Z. Phys.* 62, 1930, 253–269.
13. H. J. G. Draaisma and W. J. M. de Jonge, Surface and volume anisotropy from dipole-dipole interactions in ultrathin ferromagnetic films. *J. Appl. Phys.* 64, 1988, 3610–3613.
14. R. Jungblut, M. T. Johnson, J. aan de Stegge, A. Reinders, and F. J. A. den Broeder, Orientational and structural dependence of magnetic anisotropy of Cu/Ni/Cu sandwiches: Misfit interface anisotropy. *J. Appl. Phys.* 75, 1994, 6424–6426.
15. J. Stöhr, Exploring the microscopic origin of magnetic anisotropies with X-ray magnetic circular dichroism (XMCD) spectroscopy. *J. Magn. Magn. Mater.* 200, 1999, 470–497.
16. Y. R. Luo, *Comrehensive Handbook of Chemical Bond Energies*. CRC Press, Boca Raton, FL, 2007.
17. C. J. Lin, G. L. Gorman, C. H. Lee, R. F. C. Farrow, E. E. Marinero, H. V. Do, H. Notarys, and C. J. Chien, Magnetic and structural properties of Co/Pt multilayers. *J. Magn. Magn. Mater.* 93, 1991, 194–206.
18. P. Chowdhury, P. D. Kulkarni, M. Krishnan, Harish C. Barshilia, A. Sagdeo, S. K. Rai, G. S. Lodha, and D. V. Sridhara Rao, Effect of coherent to incoherent structural transition on magnetic anisotropy in Co/Pt multilayers. *J. Appl. Phys.* 112, 2012, 023912.
19. J. Geissler, E. Goering, M. Justen, F. Weigand, G. Schütz, J. Langer, D. Schmitz, H. Maletta, and R. Mattheis, Pt magnetization profile in a Pt/Co bilayer studied by resonant magnetic x-ray reflectometry. *Phys. Rev. B* 65, 2001, 020405.
20. T. Das, P. D. Kulkarni, S. C. Purandare, H. C. Barshilia, S. Bhattacharyya, and P. Chowdhury, Anomalous enhancement in interfacial perpendicular magnetic anisotropy through uphill diffusion. *Sci. Rep.* 4, 2014, 5328.
21. F. Yildiz, M. Przybylski, X. -D. Ma, and J. Kirschner, Strong perpendicular anisotropy in $Fe_{1-x}Co_x$ alloy films epitaxially grown on mismatching Pd(001), Ir(001), and Rh(001) substrates. *Phys. Rev. B* 80, 2009, 064415.

22. S. Ikeda, K. Miura, H. Yamamoto, K. Mizunuma, H. D. Gan, M. Endo, S. Kanai, J. Hayakawa, F. Matsukura, and H. Ohno, A perpendicular-anisotrpy CoFeB-MgO magnetic tunnel junction. *Nat. Mat.* 9, 2010, 721.

23. A. Barman, S. Wang, O Hellwig, A. Berger, E. E. Fullerton, and H. Schmidt, Ultrafast magnetization dynamics in high perpendicular anisotropy [Co/Pt]$_n$ multilayers. *J. Appl. Phys.* 101, 2007, 09D102.

24. S. Mizukami, F. Wu, A. Sakuma, J. Walowski, D. Watanabe, T. Kubota et al., Long-lived ultrafast spin precession in manganese alloys films with a large perpendicular magnetic anisotropy. *Phys. Rev. Lett.* 106, 2011, 117201.

25. S. Iihama, A. Sakuma, H. Naganuma, M. Oogane, T. Miyazaki, S. Mizukami, and Y. Ando, Low precessional damping observed for $L1_0$-ordered FePd epitaxial thin films with large perpendicular magnetic anisotropy. *Appl. Phys. Lett.* 105, 2014, 142403.

26. R. Law, R. Sbiaa, T. Liew, and T. C. Chong, Effects of Ta seed layer and annealing on magnetoresistance in CoFe/Pd-based pseudo-spin-valves with perpendicular anisotropy. *Appl. Phys. Lett.* 91, 2007, 242504.

27. V. B. Naik, H. Meng, and R. Sbiaab, Thick CoFeB with perpendicular magnetic anisotropy in CoFeB-MgO based magnetic tunnel junction. *AIP Adv.* 2, 2012, 042182.

28. B. S. Tao, D. L. Li, Z. H. Yuan, H. F. Liu, S. S. Ali, J. F. Feng et al., Perpendicular magnetic anisotropy in Ta|Co$_{40}$Fe$_{40}$B$_{20}$|MgAl$_2$O$_4$ structures and perpendicular CoFeB|MgAl$_2$O$_4$|CoFeB magnetic tunnel junction. *Appl. Phys. Lett.* 105, 2014, 102407.

29. J. Kanak, M. Czapkiewicz, T. Stobiecki, M. Kachel, I. Sveklo, A. Maziewski, and S. van Dijken, Influence of buffer layers on the texture and magnetic properties of Co/Pt multilayers with perpendicular anisotropy. *Phys. Status Solidi A* 204, 2007, 3950–3953.

30. F. J. A. den Broeder, D. Kuiper, A. P. van de Mosselaer, and W. Hoving, Perpendicular magnetic anisotropy of Co-Au multilayers induced by interface sharpening. *Phys. Rev. Lett.* 60, 1988, 2769–2772.

31. Z. Erdélyi, I. A. Szabó, and D. L. Beke, Interface sharpening instead of broadening by diffusion in ideal binary alloys. *Phys. Rev. Lett.* 89, 2002, 165901.

32. Z. Erdélyi, M. Sladecek, L.-M. Stadler, I. Zizak, G. A. Langer, M. Kis-Varga, D. L. Beke, and B. Sepiol, Transient interface sharpening in miscible alloys. *Science* 306, 2004, 1913–1915.

33. Z. Balogh, M. R. Chellali, G. -H. Greiwe, G. Schmitz, and Z. Erdélyi, Interface sharpening in miscible Ni/Cu multilayers studied by atom probe tomography. *Appl. Phys. Lett.* 99, 2011, 181902.

34. K. Yakushiji, T. Saruya, H. Kubota, A. Fukushima, T. Nagahama, S. Yuasa, and K. Ando, Ultrathin Co/Pt and Co/Pd superlattice films for MgO-based perpendicular magnetic tunnel junctions. *App. Phys. Lett.* 97, 2010, 232508.

35. K. Spörl and D. Weller, Interface anisotropy and chemistry of magnetic multilayers: Au/Co, Pt/Co and Pd/Co. *J. Magn. Magn. Mater.* 93, 1991, 379–385.

36. G. A. Bertero, R. Sinclair, C. -H. Park, and Z. X. Shen, Interface structure and perpendicular magnetic anisotropy in Pt/Co multilayers. *J. Appl. Phys.* 77, 1995, 3953–3959.

37. S. Bandiera, R. C. Sousa, B. Rodmacq, and B. Dieny, Enhancement of perpendicular magnetic anisotropy through reduction of Co-Pt interdiffusion in (Co/Pt) multilayers. *Appl. Phys. Lett.* 100, 2012, 142410.

38. G. A. Bertero and R. Sinclair, Kerr rotations and anisotropy in (Pt/Co/Pt)/X multilayers. *IEEE Trans. Mag.* 31, 1995, 3337.

39. P. C. McIntyre, D. T. Wu, and M. Nastasi, Interdiffusion in epitaxial Co/Pt multilayers. *J. Appl. Phys.* 81, 1997, 637–645.

40. G. Inden, The role of magnetism in the calculation of phase diagrams. *Physica* 103B, 1981, 82–100.

41. P. W. Rooney, A. L. Shapiro, M. Q. Tran, and F. Hellman, Evidence of a surfacemedi-ated magnetically induced miscibility gap in Co-Pt alloy thin films. *Phys. Rev. Lett.* 75, 1995, 1843.

42. W. H. Meiklejohn and C. P. Bean, New magnetic anisotropy. *Phys. Rev.* 102, 1956, 1413–1414.

43. S. Maat, K. Takano, S. S. P. Parkin, and E. E. Fullerton, Perpendicular exchange bias of Co/Pt multilayers. *Phys. Rev.* Lett. 87, 2001, 087202.

44. J. Sort, V. Baltz, F. Garcia, B. Rodmacq, and B. Dieny, Tailoring perpendicular exchange bias in [Pt/Co]-IrMn multilayers. *Phys. Rev. B* 71, 2005, 054411.

45. J. -Y. Chen, J. -F. Feng, Z. Diao, G. Feng, J. M. D. Coey, and X.-F. Han, Magnetic prop-erties of exchange-biased [Co/Pt]$_n$ multilayer with perpendicular magnetic anisotropy. *IEEE Teans. Magn.* 46, 2010, 1401.

46. N. J. Gökemeijer, T. Ambrose, and C. L. Chien, Long-range exchange bias across a spacer layer. *Phys. Rev. Lett.* 79, 1997, 4270–4273.

47. F. Torres and M. Kiwi, A quantum exchange bias model for coupling across a nonmag-netic interlayer. *IEEE* Teans. *Magn.* 50, 2014, 4800104.

48. Y. Meng, J. Li, P.-A. Glans, C. A. Jenkins, E. Arenholz, A. Tan et al., Magnetic interlayer coupling between antiferromagnetic CoO and ferromagnetic Fe across a Ag spacer layer in epitaxially grown CoO/Ag/Fe/Ag(001). *Phys. Rev. B* 85, 2012, 014425.

49. F. Garcia, J. Sort, B. Rodmacq, S. Auffret, and B. Dieny, Large anomalous enhance-ment of perpendicular exchange bias by introduction of a nonmagnetic spacer between the ferromagnetic and antiferromagnetic layers. *Appl. Phys. Lett.* 83, 2003, 3537.

50. A. Moser, K. Takano, D. T. Margulies, M. Albrecht, Y. Sonobe, Y. Ikeda, S. Sun, and E. E. Fullerton, Magnetic recording: Advancing into the future. *J. Phys. D: Appl. Phys.* 35, 2002, R157–R167.

51. B. Carvello, C. Ducruet, B. Rodmacq, S. Auffret, E. Gautier, G. Gaudin, and B. Dieny, Sizable room-temperature magnetoresistance in cobalt based magnetic tunnel junc-tions with out-of-plane anisotropy. *Appl. Phys. Lett.* 92, 2008, 102508.

52. S. Fukami, T. Suzuki, K. Nagahara, N. Ohshima, and N. Ishiwata, Large thermal sta-bility independent of critical current of domain wall motion in Co/Ni nanowires with step pinning sites. *J. Appl. Phys.* 108, 2010, 113914.

53. J. H. Park, C. Park, T. Jeong, M. T. Moneck, N. T. Nufer, and J. G. Zhu, Co/Pt multilayer based magnetic tunnel junctions using perpendicular magnetic anisotropy. *J. Appl. Phys.* 103, 2008, 07A917.

54. J. H. Park, C. Park, and J. G. Zhu, Interfacial oxidation enhanced perpendicular mag-netic anisotropy in low resistance magnetic tunnel junctions composed of Co/Pt multi-layer electrodes. *IEEE Trans. Mag.* 44, 2008, 2577.

55. Yi. Wang, W. X. Wang, H. X. Wei, B. S. Zhang, W. S. Zhan, and X. F. Han, Effect of annealing on the magnetic tunnel junction with Co/Pt perpendicular anisotropy ferromagnetic multilayers. *J. Appl. Phys.* 107, 2010, 09C711.

56. H. X. Wei, Q. H. Qin, Z. C. Wen, X. F. Han, and X.-G. Zhang, Magnetic tunnel junction sensor with Co/Pt perpendicular anisotropy ferromagnetic layer. *Appl. Phys. Lett.* 94, 2009, 172902.

57. L.-X. Ye, C.-M. Lee, J.-W. Syu, Y.-R. Wang, K.-W. Lin, Y.-H. Chang, and T.-H. Wu, Effect of annealing and barrier thickness on MgO-based Co/Pt and Co/Pd multilayered perpendicular magnetic tunnel junctions. *IEEE Trans. Mag.* 44, 2008, 3601.

58. K. Yakushiji, K. Noma, T. Saruya, H. Kubota, A. Fukushima, T. Nagahama, S. Yuasa, and K. Ando, High magnetoresistance ratio and low resistance–area product in mag-netic tunnel junctions with perpendicularly magnetized electrodes. *Appl. Phys. Express* 3, 2010, 053003.

59. A. D. Kent, Perpendicular all the way. *Nat. Mat.* 9, 2010, 699.

10 Catalytic Application of Magnetic Nanocomposites

Yinghuai Zhu

CONTENTS

ABSTRACT

Following advanced developments in synthesis and characterization technology, functionalized magnetic nanoparticles have been emerging as viable nanocomposites in various areas, including catalysis. Plenty of fascinating nanocomposites comprising of magnetic species and active centers have been prepared and used as catalysts in various organic reactions. These types of catalysts can be easily recovered from a reaction solution by application of a magnetic field. They also demonstrate reasonable, even promising in some cases, catalytic activities apart from advantages in the catalyst recycle. The benefits have been repeatedly reported in different organic transformations. Therefore, construction and examination of highly efficient magnetic nanocomposites will continue to attract increasing research interests both in academia and in industry. This chapter primarily addresses the commonly used synthetic methodology and recent successful catalytic applications of magnetic nanocomposites.

10.1   INTRODUCTION

Nanotechnology provides major advances and promising future in catalysis, which allows scientists to work at molecular levels in catalysis. It's widely recognized that catalysts always play crucial roles both in industry and in our daily life. In principle, catalysts can be classified as homogeneous and heterogeneous. Homogeneous catalysts such as transition metal complexes are soluble in reaction media to form one-phase system; all the active centers are easily accessible by reactants; while heterogeneous catalysts are insoluble in reactants, the related reactions occur in different phases. In general, homogeneous catalysts are highly active, and thus the corresponding reactions occur fast. However, recovery and reuse of homogeneous catalysts remain a big challenge. The issue is becoming more and more urgent for the pharmaceutical and fine chemistry industry due to the increasing importance of environmental benign and economical requirements [1]. To overcome the disadvantages, highly active homogeneous catalysts have been supported on various materials and examined for diverse organic reactions. In contrast, most of the heterogeneous catalysts can be recovered conveniently either by filter or by centrifugation. The drawbacks of heterogeneous catalysts include the following: (1) more critical reaction conditions such as elevated temperature or high gas pressure for the gas-involved reactions are needed to activate the reactions; (2) since the reactions occur on the surface of varying catalytic active centers by interaction of reactants, reactions using heterogeneous catalysts normally need longer time to be finished compared with one-phase homogeneous systems [1]. It's believed that increasing catalyst surface area will enhance the interaction of reactants and active centers and thus will definitely benefit the reaction procedure.

Following the advanced developments in nanotechnology, materials in nanoscale range and suitable for application in catalysis either as supports or as catalysts have been conveniently prepared and widely used [2–6]. As the size of the nanoparticles decreases, the surface-to-volume ratio of the particle increases. The ratio is significantly high for nanoparticles, and that enables a great portion of the atoms to reside on the surface. Therefore, more active centers are easily accessible for the reagents. When compared with classic bulky metal-based heterogeneous catalysts, nanoscaled metal particles-based catalysts have a higher surface-to-volume ratio, and thus enhance the interaction between catalytic active centers and reactants, and enable more reactions to occur at the same time. However, following the newly emerged nanoscale catalysts, conventional separation methods such as the filter could not meet the requirements of purifying nanocatalysts because they may block the filter paper, and thus it usually takes a long time to recycle the nanoscaled catalysts. Success of separating nanocatalysts at laboratory scale can be easily achieved by centrifugation. However, it is apparently unpractical to use centrifugation separation in tons-scale operations, which are common in industry. To overcome this limitation, nanoscaled magnetic materials have been introduced into catalysis. Recently, functionalized magnetic nanoparticles have been emerging as viable nanocomposites in various areas, including catalysis. Indeed, magnetic nanocomposite-based catalyst is becoming one of the fast-growing frontiers in catalysis. Nanoscaled magnetic nanoparticles (MNPs) are generally composed of metal oxides such as ferrites [7] and iron oxides [8]. MNPs that have at least one dimension are between approximately 1 and 100 nm and thereby contain several hundreds to hundreds of thousands of atoms. Iron oxide-based MNPs, such as magnetite (Fe_3O_4), maghemite (β-Fe_2O_3 and γ-Fe_2O_3), hematite (α-Fe_2O_3), and ϵ-Fe_2O_3, are most widely used in the catalysis area. This type of MNP has advantages of low toxicity and oxidative stability in comparison with MNPs that are made from other elements such as Co. These are often strong paramagnetic materials and thus can be attracted by an externally applied magnetic field. MNPs are classified as single-core and multicore particles. Single-core MNPs have one magnetic core coated with a matrix. Multicore MNPs have more than one core in a matrix. The magnetic properties such as the relaxation time can be changed by reducing the size of the MNPs or using different types of magnetic materials. Therefore, iron oxide-based nanoparticles have been and will continue to be useful in many biomedical applications in the areas of diagnosis, therapy, actuating, and imaging. However, it is important to note that both single-core and multicore MNPs tend to aggregate. Optimization of a matrix is essential to control the particle size and thereby makes the preparation reproducible.

Applications of magnetic nanomaterials have been widely investigated in many areas such as in biomedicine, including drug delivery and diagnostic applications [9–12]. These materials can be quickly and easily separated from a reaction mixture by application of an external magnetic field for reuse. In current catalysis, magnetic nanomaterials have been used either as support or as pristine catalysts. Furthermore, internal diffusion limitations can be avoided because all the accessible surface areas of this type of materials are external. MNPs have been functionalized by various functional organic and organometallic groups. This chapter summarizes the advanced developments of the magnetic catalyst composites, their synthesis, and their applications in organic transformations.

10.2 SYNTHESIS AND APPLICATION OF MAGNETIC NANOPARTICLES-SUPPORTED HOMOGENEOUS CATALYSTS

Magnetic nanomaterials continually attract growing research interests due to their unique physical properties such as superparamagnetism, irreversibility, saturation field, and considerable potential applications. In addition, these materials also have good wear resistance, high hardness, and outstanding mechanical properties even at high temperatures [13–15]. In industry, magnetic nanomaterials have found broad applications such as magnetic inks for bank checks, magnetic recording media, and magnetic seals in motors [6,13–15]. Magnetic nanoparticles also have potential promising applications in biomedicine such as drug delivery, and nuclear magnetic resonance (NMR) imaging for clinical diagnosis [9]. The nanocomposites discussed here are normally less than 100 nm.

10.2.1 Current Synthesis of Magnetic Nanoparticles

It has been repeatedly demonstrated that the preparation procedure of magnetic nanomaterials is crucial to applications in most of the cases. The preparation methods determine morphology such as particle size and shape as well as size distribution, surface chemistry, and magnetic properties of the produced nanoparticles. For example, ferri- and ferromagnetic nanoparticles prepared from wet chemistry and gas phase generally have a spherical shape. Preparation of magnetic nanoparticles is particularly important in catalysis applications because the preparation method determines the degree and distribution of structural defects and impurity of the particles that have crucial effects on catalyst performances [16,17].

To date, many methods have been developed to synthesize magnetic nanoparticles, which are the important composite for the type nanomaterials. Great efforts are devoted to the preparation of uniform magnetic nanoparticles both in size and in shape [6,18–20]. Iron oxides between 10 and 20 nm show the typical ferro- to superparamagnetic phase transformation [16]. The common methods include co-precipitation; thermal decomposing; sonochemical, microwave-assisted vapor deposition; and laser pyrolysis [6,18–20]. Co-precipitation is easy to perform in laboratory, and thus the method is highly preferred. On the other hand, thermal decomposition produces MNPs with a narrow range of particle size and uniform morphology [18–20]. Magnetic nanocomposites including pure metals such as Fe and Ni; metal oxides such as γ-Fe_2O_3; ferrites such as $MnFe_2O_4$; and alloys such as CoPt and FePt have been prepared [21–29]. The physical and chemical properties of the resulting magnetic nanoparticles obviously depend strongly on the fabrication conditions used in the procedures such as precursor, concentration, and pH value of solution. Other conditions such as the thermal treatment model (temperature, heating rate, etc.) also influence MNP properties. For example, when heating above 500°C, ferromagnetic γ-Fe_2O_3 can be easily transformed into the antiferromagnetic and more stable phase α-Fe_2O_3 [30]. Therefore, it is important to optimize preparation conditions to produce the desired products. Typical particle sizes for the ferro- to superparamagnetic particles are in the range of 10–20 nm for oxides and 1–3 nm for metals [16]. The addition of polymers was found to limit the particle size [31]. Ultra-small MNPs (<5 nm) with uniform size distribution were reported by using the water-in-oil microemulsion method [32].

In general, nanoscaled MNPs are very sensitive to oxidation, particularly for pure metals and metal alloys due to their extremely high surface area. In addition, MNPs tend to agglomerate to form bulky particles due to their magnetic dipole interaction. After formation of bulky particles, the composites will significantly lose their

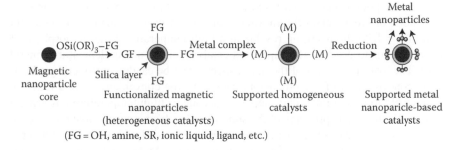

FIGURE 10.1 Methods of preparation of magnetic nanoparticles-supported catalysts.

catalytic activity or completely deactivate in some cases. Therefore, it's important to stabilize the produced MNPs to prohibit aggregation and deactivation. To protect the MNPs, they can be coated with the stabilizers such as polymers and surfactants [33,34]. Alternatively, they can be anchored in an optimized support such as polymers, inorganic metal oxides, and organic–inorganic hybrids. Various catalyst composites comprising of supports and active metal nanoparticles have been reported [35–39]. For catalyst applications, the MNPs need to meet the requirements of size uniformity, homogeneous dispersion in reaction media, and satisfied stability in the related reactions. The pristine MNPs can be functionalized either by covalent conjugation or by electrostatic adsorption [6,40–43]. The introduced functional groups may form a new layer outside of the original nanoparticle surface or simply replace the existing groups.

Magnetic nanoparticles have been used as catalyst supports with designed functionalization apart from being directly used as catalysts in some organic reactions such as oxidation. It was reported that magnetic core was coated with an amorphous silica shell to form more complicated nanoparticles, $SiO_2@Fe_2O_3$ [44]. In general, silicates such as sodium metasilicate and iron salts, for example, $FeCl_3$ and $Fe(NO_3)_3$, are used to prepare the silica-fabricated nanoparticles and magnetic nanoparticles, respectively. Functionalized magnetic nanoparticles are commonly prepared from the so-called sol–gel procedure in which tetraalkyl orthosilicate was used as a silica source [45]. The functional groups such as $-NH_2$ and $-OH$ can be introduced to the silica-coated magnetic nanoparticles in the step. The particle size was closely related to the amount of solvents trapped in the sol–gel rather than the silica source used [45]. The silica shell could be further fabricated to conjugate with varying ligands such as $-PPh_2$ due to abundant $-OH$ on the surface or pre-introduced functional groups. The supported ligands could coordinate with active metal centers or stabilize metal nanoparticles to form magnetic nanoparticles-supported heterogeneous catalysts as shown in Figure 10.1 [5,6,46].

10.2.2 CURRENT SYNTHETIC METHODOLOGIES OF MAGNETIC NANOCOMPOSITE-BASED CATALYSTS

10.2.2.1 Magnetic Nanoparticles-Supported Metal Nanoparticles

Preparation of supported metal nanoparticles-based heterogeneous catalysts normally includes the following steps: (1) impregnation of metal complexes' precursors on selected supports such as activated carbon, γ-Al₂O₃, by suspending the supports in a solution of precursors; (2) removal of volatile solvents and drying to a constant weight; (3) calcite the resulting residue at a high temperature in an atmosphere of hydrogen or air to produce the supported catalysts either metal elemental or metal oxides. The loading amount of the metal species can be easily controlled by adjusting the concentration of the precursors. However, the convenient method may not be applicable to prepare magnetic supports because a high temperature has critical effects on the magnetic property of the support. It has been reported that an increase in temperature decreases the coercive force and hysteresis loss in iron-based magnet [47]. Ferromagnetism is lost at the Curie temperature. Materials become paramagnet after the Curie temperature. Therefore, magnetic nanoparticle-supported metal particles are preferred to be prepared by a wet chemistry method in which the active metal particles are co-precipitated or postprecipitated on the supports. One widely used method to prepare metal nanoparticles is the reduction of metal salts or ligand-coordinated complexes in the presence of stabilizers. In this regard, various reducing reagents such as $NaBH_4$, N_2H_4, and HCO_2H have been used to produce supported metal nanoparticles in the presence of stabilizers [5,6]. The particle size distribution and shape of the obtained metal particles are highly dependent on the used reductants and stabilization agents such as ionic liquids, surfactants, and functional polymers [42,48,49]. We have been successfully using the reducing technology to prepare magnetic nanoparticle-supported, ultra-small Pd(0) particles (<1 nm) as catalysts. In this case, the pre-immobilized phosphate functional groups were used as stabilizing ligands (see Figure 10.2) [42].

Ionic liquid has been recognized as a green solvent and widely investigated [50,51]. The catalyst recycles can be achieved by extracting products from reaction mixtures using an optimized solvent [50,51]. Ionic liquid has also been used to stabilize metal nanoparticles such as Ru⁰ nanoparticles to produce well-distributed small catalytic systems (Figure 10.3) [5,6,49]. It is reasonable to combine advantages of both sides by forming a hybrid comprising of magnetic nanosupport and ionic liquid such as

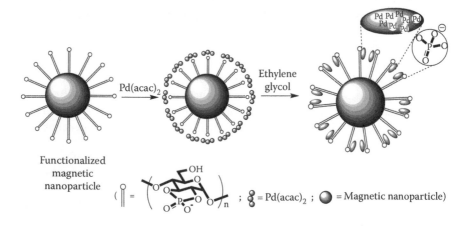

FIGURE 10.2 Synthesis of magnetic nanoparticle-supported palladium nanoparticles. (From Y. Zhu et al., *Adv. Synth. Catal.* 349, 2007, 1917.)

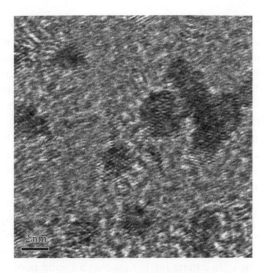

FIGURE 10.3 Transmission electron microscope (TEM) picture of Ru nanoparticles in tri-hexyltetradecylphosphonium dodecylbenzenesulfonate. (From Y. Zhu et al., *J. Am. Chem. Soc.* 129, 2007, 6507.)

an ionic liquid-coated or ionic liquid-functionalized magnetic support. Indeed, ionic liquid-functionalized magnetic nanoparticles, for example, Figure 10.4, have been widely investigated and well documented [52–56].

Metal nanoparticles can also be prepared using other methods, including UV photolysis [57], thermal decomposition [58], metal vapor deposition [59], electrochemical reduction [60], sonochemical decomposition [61], microwave irradiation [62], and rapid expansion of supercritical fluid solutions [63]. Bi- and multimetallic nanoparticles can be prepared similarly using the above-mentioned synthetic routes both for homo- and for hetero-metallic nanoparticles. For instance, different types of bimetallic nanoparticles such as Ru_5Pd, Ru_6Pd_6, and related molecules have been prepared by the thermolysis of appropriate carbonylate precursors [64].

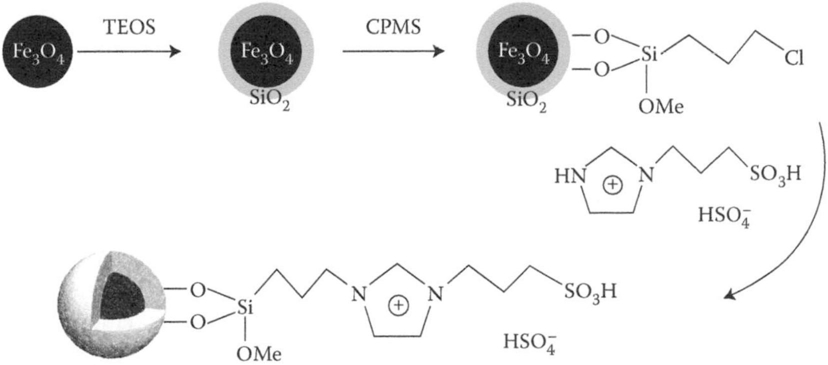

FIGURE 10.4 Synthesis route to IL-Fe_3O_4@SiO_2. (From Z. Wu et al., *Ind. Eng. Chem. Res.* 53, 2014, 3040.)

10.2.2.2 Magnetic Nanoparticles-Supported Homogeneous Catalysts

Homogeneous catalyst has been known and used for centuries. It is generally highly active and indispensable in organic synthesis. However, implementation of a homogeneous catalyst system can encounter several practical problems upon scale-up and large-scale industrial utilization, particularly in the cost of catalyst components (both metal centers and ligand) and the difficulty faced in the separation and recycling of the catalyst. Furthermore, the remaining toxic metal species may cause heavy product contamination. Therefore, it is highly expected to immobilize the active homogeneous catalysts on a support to enable their recoverability. Commonly used supports such as active carbon, carbon nanotubes, and zeolite-type materials have been employed to heterogenize organometallic catalysts or ligands [65–67]. The supported homogeneous catalyst usually does not overcome problems such as leaching of active centers and/or the lowering of activity [5,6]. In the past few years, a great amount of magnetic particles-supported organometallic catalysts have been reported [68,69]. As described in Figure 10.1, the functional groups immobilized on a magnetic core can be further used as a ligand pool to coordinate with metal complexes to form supported homogeneous catalysts. In some cases, the optimized supported functional groups can be directly used as catalysts, also called supported organic catalysts [5,6,68,69]. For instance, ionic liquid species supported on magnetic nanoparticles, as exhibited in Figure 10.4, showed good performance in esterification of oleic acid (OA) with short-chain alcohols and transesterification of soybean oil [54]. The catalyst could be easily recovered by an additional magnet with sustained catalytic activity after being reused eight times [54]. Ferrite magnets are commonly used as an additional magnetic force to attract the MNPs-based catalyst composite from a reaction mixture followed by wash with a suitable solvent to remove any remaining soluble reaction species.

10.3 CATALYTIC APPLICATION OF MAGNETIC NANOPARTICLES-SUPPORTED CATALYSTS

10.3.1 Oxidation Reactions

Oxidation reactions are well recognized as fundamental chemical transformation from the view of both academic and industry applications. In some traditional arts, excessive toxic and corrosive oxidants are used, and that has caused both environmental problems. Furthermore, they are not economic and sustainable. However, oxidation procedures accelerated by heterogeneous catalysts minimize the use of oxidants and thus are economically and environmentally benign. Various solid materials have been used for the metal-based oxidation catalysts to enable their recyclability. As a unique support, magnetic nanoparticles have attracted increasing interests in catalytic oxidation research due to their high activity and recyclability. In this part, we summarize the current applications of catalytic oxidation reactions with a specific emphasis on magnetic support-based heterogeneous catalysts. We are expecting it to serve as a reference guide for the selective oxidation reactions.

Controlling oxidation (also known as selective oxidation) of alcohols to the desired aldehydes, ketones, and carboxylic acids is an important organic reaction.

The resulting oxidation products are important intermediates or high-valued compounds in pharmaceutical and fine chemistry [70–72]. Oxidation reactions of benzyl alcohols (Ph-CHR(OH)) are commonly used to examine a catalyst in the procedure of catalyst development. Recently, both transition metal complexes and metal nanoparticles have been immobilized on magnetic nanosupports. Zhang et al. reported that magnetic nanoparticles could be functionalized with proline and further coordinated with palladium acetate (PdAc$_2$) to form supported Pd complexes, Fe$_3$O$_4$@SiO$_2$-Pro-PdAc$_2$ [73]. The catalyst composite was active for aerobic oxidation of benzyl alcohol with O$_2$ and substituted benzyl alcohols to corresponding aldehydes and ketones in 46%–97% yields as shown in Figure 10.5 [69]. For the secondary benzyl alcohols, corresponding ketone products were obtained in 71%–81% yields under the same conditions [73]. As expected, the catalyst could be separated from the reaction mixture easily by using an external magnet and reused more than eight times with sustained activity [73].

Magnetic nanoparticle-supported Co-NPs, Fe$_3$O$_4$@Co-NPs, were also prepared and used for catalytic oxidation of benzyl alcohols to corresponding carbonyl products in a yield of 79%–94% [74]. In the example, tert-butyl hydroperoxide in excess was used as oxidant, and the reaction was performed at 80°C. The catalyst could be recovered conveniently and reused for 7 runs with negligible leaching of Co (0.09%) [74]. Multiwalled carbon nanotube-supported Pd@Ni nanoparticles, Pd@Ni/MWCNTs, were also separable by using an external magnet [75]. Catalyst Pd@Ni/MWCNTs showed very high activity during the oxidation reaction of benzyl alcohol to benzaldehyde. Λ yield of 97% was reported using H$_2$O$_2$ as oxidant in water medium at 80°C for 6 h in the presence of K$_2$CO$_3$ [75]. Pd-Re NPs/Fe$_3$O$_4$@SiO$_2$ was prepared in our laboratory; the catalyst was active for selective oxidations of benzyl alcohol to benzyl aldehyde and glycerol to dihydroxyacetone in moderate yields. Oxidation of sulfides (RSR′) to sulfoxides (RSOR′) was also explored using magnetic nanoparticle-supported Mn complexes, [Br$_2$Mn(tetraphenylporphyrin)Ac]/Fe$_3$O$_4$@SiO$_2$ [76].

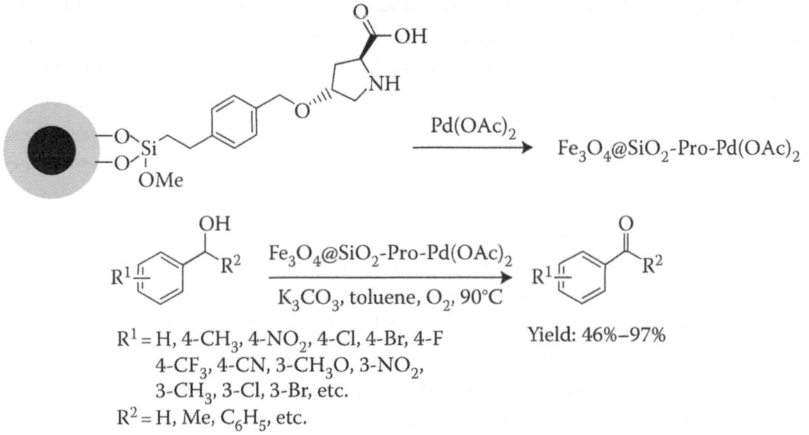

FIGURE 10.5 Oxidation of alcohols catalyzed by Fe$_3$O$_4$@SiO$_2$–Pro–Pd(OAc)$_2$. (From D. Wang and D. Astruc, *Chem. Rev.* 114, 2014, 6949.)

Sulfoxides were obtained in 60%–96% yields in 1 h. In the reaction, *tert-n*-byutyl-ammonium peroxomonosulfate was used as an oxidant [76]. The commonly used heterogeneous oxidation catalyst tungstophosphoric acid was supported on γ-Fe_2O_3@ SiO_2 to form a magnetic recyclable composite γ-Fe_2O_3@SiO_2-$H_3PW_{12}O_{40}$ [77]. The composite exhibited good activity for amine oxidation to produce corresponding nitrones in moderate to high yields. The catalyst was also recyclable with sustained activity [77]. Oxidative amidations of aromatic aldehydes with amines catalyzed by magnetic nanocatalysts, such as magnetic carbon nanotubes@SiO_2-ligand-CuI [78] and $CuFe_2O_4$ [79], were also reported. The carboxamide products were obtained in 48%–85% yields [78].

Epoxidation is another important type of oxidation reaction, which has been widely used in fine chemistry. Silver-based catalysts are significant in epoxidation of alkenes. Chen and others first applied Ag-supported magnetic nanocatalysts in the epoxidation of styrene with high activity and selectivity [80]. Other reported magnetic recyclable epoxidation catalysts include ionic liquid-type peroxotungstates such as [HDMIM][W_2O_{11}]/Fe_3O_4@SiO_2 [81], metalloporphyrin complex [Mn(TPyP) Ac]/Fe_3O_4@SiO_2 [82], and $CoFe_2O_4$ [83]. All the supported epoxidation catalysts were recyclable with sustained activities.

10.3.2 HYDROGENATION REACTIONS

10.3.2.1 Hydrogenation of Alkenes

Hydrogenation reaction of various unsaturated organic substrates such as alkenes, aldehydes, ketones and nitroaromatics, is recognized as a crucial art in organic synthesis; it is one of the most important organic transformations in industry and academia [84,85]. Apart from being widely used in fine chemistry and petro-chemistry, recently hydrogenation has shown important applications in the area of renewable and clean energy such as biomass conversion to value-added products [86,87]. Palladium-based catalysts are powerful and dominate in the hydrogenation of unsaturated substrates in existing arts and will continue to play the important role in the future due to their much higher activity [88]. Magnetic cobalt nanoparticles-fabricated ordered mesoporous carbons (OMC) produced from mesoporous silica SBA-15 templates were prepared to support Pd-based catalysts [89]. The produced catalyst Pd NPs (1 wt%)/Co-OMC is active for a continuous hydrogenation of octene to octane. This reaction was highly reproducible, and the catalyst can be conveniently recovered from the reaction solution by applying an external magnetic field [89]. Reiser and others prepared various catalysts of Pd-NPs/Co@C from $Pd_2(dba)_3$ precursor under microwave irradiation [90]. This type of catalyst is active for hydrogenation of *trans*-stilbene. It was found that the catalytic activity is highly dependent on the particle size of the supported Pd-NPs. The highest TOF of 11,095 h^{-1} has been reached by using the supported catalysts with a Pd size of around 2.7 nm [90]. However, it was also found that the smaller Pd-NPs tended to leach [90]. To well stabilize and enrich the supported Pd-NPs, Co@C was functionalized with ionic liquid to form a catalyst composite of Pd-NPs-IL/Co@C [91]. The modified catalyst exhibited a higher Pd loading amount of 34 wt% and a TOF of 50 h^{-1} in the hydrogenation of *trans*-stilbene [91]. The catalyst could be recycled more than 12 times with sustained activity [91]. Pd-NPs produced *in situ* from decomposition of Pd(dba)$_2$ were also supported on a terpyridine-fabricated Fe$_3$O$_4$@SiO$_2$ [92]. The catalyst was highly active for the hydrogenation of cyclohexene at relatively mild conditions (75°C, 6 atm H$_2$) with an initial TOF of 50,400 h^{-1} [92].

Jacinto et al. reported Rh-NPs supported by silica-coated MNPs [93]. In the procedure, the silica shell was functionalized with 3-(aminopropyl)-triethoxysilane to produce support Fe$_3$O$_4$@SiO$_2$-link-NH$_2$, and the Rh-NPs were incorporated by reducing the RhCl$_3$-loaded hybrid with H$_2$ [93]. The resulting Rh-NPs (3 ~ 5 nm) were found to be a highly active catalyst for hydrogenation of cyclohexene and benzene at 75°C and 6 atm of H$_2$ with a TOF of 40,000 and 1100 h^{-1}, respectively [93]. The catalyst was reusable for 20 cycles with negligible Rh leaching [93]. Rhodium complex [Rh(TPPTS)$_3$Cl] (TPPTS = trisodium triphenylphosphine-3,3′,3″-trisulfonate) was supported on MNPs without further surface modification using the above-described co-precipitation procedure [94]. The MNP-[Rh(TPPTS)$_3$Cl] composite successfully catalyzed 10 consecutive reaction cycles of the hydrogenation of dimethyl itaconate to dimethyl 2-methylsuccinate with 100% conversion without any loss of activity [94].

10.3.2.2 Hydrogenation of Alkynes

Platinum nanoparticles were also supported on ionic liquid-modified Fe$_3$O$_4$ nanoparticles by reducing pre-impregnated platinum complexes [52]. The catalyst was used in the selective hydrogenation of alkynes and α,β-unsaturated aldehydes. In the chemoselective hydrogenation of cinnamaldehyde, the catalyst showed high selectivity to producing 3-phenylprop-2-en-1-ol in 99% yield with a TOF of 3.3 h^{-1} as shown in Figure 10.6a [52]. In the selective hydrogenation of diphenylacetylene, the catalyst provided 95% *cis*-stilbene isomer with a TOF of 2.5 h^{-1}, and the catalyst was recyclable [52]. Pd-NPs and magnetic CuFe$_2$O$_4$ were encapsulated in a silica microsphere, SiO$_2$@CuFe$_2$O$_4$-Pd-NPs, and used for hydrogenation of phenylacetylene

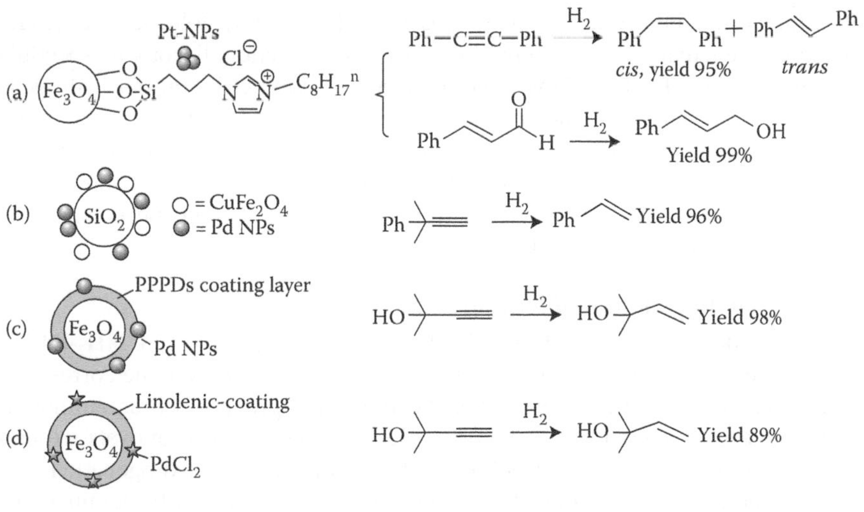

FIGURE 10.6   Selective hydrogenation of alkynes to alkenes.

(see Figure 10.6b) [95]. A high yield of over 96% was achieved in hexane media in 2.5 h by using hydrogen balloon. In comparison, SiO_2@$CoFe_2O_4$-Pd-NPs and SiO_2@ Fe_3O_4-Pd-NPs provided lower yields of 18% and 52%, respectively [95]. The SiO_2@ Pd-NPs was not active. The results suggest that magnetic species may enhance catalyst activity in this case [95]. On the other hand, polymers such as polyphenyl-enepyridyl dendrons (PPPDs) were also used to coat the magnetic particle cores to stabilize supported metal nanoparticles [96]. The embedded supported Pd-NPs exhibited high activity in selective hydrogenation of dimethylethynylcarbinol to dimethylvinylcarbinol in 98% yield as described in Figure 10.6c [96]. The catalyst was recyclable for three times. Palladium (+2) chloride was supported on linolenic-coated Fe_3O_4 nanoparticles (Figure 10.6d) [97]. The composite also showed high activity in selective hydrogenation reaction of dimethylethynylcarbinol with a high TOF of 7.9 s^{-1} [97]. The supported Pd(+2) complexes were considered to be the true active species [97].

10.3.2.3 Hydrogenation of Carbonyl Compounds

Ruthenium-based hydrogenation catalysts were supported on magnetic nanoparticles to make them recyclable. Ru-NPs were conveniently immobilized on Fe_3O_4@ SiO_2 to form catalyst Ru-NPs/Fe_3O_4@SiO_2 [98]. Ru-NPs/Fe_3O_4@SiO_2 was highly active in hydrogenation of acetophenone in 99% yield in the presence of KOH under microwave irradiation. And usually, the catalyst could be used more than 3 times with a 0.08% Ru leaching [98]. The catalyst is stable to various functional groups such as halide, NO_2, and NH_2. Chiral ligands-coordinated complexes were also supported on magnetic nanoparticles and demonstrated high activity in asymmetric hydrogenation to produce chiral alcohols. MNPs-supported [Ru(BINAP-PO_3H_2) (DPEN)Cl_2] (BINAP-PO_3H_2 = (R)-2,2′-bis(diphenylphosphino)-1,1′-binaphthyl-4-phos phonic acid, DPEN = (R,R)-1,2-diphenylethylenediamine) provided high activity and enantioselectivity in heterogeneous asymmetric hydrogenations with H_2 (Figure 10.7a) [99]. The catalyst was recyclable up to 14 times with sustained activity [99]. Chiral ruthenium complex, Ru-TsDPEN (TsDPEN = N-(p-toluenesulfonyl)-1,2-diphenylethylenediamine) was immobilized on a magnetic siliceous mesocellular foam material (Ru-TsDPEN/F(M)) and used in the asymmetric transfer hydrogenation of aromatic ketones as shown in Figure 10.7b [100]. The activity of the immobilized catalyst is comparable with its homogeneous precursor. Moreover, supported chiral rhodium complex, as detailed in Figure 10.7c was prepared and showed high asymmetric selectivity and conversion in hydrogenation of aromatic ketones with HCO_2Na as a hydrogen source [101].

10.3.2.4 Hydrogenation of Nitroaromatics

Nanocomposites Fe_2O_3@SiO_2-link-HS-Pd-NPs and Fe_2O_3@SiO_2-link-NH_2-Pd-NPs were reported as active catalysts for reductions of nitroaromatics to the corresponding anilines [102]. It was believed that the functional groups such as aminoethyl groups could bind supported Pd nanoparticles (around 2–3 nm) catalysts strongly, thus reducing aggregation and leaching of Pd-NPs [102]. The hydrogenation reactions are conducted faster with these catalysts in comparison with commercially available Pd/C catalyst. The conversion rates are 0.39, 0.12, and 0.08 μmol/s for

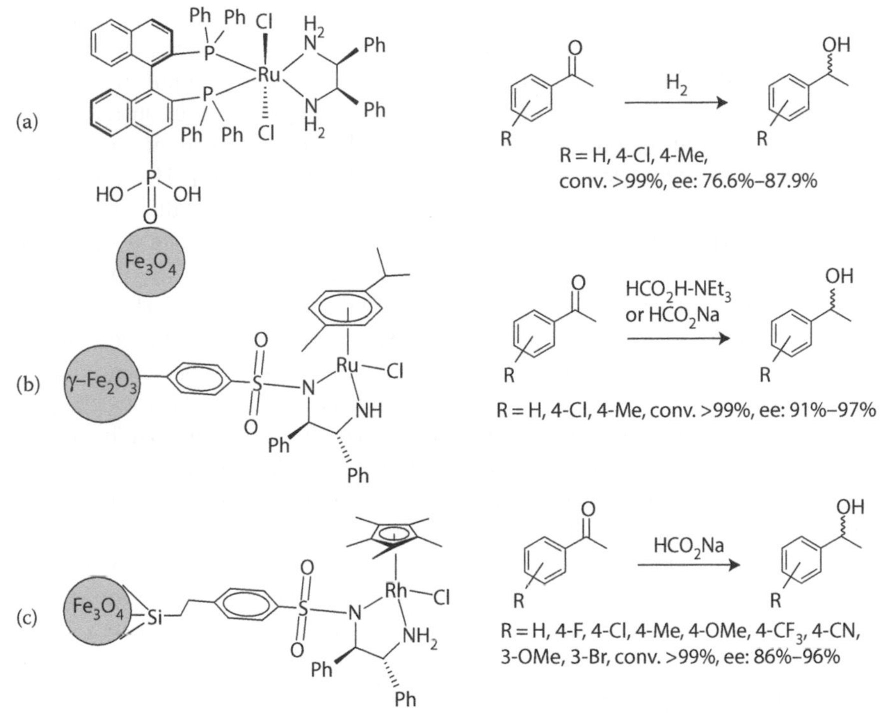

FIGURE 10.7 Enantio excess (ee %) values for hydrogenation of aromatic ketones.

catalysts Fe$_2$O$_3$@SiO$_2$-link-NH$_2$-PdNPs, Fe$_2$O$_3$@SiO$_2$-link-HS-PdNPs, and Pd/C, respectively, when the turnover number (TON) was set at 2000 for 100% conversion [102]. Furthermore, Fe$_2$O$_3$@SiO$_2$-link-NH$_2$-PdNPs also showed higher Pd dispersion and reduced agglomerate and growth of Pd nanoclusters during the hydrogenation reaction due to amine groups that have stronger affinity for Pd nanoparticles [102]. Pd-NPs supported on carbon fibrils encapsulating NiFe alloy particles also provided good activity for the hydrogenation reaction of nitrobenzene with a hydrogenation consumption of 1.5 mL/min at room temperature [103].

Recently, diverse magnetic nanoparticle-supported transition metal nanoparticles such as Pd, Au, Ag, and Fe have been prepared and used for hydrogenation of nitroaromatics [69]. Au-NPs were supported on magnetic nanoparticles with varying functional surfaces such as graphene oxides (GOs)-encapsulated polydopamine (PDA) (Au-NPs/Fe$_3$O$_4$@PDA@GOs) [104] and PDA alone (Au-NPs/Fe$_3$O$_4$@PDA) [105]. The two catalysts were examined for reduction of o-nitroaniline with NaBH$_4$. Catalyst Au-NPs/Fe$_3$O$_4$@PDA@GOs (4 min, conv. >99%) was more active than Au-NPs/Fe$_3$O$_4$@PDA (7 min, conv. 99%) under similar conditions. The catalysts were easily recovered by using an external magnet and reused more than eight runs [104,105]. Unfortunately, both catalysts showed a heavy leaching amount of Au; 30% and 18% of the loaded Au was lost for Au-NPs/Fe$_3$O$_4$@PDA@GOs and Au-NPs/Fe$_3$O$_4$@PDA after 11 and 8 runs, respectively [104,105].

Poly(N,N'-methylenebis(acrylamide)-co-poly(2-dimethylaminoethylmethacry late)-functionalized, magnetic nanoparticle-supported Au-NPs, named Au-NPs/ Fe_3O_4@SiO2@PHEMA-co-PDMAEMA, were reported with a small and narrow-distributed Au-NPs (~3.7 nm) [106]. The catalyst catalyzed reduction reaction of 4-nitrophenol to 4-aminophenol quantitatively within 15 min at room temperature. The catalyst was reused six times with sustained activity [106]. However, metal nanoparticle leaching may cause a decrease in catalyst performance due to the loss of active centers. Meanwhile, the leaching metals may also contaminate the products. To maintain the high activity, Yao et al. reported a catalyst comprising of mesoporous SiO_2-coated, magnetic nanoparticle-supported Pd-NPs (Pd-NPs/Fe_xO_y@SiO$_2$ [107]. The coating layer of mesoporous SiO_2 was slightly larger than the magnetic core, and thus the layer was movable to enhance the interaction between Pd active centers and substrates. The catalyst is highly active in reduction of 4-nitrophenol to 4-aminophenol with $NaBH_4$ and recyclable at least 10 times with sustained activity (100% conversion). The hybrid was found highly stable even at ultrasonic treatment [107]. Interestingly, pristine Fe_3O_4 magnetic nanoparticles produced *in situ* by reducing an iron complex such as $Fe(acac)_3$, $FeAc_2$, $FeCl_3 \cdot 6H_2O$, and $FeCl_2 \cdot 4H_2O$ with hydrazine in excess also showed high activity in the hydrogenation of nitroaromaticles under microwave irradiation [108,109]. With the catalysts, various functionalized aniline products were obtained in 95%–99% yields from corresponding nitroarenes using hydrazine hydrate as the reductant within 2–8 min [108,109]. The *in situ*-prepared, Fe_3O_4 nanoparticle-based catalysts were also recyclable with sustained activity. However, both commercially available Fe_3O_4 and Fe^0 powders were not active in this type of reaction [108,109]. $Fe_{25}Co_{75}$ nanoparticles were prepared and deposited onto the surface of grapheme oxide; the supported MNPs exhibited high activity for the reduction of 4-nitrophenol with $NaBH_4$ in a TOF value of 2.9×10^{16} s^{-1} [110,111]. Other active and magnetic recyclable catalysts, including Fe_3O_4-NPs/GOs [112] and Ni-NPs/GOs [113], were also reported using $NaBH_4$ as the reducing agent. It was found that the activity of these magnetic nanoparticles-based catalysts could be significantly improved by near-infrared irradiation [114]. Ma and others prepared Pt-NPs (~5 nm) catalyst supported on a carbon-coated MNPs, Pt-NPs/Fe_3O_4@C [115]. The catalyst was highly effective for substituted nitrobenzene, R-C_6H_4-NO_2 (R = H, 4-Cl, 4-Br, 4-Me, 2-NH_2, 3-NH_2, 4-NH_2, 2-OH, 4-OH, 4-CHO, 4-CH_2OH, 4-COMe, and 2-NO_2), to the corresponding anilines in 95%–99% yields with H_2 as the reductant at room temperature [115].

10.3.3 C–C Coupling

The coupling reactions such as the Miyaura–Suzuki cross-coupling and the Heck coupling reactions have been discovered a long time ago and are commonly used for organic synthesis to form new C–C bonds. The method is highly important both in academe and in industry, including fine chemistry and pharmacy. In the existing arts, palladium complexes-based homogeneous catalysts dominate the catalytic C–C coupling reactions due to their extremely high activity [116–120]. Transition metals-catalyzed C–C coupling reactions have been well investigated and documented [116–120]. However, like other homogeneous catalytic systems, removal of the catalyst species from the reaction mixture remains a big challenge. This crucial problem has to be addressed to broaden applications of this type of reaction, particularly in pharmaceuticals in which the toxic metal residues must be removed to meet a governmental requirement. To overcome this big disadvantage, homogeneous catalysts have been heterogenized, and supported catalysts have emerged [116–120]. As mentioned earlier, magnetic nanoparticles are potentially promising supports; therefore, this type of material has been employed to anchor active catalyst species in C–C coupling reactions. This part will summarize recent developments in the area.

10.3.3.1 Miyaura–Suzuki Cross-Coupling

In general, a reaction mixture of the Miyaura–Suzuki cross-coupling is comprised of substrates, typically organoboron compound and an organic halide, base, and palladium-derived catalyst [121]. Advantages of this technology include excellent product yields and the mild experimental conditions used [121–124]. Magnetic nanoparticle-supported palladium complexes such as N-heterocyclic carbene (NHC)-coordinated Pd complexes, as shown in Figure 10.8, have been prepared and used for the Miyaura–Suzuki cross-coupling reactions between aryl boronic acids and aryl bromides or aryl iodides via liquid- or solid-phase processes with isolated yields of 60%–90% [125–129].

Zhu et al. reported that ultra-fine Pd(0) nanoparticles (<1.0 nm) can also be immobilized onto magnetic nanoparticles enriched with phosphate functional groups. The catalyst showed high efficiency for both the Miyaura–Suzuki cross-coupling and the Heck coupling reactions (Figure 10.2) [42]. These supported catalysts can be well dispersed in both aqueous and organic phases to produce a quasi-homogeneous

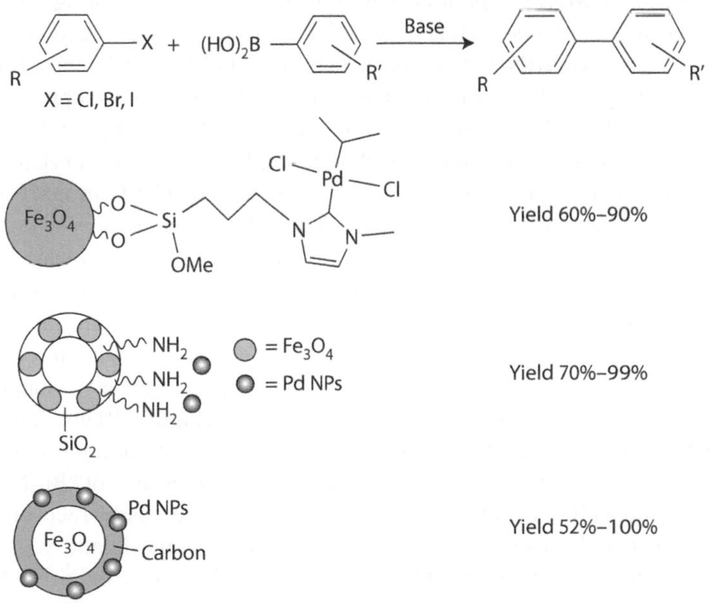

FIGURE 10.8 MNPs-supported Pd catalysts for the Miyaura–Suzuki cross-coupling reaction.

catalyst system. Similar to other MNP-supported catalysts, this catalyst could also be easily separated from the reaction medium by magnetic force. Although the aforementioned results indicate that MNP-supported catalysts could potentially find useful application in the Miyaura–Suzuki cross-coupling reactions, the method remains relatively new and has not been thoroughly investigated. More research is necessary to improve the yields and reaction conditions [42].

10.3.3.2   Heck Coupling

Catalytic reactions between aryl or vinyl halides and activated alkenes produce substituted *trans*-olefins. This type of reactions are well known as the Heck coupling and widely used for organic synthesis in the presence of palladium-based catalysts [130–137]. Again, for palladium complexes-based homogeneous catalysts, the challengeable issues of catalyst removal and recyclability remain unsolved similar to other homogeneous catalysis systems such as the Miyaura–Suzuki cross-coupling. NHC ligand-coordinated palladium catalysts were supported on magnetic nanoparticles and exhibited high activity in the Heck cross-coupling reactions between substituted aryl bromides or iodides and n-butylacrylate [125]. The reported yields were 95%–97% with good recyclability and sustained catalytic activity [125]. Small Pd-NPs of 3.6 nm were supported on magnetic carboxylated polypyrrole nanotubes, and the resulting nanocomposite provided extremely high activity for the Heck cross-coupling reactions between aryl iodides and n-butylacrylate or styrene [138]. A high yield of more than 97% with a TOF of >28.3 h^{-1} was achieved with the catalyst [138]. The catalyst composite is more active than the carbon-supported analogs due to their higher surface area [138].

Applications of MNPs-supported catalysts in the Heck coupling reaction are attracting growing interest. More works have been reported recently. Various surface-functionalized MNPs have been prepared and used in supporting palladium-based catalysts. Phosphine ligands-functionalized, MNPs-supported $PdAc_2$ ($PdAc_2/Fe_3O_4@SiO_2$-PPh_2) (Figure 10.9a) was reported [139]. The catalyst composite showed promising activity in the Heck reaction between ethyl acrylate and substituted chlorobenzenes in the presence of base K_2CO_3 at 120°C in DMF media to produce *trans*-arylated olefin in 88%–95% yields [139]. The activity is comparable with corresponding homogeneous $PdAc_2$ catalyst and is recyclable at least four times with sustained activity and negligible leaching of Pd (<1 wt% of the initial loaded Pd). The authors also examined the catalyst $PdAc_2/Fe_3O_4@SiO_2$ in the absence of phosphine groups. However, the catalyst showed very low activity in comparison with $PdAc_2/Fe_3O_4@SiO_2$-PPh_2 [139]. Another example is Pd-NPs supported on OA-functionalized MNPs (Pd-NPs/$Fe_3O_4@OA$) [140] (Figure 10.9b). The catalyst was robust and able to tolerate a wide substrate scope of aryl halides and provided various olefin products in good to excellent yields in the presence of $^{n}Bu_3N$ at 120°C in DMAc media [140]. In addition, the catalyst could be easily recovered by an external magnet and reused with sustained activity. Triazine ligand was immobilized onto the surface of $Fe_3O_4@SiO_2$ and further coordinated with Pd(EDTA) complex to heterogenize the homogeneous catalyst as shown in Figure 10.9c [141]. The catalyst showed high activity (80%–97% yields, 1.9–14×10^3 h^{-1} TOFs) in the Heck coupling reactions with a wide scope of functional aryl halides [141]. The catalyst

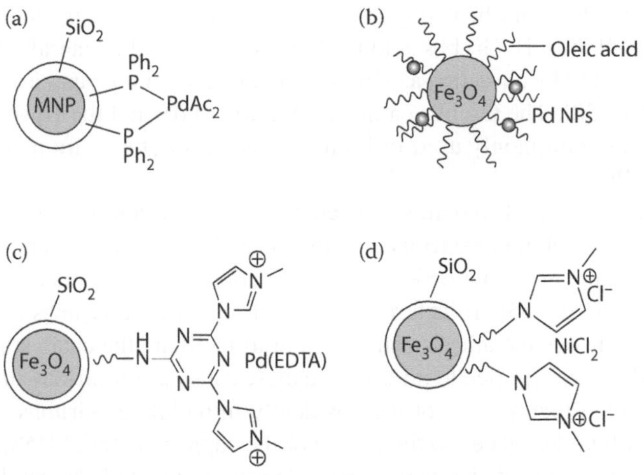

FIGURE 10.9   Magnetic nanoparticles-supported catalysts for the Heck coupling reaction.

performance could be improved by ultrasound irradiation during the reaction course with extremely high TOFs of $51–250 \times 10^3$ h^{-1} [141]. Apart from commonly used palladium nanoparticles, less expensive nickel complexes-based catalysts also exhibited activity to the Heck coupling reactions. For example, $NiCl_2$ supported on the ionic liquid-functionalized magnetic nanoparticles (Figure 10.9d) was prepared and used as a catalyst for a coupling reaction between aryl halides and ethyl but-3-enoates [142]. The Ni-based heterogeneous catalyst was more active for aryl iodides to provide olefins in 88%–98% yields in comparison with aryl bromides and aryl chlorides which gave <30% yield [142]. More research has to be done to investigate a suitable shell for the MNPs to stabilize the metal cores, as well as to reduce leaching, yet enable easy access to the substrate to the catalyst.

Magnetic nanoparticles-supported, palladium-based catalysts have also been used for a coupling reaction between aryl/vinyl halide and terminal alkynes, which is well known as the Sonogashira cross-coupling reaction. Schiff base ligands have been immobilized onto silica-coated MNPs, and used to form Schiff base-coordinated Pd complexes, $Fe_3O_4@SiO_2$-Schiff base-Pd(II), and $Fe_3O_4@SiO_2$-Schiff base-PdAc$_2$ [143,144]. The catalysts were highly active for aryl iodides and bromides in the presence of triethylamine in DMF at 90°C to produce internal alkynes in 85%–93% yields. Middle to high yield was reported for aryl chlorides under the same conditions [143,144]. The catalysts were recyclable for six times with sustained activity. The Hiyama cross-coupling reactions were also explored. The catalyst Pd-NPs/Fe_3O_4 was reported active in the Hiyama coupling between aryl bromides and aryl siloxanes in aqueous solution to form biaryl subunits in good to excellent yields [145].

10.3.4   Olefin Metathesis

Olefin metathesis is a well-explored synthetic methodology and is a powerful method to construct new C=C bonds in organic synthesis [146–150]. It is well recognized that the method is essential to transform the landscape of organic synthesis.

The 2005 Nobel Prize in Chemistry was awarded to the pioneers in olefin metathesis: Yves Chauvin, Robert H. Grubbs, and Richard R. Schrock. In general, olefin metathesis reactions could be classified as ring-closing, ring-opening, and cross-metathesis reactions [146–150]. The Grubbs catalysts (Figure 10.10) and the Hoveyda catalysts (Figure 10.9) are commonly used in homogeneous catalytic olefin metathesis reactions [146–150].

To date, catalyst recycle remains a big challenge despite continuing efforts on separation and recovery of homogeneous metathesis catalysts from reaction mixtures. The reported techniques are either toxic due to application of noxious materials or difficult to be scaled up [151–155]. To heterogenize the highly active Grubbs and Hoveyda–Grubbs homogeneous metathesis catalysts and to facilitate the catalyst recyclability, plenty of works were reported to immobilize these catalysts on various supports [156–158]. It was found that activity of the covalently immobilized Grubbs and Hoveyda catalysts is profoundly related to the nature of the support material [158].

The magnetic nanoparticles were also used to support metathesis catalysts because they could be separated conveniently by application of an external magnetic force as mentioned earlier. Zhu et al. first reported a success of immobilization of the second-generation Hoveyda–Grubbs catalyst on MNP-support as shown in Figure 10.11 [43]. A loading amount of 0.28 mmol Ru/g support was achieved according to the report. The MNPs-supported catalysts are active for both the self- and cross-metatheses of methyl oleate as shown in Figure 10.12 [43]. With the catalyst, a TOF of 12 h^{-1} was achieved in a cross-metathesis reaction between methyl oleate and methyl acrylate. The activity is much higher than that of the second-generation Grubbs catalyst which gave a TOF of less than 1 h^{-1} and comparable to the unsupported second-generation Hoveyda–Grubbs catalyst (TOF ~ 11 h^{-1}). In addition, the MNP-supported catalysts can be well dispersed in the reaction mixture, display sustained activity, and can easily be separated by external magnetic attraction [43].

The first-generation Grubbs catalyst was also immobilized by formation of a covalent bond on MNPs and used for a ring-closing metathesis (RCM) reaction [159]. It was reported that the supported catalyst exhibited promising activity, which was comparable to homogeneous and other supported catalytic systems. With this supported catalyst, the RCM of N,N-diallyl-4-methylbenzenesulfonamide was carried out in DCM at 40°C. The corresponding cyclic olefin was formed with 95%–99%

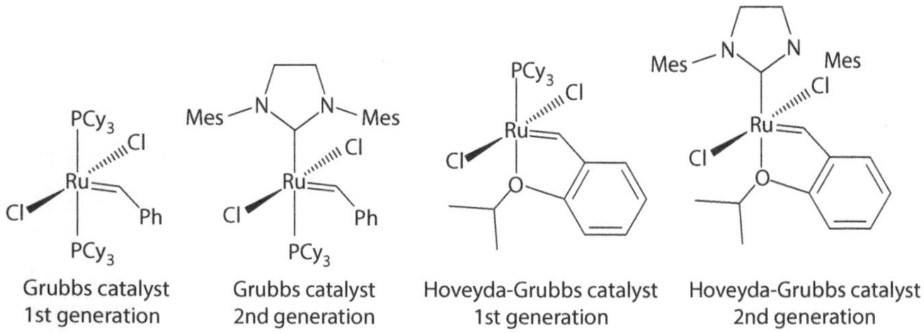

Grubbs catalyst 1st generation   Grubbs catalyst 2nd generation   Hoveyda-Grubbs catalyst 1st generation   Hoveyda-Grubbs catalyst 2nd generation

FIGURE 10.10 Metathesis catalysts.

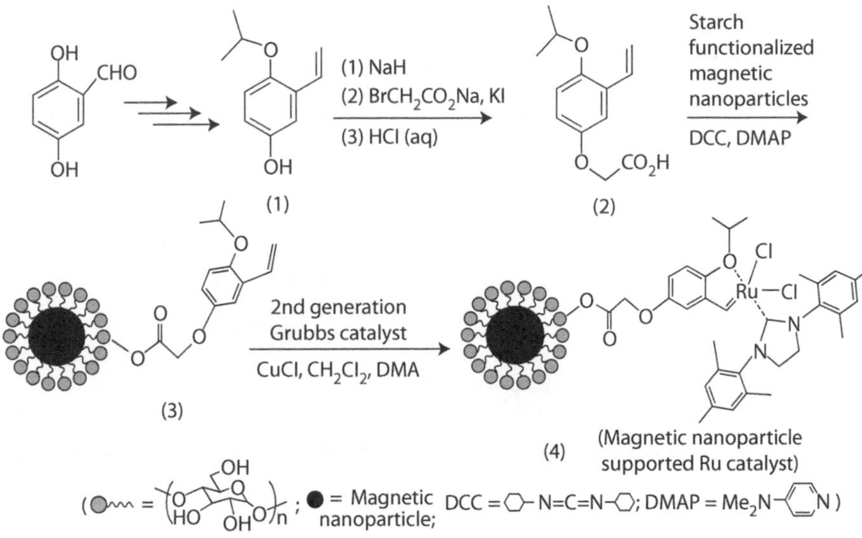

FIGURE 10.11 Synthesis of the MNPs-supported, second-generation Hoveyda–Grubbs catalyst. (From Y. Zhu et al., *Adv. Synth. Catal.* 351, 2009, 2650.)

conversions, and the catalyst was recycled up to 22 cycles without considerable loss in catalytic efficiency [159]. In addition, pyridine ligand was immobilized onto silica-coated MNPs by a "click" reaction. The supported pyridine ligand reacted with free metathesis catalyst, (1,3-bis(2,4,6-trimethylphenyl)-2-imidazolidinylidene) dichloro(phenylmethylene)(pyridine)ruthenium $(NHC)RuCl_2(Py)_2(=CHPh)$, to form a catalyst mixture of $Fe_3O_4@SiO_2$-link-Py$(NHC)RuCl_2(Py)(=CHPh)$ and $(Fe_3O_4@SiO_2$-link-Py$)_2(NHC)RuCl_2(=CHPh)$ [160]. The mixture was active for ring-opening

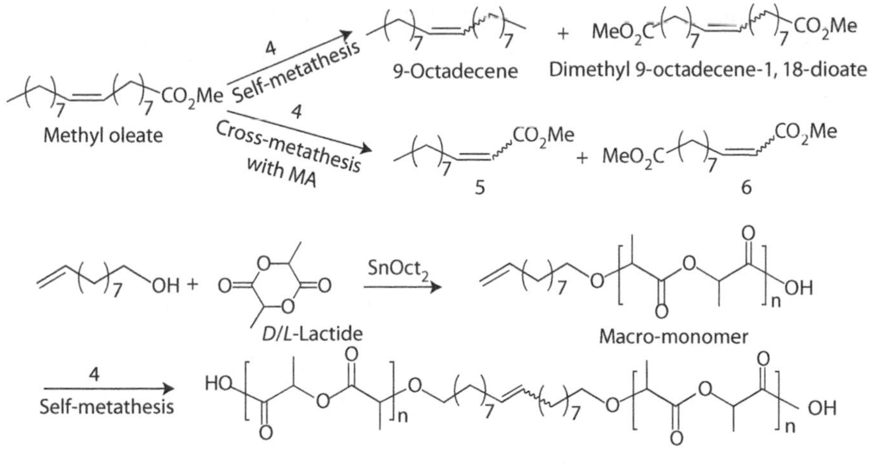

FIGURE 10.12 Metathesis reactions of methyl oleate. (From Y. Zhu et al., *Adv. Synth. Catal.* 351, 2009, 2650.)

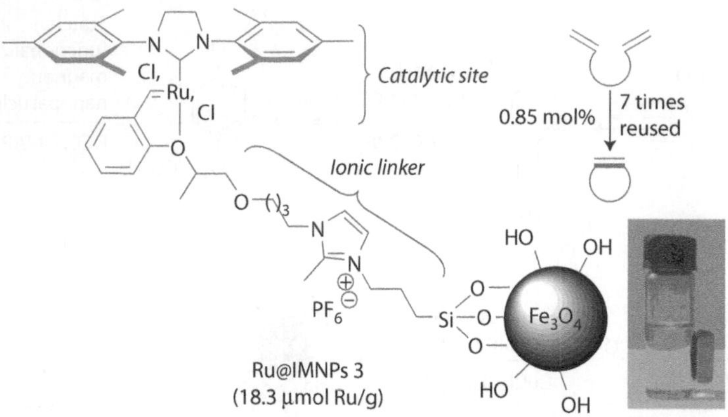

FIGURE 10.13 MNPs-supported metathesis catalysts. (From S.-W. Chen et al., *Org. Lett.* 16, 2014, 4969.)

metathesis polymerization reaction of a norbornene. However, these catalysts were reported to be less active than their corresponding unsupported counterparts due to the less efficient coordination between supported pyridine ligand and ruthenium functioning center [160]. Therefore, it is essential to optimize immobilization methodology to achieve high activity. Highly efficient straightforward procedures are always desired to heterogenize the active homogeneous catalysts.

Chen et al. reported that the MNPs-supported, second-generation Hoveyda–Grubbs catalyst was highly efficient for RCM reactions in DCM at room temperature to produce cyclic olefins as shown in Figure 10.13 [161]. In the catalyst, active Ru species were covalently attached to the MNPs by an imidazolium salt linker. The catalyst was extremely active, and the RCM reactions were carried out in less than 1 h with more than 96% conversions. The supported catalyst also demonstrated high activity in cross-metathesis reactions between methyl acrylate with 5-pentenylbezoate and 3-phenyl-1-propene providing the corresponding cross-metathesis products in excellent yield and with high E selectivity [161]. The catalysts can easily be recovered magnetically and reused up to seven times with a minimal leaching of ruthenium species [161].

10.3.5 HYDROFORMYLATION REACTIONS

Hydroformylation reaction is well known as one of the most important organic synthetic methods. It has been extensively studied in homogeneous catalytic reactions with large-scale industrial applications [162–164]. Rhodium-based catalysts dominate most of the existing arts to prepare aldehydes. Recycle of the expensive and toxic Rh catalysts from the reaction mixture is necessary from the viewpoint of economy, environmental benign, and product safety. Although Rh species could be separated from a reaction mixture by a phosphorus ligand, 3,3′,3″-phosphanetriyltris (benzenesulfonic acid) trisodium salt, due to its high water solubility [162], magnetic nanocomposite catalytic systems could also play an important role in recycling

Rh-based catalysts. In this type of reaction, regioselectivity remains a big challenge; much more effort has been spent on improving the selectivity and conversion of hydroformylations [163,164]. In comparison with initial cobalt catalyst, rhodium catalysts exhibited enhanced selectivity and thus attracted more interest. To facilitate the separation of catalyst from product mixtures, Rh-based catalysts coordinated with water-soluble ligands were developed and subsequently used in a biphasic hydroformylation of propene to produce the corresponding linear and branched aldehydes [163,164].

Magnetic nanocomposite could also be helpful in product separation in hydroformylation reactions. However, very limited works have been reported to date. The neutral and cationic Rh complexes such as $[Rh(1,5\text{-}cod)(\mu\text{-}S(CH_2)_{10}CO_2H)]_2$ (Rh-MUA; MUA = 11-mercaptoundecanoic acid; cod = 1,5-cyclooctadiene) and $[Rh(1,5\text{-}cod)(\eta^6\text{-}benzoic\ acid)]BF_4$ (Rh$^+$-BA; BA = benzoic acid) were heterogenized by immobilizing onto various supports, including poly(iodomethylstyrene-co-divinylbenzene) (PID) as a micrometer-size catalyst support and cobalt-ferrite $(CoFe_2O_4)$ nanoparticles as a nanoscaled magnetic catalyst support [165,166]. It was claimed that the carboxylic acid groups in the cationic rhodium complex could strongly bind to the surface of Co-ferrite nanoparticles [165,166]. In addition, the immobilized cationic charge of the complex could introduce strong ionic repulsion to balance magnetic attraction and thus stabilizes the nanoparticles [165,166]. Magnetic recyclable catalyst, $[Rh^+\text{-}BA]/CoFe_2O_4$, was used in hydroformylation of 4-vinylanisole and exhibited good performance (TOF > 14 h^{-1}) as shown in Figure 10.14 [166]. The activity is considered excellent compared to the catalysts immobilized on other support materials and, however, slightly lower than an unsupported counterpart (B/L = 90/10, conversion >99%, TOF = 41 h^{-1}) [166].

Alper and others functionalized MNPs $Fe_3O_4@SiO_2$ by growing polyaminoamido (PAMAM) dendrons on the surface to enhance the solubility of the catalyst support in organic solvents [167]. Three generations G(0), G(1), and G(2) that correspond to the structures $(MNP\text{-}O\text{-})_3Si\text{-}(CH_2)_3\text{-}NH_2$, $(MNP\text{-}O\text{-})_3Si\text{-}(CH_2)_3\text{-}N[(CH_2)_2\text{-}CONH(CH_2)_2NH_2]_2$, and $(MNP\text{-}O\text{-})_3Si\text{-}(CH_2)_3\text{-}N\{(CH_2)_2\text{-}CONH(CH_2)_2N[(CH_2)_2\text{-}CONH(CH_2)_2NH_2]_2\}_2$, respectively, were built by two subsequent steps of Michael-type addition of methyl acrylate followed by amidation with ethylenediamine [167]. The PAMAM layers were then phosphinized and coordinated with $[Rh(cod)Cl_2]$ to form Rh-complex-immobilized G(0)–G(2) magnetic nanocomposites [167]. Catalysts Rh-complex-immobilized G(0) and G(1) were used in the hydroformylation of 4-vinylanisole. Both catalysts exhibited high selectivity to branched

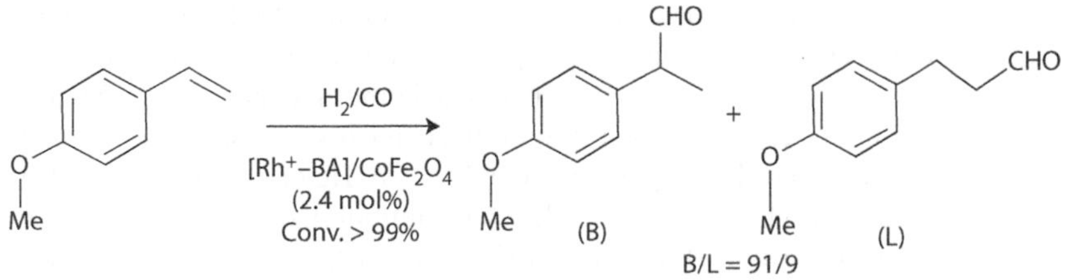

FIGURE 10.14 Hydroformylation of 4-vinylanisole.

aldehydes (B/L > 35/1) with around 100% conversions and lower TOF (TOF < 1 h^{-1}) [167]. Both Rh-G(0) and Rh-G(1) can be reused for five reaction cycles of hydroformylation of styrene without deteriorating its regioselectivity. However, on the fifth cycle, Rh-G(0) demonstrated reduced conversion yield, whereas no such deterioration was observed for Rh-G(1) [167]. Triphenylphosphine ligand was supported on the paramagnetic iron oxide nanoparticles (SPIONs, ~10 nm) by covalent bonds [168]. In the work, homogeneous catalyst RhAc$_2$ and MNPs-supported triphenylphosphine ligands were added together with substrates, and the reactions were conducted at 90°C for 9 h to reach the highest yield of 93.9% in the styrene hydroformylation reaction [168]. The MNPs-supported RhAc$_2$ was formed *in situ* and recovered for next runs with sustained activity. The selectivity of linear/branched isomers is at a range of 0.29–1.66 (l/b) [168].

10.3.6 CLICK REACTIONS

The catalytic azide (–N$_3$)-alkyne (–C ≡ CH) 1,3-dipolar cycloaddition, well known as "click" reaction, usually occurs in mild reaction conditions with high yield and good functional group tolerance. To date, click cycloaddition has been attracting growing interest from both academia and industry, particularly in synthetic chemistry and pharmaceutical chemistry, to construct a five-member heterocyclic 1,2,3-triazoles. Cu [169,170] and Ru [171] complexes-based catalysts are commonly used for this type of reaction to produce 1,4- and 1,5-substituted 1,2,3-triazoles, respectively. It is essential to recycle these active metal complexes from a reaction mixture from the view point of economy and environmental benefit as well as reducing product contaminations. MNPs-supported Cp*(PPh$_3$)$_2$Ru (II) catalyst, (γ-Fe$_2$O$_3$@-SiO$_2$-functional linker-PPh$_2$)$_2$Cp*RuCl, was prepared and evaluated for click reaction [172]. 1,5-Disubstituted 1,2,3-triazoles were obtained in around 91% yield within 3 h from the reactions between phenylacetylenes and benzyl azides [172]. The substrate scope was wide, including aryl aliphatic, and ferrocenyl acetylenes that showed good reactivity with benzyl azides in the presence of the MNPs-supported Ru catalyst. The aliphatic azides also showed good reactivity for click reactions in comparison with aryl azides which gave lower yields. The catalyst could be easily recovered by magnetic attraction and recycled at least five times with a slight loss of activity down to 77% [172].

Copper complexes are also important catalysts for cycloaddition of alkynes and azides [169,170]. Cu-catalytic systems are widely used and recognized as a model of click reaction due to their mild reaction conditions, atom economy, extremely high regioselectivity, and broad substrate scope. Similar to Ru-based catalysts, cytotoxic Cu species residue may cause products contamination, and that limits the applications of Cu-based catalysts in biomedicine and electronic. Therefore, active copper complexes have been immobilized on diverse supports to prepare corresponding Cu-based heterogeneous catalysts. *N*-Heterocarbene ligand IAD (IAD = 1,3-di(adamantyl)imidazol-2-ylidene) was immobilized onto MNPs, Fe$_3$O$_4$@SiO$_2$, and coordinated with CuI to form CuI(IAD)/Fe$_3$O$_4$@SiO$_2$ [173]. The supported catalyst exhibited good activity for the click reaction between benzyl azide and phenylacetylene in aqueous media with 92% yield. The catalyst was recycled for eight times with sustained activity. The reaction scope was broad; a variety of triazoles were isolated in the middle to high yields [173].

One-pot synthesis of triazole was developed; the process occurred through the cascade reaction of benzyl bromides, alkyne, and sodium azide. Amine ligands 3-aminopropyltrimethoxysiliane (APTS) and [3-(2-aminoethylamino)propyl]-trimethoxysilane (AAPTS) could react with MNPs $Fe_3O_4@SiO_2$ and generated amino groups-functionalized $Fe_3O_4@SiO_2$ [174]. The supported amino ligands could coordinate with CuBr to form $Fe_3O_4@SiO_2$-APTS-CuBr and $Fe_3O_4@SiO_2$-AAPTS-CuBr as shown in Figure 10.15a and b, respectively. Both catalysts were highly active for the above cascade reactions to form corresponding triazoles in good to excellent yields with 100% selectivity in a mixed medium of water and PEG 400 under microwave irradiation [174]. Very low yields were reported when aliphatic halides were used for the cascade reaction due to their poor microwave absorbing properties, base dielectric constants, and formation of 1,3-diynes byproduct [174]. Wang et al. functionalized MNPs γ-$Fe_2O_3@SiO_2$ with tris(triazolyl) groups and further metalized it with CuBr to form MNPs-supported tris(triazolyl)Cu(I) complex as presented in Figure 10.15c [175]. The complex was reported to be a highly active catalyst in alkyne azide cycloaddition reaction between benzyl azide and phenylacetylene in water media at room temperature. A high isolated yield of 97% was reached after 20 h with a Cu loading amount of 0.5 mol% [175]. The catalyst showed excellent recoverability with sustained activity and negligible leaching of Cu. Indeed, around 1.5 ppm leaching of Cu was detected based on inductively coupled plasma (ICP) mass spectrometry analysis. In addition, the catalyst was also applicable to the one-pot cascade synthesis of triazoles and dendrimer synthesis via triazolyl linkers [175].

Interestingly, Cu^0 nanoparticles produced from a reduction of $CuCl_2$ by lithium arsenide were supported onto commercially available MNPs MagSilica,

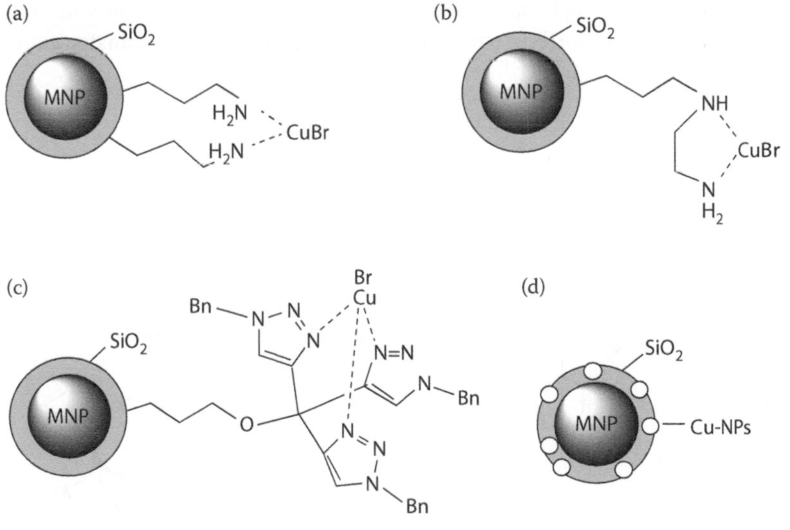

FIGURE 10.15 Magnetically recyclable Cu-based catalysts for "click" reaction.

Cu-NPs/MagSilica as shown in Figure 10.15d [176]. The spherical Cu-NPs well dispersed on the magnetic support, with a narrow size distribution and an average particle size of around 3 nm. The Cu-NPs/MagSilica catalyst was found to be very efficient in the cascade click reaction of terminal alkynes and azides generated *in situ* from sodium azide and diverse organic halides [176]. A high yield of 98% was reported in water media at 70°C within 1 h. However, low yields of triazoles were obtained when reactions were conducted in organic solvents such as THF, due to the hampering of *in situ* generation of azides [176]. In addition, magnetic nanoparticle cores were also used as catalysts and demonstrated good activity for the one-pot cascade cycloaddition reactions. It was reported that γ-Fe_2O_3 nanoparticles capped with grapheme sheets catalyzed reactions of benzyl halides, sodium azide, and aryl alkynes to produce a series of 1,4-disubstituted-1,2,3-triazoles in water media with 70%–93% isolated yields [177]. It was believed that grapheme composites play an important role in stabilization of MNPs and electron migration [177].

10.3.7 Reactions Catalyzed by Magnetic Retrievable Organocatalysts

Organocatalysts are essential in organic synthesis. In these processes, nonmetal catalysts are used. Therefore, organocatalysts make great contribution to green chemistry. For example, organic acids are used as catalysts in the cellulose modification reactions [178]. Chiral organocatalysts such as proline are used in asymmetric synthesis to make chiral products [179]. There is growing interest in development and immobilization of organocatalysts. Therefore, magnetic nanoparticles offer a few advantages to enable recyclability of organocatalysts which are difficult to be removed from a reaction mixture in a convenient method. Organocatalysts have been immobilized on MNPs, which are coated with silica or polymer, to facilitate catalyst separation and recovery [180–185].

Kawamura and Sato reported that the crown ethers were supported on the silica-coated magnetic core–shell nanostructure by forming covalent bonds as shown in Figure 10.16a. The supported crown ether was used as a phase-transfer catalyst in

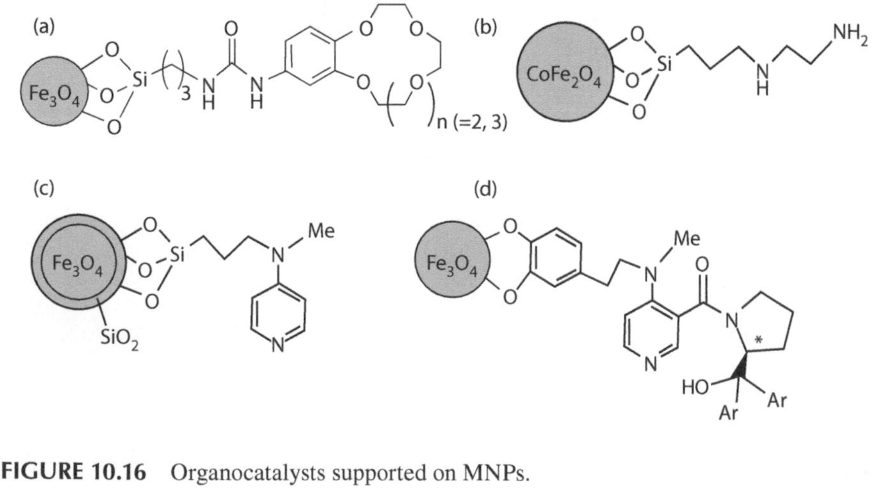

FIGURE 10.16 Organocatalysts supported on MNPs.

the nucleophilic substitution reaction between benzyl bromide and potassium acetate. The supported catalyst exhibited significantly improved activity in comparison with the corresponding parent homogeneous catalyst. It was reusable at least 8 times with a conversion of greater than 99% and a TOF of >4.1 h^{-1} for each cycle [180]. Amine is commonly used as an organic base. Diaminosilane-functionalized MNPs (CoFe$_2$O$_4$) (Figure 10.16b) were prepared and used as efficient heterogeneous base catalysts for the Knoevenagel condensation of benzaldehyde with malononitrile to produce 2-benzylidenemalononitrile [182]. The reaction rates over the MNPs-supported amine catalyst are comparable to the large pore mesoporous silicas-supported diamine and higher than the small pore MCM-48 silica with ~22 Å diameter pores-supported amines [182]. MNPs (CoFe$_2$O$_4$) were also used to support perfluorosulfonic acid groups by reacting triethoxysilylperfluorosulfonyl fluoride with MNPs, CoFe$_2$O$_4$@SiO$_2$, followed by acidifying with nitric acid [183]. The supported acids were reported to be active for the deprotection reaction of benzaldehyde dimethylacetal with water to produce benzaldehyde (Ph – CH(OMe)$_2$ + H$_2$O → Ph – CHO + 2 MeOH). The catalyst exhibited a high TOF of 54 min^{-1}, which is one-half of homogeneous catalyst triflic acid [183].

DMAP (4-N,N-dimethylaminopyridine) and a chiral analog were supported on MNPs of Fe$_3$O$_4$@SiO$_2$ (see Figure 10.16c,d) and used as the acylation catalysts [184,185]. The acylation of 1-phenylethanol with acetic anhydride was carried out with conversions of more than 98% in the presence of catalyst Fe$_3$O$_4$@SiO$_2$–linker–DMAP. The catalyst could be recovered and reused up to 30 times without any loss in activity [184]. The MNP-supported chiral DMAP (Figure 10.16d) was reported as a high efficient catalyst for the asymmetric acylation of monoprotected cis-cyclohexane-1,2-diols. The reaction is synthetically useful in kinetic resolution of sec-alcohols to prepare optically pure products. This supported chiral organocatalyst also offered promising recyclability. It could be reused for more than 32 consecutive cycles with sustained catalytic performance with a conversion of 64% and 93% enantio excess (ee) at the 32nd cycle [185]. Nasseri and others reported the synthesis of methylene dipyridine-supported Fe$_3$O$_4$@SiO$_2$ particles linked by a triethoxysilane [186]. The catalyst showed good activity and recyclability for the synthesis of biologically active pyrazolophalazinyl spirooxindoles.

TEMPO (2,2,6,6-tetramethylpiperidin-1-yl)oxyl) is a commonly used oxidation catalyst in organic synthesis to prepare carbonyl compounds from corresponding alcohols. Garrell and others supported TEMPO species on the surface of phosphonate-coated superparamagnetic iron oxide via a triazole linker [187]. The supported TEMPO catalyst provided good selectivity in controlling the oxidation of alcohols to aldehydes. The reactions were carried out in acidic environment with Mn (II) and Cu (II) co-catalysts or in basic conditions with sodium perchlorate oxidant [187]. Furthermore, TEMPO was also immobilized onto MNPs through an organosilicon linker [188]. The organosilicon-linked TEMPO exhibited good activity for the oxidation of alcohols to carbonyls. The metal-free catalyst could be recycled up to 20 runs in the oxidation of benzyl alcohol with sustained activity [188].

A group of ammonium species-supported MNPs were reported recently. The ammonium ions could be used as linker or catalytic active species as shown in Figure 10.17. Yin and others reported that chiral proline ligands were supported on MNPs

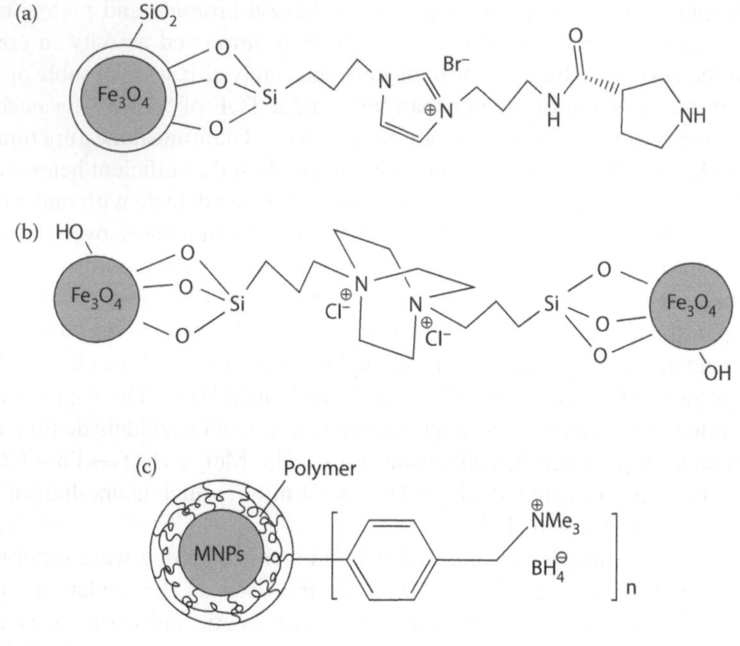

FIGURE 10.17 Examples of ammonium ion-functionalized MNPs.

through an ammonium ionic bridge as described in Figure 10.17a [189]. The magnetically retrievable catalyst was used in the asymmetric aldol reaction conducted in aqueous media and showed promising performance with high yield and enantioselectivity [189]. Interestingly, a well-ordered assembly of MNPs was achieved by application of the double-charged diazoniabicyclo[2,2,2]octane linkers attaching to the coating silica matrix (Figure 10.17b) [190]. The composite was employed as a phase-transfer catalyst in the transformations of benzyl halides to benzyl acetates and thiocyanates in high yield and chemoselectivity [190]. Reiser and others prepared magnetic exchange resins by reacting polymer-coated Co and Fe nanoparticles with trimethyl amine activated with microwave irradiation [191]. A high degree of functionalization (>95%) was achieved with a loading amount of 3.0 mmol/g. A wide scope of urea and thiourea were produced with the magnetic resin catalyst. The catalyst could be conveniently recovered and reused for next runs [191].

10.3.8 Magnetic Recyclable Biocatalysts

Biocatalysis is based on enzymes that are mainly available from Mother Nature. Both isolated and living cell-residing enzymes have been employed in organic transformations to produce desired chemicals [192]. In general, biocatalysts own advantages of high chemoselectivity, regioselectivity, and enantioselectivity. They can serve as highly efficient and specific catalysts. Indeed, biocatalyst plays important roles both in academia and in industry due to their capability to synthesize enantiopure compounds, which are essential chiral building blocks in agrochemistry and pharmacy. Application of enzyme catalysts, such as wine and cheese, has a long history, which continues even today, and will also continue in the future. To enable convenient recyclability, plenty of works have been done to heterogenize enzyme catalysts. In this regard, it is a potential alternative to immobilize enzyme catalyst on the magnetically retrievable nanoparticles. However, issues such as activity and selectivity, which are closely associated with the applications of enzyme catalysts, need to be prudent in spite of benefits facilitating retention and recyclability of the high-cost enzymes from the reaction mixture.

Surfactant-stabilized *Candida rugosa* lipase was supported on the MNPs produced from precipitation procedure in the presence of gum Arabic, and used in a multi-step synthesis of ethyl-isovalerate [193]. With this catalyst composite, a conversion rate of around 80% was achieved. It was found that across a number of cycles, the supported enzyme retained greater activity in comparison with its un-supported counterpart [193]. Vali and others supported glucose oxidases on the surface of silica-coated magnetic nanoparticles by forming a covalent chemical bond with a loading amount of 95 mg/g [194]. The MNPs were prepared via microemulsion method and showed no observed cytotoxic activity against human lung carcinoma cell and brine shrimp lethality. The nanocomposite showed high storage stability with 98% of its initial activity retained after 45 days. Furthermore, 90% of activity also remained after 12 repeated uses. The supported glucose oxidases exhibited enhanced thermal stability up to 80°C and less pH sensitivity in solution [194].

Kim et al. immobilized *Mugil cephalus* epoxide hydrolase to magnetic mesoporous silica via cross-linking method [195]. It was reported that the silica support has bottle-neck mesopores and thus effectively prohibited the leaching of catalytic species. The supported *M. cephalus* was stable and used for enantioselective hydrolysis of racemic epoxides. A 45% yield with 98%ee of (S)-styrene oxide was obtained in the presence of the catalyst. The catalyst was magnetically recoverable and reused more than 7 times with more than 50% of the initial activity retained [195]. Hydrolysis enzyme α-amylase was anchored on the amine-functionalized MNPs $Fe_3O_4@SiO_2$, via a convenient adsorptive method by Zhang et al. with a high loading of 235 mg/g [196]. This supported α-amylase was used as an efficient catalyst for in starch hydrolysis reaction. The supported enzyme showed about 80% activity in comparison with its free counterpart in the hydrolysis reaction. Nevertheless, the supported α-amylase was recyclable up to three times with sustained activity under the same conditions, and thus the total TON increased significantly [196]. Goh et al. supported hydrolysis enzyme *Amyloglucosidase (AMG)* onto magnetic single-walled carbon nanotubes (mSWCNTs) by physical adsorption and covalent immobilization [197]. The mSWCNTs were produced by incorporating iron oxide nanoparticles into SWCNTs. The supported AMG exhibited retained activity for starch hydrolysis with up to 40% catalytic efficiency after ten cycles. Furthermore, it was demonstrated that the supported enzyme could be stored at least one month at 4°C with retained activity. However, the mSWCNTs-supported AMG showed lower activity in comparison with its unsupported and SWCNTs-supported counterparts [197]. *C. rugosa* lipase was immobilized onto β-cyclodextrin-grafted Fe_3O_4 MNPs and used as catalyst for the hydrolysis of *p*-nitro-phenylpalmitate and enantioselective hydrolysis of racemic Naproxen methyl ester [198]. The supported lipase showed excellent performance providing high conversion and enantioselectivity with a 98%ee of S-Naproxen acid (Expected (E) value = 399) in comparison with free lipase, which gave an E value of 137 only. Furthermore, the supported *C. rugosa* lipase was magnetically recyclable [198].

Vinoba et al. immobilized enzyme *Bovine carbonic* anhydrase, which presented excellent activity in the catalytic conversion of CO_2 to bicarbonate, on (octa(aminophenyl)silsesquioxane)-functionalized $Fe_3O_4@SiO_2$ through the covalent chemical bonds [199]. The catalyst showed high activity in hydration and sequestration of CO_2 to $CaCO_3$ and excellent recyclability. After 30 consequent runs, activity of the supported enzyme remained high with 26-fold higher CO_2 capture efficiency than its unsupported counterpart. In addition, the catalyst provided extended storage stability, after 30 days with nearly 82% of its initial activity retained [199].

Similar to other nanocatalysts, the particle size is an essential factor for magnetically recyclable biocatalyst. Small particles own larger surface area, good distribution, and thus always exhibit higher catalyst activity. Aggregation of the nanocatalyst is not desired because the phenomenon will cause activity drop dramatically. However, recovery of a magnetic nanocatalyst comprising of small particles takes longer time than the big particles. Short separation time is necessary for supported biocatalysts. Ngo et al. reported that the alcohol dehydrogenase-supported MNPs with a small size of 65 nm could form big clusters at a pH of 8.0 and thus could be easily separated from a reaction mixture by an external magnet within only 4 s [200]. The interaction between supported biocatalysts in the clusters is weak and reversible. The clusters could be easily dissociated by simple shaking. After reaction, the clusters were regenerated by stopping shaking to enable a fast magnetic recovery. It was reported that both the pH value and enzyme modification are essential to form clusters. Pristine MNPs could not form cluster under the same conditions [200]. The supported alcohol dehydrogenase was used as catalyst in the enantioselective reduction of 7-methoxy-2-tetralone to produce (R)-7-methoxy-2-tetralol. The reaction was conducted within 60 min to obtain (R)-7-methoxy-2-tetralol in 97% yield and more than 99%ee. The catalyst retained 80% of its original activity after recycling 14 times [200].

10.3.9 MAGNETIC RETRIEVABLE PHOTOCATALYSTS

Photocatalysis has been known and used for decades. In a course of photocatalytic reaction, photons are absorbed by the catalysts to generate active free radicals, which undergo the reactants' transformation. Both homogeneous and heterogeneous photocatalysts have been investigated and well documented in the last few years [201–203]. The Fe^+ and Fe^+/H_2O_2 systems are commonly used homogeneous photocatalysts, while transition metal oxides such as TiO_2 and semiconductors are widely used heterogeneous photocatalysts. In principle, a heterogeneous photocatalyst owns a void energy region, and thus the free electrons and holes produced by photoactivation in the solid will remain and initiate the secondary reactions. Photocatalysts have plenty of important applications such as water splitting to produce hydrogen from water, oxidation of organic contaminants, and decomposition of aromatic hydrocarbons [201–203]. Following the rapid development of the economy, water pollution is becoming an increasingly serious concern, particularly in developing countries. Therefore, it is urgently needed to develop new materials and techniques to treat wastewater efficiently.

TiO_2 nanoparticles-based catalysts are widely used in the photodegradation of organic contaminants such as methylene blue (MB) dyes in wastewater. TiO_2 catalysts were supported on the $CoFe_3O_4@SiO_2$ nanoparticles [204]. The nanocatalyst showed promising activity in the degradation of MB dyes under UV irradiation. As reported, 98.3% of MB dyes were removed within 40 min [204]. Huang et al. synthesized a magnetic composite of $CoFe_2O_4/g-C_3N_4$ by a simple calculation method [205]. The catalyst demonstrated high activity in the photodegradation of MB dyes with hydrogen peroxide under visible-light irradiation. Up to 97.3% of the MB dye could be removed within 3 h under their conditions [205].

In addition, TiO_2 was supported on MNPs to form a core–shell nanocomposite, $Fe_3O_4@SiO_2@TiO_2$, via the layer-by-layer synthetic method [206]. The SiO_2 shells play important roles in preventing photodissociation and transfer of electron–holes from TiO_2 to core particle. Therefore, the photocatalyst showed high activity in the degradation of dye rhodamin B under UV illumination [206]. Core–shell magnetic nanoparticle $Fe_3O_4@TiO_2$ was modified with cyclodextrin ligand and used as a photocatalyst in the decomposition of endocrine-disrupting chemicals [207]. This nanocomposite is water dispersible and able to capture and destroy the pollutants present in water under UV illumination such as bisphenol A and dibutyl phthalate. In the component, immobilized cyclodextrin ligands are responsible for aqueous dispersibility of the composite and capture of organic pollutants. The amorphous TiO_2 layer works as a photocatalyst for the degradation of the pollutants. The magnetic nanocatalyst could be recycled up to ten times with a slight drop in photoactivity [207]. Liu and others prepared Sr^{2+}-doped MNPs $TiO_2/Ni_{0.6}Zn_{0.4}Fe_2O_4$ to enhance the photocatalytic performance and recyclability [208]. It was found that doping a low concentration of Sr^{2+} (<0.25 wt%) generated small and uniform particles with higher surface area. The catalyst exhibited a high efficiency of greater than 90% in the degradation of bisphenol A under both UV and visible-light irradiation, which is over two-times higher activity than $TiO_2/Ni_{0.6}Zn_{0.4}Fe_2O_4$, $Ni_{0.6}Zn_{0.4}Fe_2O_4$, and commercially available TiO_2. Furthermore, the hybrid demonstrated a good recyclability with sustained activity [208].

10.4   CONCLUSIONS AND OUTLOOK

Heterogeneous catalysis has been widely employed in industry for a few decades, particularly in refining processes and petrochemistry. Following advanced developments in nanotechnology, more and more nanomaterials could be designed, synthesized, and fully characterized. Recently, application of heterogeneous catalysis in fine chemistry has rapidly increased to prepare elaborated and expensive products on a small scale. Particular attention should be paid to atom economy reactions, energy saving, and to build up high-efficiency processes with high chemo-, regio-, and enantioselectivity. As mentioned earlier, nanoscaled magnetic materials with controlled particle size can be separated easily by a convenient external magnet. Application of magnetic nanoparticle catalysts has the potential of avoiding separation steps in organic syntheses and thus dramatically reduces the cost of synthesis. Therefore, magnetic nanoparticles-based material is a promising alternate to be used as catalyst support. Both high active homogeneous catalysts and transition metal nanoparticles-based nanocatalysts can be immobilized onto magnetic nanoparticles either by adsorption or by incorporation of covalent bonds.

In the past few years, tremendous progress has been made in the preparation of magnetic nanoparticle-supported catalysts and their applications in various organic conversions, including asymmetric syntheses due to the high demands in green and sustainable chemistry. However, to date, it remains a big challenge to prepare and stabilize the monodisperse magnetic nanoparticles with controllable size. The easy aggregation of magnetic nanoparticles may hinder their practical application as catalyst support. To prevent undesired aggregation and graft catalytic species, it is essential to modify magnetic nanoparticles accordingly. In this regard, magnetic nanoparticles have been functionalized with stabilizing ligands, which can coordinate with active metal centers, or coating encapsulating materials such as silica, graphene, and carbon nanotubes. Progress of the synthesizes in the area also opens up new exciting opportunities for novel organic–inorganic multifunctional hybrids, which could facilitate the conductance of a multistep catalytic transformation in one-pot reactions with magnetic recyclable catalysts as well as other applications. For example, the magnetic nanoparticles functionalized with coordinating ligands enable effective removal of toxic transition metal-based catalysts from a reaction mixture. And thus, the approach could be applicable in industrial products such as biopharmaceutical, food additives, fragrances, magnetically guided drug delivery, and magnetic resonance imaging enhancement. Although commonly used NMR spectroscopy, which may present direct evidence for the analysis of chemical structure, cannot be used to identify the immobilized functional groups due to the paramagnetic nature of MNPs support, other technologies, including photo- and thermal analyses, may be applicable in characterization of MNPs-supported catalyst.

Promising progress has been made on magnetic nanoparticle-based catalyst systems regarding activity and recyclability. In the past few years, transition metal complexes-based homogeneous catalysts, organocatalysts, and enzymes have been supported onto MNPs by a covalent or noncovalent binding process to form magnetically recyclable catalysts. These catalysts have been used and showed reasonable activity in a wide range of organic reactions such as oxidation, hydrogenation, and C–C cross-coupling reactions. Another essential development is the immobilization of enzymatic catalysts onto MNPs. The methodology is highly important for biopharmaceutical applications despite existing challenges and difficulties in the particular area. However, the unavoidable limitations related to the intrinsic instability of magnetic nanoparticles have to be addressed in order to enhance the reusability of the supported catalysts, thus minimizing the cost of the catalyst precursors.

It is reported that the pristine magnetic nanoparticles are relatively reactive. The magnetic cores may interact with strong acid and complexing ligands, which are commonly used in catalytic reactions. Therefore, stabilization and surface functionalization of magnetic nanoparticles remain an active field of research in the future. In this regard, it is urgently expected to optimize the surface functionalization procedure on a large scale to maximize the surface area for substrate access. The optimized modification procedure is also expected to reduce or eliminate the leaching of supported active species to the reaction media. It has been found that some active sites could leak from their heterogeneous supports to the reaction media, and work as real catalysts. The heterogeneous catalyst composites only act as the sources for homogeneous catalysis. Therefore, it's essential to prove that the MNPs-supported catalysts are responsible for the activity and that is exactly what has been confirmed in most of the reported results indicating very limited leaching of active centers and sustained activity after consecutive runs.

On the other hand, magnetic nanoparticles-supported catalysts have tremendous potential and advantages to be used in flow chemistry. In this case, the supported catalyst composites can be confined in a flow reactor and agitated with an external magnetic force to avoid potential clogs. Another potential application is associated to hyperthermic capability of magnetic nanoparticles. It has been found over many years that magnetic nanoparticles produce heat when subjected to an alternating magnetic field. Therefore, they can be potentially used as effective heating agents. To date, very limited works have been done to explore the potential opportunity. Nevertheless, successful applications of hyperthermic capability enable us to combine the catalysis with the confined heating area, and thus save energy and reduce costs.

ACKNOWLEDGMENTS

The author thanks the Institute of Chemical and Engineering Sciences, Agency for Science, Technology and Research, Singapore, and Singapore-MIT Alliance for Research and Technology Innovation Centre (NG120510ENG(IGN)) for their financial support.

REFERENCES

1. B. Viswanathan (ed.), Challenges in catalysis, in: *Catalysis: Selected Applications*, Alpha Science International Ltd., Oxford, UK, 2009, pp. 1.1–1.12.
2. P. Barbaro and F. Liguori (eds.), *Heterogenized Homogeneous Catalysts for Fine Chemicals Production: Materials and Processes*, Springer, Dordrecht, 2010.
3. R. Alain, S. Jürgen, and P. Henri, *Chem. Rev.* 102, 2002, 3757.
4. A. Z. Moshfegh, *J. Phy. D Appl. Phy.* 42, 2009, 233001.
5. Y. H. Zhu, C. N. Lee, R. A. Kemp, N. S. Hosmane, and J. A. Maguire, *Chem. Asian J.* 3, 2008, 650.
6. Y. Zhu, L. P. Stubbs, F. Ho, R. Liu, S. C. Peng, J. A. Maguire, and N. S. Hosmane, *ChemCatChem* 2, 2010, 365.
7. R. H. Kodama, *J. Magn. Magn. Mater.* 200, 1999, 359–372.
8. S. Laurent, D. Forge, M. Port, A. Roch, C. Robic, L. Vander Elst, and R. N. Muller, *Chem. Rev.* 108, 2008, 2064.
9. P. Tartaj, M. del P. Morales, S. Veintemillas-Verdagauer, T. Gonzalez-Carreno, and C. J. Serna, *J. Phys. D Appl. Phys.* 36, 2003, R182.
10. C. M. Niemeyer, *Angew. Chem. Int. Ed.* 40, 2001, 4128.
11. J. D. Hood, M. Bednarski, R. Frausto, S. Guccione, R. A. Reisfeld, R. Xiang, and D. A. Cheresh, *Science* 296, 2002, 2404.

12. D. W. Grainger and T. Okano, *Adv. Drug Del. Rev.* 55, 2003, 311.
13. B. M. Berkovsky, V. F. Medvedev, and M. S. Krokov, *Magnetic Fluids: Engineering Applications*, Oxford University Press, Oxford, 1993.
14. S. W. Charles and J. Popplewell, Properties and applications of magnetic liquids, in: *Hand Book of Magnetic Materials*, K. H. J. Buschow (ed.), Wiley, Colorado, 1986, 2, p. 153.
15. A. E. Merbach and E. Tóth, *The Chemistry of Contrast Agents in Medical Magnetic Resonance Imaging*, Wiley, Chichester, UK, 2001.
16. X. Batlle and A. Labarta, *J. Phys. D: Apply. Phys.* 35, 2002, R15.
17. M. P. Morales, S. Veintemillas-Verdaguer, M. I. Montero, C. J. Serna, A. Roig, L. Casas, B. Martinez, and F. Sandiumenge, *Chem. Mater.* 11, 1999, 3058.
18. E. Matijevic, *Chem. Mater.* 5, 1993, 412.
19. T. Sugimoto, *Fine Particles: Synthesis, Characterisation and Mechanism of Growth*, Marcel Dekker, New York, 2000.
20. X. Younan, B. Gates B, Y. Yin, and Y. Lu, *Adv. Mater.* 12, 2000, 693.
21. X. Sun, A. Gutierrez, M. J. Yacaman, X. Dong, and S. Jin, *Mater. Sci. Eng. A* 286, 2000, 157.
22. V. F. Puntes, K. M. Krishan, and A. P. Alivisatos, *Science* 291, 2001, 2115.
23. F. Grasset, N. Labhsetwar, D. Li, D. C. Park, N. Saito, H. Haneda et al., *Langmuir* 18, 2002, 8209.
24. S. Neveu, A. Bee, M. Robineau, and D. Talbot, *J. Colloid Interface Sci.* 255, 2002, 293.
25. S. Sun, H. Zeng, *J. Am. Chem. Soc.* 124, 2002, 8204.
26. J. Park, K. An, Y. Hwang, J.-G. Park, H.-J. Noh, J.-Y. Kim, J.-H. Park, N.-M. Hwang, and T. Hyeon, *Nat. Mater.* 3, 2004, 891.
27. J. Hu, I. M. C. Lo, and G. Chen, *Sep. Purif. Technol.* 56, 2007, 249.
28. S. Sun, C. B. Murray, D. Weller, L. Folks, and A. Moser, *Science* 287, 2000, 1989.
29. E. V. Shevchenko, D. V. Talapin, A. L. Rogach, A. Kornowski, M. Haase, and H. Weller, *J. Am. Chem. Soc.* 124, 2002, 11480.
30. D. Ortega, J. S. Garitaonandia, C. Barrera-Solano, M. Ramirez-Del-Solar, E. Blanco, and M. Dominguez, *J. Non-Cryst. Solids* 352, 2006, 2801.
31. M. A. Morales, P. V. Finotelli, J. A. H. Coaquira, M. H. M. Rocha-Leao, C. Diaz-Aguila, E. M. Baggio-Saitovitch, and A. M. Rossi, *Mater. Sci. Eng. C-Biomimetic Supramol. Syst.* 28, 2008, 253.
32. S. Santra, R. Tapec, N. Theodoropoulou, J. Dobson, A. Hebard, and W. H. Tan, *Langmuir* 17, 2001, 2900–2906.
33. Y. Lu, X. Lu, B. T. Mayers, T. Herricks, and Y. Xia, *J. Solid State Chem.* 181, 2008, 1530.
34. A. M. Schmidt, *Macromol. Rapid Commun.* 26, 2005, 93.
35. L. Sun, C. Zhang, L. Chen, J. Liu, H. Jin, H. Xu, and L. Ding, *Anal. Chim. Acta* 638, 2009, 162.
36. W. Wu, Q. He, H. Chen, J. Tang, and L. Nie, *Nanotechnology* 18, 2007, 145609.
37. Q. Sun, B. V. Reddy, M. Marquez, P. Jena, C. Gonzalez, and Q. Wang, *J. Phys. Chem. C* 111, 2007, 4159.
38. X. L. Zhao, Y. L. Shi, Y. Q. Cai, and S. F. Mou, *Environ. Sci. Technol.* 42, 2008, 1201.
39. X. Zhao, Y. Shi, T. Wang, Y. Cai, and G. Jiang, *J. Chromatogr. A* 1188, 2008, 140.
40. I. L. Medintz, H. T. Uyeda, E. R. Goldman, and H. Mattoussi, *Nat. Mater.* 4, 2005, 435.
41. X. Michalet, F. F. Pinaud, L. A. Bentolila, J. M. Tsay, S. Doose, and J. J. Li, *Science* 307, 2005, 538.
42. Y. Zhu, C. P. Ship, A. Emi, Z. Su, Monalisa, and R. A. Kemp, *Adv. Synth. Catal.* 349, 2007, 1917.
43. Y. Zhu, K. Loo, H. Ng, C. Li, P. S. Ludger, F. S. Chia, M. Tan, and C. P. Ship, *Adv. Synth. Catal.* 351, 2009, 2650.

44. R. De Palma, J. Trekker, S. Peeters, M. J. Van Bael, K. Bonroy, R. Wirix-Speetjens, G. Reekmans, W. Laureyn, G. Borghs, and G. Maes, *J. Nanosci. Nanotechnol.* 7, 2007, 4626.

45. C. Cannas, G. Concas, D. Gatteschi, A. Musinu, G. Piccaluga, and C. Sangregorio, *J. Mater. Chem.* 12, 2002, 3141.

46. Y. Ichiyanagi, S. Moritake, S. Taira, and M. Setou, *J. Magn. Magn. Mater.* 310, 2007, 2877.

47. R. M. Bozorth, *Ferromagnetism*, IEEE Press, IEEE Magnetic Society, 1951, pp. 59, 713.

48. L. S. Ott and R. G. Finke, *Coordin. Chem. Rev.* 251, 2007, 1075.

49. Y. Zhu, E. Widjaja, L. P. S. Shirley, Z. Wang, K. Carpenter, J. A. Maguire, N. S. Hosmane, and M. F. Hawthorne, *J. Am. Chem. Soc.* 129, 2007, 6507.

50. Y. Zhu, O. B. Algin Oh, S. Xiao, N. S. Hosmane, and J. A. Maguire, Ionic liquids in catalytic biomass transformation, in: *Application of Ionic Liquids in Science and Technology*, S. Handy (ed.), InTech (Open Access Publisher), Rijeka, Croatia, September, 2011.

51. Y. Zhu, Y. Karen Tang, and N. S. Hosmane, Applications of ionic liquids in lignin chemistry, in: *Ionic Liquids—New Aspects for the Future*, J.-I. Kadokawa (ed.), InTech (Open Access Publisher), Rijeka, Croatia, January, 2013.

52. R. Abu-Reziq, D. Wang, M. Post, and H. Alpera, *Adv. Synth. Catal.* 349, 2007, 2145.

53. Y. Jiang, C. Guo, H. Xia, I. Mahmood, C. Liu, and H. Liu, *J. Mol. Catal. B Enzym.* 58, 2009, 103.

54. Z. Wu, Z. Li, G. Wu, L. Wang, S. Lu, L. Wang, H. Wan, and G. Guan, *Ind. Eng. Chem. Res.* 53, 2014, 3040.

55. J. Safari and Z. Zarnegar, *New J. Chem.* 38, 2014, 358.

56. E. Santos, J. Albo, and A. Irabien, *RSC Adv.* 4, 2014, 40008.

57. C. E. Allmond, A. T. Sellinger, K. Gogick, and J. M. Fitz-gerald, *Appl. Phys. A* 86, 2007, 477.

58. D. S. Shephard, T. Maschmeyer, G. Sankar, J. M. Thomas, D. Ozkaya, B. F. G. Johnson, R. Raja, R. D. Oldroyd, and R. G. Bell, *Chem. Eur. J.* 4, 1998, 1214.

59. K. J. Klabunde, Y. X. Li, and B. J. Tan, *Chem. Mater.* 3, 1991, 30.

60. M. T. Reetz and W. J. Heibig, *J. Am. Chem. Soc.* 116, 1994, 7401.

61. K. S. Suslick, M. Fang, and T. Hyeon, *J. Am. Chem. Soc.* 118, 1996, 11960.

62. W. Tu, H. Liu, *Chem. Mater.* 12, 2000, 564.

63. Y.-P. Sun, H. W. Rollins, and R. Guduru, *Chem. Mater.* 11, 1999, 7.

64. J. M. Thomas, B. F. G Johnson, R. Raja, and P. A. Midgley, *Acc. Chem. Res.* 36, 2003, 20.

65. A.-M. Unsitalo, T. T. Pakkanen, E. I. Iskola, *J. Mol. Catal. A Chem.* 177, 2002, 179.

66. H. Schneider, G. T. Puchta, F. A. R. Kaul, G. R. Sieber, F. Lefebvre, G. Saggio, D. Mihalios, W. A. Hermann, and J. M. Basset, *J. Mol. Catal. A Chem.* 170, 2001, 127.

67. J. Wrzyszcz, M. Zawadki, A. M. Trzeciak, W. Tylus, and J. J. Ziolkowski, *Catal. Lett.* 93, 2004, 85.

68. J. Govan and Y. K. Gun'ko, *Nanomaterials* 4, 2014, 222.

69. D. Wang and D. Astruc, *Chem. Rev.* 114, 2014, 6949.

70. C. Kohlpaintner, M. Schulte, J. Falbe, P. Lappe, and J. Weber, *Aldehydes, Aliphatic in Ullmann's Encyclopedia of Industrial Chemistry*, Wiley-VCH, Weinheim, 2008.

71. H. Siegel and M. Eggersdorfer, Ketones, in: *Ullmann's Encyclopedia of Industrial Chemistry*, Wiley-VCH, Weinheim, 2005.

72. R. Wilhelm, Carboxylic acids, aliphatic, in: *Ullmann's Encyclopedia of Industrial Chemistry*, Wiley-VCH, Weinheim, 2002.

73. L. Zhang, P. Li, J. Yang, M. Wang, and L. Wang, *ChemPlusChem.* 79, 2014, 217.

74. M. B. Gawande, A. Rathi, I. D. Nogueira, C. A. A. Ghumman, N. Bundaleski, O. M. N. D. Teodoro, and P. S. Branco, *ChemPlusChem.* 77, 2012, 865.

75. M. Zhang, Q. Sun, Z. Yan, J. Jing, W. Wei, D. Jiang, J. Xie, and M. Chen, *Aust. J. Chem.* 66, 2013, 564.

76. A. Rezaeifard, M. Jafarpour, P. Farshid, and A. Naeimi, *Eur. J. Inorg. Chem.* 2012, 5515.

77. F. Nikbakht, A. Heydari, D. Saberi, and K. Aizi, *Tetrahedron Lett.* 54, 2013, 6520.

78. D. Saberi and A. Heydari, *Appl. Organometal. Chem.* 28, 2014, 101.

79. A. S. Kumar, B. Thulasiram, S. B. Laxmi, V. S. Rawat, and B. Sreedhar, *Tetrahedron* 70, 2014, 6059.

80. D. H. Zhang, G. D. Li, J. X. Lia, and J. S. Chen, *Chem. Commun.* 2008, 3414.

81. Y. Qiao, H. Li, L. Hua, L. Orzechowski, K. Yan, B. Feng, Z. Pan, N. Theyssen, W. Leitner, and Z. Hou, *ChemPlusChem* 77, 2012, 1128.

82. M. S. Saeedi, S. Tangestaninejad, M. Moghadam, V. Mirkhani, I. Mohammadpoor-Baltork, and A. R. Khosropour, *Polyhedron* 49, 2013, 158.

83. M. Kooti and M. Afshari, *Mater. Res. Bull.* 47, 2012, 3473.

84. BASF Introduces New NanoSelect Catalysts for Hydrogenation Reactions in Pharma Applications, available in: http://www.azonano.com/news.asp?newsID=7888, last reached on 18/11/2009.

85. S. W. Row, T. Y. Chae, K. S. Yoo, S. D. Lee, D. W. Lee, and Y. Shul, *Can. J. Chem. Eng.* 85, 2008, 925.

86. UK Exploring Hydrogenation for Biofuel-Diesel Mixture, available in: http://www.greencarcongress.com/2005/07/uk_exploring_hy.html, 5 July 2005, last reached on 18/11/2009.

87. R. W. Dorner, D. R. Hardy, F. W. Williams, B. H. Davis, H. D. Willauer, *Energy Fuels* 23, 2009, 4190.

88. A. H. Hoveyda, D. A. Evans, and G. C. Fu, *Chem. Rev.* 93, 1993, 1307.

89. A. H. Lu, W. Schimdt, N. Matoussevitch, H. Bönnemann, B. Spliethoff, B. Tesche, E. Bill, W. Kiefer, and F. Schüth, *Angew. Chem. Int. Ed.* 43, 2004, 4303.

90. Q. M. Kainz, R. Linhardt, R. N. Grass, G. Vilé, J. Pérez-Ramírez, W. J. Stark, and O. Reiser, *Adv. Funct. Mater.* 24, 2014, 2020.

91. R. Linhardt, Q. M. Kainz, R. N. Grass, W. J. Stark, and O. Reiser, *RSC Adv.* 4, 2014, 8541.

92. M. Guerrero, N. J. S. Costa, L. L. R Vono, L. M. Rossi, E. V. Gusevskayad, and K. J. Philippot, *Mater. Chem. A* 1, 2013, 1441.

93. M. J. Jacinto, P. K. Kiyohara, S. H. Masunaga, R. F. Jardim, and L. M. Rossi, *Appl. Catal. A* 338, 2008, 52.

94. U. Laska, C. G. Frost, P. K. Plucinski, and G. J. Price, *Catal. Lett.* 122, 2008, 68.

95. K. H. Lee, B. Lee, K. R. Lee, M. H. Yi, and N. H. Hur, *Chem. Commun.* 48, 2012, 4414.

96. N. V. Kuchkina, Y. E. Yuzik-Klimova, S. A. Sorokina, A. S. Peregudov, D. Y. Antonov, S. H. Gage et al., *Macromolecules* 46, 2013, 5890.

97. S. H. Gage, B. D. Stein, L. Z. Nikoshvili, V. G. Matveeva, M. G. Sulman, E. M. Sulman, D. G. Morgan, E. Y. Yuzik-Klimova, W. E. Mahmoud, and L. M. Bronstein, *Langmuir* 29, 2013, 466.

98. R. B. N. Baig and R. S. Varma, *ACS Sustainable Chem. Eng.* 1, 2013, 805.

99. A. Hu, G. T. Yee, and W. Lin, *J. Am. Chem. Soc.* 127, 2005, 12486.

100. J. Li, Y. Zhang, D. Han, Q. Gao, and C. Li, *J. Mol. Catal. A* 298, 2009, 31.

101. X. Gao, R. Liu, D. Zhang, M. Wu, T. Cheng, and G. Liu, *Chem. Eur. J.* 20, 2014, 1515.

102. D. K. Yi, S. S. Lee, and J. Y. Ying, *Chem. Mater.* 18, 2006, 2459.

103. T. Wendy, A. B. Ageeth, and W. G. John, *Catal. Today* 48, 1999, 329.

104. T. Zeng, X.-L. Zhang, Y.-R. Ma, H.-Y. Niu, and Y.-Q. Cai, *J. Mater. Chem.* 22, 2012, 18658.

105. T. Zeng, X.-L. Zhang, H.-Y. Niu, Y.-R. Ma, W.-H. Li, and Y.-Q. Cai, *Appl. Catal. B Environ.* 26, 2013, 134.

106. B. Liu, D. Zhang, J. Wang, C. Chen, X. Yang, and C. Li, *J. Phys. Chem. C* 117, 2013, 6363.
107. T. Yao, T. Cui, X. Fang, F. Cui, and J. Wu, *Nanoscale* 5, 2013, 5896.
108. D. Cantillo, M. Baghbanzadeh, and C. O. Kappe, *Angew. Chem. Int. Ed.* 51, 2012, 10190.
109. D. Cantillo, M. M. Moghaddam, and C. O. Kappe, *J. Org. Chem.* 78, 2013, 4530.
110. F. Chen, P. Xi, C. Ma, C. Shao, J. Wang, S. Wang, G. Liu, and Z. Zeng, *Dalton Trans.* 42, 2013, 7936.
111. X. Wang, D. Liu, S. Song, and H. Zhang, *Chem. Eur. J.* 19, 2013, 5169.
112. G. He, W. Liu, X. Sun, Q. Chen, X. Wang, and H. Chen, *Mater. Res. Bull.* 48, 2013, 1885.
113. F. Zamani and S. Kianpour, *Catal. Commun.* 45, 2014, 1.
114. B. H. Lai, Y. R. Lin, and D. H. Chen, *Chem. Eng. J.* 223, 2013, 418.
115. M. Xie, F. Zhang, Y. Long, and J. Ma, *RSC Adv.* 3, 2013, 10329.
116. Y. Zhu and N. S. Hosmane, *Coord. Chem. Rev.* 293–294, 2015, 357.
117. N. Miyaura and A. Suzuki, *Chem. Rev.* 95, 1995, 2457.
118. A. Fihri, M. Bouhrara, B. Nekoueishahraki, J.-M. Basset, and V. Polshettiwar, *Chem. Soc. Rev.* 40, 2011, 5181.
119. M. Moreno-Manas and R. Pleixats, *Acc. Chem. Res.* 36, 2003, 638.
120. R. Jana, T. P. Pathak, and M. S. Sigman, *Chem. Rev.* 111, 2011, 1417.
121. A. F. Littke, C. Dai, and G. C. Fu, *J. Am. Chem. Soc.* 122, 2000, 4020.
122. S.-Y. Liu, M. J. Choi, and G. C. Fu, *Chem. Commun.* 2001, 2408.
123. M. R. Netherton, C. Dai, K. Neuschüth, and G. C. Fu, *J. Am. Chem. Soc.* 123, 2001, 10099.
124. S. Li, Y. Lin, J. Cao, and S. Zhang, *J. Org. Chem.* 72, 2007, 4067.
125. P. D. Stevens, G. Li, J. Fan, M. Yen, and Y. Gao, *Chem. Commun.* 2005, 4435.
126. P. D. Stevens, J. Fan, H. M. R. Gardimalla, M. Yen, and Y. Gao, *Org. Lett.* 11, 2005, 2085.
127. Y. Zheng, P. D. Stevens, and Y. Gao, *J. Org. Chem.* 71, 2006, 537.
128. P. Wang, F. Zhang, Y. Long, M. Xie, R. Li, and J. Ma, *Catal. Sci. Technol.* 3, 2013, 1618.
129. M. Zhu and G. Diao, *J. Phys. Chem. C* 115, 2011, 24743.
130. R. F. Heck, *Org. React.* 27, 1982, 345.
131. J. E. Plevyak and R. F. Heck, *J. Org. Chem.* 43, 1978, 2454.
132. R. F. Heck, *Acc. Chem. Res.* 12, 1979, 146–151.
133. W. Cabri and I. Candiani, *Acc. Chem. Res.* 28, 1995, 2.
134. I. P. Belestskaya and A. V. Cheprakov, *Chem. Rev.* 100, 2000, 3009.
135. M. T. Reetz and E. Westermann, *Angew. Chem. Int. Ed.* 39, 2000, 165.
136. A. Biffis, M. Zecca, and M. Basato, *J. Mol. Catal. A.* 173, 2001, 249.
137. Z. Wang, P. Xiao, B. Shen, and N. He, *Colloids and Surfaces A* 276, 2006, 116.
138. S. Ko and J. Jang, *Angew. Chem. Int. Ed.* 45, 2006, 7564.
139. A. Khalafi-Nezhad and F. Panahi, *J. Organomet. Chem.* 741–742, 2013, 7.
140. E. Rafiee, A. Ataei, S. Nadri, M. Joshaghani, and S. Eavani, *Inorg. Chim. Acta* 409, 2014, 302.
141. M. Ghotbinejad, A. R. Khosropour, I. Mohammadpoor-Baltork, M. Moghadam, S. Tangestaninejad, and V. Mirkhani, *RSC Adv.* 4, 2014, 8590.
142. J. Safari and Z. Zarnegar, *C. R. Chim.* 16, 2013, 821.
143. M. Esmaeilpour, A. R. Sardarian, and J. Javidi, *J. Organomet. Chem.* 749, 2014, 233.
144. N. T. S. Phan and H. V. Le, *J. Mol. Catal. A Chem.* 334, 2011, 130.
145. B. Sreedhar, A. S. Kumar, and D. Yada, *Synlett* 8, 2011, 1081.
146. M. R. Buchmeiser, *Chem. Rev.* 109, 2009, 303.
147. H. Chayanant, H.-L. Su, S. B. Hassan, and E. B. David, *Org. Lett.* 11, 2009, 665.
148. B. G. Steven, S. K. Jason, L. G. Brian, and A. H. Hoveyda, *J. Am. Chem. Soc.* 122, 2000, 8168.

149. R. R. Schrock and A. H. Hoveyda, *Angew. Chem. Int. Ed.* 42, 2003, 4592.
150. S. Pawel, M. Marc, and G. Karol, *Chem. Soc. Rev.* 37, 2008, 2433.
151. H. D. Maynard and R. H. Grubbs, *Tetrahedron Lett.* 40, 1999, 4137.
152. J. H. Cho and B. M. Kim, *Org. Lett.* 5, 2003, 531.
153. S. H. Hong and R. H. Grubbs, *Org. Lett.* 9, 2007, 1955–1957.
154. B. Daniel and G. Karol, *Angew. Chem. Int. Ed.* 47, 2008, 2.
155. M. Anna, G. Lukasz, and G. Karol, *Chem. Commun.* 2006, 841.
156. Q. W. Yao, *Angew. Chem. Int. Ed.* 39, 2000, 3896.
157. Q. W. Yao and A. R. Motta, *Tetrahedron Lett.* 45, 2004, 2447.
158. F. Michalek, D. Mädge, J. Rühe, and W. Bannwarth, *J. Organometal. Chem.* 691, 2006, 5172.
159. C. Che, W. Li, S. Lin, J. Chen, J. Zhang, J. Wu, Q. Zheng, G. Zhang, Z. Yang, and B. Jiang, *Chem. Commun.* 2009, 5990.
160. D. Wong and D. Astruc, *Molecules* 19, 2014, 4635.
161. S.-W. Chen, Z.-C. Zhang, M. Ma, C.-M. Zhong, and S.-G. Lee, *Org. Lett.* 16, 2014, 4969.
162. E. Kuntz, *CHEMTECH* 17, 1987, 570.
163. F. Ungvary, *J. Organomet. Chem.* 477, 1994, 363.
164. Y. Yan, X. Zhang, and X. Zhang, *Adv. Synth. Catal.* 349, 2007, 1582.
165. T. J. Yoon, W. Lee, Y.-S. Oh, and J.-K. Lee, *New J. Chem.* 27, 2003, 227.
166. T. J. Yoon, J. I. Kim, and J. K. Lee, *Inorg. Chim. Acta.* 345, 2003, 228.
167. R. Abu-Reziq, H. Alper, D. Wang, and M. L. Post, *J. Am. Chem. Soc.* 128, 2006, 5279.
168. C. Duanmu, L. Wu, J. Gu, X. Xu, L. Feng, and X. Gu, *Catal. Commun.* 48, 2014, 45.
169. V. V. Rostovtsev, L. G. Green, V. V. Fokin, and K. B. Sharpless, *Angew. Chem. Int. Ed.* 41, 2002, 2596.
170. C. W. Tornøe, C. Christensen, and M. Meldal, *J. Org. Chem.* 67, 2002, 3057.
171. L. Zhang, X. Chen, P. Xue, H. H. Y. Sun, I. D. Williams, K. B. Sharpless, V. V. Fokin, and G. Jia, *J. Am. Chem. Soc.* 127, 2005, 15998.
172. D. Wang, L. Salmon, J. Ruiz, and D. Astruc, *Chem. Commun.* 49, 2013, 6956.
173. J.-M. Collinson, J. D. E. T. Wilton-Ely, and S. Díez-Gonza´lez, *Chem. Commun.* 49, 2013, 11358.
174. X. Xiong and L. Cai, *Catal. Sci. Technol.* 3, 2013, 1301.
175. D. Wang, L. Etienne, M. E. Igartua, S. Moya, and D. Astruc, *Chem. A Eur. J.* 20, 2014, 4047.
176. F. Nador, M. A. Volpe, F. Alonso, A. Feldhoff, A. Kirschning, and G. Radivoy, *Appl. Catal. A Gen.* 455, 2013, 39.
177. N. Salam, A. Sinha, P. Mondal, A. S. Roy, N. R. Jana, and S. M. Islam, *RSC Adv.* 3, 2013, 18087.
178. A. Córdova and J. Hafrén, WO 2006068611 A1, 2006.
179. Z. G. Hajos and D. R. Parrish, U.S. Patent 3975440, 1970.
180. M. Kawamura and K. Sato, *Chem. Commun.* 2007, 3404.
181. D. Lee, J. Lee, H. Lee, S. Jin, T. Hyeon, and B. M. Kim, *Adv. Synth. Catal.* 348, 2006, 41.
182. N. T. S. Phan and C. W. Jones, *J. Mol. Cat. A Chem.* 253, 2006, 123.
183. C. S. Gill, B. A. Price, and C. W. Jones, *J. Cat.* 251, 2007, 145.
184. C. O. Dalaigh, S. A. Corr, Y. Gun'ko, and S. J. Connon, *Angew. Chem. Int. Ed.* 46, 2007, 4329.
185. O. Gleeson, R. Tekoriute, Y. K. Gun'ko, and S. J. Connon, *Chem. A Eur. J.* 15, 2009, 5669.
186. S. M. Sadeghzadeh and M. A. Nasseri, *Catal. Today* 217, 2013, 80.
187. A. K. Tucker-Schwartz and R. L. Garrell, *Chemistry* 16, 2010, 12718.
188. B. Karimi and E. A. Farhangi, *Chemistry* 17, 2011, 6056.

189. Y. Kong, R. Tan, L. Zhao, and D. Yin, *Green Chem.* 15, 2013, 2422.
190. J. Davarpanah and A. R. Kiasat, *Catal. Commun.* 42, 2013, 98.
191. Q. M. Kainz, M. Zeltner, M. Rossier, W. J. Stark, and O. Reiser, *Chemistry* 19, 2013, 10038.
192. L. Andreas, S. Karsten, and W. Christian (eds.), *Industrial Biotransformations*, 2nd ed. John Wiley & Sons, Weinheim, 2006, p. 556.
193. I. Mahmood, I. Ahmad, G. Chen, and L. Huizhou, *Biochem. Eng. J.* 73, 2013, 72.
194. K. Ashtari, K. Khajeh, J. Fasihi, P. Ashtari, A. Ramazani, and H. Vali, *Int. J. Biol. Macromol.* 50, 2012, 1063.
195. Y. H. Kim, I. Lee, S. H. Choi, O. K. Lee, J. Shim, J. Lee, J. Kim, and E. Y. Lee, *J. Mol. Catal. B* 89, 2013, 48.
196. Q. Zhang, X. Han, and B. Tang, *RSC Adv.* 3, 2013, 9924.
197. W. J. Goh, V. S. Makam, J. Hu, L. Kang, M. Zheng, S. L. Yoong, C. N. Udalagama, and G. Pastorin, *Langmuir* 28, 2012, 16864.
198. E. Ozyilmaza, S. Sayina, M. Arslanb, and M. Yilmaz, *Colloid. Surf. B Biointerfaces* 113, 2014, 182.
199. M. Vinoba, M. Bhagiyalakshmi, S. K. Jeong, S. C. Nam, and Y. Yoon, *Chem. Eur. J.* 18, 2012, 12028.
200. T. P. N. Ngo, W. Zhang, W. Wang, and Z. Li, *Chem. Commun.* 48, 2012, 4585.
201. A. L. Linsebigler, G. Lu, and J. T. Yates, *Chem. Rev.* 95, 1995, 735.
202. M. N. Chong, B. Jin, C. W. K. Chow, and C. Saint, *Water Res.* 44, 2010, 2997.
203. A. O. Ibhadon and P. Fitzpatrick, *Catalysts* 3, 2013, 189.
204. F. A. Harraz, R. M. Mohamed, M. M. Rashad, Y. C. Wang, and W. Sigmund, *Ceram. Int.* 140, 2014, 375.
205. S. Huang, Y. Xu, M. Xie, H. Xu, M. He, J. Xia, L. Huang, and H. Li, *Coll. Surf. A Phys. Eng. Aspects* 478, 2015, 71.
206. J. P. Cheng, R. Ma, M. Li, J. S. Wu, F. Liu, and X. B. Zhang, *Chem. Eng. J.* 210, 2012, 80.
207. R. Chalasani and S. Vasudevan, *ACS Nano* 7, 2013, 4903.
208. F. Liu, Y. Xie, C. Yu, X. Liu, Y. Ma, L. Liu, and Y. Ling, *RSC Adv.* 5, 2015, 24056.

11 Magnetic Domain Walls for Memory and Logic Applications

Chandrasekhar Murapaka, Indra Purnama, and Wen Siang Lew

CONTENTS

ABSTRACT

Magnetic domain walls in ferromagnetic nanowires have attracted tremendous research interest in the recent years due to their potential applications in non-volatile magnetic memory and logic technologies. This chapter discusses in detail about the domain walls (DWs) and their dynamic behavior in ferromagnetic nanowires. The key topics described are methods for DW generation, different driving mechanisms, pinning and depinning schemes and finally detection techniques. At first, spin structures of different DWs depending on the dimensions of ferromagnetic nanostructures are discussed. Various methods to inject the DWs in the patterned nanowires by magnetic field and current are described. The DW dynamics behavior driven by magnetic field and current are presented in single nanowire and coupled two nanowire systems. Some light is shed on the effect of magnetostatic coupling on the DW dynamics. Two different kinds of remote-driving mechanisms with internal coupling field and exchange energy are described in detail. The DW pinning by notch and anti-notch structures are presented in the following section. Finally, the chapter is concluded with the insights on the DW detection methods by magnetoresistance and magnetic force microscopy imaging techniques.

11.1 INTRODUCTION OF DOMAIN WALL

Ferromagnetic materials are composed of domains of different magnetization directions, with domain walls (DWs) as the transition regions. DWs can be seen as defects in the magnetization configuration, which are created to reduce the overall energy of the system. DWs have a lot of interesting properties that make them attractive for future spintronic memory and logic applications. For instance, IBM has proposed a magnetic memory (famously known as racetrack memory) that is based on DW motion [1–3]. In such a device, the binary bits are represented by magnetic domains that are separated by DWs in a ferromagnetic nanowire. The data can then be driven to the read and the write sensors by a applying spin-polarized current to move the DWs along the racetrack [4,5]. The direction of the current dictates the motion of the DWs and data bits. The DWs are also found to have potential applications in magnetic logic [6–8]. For logic applications, the DW itself can act either as a binary bit for the Boolean logic operation or as a medium by which the bit is transferred. Hence, it is of great importance to understand the various kinds of DWs and their behavior in ferromagnetic nanostructures.

11.1.1 DWs IN PLANAR NANOWIRES

There are two kinds of DWs, Bloch and Néel, which exist in ferromagnetic materials depending on the thickness of nanostructures. The two kinds of DWs can be distinguished by the direction of spin rotation within the DW [9]. Bloch walls are stable configurations in thick nanostructures in which the spins rotate out-of-plane [10], whereas Néel walls are stable configurations in thin nanostructures in which the spins rotate in-plane [11,12]. The two DWs are shown in Figure 11.1.

In this chapter, we describe the properties and dynamics of Néel walls. According to the spin structure within the DW, a Néel wall can be differentiated into two types: transverse wall and vortex wall [13]. In the transverse DW, the spins point in a specific direction, and it possesses an asymmetric width along the y axis, which yields a triangular-shaped wall, as shown in Figure 11.2a. The transverse component of the wall also has two directions, which can be defined as the chirality of the transverse DW. If the spins point along the $+y$ direction, then the DW is said to possess an "Up" chirality, whereas in the case where the spins point along the $-y$ direction, the DW

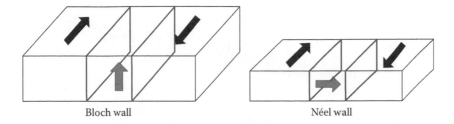

Bloch wall   Néel wall

FIGURE 11.1 Schematic illustrations of the Bloch wall in thick films and the Néel wall in thin films.

is said to possess a "Down" chirality. The vortex wall has a different magnetization configuration; the spins in the wall curl clockwise or counterclockwise around a vortex core, as shown in Figure 11.2b. The core of the vortex wall can also point either in the $+z$ or in the $-z$ direction.

The DW type in a ferromagnetic nanostructure largely depends on geometry and the material properties. The phase diagram of the energetically stable walls as a function of the nanostructure geometry has been calculated by McMichael and Donahue [14]. However, further differentiation on the walls has been proposed by Nakatani et al. [15], which adds an asymmetry to the transverse wall type. The phase diagram shown in Figure 11.2c clearly manifests the fact that vortex DWs are stable configurations in thicker and wider nanostructures whereas the transverse DWs are stable configurations in thinner and narrower nanostructures.

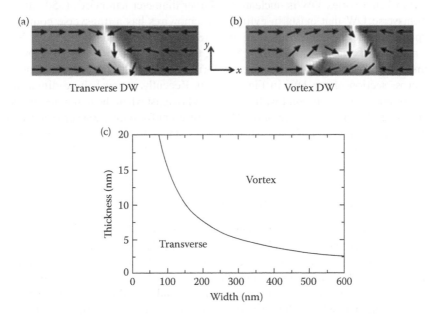

FIGURE 11.2 (a) Spin configuration of a transverse DW. (b) Spin configuration of a vortex DW. (c) Phase diagram of stability of the transverse and vortex DWs in planar nanostrips according to the thickness and width of the nanowire.

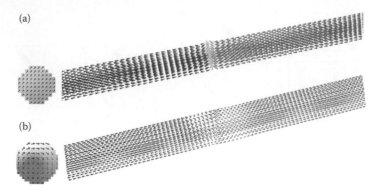

FIGURE 11.3 Transverse (a) and vortex (b) DWs that are found in cylindrical nanowires and the respective cross-sectional images.

11.1.2 DWs in Cylindrical Nanowires

The spin configuration and the magnetization reversal process of magnetic cylindrical nanowires are different from those of their planar counterparts due to the geometrical symmetry. It was predicted that two kinds of reversal modes can be observed in cylindrical nanowires, namely, transverse reversal mode and vortex reversal mode [16]. The transverse reversal mode occurs when a transverse DW is nucleated in smaller diameter nanowires (<50 nm), whereas the vortex reversal mode occurs when a vortex DW is nucleated in larger diameter nanowires (>50 nm) [17]. A transverse DW that exists in cylindrical nanowires has a transverse component perpendicular to the nanowire's long axis [18–20], as shown in Figure 11.3a. When the nanowire diameter is more than 50 nm, the demagnetization energy allows the spins to follow the cylindrical surface to form a vortex spin structure, as shown by the cross section of the DW in Figure 11.3b. Recently, a novel three-dimensional DW, so called the Bloch-point wall, is proposed to exist when the nanowire diameter is 50 nm [21,22]. The structure of the DW is different from the transverse and vortex states, as it is composed of two complex vortex states. In large diameter nanowires, similar DW structures called helical DWs were found where the DW serves as a transition between two vortex states of opposite chirality [23].

11.2 DW INJECTION IN PLANAR NANOWIRES

DW injection into a ferromagnetic nanowire is the first and vital step for the realization of DW-based memory and logic devices. Particularly for memory applications, a DW separates two bits of data while the motion of the DW leads to the reading or writing process. For logic applications, the DW acts as a medium for the logic operations. Therefore, it is of great interest to understand the DW injection methods in ferromagnetic nanowires. There are three different kinds of DW injection methods demonstrated here: (1) DW injection by the application of global external magnetic field, (2) DW injection by the application of an oblique field, and (3) DW injection using current-carrying stripe line.

11.2.1   DW Injection by Nucleation Pad

In permalloy ($Ni_{81}Fe_{19}$), due to zero crystalline anisotropy, the magnetization is always constrained to follow the shape of the nanostructure according to the shape anisotropy. Thus, permalloy is universally accepted to be considered as an ideal in-plane magnetic anisotropy material for all DW studies. A DW in a permalloy nanowire can be injected by reversing the magnetization from one end of the nanowire by the application of external magnetic field. However, a large nucleation magnetic field of around 400 Oe is required to switch the magnetization of a nanowire with dimensions of 2 μm × 100 nm × 10 nm (length, width, and thickness). To reduce the nucleation field, a diamond-shaped pad can be attached to the left edge of the nanowire. Shown in Figure 11.4 is the scanning electron microscopy (SEM) image of a nanowire with a diamond nucleation pad that is fabricated by electron beam lithography technique and Ar ion milling. The dimensions of the nucleation pad are 2 μm × 2 μm × 10 nm. The introduction of a nucleation pad results in low shape anisotropy compared to the nanowire, hence will allow a DW to be easily nucleated and injected into the track, and the magnetic field needed to reverse the magnetization in the track can be also reduced by the injection of the DW.

The magnetization configuration of the structure during the reversal process is studied using micromagnetic simulations and magnetic force microscopy (MFM) imaging. Initially, the structure is saturated with −300 G magnetic field along the x direction (nanowire long axis) and then relaxed. Afterwards, the magnetic field is reversed and slowly increased along the +x direction. Shown in Figure 11.5 are the MFM images and the corresponding micromagnetic configuration of the structure during the DW nucleation. As the shape anisotropy of the pad is lower than that of the nanowire, the magnetization configuration of the pad follows the diamond shape

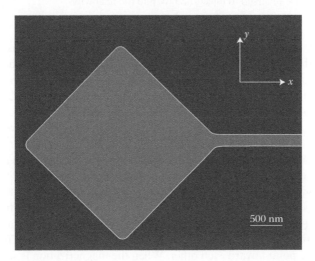

FIGURE 11.4   SEM image of the permalloy nanowire with a diamond-shaped nucleation pad for DW injection.

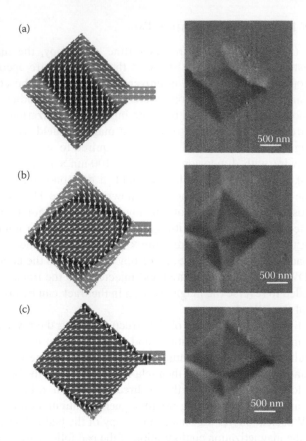

FIGURE 11.5 (a–c) Evolution of the magnetization dynamics in the nucleation pad when the magnetic field is reversed to inject the DW into the nanowire.

at remanence, as shown in Figure 11.5a. When the magnetic field is increased to +30 G, the MFM image clearly shows that the magnetization within the diamond structure switches first by forming a vortex magnetization configuration [24–27], as shown in Figure 11.5b. When the magnetic field is increased further, the vortex core travels to the $-y$ direction (transverse to the nanowire long axis). When the magnetic field is increased to 70 G, the vortex core completely moves toward the bottom of the diamond pad and a DW is pushed into the nanowire, as shown in Figure 11.5c. A small additional magnetic field can then be applied to move the DW through the nanowire. This method has an advantage of injection of multiple DWs into the nanowire, which is crucial for the memory applications [28]. Depending on the initial saturation magnetization direction, the injected DW can be of either head-to-head (HH) or tail-to-tail (TT). When the magnetization is saturated along the $-x$ direction and the field is applied in the $+x$ direction, a HH DW is injected where the spins point toward each other. When the magnetization is saturated along the $+x$ direction and the field is applied in the $-x$ direction, a TT DW is injected where the spins point away from each other.

11.2.2   DW Creation by Oblique Field

An alternative technique to create a DW is to use an L-shaped nanowire and apply an oblique magnetic field to the curvature [29–32], as shown in Figure 11.6a. The width of the nanowire here is 150 nm and its thickness is 10 nm to ensure transverse DW nucleation.

In the L-shaped nanowire, the curvature provides an avenue for the magnetization to converge or diverge such that the total energy state of the system is lowered. When a large magnetic field is applied with certain angle to the nanowire long axis, a DW is created at the curvature. Depending on the direction of magnetic field, the magnetization at the curvature converges or diverges to form either a HH DW or a TT DW, respectively. However, to move the DW from the curvature, a large magnetic field is needed to depin the DW. The MFM image in Figure 11.6b shows a transverse HH DW with a down chirality that is created in the curvature after the application of an external field at 225°. This method can then be used to select the chirality of the injected DW by controlling the orientation of the magnetic field [32].

11.2.3   DW Generation by Local Oersted Field

Among the various DW injection mechanisms, the two methods discussed previously are the simplest approaches to create a DW. However, such methods lack the

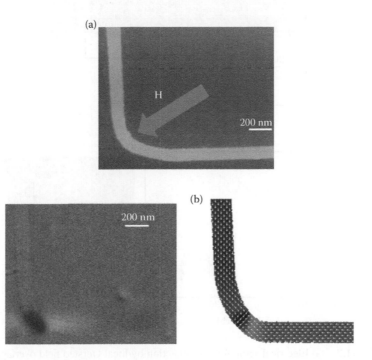

FIGURE 11.6   (a) Schematic of the direction of magnetic field applied overlapped with AFM image of the curved nanowire. (b) MFM image and the corresponding simulated magnetization of transverse HH DW.

capability of individual DW manipulation as the global magnetic field affects the whole structure. A much better option is then to use a current-carrying stripe line to generate an Oersted magnetic field to inject the DW locally [33–36]. To generate the local magnetic field, the stripe line should be placed transverse to the magnetic nanowire, as shown by the SEM image in Figure 11.7a. The stripe line is made of Cr/Au (5 nm/100 nm) with a width of 800 nm. Using a pulse generator, a voltage pulse of 2.5 V with a pulse width of 50 ns is injected into the stripe line. Due to low resistance of the stripe line, current shunting can be prevented and thus the current only passes through the stripe line. The generation of the DW in the ferromagnetic nanowire is detected by anisotropic magnetoresistance (AMR) measurements while applying a small bias current (10 µA) through the nanowire from A to B, as shown in Figure 11.7b. A drop in the resistance reading represents the generation of a pair of DWs in the nanowire.

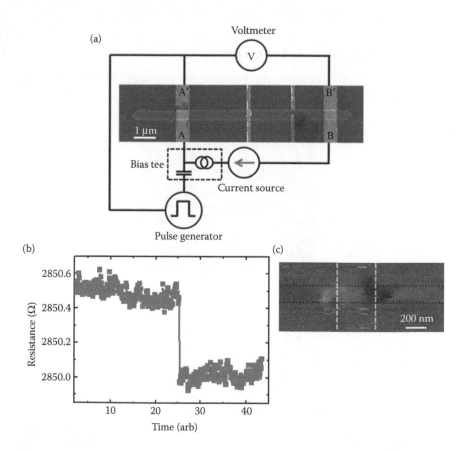

FIGURE 11.7 (a) Electrical setup for DW injection by local Oersted field overlapped with SEM image of the permalloy nanowire with nonmagnetic stripe line. (b) Resistance of the nanowire during the DW injection. (c) MFM image of the two DWs created by local Oersted field generation.

A drop in the resistance of around 0.5 Ω shows that a vortex DW is formed on each side of the stripe line. Shown in Figure 11.7c is the MFM image of the nanowire after the current pulse injection. A TT DW (bright contrast) and a HH DW (dark contrast) can be clearly seen in the MFM image. The probability of successful DW generation in 50 attempts for different pulse durations, ranging from 1 to 50 ns, is shown in Figure 11.8. The maximum probability achieved was less than 60% for 50 ns pulse duration, which shows that the DW injection by local field generation method is fast but stochastic in nature. The stochastic behavior of DW generation is due to the mutual attraction of the two DWs. The HH DW at the left of the stripe possesses a positive magnetic charge while the TT DW at the right of the stripe possesses a negative magnetic charge. The attraction between the two DWs hence causes either annihilation of the DWs or formation of bound state depending on the position of their topological charges [37]. To eliminate the stochastic behavior, one needs to avoid the mutual interaction between the two DWs. The probability of the DW generation can then be increased by the application of either a global magnetic field or long current pulses [33,38]. Application of a global magnetic field separates the two DWs far from each other to minimize the magnetostatic interaction between the two DWs; however, it may also affect the rest of the DWs in the nanowire. Injection with a longer current pulse also moves the nucleated DWs away from each other; however, it increases the power consumption significantly.

A simple yet novel design [39] to increase the probability of successful DW injection has been demonstrated recently where the edge stray magnetic field of the nanowire edge is exploited. Shown in Figure 11.9a is the SEM image of a nanowire with the stripe line placed close to the nanowire edge. When a current pulse is injected through the stripe line, the DW created at the left edge of the stripe line will be attracted to the nanowire edge due to the intrinsic stray field, which results in an annihilation. The DW nucleated at the right edge of the stripe line moves into the nanowire as it is not affected by the other DWs at this moment, resulting in a

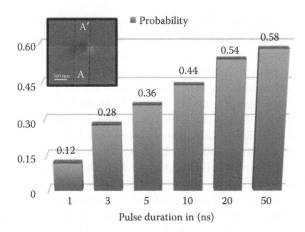

FIGURE 11.8 Histograms to show the probability of successful DW injection when two DWs are injected. Inset shows the two DWs injected.

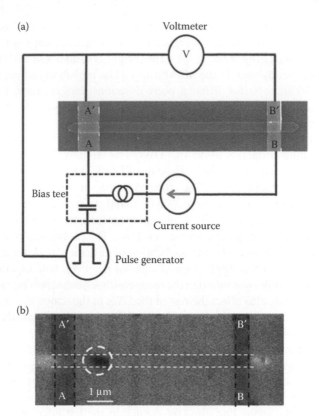

FIGURE 11.9 (a) Electrical setup for DW injection by local Oersted field overlapped with SEM image of the permalloy nanowire with nonmagnetic strip line shifted close to the edge of the nanowire. (b) MFM image of the single DW generated by local Oersted field generation.

successful DW injection. Shown in Figure 11.9b is the MFM image of a single DW that is generated by using this method. The probability of successful DW injection as a function of the pulse width at different current densities is shown in Figure 11.10. A single DW generation probability of 100% was observed for the current density of 1.1×10^{12} A/m^2 with a pulse duration of just 10 ns.

Figure 11.11 shows the spin-state evolution of DW nucleation obtained by micromagnetic simulations for different distances (d) between the edge of the nanowire and the injection strip line. The nanowire width was 300 nm with a thickness of 10 nm, while the strip line width was 1 μm with a thickness of 30 nm. A current pulse of 1×10^{12} A/m^2 with a pulse width of 1 ns was injected into the strip line. The rise and fall time of the current pulse was 300 ps. Initially, two DWs were nucleated underneath the strip line in all configurations. In particular, when the strip line is aligned with the edge of the nanowire, that is, $d = 0$, only one DW survives after the current pulse injection. The DW that was nucleated near the nanowire edge was annihilated there, while the DW at the right of the stripe line was pushed into the

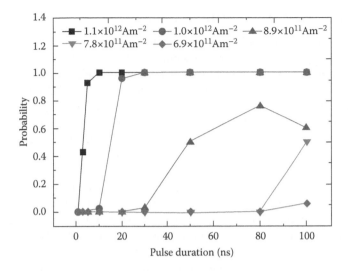

FIGURE 11.10   Successful DW injection probability as a function of current pulse duration at various current densities.

nanowire and away from the strip line. As the distance between the strip line and the edge of the nanowire was increased, that is, $d = 100$ nm and 200 nm, the stabilization of the DW took longer. At $d = 300$ nm, the DW nucleation collapsed and failed to form well-defined DWs. When the pulse duration was increased to 1.5 ns, single DW generation was successfully achieved for $d = 300$ nm. This is because when the Oersted field was sustained for a longer period of time, it allowed the DW at the right side of the stripe line to be pushed further into the nanowire.

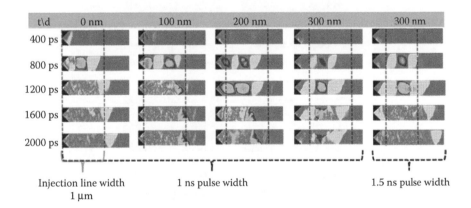

FIGURE 11.11   Simulated magnetization configurations of the nanowire during the DW generation at various instances when the stripe line is placed at different distances from the stripe line.

11.2.4 Transverse DW Chirality Selection

As shown by the magnetization configurations in Figure 11.12, four types of transverse DW configurations are possible within a nanowire with in-plane magnetization. If the transverse DW points along the +y direction (spins within the DW rotate in clockwise direction), then the DW is said to possess an "Up" chirality. When the DW transverse component points along the −y direction (spins within the DW rotate in anticlockwise direction), the DW is said to possess a "Down" chirality. In the rest of the chapter, HH DW with "Up" chirality is represented by HH-U and that with "Down" chirality is represented by HH-D. TT DW with "Up" and "Down" chiralities are represented by TT-U and TT-D, respectively.

The DW chirality plays a vital role in the DW dynamics for the memory and logic applications. For instance, the pinning potential of an artificial pinning site is influenced by the chirality of the DW [40,41]. In two-nanowire systems, the magnetostatic coupling between two DWs is a function of chirality of the DWs. The coupling strength can be tuned by varying the combinations of DWs with different chiralities [42,43], so it is of paramount interest to inject the DW with a specific chirality into the nanowire. To control the chirality of the DW that is injected into the nanowire, a chirality selector is added between the nucleation pad and the magnetic nanowire, as shown in the SEM image in Figure 11.13.

The dimensions of the chirality selector are 200 nm wide (along x axis) and 5 μm long (along y axis). The chirality selector, which has an easy axis along the ±y direction, is placed at the junction between the nucleation pad and the nanowire. The chirality selector is saturated either in the +y or in the −y direction prior to the

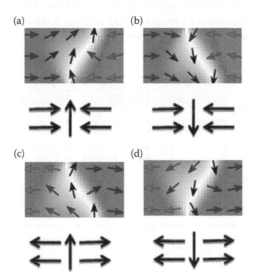

FIGURE 11.12 Magnetization configurations of four different types of transverse DWs that exist in narrow ferromagnetic nanowires. (a) Head to head DW with up chirality (HH-U), (b) head to head DW with down chirality (HH-D), (c) tail to tail DW with up chirality (TT-U), and (d) tail to tail DW with down chirality (TT-D).

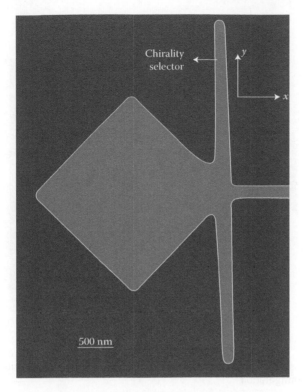

FIGURE 11.13 SEM image of the nanowire structure connected with a diamond-shaped nucleation pad and a transverse nanowire that acts as a chirality selector.

DW injection. Due to the strong shape anisotropy, the magnetization of the chirality selector is not affected by the horizontal magnetic field that is applied to nucleate the DW. As shown by the magnetization configuration in Figure 11.14, when the chirality selector is fixed in the +y direction and the nanowire is saturated in the +x direction, the transverse component of the DW aligns with the magnetization direction

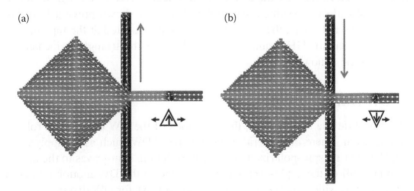

FIGURE 11.14 Magnetization configuration of the injection of (a) TT DW with "Up" chirality and (b) TT DW with "Down" chirality.

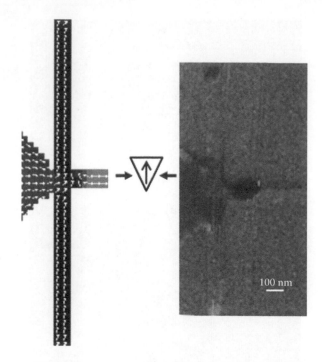

FIGURE 11.15 MFM image of a HH DW with "Up" chirality along with the simulated magnetization configuration.

of the chirality selector. This results in the injection of TT-U into the nanowire, as shown in Figure 11.14a. When the magnetization of the chirality selector is reversed to the −y direction, a TT-D is injected, as shown in Figure 11.14b.

When the chirality selector is saturated in the +y direction and the nanowire is saturated in the −x direction, a HH-U is injected, as shown in the magnetization configuration in Figure 11.15. The triangular base that is located at the top edge supports the simulation result that a HH-U is injected into the nanowire. Here, the chirality of the DW is determined by the shape of the injected DW in the MFM image, as transverse DWs always assume a triangular shape. As shown previously in Figure 11.12, for HH-U and TT-D, the base of the triangle is located at the top edge of the nanowire, whereas for HH-D and TT-U, the base of the triangle is located at the lower edge of the nanowire.

11.3 DW DRIVING

For DW-based device applications, the dynamics of the DW plays a key role in the device operation. There are several ways to drive a DW, such as by applying magnetic field, injecting spin-polarized current, introducing spin-waves to the nanowire, or combining all of these [44–48]. Here, we describe the DW motion by magnetic field and current, which are the most conventional ways for DW driving. Some light is also shed on the recent developments of DW remote driving by magnetostatic coupling and exchange coupling.

11.3.1 Field Induced DW Motion

The DW motion induced by magnetic field is well-known in magnetism as it acts as a medium for the magnetization reversal process in magnetic films. For device applications, the key parameter during the DW motion is the DW mobility, which is the rate of change of velocity as a function of the applied magnetic field. The mobility of the DW defines the limiting operating velocity of a DW-based device. Several theoretical studies predicted that the DW mobility is not linear but asymptotical with the increase in the magnetic field [49–51]. The DW mobility has two regimes with respect to the increase in the field: the linear and the precession regimes. The linear regime is encountered during the application of low field where the velocity increases with the magnetic field. In this regime, the mobility of the DW is defined by

$$\mu = \frac{\gamma\Delta}{\alpha} \tag{11.1}$$

where μ is the DW mobility, γ is the gyromagnetic ratio, Δ is the DW width, and α is the Gilbert damping constant, which is a material parameter.

The constant DW mobility holds only when the magnetic field H is below a critical field, which is also known as Walker field H_W (regime I in Figure 11.16). When the magnetic field is above the Walker field, the DW enters a precession regime. In this precession regime, the DW transverse component moves away from the DW motion, which also causes a periodic transition between the Bloch wall and the Néel wall [44]. The velocity of the DW drops greatly in this regime. The velocity of the DW drops significantly as a portion of the DW energy is used to transform the DW rather than to move it forward. The DW transformations are led by a nucleation of an out-of-plane anti-vortex core from one end of the DW, and its motion is shown in the snapshot images in Figure 11.17. The presence of the anti-vortex core here also

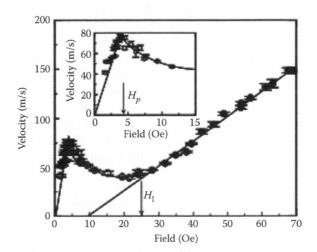

FIGURE 11.16   DW velocity as a function of magnetic field.

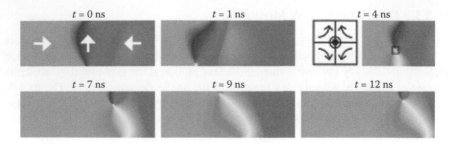

FIGURE 11.17 DW structure at various instances during the DW motion and the Walker breakdown.

shows that the DW has been changed into the Bloch-type wall. The core of the anti-vortex traverses through the width of the nanowire and annihilates at the other end to bring the DW back to the Néel-type wall. For a transverse DW, the precessional motion leads to a periodic flipping of the DW chirality. However, for vortex type, the DW orientation (either clockwise or anticlockwise) is still preserved. This phenomenon of the precessional motion of DW is called the Walker breakdown, which greatly limits the maximum velocity with which the DW can be driven. This is a major drawback of field-driven DW devices. In the presence of magnetic field, the direction of the DW motion depends on the type of the DW. HH and TT DWs always move in opposite directions when driven by the magnetic field. When multiple DWs are driven by the magnetic field, DWs with opposite charges move in opposite directions, which leads to the mutual annihilation and loss of data in a nanowire track if used as a memory device.

The motion of the DW in the presence of magnetic field can be described by using the Landau–Lifshitz–Gilbert (LLG) equation. Landau and Lifshitz [52] initially introduced the following equation:

$$\frac{\partial M}{\partial t} = -\gamma M \times H_{eff} - \frac{\lambda}{M_s} M \times (M \times H_{eff}) \qquad (11.2)$$

where $\lambda > 0$ is a constant that is unique to the material. The first term describes the Larmor precession and the second term describes the damping torque. The precessional motion of the magnetization in the presence of magnetic field under the damping torque is shown in Figure 11.18.

In 1955, Gilbert [53] reformulated the theory as he realized that the damping term introduced by Landau and Lifshitz cannot account for large damping. The magnetization dynamics equation, now referred to as the Landau–Lifshitz–Gilbert equation, becomes

$$\frac{\partial M}{\partial t} = -\gamma M \times H_{eff} + \frac{\alpha}{M_s} M \times \frac{\partial M}{\partial t} \qquad (11.3)$$

where $\alpha > 0$. The equation proposed by Landau and Lifshitz and the one reformulated by Gilbert have been proved to be equivalent.

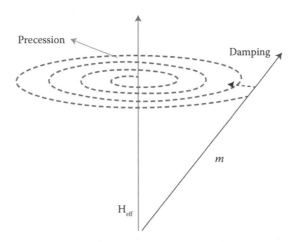

FIGURE 11.18 Sketch of the damped magnetization precession under the influence of an effective field.

11.3.2 Spin Transfer Torque Induced DW Motion

Current-induced DW motion is promising because of the unidirectional DW motion irrespective of the type of the DW. When current is applied to a ferromagnetic material, the conduction electrons become spin polarized due to spin-dependent scattering. These conduction electrons are then aligned according to the local magnetization direction because of the angular momentum transfer from local moments. To conserve the total angular momentum, the conduction electrons exert a torque on the local moments, and this phenomenon is called as spin-transfer torque (STT). The concept of STT was first proposed by Berger in 1984 [54], where he described it as an exchange interaction between the conduction electrons and local moments (DW) of the ferromagnetic material. The DW motion was demonstrated in ferromagnetic thin films using a microsecond-long current pulse. The current density required to drive the DWs is found to be quite high, and the DW displacements are in the order of micrometer [55]. The concept did not receive much attention for almost a decade. Later, Slonczewski and Berger independently predicted that when current is applied perpendicular to a giant magnetoresistance (GMR) multilayer, the STT can switch the free-layer magnetization [56]. The concept of magnetization reversal by STT was first experimentally confirmed by Tsoi et al. in point contact geometries in multilayer thin films [57]. These developments stimulated back the interest in current-induced DW motion in ferromagnetic nanowires. The DW motion induced by STT is mostly studied in nanowires of in-plane soft magnetic material such as $Ni_{81}Fe_{19}$ (permalloy). In the last decade, several studies reported the experimental observation of DW motion driven by STT. Yamaguchi et al. [58] have first shown the real-space observation of the back-and-forth motion of vortex DW by microsecond current pulses. The critical current density is found to be in the order of 10^{12} A/m² (Figure 11.19).

In several other studies in multilayer magnetic nanowires, the critical current density to move the DW is found to be much lower than that reported in a single

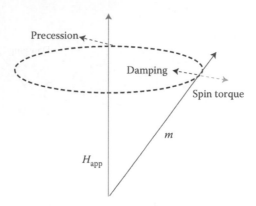

FIGURE 11.19 Schematic shows the directions of three different torques acting on the magnetization m under the application of magnetic field and current.

magnetic layer [59]. Additionally, with the developments in lithography techniques, a very high DW velocity of 200 m/s has been demonstrated by IBM [1], which makes the implementation of DW for memory and logic applications to be very promising. Besides the experimental work, the current-induced DW motion has also attracted the interests of theoretical scientists in describing the equation of motion of the DW in the presence of current [59–61]. The most general agreement among all the theoretical modeling shows that the interaction of electric currents and the local magnetizations can be explained by considering two different STTs, which are termed as adiabatic STT and nonadiabatic STT. The modified LLG equation with the addition of these two terms can be written as [59]

$$\frac{\partial M}{\partial t} = -\gamma M \times H + \frac{\alpha}{M_s} M \times \frac{\partial M}{\partial t} + STT_1 + STT_2 \qquad (11.4)$$

where M is the magnetization, M_s is the saturation magnetization of material, γ_0 is the gyromagnetic ratio, and α is the Gilbert damping constant. The first two terms account for the torque by the effective field H_{eff} and the Gilbert damping torque, parameterized by a dimensionless α as described previously. The precessional motion of the magnetization under STT is shown in Figure 11.18. Here, the STT may act against the damping to sustain the DW precession, which finds application in STT-based microwave oscillators. For permalloy, damping parameter α is in the range of 0.08–0.15 [61–63]. The last two terms on the right-hand side are the STT terms that represent the two mechanisms behind the current-induced DW motion. As proposed by Slonczewski, STT_1 describes the angular momentum transfer between the conduction electrons and the local spins of the material. In this case, it is assumed that when the electrons move across the DW, they adjust their spin orientations adiabatically to the direction of the local moments while transferring angular momentum to the DW.

The adiabatic STT_1 term can be written as

$$STT_1 = (u \cdot \nabla)M \tag{11.5}$$

where u is the conduction electron velocity quantified by magnitude of the STT given by

$$u = \frac{g\mu_B p}{2eM_s} \tag{11.6}$$

where g, μ_B, and e are the Landè factor, Bhor magnetron, and electron charge, respectively; M_s is the saturation magnetization; and p is the conduction-electron-spin polarization, with the values ranging from 0.4 to 0.7. The expression of the adiabatic STT also means that the conduction electron acquires a Berry phase as it moves within the ferromagnetic. Additionally, it is also possible to induce the inverse phenomenon, whereby an electromotive force is created across the nanowire when the magnetization configuration of the nanowire, that is, the DW, is moved using an external magnetic field [64].

The STT_2 term, which is generally called as nonadiabatic STT term, can be defined as

$$STT_2 = \frac{\beta}{M_s} M \times [(u \cdot \nabla)M] \tag{11.7}$$

where β is the non-adiabatic parameter.

It is generally believed that the nonadiabatic term originates from the spatial mistracking of the conduction electron spins and the local moments, where the spin-polarized conduction electrons are scattered (reflected) when they cannot follow the local moments. Zhang and Li in 2004 [61] proposed a model in which there is a slight mistracking between the electron spin and the local magnetization direction during the spin-transfer phenomenon, which means that the spin of the electron is not exactly aligned in the direction of the local magnetic moment when it moves along the ferromagnetic material. This mistracking generates nonequilibrium spin accumulation across the magnetic DW, and the accumulated spins are then relaxed through a spin-flip scattering process toward the local magnetization direction. The Zhang–Li model eventually then leads to both adiabatic and nonadiabatic STT terms. Many experimental works were devoted to estimate the nonadiabacity in the current-induced DW motion. Thomas et al. in 2008 estimated the β to be 0.04 [62]. Recent experimental verifications prove that β varies between 0.08 and 0.15 [63,65,66] for permalloy.

Shown in Figure 11.20a is the experimental setup to demonstrate the DW motion induced by the injection of nanosecond current pulses. A programmable pulse

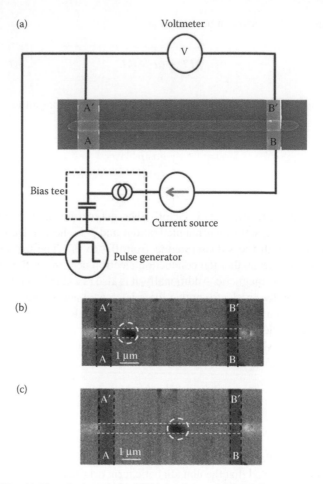

FIGURE 11.20 (a) Electrical setup for DW injection and driving by a current. (b) The DW injected and (c) the DW driven by STT.

generator (Picosecond 10300B) is connected to a radio frequency (RF) probe for injecting and driving the DW, as schematically shown in Figure 11.20. A voltmeter (Keithley 2000) and a constant current source (Keithley 2400) were used to measure the AMR of the nanowire. A bias tee was used to isolate the pulse generator from the dc source. For DW generation, the sample was initially saturated with a 1 kG magnetic field along the wire long axis. The potential difference across the nanowire (AA' to BB') was directly measured via dc probes while a constant current of 100 μA was maintained. A current density in the order of 1.1×10^{12} A/m² was injected from A to A'. The voltage reading of the voltmeter drops at that instant, which indicates DW generation. For DW driving, a current pulse was injected from BB' to AA', as shown in Figure 11.20a, which means that the electrons flow from AA' to BB'. A constant bias current of 500 μA was injected through the nanowire from BB' to AA' using a current source. The driving current density was 1.3×10^{12} A/m² with a pulse duration of 10 ns. The generated single DW shown in the MFM image of Figure

11.20b was driven with two pulses of current density, $J = 1.3 \times 10^{12}$ A/m^2 and a pulse width of 10 ns. The MFM imaging of the device confirms that the displacement of the DW from its initial position was ~2 μm, as shown in Figure 11.20c. The average velocity was estimated to be ~100 m/s.

Aside from the STT, it is possible to induce additional torques to drive the DW by creating a structure whereby the ferromagnetic layer is adjacent to a layer of non-magnetic materials with strong spin–orbit coupling [67–69] or when it is deposited on the surface of a three-dimensional topological insulator [70]. In this setup, the local magnetizations of the ferromagnet receive torques from the spin Hall effect and the Rashba effect. In the spin Hall effect, an interfacial spin accumulation is created through the imbalance of the deflections of spin-up and spin-down electrons away from the electric field direction in a nonmagnetic layer. The resulting transverse pure spin current is then injected into the adjacent magnetic layer to drive the DW. The torque from the spin Hall effect can be written as

$$\tau_{SHE} = -\frac{\gamma \hbar \theta_H J}{2eM_s t_F}(M \times \sigma \times M) \tag{11.8}$$

where θ_H is the parameter characterizing the maximum yield for the conversion of a longitudinal charge current density into a transverse spin current density, t_F is the thickness of the ferromagnetic layer, γ is the gyromagnetic ratio, J is the current density, and σ is the vector that shows the direction of the torque. When the ferromagnetic layer and the nonmagnetic layer are stacked vertically, σ is directed along the width of the nanowire.

For the Rashba effect, it is induced by the electric field gradient due to a symmetry breaking at the surface between the ferromagnetic material and the spin–orbit metal. The torque from the Rashba effect can be written as

$$H_{SHE} = -\frac{\gamma \alpha_R J P}{\mu_B M_s}(M \times \sigma) \tag{11.9}$$

where P is the polarization of carriers in the ferromagnetic layer, μ_B is the Bohr magneton, and α_R is the Rashba parameter that is averaged over the magnetic film thickness. In general, the Rashba field affects the dynamics of a current-driven DW, similar to how a constant in-plane external magnetic field does; that is, the Rashba field increases the rigidity of the DW, which allows the DW to receive and be driven with a higher current density.

11.3.3 DW Remote Driving

DW remote driving is a method to induce DW motion without the direct application of current and/or magnetic field to the respective nanowire. The DW remote driving can be induced in two different ways: by magnetostatic coupling of DWs and by exchange coupling of DWs. Here, these two concepts of DW coupling and their effect on DW motion are discussed in detail.

11.3.3.1 Magnetostatic Coupling Induced DW Motion

In ferromagnetic materials, DW carries a magnetic charge [71] and emanates stray magnetic field similar to edge of the nanowires. The magnetic volume pole that is analogous to the magnetic charge along a HH DW is plotted in Figure 11.21. The magnetic volume pole is calculated by taking the divergence of magnetization along the nanowire

$$Q = -\nabla \cdot \vec{M} \qquad (11.10)$$

where Q is the magnetic charge and $\vec{M}$ is the magnetization vector.

As shown in Figure 11.21, a HH DW possesses a positive magnetic charge due to the convergence of magnetization and similarly a TT DW is seen as a negative magnetic charge due to the divergence of magnetization. The magnetic charge allows the DW to interact with another DW in closely spaced nanowires. This magnetostatic interaction is called DW coupling. In this section, the strength of the DW coupling, effect of coupling on DW dynamics, and the coupling-induced DW motion and Walker breakdown are discussed in detail.

Initially, the interaction between two different types of transverse DWs—HH and TT DWs—relaxed in two closely spaced nanowires is discussed. The two-nanowire system is relaxed with no external field or current. The DWs (HH and TT) are attracted to each other via their stray magnetic field and move close to each other before reaching an equilibrium position where the two DWs are aligned along each other, as shown in Figure 11.22. The interaction can be seen as two magnetic charges of opposite polarities being attracted to each other. The total energy of the system is minimized in this equilibrium stable state. As many as eight chirality combinations are possible between HH and TT DWs. In all possible combinations, the total energy of the system is reduced compared to that in the case where zero interaction is present; that is, two nanowires are assumed to be placed at an infinite distance from each

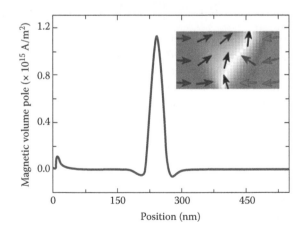

FIGURE 11.21 Magnetic volume pole calculated along the length of the nanowire with a HH DW.

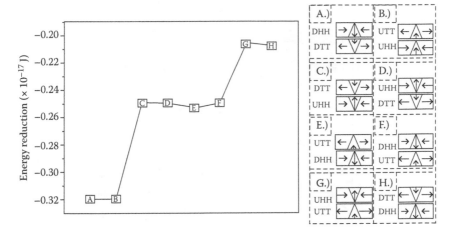

FIGURE 11.22 Energy of the system for various combinations of DW chirality to show the stability.

other. In the zero interaction case, the total energy of the system obtained from the simulation is 2.4×10^{-17} J. With respect to their energy value, the eight combinations are separated into three regimes, as shown in Figure 11.22. In A and B where the two DWs have their bases facing each other, the energy of the system is the lowest. Reversing the chirality of one of the two DWs gives four combinations. Here, the base of one DW is facing the apex of the other DW; these four combinations have the same energy, which is higher than the energy in A and B. In G and H where the apexes of the DWs are facing each other, the system has the highest total energy. The energy value of the system gives the information on the coupling strength of the combinations: lower energy value equals to stronger coupling. The DW coupling is strongest in A and B whereas in G and H, the coupling is weakest.

To study the DW remote driving induced by magnetostatic coupling, spin-polarized current is applied to a nanowire with a TT DW. The current drives the TT DW to move along the nanowire long axis. Interestingly, the HH DW in the adjacent nanowire also moves in the same direction. Similar behavior is observed when spin-polarized current is applied only to the nanowire with the HH DW. The phenomenon of DW motion in the nanowire with no current applied reveals that the coupling between the two DWs is strong enough to induce DW motion. The two DW systems can be named as a coupled DW system (CDWS).

The displacement of the CDWS as a function of time for various current densities calculated using micromagnetic simulations is shown in Figure 11.23. For current densities of $J < 2.7 \times 10^{12}$ A/m², the CDWS moves with a constant velocity along the nanowire. The velocity of the CDWS is 326 and 407.82 m/s for $J = 2.1 \times 10^{12}$ and 2.7×10^{12} A/m², respectively. The velocity of the CDWS increases linearly with increasing current density. The DWs also retain their shapes as they propagate along the nanowire. When the current density increases to $J = 3.1 \times 10^{12}$ A/m², the average velocity drops to 163 m/s. Increasing the current density beyond a certain value results in a drastic drop of the average velocity. It proves that the Walker breakdown

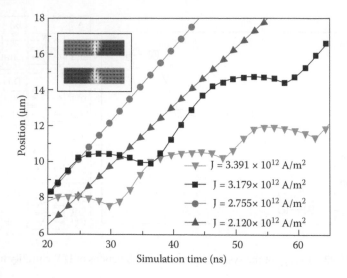

FIGURE 11.23 DW displacements as a function of simulation time at various current densities.

takes place in CDWS as well. However, the Walker breakdown limit is found to be shifted to a higher current density as compared to a single nanowire system. Shown in Figure 11.24 are the average velocities of the DW in coupled and single nanowire systems. It shows that DW chirality breaks and flips between up and down in the CDWS when the current density is above the Walker breakdown limit. Chirality flipping in CDWS is observed to occur in both of the nanowires, even though current is only applied to one of the nanowires. For the case where the current density is higher

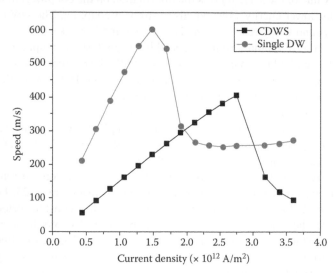

FIGURE 11.24 DW velocity as a function of current density for single and coupled DW systems.

than the Walker breakdown current density limit, the average velocity of the CDWS drops appreciably due to the chirality flipping of the DWs.

The coupling-assisted DW remote driving is also studied in the repulsion regime, where two HH DWs are relaxed in two nanowires. Figure 11.25 shows all possible four chirality combinations—Up–Down (UD), Up–Up (UU), Down–Down (DD),

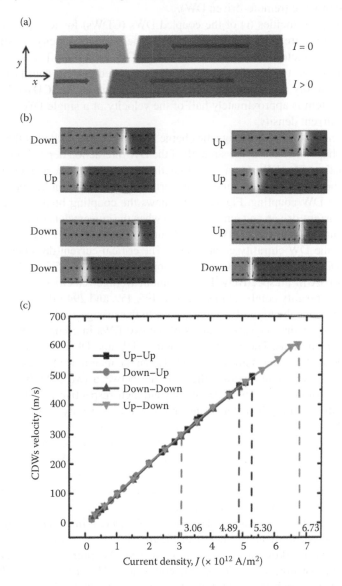

FIGURE 11.25   (a) Schematic diagram of a two-nanowire system. The current is applied to the bottom nanowire only. (b) Four possible DW chirality combinations: Up–Down (UD), Up–Up (UU), Down–Down (DD), Down–Up (DU). (c) Plot of velocities of four possible DW chirality combinations as a function of applied current density. Dotted lines represent the threshold current densities for all four chirality combinations.

and Down–Up (DU)—with the chirality of the DWs in the bottom and the upper nanowires indicated by the first and the second letters, respectively. Spin-polarized current is then applied to the bottom nanowire to drive the bottom DW (current-driven DW). However, the simulations also show that the upper DW is driven in the same direction as the bottom DW, even though there is no current that is applied to the upper nanowire (remote-driven DW).

The average velocities (v) of the coupled DWs (CDWs) for four possible chirality combinations are shown in Figure 11.25c. The average velocity is estimated by averaging the remote-driven and current-driven DW velocities. The velocity of the CDWs is observed to linearly increase with the applied current density for all possible four chirality combinations. The average velocity of the CDWs in the two-nanowire system is approximately half of the velocity of a single DW system when we fix the current density.

For a transverse DW, the magnetic charge is always concentrated at the base of its triangular shape. Therefore, the strength of the DW interaction depends on the chirality combination. The DW interaction strength does not affect the average velocity of the CDWs but determines the maximum current density that can be applied without breaking the DW coupling. Figure 11.25 shows the coupling breaking between the two DWs (current-driven and remote-driven) when the current density exceeds certain threshold current density (J_c). The critical current density for DW breaking strongly depends on the DW chirality combination. The critical current densities (J_c) for the UD, UU, DD, and DU chirality combinations are 6.7×10^{12}, 5.3×10^{12}, 4.8×10^{12}, and 3.0×10^{12} A/m², respectively. The maximum average velocities (v_{max}) of UD, UU, DD, and DU chirality combinations are 604, 495, 459, and 294 m/s, respectively. It is found that UD has the highest J_c and v_{max} due to the strongest interaction as the closest positioning of the magnetic charges of the two DWs in comparison to the other chirality combinations. The coupling strength of UU and DD chirality combinations is almost equal due to the same distance between the magnetic charges.

For all chirality combinations, the coupling between two DWs is broken when the applied current density is increased beyond their respective J_c. Interestingly, the process of the coupling is different depending on the chirality combination. Figure 11.26a and b shows the position of each DW in the system with respect to the simulation time for $J > J_c$ for UU and UD combinations, respectively. The shapes and relative positions of both the current-driven and remote-driven DWs at different stages of the simulation are also shown. For both UU and DD chirality combinations, the coupling breaking phenomenon is preceded by a change in the shape of the current-driven DW due to the Walker breakdown.

For UD chirality combination, the DW coupling breaking occurs only after two shape-change phenomena, as shown in Figure 11.26b. As shown in the inset, the first shape-change takes place at the remote-driven DW (point A in the inset), which is due to the switching of the transverse component (M_y) of the remote-driven DW from negative to positive orientation. The subsequent shape-change occurs at the current-driven DW (point B in the inset). The shape-change phenomenon at the remote-driven DW is remarkable as there is no current being applied to drive the DW. The detailed micromagnetic simulations reveal that the DW in a single nanowire of 100 nm wide and 6 nm thickness undergoes the Walker breakdown by driving

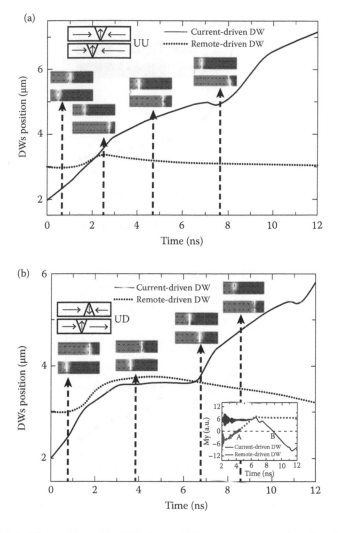

FIGURE 11.26 The position of the DWs in two different systems as a function of simulation time for current densities higher than threshold. The DW chiralities and relative positions are also shown at different times of simulation. Dotted and solid lines represent the position of remote-driven DW and current-driven DW, respectively. (a) UU and (b) UD. The inset represents the chirality change of DWs with time.

the DW with the application of an external field (H_{WB}) of 18 G. However, in a two-nanowire system, the stray magnetic field from the current-driven DW remote drives DW in the current-free nanowire. The stray field from the current-driven DW is responsible for shifting the Walker breakdown to a higher limit. This allows the driving of the remote DWs with the velocity of >620 m/s.

The distance between the current-driven lower DW and the remote-driven upper DW is varied at various current densities by keeping the interwire spacing constant at 50 nm (Figure 11.27a). The DWs come closer to each other as the applied

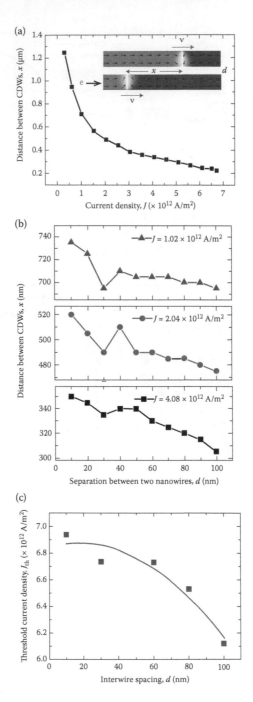

FIGURE 11.27 (a) Distance between adjacent DWs as a function of applied current density showing columbic force nature of magnetostatic forces. (b) Distance between adjacent DWs against separation between nanowires for different current densities. (c) Threshold current densities as a function of separation between nanowires.

current density is increased, proving the columbic nature of the DW interaction. In Figure 11.27b, the distance between the two DWs is plotted for varying inter-wire spacing. The results show that the distance between the two DWs decreases as the interwire separation is increased. Two different trends are observed depending on the proximity between the nanowires. When the two nanowires are placed at a separation $d < 30$ nm to each other, the distance between the two DWs decreases sharply with applied current density. This behavior can be attributed to the charge distribution of the DW. The magnetic charge distribution of a DW has a Gaussian shape: for the HH DW, a strong positive charge is spread out along the base of the triangular shape of the DW, while a weak negative charge is spread out along the apex of the triangular shape [42]. The weak negative charge at the apex of the DW affects the coupling in the two-nanowire system when the interwire spacing is small ($d < 30$ nm). Each of the DWs acts as a dipole, and the interaction between the two DWs can be seen as a dipole-to-dipole interaction. However, when the interwire spacing between the two nanowires is large ($d > 30$ nm), the effect of the weak negative charge becomes negligible, and thus the DW interaction just becomes a charge-to-charge interaction. The charge-to-charge interaction is revealed by the gradual change in the DW distance in Figure 11.27b for interwire separation $d > 30$ nm. The threshold current density is also found to decrease with increasing nanowire spacing (d), as shown in Figure 11.27c.

The interaction of the DWs in multiple nanowires was studied as well. Shown in Figure 11.28 is the comparison between the two- and three-nanowire systems for the strongest coupling strength combinations (UD and UDD, respectively) along with a single nanowire system in terms of the average velocity as a function of the applied current density. For interwire separation of 50 nm, the average DW velocity drops significantly with the increase of the total number of DWs in the system. For a particular current density, the velocity of the CDWs in the three- and two-nanowire

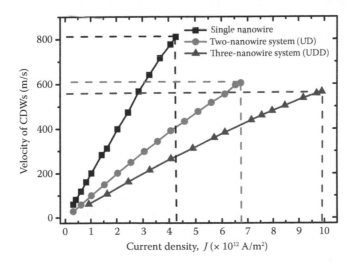

FIGURE 11.28 Plot showing the velocity variation of the CDWs in different nanowire systems as a function of applied current densities.

systems drops to about 1/3 and 1/2 compared to that in the single nanowire system, respectively. The linear reduction in the velocity with the increase in the DWs in the system is due to an increase in the inertia of the system [72]. However, the critical current density of the CDWs is found to be increased with the number of the DWs in the system. The critical current density of the three-nanowire system is twice as compared to that of the single nanowire system.

To understand the key characteristics of the coupling in the CDWS, spin-polarized current is applied to both the nanowires in opposite directions in the two-nanowire system, that is, for the TTDW along the +x direction and to the HH DW along the −x direction. The current drives the two DWs to move away from each other, while the coupling between the two DWs acts against the spin-polarized current as the two DWs attract each other. Hence, the resultant DW motion is due to the competition between the two different forces (coulomb attraction and STT). Figure 11.29a shows the separation between the two DWs along the nanowire long axis as a function of simulation time. The distance between the two DWs oscillates before the two DWs reach a certain equilibrium position. At any instance, the velocities of the two DWs are equal in magnitude but opposite in direction. Mathematically, this was described by a damped oscillation equation

$$x(t) = A e^{-\zeta \omega_0 t} \sin\left(\sqrt{1-\zeta^2}\,\omega_0 t + \varphi\right) \tag{11.11}$$

$$\ddot{x} + 2\zeta\omega_0\dot{x} + \omega_0^2 x = \frac{F(J)}{m} \tag{11.12}$$

where x is the separation between the two DWs, ω_0 is the natural frequency, and ζ is the damped oscillation's damping parameter. Here, the two DWs are assumed to be finite masses connected by a spring [73]. $F_s(J)$ is the force exerted by the spin-polarized current to the DWs. In the one-dimensional model, the force exerted by a spin-polarized current to a DW is a linear function of current density

$$F_s(J) = \gamma J \tag{11.13}$$

where γ is the constant relating the current density to the magnitude of the force. After 10 ns, the two DWs stop and are displaced by an equal distance from the initial position. Thus, by setting $\ddot{x} = 0$ and $\dot{x} = 0$ in Equation 11.4 gives us

$$x_f = \frac{\gamma}{\omega_0^2 m} J \tag{11.14}$$

The final separation (x_f) between the two DWs increases linearly with respect to linear increase of current density, as shown in Figure 11.29b. Hence, the natural frequency (ω_0) is approximated to be independent of the current density. In equilibrium, the torque from the spin-polarized current (F_s) and the columbic attraction force due to the coupling (F_c) are equal:

$$F_c = F_s \tag{11.15}$$

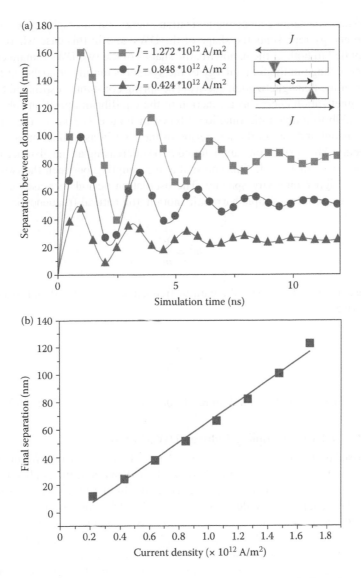

FIGURE 11.29 (a) DW separation as a function of simulation time during relaxation after the application of different current densities. (b) Final separation between the two DWs as a function of current density.

It leaves the interaction force to be a linear function of the separation between the DWs:

$$F_c = m\omega_0^2 x \tag{11.16}$$

The two-DW system is modeled as two masses connected by a spring and the spring constant is obtained from the energy evolution of the system. The total energy of the system is a sum of its demagnetization energy and its exchange energy. The exchange energy represents the shape of the DWs, and in this case where the current is applied to both of the nanowire, the shapes of the two DWs remain constant. Hence, the change in the total energy of the system is mainly contributed from the evolution of the demagnetization energy. The spring constant is estimated by fitting the demagnetization energy as a function of the equilibrium positions obtained in Figure 11.29b to a quadratic function. Shown in Figure 11.30a are the calculated spring constant and the oscillation period of the CDWS as a function of the interwire spacing. The spring constant is found to be decreasing as the distance between the nanowires is increased. This shows that the coupling between the two DWs is weaker for higher interwire spacing. The oscillation period increases as the distance is increased. In a simple harmonic motion, the natural frequency is related to the mass by

$$\omega_0^2 = \frac{k}{m} \tag{11.17}$$

We use $\omega_0 = 2\pi/T$ to relate the natural frequency to the oscillation period, and thus we have

$$m = \frac{kT^2}{4\pi^2} \tag{11.18}$$

The obtained mass shown in Figure 11.30b is in the order 10^{-24} kg [46,72].

11.3.3.2   Exchange Coupling Induced DW Motion

As discussed in the previous section, we understand that interwire DW coupling can be used as a driving force to remotely drive the DW without the application of the current to the DW in the active nanowire. Here, we discuss how we can also explore intrawire DW coupling to induce DW motion. The DWs in a same nanowire also couple with each other. However, the coupling is a combination of magnetostatic and exchange interaction. The properties of exchange coupling and its role in the remote driving are discussed in this section. Besides carrying a magnetic charge, the DWs are also considered as a composite of topological charges [74]. A vortex DW is a combination of an integer winding number and two half-integer winding numbers, whereas a transverse DW is a composite of two half-integer edge defects. The relative positions of these edge defects (winding numbers) define the chirality of the DW. When two DWs (HH and TT) are relaxed next to each other in a same nanowire, the DWs get attracted to each other. The attraction can lead to two different phenomena depending on the position of the winding numbers. The DW coupling in a intrananowire system annihilates or forms a bound state. The two possibilities, annihilation and formation of the bound state, are shown in Figure 11.31. If the two DWs have opposite winding numbers facing each other, then it leads to mutual annihilation. If DWs of same winding numbers face each other, then it leads to the formation of a bound state [37].

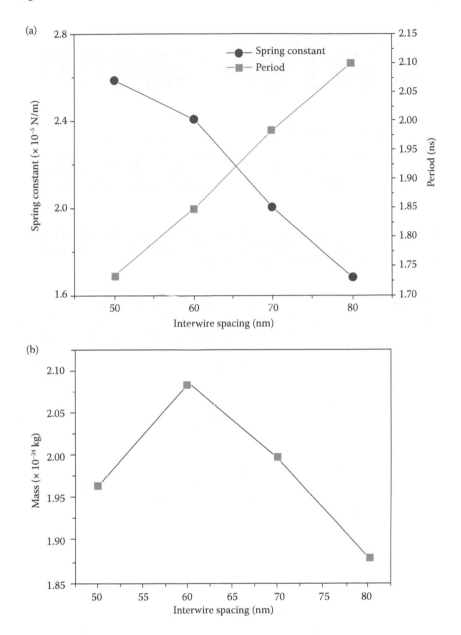

FIGURE 11.30 (a) Spring constant and oscillation period of the damped harmonic motion as a function of interwire spacing. (b) DW mass as a function of interwire spacing.

We consider the case of a HH DW and a TT DW with similar winding numbers facing each other is placed in a single nanowire to study the formation of bound state and the intra-DW interaction in the same nanowire. Shown in Figure 11.32 are the magnetization configurations of the bound state of two DWs carrying similar winding numbers facing each other. When the DWs are attracted, the DWs are

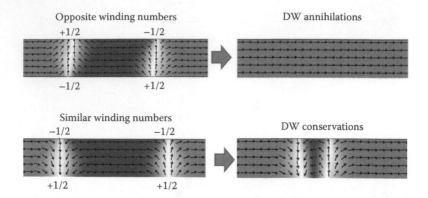

FIGURE 11.31 Schematic showing the winding numbers of transverse DWs. In the case where they have the same winding numbers, the two DWs are compressed upon collision and annihilate when opposite winding numbers are facing each other.

compressed at the edges and they collide with each other. In the close proximity of the DWs, the compression is represented as the repulsion force that they exert to each other. The magnetization components (M) of a relaxed DW can be approximated as

$$\vec{M} = M_s(\hat{x}\cos\theta + \hat{y}\sin\theta) \tag{11.19}$$

$$\cos\theta = -\tanh\left(\frac{x}{\lambda}\right) \tag{11.20}$$

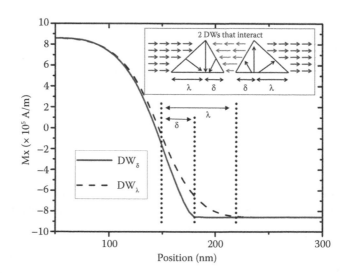

FIGURE 11.32 Magnetization component in the x direction of an unperturbed DW (DW$_\lambda$) and a compressed DW (DW$_\delta$). Inset is the illustration of the compressed DWs.

for $0 < x < 2\lambda$, with 2λ as the full width of a transverse DW [75]. The saturation magnetization of the permalloy (M_s) is 860,000 A/m. The *cosine* function is equal to a negative *tanh* function for HH DW and positive function for TT DW. The exchange energy of a single DW is therefore equal to

$$E_{exc} = Awt \int_{-\lambda}^{+\lambda} |\nabla M|^2 \, dV \approx 2\frac{Awt}{\lambda} \qquad (11.21)$$

where w and t are the width and the thickness of the nanowire, respectively. A is the exchange constant of the permalloy. The expression shows that the exchange energy increases rapidly as the half-width of the transverse DW (λ) is reduced. The preceding energy calculation is applicable when the DW has a width of 2λ; when the DW is compressed, the energy of each side of the DW is calculated separately:

$$E_{exc}(DW) = E_{exc}(left\ side) + E_{exc}(right\ side) = \frac{Awt}{\delta_{leftside}} + \frac{Awt}{\delta_{rightside}} \qquad (11.22)$$

Shown in Figure 11.33 is the exchange energy of a relaxed transverse DW (DW_λ) and a DW that is compressed at the left side (DW_δ, a snapshot of the compressed DW is included in the upper inset) as functions of the nanowire width. The calculated half-width in remanence, both the uncompressed (λ) and the compressed (δ), is shown in the inset of Figure 11.33.

The increased exchange energy of the compressed DWs as compared to the relaxed DWs is considered as the work done by the system to oppose the repulsion

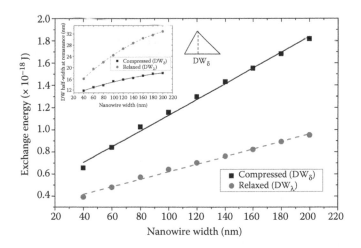

FIGURE 11.33 Exchange energy of unperturbed (E_λ) and compressed (E_δ) DWs as a function of wire width. Shown in the inset is the half-width of the unperturbed DW and the compressed DW.

force (F_{exc}) during the compression. In equilibrium, when no external field and current are applied, the repulsion force prevents the two DWs to annihilate under their own magnetostatic interaction. By taking the derivative of the exchange energy with respect to the DW compression (δ) the repulsion force is estimated as follows:

$$F_{exc} = -\frac{\partial E_{exc}(compressed\text{-}side)}{\partial \lambda}\bigg|_{\lambda=\delta} = \frac{Awt}{\delta^2} \tag{11.23}$$

For a nanowire of 100 nm width and 6 nm thickness, the DW compression (δ_c) is 15.1 nm; F_{exc} estimated from the equation is 3.42×10^{-11} kg·m/s². The magnitude of the exchange repulsion force is comparable to the driving force that is exerted by the spin-polarized and the interwire magnetostatic force [76]. When current is not applied to the system, the exchange repulsion force (F_{exc}) is balanced by the intrawire magnetostatic interaction (F_{mag}). The magnetostatic interaction when the two DWs are compressed is calculated from

$$F_{mag} = -\frac{1}{4\pi\mu_0}\frac{Q^2}{r^2} = -\frac{\mu_0}{4\pi}\frac{(CMwt(\tanh 1 + \tanh(\delta/\lambda)))^2}{(2\delta)^2} \tag{11.24}$$

where μ_0 is the vacuum permeability constant, Q is the magnetic dipole of the DW ($Q = -\mu_0 C \int_{-\lambda}^{\delta} \nabla \cdot \mathbf{M} dV$), r is the distance between the two DWs, and C is the proportionality constant that relates the DW shape and magnetic charge representation. The total force that acts on the DW as a function of the compression is then equal to

$$F_{tot} = F_{exc} - F_{mag} = \frac{Awt}{\delta^2} - \frac{\mu_0}{16\pi}\frac{(CMwt(\tanh 1 + \tanh(\delta/\lambda)))^2}{(\delta)^2} \tag{11.25}$$

At equilibrium, when no current is applied, the exchange repulsion force is balanced with the magnetostatic attraction force ($F_{exc}(\delta_{remanance}) = F_{mag}(\delta_{remanance})$), which gives us

$$C = \frac{4}{M(\tanh 1 + \tanh(\delta/\lambda))}\sqrt{\frac{\pi A}{\mu_0 wt}} \tag{11.26}$$

For a nanowire of 100 nm width and 6 nm thickness, the proportionality constant is equal to $C_{w=100,t=6} \approx 0.83$. The total force as a function of δ is shown in Figure 11.34. The graph shows that the total force that acts on the compressed DW is an attractive force until the half-width of the two DWs reaches a remanance value, and any further compression will change the total force into a repulsive force.

To explore the role of the exchange repulsion force in DW driving, the one-directional collision between two DWs is considered. In the one-directional collision, a single DW is driven by a local driving force and is allowed to collide with another DW in the same nanowire. The remote driving technique that is discussed

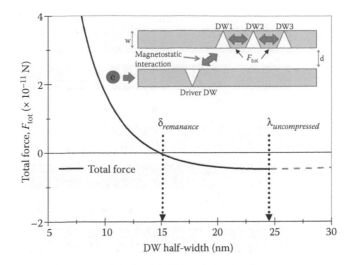

FIGURE 11.34 Calculated total force (F_{tot}) as a function of DW compression length (δ). Shown in the upper-right inset is the schematic diagram of the simulation model. The two nanowires are separated at 50 nm away.

in the earlier section is employed [78–80] as the local driving. In the remote driving technique, another nanowire with one DW is placed next to our nanowire of interest. The first nanowire, or the nanowire of interest, is called the active nanowire, while the second nanowire is called the driver nanowire. A single DW, which shall be called the driver DW, is generated in the driver nanowire to be used as a stray magnetic field generator. The driver DW and the first DW (DW1) in the active nanowire are coupled, and their dynamics are related to each other [78–81]. Shown in the upper-right inset of Figure 11.34 is the diagram of simulations. The driver DW and DW1 are oriented such that the triangle bases face each other to ensure the strongest magnetostatic attraction between them. To ensure that the topological repulsion is present, the next DWs in the active nanowire need to have similar winding numbers placement; this is achieved by aligning all of the DWs in the active nanowire such that the bases of their triangular shapes face the driver nanowire.

When spin-polarized current is applied to the driver nanowire, the driver DW moves and it exerts a repulsion force on DW1 in the active nanowire due to its stray magnetic field interaction. The DW1 is remote-driven in the same direction as the driver DW due to the magnetostatic repulsion. Eventually, DW1 collides with the next DW (DW2) in the active nanowire. While the remote driving technique serves as the driving force for DW1 in the active nanowire, it is the exchange repulsion that continuously drives DW2 ahead. Here, DW2 and DW1 are compressed further beyond the half-width remanance ($\delta_{remanance}$); the interaction between DW1 and DW2 then becomes repulsive, which then serves to reduce the velocity of DW1 and at the same time becomes the driving source for DW2.

The snapshots of the simulation showing the motion of the DWs are shown in Figure 11.35. Initially (I), the driver DW and DW1 move together with an equal

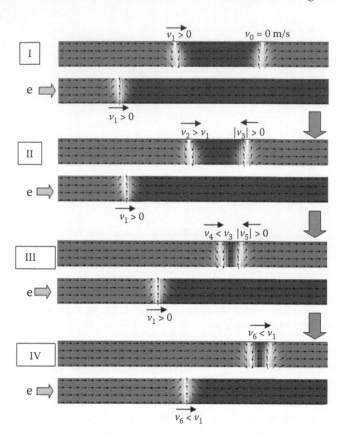

FIGURE 11.35 Snapshot of the simulations showing the collision between DW1 and DW2 in the active nanowire.

velocity of $v_1 \approx 90$ m/s. For a short period of time (II) between $t \approx 6$ and 8 ns, the magnetostatic attraction between DW1 and DW2 propels DW1 to move forward, increasing its velocity to $v_2 \approx 130$ m/s. The attraction pulls DW2 to move backward with a velocity of $v_3 \approx 40$ m/s. At $t \approx 8$ ns, the collision between the two DWs takes place and the topological repulsion starts to become more prominent. The two DWs get compressed (III). As a result, DW2 is pushed by DW1 to move in the same direction (IV) with a velocity of $v_6 \approx 50$ m/s. Figure 11.36 shows the velocities of DW1 and DW2 as a function of simulation time. From $t \approx 8$ to 14 ns, the two DWs are pushing each other until they arrive at an equal velocity. Because the original velocity of the driver DW (v_1) is higher than the velocity of the two DWs in the active nanowire, the driver DW is able to move closer to DW1. However, this results in the increase in the magnetostatic interaction. The driver DW is slowed down until it reaches to equilibrium with the same velocity as the DWs in the active nanowire (v_6).

Now we discuss the possibility of driving multiple DWs [82] in the active nanowire by utilizing mutual collision processes between them. For instance, when a third DW (DW3) is present in the active nanowire, it will collide with DW2. The collision

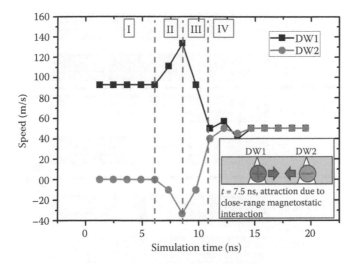

FIGURE 11.36 Velocity of the DWs as a function of simulation time. In [I], DW1 is being remote-driven and moves with the same velocity as the driver DW. In [II], DW2 moves back to approach DW1 due to the magnetostatic interaction. The magnetostatic interaction also increases the forward velocity of DW1. Shown in the inset is an illustration of the magnetostatic attraction between DWs of opposite magnetic charge. At [III], the two DWs collide and are compressed. At [IV], the topological force drives DW2 to move in front of DW1. A time span of $t \approx 6$ ns is needed for the two DWs to reach equilibrium starting from the moment of impact.

will compress both of the DWs, and as a result, DW3 will move together with DW1 and DW2. This method is extended to drive up to four DWs in the active nanowire at the same time; the DW group moves with a velocity of $v \approx 30$ m/s under applied current density of $J = 1.06 \times 10^{12}$ A/m^2.

As described earlier, the motions of DW2 and the subsequent DWs are the result of collisions with the remote-driven DW. However, as shown in Figure 11.36, the process is not instantaneous; $t \approx 6$ ns is needed for the DWs that are involved in the collisions to reach an equilibrium velocity. Therefore, it is possible for a driver DW that is moving with a very high velocity to "bypass" a group of DWs in the active nanowire when there is not enough time for the DW group to reach equilibrium. Shown in Figure 11.37 are the snapshots of the simulation with current density of 4.24×10^{12} A/m^2 applied to the driver nanowire. DW1 and DW2 are initially placed very close to each other and far away from the driver DW. Due to the high propagation velocity of the driver DW, the interaction time between the driver DW and DW1 is very short. Consequently, there is not enough time to decelerate the driver DW and accelerate DW2 to reach equal velocity. As a result, the driver DW is able to move pass DW1 and DW2 and continues its motion without any coupling to the DWs in the active nanowire. In other words, by adjusting the applied current density to the driver nanowire, we will be able to control the coupling between the DWs in the active and the driver nanowires.

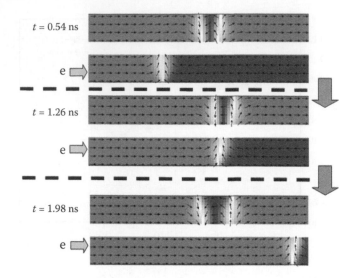

FIGURE 11.37 Snapshots of the simulated DWs when a very high current density is applied to the driver nanowire. Due to the high velocity motion of the driver DW, it is able to move pass the DW group in the active nanowire.

11.4 DW PINNING AND DEPINNING IN FERROMAGNETIC NANOWIRES

The ability to pin the DW at a specific position is essential for DW-based memory, much like in the racetrack memory where the DWs need to be pinned in order for the data to be read or written. The understanding and characterization of DW pinning in ferromagnetic nanowires are essential to the advancement of its application in spintronic devices. For the DW to be pinned, a geometrical adjustment can be made to the device as the pinning site such that the DW energy is lowered there [33,35,40,42,83]. The DW can then be driven out and depinned from the pinning site by supplying external energy to the device through the application of external field or current. Additionally, it is also possible to introduce pinning to the DW by making use of the stray field from an adjacent nanomagnet.

11.4.1 Domain Pinning at Notch Geometry in Planar Nanowires

Initial studies of DW pinning were mainly based on triangular notch with the height and width equivalent to a third of the track height [40,41,84]. The results revealed the triangular shape of transverse DWs as they were pinned by the notch. When transverse DWs with different charges and chiralities were injected and driven to the notch structure, only two out of four DWs were pinned by the same triangular notch, which is illustrated in Figure 11.38. Such a pinning/depinning trend shows that DW pinning strongly depends on the chirality of the transverse DW.

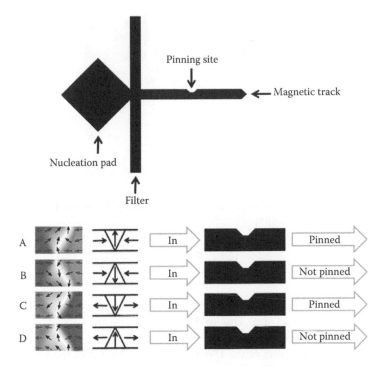

FIGURE 11.38 Schematic diagram of the notch pinning site in the magnetic track. The results of pinning for four different combinations of the DWs injected and driven in the magnetic track.

A qualitative analysis can be drawn from the concepts of topological defects proposed by Tchernyshyov and Chern [74]. As discussed in previous sections, DWs are composite objects that contain two topological defects in the magnetically ordered state of the solid. Adopting their notation as summarized in Table 11.1, each DW can then be assigned with two winding numbers of +1/2 and −1/2, as shown in Figure 11.39.

From the theoretical calculations, it was estimated that the defect with winding number +1/2 has a higher energy compared to that with the −1/2 portion of the DW.

TABLE 11.1

Magnetization Configurations and Corresponding Winding Numbers

Winding Number   Structure

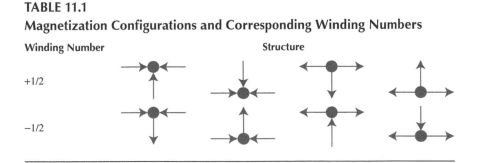

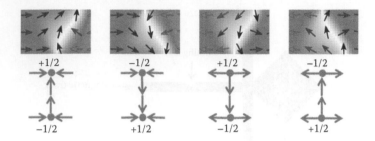

FIGURE 11.39 Position of winding numbers for four different transverse DWs.

This was observed to be consistent with the DW theory. From the DW theory, DW energy (E_{DW}) is defined as [85]

$$E_{DW} = \frac{\pi^2 A}{\delta} + K\delta \qquad (11.27)$$

where A is the exchange stiffness constant, K is the anisotropy energy density, and δ is the DW width (lateral size). Due to the dominance of the second term, the DW energy is then almost proportional to the DW width.

Figure 11.40 shows a plot of the DW energy as a function of the DW thickness, and it can be seen that a DW with a larger wall thickness possesses a higher energy. The DW thickness (δ) is a function of the lattice constant and also the number of planes over which the magnetization rotates. As such, for the transverse DW, δ varies as a function of the width of the transverse component of the DW magnetization. In other words, the energy of the DW is proportional to the width of the triangular shape of the DW. Due to the triangular shape of the transverse component of the TDW, the energy of the wall varies along the y direction, with the base of the wall (triangle) having the highest energy. From Figure 11.38, we note that pinning

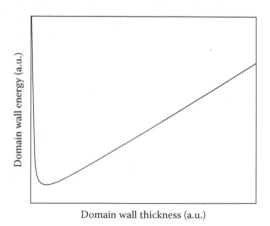

Domain wall thickness (a.u.)

FIGURE 11.40 DW energy variation along the transverse width of the nanowire.

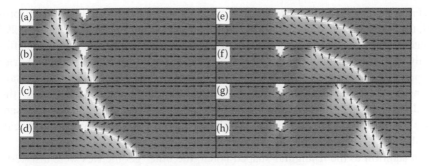

FIGURE 11.41 Snapshot images of the magnetization configurations when a TT DW with down chirality is driven through a notch structure.

is effective only when the notch provides a potential barrier to the highest energy component of the DW.

By decomposing the DW into higher and lower energy segments, we propose another qualitative description to the depinning trend observed. When we consider a TT-U DW with its triangular base (+1/2) at the bottom, the lower energy triangular apex or tip (−1/2) is "blocked" by the notch while the higher energy base is free of such constrictions (Figure 11.41a,b). The higher energy base continues to propagate through the track (since it encounters no barrier) and drag the lower energy tip along the nanowire, which can be seen from the deformation of the DW (Figure 11.41c–e). The tip then detaches itself from the notch (Figure 11.41f) and the wall regains its shape and continues propagating through the track (Figure 11.41g,h). In other words, we can say that the DW is not pinned by the notch when the notch is located at the lower energy base of the DW. In contrast, the DW is impeded by the notch when the higher energy base is located at the top while the lower energy tip is located at the bottom. In this case, even though the lower energy tip is able to avoid the notch, it does not have enough energy to drag the rest of the DW with it, resulting in a pinned DW. The external magnetic field that is needed to depin a transverse DW for top notch, bottom notch, and double notch are summarized in Table 11.2.

TABLE 11.2
DW Depinning Field for Top, Bottom, and Double Notch Structures

Notch Type	Minimum Field Needed to Depin DW from Notch (Oe)			
	TT-D	TT-U	HH-D	HH-U
	−257.5	0	0	257.5
	0	−257.5	257.5	0
	−262.5	−262.5	262.5	262.5

From the data in Table 11.2, it can be observed that when the double notch is present, the depinning field needed increases by only 5 Oe (an increase of less than 2%), suggesting that the presence of the additional notch does not have a significant effect on the depinning field of the DW. It can also be observed that there are symmetrical properties between the two different DWs. TT-D DW and HH-U DW have the same triangular wall shape and the same positioning of winding numbers; as a result, they need the same amount of external field to be depinned from the notches. The same goes for the TT-U DW and HH-D DW. This suggests that the DWs are physically similar except with an opposite magnetization.

In addition to the triangular-shaped notch, the other geometries were also tested to investigate the effect of geometry on the pinning of DWs. The various geometries tested and the depinning fields for the various geometries are also tabulated in Table 11.3.

The effect of different notch heights on the depinning field was investigated with the TT-D and TT-U DWs, and their results are presented in this section. The definition of the notch height was shown as H_N in Figure 11.42. For TT-D DW, the H_N/H_T ratio was varied in the range of 0.1–0.9 and the results for the various notch geometries were plotted in Figure 11.43. Above the threshold value of 0.25, the trend is approximately linear with only a slight increase in depinning field even as the ratio of H_N/H_T increases. This result can be explained by looking at the energy profile of the transverse DW, which shows that most of the DW energy is located at the base

TABLE 11.3
DW Depinning Field for Different Notch Geometries

Notch Shape	Mask Design	Depinning Field, H_{depin} (Oe)			
		TT-D	TT-U	HH-D	HH-U
Triangle		−157.5	0	0	157.5
Trapezium		−112.5	0	0	112.5
Semicircle		−127.5	0	0	128
Rectangle		−120	0	0	120
Inverted triangle		−47.5	0	0	47.5
Inverted trapezium		−25	0	0	22
Inverted semicircle		−69	0	0	69

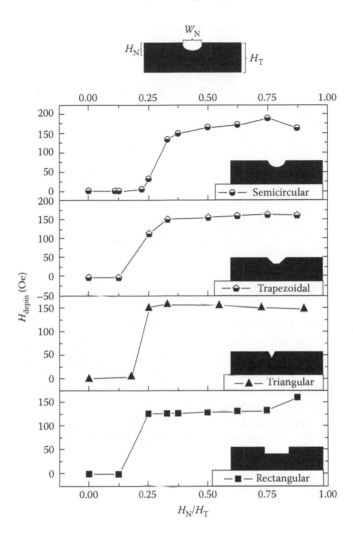

FIGURE 11.42   DW depinning field as a function of notch height.

of the wall. Hence, further increase in the notch height does not add any additional obstruction to the DW. Figure 11.44 shows the depinning field as a function of notch width (W_N). For the trends displayed by the semicircular, trapezoidal, and triangular geometries, it appears that the pinning is effective only within a range of W_N/H_T ratio from 0.1 to 1.5. Above a W_N/H_T ratio of 1.5, pinning is no longer effective and the DW moves freely across the notch [41].

11.4.2   DW PINNING BY AN ANTI-NOTCH STRUCTURE

Another possibility for geometrical pinning site in a nanowire is an anti-notch/ protrusion. Here we discuss the DW pinning by an anti-notch of a rectangular shape,

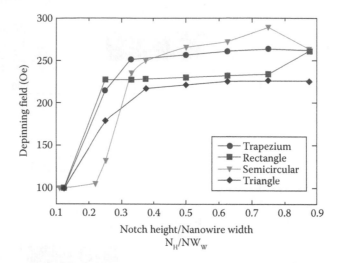

FIGURE 11.43 DW depinning field strength as a function of notch height.

as shown in Figure 11.45a. The anti-notch width is defined as W_{AN} and the height is defined as H_{AN}. The effect of the anti-notch width on the type of pinning potential experienced by HH-U DW and HH-D DW is discussed. The anti-notch width (W_{AN}) was varied from 40 to 200 nm, while the anti-notch height (H_{AN}) was kept at 200 nm. The DWs were driven from left to right along the +x direction in the nanowire. All simulations are performed at a current density $J = 1.48 \times 10^{12}$ A/m² for this wire geometry.

At first, the equilibrium DW positions for HH-U and HH-D at an anti-notch geometry with $W_{AN} = 40$ nm are calculated. For HH-U, the DW is stable beneath

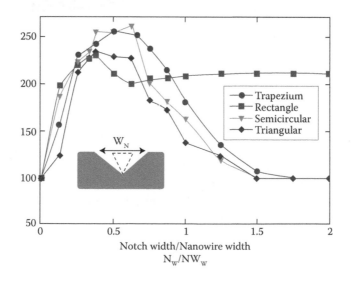

FIGURE 11.44 DW depinning field strength as a function of notch width.

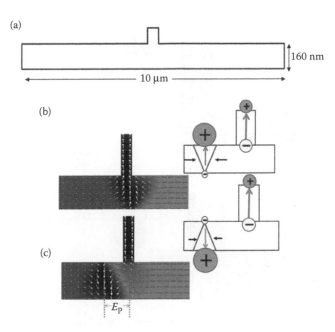

FIGURE 11.45 (a) Schematic of the anti-notch structure along the magnetic track. (b) Magnetization configuration of the DW pinning when a HH-U is injected and driven to the anti-notch width of 40 nm. (c) Magnetization configuration of the DW pinning when a HH-D is injected and driven to the anti-notch width of 40 nm.

the anti-notch structure, as shown in Figure 11.45b. For HH-D, the DW is stopped away at a distance E_P from the center of the anti-notch structure, as shown in Figure 11.45c. This difference in equilibrium position is due to the potential seen by the DW with different chiralities at the anti-notch. For HH-U, the anti-notch acts as a potential well, whereas for HH-D, the anti-notch is seen as a potential barrier. It reveals that the potential at the anti-notch as experienced by the DW is chirality dependent. Interestingly, a completely different behavior is observed when $W_{AN} = 120$ nm. The equilibrium positions of the DWs are shown in Figure 11.46. There is a change in the type of the potential disruption seen by the DW at the anti-notch. The anti-notch acts as a potential barrier for HH-U and a potential well for HH-D. This implies a change in the potential landscape of the anti-notch with varying width.

To gain a better understanding of the DW interaction with the anti-notch, the equilibrium positions of the DW at various anti-notch widths are shown in Table 11.4. The transformation in the potential landscape occurs when the anti-notch width is $W_{AN} = 100$ nm. The change in the potential polarity of the anti-notch can be explained by the orientation of the spins along the anti-notch. For the anti-notch width of $W_{AN} \leq 100$ nm, the magnetization state within the anti-notch prefers to align parallel to the y direction to minimize the demagnetization energy as induced by the shape anisotropy. This magnetic configuration within the anti-notch leads to the formation of magnetic charges oriented in the y direction, as depicted in the inset of Figure 11.45b. The presence of the magnetic charges explains the different potential

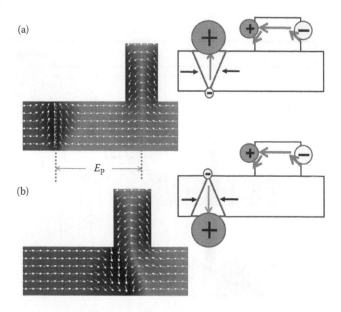

FIGURE 11.46 (a) Magnetization configuration of the DW pinning when a HH-U is injected and driven to the anti-notch width of 120 nm. (b) Magnetization configuration of the DW pinning when a HH-D is injected and driven to the anti-notch width of 120 nm.

as seen by the HH DW with different transverse components. From Figure 11.45b, it can be clearly observed that when the HH-U reaches the anti-notch, the transverse spins within the DW are aligned in the same direction as the spins in the anti-notch. The magnetic charges are of opposite polarity, as seen in the inset of Figure 11.45b, which leads to the attraction of the DW beneath the anti-notch to minimize the demagnetization energy. Conversely, for the HH-D, the transverse spins of the DW are aligned in the opposite direction with the spins of the anti-notch, leading to the magnetic charges being in the same polarity, as seen in the inset of Figure 11.45c. This results in the repulsion between the same magnetic charges, giving rise to the formation of a potential barrier for HH-D.

For $W_{AN} > 100$ nm, we observed a change in the polarity of the potential as seen by the DWs at the anti-notch. This is due to the decrease in the demagnetization factor along the x direction in the anti-notch. The magnetization direction is no longer constrained to the y direction. The spins within the anti-notch have a preferential alignment along the x direction, which is the same as the magnetization direction of the nanowire. The transverse components of the DW are now aligned orthogonally to the magnetization in the anti-notch. This leads to the repulsion of the HH-U due to the interaction of the positive charges from both the DW and the anti-notch. The DW is pushed to a distance E_p away from the center of the anti-notch. For HH-D, the opposite charge leads to the attraction of the DW within the notch, trapping the DW at the left edge of the anti-notch. Additionally, when the anti-notch acts as a barrier, the DW undergoes a damped oscillation prior to reaching the equilibrium position.

Figure 11.47 shows the variation of the equilibrium position of the HH-U from the center of the anti-notch ($W_{AN} > 100$ nm) with increasing anti-notch width. The

TABLE 11.4
Magnetization Configurations of DW Pinned for Different Widths of Anti-Notch

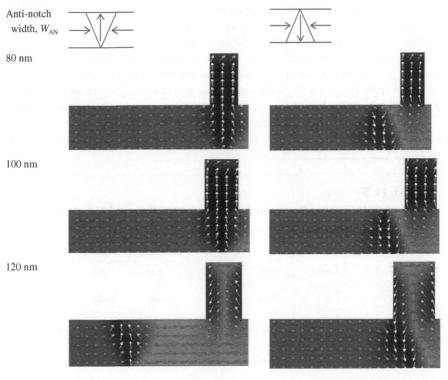

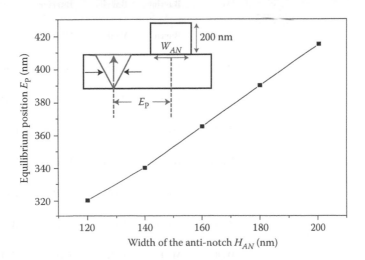

FIGURE 11.47 DW equilibrium position as a function of anti-notch width.

equilibrium position is moving farther from the anti-notch with increasing anti-notch width. The increase in the distance between the anti-notch and the HH-U is attributed to the increase in the potential of the anti-notch with the increasing width. When the anti-notch acts as a potential barrier for the HH-D ($W_{AN} \leq 100$ nm), the equilibrium position is almost stable at all widths, which is around 70 nm away from the left edge of the anti-notch. This shows that there is no significant change in the potential of the anti-notch with the variation of the width when $W_{AN} \leq 100$ nm [43].

The effect of the anti-notch height on the motion of the current-driven transverse DW in $Ni_{80}Fe_{20}$ nanowire has also been investigated. The anti-notch height was varied from $H_{AN} = 100$ to 300 nm at different widths ranging from $W_{AN} = 40$ to 160 nm. The results obtained from the micromagnetic simulations are summarized in Table 11.5. The results reveal that the anti-notch of all heights from $H_{AN} = 100$ to 300 nm acts as a potential well for HH-U and a potential barrier for HH-D at

TABLE 11.5
Type of DW Pinning Potential for Different Widths and Heights of Anti-Notch Structures

Width (nm)	Chirality	Height (nm)				
		100	150	200	250	300
40	→\↑/←	Well	Well	Well	Well	Well
	→/↓\←	Barrier	Barrier	Barrier	Barrier	Barrier
60	→\↑/←	Barrier	Well	Well	Well	Well
	→/↓\←	Well	Barrier	Barrier	Barrier	Barrier
100	→\↑/←	Barrier	Barrier	Well	Well	Well
	→/↓\←	Well	Well	Barrier	Barrier	Barrier
120	→\↑/←	Barrier	Barrier	Barrier	Well	Well
	→/↓\←	Well	Well	Well	Barrier	Barrier
140	→\↑/←	Barrier	Barrier	Barrier	Barrier	Well
	→/↓\←	Well	Well	Well	Well	Barrier
160	→\↑/←	Barrier	Barrier	Barrier	Barrier	Barrier
	→/↓\←	Well	Well	Well	Well	Well

anti-notch width $W_{AN} = 40$ nm. Conversely, it acts as a potential barrier for HH-U and a potential well for HH-D when anti-notch width $W_{AN} = 160$ nm. However, as the height of the anti-notch changes from $H_{AN} = 100$ to 300 nm, a transition in the polarity of the potential is observed at increasing widths from $W_{AN} = 60$ to 140 nm. On the basis of careful observation of the potential disruption variation with the dimensions of the anti-notch, it is shown that the transition in the potential polarity is at the anti-notch H_{AN}/W_{AN} ratio of 2.

The transition in the potential behavior of the anti-notch is due to the change in the relative orientation between the spins in the anti-notch and the nanowire. For $H_{AN}/W_{AN} < 2$, the spins in the anti-notch and the wire are almost parallel to each other. However, in the case of $H_{AN}/W_{AN} \geq 2$, the spins in the anti-notch and the wire are orthogonal to each other. When $H_{AN}/W_{AN} < 2$, the magnetic charges at the HH-U and the anti-notch are of same polarity and cause repulsion resulting in a potential barrier at the anti-notch. When $H_{AN}/W_{AN} \geq 2$, the magnetic charges at the HH-U and the anti-notch are of opposite polarity and cause attraction between them resulting in a potential well at the anti-notch. Similar but opposite behavior is observed in the case of HH-D.

The equilibrium positions from the center of the anti-notch have been calculated when the anti-notch acts as a barrier for HH-U and HH-D. The variation of the equilibrium position of HH-U with the anti-notch height is shown in Figure 11.48. The plot shows that the equilibrium position of HH-U moves away from the center of the anti-notch as the height of the anti-notch increases. This is attributed to the increase in the potential with the height of the anti-notch for HH-U when $H_{AN}/W_{AN} < 2$. However, in the case of HH-D, the equilibrium position of the DW is almost stable with varying height. This stable behavior shows that the potential barrier is constant with varying height of the anti-notch for HH-D when $H_{AN}/W_{AN} \geq 2$.

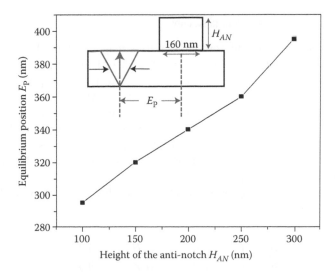

FIGURE 11.48 DW equilibrium position as a function of anti-notch height.

11.5 DW DETECTION

11.5.1 DW Resistance

When current is applied to a ferromagnetic nanowire with a DW, the conduction electron scatters at the DW due to a change in the local magnetization [86,87], which results in a change of resistance. The presence and absence of a DW in the nanowire can then be detected by this change in the nanowire resistance. The DW resistance can be understood from the AMR effect [88–90]. AMR is the change in the resistance of the magnetic material due to the change in the relative orientation between the current and the local magnetization. The origin of the AMR effect lies in the interaction between conduction s electrons and localized 3d electrons. In the absence of a magnetic field, s electrons are scattered by 3d electrons. In the presence of magnetic field, the cloud of 3d electrons deforms, which affects the s-electron scattering and changes the resistance. The change in resistance can be either positive or negative depending on the material.

AMR effect can be explained by a simplified equation

$$R(\Phi) = R(0) - \Delta R \cos^2(\Phi) \tag{11.28}$$

where $R(0)$ is the resistance at zero applied field and Φ is the relative orientation between the current and the applied magnetic field (local magnetization). Thus, the resistance is lower when the magnetization is transverse to the current. The DW has a component transverse to the nanowire length and the current; it leads to a drop in the resistance. The drop in resistance due to the presence of the DW is very small, which is typically ~0.2 Ω.

11.5.2 Magnetic Force Microscopy

MFM is a scanning probe microscopy that uses magnetic force between the cantilever tip and the surface to plot the image. In MFM imaging, a tip coated with a magnetic material is used to scan the surface to acquire the spatial variation of magnetic field instead of topographic information. MFM detects the force gradient between the tip magnetization and the local magnetization on the surface of the sample [91–93]. Due to the magnetic coating, the spatial resolution of the MFM tip is lowered compared to that of the AFM tip. Atomic forces are short-range forces while magnetostatic forces are long range, so MFM can work at a higher distance between the tip and the sample.

The force gradient acting on the cantilever tip due to the stray field from the sample is given by

$$\frac{\delta F}{\delta Z} = \int_{tip} \left(m_x \frac{\partial^2 H_x}{\partial^2 Z^2} + m_y \frac{\partial^2 H_y}{\partial^2 Z^2} + m_z \frac{\partial^2 H_z}{\partial^2 Z^2} \right) \tag{11.29}$$

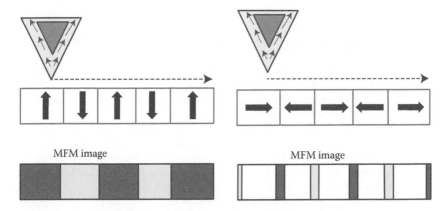

FIGURE 11.49 Schematics of MFM images for in-plane and out-of-plane nanowires.

where F is the magnetic force, Z is the coordinate perpendicular to the surface, and m_x, m_y, m_z, H_x, H_y, and H_z are magnetization components and field components at a particular point within the tip, respectively.

Due to this force gradient, the tip vibrating at its resonance frequency suffers a shift in the frequency. The measured shift can be related to the force gradient by

$$\frac{\delta F}{\delta Z} = -2K \frac{\delta f}{f} \tag{11.30}$$

where K is the spring constant of the tip material and f is the resonance frequency of the tip. The frequency shift (δf) causes the phase difference, which in turn is used for mapping the MFM image. The major problem with MFM imaging is the separation between the topological and the magnetic data. To separate the topology reading from the magnetic reading, MFM scanning is performed in the lift mode. Figure 11.49 shows two kinds of interactions between the tip and the sample for in-plane and out-of-plane nanowires with DWs. For out-of-plane structures, contrasts in the MFM images show the two possible orientations of the magnetic domains. For in-plane structures, the stray field comes only from the DW and the contrast shall show whether the DW is a HH or a TT DW.

REFERENCES

1. S. S. P. Parkin, M. Hayashi, and L. Thomas, *Science* 320, 2008, 190.
2. X. Jiang, L. Thomas, R. Moriya, M. Hayashi, B. Bergman, C. Rettner, and S. S. P. Parkin, *Nat. Commun.* 1, 2010, 25.
3. M. Hayashi, L. Thomas, R. Moriya, C. Rettner, and S. S. P. Parkin, *Science* 320, 2008, 209.
4. A. J. Annunziata, M. C. Gaidis, L. Thomas, C. W. Chien, C. C. Hung, P. Chevalier et al., in *Electron Devices Meeting, 2011 IEEE International*, IEEE, Washington, DC, 2011, pp. 24.3.1–24.3.4.
5. J. H. Franken, H. J. M. Swagten, and B. Koopmans, *Nat. Nanotechnol.* 7, 2012, 499.

6. D. A. Allwood, G. Xiong, M. D. Cooke, C. C. Faulkner, D. Atkinson, N. Vernier, and R. P. Cowburn, *Science* 296, 2002, 2003.

7. D. A. Allwood, G. Xiong, C. C. Faulkner, D. Atkinson, D. Petit, and R. P. Cowburn, *Science* 309, 2005, 1688.

8. P. Xu, K. Xia, C. Gu, L. Tang, H. Yang, and J. Li, *Nat. Nanotechnol.* 3, 2008, 97.

9. B. N. Filippov, *Phys. Solid State* 54, 2012, 2407.

10. M. Redjdal, J. Giusti, M. F. Ruane, and F. B. Humphrey, *J. Appl. Phys.* 91, 2002, 7547.

11. M. Kläui, *J. Phys. Condens. Matter* 20(31), 2008, 313001.

12. G. Catalan, J. Seidel, R. Ramesh, and J. F. Scott, *Rev. Mod. Phys.* 84, 2012, 119.

13. G. S. D. Beach, M. Tsoi, J. L. Erskine, *J. Magn. Magn. Mater.* 320, 2008, 1272.

14. R. D. McMichael and M. J. Donahue, *IEEE Trans. Magn.* 33, 1997, 4167.

15. Y. Nakatani, A. Thiaville, and J. Miltat, *J. Magn. Magn. Mater.* 290, 2005, 750.

16. R. Hertel, *J. Magn. Magn. Mater.* 249, 2002, 251.

17. R. Hertel and J. Kirschner, *Phys. B* 343, 2004, 206.

18. R. Wieser, U. Nowak, and K. D. Usadel, *Phys. Rev. B* 69, 2004, 064401.

19. R. Wieser, E. Y. Vedmedenko, P. Weinberger, and R. Weisendanger, *Phys. Rev. B* 82, 2010, 144430.

20. M. Yan, A. Kakay, S. Gliga, and R. Hertel, *Phys. Rev. Lett.* 104, 2010, 057201.

21. N. Biziere, C. Gatel, R. Lassalle-Balier, M. C. Clochard, J. E. Wegrowe, and E. Snoeck, *Nano Lett.* 13, 2013, 2053.

22. H.-G. Piao, J. H. Shim, D. Djuhana, and D. H. Kim, *Appl. Phys. Lett.* 102, 2013, 112405.

23. M. C. Sekhar, H. F. Liew, I. Purnama, W. S. Lew, M. Tran, and G. C. Han, *Appl. Phys. Lett.* 101, 2012, 152406.

24. A. Kunz and S. C. Reiff, *Appl. Phys. Lett.* 94, 2009, 192504.

25. K. He, D. J. Smith, and M. R. McCartney, *Appl. Phys. Lett.* 95, 2009, 182507.

26. S.-M. Ahn and K.-W. Moon, *Nanotechnology* 24, 2013, 105304.

27. A. Vogel, S. Wintz, J. Kimling, M. Bolte, T. Strache, M. Fritzsche, M.-Y. Im, P. Fischer, G. Meier, and J. Fassbender, *IEEE Trans. Magn.* 46, 2010, 1708.

28. A. Kunz, J. D. Priem, S. C. Reiff, *Proc. SPIE* 7760, 2010, 776005.

29. E. R. Lewis, D. Petit, A.-V. Jausovec, L. O'Brien, D. E. Read, H. T. Zeng, and R. P. Cowburn, *Phys. Rev. Lett.* 102, 2009, 057209.

30. E. R. Lewis, D. Petit, L. O'Brien, A. Fernandez-Pacheco, J. Sampaio, A.-V. Jausovec, H. T. Zeng, D. E. Read, and R. P. Cowburn, *Nat. Mater.* 9, 2010, 980.

31. E. R. Lewis, D. Petit, L. O'Brien, A.-V. Jausovec, H. T. Zeng, D. E. Read, and R. P. Cowburn, *Appl. Phys. Lett.* 98, 2011, 042502.

32. D. Petit, A.-V. Jausovec, H. T. Zeng, E. Lewis, L. O'Brien, D. Read, and R. P. Cowburn, *J. Appl. Phys.* 103, 2008, 114307.

33. M. Munoz and J. L. Prieto, *Nat. Commun.* 2, 2011, 1.

34. L. Bocklage, F. U. Stein, M. Martens, T. Matsuyama, and G. Meier, *App. Phys. Lett.* 103, 2013, 092406.

35. M. Hayashi, L. Thomas, C. Rettner, R. Moriya, and S. S. P. Parkin, *Nat. Phys.* 3, 2007, 21.

36. M. Hayashi, L. Thomas, C. Rettner, R. Moriya, X. Jiang, and S. S. P. Parkin, *Phys. Rev. Lett.* 97, 2006, 207205.

37. L. Thomas, M. Hayashi, R. Moriya, C. Rettner, and S. S. P. Parkin, *Nat. Commun.* 3, 2012, 810.

38. A. Pushp, T. Phung, C. Rettner, B. P. Hughes, S.-H. Yang, L. Thomas, and S. S. P. Parkin, *Nat. Phys.* 9, 2013, 505.

39. C. Guite, I. S. Kerk, M. C. Sekhar, M. Ramu, S. Goolaup, and W. S. Lew, *Sci. Rep.* 4, 2014, 7459.

40. D. Atkinson, D. S. Eastwood, L. K. Bogart, *Appl. Phys. Lett.* 92, 2008, 022510.

41. S. Goolaup, S. C. Low, M. Chandra Sekhar, and W. S. Lew, *J. Phys. Conf. Ser.* 266, 2011, 012079.

42. D. Petit, A.-V. Jausovec, H. T. Zeng, E. Lewis, L. O'Brien, D. Read, and R. P. Cowburn, *Phys. Rev. B* 79, 2009, 214405.
43. M. Chandra Sekhar, S. Goolaup, I. Purnama, and W. S. Lew, *J. Phys. D Appl. Phys.* 44, 2011, 235002.
44. G. S. D. Beach, C. Nistor, C. Knutson, M. Tsoi, and J. L. Erskine, *Nat. Mater.* 4, 2005, 741744.
45. Y. Nakatani, A. Thiaville, and J. Miltat, *Nat. Mater.* 2, 2003, 521.
46. M. Jamali, K.-J. Lee, and H. Yang, *Appl. Phys. Lett.* 98, 2011, 092501.
47. T. Koyama, D. Chiba, K. Ueda, K. Kondou, H. Tanigawa, S. Fukami et al., *Nat. Mater.* 10, 2011, 194.
48. S. Emori, U. Bauer, S. M. Ahn, E. Martinez, and G. S. D. Beach, *Nat. Mater.* 12, 2013, 611.
49. N. L. Schryer and L. R. Walker, *J. Appl. Phys.* 45, 1974, 5406.
50. X. R. Wang, P. Yan, J. Lu, and C. He, *Ann. Phys.* 324, 2009, 1815.
51. X. R. Wang, P. Yan, J. Lu, *Europhys. Lett.* 86, 2009, 67001.
52. L. Landau and E. Lifshitz, *Phys. Z. Sowjetunion* 8, 1935, 153.
53. T. L. Gilbert, *IEEE Trans. Magn.* 40(6), 2004, 3443.
54. L. Berger, *J. Appl. Phys.* 55, 1984, 1954.
55. L. Berger and C.-Y. Hung, *J. Appl. Phys.* 63, 1988, 4276.
56. J. C. Slonczewski, *J. Magn. Magn. Mater.* 159, 1996, L1.
57. M. Tsoi, A. G. M. Jansen, J. Bass, W. C. Chiang, M. Seck, V. Tsoi, and P. Wyder, *Phys. Rev. Lett.* 80, 1998, 4281.
58. A. Yamaguchi, T. Ono, S. Nasu, K. Miyake, K. Mibu, and T. Shinjo, *Phys. Rev. Lett.* 92, 2004, 077205.
59. A. Thiaville, Y. Nakatani, J. Miltat, and Y. Suzuki, *Europhys. Lett.* 69, 2005, 990.
60. G. Tatara and H. Kohno, *Phys. Rev. Lett.* 92, 2004, 086601.
61. S. Zhang and Z. Li, *Phys. Rev. Lett.* 93, 2004, 127204.
62. L. Thomas, M. Hayashi, X. Jiang, R. Moriya, C. Rettner, and S. S. P. Parkin, *Nature* 443, 2006, 197.
63. M. L. Schneider, T. Gerrits, A. B. Kos, and T. J. Silva, *Appl. Phys. Lett.* 87, 2005, 072509.
64. G. Tatara, H. Kohno, and J. Shibata, *Phys. Rep.* 468, 2008, 213.
65. M. Eltschka, M. Wötzel, J. Rhensius, S. Krzyk, U. Nowak, M. Kläui et al., *Phys. Rev. Lett.* 105, 2010, 056601.
66. L. Heyne, J. Rhensius, D. Ilgaz, A. Bisig, U. Rüdiger, M. Kläui et al., *Phys. Rev. Lett.* 105, 2010, 187203.
67. S.-H. Yang, K.-S. Ryu, and S. Parkin, *Nat. Nanotechnol.* 10, 2015, 221.
68. D. Bang and H. Awano, *J. Appl. Phys.* 117, 2015, 17D916.
69. K. Obata and G. Tatara, *Phys. Rev. B* 77, 2008, 214429.
70. J. Chen, M. B. A. Jalil, and S. G. Tan, *J. Phys. Soc. Jpn.* 83, 2014, 064710.
71. H. T. Zeng, D. Petit, L. O'Brien, D. Read, E. R. Lewis, and R. P. Cowburn, *J. Magn. Magn. Mater.* 322, 2010, 2010.
72. E. Saitoh, H. Miyajima, T. Yamaoka, and G. Tatara, *Nature* 432, 2004, 203.
73. V. W. Doring, Z. *Naturforsch.* A3, 1948, 373.
74. O. Tchernyshyov and G. W. Chern, *Phys. Rev. Lett.* 95, 2005, 197204.
75. L. D. Landau and E. M. Lifshitz, *Electrodynamics of Continuous Media*, Pergamon Press, 1960.
76. L. Bocklage, B. Krüger, T. Matsuyama, M. Bolte, U. Merkt, D. Pfannkuche, and G. Meier G, *Phys. Rev. Lett.* 103, 2009, 197204.
77. I. Purnama, M. Chandra Sekhar, S. Goolaup, and W. S. Lew, *Appl. Phys. Lett.* 99, 2011, 152501.
78. I. Purnama, M. Chandra Sekhar, S. Goolaup, and W. S. Lew, *IEEE Trans. Magn.* 47, 2011, 3081.

79. L. O'Brien, E. R. Lewis, A. Fernández-Pacheco, D. Petit, and R. P. Cowburn, *Phys. Rev. Lett.* 108, 2012, 187202.

80. S. Krishnia, I. Purnama, and W. S. Lew, *Appl. Phys. Lett.* 105, 2014, 042404.

81. T. J. Hayward, M. T. Bryan, P. W. Fry, P. M. Fundi1, M. R. J. Gibbs, D. A. Allwood, M.-Y. Im, and P. Fischer, *Phys. Rev. B* 81, 2010, 020410(R).

82. I. Purnama, M. C. Sekhar, W. S. Lew, and T. Ono, *Appl. Phys. Lett.* 104, 2014, 092414.

83. S.-H. Huang and C.-H. Lai, *Appl. Phys. Lett.* 95, 2009, 032505.

84. M. Klaui, H. Ehrke, U. Rüdiger, T. Kasama, R. E. Dunin-Borkowski, D. Backes et al., *Appl. Phys. Lett.* 87, 2005, 102509.

85. C. H. Marrows, *Adv. Phys.* 54(8), 2005, 585.

86. G. Tatara and H. Fukuyama, *Phys. Rev. Lett.* 78, 1997, 3773.

87. E. Simanek, *Phys. Rev. B* 63, 2001, 224412.

88. U. Ruediger, J. Yu, S. Zhang, A. D. Kent, and S. S. P. Parkin, *Phys. Rev. Lett.* 80, 1998, 5639.

89. Z. Yuan, Y. Liu, A. A. Starikov, P. J. Kelly, and A. Brataas, *Phys. Rev. Lett.* 109, 2012, 267201.

90. M. Hayashi, L. Thomas, C. Rettner, R. Moriya, and S. S. P. Parkin, *Nat. Phys.* 3(1), 2007, 21–25.

91. A. Hubert, W. Rave, and S. L. Tomlinson, *Phys. Status Solidi B* 204, 1997, 817.

92. W. Rave, L. Belliard, M. Labrune, A. Thiaville, and J. Miltat, *IEEE Trans. Magn.* 30, 1994, 4473.

93. J. M. Gracia, A. Thiaville, and J. Miltat, *J. Magn. Magn. Mater.* 249, 2002, 163.

Index